For more information, contact:

BROOKS/COLE PUBLISHING COMPANY
511 Forest Lodge Road
Pacific Grove, CA 93950
USA

International Thomson Editores
Campos Eliseos 385, Piso 7
Col. Polanco
11560 México D. F. México

International Thomson Publishing Europe
Berkshire House 168-173
High Holborn
London WC1V 7AA
England

International Thomson Publishing GmbH
Königswinterer Strasse 418
53227 Bonn
Germany

Thomas Nelson Australia
102 Dodds Street
South Melbourne, 3205
Victoria, Australia

International Thomson Publishing Asia
221 Henderson Road
#05-10 Henderson Building
Singapore 0315

Nelson Canada
1120 Birchmount Road
Scarborough, Ontario
Canada M1K 5G4

International Thomson Publishing Japan
Hirakawacho Kyowa Building 3 F
2-2-1 Hirakawacho
Chiyoda-ku, Tokyo 102
Japan

Printed in the United States of America

5 4 3 2 1

ISBN 0-534-95575-4

Complete Solutions Manual for Zill's
A First Course in Differential Equations with Modeling Applications 6th Edition

and

Complete Solutions Manual for Zill & Cullen's
Differential Equations with Boundary-Value Problems 4th Edition

Warren S. Wright
Loyola Marymount University
and
Carol D. Wright

PWS Publishing Company

Brooks/Cole Publishing Company

I⫪P® An International Thomson Publishing Company

Pacific Grove · Albany · Bonn · Boston · Cincinnati · Detroit · London · Madrid · Melbourne
Mexico City · New York · Paris · San Francisco · Singapore · Tokyo · Toronto · Washington

Table of Contents

1 Introduction to Differential Equations

Exercises 1.1

1. Second-order; linear.

2. Third-order; nonlinear because of $(dy/dx)^4$.

3. First-order; nonlinear because of yy'.

4. First-order; linear.

5. Fourth-order; linear.

6. Second-order; nonlinear because of $\sin y$.

7. Second-order; nonlinear because of $\left(d^2y/dx^2\right)^2$.

8. Second-order; nonlinear because of $1/r^2$.

9. Third-order; linear.

10. First-order; nonlinear because of y^2.

11. From $y = e^{-x/2}$ we obtain $y' = -\frac{1}{2}e^{-x/2}$. Then $2y' + y = -e^{-x/2} + e^{-x/2} = 0$.

12. From $y = 8$ we obtain $y' = 0$, so that $y' + 4y = 0 + 4(8) = 32$.

13. From $y = e^{3x} + 10e^{2x}$ we obtain $dy/dx = 3e^{3x} + 20e^{2x}$. Then

$$\frac{dy}{dx} - 2y = \left(3e^{3x} + 20e^{2x}\right) - 2\left(e^{3x} + 10e^{2x}\right) = e^{3x}.$$

14. From $y = \dfrac{6}{5} - \dfrac{6}{5}e^{-20t}$ we obtain $dy/dt = 24e^{-20t}$, so that

$$\frac{dy}{dt} + 20y = 24e^{-20t} + 20\left(\frac{6}{5} - \frac{6}{5}e^{-20t}\right) = 24.$$

15. From $y = 5\tan 5x$ we obtain $y' = 25\sec^2 5x$. Then

$$y' = 25\sec^2 5x = 25\left(1 + \tan^2 5x\right) = 25 + (5\tan 5x)^2 = 25 + y^2.$$

16. From $y = \left(\sqrt{x} + c_1\right)^2$ we obtain $y' = 2\left(\sqrt{x} + c_1\right)/2\sqrt{x}$, so that

$$y' = \frac{\sqrt{x} + c_1}{\sqrt{x}} = \sqrt{\frac{\left(\sqrt{x} + c_1\right)^2}{x}} = \sqrt{\frac{y}{x}}.$$

17. From $y = \dfrac{1}{2}\sin x - \dfrac{1}{2}\cos x + 10e^{-x}$ we obtain $y' = \dfrac{1}{2}\cos x + \dfrac{1}{2}\sin x - 10e^{-x}$. Then

$$y' + y = \left(\frac{1}{2}\cos x + \frac{1}{2}\sin x - 10e^{-x}\right) + \left(\frac{1}{2}\sin x - \frac{1}{2}\cos x + 10e^{-x}\right) = \sin x.$$

18. First write the differential equation in the form $2xy + \left(x^2 + 2y\right) y' = 0$. Implicitly differentiating $x^2 y + y^2 = c_1$ we obtain $2xy + \left(x^2 + 2y\right) y' = 0$.

19. First write the differential equation in the form $y' = -2y/x$. From $y = -1/x^2$ we obtain $y' = 2x^{-3}$, so that $-2y/x = 2x^{-3} = y'$.

20. From $y = x + 1$ we obtain $y' = 1$, so that $(y')^3 + xy' = 1 + x = y$.

21. Implicitly differentiating $y^2 = c_1 \left(x + \frac{1}{4} c_1\right)$ we obtain $y' = c_1/2y$. Then

$$2xy' + y(y')^2 = \frac{c_1 x}{y} + \frac{c_1^2}{4y} = \frac{y^2}{y} = y.$$

22. Writing $y = x|x|$ as $\quad y = \begin{cases} x^2, & \text{if } x \geq 0 \\ -x^2, & \text{if } x < 0 \end{cases} \quad$ we see that $\quad |y| = x^2, \quad -\infty < x < \infty$, and

$\sqrt{|y|} = \begin{cases} x, & \text{if } x \geq 0 \\ -x, & \text{if } x < 0 \end{cases}$. Since $\quad y' = \begin{cases} 2x, & \text{if } x \geq 0 \\ -2x, & \text{if } x < 0, \end{cases} \quad$ it is apparent that $y' = 2\sqrt{|y|}$.

23. From $y = x \ln x$ we obtain $y' = 1 + \ln x$. Then $y' - \frac{1}{x} y = 1$.

24. Differentiating $P = ac_1 e^{at} / \left(1 + bc_1 e^{at}\right)$ we obtain

$$\frac{dP}{dt} = \frac{\left(1 + bc_1 e^{at}\right) a^2 c_1 e^{at} - ac_1 e^{at} \cdot abc_1 e^{at}}{(1 + bc_1 e^{at})^2}$$

$$= \frac{ac_1 e^{at}}{1 + bc_1 e^{at}} \cdot \frac{\left[a\left(1 + bc_1 e^{at}\right) - abc_1 e^{at}\right]}{1 + bc_1 e^{at}} = P(a - bP).$$

25. Implicitly differentiating $\ln \frac{2 - X}{1 - X} = \ln(2 - X) - \ln(1 - X) = t$ we obtain

$$\frac{-1}{2 - X} \cdot \frac{dX}{dt} - \frac{-1}{1 - X} \cdot \frac{dX}{dt} = 1.$$

Then $\dfrac{dX}{dt} = (2 - X)(1 - X)$.

26. Differentiating $y = e^{-x^2} \int_0^x e^{t^2} dt + c_1 e^{-x^2}$ we obtain

$$y' = e^{-x^2} e^{x^2} - 2xe^{-x^2} \int_0^x e^{t^2} dt - 2c_1 xe^{-x^2} = 1 - 2xe^{-x^2} \int_0^x e^{t^2} dt - 2c_1 xe^{-x^2}.$$

Substituting into the differential equation, we have

$$y' + 2xy = 1 - 2xe^{-x^2} \int_0^x e^{t^2} dt - 2c_1 xe^{-x^2} + 2xe^{-x^2} \int_0^x e^{t^2} dt + 2c_1 xe^{-x^2} = 1.$$

2

27. First write the differential equation in the form $y' = \dfrac{-x^2 - y^2}{x^2 - xy}$. Then $c_1(x + y)^2 = xe^{y/x}$ implies

$c_1 = \dfrac{xe^{y/x}}{(x + y)^2}$ and implicit differentiation gives $2c_1(x + y)(1 + y') = xe^{y/x}\dfrac{xy' - y}{x^2} + e^{y/x}$. Solving

for y' we obtain

$$y' = \frac{e^{y/x} - \dfrac{y}{x}e^{y/x} - 2c_1(x + y)}{2c_1(x + y) - e^{y/x}} = \frac{1 - \dfrac{y}{x} - \dfrac{2x}{x + y}}{\dfrac{2x}{x + y} - 1} = \frac{-x^2 - y^2}{x^2 - xy}.$$

28. From $y = c_1e^{3x} + c_2e^{-4x}$ we obtain $y' = 3c_1e^{3x} - 4c_2e^{-4x}$ and $y'' = 9c_1e^{3x} + 16c_2e^{-4x}$, so that $y'' + y' - 12y = 0$.

29. From $y = e^{3x}\cos 2x$ we obtain $y' = 3e^{3x}\cos 2x - 2e^{3x}\sin 2x$ and $y'' = 5e^{3x}\cos 2x - 12e^{3x}\sin 2x$, so that $y'' - 6y' + 13y = 0$.

30. From $y = e^{2x} + xe^{2x}$ we obtain $\dfrac{dy}{dx} = 3e^{2x} + 2xe^{3x}$ and $\dfrac{d^2y}{dx^2} = 8e^{2x} + 4xe^{2x}$ so that $\dfrac{d^2y}{dx^2} - 4\dfrac{dy}{dx} + 4y = 0$.

31. From $y = \cosh x + \sinh x$ we obtain $y' = \sinh x + \cosh x$ and $y'' = \cosh x + \sinh x = y$.

32. From $y = c_1\cos 5x$ we obtain $y' = -5c_1\sin 5x$ and $y'' = -25c_1\cos 5x$, so that $y'' + 25y = 0$.

33. From $y = \ln|x + c_1| + c_2$ we obtain $y' = \dfrac{1}{x + c_1}$ and $y'' = \dfrac{-1}{(x + c_1)^2}$, so that $y'' + (y')^2 = 0$.

34. From $y = -\cos x \ln(\sec x + \tan x)$ we obtain $y' = -1 + \sin x \ln(\sec x + \tan x)$ and $y'' = \tan x + \cos x \ln(\sec x + \tan x)$. Then $y'' + y = \tan x$.

35. From $y = c_1 + c_2x^{-1}$ we obtain $y' = -c_2x^{-2}$ and $y'' = 2c_2x^{-3}$, so that $x\dfrac{d^2y}{dx^2} + 2\dfrac{dy}{dx} = 0$.

36. From $y = x\cos(\ln x)$ we obtain $y' = -\sin(\ln x) + \cos(\ln x)$ and $y'' = \dfrac{-1}{x}\cos(\ln x) - \dfrac{1}{x}\sin(\ln x)$, so that $x^2y'' - xy' + 2y = 0$.

37. From $y = x^2 + x^2\ln x$ we obtain $y' = 3x + 2x\ln x$ and $y'' = 5 + 2\ln x$ so that $x^2y'' - 3xy' + 4y = 0$.

38. From $y = c_1\sin 3x + c_2\cos 3x + 4e^x$ we obtain $y' = 3c_1\cos 3x - 3c_2\sin 3x + 4e^x$, $y'' = -9c_1\sin 3x - 9c_2\cos 3x + 4e^x$, and $y''' = -27c_1\cos 3x + 27c_2\sin 3x + 4e^x$, so that $y''' - y'' + 9y' - 9y = 0$.

39. From $y = x^2e^x$ we obtain $y' = x^2e^x + 2xe^x$, $y'' = x^2e^x + 4xe^x2e^x$, and $y''' = x^2e^x + 6xe^x + 6e^x$, so that $y''' - 3y'' + 3y' - y = 0$.

40. From $y = c_1x + c_2x\ln x + 4x^2$ we obtain $y' = c_1 + c_2 + c_2\ln x + 8x$, $y'' = c_2x^{-1} + 8$, and $y''' = -c_2x^{-2}$, so that $x^3\dfrac{d^3y}{dx^3} + 2x^2\dfrac{d^2y}{dx^2} - x\dfrac{dy}{dx} + y = 12x^2$.

41. From $y = \begin{cases} -x^2, & x < 0 \\ x^2, & x \geq 0 \end{cases}$ we obtain $y' = \begin{cases} -2x, & x < 0 \\ 2x, & x \geq 0 \end{cases}$ so that $xy' - 2y = 0$.

42. From $y = \begin{cases} 0, & x < 0 \\ x^3, & x \geq 0 \end{cases}$ we obtain $y' = \begin{cases} 0, & x < 0 \\ 3x^2, & x \geq 0 \end{cases}$ so that $(y')^2 = \begin{cases} 0, & x < 0 \\ 9x^4, & x \geq 0 \end{cases}$.

43. By inspection, $y = -1$ is a singular solution. Note that this is the "solution" obtained by computing the limit as c approaches infinity of the one-parameter family of solutions.

44. The function $y = \begin{cases} \sqrt{4 - x^2}, & -2 < x < 0 \\ -\sqrt{4 - x^2}, & 0 \leq x < 2 \end{cases}$ is not continuous at $x = 0$ (the left hand limit is 2 and the right hand limit is -2,) and hence y' does not exist at $x = 0$.

45. From $y = e^{mx}$ we obtain $y' = me^{mx}$ and $y'' = m^2 e^{mx}$. Then $y'' - 5y' + 6y = 0$ implies

$$m^2 e^{mx} - 5me^{mx} + 6e^{mx} = (m - 2)(m - 3)e^{mx} = 0.$$

Since $e^{mx} > 0$ for all x, $m = 2$ and $m = 3$. Thus $y = e^{2x}$ and $y = e^{3x}$ are solutions.

46. From $y = e^{mx}$ we obtain $y' = me^{mx}$ and $y'' = m^2 e^{mx}$. Then $y'' + 10y' + 25y = 0$ implies

$$m^2 e^{mx} + 10me^{mx} + 25e^{mx} = (m + 5)^2 e^{mx} = 0.$$

Since $e^{mx} > 0$ for all x, $m = -5$. Thus, $y = e^{-5x}$ is a solution.

47. Using $y' = mx^{m-1}$ and $y'' = m(m-1)x^{m-2}$ and substituting into the differential equation we obtain $m(m-1)x^m - x^m = \left(m^2 - m - 1\right) x^m = 0$. Solving $m^2 - m - 1 = 0$ we obtain $m = \left(1 \pm \sqrt{5}\right)/2$. Thus, two solutions of the differential equation on the interval $0 < x < \infty$ are $y = x^{(1+\sqrt{5})/2}$ and $y = x^{(1-\sqrt{5})/2}$.

48. Using $y' = mx^{m-1}$ and $y'' = m(m-1)x^{m-2}$ and substituting into the differential equation we obtain $x^2 y'' + 6xy' + 4y = [m(m - 1) + 6m + 4]x^m$. The right side will be zero provided m satisfies

$$m(m - 1) + 6m + 4 = m^2 + 5m + 4 = (m + 4)(m + 1) = 0.$$

Thus, $m = -4, -1$ and two solutions of the differential equation on the interval $0 < x < \infty$ are $y = x^{-4}$ and $y = x^{-1}$.

49. From $x = e^{-2t} + 3e^{6t}$ and $y = -e^{-2t} + 5e^{6t}$ we obtain

$$\frac{dx}{dt} = -2e^{-2t} + 18e^{6t} \quad \text{and} \quad \frac{dy}{dt} = 2e^{-2t} + 30e^{6t}.$$

Then

$$x + 3y = (e^{-2t} + 3e^{6t}) + 3(-e^{-2t} + 5e^{6t})$$

$$= -2e^{-2t} + 18e^{6t} = \frac{dx}{dt}$$

and

$$5x + 3y = 5(e^{-2t} + 3e^{6t}) + 3(-e^{-2t} + 5e^{6t})$$

$$= 2e^{-2t} + 30e^{6t} = \frac{dy}{dt} \, .$$

50. From $x = \cos 2t + \sin 2t + \frac{1}{5}e^t$ and $y = -\cos 2t - \sin 2t - \frac{1}{5}e^t$ we obtain

$$\frac{dx}{dt} = -2\sin 2t + 2\cos 2t + \frac{1}{5}e^t \quad \text{and} \quad \frac{dy}{dt} = 2\sin 2t - 2\cos 2t - \frac{1}{5}e^t$$

and

$$\frac{d^2x}{dt^2} = -4\cos 2t - 4\sin 2t + \frac{1}{5}e^t \quad \text{and} \quad \frac{d^2y}{dt^2} = 4\cos 2t + 4\sin 2t - \frac{1}{5}e^t.$$

Then

$$4y + e^t = 4(-\cos 2t - \sin 2t - \frac{1}{5}e^t) + e^t$$

$$= -4\cos 2t - 4\sin 2t + \frac{1}{5}e^t = \frac{d^2x}{dt^2}$$

and

$$4x - e^t = 4(\cos 2t + \sin 2t + \frac{1}{5}e^t) - e^t$$

$$= 4\cos 2t + 4\sin 2t - \frac{1}{5}e^t = \frac{d^2y}{dt^2} \, .$$

51. (a) The differential equations

$$\left(\frac{dy}{dx}\right)^2 + 1 = 0 \quad \text{and} \quad (y')^2 + y^2 + 4 = 0$$

have no real solutions.

(b) The differential equations

$$|y'| + |y| = 0 \quad \text{and} \quad (y')^2 + y^2 = 0$$

have the trivial solution $y = 0$, but no other solutions.

52. Since the nth derivative of $\phi(x)$ must exist if $\phi(x)$ is a solution of the nth order differential equation, all lower-order derivatives of $\phi(x)$ must exist and be continuous. [Recall that a differentiable function is continuous.]

53. (a) $y = 0$ and $y = a/b$.

(b) Since $dy/dx = y(a - by) > 0$ for $0 < y < a/b$, $y = \phi(x)$ is increasing on this interval. Since $dy/dx < 0$ for $y < 0$ or $y > a/b$, $y = \phi(x)$ is decreasing on these intervals.

(c) Using implicit differentiation we compute

$$\frac{d^2y}{dx^2} = y(-by') + y'(a - by) = y'(a - 2by).$$

5

Solving $d^2y/dx^2 = 0$ we obtain $y = a/2b$. Since $d^2y/dx^2 > 0$ for $0 < y < a/2b$ and $d^2y/dx^2 < 0$ for $a/2b < y < a/b$, the graph of $y = \phi(x)$ has a point of inflection at $y = a/2b$.

(d)

54. (a) Differentiating we obtain

$$y' = xy'' + y' + f'(y')y''$$
$$y''(x + f'(y')) = 0.$$

If $y'' = 0$ then $y' = c$. Substituting into Clairaut's equation gives $y = cx + f(c)$, a straight line.

(b) If $x + f'(y') = 0$ then $x = -f'(t)$ where $t = y'$, and Clairaut's equation gives $y = -tf'(t) + f(t)$. The two equations for x and y are parametric equations of a singular solution.

(c) Identifying $f(t) = t^2$ and $f'(t) = 2t$ we see that a one-parameter family of solutions is $y = cx + c^2$ and a singular solution is determined by

$$x = -2t$$

$$y = -2t^2 + t^2 = -t^2.$$

Eliminating the parameter we have $y = -\frac{1}{4}x^2$.

——— Exercises 1.2 ———

1. For $f(x, y) = y^{2/3}$ we have $\dfrac{\partial f}{\partial y} = \dfrac{2}{3}y^{-1/3}$. Thus the differential equation will have a unique solution in any rectangular region of the plane where $y \neq 0$.

2. For $f(x, y) = \sqrt{xy}$ we have $\dfrac{\partial f}{\partial y} = \dfrac{1}{2}\sqrt{\dfrac{x}{y}}$. Thus the differential equation will have a unique solution in any region where $x > 0$ and $y > 0$ or where $x < 0$ and $y < 0$.

3. For $f(x, y) = \dfrac{y}{x}$ we have $\dfrac{\partial f}{\partial y} = \dfrac{1}{x}$. Thus the differential equation will have a unique solution in any region where $x \neq 0$.

4. For $f(x, y) = x + y$ we have $\dfrac{\partial f}{\partial y} = 1$. Thus the differential equation will have a unique solution in the entire plane.

5. For $f(x, y) = \dfrac{x^2}{4 - y^2}$ we have $\dfrac{\partial f}{\partial y} = \dfrac{2x^2 y}{(4 - y^2)^2}$. Thus the differential equation will have a unique solution in any region where $y < -2$, $-2 < y < 2$, or $y > 2$.

6. For $f(x, y) = \dfrac{x^2}{1 + y^3}$ we have $\dfrac{\partial f}{\partial y} = \dfrac{-3x^2 y^2}{(1 + y^3)^2}$. Thus the differential equation will have a unique solution in any region where $y \neq -1$.

7. For $f(x, y) = \dfrac{y^2}{x^2 + y^2}$ we have $\dfrac{\partial f}{\partial y} = \dfrac{2x^2 y}{(x^2 + y^2)^2}$. Thus the differential equation will have a unique solution in any region not containing $(0, 0)$.

8. For $f(x, y) = \dfrac{y + x}{y - x}$ we have $\dfrac{\partial f}{\partial y} = \dfrac{-2x}{(y - x)^2}$. Thus the differential equation will have a unique solution in any region where $y < x$ or where $y > x$.

9. For $f(x, y) = x^3 \cos y$ we have $\dfrac{\partial f}{\partial y} = -x^2 \sin y$. Thus the differential equation will have a unique solution in the entire plane.

10. For $f(x, y) = (x - 1)e^{y/(x-1)}$ we have $\dfrac{\partial f}{\partial y} = e^{y/(x-1)}$. Thus the differential equation will have a unique solution in any region where $x \neq 1$.

11. Two solutions are $y = 0$ and $y = x^3$.

12. Two solutions are $y = 0$ and $y = x^2$. (Also, any constant multiple of x^2 is a solution.)

For Problems 13–16 we identify $f(x, y) = \sqrt{y^2 - 9}$ and $\partial f / \partial y = y^2 / \sqrt{y^2 - 9}$. We further note that $f(x, y)$ is discontinuous for $|y| < 3$ and that $\partial f / \partial y$ is discontinuous for $|y| < 3$. We then apply Theorem 1.1.

13. The differential equation has a unique solution at $(1, 4)$.

14. The differential equation is not guaranteed to have a unique solution at $(5, 3)$.

15. The differential equation is not guaranteed to have a unique solution at $(2, -3)$.

16. The differential equation is not guaranteed to have a unique solution at $(-1, 1)$.

17. (a) A one-parameter family of solutions is $y = cx$. Since $y' = c$, $xy' = xc = y$ and $y(0) = c \cdot 0 = 0$.

(b) Writing the equation in the form $y' = y/x$ we see that R cannot contain any point on the y-axis. Thus, any rectangular region disjoint from the y-axis and containing (x_0, y_0) will determine an interval around x_0 and a unique solution through (x_0, y_0). Since $x_0 = 0$ in part (a) we are not guaranteed a unique solution through $(0, 0)$.

(c) The piecewise-defined function which satisfies $y(0) = 0$ is not a solution since it is not differentiable at $x = 0$.

18. (a) Since $1 + y^2$ and its partial derivative with respect to y are continuous everywhere in the plane, the differential equation has a unique solution through every point in the plane.

(b) Since $\dfrac{d}{dx}(\tan x) = \sec^2 x = 1 + \tan^2 x$ and $\tan 0 = 0$, $y = \tan x$ satisfies the differential equation and the initial condition. Since $-2 < \pi/2 < 2$ and $\tan x$ is undefined for $x = \pi/2$, $y = \tan x$ is not a solution on the interval $-2 < x < 2$.

(c) Since $\tan x$ is continuous and differentiable on $(-\pi/2, \pi/2)$ and not defined at the endpoints of this interval, this is the largest interval of validity for which $y = \tan x$ is a solution of $y' = 1 + y^2$, $y(0) = 0$.

19. (a) We identify $f(x, y) = y^2$ and note that f and $\partial f/\partial y$ are continuous in the plane, so by Theorem 1.1 the differential equation has a unique solution through any point in the plane.

(b)

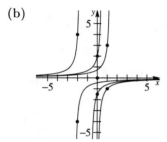

(c) $(0, 0)$: $(-\infty, \infty)$; $(0, 2)$: $(-\infty, 1/2)$; $(1, 3)$: $(-\infty, -4/3)$; $(-2, 4)$: $(-\infty, -2)$; $(-2, -4)$: $(-2, \infty)$; $(0, -1.5)$: $(-1/2, \infty)$; $(1, -1)$: $(0, \infty)$

20. (a) Identifying $f(x, y) = x/y$ we see that f and $\partial f/\partial y$ are continuous and differentiable for $y > 0$. We thus take R to be the upper half-plane.

(b)

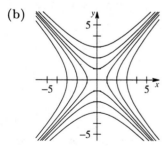

(c) The graph looks like a family of hyperbolas of the form $y^2 - x^2 = c^2$.

21. Setting $x = 0$ and $y = -1/3$ we have $1/(1 + c_1) = -1/3$ so $c_1 = -4$. A solution of the initial-value problem is $y = 1/(1 - 4e^{-x})$.

22. Setting $x = -1$ and $y = 2$ we have $1/(1 + c_1 e) = 2$ so $c_1 = -1/2e$. A solution of the inital-value problem is $y = 1/(1 - \frac{1}{2}e^{-x-1})$.

23. From the initial conditions we obtain the system

$$c_1 + c_2 = 1$$

$$c_1 - c_2 = 2.$$

Solving we get $c_1 = \frac{3}{2}$ and $c_2 = -\frac{1}{2}$. A solution of the initial-value problem is $y = \frac{3}{2}e^x - \frac{1}{2}e^{-x}$.

24. From the initial conditions we obtain the system

$$c_1 e + c_2 e^{-1} = 0$$

$$c_1 e - c_2 e^{-1} = e.$$

Solving we get $c_1 = \frac{1}{2}$ and $c_2 = -\frac{1}{2}e^2$. A solution of the initial-value problem is $y = \frac{1}{2}e^x - \frac{1}{2}e^{2-x}$.

25. From the initial conditions we obtain

$$c_1 e^{-1} + c_2 e = 5$$

$$c_1 e^{-1} - c_2 e = -5.$$

Solving we get $c_1 = 0$ and $c_2 = 5e^{-1}$. A solution of the initial-value problem is $y = 5e^{-x-1}$.

26. From the initial conditions we obtain

$$c_1 + c_2 = 0$$

$$c_1 - c_2 = 0.$$

Solving we get $c_1 = c_2 = 0$. A solution of the initial-value problem is $y = 0$.

27. The theorem guarantees a unique (meaning single) solution through any point. Thus, there cannot be two distinct solutions through any point.

28. Setting $x = \pi$ and $y = 1/8$ we have $1/c^3 = 1/8$ so $c = 2$. A solution of the initial-value problem is $y = 1/(2 - \sin x)^3$. Setting $x = \pi$ and $y = 8$ we have $1/c^3 = 8$ so $c = 1/2$. A solution of the initial-value problem is $y = 1/(1/2 - \sin x)^3$.

When $y(\pi) = y_0$ is the initial condition, $c = 1/y_0^{1/3}$ and the solution is $y = 1/(1/y_0^{1/3} - \sin x)^3$. This solution will exist on a finite interval when $|1/y_0^{1/3}| \leq 1$ or $|y_0| \geq 1$. It will exist on the entire real line when $|y_0| < 1$, $y_0 \neq 0$.

_____ **Exercises 1.3** _____

1. $\dfrac{dP}{dt} = kP + r$

2. Let b be the rate of births and d the rate of deaths. Then $b = k_1 P$ and $d = k_2 P$. Since $dP/dt = b - d$, the differential equation is

$$\frac{dP}{dt} = k_1 P - k_2 P.$$

3. The differential equation is $x'(t) = r - kx(t)$ where $k > 0$.

4. By analogy with differential equation modeling the spread of a disease we assume that the rate at which the technological innovation is adopted is proportional to the number of people who have adopted the innovation and also to the number of people, $y(t)$, who have not yet adopted it. If one person who has adopted the innovation is introduced into the population then $x + y = n + 1$ and

$$\frac{dx}{dt} = kx(n + 1 - x), \quad x(0) = 1.$$

5. The rate at which salt is leaving the tank is

$$(3 \text{ gal/min}) \cdot \left(\frac{A}{300} \text{ lb/gal}\right) = \frac{A}{100} \text{ lb/min}.$$

Thus $dA/dt = A/100$.

6. The rate at which salt is entering the tank is

$$R_1 = (3 \text{ gal/min}) \cdot (2 \text{ lb/gal}) = 6 \text{ lb/min}.$$

Since the solution is pumped out at a slower rate, it is accumulating at the rate of $(3 - 2)\text{gal/min} = 1 \text{ gal/min}$. After t minutes there are $300 + t$ gallons of brine in the tank. The rate at which salt is leaving is

$$R_2 = (2 \text{ gal/min}) \cdot \left(\frac{A}{300 + t} \text{ lb/gal}\right) = \frac{2A}{300 + t} \text{ lb/min}.$$

The differential equation is

$$\frac{dA}{dt} = 6 - \frac{2A}{300 + t}.$$

7. The volume of water in the tank at time t is $V = A_w h$. The differential equation is then

$$\frac{dh}{dt} = \frac{1}{A_w} \frac{dV}{dt} = \frac{1}{A_w}\left(-cA_0 \sqrt{2gh}\right) = -\frac{cA_0}{A_w}\sqrt{2gh}.$$

Using $A_0 = \pi \left(\dfrac{2}{12}\right)^2 = \dfrac{\pi}{36}$, $A_w = 10^2 = 100$, and $g = 32$, this becomes

$$\frac{dh}{dt} = -\frac{c\pi/36}{100}\sqrt{64h} = -\frac{c\pi}{450}\sqrt{h}.$$

10

8. The differential equation is

$$\frac{dh}{dt} = -\frac{cA_0}{A_w}\sqrt{2gh}.$$

Using $A_0 = \pi(1/24)^2 = \pi/576$, $A_w = \pi(2)^2 = 4\pi$, and $g = 32$, this becomes

$$\frac{dh}{dt} = -\frac{c\pi/576}{4\pi}\sqrt{64h} = \frac{c}{288}\sqrt{h}.$$

9. Since $i = \frac{dq}{dt}$ and $L\frac{d^2q}{dt^2} + R\frac{dq}{dt} = E(t)$ we obtain $L\frac{di}{dt} + Ri = E(t)$.

10. By Kirchoff's second law we obtain $R\frac{dq}{dt} + \frac{1}{C}q = E(t)$.

11. The differential equation is $\frac{dA}{dt} = k(M - A)$.

12. The differential equation is $\frac{dA}{dt} = k_1(M - A) - k_2A$.

13. By the Pythagorean Theorem the slope of the tangent line is $y' = \dfrac{-y}{\sqrt{s^2 - y^2}}$.

14. From Newton's second law we obtain $m\frac{dv}{dt} = -kv^2 + mg$.

15. The net force acting on the mass is

$$F = ma = m\frac{d^2x}{dt^2} = -k(s + x) + mg = -kx + mg - ks.$$

Since the condition of equilibrium is $mg = ks$, the differential equation is

$$m\frac{d^2x}{dt^2} = -ks.$$

16. We have from Archimedes' principle

upward force of water on barrel = weight of water displaced

$$= (62.4) \times (\text{volume of water displaced})$$

$$= (62.4)\pi(s/2)^2 y = 15.6\pi s^2 y.$$

It then follows from Newton's second law that $\dfrac{w}{g}\dfrac{d^2y}{dt^2} = -15.6\pi s^2 y$ or $\dfrac{d^2y}{dt^2} + \dfrac{15.6\pi s^2 g}{w}y = 0$, where $g = 32$ and w is the weight of the barrel in pounds.

17. We see from the figure that $\alpha + \phi = \pi = 2\theta + \alpha$ so that $\phi = 2\theta$.

Now the slope of the tangent line is

$$\frac{dy}{dx} = \tan\theta = \tan\frac{\phi}{2} = \frac{\sin\phi}{1+\cos\phi} = \frac{\sin(\pi-\alpha)}{1+\cos(\pi-\alpha)}$$

$$= \frac{\sin\alpha}{1-\cos\alpha} = \frac{y/\sqrt{x^2+y^2}}{1-x/\sqrt{x^2+y^2}} = \frac{y}{\sqrt{x^2+y^2}-x}.$$

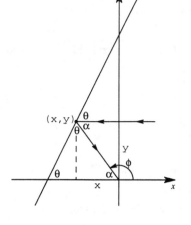

Then

$$\left(\sqrt{x^2+y^2} - x\right)\frac{dy}{dx} = y,$$

$$\sqrt{x^2+y^2}\,\frac{dy}{dx} = y + x\frac{dy}{dx}$$

and

$$(x^2+y^2)\left(\frac{dy}{dx}\right)^2 = y^2 + 2xy\frac{dy}{dx} + x^2\left(\frac{dy}{dx}\right)^2.$$

Simplifying we obtain

$$y\left(\frac{dy}{dx}\right)^2 = y + 2x\frac{dy}{dx}$$

or

$$y\left(\frac{dy}{dx}\right)^2 - 2x\frac{dy}{dx} = y.$$

18. This differential equation could describe a population that undergoes periodic fluctuations.

19. The radius, volume, and surface area of the snowball change with time. As the snowball melts the volume decreases. It seems reasonable to assume that the rate at which the snowball melts is proportional to the surface area. That is, $dV/dt = -kS$ where k is positive. Now $V = 4\pi r^3/3$ and $S = 4\pi r^2$, so

$$\frac{dV}{dt} = 4\pi r^2\frac{dr}{dt} = S\frac{dr}{dt} = -kS.$$

Thus, $dr/dt = -k$.

20. We assume that the plow clears snow at a constant rate of k cubic miles per hour. Let t be the time in hours after noon, $x(t)$ the depth in miles of the snow at time t, and $y(t)$ the distance the plow has moved in t hours. Then dy/dt is the velocity of the plow and the assumption gives

$$wx\frac{dy}{dt} = k$$

where w is the width of the plow. Each side of this equation simply represents the volume of snow plowed in one hour. Now let t_0 be the number of hours before noon when it started snowing and let s be the constant rate in miles per hour at which x increases. Then for $t > -t_0$, $x = s(t + t_0)$. The differential equation then becomes

$$\frac{dy}{dt} = \frac{k}{ws}\frac{1}{t + t_0}.$$

Integrating we obtain

$$y = \frac{k}{ws}\left[\ln(t + t_0) + c\right]$$

where c is a constant. Now when $t = 0$, $y = 0$ so $c = -\ln t_0$ and

$$y = \frac{k}{ws}\ln\left(1 + \frac{t}{t_0}\right).$$

Finally, from the fact that when $t = 1$, $y = 2$ and when $t = 2$, $y = 3$, we obtain

$$\left(1 + \frac{2}{t_0}\right)^2 = \left(1 + \frac{1}{t_0}\right)^3.$$

Expanding and simplifying gives $t_0^2 + t_0 - 1 = 0$. Since $t_0 > 0$, we find $t_0 \approx 0.618$ hours $\approx$ 37 minutes. Thus it started snowing at about 11:23 in the morning.

21. A possible motion for the mass is that it oscillates back and forth around the center of the earth. The gravitational force on m is $F = -kM_r m/r^2$. Since $M_r = 4\pi\delta r^3/3$ and $M = 4\pi\delta R^3/3$ we have $M_r = r^3 M/R^3$ and

$$F = -k\frac{M_r m}{r^2} = -k\frac{r^3 M m/R^3}{r^2} = -k\frac{mM}{R^3}r.$$

Now from $F = ma = d^2r/dt^2$ we have

$$m\frac{d^2r}{dt^2} = -k\frac{mM}{R^3}r \quad \text{or} \quad \frac{d^2r}{dt^2} = -\frac{kM}{R^3}r.$$

22. From the graph we estimate $T_0 = 180°$ and $T_m = 75°$. We observe that when $T = 85$, $dT/dt \approx -1$. From the differential equation we then have

$$k = \frac{dT/dt}{T - T_m} = \frac{-1}{85 - 75} = -0.1.$$

Chapter 1 Review Exercises

1. For $f(x, y) = \left(25 - x^2 - y^2\right)^{-1}$ we obtain $f_y(x, y) = 2y\left(25 - x^2 - y^2\right)^{-2}$ so there will be a unique solution for any point (x_0, y_0) in the region $x^2 + y^2 < 25$ or $x^2 + y^2 > 25$.

2. $y = 0$

3. False; since $y = 0$ is a solution.

4. True; since $f(x, y) = (y - 1)^3$ and $f_y(x, y) = 3(y - 1)^2$ are continuous everywhere in the plane.

5. First-order; ordinary; nonlinear because of y^2.

6. Third-order; ordinary; nonlinear because of $\sin xy$.

7. Second-order; partial.

8. Second-order; ordinary; linear.

9. From $y = x + \tan x$ we obtain $y' = 1 + \sec^2 x$, and $y'' = 2\sec^2 x \tan x$. Using $1 + \tan^2 x = \sec^2 x$ we have $y' + 2xy = 2 + x^2 + y^2$.

10. From $y = c_1 \cos(\ln x) + c_2 \sin(\ln x)$ we obtain $y' = \dfrac{1}{x}\left[c_2 \cos(\ln x) - c_1 \sin(\ln x)\right]$ and

$$y'' = \frac{-1}{x^2}\left[c_1 \cos(\ln x) + c_2 \sin(\ln x) + c_2 \cos(\ln x) - c_1 \sin(\ln x)\right]$$

so that $x^2 y'' + xy' + y = 0$.

11. From $y = c_1 e^x + c_2 e^{-x} + c_3 e^{2x} + 3$ we obtain $y' = c_1 e^x - c_2 e^{-x} + 2c_3 e^{2x}$, $y'' = c_1 e^x + c_2 e^{-x} + 4c_3 e^{2x}$, and $y''' = c_1 e^x - c_2 e^{-x} + 8c_3 e^{2x}$ so that $y''' - 2y'' - y' + 2y = 6$.

12. From $y = \sin 2x + \cosh 2x$ we obtain $y^{(4)} = 16 \sin 2x + 16 \cosh 2x$ so that $y^{(4)} - 16y = 0$.

13. $y = x^2$

14. $y = e^{5x}$

15. $y = \frac{1}{2}x^2$

16. $y = 2$

17. $y = e^x$, $y = 0$

18. $y = \sqrt{x}$

19. $y = \sin x$, $y = \cos x$, $y = 0$

20. $y = e^x$

21. For all values of y, $y^2 - 2y \geq -1$. Avoiding left– and right–hand derivatives, we then must have $x^2 - x - 1 > -1$. That is, $x < 0$ or $x > 1$.

22. If $|x| < 2$ and $|y| > 2$, then $(dy/dx)^2 < 0$ and the differential equation has no real solutions. This is also true for $|x| > 2$ and $|y| < 2$.

23. The differential equation is $\dfrac{dh}{dt} = -\dfrac{A_0}{A_w}\sqrt{2gh}$. We have $A_0 = \dfrac{1}{4}$. To find A_w we note that the radius

r corresponding to A_w satisfies $\dfrac{r}{h} = \dfrac{8}{20}$. Thus $r = \dfrac{2h}{5}$ and $A_w = \dfrac{4\pi h^2}{25}$. Then

$$\frac{dh}{dt} = -\frac{1/4}{4\pi h^2/25}\sqrt{2gh} = -\frac{25\sqrt{2gh}}{16\pi h^{3/2}}.$$

24. From Newton's second law we obtain $m\dfrac{dv}{dt} = \dfrac{1}{2}mg - \mu\dfrac{\sqrt{3}}{2}mg$ or $\dfrac{dv}{dt} = 16\left(1 - \sqrt{3}\,\mu\right)$.

25. (a) From $g = k/R^2$ we find $k = gR^2$.

(b) Using $a = \dfrac{d^2r}{dt^2}$ and part (a) we obtain $\dfrac{d^2r}{dt^2} = a = \dfrac{k}{r^2} = \dfrac{gR^2}{r^2}$ or $\dfrac{d^2r}{dt^2} - \dfrac{gR^2}{r^2} = 0.$

(c) Part (b) becomes $\dfrac{dv}{dr}\dfrac{dr}{dt} - \dfrac{gR^2}{r^2} = 0$ or $v\dfrac{dv}{dr} - \dfrac{gR^2}{r^2} = 0.$

2 First-Order Differential Equations

_____ **Exercises 2.1** _____

In many of the following problems we will encounter an expression of the form $\ln|g(y)| = f(x) + c$. To solve for $g(y)$ we exponentiate both sides of the equation. This yields $|g(y)| = e^{f(x)+c} = e^c e^{f(x)}$ which implies $g(y) = \pm e^c e^{f(x)}$. Letting $c_1 = \pm e^c$ we obtain $g(y) = c_1 e^{f(x)}$.

1. From $dy = \sin 5x\, dx$ we obtain $y = -\dfrac{1}{5}\cos 5x + c$.

2. From $dy = (x+1)^2\, dx$ we obtain $y = \dfrac{1}{3}(x+1)^3 + c$.

3. From $dy = -e^{-3x}\, dx$ we obtain $y = \dfrac{1}{3}e^{-3x} + c$.

4. From $dy = \dfrac{1}{x^2}\, dx$ we obtain $y = \dfrac{-1}{x} + c$.

5. From $dy = \dfrac{x+6}{x+1}\, dx = \left(1 + \dfrac{5}{x+1}\right) dx$ we obtain $y = x + 5\ln|x+1| + c$.

6. From $dy = 2xe^{-x}dx$ we obtain $y = -2xe^{-x} - 2e^{-x} + c$.

7. From $\dfrac{1}{y}\, dy = \dfrac{4}{x}\, dx$ we obtain $\ln|y| = 4\ln|x| + c$ or $y = c_1 x^4$.

8. From $\dfrac{1}{y}\, dy = -2x\, dx$ we obtain $\ln|y| = -x^2 + c$ or $y = c_1 e^{-x^2}$.

9. From $\dfrac{1}{y^3}\, dy = \dfrac{1}{x^2}\, dx$ we obtain $y^{-2} = \dfrac{2}{x} + c$.

10. From $\dfrac{1}{y+1}\, dy = \dfrac{1}{x}\, dx$ we obtain $\ln|y+1| = \ln|x| + c$ or $y + 1 = c_1 x$.

11. From $y^2\, dy = \left(\dfrac{1}{x^2} + \dfrac{1}{x}\right) dx$ we obtain $\dfrac{1}{3}y^3 = \dfrac{-1}{x} + \ln|x| + c$ or $xy^3 = -3 + 3x\ln|x| + c_1 x$.

12. From $\left(\dfrac{1}{y} + 2y\right) dy = \sin x\, dx$ we obtain $\ln|y| + y^2 = -\cos x + c$.

13. From $e^{-2y}dy = e^{3x}dx$ we obtain $3e^{-2y} + 2e^{3x} = c$.

14. From $ye^y dy = \left(e^{-x} + e^{-3x}\right) dx$ we obtain $ye^y - e^y + e^{-x} + \dfrac{1}{3}e^{-3x} = c$.

15. From $\dfrac{y}{2+y^2}\, dy = \dfrac{x}{4+x^2}\, dx$ we obtain $\ln|2+y^2| = \ln|4+x^2| + c$ or $2 + y^2 = c_1\left(4 + x^2\right)$.

16. From $\left(\dfrac{1}{y^2} + 1\right) dy = \dfrac{1}{1+x^2}\, dx$ we obtain $\dfrac{-1}{y} + y = \tan^{-1} x + c$.

16

17. From $2y\,dy = \dfrac{x}{x+1}\,dx$ we obtain $y^2 = x - \ln|x+1| + c.$

18. From $\dfrac{y^2}{y+1}\,dy = \dfrac{1}{x^2}\,dx$ we obtain $\dfrac{1}{2}y^2 - y + \ln|y+1| = -\dfrac{1}{x} + c$ or $\dfrac{1}{2}y^2 - y + \ln|y+1| = -\dfrac{1}{x} + c_1.$

19. From $\left(y + 2 + \dfrac{1}{y}\right)dy = x^2 \ln x\,dx$ we obtain $\dfrac{y^2}{2} + 2y + \ln|y| = \dfrac{x^3}{3}\ln|x| - \dfrac{1}{9}x^3 + c.$

20. From $\dfrac{1}{(2y+3)^2}\,dy = \dfrac{1}{(4x+5)^2}\,dx$ we obtain $\dfrac{2}{2y+3} = \dfrac{1}{4x+5} + c.$

21. From $\dfrac{1}{S}\,dS = k\,dr$ we obtain $S = ce^{kr}.$

22. From $\dfrac{1}{Q-70}\,dQ = k\,dt$ we obtain $\ln|Q-70| = kt + c$ or $Q - 70 = c_1 e^{kt}.$

23. From $\dfrac{1}{P-P^2}\,dP = \left(\dfrac{1}{P} + \dfrac{1}{1-P}\right)dP = dt$ we obtain $\ln|P| - \ln|1-P| = t + c$ so that $\ln\dfrac{P}{1-P} = t + c$

or $\dfrac{P}{1-P} = c_1 e^t.$ Solving for P we have $P = \dfrac{c_1 e^t}{1 + c_1 e^t}.$

24. From $\dfrac{1}{N}\,dN = \left(te^{t+2} - 1\right)dt$ we obtain $\ln|N| = te^{t+2} - e^{t+2} - t + c.$

25. From $\dfrac{1}{\csc y}\,dy = -\dfrac{1}{\sec^2 x}\,dx$ or $\sin y\,dy = -\cos^2 x\,dx = -\dfrac{1}{2}(1 + \cos 2x)\,dx$ we obtain

$-\cos y = -\dfrac{1}{2}x - \dfrac{1}{4}\sin 2x + c$ or $4\cos y = 2x + \sin 2x + c_1.$

26. From $2y\,dy = -\dfrac{\sin 3x}{\cos^3 3x}\,dx = -\tan 3x \sec^2 3x\,dx$ we obtain $y^2 = -\dfrac{1}{6}\sec^2 3x + c.$

27. From $\dfrac{e^{2y} - y}{e^y}\,dy = -\dfrac{\sin 2x}{\cos x}\,dx = -\dfrac{2\sin x \cos x}{\cos x}\,dx$ or $\left(e^y - ye^{-y}\right)dy = -2\sin x\,dx$ we obtain

$e^y + ye^{-y} + e^{-y} = 2\cos x + c.$

28. From $\tan y\,dy = x\cos x\,dx$ we obtain $\ln|\sec y| = x\sin x + \cos x + c.$

29. From $\dfrac{e^y}{(e^y+1)^2}\,dy = \dfrac{-e^x}{(e^x+1)^3}\,dx$ we obtain $-(e^y+1)^{-1} - \dfrac{1}{2}(e^x+1)^{-2} + c.$

30. From $\dfrac{y}{(1+y^2)^{1/2}}\,dy = \dfrac{x}{(1+x^2)^{1/2}}\,dx$ we obtain $\left(1+y^2\right)^{1/2} = \left(1+x^2\right)^{1/2} + c.$

31. From $\dfrac{y}{(y+1)^2}\,dy = \dfrac{1}{1-x^2}\,dx$ or $\left(\dfrac{1}{y+1} - \dfrac{1}{(y+1)^2}\right)dy = \left(\dfrac{1/2}{1+x} + \dfrac{1/2}{1-x}\right)dx$ we obtain

$$\ln|y+1| + \dfrac{1}{y+1} = \dfrac{1}{2}\ln|1+x| - \dfrac{1}{2}\ln|1-x| + c.$$

32. From $2y\,dy = (2x+1)\,dx$ we obtain $y^2 = x^2 + x + c.$

33. From $\dfrac{y-2}{y+3}\,dy = \dfrac{x-1}{x+4}\,dx$ or $\left(1 - \dfrac{5}{y-3}\right)dy = \left(1 - \dfrac{5}{x+4}\right)dx$ we obtain

$$y - 5\ln|y-3| = x - 5\ln|x+4| + c \quad\text{or}\quad \left(\dfrac{x+4}{y-3}\right)^5 = c_1 e^{x-y}.$$

34. From $\dfrac{y+1}{y-1}\,dy = \dfrac{x+2}{x-3}\,dx$ or $\left(1 + \dfrac{2}{y-1}\right)dy = \left(1 + \dfrac{5}{x-3}\right)dx$ we obtain

$$y + 2\ln|y-1| = x + 5\ln|x-3| + c \quad\text{or}\quad \dfrac{(y-1)^2}{(x-3)^5} = c_1 e^{x-y}.$$

35. From $\dfrac{1}{(2\cos^2 y - 1) - \cos^2 y}\,dy = \sin x\,dx$ or $\dfrac{1}{\cos^2 y - 1}\,dy = -\csc^2 y\,dy = \sin x\,dx$ we obtain

$\cot y = -\cos x + c.$

36. From $\sec y\dfrac{dy}{dx} + \sin x\cos y - \cos x\sin y = \sin x\cos y + \cos x\sin y$ we find $\sec y\,dy = 2\sin y\cos x\,dx$ or

$\dfrac{1}{2\sin y\cos y}\,dy = \csc 2y\,dy = \cos x\,dx.$ Then $\dfrac{1}{2}\ln|\csc 2y - \cot 2y| = \sin x + c.$

37. From $x\,dx = \dfrac{1}{\sqrt{1-y^2}}\,dy$ we obtain $\dfrac{1}{2}x^2 = \sin^{-1} y + c$ or $y = \sin\left(\dfrac{x^2}{2} + c_1\right).$

38. From $\dfrac{y}{\sqrt{4+y^2}}\,dy = \dfrac{1}{\sqrt{4-x^2}}\,dx$ we obtain $\sqrt{4+y^2} = \sin^{-1}\dfrac{x}{2} + c.$

39. From $\dfrac{1}{y^2}\,dy = \dfrac{1}{e^x + e^{-x}}\,dx = \dfrac{e^x}{(e^x)^2 + 1}\,dx$ we obtain $-\dfrac{1}{y} = \tan^{-1} e^x + c.$

40. To integrate $dx/(x + \sqrt{x})$ make the substitution $u^2 = x.$ Then $2u\,du = dx$ and

$$\int \dfrac{dx}{x + \sqrt{x}} = \int \dfrac{2u\,du}{u^2 + u} = \int \dfrac{2\,du}{u+1} = 2\ln|u+1| + c = 2\ln\left(\sqrt{x} + 1\right) + c.$$

Thus, from $\dfrac{1}{y + \sqrt{y}}\,dy = \dfrac{1}{x + \sqrt{x}}\,dx$ we obtain $2\ln\left(\sqrt{y} + 1\right) = 2\ln\left(\sqrt{x} + 1\right) + c$ or

$\sqrt{y} + 1 = c_1\left(\sqrt{x} + 1\right).$

41. From $\dfrac{\sin x}{1 + \cos x}\,dx = \dfrac{1}{e^{-y} + 1}\,dy = \dfrac{e^y}{1 + e^y}\,dy$ we obtain $-\ln(1 + \cos x) = \ln(1 + e^y) + c$ or

$(1 + e^y)(1 + \cos x) = c_1.$ Using $y(0) = 0$ we find $c_1 = 4.$ The solution of the initial-value problem

is $(1 + e^y)(1 + \cos x) = 4.$

42. From $\dfrac{1}{1 + (2y)^2}\,dy = \dfrac{-x}{1 + (x^2)^2}\,dx$ we obtain

$$\dfrac{1}{2}\tan^{-1} 2y = -\dfrac{1}{2}\tan^{-1} x^2 + c \quad\text{or}\quad \tan^{-1} 2y + \tan^{-1} x^2 = c_1.$$

Using $y(1) = 0$ we find $c_1 = \pi/4$. The solution of the initial-value problem is

$$\tan^{-1} 2y + \tan^{-1} x^2 = \frac{\pi}{4}.$$

43. From $\dfrac{y}{\sqrt{y^2 + 1}} \, dy = 4x \, dx$ we obtain $\sqrt{y^2 + 1} = 2x^2 + c$. Using $y(0) = 1$ we find $c = \sqrt{2}$. The

solution of the initial-value problem is $\sqrt{y^2 + 1} = 2x^2 + \sqrt{2}$.

44. From $\dfrac{1}{y} \, dy = (1 - t) \, dt$ we obtain $\ln|y| = t - \frac{1}{2}t^2 + c$ or $y = c_1 e^{t - t^2/2}$. Using $y(1) = 3$ we find

$c = 3e^{-1/2}$. The solution of the initial-value problem is $y = 3e^{t - t^2/2 - 1/2} = 3e^{-(t-1)^2/2}$.

45. From $\dfrac{1}{x^2 + 1} \, dx = 4 \, dy$ we obtain $\tan^{-1} x = 4y + c$. Using $x(\pi/4) = 1$ we find $c = -3\pi/4$. The

solution of the initial-value problem is $\tan^{-1} x = 4y - \dfrac{3\pi}{4}$ or $x = \tan\left(4y - \dfrac{3\pi}{4}\right)$.

46. From $\dfrac{1}{y^2 - 1} \, dy = \dfrac{1}{x^2 - 1} \, dx$ or $\dfrac{1}{2}\left(\dfrac{1}{y - 1} - \dfrac{1}{y + 1}\right) dy = \dfrac{1}{2}\left(\dfrac{1}{x - 1} - \dfrac{1}{x + 1}\right) dx$ we obtain

$\ln|y - 1| - \ln|y + 1| = \ln|x - 1| - \ln|x + 1| + c$ or $\dfrac{y - 1}{y + 1} = \dfrac{x - 1}{x + 1} + c$. Using $y(2) = 2$ we find $c = 0$.

The solution of the initial-value problem is $\dfrac{y - 1}{y + 1} = \dfrac{x - 1}{x + 1}$ or $y = x$.

47. From $\dfrac{1}{y} \, dy = \dfrac{1 - x}{x^2} \, dx = \left(\dfrac{1}{x^2} - \dfrac{1}{x}\right) dx$ we obtain $\ln|y| = -\dfrac{1}{x} - \ln|x| = c$ or $xy = c_1 e^{-1/x}$. Using

$y(-1) = -1$ we find $c_1 = e^{-1}$. The solution of the initial-value problem is $xy = e^{-1 - 1/x}$.

48. From $\dfrac{1}{1 - 2y} \, dy = dx$ we obtain $-\dfrac{1}{2}\ln|1 - 2y| = x + c$ or $1 - 2y = c_1 e^{-2x}$. Using $y(0) = 5/2$ we

find $c_1 = -4$. The solution of the initial-value problem is $1 - 2y = -4e^{-2x}$ or $y = 2e^{-2x} + \dfrac{1}{2}$.

49. From $\left(\dfrac{-1/6}{y + 3} + \dfrac{1/6}{y - 3}\right) dy = dx$ we obtain $\dfrac{y - 3}{y + 3} = ce^{6x}$.

    (a) If $y(0) = 0$ then $y = 3 \dfrac{1 - e^{6x}}{1 + e^{6x}}$.

    (b) If $y(0) = 3$ then $y = 3$.

    (c) If $y(1/3) = 1$ then $y = 3 \dfrac{2 - e^{6x - 2}}{2 + e^{6x - 2}}$.

50. From $\left(\dfrac{1}{y - 1} + \dfrac{-1}{y}\right) dy = \dfrac{1}{x} \, dx$ we obtain $\ln|y - 1| - \ln|y| = \ln|x| + c$ or $y = \dfrac{1}{1 - c_1 x}$. Another

solution is $y = 0$.

    (a) If $y(0) = 1$ then $y = 1$.

    (b) If $y(0) = 0$ then $y = 0$.

(c) If $y(1/2) = 1/2$ then $y = \dfrac{1}{1 + 2x}$.

51. By inspection a singular solution is $y = 1$.

52. By inspection a singular solution is $y = 0$.

53. The singular solution $y = 1$ satisfies the initial-value problem.

54. Separating variables we obtain $\dfrac{dy}{(y-1)^2} = dx$. Then $-\dfrac{1}{y-1} = x + c$ and $y = \dfrac{x+c-1}{x+c}$. Setting $x = 0$ and $y = 1.01$ we obtain $c = -100$. The solution is $y = \dfrac{x-101}{x-100}$.

55. Separating variables we obtain $\dfrac{dy}{(y-1)^2 + 0.01} = dx$. Then $10 \tan^{-1} 10(y-1) = x + c$ and $y = 1 + \dfrac{1}{10} \tan \dfrac{x+c}{10}$. Setting $x = 0$ and $y = 1$ we obtain $c = 0$. The solution is $y = 1 + \dfrac{1}{10} \tan \dfrac{x}{10}$.

56. Separating variables we obtain $\dfrac{dy}{(y-1)^2 - 0.01} = dx$. Then $5 \ln \left| \dfrac{10y-11}{10y-9} \right| = x + c$. Setting $x = 0$ and $y = 1$ we obtain $c = 5 \ln 1 = 0$. The solution is $5 \ln \left| \dfrac{10y-11}{10y-9} \right| = x$.

57. (a) From Example 2 the solution is $x^2 + y^2 = 25$. Solving for y we get $y = \sqrt{25 - x^2}$ where the positive square root is used because the graph of the solution passes through $(4, 3)$. This solution is valid on the interval $(-5, 5)$. It is not valid at the endpoints because it is not differentiable there.

(b) Since $dy/dx = -x/y$ is not defined when $y = 0$, the initial-value problem has no solution.

58. When $y = r$, $dy/dx = 0$ and $g(x)h(y) = g(x)h(r) = 0$, so $y = r$ is a solution of the differential equation.

Exercises 2.2

1. Let $M = 2x - 1$ and $N = 3y + 7$ so that $M_y = 0 = N_x$. From $f_x = 2x - 1$ we obtain $f = x^2 - x + h(y)$, $h'(y) = 3y + 7$, and $h(y) = \dfrac{3}{2}y^2 + 7y$. The solution is $x^2 - x + \dfrac{3}{2}y^2 + 7y = c$.

2. Let $M = 2x + y$ and $N = -x - 6y$. Then $M_y = 1$ and $N_x = -1$, so the equation is not exact.

3. Let $M = 5x + 4y$ and $N = 4x - 8y^3$ so that $M_y = 4 = N_x$. From $f_x = 5x + 4y$ we obtain $f = \dfrac{5}{2}x^2 + 4xy + h(y)$, $h'(y) = -8y^3$, and $h(y) = -2y^4$. The solution is $\dfrac{5}{2}x^2 + 4xy - 2y^4 = c$.

4. Let $M = \sin y - y \sin x$ and $N = \cos x + x \cos y - y$ so that $M_y = \cos y - \sin x = N_x$. From $f_x = \sin y - y \sin x$ we obtain $f = x \sin y + y \cos x + h(y)$, $h'(y) = -y$, and $h(y) = \frac{1}{2}y^2$. The solution is $x \sin y + y \cos x - \frac{1}{2}y^2 = c$.

5. Let $M = 2y^2 x - 3$ and $N = 2yx^2 + 4$ so that $M_y = 4xy = N_x$. From $f_x = 2y^2 x - 3$ we obtain $f = x^2 y^2 - 3x + h(y)$, $h'(y) = 4$, and $h(y) = 4y$. The solution is $x^2 y^2 - 3x + 4y = c$.

6. Let $M = 4x^3 - 3y \sin 3x - y/x^2$ and $N = 2y - 1/x + \cos 3x$ so that $M_y = -3 \sin 3x - 1/x^2$ and $N_x = 1/x^2 - 3 \sin 3x$. The equation is not exact.

7. Let $M = x^2 - y^2$ and $N = x^2 - 2xy$ so that $M_y = -2y$ and $N_x = 2x - 2y$. The equation is not exact.

8. Let $M = 1 + \ln x + y/x$ and $N = -1 + \ln x$ so that $M_y = 1/x = N_x$. From $f_y = -1 + \ln x$ we obtain $f = -y + y \ln x + h(y)$, $h'(x) = 1 + \ln x$, and $h(y) = x \ln x$. The solution is $-y + y \ln x + x \ln x = c$.

9. Let $M = y^3 - y^2 \sin x - x$ and $N = 3xy^2 + 2y \cos x$ so that $M_y = 3y^2 - 2y \sin x = N_x$. From $f_x = y^3 - y^2 \sin x - x$ we obtain $f = xy^3 + y^2 \cos x - \frac{1}{2}x^2 + h(y)$, $h'(y) = 0$, and $h(y) = 0$. The solution is $xy^3 + y^2 \cos x - \frac{1}{2}x^2 = c$.

10. Let $M = x^3 + y^3$ and $N = 3xy^2$ so that $M_y = 3y^2 = N_x$. From $f_x = x^3 + y^3$ we obtain $f = \frac{-1}{4}x^4 + xy^3 + h(y)$, $h'(y) = 0$, and $h(y) = 0$. The solution is $\frac{1}{4}x^4 + xy^3 = c$.

11. Let $M = y \ln y - e^{-xy}$ and $N = 1/y + x \ln y$ so that $M_y = 1 + \ln y + ye^{-xy}$ and $N_x = \ln y$. The equation is not exact.

12. Let $M = 2x/y$ and $N = -x^2/y^2$ so that $M_y = -2x/y^2 = N_x$. From $f_x = 2x/y$ we obtain $f = \frac{x^2}{y} + h(y)$, $h'(y) = 0$, and $h(y) = 0$. The solution is $x^2 = cy$.

13. Let $M = y - 6x^2 - 2xe^x$ and $N = x$ so that $M_y = 1 = N_x$. From $f_x = y - 6x^2 - 2xe^x$ we obtain $f = xy - 2x^3 - 2xe^x + 2e^x + h(y)$, $h'(y) = 0$, and $h(y) = 0$. The solution is $xy - 2x^3 - 2xe^x + 2e^x = c$.

14. Let $M = 3x^2 y + e^y$ and $N = x^3 + xe^y - 2y$ so that $M_y = 3x^2 + e^y = N_x$. From $f_x = 3x^2 y + e^y$ we obtain $f = x^3 y + xe^y + h(y)$, $h'(y) = -2y$, and $h(y) = -y^2$. The solution is $x^3 y + xe^y - y^2 = c$.

15. Let $M = 1 - 3/x + y$ and $N = 1 - 3/y + x$ so that $M_y = 1 = N_x$. From $f_x = 1 - 3/x + y$ we obtain $f = x - 3 \ln|x| + xy + h(y)$, $h'(y) = 1 - \frac{3}{y}$, and $h(y) = y - 3 \ln|y|$. The solution is $x + y + xy - 3 \ln|xy| = c$.

16. Let $M = xy^2 \sinh x + y^2 \cosh x$ and $N = e^y + 2xy \cosh x$ so that $M_y = 2xy \sinh x + 2y \cosh x = N_x$. From $f_y = e^y + 2xy \cosh x$ we obtain $f = e^y + xy^2 \cosh x + h(y)$, $h'(y) = 0$, and $h(y) = 0$. The solution is $e^y + xy^2 \cosh x = c$.

21

17. Let $M = x^2 y^3 - 1/(1 + 9x^2)$ and $N = x^3 y^2$ so that $M_y = 3x^2 y^2 = N_x$. From

$f_x = x^2 y^3 - 1/(1 + 9x^2)$ we obtain $f = \frac{1}{3} x^3 y^3 - \frac{1}{3} \arctan(3x) + h(y)$, $h'(y) = 0$, and $h(y) = 0$. The solution is $x^3 y^3 - \arctan(3x) = c$.

18. Let $M = -2y$ and $N = 5y - 2x$ so that $M_y = -2 = N_x$. From $f_x = -2y$ we obtain $f = -2xy + h(y)$, $h'(y) = 5y$, and $h(y) = \frac{5}{2} y^2$. The solution is $-2xy + \frac{5}{2} y^2 = c$.

19. Let $M = \tan x - \sin x \sin y$ and $N = \cos x \cos y$ so that $M_y = -\sin x \cos y = N_x$. From $f_x = \tan x - \sin x \sin y$ we obtain $f = \ln|\sec x| + \cos x \sin y + h(y)$, $h'(y) = 0$, and $h(y) = 0$. The solution is $\ln|\sec x| + \cos x \sin y = c$.

20. Let $M = 3x \cos 3x + \sin 3x - 3$ and $N = 2y + 5$ so that $M_y = 0 = N_x$. From $f_x = 3x \cos 3x + \sin 3x - 3$ we obtain $f = x \sin 3x - 3x + h(y)$, $h'(y) = 2y + 5$, and $h(y) = y^2 + 5y$. The solution is $x \sin 3x - 3x + y^2 + 5y = c$.

21. Let $M = 4x^3 + 4xy$ and $N = 2x^2 + 2y - 1$ so that $M_y = 4x = N_x$. From $f_x = 4x^3 + 4xy$ we obtain $f = x^4 + 2x^2 y + h(y)$, $h'(y) = 2y - 1$, and $h(y) = y^2 - y$. The solution is $x^4 + 2x^2 y + y^2 - y = c$.

22. Let $M = 2y \sin x \cos x - y + 2y^2 e^{xy^2}$ and $N = -x + \sin^2 x + 4xye^{xy^2}$ so that

$$M_y = 2 \sin x \cos x - 1 + 4xy^3 e^{xy^2} + 4ye^{xy^2} = N_x.$$

From $f_x = 2y \sin x \cos x - y + 2y^2 e^{xy^2}$ we obtain $f = y \sin^2 x - xy + 2e^{xy^2} + h(y)$, $h'(y) = 0$, and $h(y) = 0$. The solution is $y \sin^2 x - xy + 2e^{xy^2} = c$.

23. Let $M = 4x^3 y - 15x^2 - y$ and $N = x^4 + 3y^2 - x$ so that $M_y = 4x^3 - 1 = N_x$. From $f_x = 4x^3 y - 15x^2 - y$ we obtain $f = x^4 y - 5x^3 - xy + h(y)$, $h'(y) = 3y^2$, and $h(y) = y^3$. The solution is $x^4 y - 5x^3 - xy + y^3 = c$.

24. Let $M = 1/x + 1/x^2 - y/(x^2 + y^2)$ and $N = ye^y + x/(x^2 + y^2)$ so that

$M_y = (y^2 - x^2)/(x^2 + y^2)^2 = N_x$. From $f_x = 1/x + 1/x^2 - y/(x^2 + y^2)$ we obtain

$f = \ln|x| - \frac{1}{x} - \arctan\left(\frac{x}{y}\right) + h(y)$, $h'(y) = ye^y$, and $h(y) = ye^y - e^y$. The solution is

$$\ln|x| - \frac{1}{x} - \arctan\left(\frac{x}{y}\right) + ye^y - e^y = c.$$

25. Let $M = x^2 + 2xy + y^2$ and $N = 2xy + x^2 - 1$ so that $M_y = 2(x + y) = N_x$. From $f_x = x^2 + 2xy + y^2$ we obtain $f = \frac{1}{3} x^3 + x^2 y + xy^2 + h(y)$, $h'(y) = -1$, and $h(y) = -y$. The general solution is $\frac{1}{3} x^3 + x^2 y + xy^2 - y = c$. If $y(1) = 1$ then $c = 4/3$ and the solution of the initial-value problem is $\frac{1}{3} x^3 + x^2 y + xy^2 - y = \frac{4}{3}$.

26. Let $M = e^x + y$ and $N = 2 + x + ye^y$ so that $M_y = 1 = N_x$. From $f_x = e^x + y$ we obtain $f = e^x + xy + h(y)$, $h'(y) = 2 + ye^y$, and $h(y) = 2y + ye^y - y$. The general solution is $e^x + xy + 2y + ye^y - e^y = c$. If $y(0) = 1$ then $c = 3$ and the solution of the initial-value problem is $e^x + xy + 2y + ye^y - e^y = 3$.

27. Let $M = 4y + 2x - 5$ and $N = 6y + 4x - 1$ so that $M_y = 4 = N_x$. From $f_x = 4y + 2x - 5$ we obtain $f = 4xy + x^2 - 5x + h(y)$, $h'(y) = 6y - 1$, and $h(y) = 3y^2 - y$. The general solution is $4xy + x^2 - 5x + 3y^2 - y = c$. If $y(-1) = 2$ then $c = 8$ and the solution of the initial-value problem is $4xy + x^2 - 5x + 3y^2 - y = 8$.

28. Let $M = x/2y^4$ and $N = \left(3y^2 - x^2\right)/y^5$ so that $M_y = -2x/y^5 = N_x$. From $f_x = x/2y^4$ we obtain $f = \dfrac{x^2}{4y^4} + h(y)$, $h'(y) = \dfrac{3}{y^3}$, and $h(y) = -\dfrac{3}{2y^2}$. The general solution is $\dfrac{x^2}{4y^4} - \dfrac{3}{2y^2} = c$. If $y(1) = 1$ then $c = -5/4$ and the solution of the initial-value problem is $\dfrac{x^2}{4y^4} - \dfrac{3}{2y^2} = -\dfrac{5}{4}$.

29. Let $M = y^2 \cos x - 3x^2 y - 2x$ and $N = 2y \sin x - x^3 + \ln y$ so that $M_y = 2y \cos x - 3x^2 = N_x$. From $f_x = y^2 \cos x - 3x^2 y - 2x$ we obtain $f = y^2 \sin x - x^3 y - x^2 + h(y)$, $h'(y) = \ln y$, and $h(y) = y \ln y - y$. The general solution is $y^2 \sin x - x^3 y - x^2 + y \ln y - y = c$. If $y(0) = e$ then $c = 0$ and the solution of the initial-value problem is $y^2 \sin x - x^3 y - x^2 + y \ln y - y = 0$.

30. Let $M = y^2 + y \sin x$ and $N = 2xy - \cos x - 1/\left(1 + y^2\right)$ so that $M_y = 2y + \sin x = N_x$. From $f_x = y^2 + y \sin x$ we obtain $f = xy^2 - y \cos x + h(y)$, $h'(y) = \dfrac{-1}{1 + y^2}$, and $h(y) = -\tan^{-1} y$. The general solution is $xy^2 - y \cos x - \tan^{-1} y = c$. If $y(0) = 1$ then $c = -1 - \pi/4$ and the solution of the initial-value problem is $xy^2 - y \cos x - \tan^{-1} y = -1 - \dfrac{\pi}{4}$.

31. Equating $M_y = 3y^2 + 4kxy^3$ and $N_x = 3y^2 + 40xy^3$ we obtain $k = 10$.

32. Equating $M_y = -xy \cos xy - \sin xy + 4ky^3$ and $N_x = -20y^3 - xy \cos xy - \sin xy$ we obtain $k = -5$.

33. Equating $M_y = 4xy + e^x$ and $N_x = 4xy + ke^x$ we obtain $k = 1$.

34. Equating $M_y = 18xy^2 - \sin y$ and $N_x = 2kxy^2 - \sin y$ we obtain $k = 9$.

35. Since $f_y = N(x, y) = xe^{xy} + 2xy + 1/x$ we obtain $f = e^{xy} + xy^2 + \dfrac{y}{x} + h(x)$ so that $f_x = ye^{xy} + y^2 - \dfrac{y}{x^2} + h'(x)$. Let $M(x, y) = yc^{xy} + y^2 - \dfrac{y}{x^2}$.

36. Since $f_x = M(x, y) = y^{1/2} x^{-1/2} + x\left(x^2 + y\right)^{-1}$ we obtain $f = 2y^{1/2} x^{1/2} + \dfrac{1}{2} \ln\left|x^2 + y\right| + h(x)$ so that $f_y = y^{-1/2} x^{1/2} + \dfrac{1}{2}\left(x^2 + y\right)^{-1} + h'(x)$. Let $N(x, y) = y^{-1/2} x^{1/2} + \dfrac{1}{2}\left(x^2 + y\right)^{-1}$.

37. Let $M = 6xy^3$ and $N = 4y^3 + 9x^2 y^2$ so that $M_y = 18xy^2 = N_x$. From $f_x = 6xy^3$ we obtain $f = 3x^2 y^3 + h(y)$, $h'(y) = 4y^3$, and $h(y) = y^4$. The solution of the differential equation is

$3x^2y^3 + y^4 = c$.

38. Let $M = -y/x^2$ and $N = 1/y + 1/x$ so that $M_y = -1/x^2 = N_x$. From $f_x = -y/x^2$ we obtain $f = \dfrac{y}{x} + h(y)$, $h'(y) = \dfrac{1}{y}$, and $h(y) = \ln y$. The solution of the differential equation is $\dfrac{y}{x} + \ln|y| = c$.

39. Let $M = -x^2y^2 \sin x + 2xy^2 \cos x$ and $N = 2x^2y \cos x$ so that $M_y = -2x^2y \sin x + 4xy \cos x = N_x$. From $f_y = 2x^2y \cos x$ we obtain $f = x^2y^2 \cos x + h(y)$, $h'(y) = 0$, and $h(y) = 0$. The solution of the differential equation is $x^2y^2 \cos x = c$.

40. Let $M = xye^x + y^2e^x + ye^x$ and $N = xe^x + 2ye^x$ so that $M_y = xe^x + 2ye^x + e^x = N_x$. From $f_y = xe^x + 2ye^x$ we obtain $f = xye^x + y^2e^x + h(x)$, $h'(y) = 0$, and $h(y) = 0$. The solution of the differential equation is $xye^x + y^2e^x = c$.

41. Let $M = 2xy^2 + 3x^2$ and $N = 2x^2y$ so that $M_y = 4xy = N_x$. From $f_x = 2xy^2 + 3x^2$ we obtain $f = x^2y^2 + x^3 + h(y)$, $h'(y) = 0$, and $h(y) = 0$. The solution of the differential equation is $x^2y^2 + x^3 = c$.

42. Let $M = \left(x^2 + 2xy - y^2\right) / \left(x^2 + 2xy + y^2\right)$ and $N = \left(y^2 + 2xy - x^2\right) / \left(y^2 + 2xy + x^2\right)$ so that $M_y = -4xy/(x+y)^3 = N_x$. From $f_x = \left(x^2 + 2xy + y^2 - 2y^2\right)/(x+y)^2$ we obtain

$f = x + \dfrac{2y^2}{x+y} + h(y)$, $h'(y) = -1$, and $h(y) = -y$. The solution of the differential equation is $x^2 + y^2 = c(x+y)$.

43. To see that the equations are not equivalent consider $dx = (x/y)dy = 0$. An integrating factor is $\mu(x,y) = y$ resulting in $y\,dx + x\,dy = 0$. A solution of the latter equation is $y = 0$, but this is not a solution of the original equation.

Exercises 2.3

1. For $y' - 5y = 0$ an integrating factor is $e^{-\int 5dx} = e^{-5x}$ so that $\dfrac{d}{dx}\left[e^{-5x}y\right] = 0$ and $y = ce^{5x}$ for $-\infty < x < \infty$.

2. For $y' + 2y = 0$ an integrating factor is $e^{\int 2dx} = e^{2x}$ so that $\dfrac{d}{dx}\left[e^{2x}y\right] = 0$ and $y = ce^{-2x}$ for $-\infty < x < \infty$.

3. For $y' + 4y = \dfrac{4}{3}$ an integrating factor is $e^{\int 4dx} = e^{4x}$ so that $\dfrac{d}{dx}\left[e^{4x}y\right] = \dfrac{4}{3}e^{4x}$ and $y = \dfrac{1}{3} + ce^{-4x}$ for $-\infty < x < \infty$.

4. For $y' + \dfrac{2}{x}y = 3/x$ an integrating factor is $e^{\int (2/x)dx} = x^2$ so that $\dfrac{d}{dx}\left[x^2y\right] = 3x$ and $y = \dfrac{3}{2} + cx^{-2}$ for $0 < x < \infty$.

5. For $y' + y = e^{3x}$ an integrating factor is $e^{\int dx} = e^x$ so that $\dfrac{d}{dx}\left[e^x y\right] = e^{4x}$ and $y = \dfrac{1}{4}e^{3x} + ce^{-x}$ for $-\infty < x < \infty$.

6. For $y' - y = e^x$ an integrating factor is $e^{-\int dx} = e^{-x}$ so that $\dfrac{d}{dx}\left[e^{-x} y\right] = 1$ and $y = xe^x + ce^x$ for $-\infty < x < \infty$.

7. For $y' + 3x^2 y = x^2$ an integrating factor is $e^{\int 3x^2 dx} = e^{x^3}$ so that $\dfrac{d}{dx}\left[e^{x^3} y\right] = x^2 e^{x^3}$ and $y = \dfrac{1}{3} + ce^{-x^3}$ for $-\infty < x < \infty$.

8. For $y' + 2xy = x^3$ an integrating factor is $e^{\int 2x \, dx} = e^{x^2}$ so that $\dfrac{d}{dx}\left[e^{x^2} y\right] = x^3 e^{x^2}$ and
$y = \dfrac{1}{2}x^2 - \dfrac{1}{2} + ce^{-x^2}$ for $-\infty < x < \infty$.

9. For $y' + \dfrac{1}{x} y = \dfrac{1}{x^2}$ an integrating factor is $e^{\int (1/x)dx} = x$ so that $\dfrac{d}{dx}\left[xy\right] = \dfrac{1}{x}$ and $y = \dfrac{1}{x}\ln x + \dfrac{c}{x}$ for $0 < x < \infty$.

10. For $y' - 2y = x^2 + 5$ an integrating factor is $e^{-\int 2dx} = e^{-2x}$ so that $\dfrac{d}{dx}\left[e^{-2x} y\right] = x^2 e^{-2x} + 5e^{-2x}$
and $y = -\dfrac{1}{2}x^2 - \dfrac{1}{2}x - \dfrac{11}{4} + ce^{2x}$ for $-\infty < x < \infty$.

11. For $\dfrac{dx}{dy} + \dfrac{1}{2y} x = -2y$ an integrating factor is $e^{\int (1/2y)dy} = y^{1/2}$ so that $\dfrac{d}{dy}\left[y^{1/2} x\right] = -2y^{3/2}$ and
$x = -\dfrac{4}{5}y^2 + cy^{-1/2}$ for $0 < y < \infty$.

12. For $\dfrac{dx}{dy} - x = y$ an integrating factor is $e^{-\int dy} = e^{-y}$ so that $\dfrac{d}{dy}\left[e^{-y} x\right] = ye^{-y}$ and $x = -y - 1 + ce^y$
for $-\infty < y < \infty$.

13. For $y' + \dfrac{1}{x} y = \sin x$ an integrating factor is $e^{\int (1/x)dx} = x$ so that $\dfrac{d}{dx}\left[xy\right] = x \sin x$ and
$y = \dfrac{\sin x}{x} - \cos x + \dfrac{c}{x}$ for $0 < x < \infty$.

14. For $y' + \dfrac{x}{1 + x^2} y = -x$ an integrating factor is $e^{\int [x/(1+x^2)]dx} = \sqrt{1 + x^2}$ so that
$\dfrac{d}{dx}\left[\sqrt{1 + x^2} y\right] = -x\sqrt{1 + x^2}$ and $y = -\dfrac{1}{3}\left(1 + x^2\right) + c\left(1 + x^2\right)^{-1/2}$ for $-\infty < x < \infty$.

15. For $y' + \dfrac{e^x}{1 + e^x} y = 0$ an integrating factor is $e^{\int [e^x/(1+e^x)]dx} = 1 + e^x$ so that $\dfrac{d}{dx}\left[1 + e^x y\right] = 0$ and
$y = \dfrac{c}{1 + e^x}$ for $-\infty < x < \infty$.

16. For $y' + \dfrac{3x^2}{x^3 - 1} y = 0$ an integrating factor is $e^{\int [3x^2/(x^3-1)]dx} = x^3 - 1$ so that $\dfrac{d}{dx}\left[\left(x^3 - 1\right) y\right] = 0$
and $y = \dfrac{c}{x^3 - 1}$ for $1 < x < \infty$.

Exercises 2.3

17. For $y' + (\tan x)y = \sec x$ an integrating factor is $e^{\int \tan x\,dx} = \sec x$ so that $\dfrac{d}{dx}\left[(\sec x)\,y\right] = \sec^2 x$ and $y = \sin x + c\cos x$ for $-\pi/2 < x < \pi/2$.

18. For $y' + (\cot x)y = 2\cos x$ an integrating factor is $e^{\int \cot x\,dx} = \sin x$ so that $\dfrac{d}{dx}\left[(\sin x)\,y\right] = 2\sin x\cos x$ and $y = \sin x + c\csc x$ for $0 < x < \pi$.

19. For $y' + \dfrac{4}{x}y = x^2 - 1$ an integrating factor is $e^{\int(4/x)\,dx} = x^4$ so that $\dfrac{d}{dx}\left[x^4 y\right] = x^6 - x^4$ and

$y = \dfrac{1}{7}x^3 - \dfrac{1}{5}x + cx^{-4}$ for $0 < x < \infty$.

20. For $y' - \dfrac{x}{(1+x)}y = x$ an integrating factor is $e^{-\int[x/(1+x)]\,dx} = (x+1)e^{-x}$ so that

$\dfrac{d}{dx}\left[(x+1)e^{-x}y\right] = x(x+1)e^{-x}$ and $y = -x - \dfrac{2x+3}{x+1} + \dfrac{ce^x}{x+1}$ for $-1 < x < \infty$

21. For $y' + \left(1 + \dfrac{2}{x}\right)y = \dfrac{e^x}{x^2}$ an integrating factor is $e^{\int[1+(2/x)]\,dx} = x^2 e^x$ so that $\dfrac{d}{dx}\left[x^2 e^x y\right] = e^{2x}$ and

$y = \dfrac{1}{2}\dfrac{e^x}{x^2} + \dfrac{ce^{-x}}{x^2}$ for $0 < x < \infty$.

22. For $y' + \left(1 + \dfrac{1}{x}\right)y = \dfrac{1}{x}e^{-x}\sin 2x$ an integrating factor is $e^{\int[1+(1/x)]\,dx} = xe^x$ so that

$\dfrac{d}{dx}\left[xe^x y\right] = \sin 2x$ and $xe^x y = -\dfrac{1}{2}\cos 2x + c$ for $0 < x < \infty$.

23. For $y' + (\cot x)y = \sec^2 x\csc x$ an integrating factor is $e^{\int \cot x\,dx} = \sin x$ so that $\dfrac{d}{dx}\left[(\sin x)\,y\right] = \sec^2 x$ and $y = \sec x + c\csc x$ for $0 < x < \pi/2$.

24. For $y' + \dfrac{2\sin x}{(1-\cos x)}y = \tan x(1 - \cos x)$ an integrating factor is $e^{\int[2\sin x/(1-\cos x)]\,dx} = (1 - \cos x)^2$ so that $\dfrac{d}{dx}\left[(1 - \cos x)^2 y\right] = \tan x - \sin x$ and $y(1 - \cos x)^2 = \ln|\sec x| + \cos x + c$ for $0 < x < \pi/2$.

25. For $\dfrac{dx}{dy} + \left(1 + \dfrac{2}{y}\right)x = e^y$ an integrating factor is $e^{\int[1+(2/y)]\,dy} = y^2 e^y$ so that $\dfrac{d}{dy}\left[y^2 e^y x\right] = y^2 e^{2y}$

and $x = \dfrac{1}{2}e^y - \dfrac{1}{2}\dfrac{e^y}{y} + \dfrac{1}{4}\dfrac{e^y}{y^2} + \dfrac{ce^{-y}}{y^2}$ for $0 < y < \infty$.

26. For $y' - \dfrac{3}{x}y = \dfrac{x^4}{x+1}$ an integrating factor is $e^{-\int(3/x)\,dx} = x^{-3}$ so that $\dfrac{d}{dx}\left[x^{-3}y\right] = \dfrac{x}{x+1}$ and

$y = x^4 - x^3\ln|x+1| + cx^3$ for $-1 < x < \infty$.

27. For $y' + \left(3 + \dfrac{1}{x}\right)y = \dfrac{e^{-3x}}{x}$ an integrating factor is $e^{\int[3+(1/x)]\,dx} = xe^{3x}$ so that $\dfrac{d}{dx}\left[xe^{3x}y\right] = 1$ and

$y = e^{-3x} + \dfrac{ce^{-3x}}{x}$ for $0 < x < \infty$.

28. For $y' + \dfrac{x+2}{x+1}y = \dfrac{2xe^{-x}}{x+1}$ an integrating factor is $e^{\int [(x+2)/(x+1)]dx} = (x+1)e^x$ so that

$\dfrac{d}{dx}[(x+1)e^x y] = 2x$ and $(x+1)e^x y = x^2 + c$ for $-1 < x < \infty$.

29. For $\dfrac{dx}{dy} - \dfrac{4}{y}x = 4y^5$ an integrating factor is $e^{-\int(4/y)dy} = y^{-4}$ so that $\dfrac{d}{dy}\left[y^{-4}x\right] = 4y$ and

$x = 2y^6 + cy^4$ for $0 < y < \infty$.

30. For $y' + \dfrac{2}{x}y = \dfrac{1}{x}(e^x + \ln x)$ an integrating factor is $e^{\int (2/x)dx} = x^2$ so that $\dfrac{d}{dx}\left[x^2 y\right] = xe^x + x\ln x$

and $x^2 y = xe^x - e^x + \dfrac{x^2}{2}\ln x - \dfrac{1}{4}x^2 + c$ for $0 < x < \infty$.

31. For $y' + y = \dfrac{1 - e^{-2x}}{e^x + e^{-x}}$ an integrating factor is $e^{\int dx} = e^x$ so that $\dfrac{d}{dx}[e^x y] = \dfrac{e^x - e^{-x}}{e^x + e^{-x}}$ and

$y = e^{-x}\ln(e^x + e^{-x}) + ce^{-x}$ for $-\infty < x < \infty$.

32. For $y' - y = \sinh x$ an integrating factor is $e^{-\int dx} = e^{-x}$ so that $\dfrac{d}{dx}\left[e^{-x}y\right] = \dfrac{1}{2}\left(1 - e^{-2x}\right)$ and

$y = \dfrac{1}{2}xe^x + \dfrac{1}{4}e^{-x} + ce^x$ for $-\infty < x < \infty$.

33. For $\dfrac{dx}{dy} + \left(2y + \dfrac{1}{y}\right)x = 2$ an integrating factor is $e^{\int [2y+(1/y)]dy} = ye^{y^2}$ so that $\dfrac{d}{dy}\left[ye^{y^2}x\right] = 2ye^{y^2}$

and $x = \dfrac{1}{y} + \dfrac{1}{y}ce^{-y^2}$ for $0 < y < \infty$.

34. For $\dfrac{dx}{dy} + \dfrac{2}{y}x = e^y$ an integrating factor is $e^{\int (2/y)dy} = y^2$ so that $\dfrac{d}{dy}\left[y^2 x\right] = y^2 e^y$ and

$x = e^y - \dfrac{2}{y}e^y + \dfrac{2}{y^2} + \dfrac{c}{y^2}$ for $0 < y < \infty$.

35. For $\dfrac{dr}{d\theta} + r\sec\theta = \cos\theta$ an integrating factor is $e^{\int \sec\theta\, d\theta} = \sec\theta + \tan\theta$ so that

$\dfrac{d}{d\theta}[r(\sec\theta + \tan\theta)] = 1 + \sin\theta$ and $r(\sec\theta + \tan\theta) = \theta - \cos\theta + c$ for $-\pi/2 < \theta < \pi/2$.

36. For $\dfrac{dP}{dt} + (2t - 1)P = 4t - 2$ an integrating factor is $e^{\int (2t-1)\,dt} = e^{t^2 - t}$ so that

$\dfrac{d}{dt}\left[Pe^{t^2 - t}\right] = (4t - 2)e^{t^2 - t}$ and $P = 2 + ce^{t - t^2}$ for $-\infty < t < \infty$.

37. For $y' + \dfrac{4}{x+2}y = \dfrac{5}{(x+2)^2}$ an integrating factor is $e^{\int [4/(x+2)]dx} = (x+2)^4$ so that

$\dfrac{d}{dx}\left[(x+2)^4 y\right] = 5(x+2)^2$ and $y = \dfrac{5}{3}(x+2)^{-1} + c(x+2)^{-4}$ for $-2 < x < \infty$.

38. For $y' + \dfrac{2}{x^2 - 1} y = \dfrac{x+1}{x-1}$ an integrating factor is $e^{\int [2/(x^2-1)]dx} = \dfrac{x-1}{x+1}$ so that $\dfrac{d}{dx}\left[\dfrac{x-1}{x+1}y\right] = 1$ and $(x-1)y = x(x+1) + c(x+1)$ for $-1 < x < 1$.

39. For $y' + (\cosh x)y = 10 \cosh x$ an integrating factor is $e^{\int \cosh x\, dx} = e^{\sinh x}$ so that $\dfrac{d}{dx}\left[e^{\sinh x}y\right] = 10(\cosh x)e^{\sinh x}$ and $y = 10 + ce^{-\sinh x}$ for $-\infty < x < \infty$.

40. For $\dfrac{dx}{dy} + 2x = 3e^y$ an integrating factor is $e^{\int 2dy} = e^{2y}$ so that $\dfrac{d}{dy}\left[e^{2y}x\right] = 3e^{3y}$ and $x = e^y + ce^{-2y}$ for $-\infty < y < \infty$.

41. For $y' + 5y = 20$ an integrating factor is $e^{\int 5dx} = e^{5x}$ so that $\dfrac{d}{dx}\left[e^{5x}y\right] = 20e^{5x}$ and $y = 4 + ce^{-5x}$ for $-\infty < x < \infty$. If $y(0) = 2$ then $c = -2$ and $y = 4 - 2e^{-5x}$.

42. For $y' - 2y = x\left(e^{3x} - e^{2x}\right)$ an integrating factor is $e^{-\int 2dx} = e^{-2x}$ so that $\dfrac{d}{dx}\left[e^{-2x}y\right] = xe^x - x$ and $y = xe^{3x} - e^{3x} - \frac{1}{2}x^2e^{2x} + ce^{2x}$ for $-\infty < x < \infty$. If $y(0) = 2$ then $c = 3$ and $y = xe^{3x} - e^{3x} - \dfrac{1}{2}x^2e^{2x} + 3e^{2x}$.

43. For $\dfrac{di}{dt} + \dfrac{R}{L}i = \dfrac{E}{L}$ an integrating factor is $e^{\int (R/L)\,dt} = e^{Rt/L}$ so that $\dfrac{d}{dt}\left[ie^{Rt/L}\right] = \dfrac{E}{L}e^{Rt/L}$ and $i = \dfrac{E}{R} + ce^{-Rt/L}$ for $-\infty < t < \infty$. If $i(0) = i_0$ then $c = i_0 - E/R$ and $i = \dfrac{E}{R} + \left(i_0 - \dfrac{E}{R}\right)e^{-Rt/L}$.

44. For $\dfrac{dx}{dy} - \dfrac{1}{y}x = 2y$ an integrating factor is $e^{-\int (1/y)dy} = \dfrac{1}{y}$ so that $\dfrac{d}{dy}\left[\dfrac{1}{y}x\right] = 2$ and $x = 2y^2 + cy$ for $0 < y < \infty$. If $y(1) = 5$ then $c = -49/5$ and $x = 2y^2 - \dfrac{49}{5}y$.

45. For $y' + (\tan x)y = \cos^2 x$ an integrating factor is $e^{\int \tan x\, dx} = \sec x$ so that $\dfrac{d}{dx}[(\sec x)\, y] = \cos x$ and $y = \sin x \cos x + c \cos x$ for $-\pi/2 < x < \pi/2$. If $y(0) = -1$ then $c = -1$ and $y = \sin x \cos x - \cos x$.

46. For $\dfrac{dQ}{dx} - 5x^4 Q = 0$ an integrating factor is $e^{-\int 5x^4\, dx} = e^{-x^5}$ so that $\dfrac{d}{dx}\left[e^{-x^5}Q\right] = 0$ and $Q = ce^{x^5}$ for $-\infty < x < \infty$. If $Q(0) = -7$ then $c = -7$ and $Q = -7e^{x^5}$.

47. For $\dfrac{dT}{dt} - kT = -50k$ an integrating factor is $e^{\int (-k)\, dt} = e^{-kt}$ so that $\dfrac{d}{dt}\left[Te^{-kt}\right] = -50ke^{-kt}$ and $T = 50 + ce^{kt}$ for $-\infty < t < \infty$. If $T(0) = 150$ then $c = 150$ and $T = 50 + 150e^{kt}$.

48. For $y' + \left(1 + \dfrac{2}{x}\right)y = \dfrac{2}{x}e^{-x}$ an integrating factor is $e^{\int(1 + 2/x)dx} = x^2e^x$ so that $\dfrac{d}{dx}\left[x^2e^xy\right] = 2x$ and $y = e^{-x} + \dfrac{c}{x^2}e^{-x}$ for $0 < x < \infty$. If $y(1) = 0$ then $c = -1$ and $y = e^{-x} - \dfrac{1}{x^2}e^{-x}$.

49. For $y' + \dfrac{1}{x+1}y = \dfrac{\ln x}{x+1}$ an integrating factor is $e^{\int \frac{1}{x+1}dx} = x = 1$ so that $\dfrac{d}{dx}[(x+1)y] = \ln x$ and

$$y = \dfrac{x}{x+1}\ln x - \dfrac{x}{x+1} + \dfrac{c}{x+1} \quad \text{for } 0 < x < \infty. \text{ If } y(1) = 10 \text{ then } c = 21 \text{ and}$$

$$y = \dfrac{x}{x+1}\ln x - \dfrac{x}{x+1} + \dfrac{21}{x+1}.$$

50. For $y' + \dfrac{1}{x}y = \dfrac{1}{x}e^x$ an integrating factor is $e^{\int (1/x)dx} = x$ so that $\dfrac{d}{dx}[xy] = e^x$ and $y = \dfrac{1}{x}e^x + \dfrac{c}{x}$
for $0 < x < \infty$. If $y(1) = 2$ then $c = 2 - e$ and $y = \dfrac{1}{x}e^x + \dfrac{2-e}{x}$.

51. For $y' + 2y = f(x)$ an integrating factor is e^{2x} so that

$$ye^{2x} = \begin{cases} \frac{1}{2}e^{2x} + c_1, & 0 \le x \le 3; \\ c_2, & x > 3. \end{cases}$$

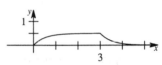

If $y(0) = 0$ then $c_1 = -1/2$ and for continuity we must have $c_2 = \frac{1}{2}e^6 - \frac{1}{2}$ so that

$$y = \begin{cases} \frac{1}{2}\left(1 - e^{-2x}\right), & 0 \le x \le 3; \\ \frac{1}{2}\left(e^6 - 1\right)e^{-2x}, & x > 3. \end{cases}$$

52. For $y' + y = f(x)$ an integrating factor is e^x so that

$$ye^x = \begin{cases} e^x + c_1, & 0 \le x \le 1; \\ -e^x + c_2, & x > 1. \end{cases}$$

If $y(0) = 1$ then $c_1 = 0$ and for continuity we must have $c_2 = 2e$
so that

$$y = \begin{cases} 1, & 0 \le x \le 1; \\ 2e^{1-x} - 1, & x > 1. \end{cases}$$

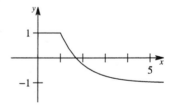

53. For $y' + 2xy = f(x)$ an integrating factor is e^{x^2} so that

$$ye^{x^2} = \begin{cases} \frac{1}{2}e^{x^2} + c_1, & 0 \le x < 1; \\ c_2, & x \ge 1. \end{cases}$$

If $y(0) = 2$ then $c_1 = 3/2$ and for continuity we must have
$c_2 = \frac{1}{2}e + \frac{3}{2}$ so that

$$y = \begin{cases} \frac{1}{2} + \frac{3}{2}e^{-x^2}, & 0 \le x < 1; \\ \left(\frac{1}{2}e + \frac{3}{2}\right)e^{-x^2}, & x \ge 1. \end{cases}$$

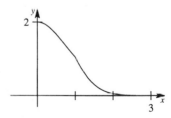

54. For

$$y' + \frac{2x}{1+x^2} y = \begin{cases} \dfrac{x}{1+x^2}, & 0 \le x < 1; \\[2mm] \dfrac{-x}{1+x^2}, & x \ge 1 \end{cases}$$

an integrating factor is $1 + x^2$ so that

$$\left(1 + x^2\right) y = \begin{cases} \frac{1}{2}x^2 + c_1, & 0 \le x < 1; \\[2mm] -\frac{1}{2}x^2 + c_2, & x \ge 1. \end{cases}$$

If $y(0) = 0$ then $c_1 = 0$ and for continuity we must have $c_2 = 1$ so that

$$y = \begin{cases} \frac{1}{2} - \dfrac{1}{2(1+x^2)}, & 0 \le x < 1; \\[2mm] \dfrac{3}{2(1+x^2)} - \frac{1}{2}, & x \ge 1. \end{cases}$$

55. An integrating factor for

$$y' + \frac{2}{x} y = \frac{10 \sin x}{x^3}$$

is x^2. Thus

$$\frac{d}{dx}[x^2 y] = 10 \frac{\sin x}{x}$$

$$x^2 y = 10 \int_0^x \frac{\sin t}{t} \, dt + c$$

$$y = 10x^{-2}\mathrm{Si}(x) + cx^{-2}.$$

From $y(1) = 0$ we get $c = -10\mathrm{Si}(1)$. Thus

$$y = 10x^{-2}\mathrm{Si}(x) - 10x^{-2}\mathrm{Si}(1) = 10x^{-2}(\mathrm{Si}(x) - \mathrm{Si}(1)).$$

Using *Mathematica* we find $y(2) \approx 1.64832$.

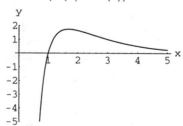

56. An integrating factor for $y' - 2xy = -1$ is e^{-x^2}. Thus

$$\frac{d}{dx}[e^{-x^2} y] = -e^{-x^2}$$

$$e^{-x^2} y = -\int_0^x e^{-t^2} \, dt = -\frac{\sqrt{\pi}}{2} \, \mathrm{erf}(x) + c.$$

From $y(0) = \sqrt{\pi}/2$, and noting that $\text{erf}(0) = 0$, we get $c = \sqrt{\pi}/2$. Thus

$$y = e^{x^2}\left(-\frac{\sqrt{\pi}}{2}\,\text{erf}(x) + \frac{\sqrt{\pi}}{2}\right) = \frac{\sqrt{\pi}}{2}\,e^{x^2}(1 - \text{erf}(x))$$

$$= \frac{\sqrt{\pi}}{2}\,e^{x^2}\,\text{erfc}(x).$$

Using *Mathematica* we find $y(2) \approx 0.226339$.

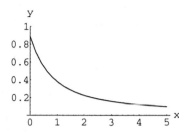

57. An integrating factor for $y' - 2xy = 1$ is e^{-x^2}. Thus

$$\frac{d}{dx}\left[e^{-x^2}y\right] = e^{-x^2}$$

$$e^{-x^2}y = \int_0^x e^{-t^2}\,dt = \text{erf}(x) + c$$

and

$$y = e^{x^2}\,\text{erf}(x) + ce^{x^2}.$$

From $y(1) = 1$ we get $1 = e\,\text{erf}(1) + ce$, so that $c = e^{-1} - \text{erf}(1)$. Thus

$$y = e^{x^2}\,\text{erf}(x) + (e^{-1} - \text{erf}(1))e^{x^2}$$

$$= e^{x^2-1} + e^{x^2}(\text{erf}(x) - \text{erf}(1)).$$

58.

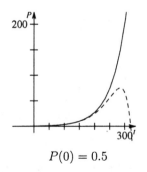

$P(0) = 0.5$

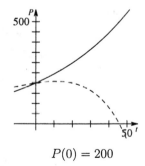

$P(0) = 2$

$P(0) = 200$

The solutions of the linearized equation are shown as solid curves, while the solutions of the nonlinear equation are shown as dashed curves. We see that the approximation remains close to the exact solution for a longer time when the initial population is smaller.

59. The first equation can be solved by separation of variables. We obtain $x = c_1 e^{-\lambda_1 t}$. From $x(0) = x_0$ we obtain $c_1 = x_0$ and so $x = x_0 e^{-\lambda_1 t}$. The second equation then becomes

$$\frac{dy}{dt} = x_0 \lambda_1 e^{-\lambda_1 t} - \lambda_2 y \qquad \text{or} \qquad \frac{dy}{dt} + \lambda_2 y = x_0 \lambda_1 e^{-\lambda_1 t}$$

which is linear. An integrating factor is $e^{\lambda_2 t}$. Thus

$$\frac{d}{dt}\left[e^{\lambda_2 t} y\right] = x_0 \lambda_1 e^{-\lambda_1 t} e^{\lambda_2 t} = x_0 \lambda_1 e^{(\lambda_2 - \lambda_1) t}$$

$$e^{\lambda_2 t} y = \frac{x_0 \lambda_1}{\lambda_2 - \lambda_1} e^{(\lambda_2 - \lambda_1) t} + c_2$$

$$y = \frac{x_0 \lambda_1}{\lambda_2 - \lambda_1} e^{-\lambda_1 t} + c_2 e^{-\lambda_2 t}.$$

From $y(0) = y_0$ we obtain $c_2 = (y_0 \lambda_2 - y_0 \lambda_1 - x_0 \lambda_1)/(\lambda_2 - \lambda_1)$. The solution is

$$y = \frac{x_0 \lambda_1}{\lambda_2 - \lambda_1} e^{-\lambda_1 t} + \frac{y_0 \lambda_2 - y_0 \lambda_1 - x_0 \lambda_1}{\lambda_2 - \lambda_1} e^{-\lambda_2 t}.$$

60. (a) Letting $y = y_1 + y_2$ we have $y(x_0) = y_1(x_0) + y_2(x_0) = \alpha + 0 = \alpha$ and

$$y' + P(x)y = (y_1 + y_2)' + P(x)(y_1 + y_2)$$

$$= (y_1' + P(x)y_1) + (y_2' + P(x)y_2) = 0 + f(x) = f(x).$$

Thus $y = y_1 + y_2$ is a solution of the initial-value problem $y' + P(x)y = f(x)$, $y(x_0) = \alpha$.

(b) By Theorem 1.1 the initial-value problem

$$y' + P(x)y = f_1(x) + f_2(x), \qquad y(x_0) = \alpha + \beta \tag{1}$$

has a unique solution. Consider $y = y_1 + y_2$. Since $y(x_0) = y_1(x_0) + y_2(x_0) = \alpha + \beta$, y satisfies the initial condition of (1). Also

$$y' + P(x)y = (y_1 + y_2)' + P(x)(y_1 + y_2)$$

$$= (y_1' + P(x)y_1) + (y_2' + P(x)y_2) = f_1(x) + f_2(x),$$

so y satisfies the differential equation in (1). Thus $y = y_1 + y_2$ is the unique solution of (1) and $y(x_0) = \alpha + \beta$.

Since $c_1 y_1$ is a solution of $y' + P(x)y = c_1 f_1(x)$, $y(x_0) = c_1 \alpha$, and $c_2 y_2$ is a solution of $y' + P(x)y = c_2 f_2(x)$, $y(x_0) = c_2 \beta$, we see by the above argument that $y(x_0) = c_1 \alpha + c_2 \beta$.

_____ **Exercises 2.4** _____

1. Letting $y = ux$ we have

$$(x - ux)\, dx + x(u\, dx + x\, du) = 0$$

$$dx + x\, du = 0$$

$$\frac{dx}{x} + du = 0$$

$$\ln |x| + u = c$$

$$x \ln |x| + y = cx.$$

$$2\, \ell n\, \mathrm{(a)}\; +l = C\, \mathrm{a}$$

2. Letting $y = ux$ we have

$$(x + ux)\, dx + x(u\, dx + x\, du) = 0$$

$$(1 + 2u)\, dx + x\, du = 0$$

$$\frac{dx}{x} + \frac{du}{1 + 2u} = 0$$

$$\ln |x| + \frac{1}{2} \ln |1 + 2u| = c$$

$$x^2 \left(1 + 2\frac{y}{x}\right) = c_1$$

$$x^2 + 2xy = c_1.$$

3. Letting $x = vy$ we have

$$vy(v\, dy + y\, dv) + (y - 2vy)\, dy = 0$$

$$vy\, dv + \left(v^2 - 2v + 1\right) dy = 0$$

$$\frac{v\, dv}{(v - 1)^2} + \frac{dy}{y} = 0$$

$$\ln |v - 1| - \frac{1}{v - 1} + \ln |y| = c$$

$$\ln \left|\frac{x}{y} - 1\right| - \frac{1}{x/y - 1} + \ln y = c$$

$$(x - y) \ln |x - y| - y = c(x - y).$$

33

4. Letting $x = vy$ we have

$$y(v\,dy + y\,dv) - 2(vy + y)\,dy = 0$$

$$y\,dv - (v + 2)\,dy = 0$$

$$\frac{dv}{v + 2} - \frac{dy}{y} = 0$$

$$\ln|v + 2| - \ln|y| = c$$

$$\ln\left|\frac{x}{y} + 2\right| - \ln|y| = c$$

$$x + 2y = c_1 y^2.$$

5. Letting $y = ux$ we have

$$\left(u^2 x^2 + ux^2\right)dx - x^2(u\,dx + x\,du) = 0$$

$$u^2\,dx - x\,du = 0$$

$$\frac{dx}{x} - \frac{du}{u^2} = 0$$

$$\ln|x| + \frac{1}{u} = c$$

$$\ln|x| + \frac{x}{y} = c$$

$$y\ln|x| + x = cy.$$

6. Letting $y = ux$ we have

$$\left(u^2 x^2 + ux^2\right)dx + x^2(u\,dx + x\,du) = 0$$

$$\left(u^2 + 2u\right)dx + x\,du = 0$$

$$\frac{dx}{x} + \frac{du}{u(u + 2)} = 0$$

$$\ln|x| + \frac{1}{2}\ln|u| - \frac{1}{2}\ln|u + 2| = c$$

$$\frac{x^2 u}{u + 2} = c_1$$

$$x^2 \frac{y}{x} = c_1\left(\frac{y}{x} + 2\right)$$

$$x^2 y = c_1(y + 2x).$$

7. Letting $y = ux$ we have

$$(ux - x)\,dx - (ux + x)(u\,dx + x\,du) = 0$$

$$\left(u^2 + 1\right)dx + x(u + 1)\,du = 0$$

$$\frac{dx}{x} + \frac{u + 1}{u^2 + 1}\,du = 0$$

$$\ln|x| + \frac{1}{2}\ln\left(u^2 + 1\right) + \tan^{-1}u = c$$

$$\ln x^2\left(\frac{y^2}{x^2} + 1\right) + 2\tan^{-1}\frac{y}{x} = c_1$$

$$\ln\left(x^2 + y^2\right) + 2\tan^{-1}\frac{y}{x} = c_1.$$

8. Letting $y = ux$ we have

$$(x + 3ux)\,dx - (3x + ux)(u\,dx + x\,du) = 0$$

$$\left(u^2 - 1\right)dx + x(u + 3)\,du = 0$$

$$\frac{dx}{x} + \frac{u + 3}{(u - 1)(u + 1)}\,du = 0$$

$$\ln|x| + 2\ln|u - 1| - \ln|u + 1| = c$$

$$\frac{x(u - 1)^2}{u + 1} = c_1$$

$$x\left(\frac{y}{x} - 1\right)^2 = c_1\left(\frac{y}{x} + 1\right)$$

$$(y - x)^2 = c_1(y + x).$$

9. Letting $y = ux$ we have

$$-ux\,dx + (x + \sqrt{u}\,x)(u\,dx + x\,du) = 0$$

$$(x + x\sqrt{u}\,)\,du + u^{3/2}\,dx = 0$$

$$\left(u^{-3/2} + \frac{1}{u}\right)du + \frac{dx}{x} = 0$$

$$-2u^{-1/2} + \ln|u| + \ln|x| = c$$

$$\ln|y/x| + \ln|x| = 2\sqrt{x/y} + c$$

$$y(\ln|y| - c)^2 = 4x.$$

35

10. Letting $y = ux$ we have

$$\left(ux + \sqrt{x^2 + u^2 x^2} \right) dx - x(u\,dx + x\,du) = 0$$

$$x\sqrt{1 + u^2}\,dx - x^2\,du = 0$$

$$\frac{dx}{x} - \frac{du}{\sqrt{1 + u^2}} = 0$$

$$\ln|x| - \ln\left| u + \sqrt{1 + u^2} \right| = c$$

$$u + \sqrt{1 + u^2} = c_1 x$$

$$y + \sqrt{y^2 + x^2} = c_1 x^2.$$

11. Letting $y = ux$ we have

$$\left(x^3 - u^3 x^3 \right) dx + u^2 x^3 (u\,dx + x\,du) = 0$$

$$dx + u^2 x\,du = 0$$

$$\frac{dx}{x} + u^2\,du = 0$$

$$\ln|x| + \frac{1}{3} u^3 = c$$

$$3x^3 \ln|x| + y^3 = c_1 x^3.$$

Using $y(1) = 2$ we find $c_1 = 8$. The solution of the initial-value problem is $3x^3 \ln|x| + y^3 = 8x^3$.

12. Letting $y = ux$ we have

$$\left(x^2 + 2u^2 x^2 \right) dx - ux^2 (u\,dx + x\,du) = 0$$

$$\left(1 + u^2 \right) dx - ux\,du = 0$$

$$\frac{dx}{x} - \frac{u\,du}{1 + u^2} = 0$$

$$\ln|x| - \frac{1}{2} \ln\left(1 + u^2 \right) = c$$

$$\frac{x^2}{1 + u^2} = c_1$$

$$x^4 = c_1 \left(y^2 + x^2 \right).$$

Using $y(-1) = 1$ we find $c_1 = 1/2$. The solution of the initial-value problem is $2x^4 = y^2 + x^2$.

13. Letting $y = ux$ we have

$$(x + uxe^u)\, dx - xe^u(u\, dx + x\, du) = 0$$

$$dx - xe^u\, du = 0$$

$$\frac{dx}{x} - e^u\, du = 0$$

$$\ln|x| - e^u = c$$

$$\ln|x| - e^{y/x} = c.$$

Using $y(1) = 0$ we find $c = -1$. The solution of the initial-value problem is $\ln|x| = e^{y/x} - 1$.

14. Letting $x = vy$ we have

$$y(v\, dy + y\, dv) + vy(\ln vy - \ln y - 1)\, dy = 0$$

$$y\, dv + v \ln v\, dy = 0$$

$$\frac{dv}{v \ln v} + \frac{dy}{y} = 0$$

$$\ln|\ln|v|| + \ln|y| = c$$

$$y \ln\left|\frac{x}{y}\right| = c_1.$$

Using $y(1) = e$ we find $c_1 = -e$. The solution of the initial-value problem is $y \ln\left|\frac{x}{y}\right| = -e$.

15. From $y' + \frac{1}{x}y = \frac{1}{x}y^{-2}$ and $w = y^3$ we obtain $\frac{dw}{dx} + \frac{3}{x}w = \frac{3}{x}$. An integrating factor is x^3 so that $x^3w = x^3 + c$ or $y^3 = 1 + cx^{-3}$.

16. From $y' - y = e^x y^2$ and $w = y^{-1}$ we obtain $\frac{dw}{dx} + w = -e^x$. An integrating factor is e^x so that $e^x w = -\frac{1}{2}e^{2x} + c$ or $y^{-1} = -\frac{1}{2}e^x + ce^{-x}$.

17. From $y' + y = xy^4$ and $w = y^{-3}$ we obtain $\frac{dw}{dx} - 3w = -3x$. An integrating factor is e^{-3x} so that $e^{-3x} = xe^{-3x} + \frac{1}{3}e^{-3x} + c$ or $y^{-3} = x + \frac{1}{3} + ce^{3x}$.

18. From $y' - \left(1 + \frac{1}{x}\right)y = y^2$ and $w = y^{-1}$ we obtain $\frac{dw}{dx} + \left(1 + \frac{1}{x}\right)w = -1$. An integrating factor is xe^x so that $xe^x w = -xe^x + e^x + c$ or $y^{-1} = -1 + \frac{1}{x} + \frac{c}{x}e^{-x}$.

$$\times \ \frac{dy}{dx} - (1+x)y = xy^2$$ **37**

19. From $y' - \frac{1}{x}y = -\frac{1}{x^2}y^2$ and $w = y^{-1}$ we obtain $\frac{dw}{dx} + \frac{1}{x}w = \frac{1}{x^2}$. An integrating factor is x so that

$xw = \ln x + c$ or $y^{-1} = \frac{1}{x}\ln x + \frac{c}{x}$.

20. From $y' + \frac{2}{3(1+x^2)}y = \frac{2x}{3(1+x^2)}y^4$ and $w = y^{-3}$ we obtain $\frac{dw}{dx} - \frac{2x}{1+x^2}w = \frac{-2x}{1+x^2}$. An

integrating factor is $\frac{1}{1+x^2}$ so that $\frac{w}{1+x^2} = \frac{1}{1+x^2} + c$ or $y^{-3} = 1 + c\left(1+x^2\right)$.

21. From $y' - \frac{2}{x}y = \frac{3}{x^2}y^4$ and $w = y^{-3}$ we obtain $\frac{dw}{dx} + \frac{6}{x}w = -\frac{9}{x^2}$. An integrating factor is x^6 so that

$x^6w = -\frac{9}{5}x^5 + c$ or $y^{-3} = -\frac{9}{5}x^{-1} + cx^{-6}$. If $y(1) = \frac{1}{2}$ then $c = \frac{49}{5}$ and $y^{-3} = -\frac{9}{5}x^{-1} + \frac{49}{5}x^{-6}$.

22. From $y' + y = y^{-1/2}$ and $w = y^{3/2}$ we obtain $\frac{dw}{dx} + \frac{3}{2}w = \frac{3}{2}$. An integrating factor is $e^{3x/2}$ so that

$e^{3x/2}w = e^{3x/2} + c$ or $y^{3/2} = 1 + ce^{-3x/2}$. If $y(0) = 4$ then $c = 7$ and $y^{3/2} = 1 + 7e^{-3x/2}$.

23. Let $u = x + y + 1$ so that $du/dx = 1 + dy/dx$. Then $\frac{du}{dx} - 1 = u^2$ or $\frac{1}{1+u^2}du = dx$. Thus

$\tan^{-1}u = x + c$ or $u = \tan(x+c)$, and $x + y + 1 = \tan(x+c)$ or $y = \tan(x+c) - x - 1$.

24. Let $u = x + y$ so that $du/dx = 1 + dy/dx$. Then $\frac{du}{dx} - 1 = \frac{1-u}{u}$ or $u\,du = dx$. Thus $\frac{1}{2}u^2 = x + c$

or $u^2 = 2x + c_1$, and $(x+y)^2 = 2x + c_1$.

25. Let $u = x + y$ so that $du/dx = 1 + dy/dx$. Then $\frac{du}{dx} - 1 = \tan^2 u$ or $\cos^2 u\,du = dx$. Thus

$\frac{1}{2}u + \frac{1}{4}\sin 2u = x + c$ or $2u + \sin 2u = 4x + c_1$, and $2(x+y) + \sin 2(x+y) = 4x + c_1$ or

$2y + \sin 2(x+y) = 2x + c_1$.

26. Let $u = x + y$ so that $du/dx = 1 + dy/dx$. Then $\frac{du}{dx} - 1 = \sin u$ or $\frac{1}{1+\sin u}du = dx$. Multiplying

by $(1 - \sin u)/(1 - \sin u)$ we have $\frac{1-\sin u}{\cos^2 u}du = dx$ or $\left(\sec^2 u - \tan u\sec u\right)du = dx$. Thus

$\tan u - \sec u = x + c$ or $\tan(x+y) - \sec(x+y) = x + c$.

27. Let $u = y - 2x + 3$ so that $du/dx = dy/dx - 2$. Then $\frac{du}{dx} + 2 = 2 + \sqrt{u}$ or $\frac{1}{\sqrt{u}}du = dx$. Thus

$2\sqrt{u} = x + c$ and $2\sqrt{y - 2x + 3} = x + c$.

28. Let $u = y - x + 5$ so that $du/dx = dy/dx - 1$. Then $\frac{du}{dx} + 1 = 1 + e^u$ or $e^{-u}du = dx$. Thus

$-e^{-u} = x + c$ and $-e^{y-x+5} = x + c$.

29. Let $u = x + y$ so that $du/dx = 1 + dy/dx$. Then $\frac{du}{dx} - 1 = \cos u$ and $\frac{1}{1+\cos u}du = dx$. Now

$$\frac{1}{1+\cos u} = \frac{1-\cos u}{1-\cos^2 u} = \frac{1-\cos u}{\sin^2 u} = \csc^2 u - \csc u\cot u$$

so we have $\int(\csc^2 u - \csc u \cot u)\,du = \int dx$ and $-\cot u + \csc u = x + c$. Thus $-\cot(x+y) + \csc(x+y) = x + c$. Setting $x = 0$ and $y = \pi/4$ we obtain $c = \sqrt{2} - 1$. The solution is

$$\csc(x + y) - \cot(x + y) = x + \sqrt{2} - 1.$$

30. Let $u = 3x + 2y$ so that $du/dx = 3 + 2\,dy/dx$. Then $\dfrac{du}{dx} = 3 + \dfrac{2u}{u+2} = \dfrac{5u+6}{u+2}$ and $\dfrac{u+2}{5u+6}\,du = dx$.

Now

$$\frac{u+2}{5u+6} = \frac{1}{5} + \frac{4}{25u+30}$$

so we have

$$\int\left(\frac{1}{5} + \frac{4}{25u+30}\right)du = dx$$

and $\dfrac{1}{5}u + \dfrac{4}{25}\ln|25u + 30| = x + c$. Thus

$$\frac{1}{5}(3x + 2y) + \frac{4}{25}\ln|75x + 50y + 30| = x + c.$$

Setting $x = -1$ and $y = -1$ we obtain $c = \frac{4}{5}\ln 95$. The solution is

$$\frac{1}{5}(3x + 2y) + \frac{4}{25}\ln|75x + 50y + 30| = x + \frac{4}{5}\ln 95$$

or
$$5y - 5x + 2\ln|75x + 50y + 30| = 10\ln 95.$$

31. We write the differential equation $M(x,y)dx + N(x,y)dy = 0$ as $dy/dx = f(x,y)$ where

$$f(x, y) = -\frac{M(x,y)}{N(x,y)}.$$

The function $f(x,y)$ must necessarily be homogeneous of degree 0 when M and N are homogeneous of degree n. Since M is homogeneous of degree n, $M(tx, ty) = t^n M(x,y)$, and letting $t = 1/x$ we have

$$M(1, y/x) = \frac{1}{x^n} M(x,y) \quad\text{or}\quad M(x,y) = x^n M(1, y/x).$$

Thus

$$\frac{dy}{dx} = f(x,y) = -\frac{x^n M(1, y/x)}{x^n N(1, y/x)} = -\frac{M(1, y/x)}{N(1, y/x)} = F\left(\frac{y}{x}\right).$$

To show that the differential equation also has the form

$$\frac{dy}{dx} = G\left(\frac{x}{y}\right)$$

we use the fact that $M(x,y) = y^n M(x/y, 1)$. The forms $F(y/x)$ and $G(x/y)$ suggest, respectively, the substitutions $u = y/x$ or $y = ux$ and $v = x/y$ or $x = vy$.

32. (a) The substitutions $y = y_1 + u$ and

$$\frac{dy}{dx} = \frac{dy_1}{dx} + \frac{du}{dx}$$

lead to

$$\frac{dy_1}{dx} + \frac{du}{dx} = P + Q(y_1 + u) + R(y_1 + u)^2$$

$$= P + Qy_1 + Ry_1^2 + Qu + 2y_1Ru + Ru^2$$

or

$$\frac{du}{dx} - (Q + 2y_1R)u = Ru^2.$$

This is a Bernoulli equation with $n = 2$ which can be reduced to the linear equation

$$\frac{dw}{dx} + (Q + 2y_1R)w = -R$$

by the substitution $w = u^{-1}$.

(b) Identify $P(x) = -4/x^2$, $Q(x) = -1/x$, and $R(x) = 1$. Then $\dfrac{dw}{dx} + \left(-\dfrac{1}{x} + \dfrac{4}{x}\right)w = -1$. An integrating factor is x^3 so that $x^3w = -\dfrac{1}{4}x^4 + c$ or $u = \left[-\dfrac{1}{4}x + cx^{-3}\right]^{-1}$. Thus, $y = \dfrac{2}{x} + u$.

Chapter 2 Review Exercises

1. linear in y, homogeneous, exact

2. linear in x

3. separable, exact, linear in x and y

4. Bernoulli in x

5. separable

6. separable, linear in x, Bernoulli

7. linear in x

8. homogeneous

9. Bernoulli

10. homogeneous, exact, Bernoulli

11. linear in x and y, exact, separable, homogeneous

12. exact, linear in y

13. homogeneous

14. separable

15. Separating variables we obtain

$$\cos^2 x \, dx = \frac{y}{y^2+1} \, dy \implies \frac{1}{2}x + \frac{1}{4}\sin 2x = \frac{1}{2}\ln\left(y^2+1\right) + c \implies 2x + \sin 2x = 2\ln\left(y^2+1\right) + c.$$

16. Write the differential equation in the form $y \ln \frac{x}{y} \, dx = \left(x \ln \frac{x}{y} - y\right) dy$. This is a homogeneous equation, so let $x = uy$. Then $dx = u \, dy + y \, du$ and the differential equation becomes

$$y \ln u(u \, dy + y \, du) = (uy \ln u - y) \, dy \quad \text{or} \quad y \ln u \, du = -dy.$$

Separating variables we obtain

$$\ln u \, du = -\frac{dy}{y} \implies u \ln|u| - u = -\ln|y| + c \implies \frac{x}{y}\ln\left|\frac{x}{y}\right| - \frac{x}{y} = -\ln|y| + c$$

$$\implies x(\ln x - \ln y) - x = -y \ln|y| + cy.$$

17. The differential equation $\dfrac{dy}{dx} + \dfrac{2}{6x+1}y = -\dfrac{3x^2}{6x+1}y^{-2}$ is Bernoulli. Using $w = y^3$ we obtain

$\dfrac{dw}{dx} + \dfrac{6}{6x+1}w = -\dfrac{9x^2}{6x+1}$. An integrating factor is $6x + 1$, so

$$\frac{d}{dx}[(6x+1)w] = -9x^2 \implies w = -\frac{3x^3}{6x+1} + \frac{c}{6x+1} \implies (6x+1)y^3 = -3x^3 + c.$$

(Note: The differential equation is also exact.)

18. Write the differential equation in the form $(3y^2 + 2x)dx + (4y^2 + 6xy)dy = 0$. Letting $M = 3y^2 + 2x$ and $N = 4y^2 + 6xy$ we see that $M_y = 6y = N_x$ so the differential equation is exact. From $f_x = 3y^2 + 2x$ we obtain $f = 3xy^2 + x^2 + h(y)$. Then $f_y = 6xy + h'(y) = 4y^2 + 6xy$ and $h'(y) = 4y^2$ so $h(y) = \frac{4}{3}y^3$. The general solution is

$$3xy^2 + x^2 + \frac{4}{3}y^3 = c.$$

19. Write the equation in the form

$$\frac{dQ}{dt} + \frac{1}{t}Q = t^3 \ln t.$$

An integrating factor is $e^{\ln t} = t$, so

$$\frac{d}{dt}[tQ] = t^4 \ln t \implies tQ = -\frac{1}{25}t^5 + \frac{1}{5}t^5 \ln t + c$$

$$\implies Q = -\frac{1}{25}t^4 + \frac{1}{5}t^4 \ln t + \frac{c}{t}.$$

41

20. Letting $u = 2x + y + 1$ we have

$$\frac{du}{dx} = 2 + \frac{dy}{dx},$$

and so the given differential equation is transformed into

$$u\left(\frac{du}{dx} - 2\right) = 1 \quad \text{or} \quad \frac{du}{dx} = \frac{2u + 1}{u}.$$

Separating variables and integrating we get

$$\frac{u}{2u + 1}\, du = dx$$

$$\left(\frac{1}{2} - \frac{1}{2}\frac{1}{2u + 1}\right) du = dx$$

$$\frac{1}{2}u - \frac{1}{4}\ln|2u + 1| = x + c$$

$$2u - \ln|2u + 1| = 2x + c_1.$$

Resubstituting for u gives the solution

$$4x + 2y + 2 - \ln|4x + 2y + 3| = 2x + c_1$$

or

$$2x + 2y + 2 - \ln|4x + 2y + 3| = c_1.$$

21. Separating variables we obtain

$$y \ln y\, dy = te^t dt \implies \frac{1}{2}y^2 \ln|y| - \frac{1}{4}y^2 = te^t - e^t + c.$$

If $y(1) = 1$, $c = -1/4$. The solution is $2y^2 \ln|y| - y^2 = 4te^t - 4e^t - 1$.

22. The equation is homogeneous, so let $x = ut$. Then $dx = u\, dt + t\, du$ and the differential equation becomes $ut^2(u\, dt + t\, du) = \left(3u^2t^2 + t^2\right) dt$ or $ut\, du = \left(2u^2 + 1\right) dt$. Separating variables we obtain

$$\frac{u}{2u^2 + 1}\, du = \frac{dt}{t} \implies \frac{1}{4}\ln\left(2u^2 + 1\right) = \ln t + c \implies 2u^2 + 1 = c_1 t^4$$

$$\implies 2\frac{x^2}{t^2} + 1 = c_1 t^4 \implies 2x^2 + t^2 = c_1 t^6.$$

If $x(-1) = 2$ then $c_1 = 9$ and the solution of the initial-value problem is $2x^2 + t^2 = 9t^6$.

23. Write the equation in the form $\dfrac{dy}{dx} + \dfrac{8x}{x^2 + 4}y = \dfrac{2x}{x^2 + 4}$. An integrating factor is $\left(x^2 + 4\right)^4$, so

$$\frac{d}{dx}\left[\left(x^2 + 4\right)^4 y\right] = 2x\left(x^2 + 4\right)^3 \implies \left(x^2 + 4\right)^4 y = \frac{1}{4}\left(x^2 + 4\right)^4 + c \implies y = \frac{1}{4} + c\left(x^2 + 4\right)^{-4}.$$

If $y(0) = -1$ then $c = -320$ and $y = \dfrac{1}{4} - 320\left(x^2 + 4\right)^{-4}$.

24. The differential equation is Bernoulli. Using $w = y^{-1}$ we obtain $-xy^2 \dfrac{dw}{dx} + 4y = x^4 y^2$ or

$\dfrac{dw}{dx} - \dfrac{4}{x} w = -x^3$. An integrating factor is x^{-4}, so

$$\frac{d}{dx}\left[x^{-4}w\right] = -\frac{1}{x} \implies x^{-4}w = -\ln x + c \implies w = -x^4 \ln x + cx^4 \implies y = \left(cx^4 - x^4 \ln x\right)^{-1}.$$

If $y(1) = 1$ then $c = 1$ and $y = \left(x^4 - x^4 \ln x\right)^{-1}$.

25. Separating variables and integrating we obtain

$$e^{-2y}\, dy = e^{-x}\, dx$$

$$-\frac{1}{2}e^{-2y} = -e^{-x} + c$$

$$e^{-2y} = 2e^{-x} + c_1.$$

If $y(0) = 0$ then $c_1 = -1$ and the solution is $e^{-2y} = 2e^{-x} - 1$. Now

$$\frac{1}{e^{2y}} = \frac{2}{e^x} - 1 \qquad \text{so} \qquad e^x = 2e^{2y} - e^{2y+x}.$$

26. Letting $M = 2r^2 \cos\theta \sin\theta + r\cos\theta$ and $N = 4r + \sin\theta - 2r\cos^2\theta$ we see that $M_r = 4r\cos\theta \sin\theta + \cos\theta = N_\theta$ so the differential equation is exact. From $f_\theta = 2r^2 \cos\theta \sin\theta + r\cos\theta$ we obtain $f = -r^2 \cos^2\theta + r\sin\theta + h(r)$. Then $f_r = -2r\cos^2\theta + \sin\theta + h'(r) = 4r + \sin\theta - 2r\cos^2\theta$ and $h'(r) = 4r$ so $h(r) = 2r^2$. The general solution is

$$-r^2 \cos^2\theta + r\sin\theta + 2r^2 = c.$$

If $r(\pi/2) = 2$ then $r = 10$ and the solution of the initial-value problem is

$$-r^2 \cos^2\theta + r\sin\theta + 2r^2 = 10.$$

3 Modeling with First-Order Differential Equations

_____ **Exercises 3.1** _____

1. Let $P = P(t)$ be the population at time t, and P_0 the initial population. From $dP/dt = kP$ we obtain $P = P_0 e^{kt}$. Using $P(5) = 2P_0$ we find $k = \frac{1}{5} \ln 2$ and $P = P_0 e^{(\ln 2)t/5}$. Setting $P(t) = 3P_0$ we have

$$3 = e^{(\ln 2)t/5} \implies \ln 3 = \frac{(\ln 2)t}{5} \implies t = \frac{5 \ln 3}{\ln 2} \approx 7.9 \text{ years.}$$

Setting $P(t) = 4P_0$ we have

$$4 = e^{(\ln 2)t/5} \implies \ln 4 = \frac{(\ln 2)t}{5} \implies t = 10 \text{ years.}$$

2. Setting $P = 10{,}000$ and $t = 3$ in Problem 1 we obtain

$$10{,}000 = P_0 e^{(\ln 2)3/5} \implies P_0 = 10{,}000 e^{-0.6 \ln 2} \approx 6597.5.$$

Then $P(10) = P_0 e^{2 \ln 2} = 4P_0 \approx 26{,}390.$

3. Let $P = P(t)$ be the population at time t. From $dP/dt = kt$ and $P(0) = P_0 = 500$ we obtain $P = 500 e^{kt}$. Using $P(10) = 575$ we find $k = \frac{1}{10} \ln 1.15$. Then $P(30) = 500 e^{3 \ln 1.15} \approx 760$ years.

4. Let $N = N(t)$ be the number of bacteria at time t and N_0 the initial number. From $dN/dt = kN$ we obtain $N = N_0 e^{kt}$. Using $N(3) = 400$ and $N(10) = 2000$ we find $400 = N_0 e^{3k}$ or $e^k = (400/N_0)^{1/3}$. From $N(10) = 2000$ we then have

$$2000 = N_0 e^{10k} = N_0 \left(\frac{400}{N_0}\right)^{10/3} \implies \frac{2000}{400^{10/3}} = N_0^{-7/3} \implies N_0 = \left(\frac{2000}{400^{10/3}}\right)^{-3/7} \approx 201.$$

5. Let $N = N(t)$ be the amount of lead at time t. From $dN/dt = kN$ and $N(0) = 1$ we obtain $N = e^{kt}$. Using $N(3.3) = 1/2$ we find $k = \frac{1}{3.3} \ln 1/2$. When 90% of the lead has decayed, 0.1 grams will remain. Setting $N(t) = 0.1$ we have

$$e^{t(1/3.3) \ln(1/2)} = 0.1 \implies \frac{t}{3.3} \ln \frac{1}{2} = \ln 0.1 \implies t = \frac{3.3 \ln 0.1}{\ln 1/2} \approx 10.96 \text{ hours.}$$

6. Let $N = N(t)$ be the amount at time t. From $dN/dt = kt$ and $N(0) = 100$ we obtain $N = 100 e^{kt}$. Using $N(6) = 97$ we find $k = \frac{1}{6} \ln 0.97$. Then $N(24) = 100 e^{(1/6)(\ln 0.97)24} = 100(0.97)^4 \approx 88.5$ mg.

7. Setting $N(t) = 50$ in Problem 6 we obtain

$$50 = 100e^{kt} \implies kt = \ln\frac{1}{2} \implies t = \frac{\ln 1/2}{(1/6)\ln 0.97} \approx 136.5 \text{ hours.}$$

8. **(a)** The solution of $dA/dt = kA$ is $A(t) = A_0 e^{kt}$. Letting $A = \frac{1}{2}A_0$ and solving for t we obtain the half-life $T = -(\ln 2)/k$.

(b) Since $k = -(\ln 2)/T$ we have

$$A(t) = A_0 e^{-(\ln 2)t/T} = A_0 2^{-t/T}.$$

9. Let $I = I(t)$ be the intensity, t the thickness, and $I(0) = I_0$. If $dI/dt = kI$ and $I(3) = 0.25I_0$ then $I = I_0 e^{kt}$, $k = \frac{1}{3}\ln 0.25$, and $I(15) = 0.00098I_0$.

10. From $dS/dt = rS$ we obtain $S = S_0 e^{rt}$ where $S(0) = S_0$.

(a) If $S_0 = \$5000$ and $r = 5.75\%$ then $S(5) = \$6665.45$.

(b) If $S(t) = \$10,000$ then $t = 12$ years.

(c) $S \approx \$6651.82$

11. Assume that $A = A_0 e^{kt}$ and $k = -0.00012378$. If $A(t) = 0.145A_0$ then $t \approx 15,600$ years.

12. Assume that $dT/dt = k(T - 5)$ so that $T = 5 + ce^{kt}$. If $T(1) = 55°$ and $T(5) = 30°$ then $k = -\frac{1}{4}\ln 2$ and $c = 59.4611$ so that $T(0) = 64.4611°$.

13. Assume that $dT/dt = k(T - 10)$ so that $T = 10 + ce^{kt}$. If $T(0) = 70°$ and $T(1/2) = 50°$ then $c = 60$ and $k = 2\ln(2/3)$ so that $T(1) = 36.67°$. If $T(t) = 15°$ then $t = 3.06$ minutes.

14. Assume that $dT/dt = k(T - 100)$ so that $T = 100 + ce^{kt}$. If $T(0) = 20°$ and $T(1) = 22°$ then $c = -80$ and $k = \ln(39/40)$ so that $T(t) = 90°$ implies $t = 82.1$ seconds. If $T(t) = 98°$ then $t = 145.7$ seconds.

15. Assume $L\,di/dt + Ri = E(t)$, $L = 0.1$, $R = 50$, and $E(t) = 50$ so that $i = \frac{3}{5} + ce^{-500t}$. If $i(0) = 0$ then $c = -3/5$ and $\lim_{t\to\infty} i(t) = 3/5$.

16. Assume $L\,di/dt + Ri = E(t)$, $E(t) = E_0 \sin \omega t$, and $i(0) = i_0$ so that

$$i = \frac{E_0 R}{L^2\omega^2 + R^2}\sin \omega t - \frac{E_0 L\omega}{L^2\omega^2 + R^2}\cos \omega t + ce^{-Rt/L}.$$

Since $i(0) = i_0$ we obtain $c = i_0 + \dfrac{E_0 L\omega}{L^2\omega^2 + R^2}$.

17. Assume $R\,dq/dt + (1/c)q = E(t)$, $R = 200$, $C = 10^{-4}$, and $E(t) = 100$ so that $q = 1/100 + ce^{-50t}$. If $q(0) = 0$ then $c = -1/100$ and $i = \frac{1}{2}e^{-50t}$.

18. Assume $R\,dq/dt + (1/c)q = E(t)$, $R = 1000$, $C = 5 \times 10^{-6}$, and $E(t) = 200$. Then $q = \frac{1}{1000} + ce^{-200t}$ and $i = -200ce^{-200t}$. If $i(0) = 0.4$ then $c = -\frac{1}{500}$, $q(0.005) = 0.003$ coulombs, and $i(0.005) = 0.1472$ amps. As $t \to \infty$ we have $q \to \frac{1}{1000}$.

19. For $0 \le t \le 20$ the differential equation is $20\, di/dt + 2i = 120$. An integrating factor is $e^{t/10}$, so

$$\frac{d}{dt}\left[e^{t/10}i\right] = 6e^{t/10} \text{ and } i = 60 + c_1 e^{-t/10}. \text{ If } i(0) = 0 \text{ then } c_1 = -60 \text{ and } i = 60 - 60e^{-t/10}.$$

For $t > 20$ the differential equation is $20\, di/dt + 2i = 0$ and $i = c_2 e^{-t/10}$.

At $t = 20$ we want $c_2 e^{-2} = 60 - 60e^{-2}$ so that $c_2 = 60\left(e^2 - 1\right)$. Thus

$$i(t) = \begin{cases} 60 - 60e^{-t/10}, & 0 \le t \le 20; \\ 60\left(e^2 - 1\right)e^{-t/10}, & t > 20. \end{cases}$$

20. Separating variables we obtain

$$\frac{dq}{E_0 - q/C} = \frac{dt}{k_1 + k_2 t} \implies -C \ln\left|E_0 - \frac{q}{C}\right| = \frac{1}{k_2}\ln|k_1 + k_2 t| + c_1 \implies \frac{(E_0 - q/C)^{-C}}{(k_1 + k_2 t)^{1/k_2}} = c_2.$$

Setting $q(0) = q_0$ we find $c_2 = \dfrac{(E_0 - q_0/C)^{-C}}{k_1^{1/k_2}}$, so

$$\frac{(E_0 - q/C)^{-C}}{(k_1 + k_2 t)^{1/k_2}} = \frac{(E_0 - q_0/C)^{-C}}{k_1^{1/k_2}} \implies \left(E_0 - \frac{q}{C}\right)^{-C} = \left(E_0 - \frac{q_0}{C}\right)^{-C}\left(\frac{k_1}{k + k_2 t}\right)^{-1/k_2}$$

$$\implies E_0 - \frac{q}{C} = \left(E_0 - \frac{q_0}{C}\right)\left(\frac{k_1}{k + k_2 t}\right)^{1/Ck_2}$$

$$\implies q = E_0 C + (q_0 - E_0 C)\left(\frac{k_1}{k + k_2 t}\right)^{1/Ck_2}.$$

21. From $dA/dt = 4 - A/50$ we obtain $A = 200 + ce^{-t/50}$. If $A(0) = 30$ then $c = -170$ and $A = 200 - 170e^{-t/50}$.

22. From $dA/dt = 0 - A/50$ we obtain $A = ce^{-t/50}$. If $A(0) = 30$ then $c = 30$ and $A = 30e^{-t/50}$.

23. From $dA/dt = 10 - A/100$ we obtain $A = 1000 + ce^{-t/100}$. If $A(0) = 0$ then $c = -1000$ and $A = 1000 - 1000e^{-t/100}$.

24. From $\dfrac{dA}{dt} = 10 - \dfrac{10A}{500 - (10 - 5)t} = 10 - \dfrac{2A}{100 - t}$ we obtain $A = 1000 - 10t + c(100 - t)^2$. If $A(0) = 0$ then $c = -\dfrac{1}{10}$. The tank is empty in 100 minutes.

25. From $\dfrac{dA}{dt} = 3 - \dfrac{4A}{100 + (6 - 4)t} = 3 - \dfrac{2A}{50 + t}$ we obtain $A = 50 + t + c(50 + t)^{-2}$. If $A(0) = 10$ then $c = -100{,}000$ and $A(30) = 64.38$ pounds.

26. (a) Initially the tank contains 300 gallons of solution. Since brine is pumped in at a rate of 3 gal/min and the solution is pumped out at a rate of 2 gal/min, the net change is an increase of 1 gal/min. Thus, in 100 minutes the tank will contain its capacity of 400 gallons.

(b) The differential equation describing the amount of salt in the tank is $A'(t) = 6 - 2A/(300 + t)$ with solution

$$A(t) = 600 + 2t - (4.95 \times 10^7)(300 + t)^{-2}, \qquad 0 \leq t \leq 100,$$

as noted in the discussion following Example 5 in the text. Thus, the amount of salt in the tank when it overflows is

$$A(100) = 800 - (4.95 \times 10^7)(400)^{-2} = 490.625 \text{ lbs.}$$

(c) When the tank is overflowing the amount of salt in the tank is governed by the differential equation

$$\frac{dA}{dt} = (3 \text{ gal/min})(2 \text{ lb/gal}) - (\frac{A}{400} \text{ lb/gal})(3 \text{ gal/min})$$

$$= 6 - \frac{3A}{400}, \qquad A(100) = 490.625.$$

Solving the equation we obtain $A(t) = 800 + ce^{-3t/400}$. The initial condition yields $c = -654.947$, so that

$$A(t) = 800 - 654.947e^{-3t/400}.$$

When $t = 150$, $A(150) = 587.37$ lbs.

(d) As $t \to \infty$, the amount of salt is 800 lbs, which is to be expected since
(400 gal)(2 lbs/gal)= 800 lbs.

(e)

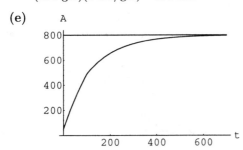

27. (a) From $m\, dv/dt = mg - kv$ we obtain $v = gm/k + ce^{-kt/m}$. If $v(0) = v_0$ then $c = v_0 - gm/k$ and the solution of the initial-value problem is

$$v = \frac{gm}{k} + \left(v_0 - \frac{gm}{k}\right)e^{-kt/m}.$$

(b) As $t \to \infty$ the limiting velocity is gm/k.

(c) From $ds/dt = v$ and $s(0) = s_0$ we obtain

$$s = \frac{gm}{k}t - \frac{m}{k}\left(v_0 - \frac{gm}{k}\right)e^{-kt/m} + s_0 + \frac{m}{k}\left(v_0 - \frac{gm}{k}\right).$$

28. From $dx/dt = r - kx$ and $x(0) = 0$ we obtain $x = r/k - (r/k)e^{-kt}$ so that $x \to r/k$ as $t \to \infty$. If $x(T) = r/2k$ then $T = (\ln 2)/k$.

29. (a) From $dP/dt = (k_1 - k_2)P$ we obtain $P = P_0 e^{(k_1 - k_2)t}$ where $P_0 = P(0)$.

(b) If $k_1 > k_2$ then $P \to \infty$ as $t \to \infty$. If $k_1 = k_2$ then $P = P_0$ for every t. If $k_1 < k_2$ then $P \to 0$ as $t \to \infty$.

30. Separating variables we obtain

$$\frac{dP}{P} = k \cos t \, dt \implies \ln|P| = k \sin t + c \implies P = c_1 e^{k \sin t}.$$

If $P(0) = P_0$ then $c_1 = P_0$ and $P = P_0 e^{k \sin t}$.

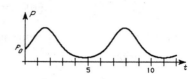

31. Write the differential equation in the form $dA/dt + (k_1 + k_2)A = k_1 M$. Then an integrating factor is $e^{(k_1 + k_2)t}$, and

$$\frac{d}{dt}\left[e^{(k_1 + k_2)t} A\right] = k_1 M e^{(k_1 + k_2)t} \implies e^{(k_1 + k_2)t} A = \frac{k_1 M}{k_1 + k_2} e^{(k_1 + k_2)t} + c$$

$$\implies A = \frac{k_1 M}{k_1 + k_2} + ce^{-(k_1 + k_2)t}.$$

Using $A(0) = 0$ we find $c = -\dfrac{k_1 M}{k_1 + k_2}$ and $A = \dfrac{k_1 M}{k_1 + k_2}\left(1 - e^{-(k_1 + k_2)t}\right)$. As $t \to \infty$, $A \to \dfrac{k_1 M}{k_1 + k_2}$.

If $k_2 > 0$, the material will never be completely memorized.

32. (a) From $y = -x - 1 + c_1 e^x$ we obtain $y' = y + x$ so that the differential equation of the orthogonal family is $\dfrac{dy}{dx} = -\dfrac{1}{y + x}$ or $\dfrac{dx}{dy} + x = -y$.

(b) An integrating factor is e^y, so

$$\frac{d}{dy}[e^y x] = -ye^y \implies e^y x = -ye^y + e^y + c \implies x = -y + 1 + ce^{-y}.$$

33. (a) Letting $t = 0$ correspond to 1790 we have $P(0) = 3.929$ and $P(t) = 3.929e^{kt}$. Using $t = 10$, which corresponds to 1800, we have

$$5.308 = P(10) = 3.929e^{10k}.$$

This implies that $k = 0.03$, so that

$$P(t) = 3.929e^{0.03t}.$$

(b)

Year	Census Population	Predicted Population	Error	% Error
1790	3.929	3.929	0.000	0.00
1800	5.308	5.308	0.000	0.00
1810	7.240	7.171	0.069	0.95
1820	9.638	9.688	-0.050	-0.52
1830	12.866	13.088	-0.222	-1.73
1840	17.069	17.682	-0.613	-3.59
1850	23.192	23.888	-0.696	-3.00
1860	31.433	32.272	-0.839	-2.67
1870	38.558	43.599	-5.041	-13.07
1880	50.156	58.901	-8.745	-17.44
1890	62.948	79.574	-16.626	-26.41
1900	75.996	107.503	-31.507	-41.46
1910	91.972	145.234	-53.262	-57.91
1920	105.711	196.208	-90.497	-85.61
1930	122.775	265.074	-142.299	-115.90
1940	131.669	358.109	-226.440	-171.98
1950	150.697	483.798	-333.101	-221.04

34. It is necessary to know the air temperature from the time of death until the medical examiner arrives. We will assume that the temperature of the air is a constant $65°F$. By Newton's law of cooling we then have

$$\frac{dT}{dt} = k(T - 65), \qquad T(0) = 82.$$

Using linearity or separation of variables we obtain $T = 65 + ce^{kt}$. From $T(0) = 82$ we find $c = 17$, so that $T = 65 + 17e^{kt}$. To find k we need more information so we assume that the body temperature at $t = 2$ hours was $75°F$. Then $75 = 65 + 17e^{2k}$ and $k = -0.2653$ and

$$T(t) = 65 + 17e^{-0.2653t}.$$

At the time of death, t_0, $T(t_0) = 98.6°F$, so $98.6 = 65 + 17e^{-0.2653t}$, which gives $t = -2.568$. Thus, the murder took place about 2.568 hours prior to the discovery of the body.

35. We will assume that the temperature of both the room and the cream is $72°F$, and that the temperature of the coffee when it is first put on the table is $175°F$. If we let $T_1(t)$ represent the temperature of the coffee in Mr. Jones' cup at time t, then

$$\frac{dT_1}{dt} = k(T_1 - 72),$$

which implies $T_1 = 72 + c_1 e^{kt}$. At time $t = 0$ Mr. Jones adds cream to his coffee which immediately reduces its temperature by an amount α, so that $T_1(0) = 175 - \alpha$. Thus $175 - \alpha = T_1(0) = 72 + c_1$, which implies $c_1 = 103 - \alpha$, so that $T_1(t) = 72 + (103 - \alpha)e^{kt}$. At $t = 5$, $T_1(5) = 72 + (103 - \alpha)e^{5k}$. Now we let $T_2(t)$ represent the temperature of the coffee in Mrs. Jones' cup. From $T_2 = 72 + c_2 e^{kt}$ and $T_2(0) = 175$ we obtain $c_2 = 103$, so that $T_2(t) = 72 + 103e^{kt}$. At $t = 5$, $T_2(5) = 72 + 103e^{5k}$. When cream is added to Mrs. Jones' coffee the temperature is reduced by an amount α. Using the

49

fact that $k < 0$ we have

$$T_2(5) - \alpha = 72 + 103e^{5k} - \alpha < 72 + 103e^{5k} - \alpha e^{5k}$$

$$= 72 + (103 - \alpha)e^{5k} = T_1(5).$$

Thus, the temperature of the coffee in Mr. Jones' cup is hotter.

36. Separating variables and integrating, we have

$$\frac{dx}{n - x} = r\,dt$$

$$- \ln|n - x| = rt + c$$

$$n - x = c_1 e^{-rt}$$

$$x = n - c_1 e^{-rt}.$$

If $t = 0$, $x = x_0$ so $c_1 = n - x_0$ and $x(t) = n - (n - x_0)e^{-rt}$. As $t \to \infty$, $e^{-rt} \to 0$ and $x(t) \to n$. That is, eventually everyone will contract the disease. Setting $x = n - 1$ and solving for t we get $t = \dfrac{1}{r}\ln(n - x_0)$. This is about how long it will take for the epidemic to run its course.

Exercises 3.2

1. From $\dfrac{dC}{dt} = C(1 - 0.0005C)$ and $C(0) = 1$ we obtain $\left(\dfrac{1}{C} + \dfrac{0.0005}{1 - 0.0005C}\right) dC = dt$ and

$C = \dfrac{1.0005e^t}{1 + 0.0005e^t}$. Then $C(10) = 1834$ supermarkets, and $C \to 2000$ as $t \to \infty$.

2. From $\dfrac{dN}{dt} = N(a - bN)$ and $N(0) = 500$ we obtain $N = \dfrac{500a}{500b + (a - 500b)e^{-at}}$. Since

$\lim\limits_{t\to\infty} N = \dfrac{a}{b} = 50{,}000$ and $N(1) = 1000$ we have $a = 0.7033$, $b = 0.00014$, and $N = \dfrac{50{,}000}{1 + 99e^{-0.7033t}}$.

3. From $\dfrac{dP}{dt} = P\left(10^{-1} - 10^{-7}P\right)$ and $P(0) = 5000$ we obtain $P = \dfrac{500}{0.0005 + 0.0995e^{-0.1t}}$ so that

$P \to 1{,}000{,}000$ as $t \to \infty$. If $P(t) = 500{,}000$ then $t = 52.9$ months.

4. From $\dfrac{dP}{dt} = P(a - bP)\left(1 - cP^{-1}\right)$ we obtain $\left(\dfrac{b/(a - bc)}{a - bP} + \dfrac{1/(a - bc)}{P - c}\right) dP = dt$ and

$P = \dfrac{c + aEe^{(a-bc)t}}{1 + bEe^{(a-bc)t}}$ where E is an arbitrary constant.

5. (a) From $\dfrac{dP}{dt} = P(a - b\ln P)$ we obtain $\dfrac{-1}{b}\ln|a - b\ln P| = t + c_1$ so that $P = e^{a/b}e^{-ce^{-bt}}$.

(b) If $P(0) = P_0$ then $c = \dfrac{a}{b} - \ln P_0$.

6. From Problem 5 we have $P = e^{a/b}e^{-ce^{-bt}}$ so that

$$\frac{dP}{dt} = bce^{a/b-bt}e^{-ce^{-bt}} \quad \text{and} \quad \frac{d^2P}{dt^2} = b^2ce^{a/b-bt}e^{-ce^{-bt}}\left(ce^{-bt} - 1\right).$$

Setting $d^2P/dt^2 = 0$ and using $c = a/b - \ln P_0$ we obtain $t = (1/b)\ln(a/b - \ln P_0)$ and $P = e^{a/b-1}$.

7. Let $X = X(t)$ be the amount of C at time t and $\dfrac{dX}{dt} = k(120 - 2X)(150 - X)$. If $X(0) = 0$ and

$X(5) = 10$ then $X = \dfrac{150 - 150e^{180kt}}{1 - 2.5e^{180kt}}$ where $k = .0001259$, and $X(20) = 29.3\,\text{grams}$. Now $X \to 60$

as $t \to \infty$, so that the amount of $A \to 0$ and the amount of $B \to 30$ as $t \to \infty$.

8. From $\dfrac{dX}{dt} = k(150 - X)^2$, $X(0) = 0$, and $X(5) = 10$ we obtain $X = 150 - \dfrac{150}{150kt + 1}$ where

$k = .000095238$. Then $X(20) = 33.3\,\text{grams}$ and $X \to 150$ as $t \to \infty$ so that the amount of $A \to 0$

and the amount of $B \to 0$ as $t \to \infty$. If $X(t) = 75$ then $t = 70\,\text{minutes}$.

9. If $\alpha \neq \beta$, $\dfrac{dX}{dt} = k(\alpha - X)(\beta - X)$, and $X(0) = 0$ then $\left(\dfrac{1/(\beta - \alpha)}{\alpha - X} + \dfrac{1/(\alpha - \beta)}{\beta - X}\right)dX = k\,dt$ so that

$X = \dfrac{\alpha\beta - \alpha\beta e^{(\alpha-\beta)kt}}{\beta - \alpha e^{(\alpha-\beta)kt}}$. If $\alpha = \beta$ then $\dfrac{1}{(\alpha - X)^2}\,dX = k\,dt$ and $X = \alpha - \dfrac{1}{kt + c}$.

10. From $\dfrac{dX}{dt} = k(\alpha - X)(\beta - X)(\gamma - X)$ we obtain

$$\left(\frac{1}{(\beta - \alpha)(\gamma - \alpha)} \cdot \frac{1}{\alpha - X} + \frac{1}{(\alpha - \beta)(\gamma - \beta)} \cdot \frac{1}{\beta - X} + \frac{1}{(\alpha - \gamma)(\beta - \gamma)} \cdot \frac{1}{\gamma - X}\right)dX = k\,dt$$

so that

$$\frac{-1}{((\beta - \alpha)(\gamma - \alpha)}\ln|\alpha - X| + \frac{-1}{(\alpha - \beta)(\gamma - \beta)}\ln|\beta - X| + \frac{-1}{(\alpha - \gamma)(\beta - \gamma)}\ln|\gamma - X| = kt + c.$$

11. From $\dfrac{dh}{dt} = -\dfrac{\sqrt{h}}{25}$ and $h(0) = 20$ we obtain $h = \left(\sqrt{20} - \dfrac{t}{50}\right)^2$. If $h(t) = 0$ then $t = 50\sqrt{20}\,\text{seconds}$.

12. In this case the differential equation is

$$\frac{dh}{dt} = -\frac{0.6}{25}\sqrt{h} = -\frac{3}{125}\sqrt{h}.$$

Separating variables and integrating we have $2h^{1/2} = -\dfrac{3}{125}t + c_1$ or $h^{1/2} = -\dfrac{3}{250}t + c_2$. From

$h(0) = 20$ we find $c_2 = \sqrt{20}$, so $h = \left(\sqrt{20} - \dfrac{3}{250}t\right)^2$. Solving $h(t) = 0$ for t we find that the tank

empties in $\dfrac{250}{3}\sqrt{20}\,\text{s} \approx 6.2\,\text{min}$.

13. From $\dfrac{dy}{dx} = \dfrac{-y}{\sqrt{100 - y^2}}$ and $h(0) = 10$ we obtain $10 \displaystyle\int (\csc\theta - \sin\theta)\, d\theta = -\,dx$ where $y = 10\sin\theta$.

Then
$$10\ln\left|\frac{10}{y} + \frac{\sqrt{100 - y^2}}{y}\right| - \sqrt{100 - y^2} = x.$$

14. From $\dfrac{dT}{dt} = k\left(T^4 - T_m^4\right)$ we obtain

$$\left[\frac{1/\left(4T_m^3\right)}{T - T_m} - \frac{1/\left(4T_m^3\right)}{T + T_m} - \frac{1/\left(2T_m^2\right)}{T^2 + T_m^2}\right] dT = k\,dt$$

so that
$$\ln\left|\frac{T - T_m}{T + T_m}\right| - 2\tan^{-1}\frac{T}{T_m} = 4T_m^3 kt + c.$$

15. (a) Separating variables we obtain

$$\frac{m\,dv}{mg - kv^2} = dt$$

$$\frac{1}{g}\frac{dv}{1 - (kv/mg)^2} = dt$$

$$\frac{\sqrt{mg}}{\sqrt{k}\,g}\frac{\sqrt{k/mg}\,dv}{1 - (\sqrt{k}\,v/\sqrt{mg})^2} = dt$$

$$\sqrt{\frac{m}{kg}}\,\tanh^{-1}\frac{\sqrt{k}\,v}{\sqrt{mg}} = t + c$$

$$\tanh^{-1}\frac{\sqrt{k}\,v}{\sqrt{mg}} = \sqrt{\frac{kg}{m}}\,t + c_1.$$

Thus the velocity at time t is

$$v(t) = \sqrt{\frac{mg}{k}}\,\tanh\left(\sqrt{\frac{kg}{m}}\,t + c_1\right).$$

Setting $t = 0$ and $v = v_0$ we find $c_1 = \tanh^{-1}(\sqrt{k}\,v_0/\sqrt{mg}\,)$.

(b) Since $\tanh t \to 1$ as $t \to \infty$, we have $v \to \sqrt{mg/k}$ as $t \to \infty$.

(c) Integrating the expression for $v(t)$ in part (a) we obtain

$$s(t) = \sqrt{\frac{mg}{k}}\int \tanh\left(\sqrt{\frac{kg}{m}}\,t + c_1\right) dt = \frac{m}{k}\ln\left[\cosh\left(\sqrt{\frac{kg}{m}}\,t + c_1\right)\right] + c_2.$$

Setting $t = 0$ and $s = s_0$ we find $c_2 = s_0 - \ln\cosh c_1$.

16. (a) Let ρ be the weight density of the water and V the volume of the object. Archimedes' principle states that the upward buoyant force has magnitude equal to the weight of the water displaced. Taking the positive direction to be down, the differential equation is

$$m\frac{dv}{dt} = mg - kv^2 - \rho V.$$

(b) Using separation of variables we have

$$\frac{m\,dv}{(mg - \rho V) - kv^2} = dt$$

$$\frac{m}{\sqrt{k}}\frac{\sqrt{k}\,dv}{(\sqrt{mg - \rho V})^2 - (\sqrt{k}\,v)^2} = dt$$

$$\frac{m}{\sqrt{k}}\frac{1}{\sqrt{mg - \rho V}}\tanh^{-1}\frac{\sqrt{k}\,v}{\sqrt{mg - \rho V}} = t + c.$$

Thus

$$v(t) = \sqrt{\frac{mg - \rho V}{k}}\tanh\left(\frac{\sqrt{kmg - k\rho V}}{m}t + c_1\right).$$

(c) Since $\tanh t \to 1$ as $t \to \infty$, the terminal velocity is $\sqrt{(mg - \rho V)/k}$.

17. (a) The differential equation is

$$\frac{dP}{dt} = P(5 - P) - 4 = -(P^2 - 5P + 4) = -(P - 4)(P - 1).$$

Separating variables and integrating, we obtain

$$\frac{dP}{(P - 4)(P - 1)} = -dt$$

$$\left(\frac{1/3}{P - 4} - \frac{1/3}{P - 1}\right)dP = -dt$$

$$\frac{1}{3}\ln\left|\frac{P - 4}{P - 1}\right| = -t + c$$

$$\frac{P - 4}{P - 1} = c_1 e^{-3t}.$$

Setting $t = 0$ and $P = P_0$ we find $c_1 = (P_0 - 4)/(P_0 - 1)$. Solving for P we obtain

$$P(t) = \frac{4(P_0 - 1) - (P_0 - 4)e^{-3t}}{(P_0 - 1) - (P_0 - 4)e^{-3t}}.$$

53

(b) Graphing solutions obtained by letting $P_0 = 5$, $P_0 = 2$, and $P_0 = \frac{3}{4}$ we observe that for $P_0 > 4$ and $1 < P_0 < 4$ the population approaches 4 as t becomes infinite. We see that for $0 < P_0 < 1$ the population decreases to 0 in finite time. (The graph continues below the t-axis, but since P represents population we are only concerned with that part of the graph that lies in the first quadrant.)

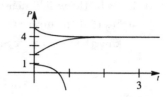

(c) To find when the population becomes extinct in the case $0 < P_0 < 1$ we set $P = 0$ in

$$\frac{P-4}{P-1} = \frac{P_0 - 4}{P_0 - 1}e^{-3t}$$

from part (a) and solve for t. This gives the time of extinction

$$t = -\frac{1}{3}\ln\frac{4(P_0 - 1)}{P_0 - 4}.$$

18. (a) We have $dP/dt = P(a - bP)$ with $P(0) = 3.929$ million. Using separation of variables we obtain

$$P(t) = \frac{3.929a}{3.929b + (a - 3.929b)e^{-at}} = \frac{a/b}{1 + (a/3.929b - 1)e^{-at}}$$

$$= \frac{c}{1 + (c/3.929 - 1)e^{-at}}.$$

At $t = 60(1850)$ the population is 23.192 million, so

$$23.192 = \frac{c}{1 + (c/3.929 - 1)e^{-60a}}$$

or $c = 23.192 + 23.192(c/3.929 - 1)e^{-60a}$. At $t = 120(1910)$

$$91.972 = \frac{c}{1 + (c/3.929 - 1)e^{-120a}}$$

or $c = 91.972 + 91.972(c/3.929 - 1)(e^{-60a})^2$. Combining the two equations for c we get

$$\left(\frac{(c - 23.192)/23.192}{c/3.929 - 1}\right)^2\left(\frac{c}{3.929} - 1\right) = \frac{c - 91.972}{91.972}$$

or

$$91.972(3.929)(c - 23.192)^2 = (23.192)^2(c - 91.972)(c - 3.929).$$

The solution of this quadratic equation is $c = 197.274$. This in turn gives $a = 0.0313$. Therefore

$$P(t) = \frac{197.274}{1 + 49.21e^{-0.0313t}}.$$

(b)

Year	Census Population	Predicted Population	Error	% Error
1790	3.929	3.929	0.000	0.00
1800	5.308	5.334	-0.026	-0.49
1810	7.240	7.222	0.018	0.24
1820	9.638	9.746	-0.108	-1.12
1830	12.866	13.090	-0.224	-1.74
1840	17.069	17.475	-0.406	-2.38
1850	23.192	23.143	0.049	0.21
1860	31.433	30.341	1.092	3.47
1870	38.558	39.272	-0.714	-1.85
1880	50.156	50.044	0.112	0.22
1890	62.948	62.600	0.348	0.55
1900	75.996	76.666	-0.670	-0.88
1910	91.972	91.739	0.233	0.25
1920	105.711	107.143	-1.432	-1.35
1930	122.775	122.140	0.635	0.52
1940	131.669	136.068	-4.399	-3.34
1950	150.697	148.445	2.252	1.49

19. Differentiating y we obtain

$$y' = -\frac{1}{(x + c_1)^2} = -y^2.$$

The differential equation of the family of orthogonal trajectories is $y' = 1/y^2$. Solving by separation of variables we obtain $y^3 = 3x + c_2$.

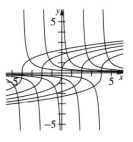

20. (a) From $V = \frac{4}{3}\pi r^3$ we have $r = (3V/4\pi)^{1/3}$. Then

$$\frac{dV}{dt} = kS = 4\pi r^2 k = 4\pi k \left(\frac{3V}{4\pi}\right)^{2/3}.$$

(b) Separating variables and integrating we have

$$\frac{dV}{V^{2/3}} = k(4\pi)^{1/3}3^{2/3}dt$$

$$3V^{1/3} - k(4\pi)^{1/3}3^{2/3}t + c$$

$$V = \left[k\left(\frac{4\pi}{3}\right)^{1/3}t + c_1\right]^3.$$

Using $V(0) = V_0$ we obtain $c_1 = V_0^{1/3}$, so

$$V(t) = \left[k\left(\frac{4\pi}{3}\right)^{1/3}t + V_0^{1/3}\right]^3.$$

(c) Differentiating $V = \frac{4}{3}\pi r^3$ with respect to t we obtain $dV/dt = 4\pi r^2\, dr/dt$. But we saw in part

(a) that $dV/dt = 4\pi r^2 k$. Thus $dr/dt = k$ and $r = kt + c$. When $t = 0$, $r = r_0$ so $r(t) = kt + r_0$. The snowball disappears when $r = 0$ or at time $t = -r_0/k$.

21. (a) Writing the equation in the form $(x - \sqrt{x^2 + y^2}\,)dx + y\,dy$ we identify $M = x - \sqrt{x^2 + y^2}$ and $N = y$. Since M and N are both homogeneous of degree 1 we use the substitution $y = ux$. It follows that

$$\left(x - \sqrt{x^2 + u^2 x^2}\,\right)dx + ux(u\,dx + x\,du) = 0$$

$$x\left[\left(1 - \sqrt{1 + u^2}\,\right) + u^2\right]dx + x^2 u\,du = 0$$

$$-\frac{u\,du}{1 + u^2 - \sqrt{1 + u^2}} = \frac{dx}{x}$$

$$\frac{u\,du}{\sqrt{1 + u^2}\,(1 - \sqrt{1 + u^2}\,)} = \frac{dx}{x}.$$

Letting $w = 1 - \sqrt{1 + u^2}$ we have $dw = -u\,du/\sqrt{1 + u^2}$ so that

$$-\ln\left(1 - \sqrt{1 + u^2}\,\right) = \ln x + c$$

$$\frac{1}{1 - \sqrt{1 + u^2}} = c_1 x$$

$$1 - \sqrt{1 + u^2} = -\frac{c_2}{x} \qquad (-c_2 = 1/c_1)$$

$$1 + \frac{c_2}{x} = \sqrt{1 + \frac{y^2}{x^2}}$$

$$1 + \frac{2c_2}{x} + \frac{c_2^2}{x^2} = 1 + \frac{y^2}{x^2}.$$

Solving for y^2 we have

$$y^2 = 2c_2 x + c_2^2 = 4\left(\frac{c_2}{2}\right)\left(x + \frac{c_2}{2}\right)$$

which is a family of parabolas symmetric with respect to the x-axis with vertex at $(-c_2/2, 0)$ and focus at the origin.

(b) Writing the differential equation as $yy' + x = \sqrt{x^2 + y^2}$ and then squaring and simplifying we obtain $y = 2xy' + y(y')^2$. Let $w = y^2$ and write the differential equation as $y^2 = 2xyy' + y^2(y')^2$. Now $dw/dx = 2yy'$ so $w = xw' + \frac{1}{4}(w')^2$. Using Problem 54 in Exercises 1.1 we obtain $w = cx + \frac{1}{4}c^2$ or $y^2 = cx + (c/2)^2$. Letting $c/2 = c_2$ we have $y^2 = 2c_2 x + c_2^2$, which is the solution obtained in part (a).

(c) Let $u = x^2 + y^2$ so that

$$\frac{du}{dx} = 2x + 2y \frac{dy}{dx} .$$

Then

$$y \frac{dy}{dx} = \frac{1}{2} \frac{du}{dx} - x$$

and the differential equation can be written in the form

$$\frac{1}{2} \frac{du}{dx} - x = -x + \sqrt{u} \quad \text{or} \quad \frac{1}{2} \frac{du}{dx} = \sqrt{u} .$$

Separating variables and integrating we have

$$\frac{du}{2\sqrt{u}} = dx$$

$$\sqrt{u} = x + c$$

$$u = x^2 + 2cx + c^2$$

$$x^2 + y^2 = x^2 + 2cx + c^2$$

$$y^2 = 2cx + c^2 .$$

22. (a) From $2W^2 - W^3 = W^2(2 - W) = 0$ we see that $W = 0$ and $W = 2$ are constant solutions.

(b) Using the DSolve function in *Mathematica* on $W' = \sqrt{4W^2 - 2W^3}$ we obtain $W(x) = 2(1 - \tanh^2 x)$ which can be written as $W(x) = 2\,\text{sech}^2 x$. The same result is obtained when the differential equation is written as $W' = -\sqrt{4W^2 - 2W^3}$.

(c)

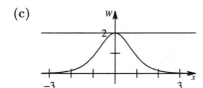

23. (a) For $0 \le t \le 15$ we use separation of variables:

$$\frac{m\,dv}{mg - kv^2} = dt$$

$$\frac{m}{\sqrt{mgk}} \tanh^{-1} \frac{\sqrt{k}\,v}{\sqrt{mg}} = t + c$$

$$\frac{\sqrt{k}\,v}{\sqrt{mg}} = \tanh\left(\sqrt{\frac{gk}{m}}\,t + c_1\right)$$

$$v = \sqrt{\frac{mg}{k}} \tanh\left(\sqrt{\frac{gk}{m}}\,t + c_1\right) .$$

From the initial condition $v(0) = 0$ we get $c_1 = 0$. This gives

$$v_1(t) = \sqrt{\frac{mg}{k}} \tanh\left(\sqrt{\frac{gk}{m}}\, t\right).$$

Integrating, we obtain

$$s_1(t) = \frac{m}{k} \ln\left(\cosh \sqrt{\frac{gk}{m}}\, t\right) + c.$$

Using $s_1(0) = 0$ we find $c = 0$ and

$$s_1(t) = \frac{m}{k} \ln\left(\cosh \sqrt{\frac{gk}{m}}\, t\right).$$

Thus, $v_1(15) \approx 172.4$ ft/s and $s_1(15) \approx 1956.1$ ft. When the parachute opens we use the solution of Problem 27, part(a) in Exercises 3.1, with $v_2(0) = v_0 = v_1(15) = 172.4$ and $s_2(0) = 0$. This gives

$$v_2(t) = \frac{mg}{k} + \left(172.4 - \frac{mg}{k}\right) e^{-kt/m}.$$

Integrating, we obtain

$$s_2(t) = \frac{mg}{k} t - \frac{m}{k}\left(172.4 - \frac{mg}{k}\right) e^{-kt/m} + c_1.$$

Using $s_2(0) = 0$ we find $c_1 = (m/k)(172.4 - mg/k)$ and

$$s_2(t) = \frac{mg}{k} t - \frac{m}{k}\left(172.4 - \frac{mg}{k}\right) e^{-kt/m} + \frac{m}{k}\left(172.4 - \frac{mg}{k}\right).$$

To find the time the skydiver reaches the ground starting from $s_2(0) = 0$ we solve

$$s_2(t) = 12,000 - s_1(15) = 12,000 - 1956.1 = 10,043.9$$

or

$$\frac{160}{7.857} t - \frac{5}{7.857}\left(172.4 - \frac{160}{7.857}\right) e^{-7.857t/5} + \frac{5}{7.857}\left(172.4 - \frac{160}{7.857}\right) = 10.043.9$$

which simplifies to

$$20.4t - 9.68e^{-1.5714t} - 9947.1 = 0.$$

Since we expect t to be relatively large, $e^{-1.5714t} \approx 0$, and we need only solve $20.4t = 9947.1$. This gives $t = 487.6$ seconds. Therefore the total time to reach the ground is $15 + 487.6 = 502.6$ seconds or 8.4 minutes. The impact velocity will be the terminal velocity

$$v_t = mg/k = 160/7.857 \approx 20.4 \text{ ft/s} = 13.9 \text{ mi/h}.$$

_____ Exercises 3.3 _____

1. The equation $dx/dt = -\lambda_1 x$ can be solved by separation of variables. Integrating both sides of $dx/x = -\lambda_1 dt$ we obtain $\ln|x| = -\lambda_1 t + c$ from which we get $x = c_1 e^{-\lambda_1 t}$. Using $x(0) = x_0$ we find $c_1 x_0$ so that $x = x_0 e^{-\lambda_1 t}$. Substituting this result into the second differential equation we have

$$\frac{dy}{dt} + \lambda_2 y = \lambda_1 x_0 e^{-\lambda_1 t}$$

which is linear. An integrating factor is $e^{\lambda_2 t}$ so that

$$\frac{d}{dt}\left[e^{\lambda_2 t} y\right] = \lambda_1 x_0 e^{(\lambda_2 - \lambda_1)t} + c_2$$

$$y = \frac{\lambda_1 x_0}{\lambda_2 - \lambda_1} e^{(\lambda_2 - \lambda_1)t} e^{-\lambda_2 t} + c_2 e^{-\lambda_2 t} = \frac{\lambda_1 x_0}{\lambda_2 - \lambda_1} e^{-\lambda_1 t} + c_2 e^{-\lambda_2 t}.$$

Using $y(0) = 0$ we find $c_2 = -\lambda_1 x_0/(\lambda_2 - \lambda_1)$. Thus

$$y = \frac{\lambda_1 x_0}{\lambda_2 - \lambda_1}\left(e^{-\lambda_1 t} - e^{-\lambda_2 t}\right).$$

Substituting this result into the third differential equation we have

$$\frac{dz}{dt} = \frac{\lambda_1 \lambda_2 x_0}{\lambda_2 - \lambda_1}\left(e^{-\lambda_1 t} - e^{-\lambda_2 t}\right).$$

Integrating we find

$$z = -\frac{\lambda_2 x_0}{\lambda_2 - \lambda_1} e^{-\lambda_1 t} + \frac{\lambda_1 x_0}{\lambda_2 - \lambda_1} e^{-\lambda_2 t} + c_3.$$

Using $z(0) = 0$ we find $c_3 = x_0$. Thus

$$z = x\left(1 - \frac{\lambda_2}{\lambda_2 - \lambda_1} e^{-\lambda_1 t} + \frac{\lambda_1}{\lambda_2 - \lambda_1} e^{-\lambda_2 t}\right).$$

2. We see from the graph that the half-life of A is approximately 4.7 days. To determine the half-life of B we use $t = 50$ as a base, since at this time the amount of substance A is so small that it contributes very little to substance B. Now we see from the graph that $y(50) \approx 16.2$ and $y(191) \approx 8.1$. Thus, the half-life of B is approximately 141 days.

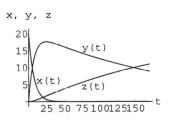

3. The amounts of x and y are the same at about $t = 5$ days. The amounts of x and z are the same at about $t = 20$ days. The amounts of y and z are the same at about $t = 147$ days. The time when y and z are the same makes sense because most of A and half of B are gone, so half of C should have been formed.

Exercises 3.3

4. Suppose that the series is described schematically by $W \xrightarrow{-\lambda_1} X \xrightarrow{-\lambda_2} Y \xrightarrow{-\lambda_3} Z$ where $-\lambda_1$, $-\lambda_2$, and $-\lambda_3$ are the decay constants for W, X and Y, respectively, and Z is a stable element. Let $w(t)$, $x(t)$, $y(t)$, and $z(t)$ denote the amounts of substances W, X, Y, and Z, respectively. A model for the radioactive series is

$$\frac{dw}{dt} = -\lambda_1 w$$

$$\frac{dx}{dt} = \lambda_1 w - \lambda_2 x$$

$$\frac{dy}{dt} = \lambda_2 x - \lambda_3 y$$

$$\frac{dz}{dt} = \lambda_3 y.$$

5. The system is

$$x_1' = 2 \cdot 3 + \frac{1}{50}x_2 - \frac{1}{50}x_1 \cdot 4 = -\frac{2}{25}x_1 + \frac{1}{50}x_2 - 6$$

$$x_2' = \frac{1}{50}x_1 \cdot 4 - \frac{1}{50}x_2 - \frac{1}{50}x_2 \cdot 3 = \frac{2}{25}x_1 - \frac{2}{25}x_2.$$

6. Let x_1, x_2, and x_3 be the amounts of salt in tanks A, B, and C, respectively, so that

$$x_1' = \frac{1}{100}x_2 \cdot 2 - \frac{1}{100}x_1 \cdot 6 = \frac{1}{50}x_2 - \frac{3}{50}x_1$$

$$x_2' = \frac{1}{100}x_1 \cdot 6 + \frac{1}{100}x_3 - \frac{1}{100}x_2 \cdot 2 - \frac{1}{100}x_2 \cdot 5 = \frac{3}{50}x_1 - \frac{7}{100}x_2 + \frac{1}{100}x_3$$

$$x_3' = \frac{1}{100}x_2 \cdot 5 - \frac{1}{100}x_3 - \frac{1}{100}x_3 \cdot 4 = \frac{1}{20}x_2 - \frac{1}{20}x_3.$$

7. A model is

$$\frac{dx_1}{dt} = 3 \cdot \frac{x_2}{100 - t} - 2 \cdot \frac{x_1}{100 + t}, \qquad x_1(0) = 100$$

$$\frac{dx_2}{dt} = 2 \cdot \frac{x_1}{100 + t} - 3 \cdot \frac{x_2}{100 - t}, \qquad x_2(0) = 50.$$

8. Since the system is closed, no salt enters or leaves the system and $x_1(t) + x_2(t) = 100 + 50 = 150$ for all time. Thus $x_1 = 150 - x_2$ and the second equation in Problem 7 becomes

$$\frac{dx_2}{dt} = \frac{2(150 - x_2)}{100 + t} - \frac{3x_2}{100 - t} = \frac{300}{100 + t} - \frac{2x_2}{100 + t} - \frac{3x_2}{100 - t}$$

or

$$\frac{dx_2}{dt} + \left(\frac{2}{100 + t} + \frac{3}{100 - t} \right) x_2 = \frac{300}{100 + t},$$

which is linear in x_2. An integrating factor is

$$e^{2\ln(100+t) - 3\ln(100-t)} = (100 + t)^2 (100 - t)^{-3}$$

so
$$\frac{d}{dt}[(100+t)^2(100-t)^{-3}x_2] = 300(100+t)(100-t)^{-3}.$$

Using integration by parts, we obtain

$$(100+t)^2(100-t)^{-3}x_2 = 300\left[\frac{1}{2}(100+t)(100-t)^{-2} - \frac{1}{2}(100-t)^{-1} + c\right].$$

Thus

$$x_2 = \frac{300}{(100+t)^2}\left[c(100-t)^3 - \frac{1}{2}(100-t)^2 + \frac{1}{2}(100+t)(100-t)\right]$$

$$= \frac{300}{(100+t)^2}[c(100-t)^3 + t(100-t)].$$

Using $x_2(0) = 50$ we find $c = 5/3000$. At $t = 30$, $x_2 = (300/130^2)(70^3c + 30 \cdot 70) \approx 47.4$lbs.

9. From the graph we see that the populations are first equal at about $t = 5.6$. The approximate periods of x and y are both 45.

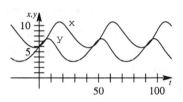

10. (a) The population $y(t)$ approaches 10,000, while the population $x(t)$ approaches extinction.

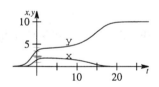

(b) The population $x(t)$ approaches 5,000, while the population $y(t)$ approaches extinction.

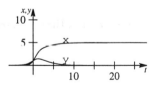

(c) The population $y(t)$ approaches 10,000, while the population $x(t)$ approaches extinction.

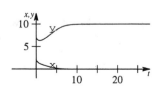

61

Exercises 3.3

(d) The population $x(t)$ approaches 5,000, while the population $y(t)$ approaches extinction.

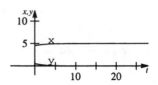

11. (a)

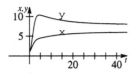

(b)

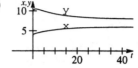

(c)

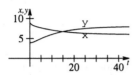

(d)

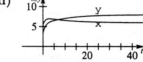

In each case the population $x(t)$ approaches 6,000, while the population $y(t)$ approaches 8,000.

12. By Kirchoff's first law we have $i_1 = i_2 + i_3$. By Kirchoff's second law, on each loop we have $E(t) = Li'_1 + R_1 i_2$ and $E(t) = Li'_1 + R_2 i_3 + \frac{1}{C}q$ so that $q = CR_1 i_2 - CR_2 i_3$. Then $i_3 = q' = CR_1 i'_2 - CR_2 i_3$ so that the system is

$$Li'_2 + Li'_3 + R_1 i_2 = E(t)$$

$$-R_1 i'_2 + R_2 i'_3 + \frac{1}{C} i_3 = 0.$$

13. By Kirchoff's first law we have $i_1 = i_2 + i_3$. Applying Kirchoff's second law to each loop we obtain

$$E(t) = i_1 R_1 + L_1 \frac{di_2}{dt} + i_2 R_2$$

and

$$E(t) = i_1 R_1 + L_2 \frac{di_3}{dt} + i_3 R_3.$$

Combining the three equations, we obtain the system

$$L_1 \frac{di_2}{dt} + (R_1 + R_2)i_2 + R_1 i_3 = E$$

$$L_2 \frac{di_3}{dt} + R_1 i_2 + (R_1 + R_3)i_3 = E.$$

14. By Kirchoff's first law we have $i_1 = i_2 + i_3$. By Kirchoff's second law, on each loop we have $E(t) = Li'_1 + Ri_2$ and $E(t) = Li'_1 + \frac{1}{C}q$ so that $q = CRi_2$. Then $i_3 = q' = CRi'_2$ so that system is

$$Li' + Ri_2 = E(t)$$

$$CRi'_2 + i_2 - i_1 = 0.$$

15. We first note that $s(t) + i(t) + r(t) = n$. Now the rate of change of the number of susceptible persons, $s(t)$, is proportional to the number of contacts between the number of people infected and the number who are susceptible; that is, $ds/dt = -k_1 s_i$. We use $-k_1$ because $s(t)$ is decreasing. Next, the rate of change of the number of persons who have recovered is proportional to the number infected; that is, $dr/dt = k_2 i$ where k_2 is positive since r is increasing. Finally, to obtain di/dt we use

$$\frac{d}{dt}(s + i + r) = \frac{d}{dt}n = 0.$$

This gives

$$\frac{di}{dt} = -\frac{dr}{dt} - \frac{ds}{dt} = -k_2 i + k_1 s i.$$

The system of equations is then

$$\frac{ds}{dt} = -k_1 s i$$

$$\frac{di}{dt} = -k_2 i + k_1 s i$$

$$\frac{dr}{dt} = k_2 i.$$

A reasonable set of initial conditions is $i(0) = i_0$, the number of infected people at time 0, $s(0) = n - i_0$, and $r(0) = 0$.

16. (a) If we know $s(t)$ and $i(t)$ then we can determine $r(t)$ from $s + i + r = n$.

(b) In this case the system is

$$\frac{ds}{dt} = -0.2 s i$$

$$\frac{di}{dt} = -0.7 i + 0.2 s i.$$

We also note that when $i(0) = i_0$, $s(0) = 10 - i_0$ since $r(0) = 0$ and $i(t) + s(t) + r(t) = 0$ for all values of t. Now $k_2/k_1 = 0.7/0.2 = 3.5$, so we consider initial conditions $s(0) = 2$, $i(0) = 8$; $s(0) = 3.4$, $i(0) = 6.6$; $s(0) = 7$, $i(0) = 3$; and $s(0) = 9$, $i(0) = 1$.

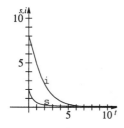
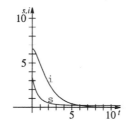
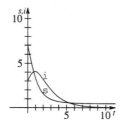
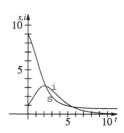

We see that an initial susceptible population greater than k_2/k_1 results in an epidemic in the sense that the number of infected persons increases to a maximum before decreasing to 0. On the other hand, when $s(0) < k_2/k_1$, the number of infected persons decreases from the start and there is no epidemic.

17. Since $x_0 > y_0 > 0$ we have $x(t) > y(t)$ and $y-x < 0$. Thus $dx/dt < 0$ and $dy/dt > 0$. We conclude that $x(t)$ is decreasing and $y(t)$ is increasing. As $t \to \infty$ we expect that $x(t) \to C$ and $y(t) \to C$, where C is a constant common equilibrium concentration.

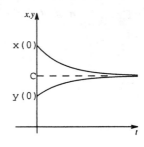

18. We write the system in the form

$$\frac{dx}{dt} = k_1(y - x)$$

$$\frac{dy}{dt} = k_2(x - y),$$

where $k_1 = \kappa/V_A$ and $k_2 = \kappa/V_B$. Letting $z(t) = x(t) - y(t)$ we have

$$\frac{dx}{dt} - \frac{dy}{dt} = k_1(y - x) - k_2(x - y)$$

$$\frac{dz}{dt} = k_1(-z) - k_2 z$$

$$\frac{dz}{dt} + (k_1 + k_2)z = 0.$$

This is a first-order linear differential equation with solution $z(t) = c_1 e^{-(k_1+k_2)t}$. Now

$$\frac{dx}{dt} = -k_1(y - x) = -k_1 z = -k_1 c_1 e^{-(k_1+k_2)t}$$

and

$$x(t) = c_1 \frac{k_1}{k_1 + k_2} e^{-(k_1+k_2)t} + c_2.$$

Since $y(t) = x(t) - z(t)$ we have

$$y(t) = -c_1 \frac{k_2}{k_1 + k_2} e^{-(k_1+k_2)t} + c_2.$$

The initial conditions $x(0) = x_0$ and $y(0) = y_0$ imply

$$c_1 = x_0 - y_0 \quad \text{and} \quad c_2 = \frac{x_0 k_2 + y_0 k_1}{k_1 + k_2}.$$

The solution of the system is

$$x(t) = \frac{(x_0 - y_0)k_1}{k_1 + k_2}e^{-(k_1+k_2)t} + \frac{x_0 k_2 + y_0 k_1}{k_1 + k_2}$$

$$y(t) = \frac{(y_0 - x_0)k_2}{k_1 + k_2}e^{-(k_1+k_2)t} + \frac{x_0 k_2 + y_0 k_1}{k_1 + k_2}.$$

As $t \to \infty$, $x(t)$ and $y(t)$ approach the common limit

$$\frac{x_0 k_2 + y_0 k_1}{k_1 + k_2} = \frac{x_0 \kappa/V_B + y_0 \kappa/V_A}{\kappa/V_A + \kappa/V_B} = \frac{x_0 V_A + y_0 V_B}{V_A + V_B}$$

$$= x_0 \frac{V_A}{V_A + V_B} + y_0 \frac{V_B}{V_A + V_B}.$$

This makes intuitive sense because the limiting concentration is seen to be a weighted average of the two initial concentrations.

19. Since there are initially 25 pounds of salt in tank A and none in tank B, and since furthermore only pure water is being pumped into tank A, we would expect that $x_1(t)$ would steadily decrease over time. On the other hand, since salt is being added to tank B from tank A, we would expect $x_2(t)$ to increase over time. However, since pure water is being added to the system at a constant rate and a mixed solution is being pumped out of the system, it makes sense that the amount of salt in both tanks would approach 0 over time.

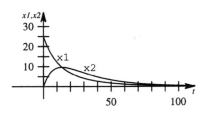

Chapter 3 Review Exercises

1. From $\dfrac{dP}{dt} = 0.018P$ and $P(0) = 4$ billion we obtain $P = 4e^{0.018t}$ so that $P(45) = 8.99$ billion.

2. Let $A = A(t)$ be the volume of CO_2 at time t. From $\dfrac{dA}{dt} = 1.2 - \dfrac{A}{4}$ and $A(0) = 16\,\text{ft}^3$ we obtain $A = 4.8 + 11.2e^{-t/4}$. Since $A(10) = 5.7\,\text{ft}^3$, the concentration is 0.017%. As $t \to \infty$ we have $A \to 4.8\,\text{ft}^3$ or 0.06%.

3. From $dE/dt = -E/RC$ and $E(t_1) = E_0$ we obtain $E = E_0 e^{(t_1-t)/RC}$.

4. From $V\,dC/dt = kA(C_s - C)$ and $C(0) = C_0$ we obtain $C = C_s + (C_0 - C_s)e^{-kAt/V}$.

5. (a) The differential equation is

$$\frac{dT}{dt} = k[T - T_2 - B(T_1 - T)] = k[(1 + B)T - (BT_1 + T_2)].$$

Separating variables we obtain $\dfrac{dT}{(1+B)T - (BT_1 + T_2)} = k\,dt$. Then

$$\frac{1}{1+B}\ln|(1+B)T - (BT_1 + T_2)| = kt + c \quad \text{and} \quad T(t) = \frac{BT_1 + T_2}{1+B} + c_3 e^{k(1+B)t}.$$

Since $T(0) = T_1$ we must have $c_3 = \dfrac{T_1 - T_2}{1+B}$ and so

$$T(t) = \frac{BT_1 + T_2}{1+B} + \frac{T_1 - T_2}{1+B}e^{k(1+B)t}.$$

(b) Since $k < 0$, $\displaystyle\lim_{t\to\infty} e^{k(1+B)t} = 0$ and $\displaystyle\lim_{t\to\infty} T(t) = \frac{BT_1 + T_2}{1+B}$.

(c) Since $T_s = T_2 + B(T_1 - T)$, $\displaystyle\lim_{t\to\infty} T_s = T_2 + BT_1 - B\left(\frac{BT_1 + T_2}{1+B}\right) = \frac{BT_1 + T_2}{1+B}$.

6. We first solve $\left(1 - \dfrac{t}{10}\right)\dfrac{di}{dt} + 0.2i = 4$. Separating variables we obtain

$$\frac{di}{40 - 2i} = \frac{dt}{10 - t}. \text{ Then}$$

$$-\frac{1}{2}\ln|40 - 2i| = -\ln|10 - t| + c \quad \text{or} \quad \sqrt{40 - 2i} = c_1(10 - t).$$

Since $i(0) = 0$ we must have $c_1 = 2/\sqrt{10}$. Solving for i we get $i(t) = 4t - \frac{1}{5}t^2$, $0 \le t < 10$. For $t \ge 10$ the equation for the current becomes $0.2i = 4$ or $i = 20$. Thus

$$i(t) = \begin{cases} 4t - \frac{1}{5}t^2, & 0 \le t < 10 \\ 20, & t \ge 10 \end{cases}.$$

7. From $y\left[1 + (y')^2\right] = k$ we obtain $dx = \dfrac{\sqrt{y}}{\sqrt{k - y}}\,dy$. If $y = k\sin^2\theta$ then

$$dy = 2k\sin\theta\cos\theta\,d\theta, \quad dx = 2k\left(\frac{1}{2} - \frac{1}{2}\cos 2\theta\right)d\theta, \quad \text{and} \quad x = k\theta - \frac{k}{2}\sin 2\theta + c.$$

If $x = 0$ when $\theta = 0$ then $c = 0$.

8. Setting $\dfrac{dv}{dt} = \dfrac{dy}{dt}\dfrac{dv}{dy} = v\dfrac{dv}{dt}$ we see that the first equation can be written as $mv\dfrac{dv}{dy} = -mg - kv^2$. Separating variables and integrating we have

$$\int \frac{mv}{mg + kv^2}\,dv = -\int dy$$

$$\frac{m}{2k}\ln(mg + kv^2) = -y + c_1$$

$$mg + kv^2 = c_2 e^{-2ky/m}$$

$$v^2 = c_3 e^{-2ky/m} - \frac{mg}{k}.$$

Using $y(0) = 0$ and $v(0) = v_0$ we see that $v = v_0$ when $y = 0$ so that $v_0^2 = c_3 - mg/k$ and $c_3 = v_0^2 + mg/k$. Thus

$$v^2 = \frac{kv_0^2 + mg}{k} e^{-2ky/m} - \frac{mg}{k}.$$

Setting $v = 0$ we see that the maximum height is

$$h = \frac{m}{2k} \ln \frac{kv_0^2 + mg}{mg}.$$

Now, setting $\dfrac{dv}{dt} = v \dfrac{dv}{dt}$ in the second equation, we have $mv \dfrac{dv}{dy} = mg - kv^2$. Separating variables and integrating we find

$$\int \frac{mv}{mg - kv^2} \, dv = \int dy$$

$$-\frac{m}{2k} \ln|mg - kv^2| = y + c_1$$

$$mg - kv^2 = c_2 e^{-2ky/m}$$

$$v^2 = \frac{mg}{k}(1 - c_3 e^{-2ky/m}).$$

In this case $v = 0$ when $y = 0$ so $c_3 = 1$ and

$$v^2 = \frac{mg}{k}(1 - e^{-2ky/m}).$$

Setting $y = h = \dfrac{m}{2k} \ln \dfrac{kv_0^2 + mg}{mg}$ and solving for v we see that the impact velocity is

$$v_i = \frac{v_0}{\sqrt{1 + kv_0^2/mg}} < v_0.$$

9. From $\dfrac{dx}{dt} = k_1 x(\alpha - x)$ we obtain $\left(\dfrac{1/\alpha}{x} + \dfrac{1/\alpha}{\alpha - x}\right) dx = k_1 \, dt$ so that $x = \dfrac{\alpha c_1 e^{\alpha k_1 t}}{1 + c_1 e^{\alpha k_1 t}}$. From $\dfrac{dy}{dt} = k_2 xy$ we obtain

$$\ln|y| = \frac{k_2}{k_1} \ln\left|1 + c_1 e^{\alpha k_1 t}\right| + c \quad \text{or} \quad y = c_2 \left(1 + c_1 e^{\alpha k_1 t}\right)^{k_2/k_1}.$$

10. In tank A the salt input is

$$\left(7 \frac{\text{gal}}{\text{min}}\right)\left(2 \frac{\text{lb}}{\text{gal}}\right) + \left(1 \frac{\text{gal}}{\text{min}}\right)\left(\frac{x_2}{100} \frac{\text{lb}}{\text{gal}}\right) = \left(14 + \frac{1}{100} x_2\right) \frac{\text{lb}}{\text{min}}.$$

The salt output is

$$\left(3 \frac{\text{gal}}{\text{min}}\right)\left(\frac{x_1}{100} \frac{\text{lb}}{\text{gal}}\right) + \left(5 \frac{\text{gal}}{\text{min}}\right)\left(\frac{x_1}{100} \frac{\text{lb}}{\text{gal}}\right) = \frac{2}{25} x_1 \frac{\text{lb}}{\text{min}}.$$

In tank B the salt input is

$$\left(5\,\frac{\text{gal}}{\text{min}}\right)\left(\frac{x_1}{100}\,\frac{\text{lb}}{\text{gal}}\right)=\frac{1}{20}x_1\,\frac{\text{lb}}{\text{min}}\,.$$

The salt output is

$$\left(1\,\frac{\text{gal}}{\text{min}}\right)\left(\frac{x_2}{100}\,\frac{\text{lb}}{\text{gal}}\right)+\left(4\,\frac{\text{gal}}{\text{min}}\right)\left(\frac{x_2}{100}\,\frac{\text{lb}}{\text{gal}}\right)=\frac{1}{20}x_2\,\frac{\text{lb}}{\text{min}}\,.$$

The system of differential equations is then

$$\frac{dx_1}{dt}=14+\frac{1}{100}x_2-\frac{2}{25}x_1$$

$$\frac{dx_2}{dt}=\frac{1}{20}x_1-\frac{1}{20}x_2.$$

4 Differential Equations of Higher-Order

Exercises 4.1

1. From $y = c_1 e^x + c_2 e^{-x}$ we find $y' = c_1 e^x - c_2 e^{-x}$. Then $y(0) = c_1 + c_2 = 0$, $y'(0) = c_1 - c_2 = 1$ so that $c_1 = 1/2$ and $c_2 = -1/2$. The solution is $y = \frac{1}{2} e^x - \frac{1}{2} e^{-x}$.

2. We have $y(0) = c_1 + c_2 = 0$, $y'(1) = c_1 e + c_2 e^{-1} = 1$ so that $c_1 = e/\left(e^2 - 1\right)$ and $c_2 = -e/\left(e^2 - 1\right)$. The solution is $y = e\left(e^x - e^{-x}\right)/\left(e^2 - 1\right)$.

3. From $y = c_1 e^{4x} + c_2 e^{-x}$ we find $y' = 4c_1 e^{4x} - c_2 e^{-x}$. Then $y(0) = c_1 + c_2 = 1$, $y'(0) = 4c_1 - c_2 = 2$ so that $c_1 = 3/5$ and $c_2 = 2/5$. The solution is $y = \frac{3}{5} e^{4x} + \frac{2}{5} e^{-x}$.

4. From $y = c_1 + c_2 \cos x + c_3 \sin x$ we find $y' = -c_2 \sin x + c_3 \cos x$ and $y'' = -c_2 \cos x - c_3 \sin x$. Then $y(\pi) = c_1 - c_2 = 0$, $y'(\pi) = -c_3 = 2$, $y''(\pi) = c_2 = -1$ so that $c_1 = -1$, $c_2 = -1$, and $c_3 = -2$. The solution is $y = -1 - \cos x - 2 \sin x$.

5. From $y = c_1 x + c_2 x \ln x$ we find $y' = c_1 + c_2 (1 + \ln x)$. Then $y(1) = c_1 = 3$, $y'(1) = c_1 + c_2 = -1$ so that $c_1 = 3$ and $c_2 = -4$. The solution is $y = 3x - 4x \ln x$.

6. From $y = c_1 + c_2 x^2$ we find $y' = 2c_2 x$. Then $y(0) = c_1 = 0$, $y'(0) = 2c_2 \cdot 0 = 0$ and $y'(0) = 1$ is not possible. Since $a_2(x) = x$ is 0 at $x = 0$, Theorem 4.1 is not violated.

7. In this case we have $y(0) = c_1 = 0$, $y'(0) = 2c_2 \cdot 0 = 0$ so $c_1 = 0$ and c_2 is arbitrary. Two solutions are $y = x^2$ and $y = 2x^2$.

8. In this case we have $y(0) = c_1 = 1$, $y'(1) = 2c_2 = 6$ so that $c_1 = 1$ and $c_2 = 3$. The solution is $y = 1 + 3x^2$. Theorem 4.1 does not apply because y and y' are evaluated at different points.

9. From $y = c_1 e^x \cos x + c_2 e^x \sin x$ we find $y' = c_1 e^x (-\sin x + \cos x) + c_2 e^x (\cos x + \sin x)$.
   (a) We have $y(0) = c_1 = 1$, $y'(0) = c_1 + c_2 = 0$ so that $c_1 = 1$ and $c_2 = -1$. The solution is $y = e^x \cos x - e^x \sin x$.
   (b) We have $y(0) = c_1 = 1$, $y(\pi) = -c_1 e^\pi = -1$, which is not possible.
   (c) We have $y(0) = c_1 = 1$, $y(\pi/2) = c_2 e^{\pi/2} = 1$ so that $c_1 = 1$ and $c_2 = e^{-\pi/2}$. The solution is $y = e^x \cos x + e^{-\pi/2} e^x \sin x$.
   (d) We have $y(0) = c_1 = 0$, $y(\pi) = -c_1 e^\pi = 0$ so that $c_1 = 0$ and c_2 is arbitrary. Solutions are $y = c_2 e^x \sin x$, for any real numbers c_2.

10. (a) We have $y(-1) = c_1 + c_2 + 3 = 0$, $y(1) = c_1 + c_2 + 3 = 4$, which is not possible.
    (b) We have $y(0) = c_1 \cdot 0 + c_2 \cdot 0 + 3 = 1$, which is not possible.

(c) We have $y(0) = c_1 \cdot 0 + c_2 \cdot 0 + 3 = 3$, $y(1) = c_1 + c_2 + 3 = 0$ so that c_1 is arbitrary and $c_2 = -3 - c_1$. Solutions are $y = c_1 x^2 - (c_1 + 3)x^4 + 3$.

(d) We have $y(1) = c_1 + c_2 + 3 = 3$, $y(2) = 4c_1 + 16c_2 + 3 = 15$ so that $c_1 = -1$ and $c_2 = 1$. The solution is $y = -x^2 + x^4 + 3$.

11. Since $a_2(x) = x - 2$ and $x_0 = 0$ the problem has a unique solution for $-\infty < x < 2$.

12. Since $a_0(x) = \tan x$ and $x_0 = 0$ the problem has a unique solution for $-\pi/2 < x < \pi/2$.

13. From $x(0) = x_0 = c_1$ we see that $x(t) = x_0 \cos \omega t + c_2 \sin \omega t$ and $x'(t) = -x_0 \sin \omega t + c_2 \omega \cos \omega t$. Then $x'(0) = x_1 = c_2 \omega$ implies $c_2 = x_1/\omega$. Thus

$$x(t) = x_0 \cos \omega t + \frac{x_1}{\omega} \sin \omega t.$$

14. Solving the system

$$x(t_0) = c_1 \cos \omega t_0 + c_2 \sin \omega t_0 = x$$

$$x'(t_0) = -c_1 \omega \sin \omega t_0 + c_2 \omega \cos \omega t_0 = x_1$$

for c_1 and c_2 gives

$$c_1 = \frac{\omega x_0 \cos \omega t_0 - x_1 \sin \omega t_0}{\omega} \quad \text{and} \quad c_2 = \frac{x_1 \cos \omega t_0 + \omega x_0 \sin \omega t_0}{\omega}.$$

Thus

$$x(t) = \frac{\omega x_0 \cos \omega t_0 - x_1 \sin \omega t_0}{\omega} \cos \omega t + \frac{x_1 \cos \omega t_0 + \omega x_0 \sin \omega t_0}{\omega} \sin \omega t$$

$$= x_0(\cos \omega t \cos \omega t_0 + \sin \omega t \sin \omega t_0) + \frac{x_1}{\omega}(\sin \omega t \cos \omega t_0 - \cos \omega t \sin \omega t_0)$$

$$= x_0 \cos \omega(t - t_0) + \frac{x_1}{\omega} \sin \omega(t - t_0).$$

15. Since $(-4)x + (3)x^2 + (1)(4x - 3x^2) = 0$ the functions are linearly dependent.

16. Since $(1)0 + (0)x + (0)e^x = 0$ the functions are linearly dependent. A similar argument shows that any set of functions containing $f(x) = 0$ will be linearly dependent.

17. Since $(-1/5)5 + (1)\cos^2 x + (1)\sin^2 x = 0$ the functions are linearly dependent.

18. Since $(1)\cos 2x + (1)1 + (-2)\cos^2 x = 0$ the functions are linearly dependent.

19. Since $(-4)x + (3)(x - 1) + (1)(x + 3) = 0$ the functions are linearly dependent.

20. From the graphs of $f_1(x) = 2 + x$ and $f_2(x) = 2 + |x|$
we see that the functions are linearly independent since
they cannot be multiples of each other.

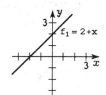

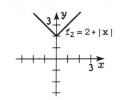

21. The functions are linearly independent since $W\left(1 + x, x, x^2\right) = \begin{vmatrix} 1+x & x & x^2 \\ 1 & 1 & 2x \\ 0 & 0 & 2 \end{vmatrix} = 2 \neq 0.$

22. Since $(-1/2)e^x + (1/2)e^{-x} + (1)\sinh x = 0$ the functions are linearly dependent.

23. The functions satisfy the differential equation and are linearly independent since
$$W\left(e^{-3x}, e^{4x}\right) = 7e^x \neq 0$$
for $-\infty < x < \infty$. The general solution is
$$y = c_1 e^{-3x} + c_2 e^{4x}.$$

24. The functions satisfy the differential equation and are linearly independent since
$$W(\cosh 2x, \sinh 2x) = 2$$
for $-\infty < x < \infty$. The general solution is
$$y = c_1 \cosh 2x + c_2 \sinh 2x.$$

25. The functions satisfy the differential equation and are linearly independent since
$$W\left(e^x \cos 2x, e^x \sin 2x\right) = 2e^{2x} \neq 0$$
for $-\infty < x < \infty$. The general solution is $y = c_1 e^x \cos 2x + c_2 e^x \sin 2x.$

26. The functions satisfy the differential equation and are linearly independent since
$$W\left(e^{x/2}, xe^{x/2}\right) = e^x \neq 0$$
for $-\infty < x < \infty$. The general solution is
$$y = c_1 e^{x/2} + c_2 x e^{x/2}.$$

27. The functions satisfy the differential equation and are linearly independent since
$$W\left(x^3, x^4\right) = x^6 \neq 0$$
for $0 < x < \infty$. The general solution is
$$y = c_1 x^3 + c_2 x^4.$$

28. The functions satisfy the differential equation and are linearly independent since

$$W\left(\cos(\ln x), \sin(\ln x)\right) = 1/x \neq 0$$

for $0 < x < \infty$. The general solution is

$$y = c_1 \cos(\ln x) + c_2 \sin(\ln x).$$

29. The functions satisfy the differential equation and are linearly independent since

$$W\left(x, x^{-2}, x^{-2}\ln x\right) = 9x^{-6} \neq 0$$

for $0 < x < \infty$. The general solution is

$$y = c_1 x + c_2 x^{-2} + c_3 x^{-2} \ln x.$$

30. The functions satisfy the differential equation and are linearly independent since

$$W(1, x, \cos x, \sin x) = 1$$

for $-\infty < x < \infty$. The general solution is

$$y = c_1 + c_2 x + c_3 \cos x + c_4 \sin x.$$

31. (a) From the graphs of $y_1 = x^3$ and $y_2 = |x|^3$ we see that the functions are linearly independent since they cannot be multiples of each other. It is easily shown that $y_1 = x^3$ solves $x^2 y'' - 4xy' + 6y = 0$. To show that $y_2 = |x|^3$ is a solution let $y_2 = x^3$ for $x \geq 0$ and let $y_2 = -x^3$ for $x < 0$.

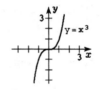 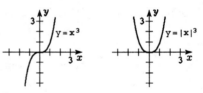

(b) If $x \geq 0$ then $y_2 = x^3$ and $W(y_1, y_2) = \begin{vmatrix} x^3 & x^3 \\ 3x^2 & 3x^2 \end{vmatrix} = 0$. If $x < 0$ then $y_2 = -x^3$ and

$$W(y_1, y_2) = \begin{vmatrix} x^3 & -x^3 \\ 3x^2 & -3x^2 \end{vmatrix} = 0.$$

This does not violate Theorem 3.3 since $a_2(x) = x^2$ is zero at $x = 0$.

(c) The functions $Y_1 = x^3$ and $Y_2 = x^2$ are solutions of $x^2 y'' - 4xy' + 6y = 0$. They are linearly independent since $W\left(x^3, x^2\right) = x^4 \neq 0$ for $-\infty < x < \infty$.

(d) The function $y = x^3$ satisfies $y(0) = 0$ and $y'(0) = 0$.

(e) Neither is the general solution since we form a general solution on an interval for which $a_2(x) \neq 0$ for every x in the interval.

32. By the superposition principle, if $y_1 = e^x$ and $y_2 = e^{-x}$ are both solutions of a homogeneous linear differential equation, then so are

$$\frac{1}{2}(y_1 + y_2) = \frac{e^x + e^{-x}}{2} = \cosh x \quad \text{and} \quad \frac{1}{2}(y_1 - y_2) = \frac{e^x - e^{-x}}{2} = \sinh x.$$

33. The functions $y_1 = e^{2x}$ and $y_2 = e^{5x}$ form a fundamental set of solutions of the homogeneous equation, and $y_p = 6e^x$ is a particular solution of the nonhomogeneous equation.

34. The functions $y_1 = \cos x$ and $y_2 = \sin x$ form a fundamental set of solutions of the homogeneous equation, and $y_p = x \sin x + (\cos x) \ln(\cos x)$ is a particular solution of the nonhomogeneous equation.

35. The functions $y_1 = e^{2x}$ and $y_2 = xe^{2x}$ form a fundamental set of solutions of the homogeneous equation, and $y_p = x^2 e^{2x} + x - 2$ is a particular solution of the nonhomogeneous equation.

36. The functions $y_1 = x^{-1/2}$ and $y_2 = x^{-1}$ form a fundamental set of solutions of the homogeneous equation, and $y_p = \frac{1}{15}x^2 - \frac{1}{6}x$ is a particular solution of the nonhomogeneous equation.

37. By the superposition principle for nonhomogeneous equations a particular solution of
$y'' - 6y' + 5y = 5x^2 + 3x - 16 - 9e^{2x}$ is $y_p = x^2 + 3x + 3e^{2x}$. A particular solution of the second equation is

$$y_p = -2y_{p_2} - \frac{1}{9}y_{p_1} = -2x^2 - 6x - \frac{1}{3}e^{2x}.$$

38. (a) $y_{p_1} = 5$

    (b) $y_{p_2} = -2x$

    (c) $y_p = y_{p_1} + y_{p_2} = 5 - 2x$

    (d) $y_p = \frac{1}{2}y_{p_1} - 2y_{p_2} = \frac{5}{2} + 4x$

Exercises 4.2

In Problems 1-10 we use reduction of order to find a second solution. In Problems 11-24 we use formula (5) from the text.

1. Define $y = u(x) \cdot 1$ so

$$y' = u', \quad y'' = u'', \quad \text{and} \quad y'' + 5y' = u'' + 5u' = 0.$$

If $w = u'$ we obtain the first-order equation $w' + 5w = 0$ which has the integrating factor $e^{5\int dx} = e^{5x}$.

Now
$$\frac{d}{dx}[e^{5x}w] = 0 \quad \text{gives} \quad e^{5x}w = c.$$

Therefore $w = u' = ce^{-5x}$ and $u = c_1 e^{-5x}$. A second solution is $y_2 = e^{-5x}$.

2. Define $y = u(x) \cdot 1$ so

$$y' = u', \quad y'' = u'', \quad \text{and} \quad y'' - y' = u'' - u' = 0.$$

If $w = u'$ we obtain the first-order equation $w' - w = 0$ which has the integrating factor $e^{-\int dx} = e^{-x}$.

Now
$$\frac{d}{dx}[e^{-x}w] = 0 \quad \text{gives} \quad e^{-x}w = c.$$

Therefore $w = u' = ce^x$ and $u = ce^x$. A second solution is $y_2 = e^x$.

3. Define $y = u(x)e^{2x}$ so

$$y' = 2ue^{2x} + u'e^{2x}, \quad y'' = e^{2x}u'' + 4e^{2x}u' + 4e^{2x}u, \quad \text{and} \quad y'' - 4y' + 4y = 4e^{2x}u'' = 0.$$

Therefore $u'' = 0$ and $u = c_1 x + c_2$. Taking $c_1 = 1$ and $c_2 = 0$ we see that a second solution is $y_2 = xe^{2x}$.

4. Defing $y = u(x)xe^{-x}$ so

$$y' = (1 - x)e^{-x}u + xe^{-x}u', \quad y'' = xe^{-x}u'' + 2(1 - x)e^{-x}u' - (2 - x)e^{-x}u,$$

and

$$y'' + 2y' + y = e^{-x}(xu'' + 2u') = 0 \quad \text{or} \quad u'' + \frac{2}{x}u' = 0.$$

If $w = u'$ we obtain the first-order equation $w' + \dfrac{2}{x}w = 0$ which has the integrating factor $e^{2 \int dx/x} = x^2$. Now

$$\frac{d}{dx}[x^2 w] = 0 \quad \text{gives} \quad x^2 w = c.$$

Therefore $w = u' = c/x^2$ and $u = c_1/x$. A second solution is $y_2 = \dfrac{1}{x}xe^{-x} = e^{-x}$.

5. Define $y = u(x)\cos 4x$ so

$$y' = -4u\sin 4x + u'\cos 4x, \quad y'' = u''\cos 4x - 8u'\sin 4x - 16u\cos 4x$$

and

$$y'' + 16y = (\cos 4x)u'' - 8(\sin 4x)u' = 0 \quad \text{or} \quad u'' - 8(\tan 4x)u' = 0.$$

If $w = u'$ we obtain the first-order equation $w' - 8(\tan 4x)w = 0$ which has the integrating factor $e^{-8 \int \tan 4x \, dx} = \cos^2 4x$. Now

$$\frac{d}{dx}[(\cos^2 4x)w] = 0 \quad \text{gives} \quad (\cos^2 4x)w = c.$$

Therefore $w = u' = c\sec^2 4x$ and $u = c_1 \tan 4x$. A second solution is $y_2 = \tan 4x \cos 4x = \sin 4x$.

6. Define $y = u(x)\sin 3x$ so

$$y' = 3u\cos 3x + u'\sin 3x, \quad y'' = u''\sin 3x + 6u'\cos 3x - 9u\sin 3x,$$

and

$$y'' + 9y = (\sin 3x)u'' + 6(\cos 3x)u' = 0 \quad \text{or} \quad u'' + 6(\cot 3x)u' = 0.$$

If $w = u'$ we obtain the first-order equation $w' + 6(\cot 3x)w = 0$ which has the integrating factor $e^{6 \int \cot 3x \, dx} = \sin^2 3x$. Now

$$\frac{d}{dx}[(\sin^2 3x)w] = 0 \quad \text{gives} \quad (\sin^2 3x)w = c.$$

Therefore $w = u' = c \csc^2 3x$ and $u = c_1 \cot 3x$. A second solution is $y_2 = \cot 3x \sin 3x = \cos 3x$.

7. Define $y = u(x) \cosh x$ so

$$y' = u \sinh x + u' \cosh x, \quad y'' = u'' \cosh x + 2u' \sinh x + u \cosh x$$

and

$$y'' - y = (\cosh x)u'' + 2(\sinh x)u' = 0 \quad \text{or} \quad u'' + 2(\tanh x)u' = 0.$$

If $w = u'$ we obtain the first-order equation $w' + 2(\tanh x)w = 0$ which has the integrating factor $e^{2 \int \tanh x \, dx} = \cosh^2 x$. Now

$$\frac{d}{dx} [(\cosh^2 x)w] = 0 \quad \text{gives} \quad (\cosh^2 x)w = c.$$

Therefore $w = u' = c \operatorname{sech}^2 x$ and $u = c_1 \tanh x$. A second solution is $y_2 = \tanh x \cosh x = \sinh x$.

8. Define $y = u(x)e^{5x}$ so

$$y' = 5e^{5x}u + e^{5x}u', \quad y'' = e^{5x}u'' + 10e^{5x}u' + 25e^{5x}u$$

and

$$y'' - 25y = e^{5x}(u'' + 10u') = 0 \quad \text{or} \quad u'' + 10u' = 0.$$

If $w = u'$ we obtain the first-order equation $w' + 10w = 0$ which has the integrating factor $e^{10 \int dx} = e^{10x}$. Now

$$\frac{d}{dx} [e^{10x}w] = 0 \quad \text{gives} \quad e^{10x}w = c.$$

Therefore $w = u' = ce^{-10x}$ and $u = c_1 e^{-10x}$. A second solution is $y_2 = e^{-10x}e^{5x} = e^{-5x}$.

9. Define $y = u(x)e^{2x/3}$ so

$$y' = \frac{2}{3}e^{2x/3}u + e^{2x/3}u', \quad y'' = e^{2x/3}u'' + \frac{4}{3}e^{2x/3}u' + \frac{4}{9}e^{2x/3}u$$

and

$$9y'' - 12y' + 4y = 9e^{2x/3}u'' = 0.$$

Therefore $u'' = 0$ and $u = c_1 x + c_2$. Taking $c_1 = 1$ and $c_2 = 0$ we see that a second solution is $y_2 = xe^{2x/3}$.

10. Define $y = u(x)e^{x/3}$ so

$$y' = \frac{1}{3}e^{x/3}u + e^{x/3}u', \quad y'' = e^{x/3}u'' + \frac{2}{3}e^{x/3}u' + \frac{1}{9}e^{x/3}u$$

and

$$6y'' + y' - y = e^{x/3}(6u'' + 5u') = 0 \quad \text{or} \quad u'' + \frac{5}{6}u' = 0.$$

If $w = u'$ we obtain the first-order equation $w' + \frac{5}{6}w = 0$ which has the integrating factor

$e^{(5/6)\int dx} = e^{5x/6}$. Now

$$\frac{d}{dx}[e^{5x/6}w] = 0 \quad \text{gives} \quad e^{5x/6}w = c.$$

Therefore $w = u' = ce^{-5x/6}$ and $u = c_1 e^{-5x/6}$. A second solution is $y_2 = e^{-5x/6}e^{x/3} = e^{-x/2}$.

11. Identifying $P(x) = -7/x$ we have

$$y_2 = x^4 \int \frac{e^{-\int -(7/x)\,dx}}{x^8}\,dx = x^4 \int \frac{1}{x}\,dx = x^4 \ln|x|.$$

A second solution is $y_2 = x^4 \ln|x|$.

12. Identifying $P(x) = 2/x$ we have

$$y_2 = x^2 \int \frac{e^{-\int (2/x)\,dx}}{x^4}\,dx = x^2 \int x^{-6}\,dx = -\frac{1}{5}x^{-3}.$$

A second solution is $y_2 = x^{-3}$.

13. Identifying $P(x) = 1/x$ we have

$$y_2 = \ln x \int \frac{e^{-\int dx/x}}{(\ln x)^2}\,dx = \ln x \int \frac{dx}{x(\ln x)^2} = \ln x \left(-\frac{1}{\ln x}\right) = -1.$$

A second solution is $y_2 = 1$.

14. Identifying $P(x) = 0$ we have

$$y_2 = x^{1/2}\ln x \int \frac{e^{-\int 0\,dx}}{x(\ln x)^2} = x^{1/2}\ln x \left(-\frac{1}{\ln x}\right) = -x^{1/2}.$$

A second solution is $y_2 = x^{1/2}$.

15. Identifying $P(x) = 2(1+x)/\left(1 - 2x - x^2\right)$ we have

$$y_2 = (x+1)\int \frac{e^{-\int 2(1+x)dx/(1-2x-x^2)}}{(x+1)^2}\,dx = (x+1)\int \frac{e^{\ln(1-2x-x^2)}}{(x+1)^2}\,dx$$

$$= (x+1)\int \frac{1-2x-x^2}{(x+1)^2}\,dx = (x+1)\int \left[\frac{2}{(x+1)^2} - 1\right] dx$$

$$= (x+1)\left[-\frac{2}{x+1} - x\right] = -2 - x^2 - x.$$

A second solution is $y_2 = x^2 + x + 2$.

16. Identifying $P(x) = -2x/\left(1 - x^2\right)$ we have

$$y_2 = \int e^{-\int -2x\,dx/(1-x^2)}\,dx = \int e^{-\ln(1-x^2)}\,dx = \int \frac{1}{1-x^2}\,dx = \frac{1}{2}\ln\left|\frac{1+x}{1-x}\right|.$$

A second solution is $y_2 = \ln|(1+x)/(1-x)|$.

17. Identifying $P(x) = -1/x$ we have

$$y_2 = x\sin(\ln x)\int \frac{e^{-\int -dx/x}}{x^2\sin^2(\ln x)}\,dx = x\sin(\ln x)\int \frac{x}{x^2\sin^2(\ln x)}\,dx$$

$$= [x\sin(\ln x)][-\cot(\ln x)] = -x\cos(\ln x).$$

A second solution is $y_2 = x\cos(\ln x)$.

18. Identifying $P(x) = -3/x$ we have

$$y_2 = x^2\cos(\ln x)\int \frac{e^{-\int -3\,dx/x}}{x^4\cos^2(\ln x)}\,dx = x^2\cos(\ln x)\int \frac{x^3}{x^4\cos^2(\ln x)}\,dx$$

$$= x^2\cos(\ln x)\tan(\ln x) = x^2\sin(\ln x).$$

A second solution is $y_2 = x^2\sin(\ln x)$.

19. Identifying $P(x) = 4x/(1+2x)$ we have

$$y_2 = e^{-2x}\int \frac{e^{-\int 4x\,dx/(1+2x)}}{e^{-4x}}\,dx = e^{-2x}\int \frac{e^{-2x+\ln(1+2x)}}{e^{-4x}}\,dx$$

$$= e^{-2x}\int(1+2x)e^{2x}\,dx = e^{-2x}\left[\frac{1}{2}e^{2x} + xe^{2x} - \frac{1}{2}e^{2x}\right] = x.$$

A second solution is $y_2 = x$.

20. Identifying $P(x) = x/(1+x)$ we have

$$y_2 = x\int \frac{e^{-\int x\,dx/(1+x)}}{x^2}\,dx = x\int \frac{e^{-x+\ln(1+x)}}{x^2}\,dx = x\int \frac{(1+x)e^{-x}}{x^2}\,dx = x\int\left(\frac{e^{-x}}{x^2} + \frac{e^{-x}}{x}\right)\,dx$$

$$= x\int \frac{e^{-x}}{x^2}\,dx + x\int \frac{e^{-x}}{x}\,dx \qquad \boxed{u = e^{-x},\ du = -e^{-x}\,dx,\ dv = \frac{1}{x^2}\,dx,\ v = -\frac{1}{x}}$$

$$= x\left(-\frac{1}{x}e^{-x} - \int \frac{e^{-x}}{x}\,dx\right) + x\int \frac{e^{-x}}{x}\,dx = -e^{-x}.$$

A second solution is $y_2 = e^{-x}$.

21. Identifying $P(x) = -1/x$ we have

$$y_2 = x\int \frac{e^{-\int -dx/x}}{x^2}\,dx = x\int \frac{dx}{x} = x\ln|x|.$$

A second solution is $y_2 = x\ln|x|$.

22. Identifying $P(x) = 0$ we have

$$y_2 = x^{-4}\int \frac{e^{-\int 0\,dx}}{x^{-8}}\,dx = x^{-4}\left(\frac{1}{9}x^9\right) = \frac{1}{9}x^5.$$

A second solution is $y_2 = x^5$.

23. Identifying $P(x) = -5/x$ we have

$$y_2 = x^3 \ln x \int \frac{e^{-\int -5\,dx/x}}{x^6 (\ln x)^2}\,dx = x^3 \ln x \int \frac{x^5}{x^6 (\ln x)^2}\,dx = x^3 \ln x \left(-\frac{1}{\ln x}\right) = -x^3.$$

A second solution is $y_2 = x^3$.

24. Identifying $P(x) = 1/x$ we have

$$y_2 = \cos(\ln x) \int \frac{e^{-\int dx/x}}{\cos^2(\ln x)}\,dx = \cos(\ln x) \int \frac{1/x}{\cos^2(\ln x)}\,dx = \cos(\ln x)\tan(\ln x) = \sin(\ln x).$$

A second solution is $y_2 = \sin(\ln x)$.

25. Define $y = u(x)e^{-2x}$ so

$$y' = -2ue^{-2x} + u'e^{-2x}, \quad y'' = u''e^{-2x} - 4u'e^{-2x} + 4ue^{-2x}$$

and

$$y'' - 4y = e^{-2x}u'' - 4e^{-2x}u' = 0 \quad \text{or} \quad u'' - 4u' = 0.$$

If $w = u'$ we obtain the first order equation $w' - 4w = 0$ which has the integrating factor $e^{-4\int dx} = e^{-4x}$. Now

$$\frac{d}{dx}[e^{-4x}w] = 0 \quad \text{gives} \quad e^{-4x}w = c.$$

Therefore $w = u' = ce^{4x}$ and $u = c_1 e^{4x}$. A second solution is $y_2 = e^{-2x}e^{4x} = e^{2x}$. We see by observation that a particular solution is $y_p = -1/2$. The general solution is

$$y = c_1 e^{-2x} + c_2 e^{2x} - \frac{1}{2}.$$

26. Define $y = u(x) \cdot 1$ so

$$y' = u', \quad y'' = u'' \quad \text{and} \quad y'' + y' = u'' + u' = 0.$$

If $w = u'$ we obtain the first order equation $w' + w = 0$ which has the integrating factor $e^{\int dx} = e^x$. Now

$$\frac{d}{dx}[e^x w] = 0 \quad \text{gives} \quad e^x w = c.$$

Therefore $w = u' = ce^{-x}$ and $u = c_1 e^{-x}$. A second solution is $y_2 = 1 \cdot e^{-x} = e^{-x}$. We see by observation that a particular solution is $y_p = x$. The general solution is

$$y = c_1 + c_2 e^{-x} + x.$$

27. Define $y = u(x)e^x$ so

$$y' = ue^x + u'e^x, \quad y'' = u''e^x + 2u'e^x + ue^x$$

78

and

$$y'' - 3y' + 2y = e^x u'' - e^x u' = 0 \quad \text{or} \quad u'' - u' = 0.$$

If $w = u'$ we obtain the first order equation $w' - w = 0$ which has the integrating factor $e^{-\int dx} = e^{-x}$. Now

$$\frac{d}{dx}[e^{-x}w] = 0 \quad \text{gives} \quad e^{-x}w = c.$$

Therefore $w = u' = ce^x$ and $u = ce^x$. A second solution is $y_2 = e^x e^x = e^{2x}$. To find a particular solution we try $y_p = Ae^{3x}$. Then $y' = 3Ae^{3x}$, $y'' = 9Ae^{3x}$, and $9Ae^{3x} - 3\left(3Ae^{3x}\right) + 2Ae^{3x} = 5e^{3x}$. Thus $A = 5/2$ and $y_p = \frac{5}{2}e^{3x}$. The general solution is

$$y = c_1 e^x + c_2 e^{2x} + \frac{5}{2}e^{3x}.$$

28. Define $y = u(x)e^x$ so

$$y' = ue^x + u'e^x, \quad y'' = u''e^x + 2u'e^x + ue^x$$

and

$$y'' - 4y' + 3y = e^x u'' - 2e^x u' = 0 \quad \text{or} \quad u'' - 2u' = 0.$$

If $w = u'$ we obtain the first order equation $w' - 2w = 0$ which has the integrating factor $e^{-2\int dx} = e^{-2x}$. Now

$$\frac{d}{dx}[e^{-2x}w] = 0 \quad \text{gives} \quad e^{-2x}w = c.$$

Therefore $w = u' = ce^{2x}$ and $u = c_1 e^{2x}$. A second solution is $y_2 = e^x e^{2x} = e^{3x}$. To find a particular solution we try $y_p = ax + b$. Then $y_p' = a$, $y_p'' = 0$, and $0 - 4a + 3(ax + b) = 3ax - 4a + 3b = x$. Then $3a = 1$ and $-4a + 3b = 0$ so $a = 1/3$ and $b = 4/9$. A particular solution is $y_p = \frac{1}{3}x + \frac{4}{9}$ and the general solution is

$$y = c_1 e^x + c_2 e^{3x} + \frac{1}{3}x + \frac{4}{9}.$$

29. (a) If $y_2 = y_1 \int \dfrac{e^{-\int P\,dx}}{y_1^2}\,dx$ then

$$y_2' = \frac{1}{y_1}e^{-\int P\,dx} + y_1' \int \frac{e^{-\int P\,dx}}{y_1^2}\,dx$$

and

$$y_2'' = \frac{-P}{y_1}e^{-\int P\,dx} - \frac{y_1'}{y_1^2}e^{-\int P\,dx} + \frac{y_1'}{y_1^2}e^{-\int P\,dx} + y_1'' \int \frac{e^{-\int P\,dx}}{y_1^2}\,dx$$

so that

$$y_2'' + Py_2' + Qy_2 = \left(y_1'' + Py_1' + Qy_1\right)\int \frac{e^{-\int P\,dx}}{y_1^2}\,dx = 0.$$

(b) $W(y_1, y_2) = \begin{vmatrix} y_1 & uy_1 \\ y_1' & uy_1' + u'y_1 \end{vmatrix} = u'y_1^2 = \dfrac{e^{-\int P\,dx}}{y_1^2}\,y_1^2 = e^{-\int P\,dx}$

30. (a) For m_1 constant, let $y_1 = e^{m_1 x}$. Then $y_1' = m_1 e^{m_1 x}$ and $y_1'' = m_1^2 e^{m_1 x}$. Substituting into the differential equation we obtain

$$ay_1'' + by_1' + cy_1 = am_1^2 e^{m_1 x} + bm_1 e^{m_1 x} + ce^{m_1 x}$$

$$= e^{m_1 x}(am_1^2 + bm_1 + c) = 0.$$

Thus, $y_1 = e^{m_1 x}$ will be a solution of the differential equation whenever $am_1^2 + bm_1 + c = 0$. Since a quadratic equation always has at least one real or complex root, the differential equation must have a solution of the form $y_1 = e^{m_1 x}$.

(b) Write the differential equation in the form

$$y'' + \frac{b}{a}y' + \frac{c}{a}y = 0,$$

and let $y_1 = e^{m_1 x}$ be a solution. Then a second solution is given by

$$y_2 = e^{m_1 x} \int \frac{e^{-bx/a}}{e^{2m_1 x}}\,dx$$

$$= e^{m_1 x} \int e^{-(b/a + 2m_1)x}\,dx$$

$$= -\frac{1}{b/a + 2m_1}\,e^{m_1 x}e^{-(b/a + 2m_1)} \qquad (m_1 \neq -b/2a)$$

$$= -\frac{1}{b/a + 2m_1}\,e^{-(b/a + m_1)}.$$

Thus, when $m_1 \neq -b/2a$, a second solution is given by $y_2 = e^{m_2 x}$ where $m_2 = -b/a - m_1$. When $m_1 = -b/2a$ a second solution is given by

$$y_2 = e^{m_1 x} \int dx = xe^{m_1 x}.$$

(c) The functions

$$\sin x = \frac{1}{2i}(e^{ix} - e^{-ix}) \qquad \cos x = \frac{1}{2}(e^{ix} + e^{-ix})$$

$$\sinh x = \frac{1}{2}(e^x - e^{-x}) \qquad \cosh x = \frac{1}{2}(e^x + e^{-x})$$

are all expressible in terms of exponential functions.

Exercises 4.3

1. From $4m^2 + m = 0$ we obtain $m = 0$ and $m = -1/4$ so that $y = c_1 + c_2 e^{-x/4}$.

2. From $2m^2 - 5m = 0$ we obtain $m = 0$ and $m = 5/2$ so that $y = c_1 + c_2 e^{5x/2}$.

3. From $m^2 - 36 = 0$ we obtain $m = 6$ and $m = -6$ so that $y = c_1 e^{6x} + c_2 e^{-6x}$.

4. From $m^2 - 8 = 0$ we obtain $m = 2\sqrt{2}$ and $m = -2\sqrt{2}$ so that $y = c_1 e^{2\sqrt{2}x} + c_2 e^{-2\sqrt{2}x}$.

5. From $m^2 + 9 = 0$ we obtain $m = 3i$ and $m = -3i$ so that $y = c_1 \cos 3x + c_2 \sin 3x$.

6. From $3m^2 + 1 = 0$ we obtain $m = i/\sqrt{3}$ and $m = -i/\sqrt{3}$ so that $y = c_1 \cos x/\sqrt{3} + c_2 \sin x/\sqrt{3}$.

7. From $m^2 - m - 6 = 0$ we obtain $m = 3$ and $m = -2$ so that $y = c_1 e^{3x} + c_2 e^{-2x}$.

8. From $m^2 - 3m + 2 = 0$ we obtain $m = 1$ and $m = 2$ so that $y = c_1 e^x + c_2 e^{2x}$.

9. From $m^2 + 8m + 16 = 0$ we obtain $m = -4$ and $m = -4$ so that $y = c_1 e^{-4x} + c_2 x e^{-4x}$.

10. From $m^2 - 10m + 25 = 0$ we obtain $m = 5$ and $m = 5$ so that $y = c_1 e^{5x} + c_2 x e^{5x}$.

11. From $m^2 + 3m - 5 = 0$ we obtain $m = -3/2 \pm \sqrt{29}/2$ so that $y = c_1 e^{(-3+\sqrt{29})x/2} + c_2 e^{(-3-\sqrt{29})x/2}$.

12. From $m^2 + 4m - 1 = 0$ we obtain $m = -2 \pm \sqrt{5}$ so that $y = c_1 e^{(-2+\sqrt{5})x} + c_2 e^{(-2-\sqrt{5})x}$.

13. From $12m^2 - 5m - 2 = 0$ we obtain $m = -1/4$ and $m = 2/3$ so that $y = c_1 e^{-x/4} + c_2 e^{2x/3}$.

14. From $8m^2 + 2m - 1 = 0$ we obtain $m = 1/4$ and $m = -1/2$ so that $y = c_1 e^{x/4} + c_2 e^{-x/2}$.

15. From $m^2 - 4m + 5 = 0$ we obtain $m = 2 \pm i$ so that $y = e^{2x}(c_1 \cos x + c_2 \sin x)$.

16. From $2m^2 - 3m + 4 = 0$ we obtain $m = 3/4 \pm \sqrt{23}\,i/4$ so that
$$y = e^{3x/4}\left(c_1 \cos \sqrt{23}\,x/4 + c_2 \sin \sqrt{23}\,x/4\right).$$

17. From $3m^2 + 2m + 1 = 0$ we obtain $m = -1/3 \pm \sqrt{2}\,i/3$ so that
$$y = e^{-x/3}\left(c_1 \cos \sqrt{2}\,x/3 + c_2 \sin \sqrt{2}\,x/3\right).$$

18. From $2m^2 + 2m + 1 = 0$ we obtain $m = -1/2 \pm i/2$ so that
$$y = e^{-x/2}(c_1 \cos x/2 + c_2 \sin x/2).$$

19. From $m^3 - 4m^2 - 5m = 0$ we obtain $m = 0$, $m = 5$, and $m = -1$ so that
$$y = c_1 + c_2 e^{5x} + c_3 e^{-x}.$$

20. From $4m^3 + 4m^2 + m = 0$ we obtain $m = 0$, $m = -1/2$, and $m = -1/2$ so that
$$y = c_1 + c_2 e^{-x/2} + c_3 x e^{-x/2}.$$

21. From $m^3 - 1 = 0$ we obtain $m = 1$ and $m = -1/2 \pm \sqrt{3}\,i/2$ so that
$$y = c_1 e^x + e^{-x/2}\left(c_2 \cos \sqrt{3}\,x/2 + c_3 \sin \sqrt{3}\,x/2\right).$$

22. From $m^3 + 5m^2 = 0$ we obtain $m = 0$, $m = 0$, and $m = -5$ so that

$$y = c_1 + c_2 x + c_3 e^{-5x}.$$

23. From $m^3 - 5m^2 + 3m + 9 = 0$ we obtain $m = -1$, $m = 3$, and $m = 3$ so that

$$y = c_1 e^{-x} + c_2 e^{3x} + c_3 x e^{3x}.$$

24. From $m^3 + 3m^2 - 4m - 12 = 0$ we obtain $m = -2$, $m = 2$, and $m = -3$ so that

$$y = c_1 e^{-2x} + c_2 e^{2x} + c_3 e^{-3x}.$$

25. From $m^3 + m^2 - 2 = 0$ we obtain $m = 1$ and $m = -1 \pm i$ so that

$$y = c_1 e^x + e^{-x}(c_2 \cos x + c_3 \sin x).$$

26. From $m^3 - m^2 - 4 = 0$ we obtain $m = 2$ and $m = -1/2 \pm \sqrt{7}\,i/2$ so that

$$y = c_1 e^{2x} + e^{-x/2}\left(c_2 \cos \sqrt{7}\,x/2 + c_3 \sin \sqrt{7}\,x/2\right).$$

27. From $m^3 + 3m^2 + 3m + 1 = 0$ we obtain $m = -1$, $m = -1$, and $m = -1$ so that

$$y = c_1 e^{-x} + c_2 x e^{-x} + c_3 x^2 e^{-x}.$$

28. From $m^3 - 6m^2 + 12m - 8 = 0$ we obtain $m = 2$, $m = 2$, and $m = 2$ so that

$$y = c_1 e^{2x} + c_2 x e^{2x} + c_3 x^2 e^{2x}.$$

29. From $m^4 + m^3 + m^2 = 0$ we obtain $m = 0$, $m = 0$, and $m = -1/2 \pm \sqrt{3}\,i/2$ so that

$$y = c_1 + c_2 x + e^{-x/2}\left(c_3 \cos \sqrt{3}\,x/2 + c_4 \sin \sqrt{3}\,x/2\right).$$

30. From $m^4 - 2m^2 + 1 = 0$ we obtain $m = 1$, $m = 1$, $m = -1$, and $m = -1$ so that

$$y = c_1 e^x + c_2 x e^x + c_3 e^{-x} + c_4 x e^{-x}.$$

31. From $16m^4 + 24m^2 + 9 = 0$ we obtain $m = \pm\sqrt{3}\,i/2$ and $m = \pm\sqrt{3}\,i/2$ so that

$$y = c_1 \cos \sqrt{3}\,x/2 + c_2 \sin \sqrt{3}\,x/2 + c_3 x \cos \sqrt{3}\,x/2 + c_4 x \sin \sqrt{3}\,x/2.$$

32. From $m^4 - 7m^2 - 18 = 0$ we obtain $m = 3$, $m = -3$, and $m = \pm\sqrt{2}\,i$ so that

$$y = c_1 e^{3x} + c_2 e^{-3x} + c_3 \cos \sqrt{2}\,x + c_4 \sin \sqrt{2}\,x.$$

33. From $m^5 - 16m = 0$ we obtain $m = 0$, $m = 2$, $m = -2$, and $m = \pm 2i$ so that

$$y = c_1 + c_2 e^{2x} + c_3 e^{-2x} + c_4 \cos 2x + c_5 \sin 2x.$$

34. From $m^5 - 2m^4 + 17m^3 = 0$ we obtain $m = 0$, $m = 0$, $m = 0$, and $m = 1 \pm 4i$ so that

$$y = c_1 + c_2 x + c_3 x^2 + e^x(c_4 \cos 4x + c_5 \sin 4x).$$

35. From $m^5 + 5m^4 - 2m^3 - 10m^2 + m + 5 = 0$ we obtain $m = -1$, $m = -1$, $m = 1$, and $m = 1$, and $m = -5$ so that

$$y = c_1 e^{-x} + c_2 x e^{-x} + c_3 e^x + c_4 x e^x + c_5 e^{-5x}.$$

36. From $2m^5 - 7m^4 + 12m^3 + 8m^2 = 0$ we obtain $m = 0$, $m = 0$, $m = -1/2$, and $m = 2 \pm 2i$ so that

$$y = c_1 + c_2 x + c_3 e^{-x/2} + e^{2x}(c_4 \cos 2x + c_5 \sin 2x).$$

37. From $m^2 + 16 = 0$ we obtain $m = \pm 4i$ so that $y = c_1 \cos 4x + c_2 \sin 4x$. If $y(0) = 2$ and $y'(0) = -2$ then $c_1 = 2$, $c_2 = -1/2$, and $y = 2 \cos 4x - \frac{1}{2} \sin 4x$.

38. From $m^2 - 1 = 0$ we obtain $m = 1$ and $m = -1$ so that $y = c_1 e^x + c_2 e^{-x}$. If $y(0) = 1$ and $y'(0) = 1$ then $c_1 + c_2 =$, $c_1 - c_2 = 1$, so $c_1 = 1$, $c_2 = 0$, and $y = e^x$.

39. From $m^2 + 6m + 5 = 0$ we obtain $m = -1$ and $m = -5$ so that $y = c_1 e^{-x} + c_2 e^{-5x}$. If $y(0) = 0$ and $y'(0) = 3$ then $c_1 + c_2 = 0$, $-c_1 - 5c_2 = 3$, so $c_1 = 3/4$, $c_2 = -3/4$, and $y = \frac{3}{4} e^{-x} - \frac{3}{4} e^{-5x}$.

40. From $m^2 - 8m + 17 = 0$ we obtain $m = 4 \pm i$ so that $y = e^{4x}(c_1 \cos x + c_2 \sin x)$. If $y(0) = 4$ and $y'(0) = -1$ then $c_1 = 4$, $4c_1 + c_2 = -1$, so $c_1 = 4$, $c_2 = -17$, and $y = e^{4x}(4 \cos x - 17 \sin x)$.

41. From $2m^2 - 2m + 1 = 0$ we obtain $m = 1/2 \pm i/2$ so that $y = e^{x/2}(c_1 \cos x/2 + c_2 \sin x/2)$. If $y(0) = -1$ and $y'(0) = 0$ then $c_1 = -1$, $\frac{1}{2}c_1 + \frac{1}{2}c_2 = 0$, so $c_1 = -1$, $c_2 = 1$, and $y = e^{x/2}\left(\sin \frac{1}{2}x - \cos \frac{1}{2}x\right)$.

42. From $m^2 - 2m + 1 = 0$ we obtain $m = 1$ and $m = 1$ so that $y = c_1 e^x + c_2 x e^x$. If $y(0) = 5$ and $y'(0) = 10$ then $c_1 = 5$, $c_1 + c_2 = 10$ so $c_1 = 5$, $c_2 = 5$, and $y = 5e^x + 5x e^x$.

43. From $m^2 + m + 2 = 0$ we obtain $m = -1/2 \pm \sqrt{7}\, i/2$ so that $y = e^{-x/2}\left(c_1 \cos \sqrt{7}\, x/2 + c_2 \sin \sqrt{7}\, x/2\right)$. If $y(0) = 0$ and $y'(0) = 0$ then $c_1 = 0$ and $c_2 = 0$ so that $y = 0$.

44. From $4m^2 - 4m - 3 = 0$ we obtain $m = -1/2$ and $m = 3/2$ so that $y = c_1 e^{-x/2} + c_2 e^{3x/2}$. If $y(0) = 1$ and $y'(0) = 5$ then $c_1 + c_2 = 1$, $-\frac{1}{2}c_1 + \frac{3}{2}c_2 = 5$, so $c_1 = -7/4$, $c_2 = 11/4$, and $y = -\frac{7}{4}e^{-x/2} + \frac{11}{4}e^{3x/2}$.

45. From $m^2 - 3m + 2 = 0$ we obtain $m = 1$ and $m = 2$ so that $y = c_1 e^x + c_2 e^{2x}$. If $y(1) = 0$ and $y'(1) = 1$ then $c_1 e + c_2 e^2 = 0$, $c_1 e + 2c_2 e^2 = 0$ so $c_1 = -e^{-1}$, $c_2 = e^{-2}$, and $y = -e^{x-1} + e^{2x-2}$.

46. From $m^2 + 1 = 0$ we obtain $m = \pm i$ so that $y = c_1 \cos x + c_2 \sin x$. If $y(\pi/3) = 0$ and $y'(\pi/3) = 2$ then $\frac{1}{2}c_1 + \frac{\sqrt{3}}{2}c_2 = 0$, $-\frac{\sqrt{3}}{2}c_1 + \frac{1}{2}c_2 = 2$, so $c_1 = -\sqrt{3}$, $c_2 = 1$, and $y = -\sqrt{3} \cos x + \sin x$.

47. From $m^3 + 12m^2 + 36m = 0$ we obtain $m = 0$, $m = -6$, and $m = -6$ so that $y = c_1 + c_2 e^{-6x} + c_3 x e^{-6x}$. If $y(0) = 0$, $y'(0) = 1$, and $y''(0) = -7$ then

$$c_1 + c_2 = 0, \quad -6c_2 + c_3 = 1, \quad 36c_2 - 12c_3 = -7,$$

so $c_1 = 5/36$, $c_2 = -5/36$, $c_3 = 1/6$, and $y = \frac{5}{36} - \frac{5}{36}e^{-6x} + \frac{1}{6}x e^{-6x}$.

48. From $m^3 + 2m^2 - 5m - 6 = 0$ we obtain $m = -1$, $m = 2$, and $m = -3$ so that

$$y = c_1 e^{-x} + c_2 e^{2x} + c_3 e^{-3x}.$$

If $y(0) = 0$, $y'(0) = 0$, and $y''(0) = 1$ then

$$c_1 + c_2 + c_3 = 0, \quad -c_1 + 2c_2 - 3c_3 = 0, \quad c_1 + 4c_2 + 9c_3 = 1,$$

so $c_1 = -1/6$, $c_2 = 1/15$, $c_3 = 1/10$, and

$$y = -\frac{1}{6}e^{-x} + \frac{1}{15}e^{2x} + \frac{1}{10}e^{-3x}.$$

49. From $m^3 - 8 = 0$ we obtain $m = 2$ and $m = -1 \pm \sqrt{3}\,i$ so that

$$y = c_1 e^{2x} + e^{-x}\left(c_2 \cos \sqrt{3}\,x + c_3 \sin \sqrt{3}\,x\right).$$

If $y(0) = 0$ and $y'(0) = -1$, and $y''(0) = 0$ then

$$c_1 + c_2 = 0, \quad 2c_1 - c_2 + \sqrt{3}\,c_3 = -1, \quad 4c_1 - 2c_2 - 2\sqrt{3}\,c_3 = 0,$$

so $c_1 = -1/6$, $c_2 = 1/6$, $c_3 = -1/2\sqrt{3}$, and

$$y = -\frac{1}{6}e^{2x} + e^{-x}\left(\frac{1}{6}\cos \sqrt{3}\,x - \frac{1}{2\sqrt{3}}\sin \sqrt{3}\,x\right).$$

50. From $m^4 = 0$ we obtain $y = c_1 + c_2 x + c_3 x^2 + c_4 x^3$. If $y(0) = 2$, $y'(0) = 3$, $y''(0) = 4$, and $y'''(0) = 5$ then $c_1 = 2$, $c_2 = 3$, $2c_3 = 4$, $6c_4 = 5$, and

$$y = 2 + 3x + 2x^2 + \frac{5}{6}x^3.$$

51. From $m^4 - 3m^3 + 3m^2 - m = 0$ we obtain $m = 0$, $m = 1$, $m = 1$, and $m = 1$ so that $y = c_1 + c_2 e^x + c_3 x e^x + c_4 x^2 e^x$. If $y(0) = 0$, $y'(0) = 0$, $y''(0) = 1$, and $y'''(0) = 1$ then

$$c_1 + c_2 = 0, \quad c_2 + c_3 = 0, \quad c_2 + 2c_3 + 2c_4 = 1, \quad c_2 + 3c_3 + 6c_4 = 1,$$

so $c_1 = 2$, $c_2 = -2$, $c_3 = 2$, $c_4 = -1/2$, and

$$y = 2 - 2e^x + 2x e^x - \frac{1}{2}x^2 e^x.$$

52. From $m^4 - 1 = 0$ we obtain $m = 1$, $m = -1$, and $m = \pm i$ so that $y = c_1 e^x + c_2 e^{-x} + c_3 \cos x + c_4 \sin x$. If $y(0) = 0$, $y'(0) = 0$, $y''(0) = 0$, and $y'''(0) = 1$ then

$$c_1 + c_2 + c_3 = 0, \quad c_1 - c_2 + c_4 = 0, \quad c_1 + c_2 - c_3 = 0, \quad c_1 - c_2 - c_4 = 1,$$

so $c_1 = 1/4$, $c_2 = -1/4$, $c_3 = 0$, $c_4 = -1/2$, and

$$y = \frac{1}{4}e^x - \frac{1}{4}e^{-x} - \frac{1}{2}\sin x.$$

53. From $m^2 - 10m + 25 = 0$ we obtain $m = 5$ and $m = 5$ so that $y = c_1 e^{5x} + c_2 x e^{5x}$. If $y(0) = 1$ and $y(1) = 0$ then $c_1 = 1$, $c_1 e^5 + c_2 e^5 = 0$, so $c_1 = 1$, $c_2 = -1$, and $y = e^{5x} - x e^{5x}$.

54. From $m^2 + 4 = 0$ we obtain $m = \pm 2i$ so that $y = c_1 \cos 2x + c_2 \sin 2x$. If $y(0) = 0$ and $y(\pi) = 0$ then $c_1 = 0$ and $y = c_2 \sin 2x$.

55. From $m^2 + 1 = 0$ we obtain $m = \pm i$ so that $y = c_1 \cos x + c_2 \sin x$. If $y'(0) = 0$ and $y'(\pi/2) = 2$ then $c_1 = -2$, $c_2 = 0$ and $y = -2 \cos x$.

56. From $m^2 - 1 = 0$ we obtain $m = 1$ and $m = -1$ so that $y = c_1 e^x + c_2 e^{-x}$ or $y = c_3 \cosh x + c_4 \sinh x$. If $y(0) = 1$ and $y'(1) = 0$ then $c_1 = 1$, $c_1 \sinh 1 + c_2 \cosh 1 = 0$, so $c_1 = 1$, $c_2 = -\sinh 1/\cosh 1$ and

$$y = \cosh x - \frac{\sinh 1}{\cosh 1} \sinh x = \frac{\cosh x \cosh 1 - \sinh x \sinh 1}{\cosh 1} = \frac{\cosh(x-1)}{\cosh 1}.$$

57. Using a CAS to solve the auxiliary equation $m^3 - 6m^2 + 2m + 1$ we find $m_1 = -0.270534$, $m_2 = 0.658675$, and $m_3 = 5.61186$. The general solution is

$$y = c_1 e^{-0.270534x} + c_2 e^{0.658675x} + c_3 e^{5.61186x}.$$

58. Using a CAS to solve the auxiliary equation $6.11m^3 + 8.59m^2 + 7.93m + 0.778 = 0$ we find $m_1 = -0.110241$, $m_2 = -0.647826 + 0.857532i$, and $m_3 = -0.647826 - 0.857532i$. The general solution is

$$y = c_1 e^{-0.110241x} + e^{-0.647826x}(c_2 \cos 0.857532x + c_3 \sin 0.857532x).$$

59. Using a CAS to solve the auxiliary equation $3.15m^4 - 5.34m^2 + 6.33m - 2.03 = 0$ we find $m_1 = -1.74806$, $m_2 = 0.501219$, $m_3 = 0.62342 + 0.588965i$, and $m_4 = 0.62342 - 0.588965i$. The general solution is

$$y = c_1 e^{-1.74806x} + c_2 e^{0.501219x} + e^{0.62342x}(c_3 \cos 0.588965x + c_4 \sin 0.588965x).$$

60. Using a CAS to solve the auxiliary equation $m^4 + 2m^2 - m + 2 = 0$ we find $m_1 = 1/2 + \sqrt{3}\,i/2$, $m_2 = 1/2 - \sqrt{3}\,i/2$, $m_3 = -1/2 + \sqrt{7}\,i/2$, and $m_4 = -1/2 - \sqrt{7}\,i/2$. The general solution is

$$y = e^{x/2}\left(c_1 \cos \frac{\sqrt{3}}{2}x + c_2 \sin \frac{\sqrt{3}}{2}x\right) + e^{-x/2}\left(c_3 \cos \frac{\sqrt{7}}{2}x + c_4 \sin \frac{\sqrt{7}}{2}x\right).$$

61. (a) Since $(m-4)(m+5)^2 = m^3 + 6m^2 - 15m - 100$ the differential equation is

$$y''' + 6y'' - 15y' - 100y = 0.$$

(b) Since $\left(m + \frac{1}{2}\right)\left(m^2 - 6m + 10\right) = m^3 - \frac{11}{2}m^2 + 7m + 5$ the differential equation is

$$y''' - \frac{11}{2}y'' + 7y' + 5y = 0.$$

(c) From the solution $y_1 = e^{-4x}\cos x$ we conclude that $m_1 = -4 + i$ and $m_2 = -4 - i$ are roots of the auxiliary equation. Hence another solution must be $y_2 = e^{-4x}\sin x$. Now dividing the polynomial $m^3 + 6m^2 + m - 34$ by $[m - (-4 + i)][m - (-4 - i)] = m^2 + 8m + 17$ gives $m - 2$.

Therefore $m_3 = 2$ is the third root of the auxiliary equation, and the general solution of the differential equation is

$$y = c_1 e^{-4x} \cos x + c_2 e^{-4x} \sin x + c_3 e^{2x}.$$

62. If $a < 0$, then the equivalent differential equation $-ay'' - by' - c = 0$ has a positive lead coefficient. Thus, we assume without loss of generality that $a > 0$. We consider three cases.

Case I: $b^2 - 4ac = 0$. The roots of the auxiliary equation are both $-b/2a$ and the general solution is $y = (c_1 + c_2 x)e^{-bx/2a}$. This solution will be bounded if and only if $-b/2a < 0$. Since $a > 0$, this implies that $b > 0$. Also, since $4ac = b^2 > 0$ we see that $c > 0$. Thus, when $b^2 - 4ac = 0$, the solutions will be bounded when all coefficients are positive.

Case II: $b^2 - 4ac < 0$. The roots of the auxiliary equation are $\alpha \pm \beta i$ where $\alpha = -b/2a$ and $\beta = \sqrt{b^2 - 4ac}/2a$. The general solution of the differential equation is $y = e^{-bx/2a}(c_1 \cos \beta x + c_2 \sin \beta x)$, which will be bounded if and only if $-b/2a < 0$. Since $a > 0$, this implies that $b > 0$. Also, since $4ac > b^2 \geq 0$, we see that $c > 0$. Thus, when $b^2 - 4ac < 0$, the solutions will be bounded when all coefficients are positive.

Case III: $b^2 - 4ac > 0$. The roots of the auxiliary equation are $(-b \pm \sqrt{b^2 - 4ac})/2a$ and the solution of the differential equation will be bounded if and only if both of these roots are negative. Since $a > 0$, this implies that $-b \pm \sqrt{b^2 - 4ac} < 0$ or $\pm\sqrt{b^2 - 4ac} < b$. In particular we must have $0 < \sqrt{b^2 - 4ac} < b$ and $b^2 - 4ac < b^2$ or $-4ac < 0$, which implies $c > 0$.

We see in every case then, that for the solutions of the differential equation to be bounded all coefficients must be positive. Because of the statement at the start of this solution, the solutions of the differential equation will also be bounded if all coefficients are negative.

63. Since $1/x \to 0$ as $x \to \infty$, we would expect the solutions of $y'' + (1/x)y' + y = 0$ to behave similar to the solutions of $y'' + y = 0$; that is, like $\sin x$ and $\cos x$ for large values of x. Solutions of $xy'' + y' + xy = 0$ are obtained using an ODE solver and are shown below with the indicated initial conditions.

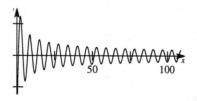

$$y(1) = 0, \quad y'(1) = 2$$

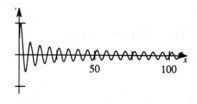

$$y(1) = 2, \quad y'(1) = 0$$

Exercises 4.4

1. From $m^2 + 3m + 2 = 0$ we find $m_1 = -1$ and $m_2 = -2$. Then $y_c = c_1 e^{-x} + c_2 e^{-2x}$ and we assume $y_p = A$. Substituting into the differential equation we obtain $2A = 6$. Then $A = 3$, $y_p = 3$ and

$$y = c_1 e^{-x} + c_2 e^{-2x} + 3.$$

2. From $4m^2 + 9 = 0$ we find $m_1 = -\frac{3}{2}i$ and $m_2 = \frac{3}{2}i$. Then $y_c = c_1 \cos \frac{3}{2}x + c_2 \sin \frac{3}{2}x$ and we assume $y_p = A$. Substituting into the differential equation we obtain $9A = 15$. Then $A = \frac{5}{3}$, $y_p = \frac{5}{3}$ and

$$y = c_1 \cos \frac{3}{2}x + c_2 \sin \frac{3}{2}x + \frac{5}{3}.$$

3. From $m^2 - 10m + 25 = 0$ we find $m_1 = m_2 = 5$. Then $y_c = c_1 e^{5x} + c_2 x e^{5x}$ and we assume $y_p = Ax + B$. Substituting into the differential equation we obtain $25A = 30$ and $-10A + 25B = 3$. Then $A = \frac{6}{5}$, $B = \frac{6}{5}$, $y_p = \frac{6}{5}x + \frac{6}{5}$, and

$$y = c_1 e^{5x} + c_2 x e^{5x} + \frac{6}{5}x + \frac{6}{5}.$$

4. From $m^2 + m - 6 = 0$ we find $m_1 = -3$ and $m_2 = 2$. Then $y_c = c_1 e^{-3x} + c_2 e^{2x}$ and we assume $y_p = Ax + B$. Substituting into the differential equation we obtain $-6A = 2$ and $A - 6B = 0$. Then $A = -\frac{1}{3}$, $B = -\frac{1}{18}$, $y_p = -\frac{1}{3}x - \frac{1}{18}$, and

$$y = c_1 e^{-3x} + c_2 e^{2x} - \frac{1}{3}x - \frac{1}{18}.$$

5. From $\frac{1}{4}m^2 + m + 1 = 0$ we find $m_1 = m_2 = 0$. Then $y_c = c_1 e^{-2x} + c_2 x e^{-2x}$ and we assume $y_p = Ax^2 + Bx + C$. Substituting into the differential equation we obtain $A = 1$, $2A + B = -2$, and $\frac{1}{2}A + B + C = 0$. Then $A = 1$, $B = -4$, $C = \frac{7}{2}$, $y_p = x^2 - 4x + \frac{7}{2}$, and

$$y = c_1 e^{-2x} + c_2 x e^{-2x} + x^2 - 4x + \frac{7}{2}.$$

6. From $m^2 - 8m + 20 = 0$ we find $m_1 = 2 + 4i$ and $m_2 = 2 - 4i$. Then $y_c = e^{2x}(c_1 \cos 4x + c_2 \sin 4x)$ and we assume $y_p = Ax^2 + Bx + C + (Dx + E)e^x$. Substituting into the differential equation we obtain

$$2A - 8B + 20C = 0$$

$$-6D + 13E = 0$$

$$-16A + 20B = 0$$

$$13D = -26$$

$$20A = 100.$$

87

Then $A = 5$, $B = 4$, $C = \frac{11}{10}$, $D = -2$, $E = -\frac{12}{13}$, $y_p = 5x^2 + 4x + \frac{11}{10} + \left(-2x - \frac{12}{13}\right)e^x$ and

$$y = e^{2x}(c_1 \cos 4x + c_2 \sin 4x) + 5x^2 + 4x + \frac{11}{10} + \left(-2x - \frac{12}{13}\right)e^x.$$

7. From $m^2 + 3 = 0$ we find $m_1 = \sqrt{3}\,i$ and $m_2 = -\sqrt{3}\,i$. Then $y_c = c_1 \cos \sqrt{3}\,x + c_2 \sin \sqrt{3}\,x$ and we assume $y_p = (Ax^2 + Bx + C)e^{3x}$. Substituting into the differential equation we obtain $2A + 6B + 12C = 0$, $12A + 12B = 0$, and $12A = -48$. Then $A = -4$, $B = 4$, $C = -\frac{4}{3}$, $y_p = \left(-4x^2 + 4x - \frac{4}{3}\right)e^{3x}$ and

$$y = c_1 \cos \sqrt{3}\,x + c_2 \sin \sqrt{3}\,x + \left(-4x^2 + 4x - \frac{4}{3}\right)e^{3x}.$$

8. From $4m^2 - 4m - 3 = 0$ we find $m_1 = \frac{3}{2}$ and $m_2 = -\frac{1}{2}$. Then $y_c = c_1 e^{3x/2} + c_2 e^{-x/2}$ and we assume $y_p = A\cos 2x + B\sin 2x$. Substituting into the differential equation we obtain $-19 - 8B = 1$ and $8A - 19B = 0$. Then $A = -\frac{19}{425}$, $B = -\frac{8}{425}$, $y_p = -\frac{19}{425}\cos 2x - \frac{8}{425}\sin 2x$, and

$$y = c_1 e^{3x/2} + c_2 e^{-x/2} - \frac{19}{425}\cos 2x - \frac{8}{425}\sin 2x.$$

9. From $m^2 - m = 0$ we find $m_1 = 1$ and $m_2 = 0$. Then $y_c = c_1 e^x + c_2$ and we assume $y_p = Ax$. Substituting into the differential equation we obtain $-A = -3$. Then $A = 3$, $y_p = 3x$ and $y = c_1 e^x + c_2 + 3x$.

10. From $m^2 + 2m = 0$ we find $m_1 = -2$ and $m_2 = 0$. Then $y_c = c_1 e^{-2x} + c_2$ and we assume $y_p = Ax^2 + Bx + Cxe^{-2x}$. Substituting into the differential equation we obtain $2A + 2B = 5$, $4A = 2$, and $-2C = -1$. Then $A = \frac{1}{2}$, $B = 2$, $C = \frac{1}{2}$, $y_p = \frac{1}{2}x^2 + 2x + \frac{1}{2}xe^{-2x}$, and

$$y = c_1 e^{-2x} + c_2 + \frac{1}{2}x^2 + 2x + \frac{1}{2}xe^{-2x}.$$

11. From $m^2 - m + \frac{1}{4} = 0$ we find $m_1 = m_2 = \frac{1}{2}$. Then $y_c = c_1 e^{x/2} + c_2 xe^{x/2}$ and we assume $y_p = A + Bx^2 e^{x/2}$. Substituting into the differential equation we obtain $\frac{1}{4}A = 3$ and $2B = 1$. Then $A = 12$, $B = \frac{1}{2}$, $y_p = 12 + \frac{1}{2}x^2 e^{x/2}$, and

$$y = c_1 e^{x/2} + c_2 xe^{x/2} + 12 + \frac{1}{2}x^2 e^{x/2}.$$

12. From $m^2 - 16 = 0$ we find $m_1 = 4$ and $m_2 = -4$. Then $y_c = c_1 e^{4x} + c_2 e^{-4x}$ and we assume $y_p = Axe^{4x}$. Substituting into the differential equation we obtain $8A = 2$. Then $A = \frac{1}{4}$, $y_p = \frac{1}{4}xe^{4x}$ and

$$y = c_1 e^{4x} + c_2 e^{-4x} + \frac{1}{4}xe^{4x}.$$

13. From $m^2 + 4 = 0$ we find $m_1 = 2i$ and $m_2 = -2i$. Then $y_c = c_1 \cos 2x + c_2 \sin 2x$ and we assume $y_p = Ax\cos 2x + Bx\sin 2x$. Substituting into the differential equation we obtain $4B = 0$ and

$-4A = 3$. Then $A = -\frac{3}{4}$, $B = 0$, $y_p = -\frac{3}{4}x\cos 2x$, and

$$y = c_1\cos 2x + c_2\sin 2x - \frac{3}{4}x\cos 2x.$$

14. From $m^2 + 4 = 0$ we find $m_1 = 2i$ and $m_2 = -2i$. Then $y_c = c_1\cos 2x + c_2\sin 2x$ and we assume $y_p = (Ax^3 + Bx^2 + Cx)\cos 2x + (Dx^3 + Ex^2 + Fx)\sin 2x$. Substituting into the differential equation we obtain

$$2B + 4F = 0$$

$$6A + 8E = 0$$

$$12D = 0$$

$$-4C + 2E = -3$$

$$-8B + 6D = 0$$

$$-12A = 1.$$

Then $A = -\frac{1}{12}$, $B = 0$, $C = \frac{25}{32}$, $D = 0$, $E = \frac{1}{16}$, $F = 0$, $y_p = \left(-\frac{1}{12}x^3 + \frac{25}{32}x\right)\cos 2x + \frac{1}{16}x^2\sin 2x$, and

$$y = c_1\cos 2x + c_2\sin 2x + \left(-\frac{1}{12}x^3 + \frac{25}{32}x\right)\cos 2x + \frac{1}{16}x^2\sin 2x.$$

15. From $m^2 + 1 = 0$ we find $m_1 = i$ and $m_2 = -i$. Then $y_c = c_1\cos x + c_2\sin x$ and we assume $y_p = (Ax^2 + Bx)\cos x + (Cx^2 + Dx)\sin x$. Substituting into the differential equation we obtain $4C = 0$, $2A + 2D = 0$, $-4A = 2$, and $-2B + 2C = 0$. Then $A = -\frac{1}{2}$, $B = 0$, $C = 0$, $D = \frac{1}{2}$, $y_p = -\frac{1}{2}x^2\cos x + \frac{1}{2}x\sin x$, and

$$y = c_1\cos x + c_2\sin x - \frac{1}{2}x^2\cos x + \frac{1}{2}x\sin x.$$

16. From $m^2 - 5m = 0$ we find $m_1 = 5$ and $m_2 = 0$. Then $y_c = c_1e^{5x} + c_2$ and we assume $y_p = Ax^4 + Bx^3 + Cx^2 + Dx$. Substituting into the differential equation we obtain $-20A = 2$, $12A - 15B = -4$, $6B - 10C = -1$, and $2C - 5D = 6$. Then $A = -\frac{1}{10}$, $B = \frac{14}{75}$, $C = \frac{53}{250}$, $D = -\frac{697}{625}$, $y_p = -\frac{1}{10}x^4 + \frac{14}{75}x^3 + \frac{53}{250}x^2 - \frac{697}{625}x$, and

$$y = c_1e^{5x} + c_2 - \frac{1}{10}x^4 + \frac{14}{75}x^3 + \frac{53}{250}x^2 - \frac{697}{625}x.$$

17. From $m^2 - 2m + 5 = 0$ we find $m_1 = 1 + 2i$ and $m_2 = 1 - 2i$. Then $y_c = e^x(c_1\cos 2x + c_2\sin 2x)$ and we assume $y_p = Axe^x\cos 2x + Bxe^x\sin 2x$. Substituting into the differential equation we obtain $4B = 1$ and $-4A = 0$. Then $A = 0$, $B = \frac{1}{4}$, $y_p = \frac{1}{4}xe^x\sin 2x$, and

$$y = e^x(c_1\cos 2x + c_2\sin 2x) + \frac{1}{4}xe^x\sin 2x.$$

18. From $m^2 - 2m + 2 = 0$ we find $m_1 = 1 + i$ and $m_2 = 1 - i$. Then $y_c = e^x(c_1 \cos x + c_2 \sin x)$ and we assume $y_p = Ae^{2x} \cos x + Be^{2x} \sin x$. Substituting into the differential equation we obtain $A + 2B = 1$ and $-2A + B = -3$. Then $A = \frac{7}{5}$, $B = -\frac{1}{5}$, $y_p = \frac{7}{5}e^{2x} \cos x - \frac{1}{5}e^{2x} \sin x$ and

$$y = e^x(c_1 \cos x + c_2 \sin x) + \frac{7}{5}e^{2x} \cos x - \frac{1}{5}e^{2x} \sin x.$$

19. From $m^2 + 2m + 1 = 0$ we find $m_1 = m_2 = -1$. Then $y_c = c_1 e^{-x} + c_2 x e^{-x}$ and we assume $y_p = A \cos x + B \sin x + C \cos 2x + D \sin 2x$. Substituting into the differential equation we obtain $2B = 0$, $-2A = 1$, $-3C + 4D = 3$, and $-4C - 3D = 0$. Then $A = -\frac{1}{2}$, $B = 0$, $C = -\frac{9}{25}$, $D = \frac{12}{25}$, $y_p = -\frac{1}{2} \cos x - \frac{9}{25} \cos 2x + \frac{12}{25} \sin 2x$, and

$$y = c_1 e^{-x} + c_2 x e^{-x} - \frac{1}{2} \cos x - \frac{9}{25} \cos 2x + \frac{12}{25} \sin 2x.$$

20. From $m^2 + 2m - 24 = 0$ we find $m_1 = -6$ and $m_2 = 4$. Then $y_c = c_1 e^{-6x} + c_2 e^{4x}$ and we assume $y_p = A + (Bx^2 + Cx)e^{4x}$. Substituting into the differential equation we obtain $-24A = 16$, $2B + 10C = -2$, and $20B = -1$. Then $A = -\frac{2}{3}$, $B = -\frac{1}{20}$, $C = -\frac{19}{100}$, $y_p = -\frac{2}{3} - \left(\frac{1}{20}x^2 + \frac{19}{100}x\right)e^{4x}$, and

$$y = c_1 e^{-6x} + c_2 e^{4x} - \frac{2}{3} - \left(\frac{1}{20}x^2 + \frac{19}{100}x\right)e^{4x}.$$

21. From $m^3 - 6m^2 = 0$ we find $m_1 = m_2 = 0$ and $m_3 = 6$. Then $y_c = c_1 + c_2 x + c_3 e^{6x}$ and we assume $y_p = Ax^2 + B \cos x + C \sin x$. Substituting into the differential equation we obtain $-12A = 3$, $6B - C = -1$, and $B + 6C = 0$. Then $A = -\frac{1}{4}$, $B = -\frac{6}{37}$, $C = \frac{1}{37}$, $y_p = -\frac{1}{4}x^2 - \frac{6}{37} \cos x + \frac{1}{37} \sin x$, and

$$y = c_1 + c_2 x + c_3 e^{6x} - \frac{1}{4}x^2 - \frac{6}{37} \cos x + \frac{1}{37} \sin x.$$

22. From $m^3 - 2m^2 - 4m + 8 = 0$ we find $m_1 = m_2 = 2$ and $m_3 = -2$. Then $y_c = c_1 e^{2x} + c_2 x e^{2x} + c_3 e^{-2x}$ and we assume $y_p = (Ax^3 + Bx^2)e^{2x}$. Substituting into the differential equation we obtain $24A = 6$ and $6A + 8B = 0$. Then $A = \frac{1}{4}$, $B = -\frac{3}{16}$, $y_p = \left(\frac{1}{4}x^3 - \frac{3}{16}x^2\right)e^{2x}$, and

$$y = c_1 e^{2x} + c_2 x e^{2x} + c_3 e^{-2x} + \left(\frac{1}{4}x^3 - \frac{3}{16}x^2\right)e^{2x}.$$

23. From $m^3 - 3m^2 + 3m - 1 = 0$ we find $m_1 = m_2 = m_3 = 1$. Then $y_c = c_1 e^x + c_2 x e^x + c_3 x^2 e^x$ and we assume $y_p = Ax + B + Cx^3 e^x$. Substituting into the differential equation we obtain $-A = 1$, $3A - B = 0$, and $6C = -4$. Then $A = -1$, $B = -3$, $C = -\frac{2}{3}$, $y_p = -x - 3 - \frac{2}{3}x^3 e^x$, and

$$y = c_1 e^x + c_2 x e^x + c_3 x^2 e^x - x - 3 - \frac{2}{3}x^3 e^x.$$

24. From $m^3 - m^2 - 4m + 4 = 0$ we find $m_1 = 1$, $m_2 = 2$, and $m_3 = -2$. Then $y_c = c_1 e^x + c_2 e^{2x} + c_3 e^{-2x}$ and we assume $y_p = A + Bxe^x + Cxe^{2x}$. Substituting into the differential equation we obtain $4A = 5$,

$-3B = -1$, and $4C = 1$. Then $A = \frac{5}{4}$, $B = \frac{1}{3}$, $C = \frac{1}{4}$, $y_p = \frac{5}{4} + \frac{1}{3}xe^x + \frac{1}{4}xe^{2x}$, and

$$y = c_1 e^x + c_2 e^{2x} + c_3 e^{-2x} + \frac{5}{4} + \frac{1}{3}xe^x + \frac{1}{4}xe^{2x}.$$

25. From $m^4 + 2m^2 + 1 = 0$ we find $m_1 = m_3 = i$ and $m_2 = m_4 = -i$. Then $y_c = c_1 \cos x + c_2 \sin x + c_3 x \cos x + c_4 x \sin x$ and we assume $y_p = Ax^2 + Bx + C$. Substituting into the differential equation we obtain $A = 1$, $B = -2$, and $4A + C = 1$. Then $A = 1$, $B = -2$, $C = -3$, $y_p = x^2 - 2x - 3$, and

$$y = c_1 \cos x + c_2 \sin x + c_3 x \cos x + c_4 x \sin x + x^2 - 2x - 3.$$

26. From $m^4 - m^2 = 0$ we find $m_1 = m_2 = 0$, $m_3 = 1$, and $m_4 = -1$. Then $y_c = c_1 + c_2 x + c_3 e^x + c_4 e^{-x}$ and we assume $y_p = Ax^3 + Bx^2 + (Cx^2 + Dx)e^{-x}$. Substituting into the differential equation we obtain $-6A = 4$, $-2B = 0$, $10C - 2D = 0$, and $-4C = 2$. Then $A = -\frac{2}{3}$, $B = 0$, $C = -\frac{1}{2}$, $D = -\frac{5}{2}$, $y_p = -\frac{2}{3}x^3 - \left(\frac{1}{2}x^2 + \frac{5}{2}x\right)e^{-x}$, and

$$y = c_1 + c_2 x + c_3 e^x + c_4 e^{-x} - \frac{2}{3}x^3 - \left(\frac{1}{2}x^2 + \frac{5}{2}x\right)e^{-x}.$$

27. We have $y_c = c_1 \cos 2x + c_2 \sin 2x$ and we assume $y_p = A$. Substituting into the differential equation we find $A = -\frac{1}{2}$. Thus $y = c_1 \cos 2x + c_2 \sin 2x - \frac{1}{2}$. From the initial conditions we obtain $c_1 = 0$ and $c_2 = \sqrt{2}$, so $y = \sqrt{2} \sin 2x - \frac{1}{2}$.

28. We have $y_c = c_1 e^{-2x} + c_2 e^{x/2}$ and we assume $y_p = Ax^2 + Bx + C$. Substituting into the differential equation we find $A = -7$, $B = -19$, and $C = -37$. Thus $y = c_1 e^{-2x} + c_2 e^{x/2} - 7x^2 - 19x - 37$. From the initial conditions we obtain $c_1 = -\frac{1}{5}$ and $c_2 = \frac{186}{5}$, so

$$y = -\frac{1}{5}e^{-2x} + \frac{186}{5}e^{x/2} - 7x^2 - 19x - 37.$$

29. We have $y_c = c_1 e^{-x/5} + c_2$ and we assume $y_p = Ax^2 + Bx$. Substituting into the differential equation we find $A = -3$ and $B = 30$. Thus $y = c_1 e^{-x/5} + c_2 - 3x^2 + 30x$. From the initial conditions we obtain $c_1 = 200$ and $c_2 = -200$, so

$$y = 200^{-x/5} - 200 - 3x^2 + 30x.$$

30. We have $y_c = c_1 e^{-2x} + c_2 x e^{-2x}$ and we assume $y_p = (Ax^3 + Bx^2)e^{-2x}$. Substituting into the differential equation we find $A = \frac{1}{6}$ and $B = \frac{3}{2}$. Thus $y = c_1 e^{-2x} + c_2 x e^{-2x} + \left(\frac{1}{6}x^3 + \frac{3}{2}x^2\right)e^{-2x}$. From the initial conditions we obtain $c_1 = 2$ and $c_2 = 9$, so

$$y = 2e^{-2x} + 9x e^{-2x} + \left(\frac{1}{6}x^3 + \frac{3}{2}x^2\right)e^{-2x}.$$

31. We have $y_c = e^{-2x}(c_1 \cos x + c_2 \sin x)$ and we assume $y_p = Ae^{-4x}$. Substituting into the differential equation we find $A = 5$. Thus $y = e^{-2x}(c_1 \cos x + c_2 \sin x) + 7e^{-4x}$. From the initial conditions we obtain $c_1 = -10$ and $c_2 = 9$, so

$$y = e^{-2x}(-10 \cos x + 9 \sin x + 7e^{-4x}).$$

Exercises 4.4

32. We have $y_c = c_1 \cosh x + c_2 \sinh x$ and we assume $y_p = Ax \cosh x + Bx \sinh x$. Substituting into the differential equation we find $A = 0$ and $B = \frac{1}{2}$. Thus

$$y = c_1 \cosh x + c_2 \sinh x + \frac{1}{2}x \sinh x.$$

From the initial conditions we obtain $c_1 = 2$ and $c_2 = 12$, so

$$y = 2 \cosh x + 12 \sinh x + \frac{1}{2}x \sinh x.$$

33. We have $x_c = c_1 \cos \omega t + c_2 \sin \omega t$ and we assume $x_p = At \cos \omega t + Bt \sin \omega t$. Substituting into the differential equation we find $A = -F_0/2\omega$ and $B = 0$. Thus $x = c_1 \cos \omega t + c_2 \sin \omega t - (F_0/2\omega)t \cos \omega t$. From the initial conditions we obtain $c_1 = 0$ and $c_2 = F_0/2\omega^2$, so

$$x = (F_0/2\omega^2) \sin \omega t - (F_0/2\omega)t \cos \omega t.$$

34. We have $x_c = c_1 \cos \omega t + c_2 \sin \omega t$ and we assume $x_p = A \cos \gamma t + B \sin \gamma t$, where $\gamma \neq \omega$. Substituting into the differential equation we find $A = F_0/(\omega^2 - \gamma^2)$ and $B = 0$. Thus

$$x = c_1 \cos \omega t + c_2 \sin \omega t + \frac{F_0}{(\omega^2 - \gamma^2)} \cos \gamma t.$$

From the initial conditions we obtain $c_1 = F_0/(\omega^2 - \gamma^2)$ and $c_2 = 0$, so

$$x = \frac{F_0}{(\omega^2 - \gamma^2)} \cos \omega t + \frac{F_0}{(\omega^2 - \gamma^2)} \cos \gamma t.$$

35. We have $y_c = c_1 + c_2 e^x + c_3 x e^x$ and we assume $y_p = Ax + Bx^2 e^x + Ce^{5x}$. Substituting into the differential equation we find $A = 2$, $B = -12$, and $C = \frac{1}{2}$. Thus

$$y = c_1 + c_2 e^x + c_3 x e^x + 2x - 12x^2 e^x + \frac{1}{2}e^{5x}.$$

From the initial conditions we obtain $c_1 = 11$, $c_2 = -11$, and $c_3 = 9$, so

$$y = 11 - 11e^x + 9x e^x + 2x - 12x^2 e^x + \frac{1}{2}e^{5x}.$$

36. We have $y_c = c_1 e^{-2x} + e^x(c_2 \cos \sqrt{3}\, x + c_3 \sin \sqrt{3}\, x)$ and we assume $y_p = Ax + B + Cxe^{-2x}$. Substituting into the differential equation we find $A = \frac{1}{4}$, $B = -\frac{5}{8}$, and $C = \frac{2}{3}$. Thus

$$y = c_1 e^{-2x} + e^x(c_2 \cos \sqrt{3}\, x + c_3 \sin \sqrt{3}\, x) + \frac{1}{4}x - \frac{5}{8} + \frac{2}{3}xe^{-2x}.$$

From the initial conditions we obtain $c_1 = -\frac{23}{12}$, $c_2 = -\frac{59}{24}$, and $c_3 = \frac{17}{72}\sqrt{3}$, so

$$y = -\frac{23}{12}e^{-2x} + e^x\left(-\frac{59}{24}\cos \sqrt{3}\, x + \frac{17}{72}\sqrt{3}\sin \sqrt{3}\, x\right) + \frac{1}{4}x - \frac{5}{8} + \frac{2}{3}xe^{-2x}.$$

37. We have $y_c = c_1 \cos x + c_2 \sin x$ and we assume $y_p = A^2 + Bx + C$. Substituting into the differential equation we find $A = 1$, $B = 0$, and $C = -1$. Thus $y = c_1 \cos x + c_2 \sin x + x^2 - 1$. From $y(0) = 5$ and $y(1) = 0$ we obtain

$$c_1 - 1 = 5$$

$$(\cos 1)c_1 + \sin(1)c_2 = 0.$$

Solving this system we find $c_1 = 6$ and $c_2 = -6 \cot 1$. The solution of the boundary-value problem is

$$y = 6 \cos x - 6(\cot 1) \sin x + x^2 - 1.$$

38. We have $y_c = e^x(c_1 \cos x + c_2 \sin x)$ and we assume $y_p = Ax + B$. Substituting into the differential equation we find $A = 1$ and $B = 0$. Thus $y = e^x(c_1 \cos x + c_2 \sin x) + x$. From $y(0) = 0$ and $y(\pi) = \pi$ we obtain

$$c_1 = 0$$

$$\pi - e^\pi c_1 = \pi.$$

Solving this system we find $c_1 = 0$ and c_2 is any real number. The solution of the boundary-value problem is

$$y = c_2 e^x \sin x + x.$$

39. We have $y_c = c_1 \cos 2x + c_2 \sin 2x$ and we assume $y_p = A \cos x + B \sin x$ on $[0, \pi/2]$. Substituting into the differential equation we find $A = 0$ and $B = \frac{1}{3}$. Thus $y = c_1 \cos 2x + c_2 \sin 2x + \frac{1}{3} \sin x$ on $[0, \pi/2]$. On $(\pi/2, \infty)$ we have $y = c_3 \cos 2x + c_4 \sin 2x$. From $y(0) = 1$ and $y'(0) = 2$ we obtain

$$c_1 = 1$$

$$\frac{1}{3} + 2c_2 = 2.$$

Solving this system we find $c_1 = 1$ and $c_2 = \frac{5}{6}$. Thus $y = \cos 2x + \frac{5}{6} \sin 2x + \frac{1}{3} \sin x$ on $[0, \pi/2]$. Now continuity of y at $x = \pi/2$ implies

$$\cos \pi + \frac{5}{6} \sin \pi + \frac{1}{3} \sin \frac{\pi}{2} = c_3 \cos \pi + c_4 \sin \pi$$

or $-1 + \frac{1}{3} = -c_3$. Hence $c_3 = \frac{2}{3}$. Continuity of y' at $x = \pi/2$ implies

$$-2 \sin \pi + \frac{5}{3} \cos \pi + \frac{1}{3} \cos \frac{\pi}{2} = -2c_3 \sin \pi + 2c_4 \cos \pi$$

or $-\frac{5}{3} = -2c_4$. Then $c_4 = \frac{5}{6}$ and the solution of the boundary-value problem is

$$y(x) = \begin{cases} \cos 2x + \frac{5}{6} \sin 2x + \frac{1}{3} \sin x, & 0 \le x \le \pi/2 \\ \frac{2}{3} \cos 2x + \frac{5}{6} \sin 2x, & x > \pi/2 \end{cases}$$

40. (a) Writing the differential equation in the form

$$\frac{d}{dx}(ay' + by) = g(x)$$

and integrating we obtain $ay' + by = G(x) + c$, where $G(x)$ is an antiderivative of $g(x)$. This is a linear first-order differential equation which we can solve using techniques discussed in Chapter 2.

(b) We have $y' + y = x^2 + e^{-x} + c_1$. An integrating factor is $e^{\int dx} = e^x$ so that

$$\frac{d}{dx}[e^x y] = x^2 e^x + 1 + c_1 e^x.$$

Then $e^x y = x^2 e^x - 2xe^x + 2e^x + x + c_1 e^x + c_2$ and

$$y = x^2 - 2x + 2 + xe^{-x} + c_1 + c_2 e^{-x}$$

$$= x^2 - 2x + xe^{-x} + c_3 + c_2 e^{-x}.$$

(c) To apply the method to, say, third-order differential equations, the y and y' terms should be missing, and so on.

41. Using the double-angle formula for the cosine, we have

$$\sin x \cos 2x = \sin x(\cos^2 x - \sin^2 x) = \sin x(1 - 2\sin^2 x) = \sin x - 2\sin^3 x.$$

Since $\sin x$ is a solution of the related homogeneous differential equation we look for a particular solution of the form $y_p = Ax \sin x + Bx \cos x + C \sin^3 x$. Substituting into the differential equation we obtain

$$2a \cos x + (6c - 2b)\sin x - 8c \sin^3 x = \sin x - 2\sin^3 x.$$

Equating coefficients we find $a = 0$, $c = \frac{1}{4}$, and $b = \frac{1}{4}$. Thus, a particular solution is

$$y_p = \frac{1}{4}x \cos x + \frac{1}{4}\sin^3 x.$$

Exercises 4.5

1. $(9D^2 - 4)y = (3D - 2)(3D + 2)y = \sin x$
2. $(D^2 - 5)y = (D - \sqrt{5})(D + \sqrt{5})y = x^2 - 2x$
3. $(D^2 - 4D - 12)y = (D - 6)(D + 2)y = x - 6$
4. $(2D^2 - 3D - 2)y = (2D + 1)(D - 2)y = 1$
5. $(D^3 + 10D^2 + 25D)y = D(D + 5)^2 y = e^x$
6. $(D^3 + 4D)y = D(D^2 + 4)y = e^x \cos 2x$

7. $(D^3 + 2D^2 - 13D + 10)y = (D-1)(D-2)(D+5)y = xe^{-x}$

8. $(D^3 + 4D^2 + 3D)y = D(D+1)(D+3)y = x^2 \cos x - 3x$

9. $(D^4 + 8D)y = D(D+2)(D^2 - 2D + 4)y = 4$

10. $(D^4 - 8D^2 + 16)y = (D-2)^2(D+2)^2 y = (x^3 - 2x)e^{4x}$

11. $D^4 y = D^4(10x^3 - 2x) = D^3(30x^2 - 2) = D^2(60x) = D(60) = 0$

12. $(2D-1)y = (2D-1)4e^{x/2} = 8De^{x/2} - 4e^{x/2} = 4e^{x/2} - 4e^{x/2} = 0$

13. $(D-2)(D+5)(e^{2x} + 3e^{-5x}) = (D-2)(2e^{2x} - 15e^{-5x} + 5e^{2x} + 15e^{-5x}) = (D-2)7e^{2x} = 14e^{2x} - 14e^{2x} = 0$

14. $(D^2 + 64)(2\cos 8x - 5\sin 8x) = D(-16\sin 8x - 40\cos 8x) + 64(2\cos 8x - 5\sin 8x)$

$$= -128\cos 8x + 320\sin 8x + 128\cos 8x - 320\sin 8x = 0$$

15. D^4 because of x^3

16. D^5 because of x^4

17. $D(D-2)$ because of 1 and e^{2x}

18. $D^2(D-6)^2$ because of x and xe^{6x}

19. $D^2 + 4$ because of $\cos 2x$

20. $D(D^2 + 1)$ because of 1 and $\sin x$

21. $D^3(D^2 + 16)$ because of x^2 and $\sin 4x$

22. $D^2(D^2 + 1)(D^2 + 25)$ because of x, $\sin x$, and $\cos 5x$

23. $(D+1)(D-1)^3$ because of e^{-x} and $x^2 e^x$

24. $D(D-1)(D-2)$ because of 1, e^x, and e^{2x}

25. $D(D^2 - 2D + 5)$ because of 1 and $e^x \cos 2x$

26. $(D^2 + 2D + 2)(D^2 - 4D + 5)$ because of $e^{-x}\sin x$ and $e^{2x}\cos x$

27. $1, x, x^2, x^3, x^4$

28. $D^2 + 4D = D(D+4); \quad 1, e^{-4x}$

29. $e^{6x}, e^{-3x/2}$

30. $D^2 - 9D - 36 = (D-12)(D+3); \quad e^{12x}, e^{-3x}$

31. $\cos\sqrt{5}\,x, \sin\sqrt{5}\,x$

32. $D^2 - 6D + 10 = D^2 - 2(3)D + (3^2 + 1^2); \quad e^{3x}\cos x, e^{3x}\sin x$

33. $D^3 - 10D^2 + 25D = D(D-5)^2; \quad 1, e^{5x}, xe^{5x}$

34. $1, x, e^{5x}, e^{7x}$

35. Applying D to the differential equation we obtain

$$D(D^2 - 9)y = 0.$$

Then

$$y = \underbrace{c_1 e^{3x} + c_2 e^{-3x}}_{y_c} + c_3$$

and $y_p = A$. Substituting y_p into the differential equation yields $-9A = 54$ or $A = -6$. The general solution is

$$y = c_1 e^{3x} + c_2 e^{-3x} - 6.$$

36. Applying D to the differential equation we obtain

$$D(2D^2 - 7D + 5)y = 0.$$

Then

$$y = \underbrace{c_1 e^{5x/2} + c_2 e^x}_{y_c} + c_3$$

and $y_p = A$. Substituting y_p into the differential equation yields $5A = -29$ or $A = -29/5$. The general solution is

$$y = c_1 e^{5x/2} + c_2 e^x - \frac{29}{5}.$$

37. Applying D to the differential equation we obtain

$$D(D^2 + D)y = D^2(D + 1)y = 0.$$

Then

$$y = \underbrace{c_1 + c_2 e^{-x}}_{y_c} + c_3 x$$

and $y_p = Ax$. Substituting y_p into the differential equation yields $A = 3$. The general solution is

$$y = c_1 + c_2 e^{-3x} + 3x.$$

38. Applying D to the differential equation we obtain

$$D(D^3 + 2D^2 + D)y = D^2(D + 1)^2 y = 0.$$

Then

$$y = \underbrace{c_1 + c_2 e^{-x} + c_3 x e^{-x}}_{y_c} + c_4 x$$

and $y_p = Ax$. Substituting y_p into the differential equation yields $A = 10$. The general solution is

$$y = c_1 + c_2 e^{-x} + c_3 x e^{-x} + 10x.$$

39. Applying D^2 to the differential equation we obtain

$$D^2(D^2 + 4D + 4)y = D^2(D + 2)^2 y = 0.$$

Then

$$y = \underbrace{c_1 e^{-2x} + c_2 x e^{-2x}}_{y_c} + c_3 + c_4 x$$

and $y_p = Ax + B$. Substituting y_p into the differential equation yields $4Ax + (4A + 4B) = 2x + 6$. Equating coefficients gives

$$4A = 2$$

$$4A + 4B = 6.$$

Then $A = 1/2$, $B = 1$, and the general solution is

$$y = c_1 e^{-2x} + c_2 x e^{-2x} + \frac{1}{2}x + 1.$$

40. Applying D^2 to the differential equation we obtain

$$D^2(D^2 + 3D)y = D^3(D + 3)y = 0.$$

Then

$$y = \underbrace{c_1 + c_2 e^{-3x}}_{y_c} + c_3 x^2 + c_4 x$$

and $y_p = Ax^2 + Bx$. Substituting y_p into the differential equation yields $6Ax + (2A + 3B) = 4x - 5$. Equating coefficients gives

$$6A = 4$$

$$2A + 3B = -5.$$

Then $A = 2/3$, $B = -19/9$, and the general solution is

$$y = c_1 + c_2 e^{-3x} + \frac{2}{3}x^2 - \frac{19}{9}x.$$

41. Applying D^3 to the differential equation we obtain

$$D^3(D^3 + D^2)y = D^5(D + 1)y = 0.$$

Then

$$y = \underbrace{c_1 + c_2 x + c_3 e^{-x}}_{y_c} + c_4 x^4 + c_5 x^3 + c_6 x^2$$

and $y_p = Ax^4 + Bx^3 + Cx^2$. Substituting y_p into the differential equation yields $12Ax^2 + (24A + 6B)x + (6B + 2C) = 8x^2$. Equating coefficients gives

$$12A = 8$$

$$24A + 6B = 0$$

$$6B + 2C = 0.$$

Then $A = 2/3$, $B = -8/3$, $C = 8$, and the general solution is

$$y = c_1 + c_2 x + c_3 e^{-x} + \frac{2}{3} x^4 - \frac{8}{3} x^3 + 8 x^2.$$

42. Applying D^4 to the differential equation we obtain

$$D^4(D^2 - 2D + 1)y = D^4(D - 1)^2 y = 0.$$

Then

$$y = \underbrace{c_1 e^x + c_2 x e^x}_{y_c} + c_3 x^3 + c_4 x^2 + c_5 x + c_6$$

and $y_p = A x^3 + B x^2 + C x + D$. Substituting y_p into the differential equation yields
$A x^3 + (B - 6A)x^2 + (6A - 4B + C)x + (2B - 2C + D) = x^3 + 4x$. Equating coefficients gives

$$A = 1$$

$$B - 6A = 0$$

$$6A - 4B + C = 4$$

$$2B - 2C + D = 0.$$

Then $A = 1$, $B = 6$, $C = 22$, $D = 32$, and the general solution is

$$y = c_1 e^x + c_2 x e^x + x^3 + 6x^2 + 22x + 32.$$

43. Applying $D - 4$ to the differential equation we obtain

$$(D - 4)(D^2 - D - 12)y = (D - 4)^2(D + 3)y = 0.$$

Then

$$y = \underbrace{c_1 e^{4x} + c_2 e^{-3x}}_{y_c} + c_3 x e^{4x}$$

and $y_p = A x e^{4x}$. Substituting y_p into the differential equation yields $7A e^{4x} = e^{4x}$. Equating
coefficients gives $A = 1/7$. The general solution is

$$y = c_1 e^{4x} + c_2 e^{-3x} + \frac{1}{7} x e^{4x}.$$

44. Applying $D - 6$ to the differential equation we obtain

$$(D - 6)(D^2 + 2D + 2)y = 0.$$

Then

$$y = \underbrace{e^{-x}(c_1 \cos x + c_2 \sin x)}_{y_c} + c_3 e^{6x}$$

and $y_p = Ae^{6x}$. Substituting y_p into the differential equation yields $50Ae^{6x} = 5e^{6x}$. Equating coefficients gives $A = 1/10$. The general solution is

$$y = e^{-x}(c_1 \cos x + c_2 \sin x) + \frac{1}{10}e^{6x}.$$

45. Applying $D(D-1)$ to the differential equation we obtain

$$D(D-1)(D^2 - 2D - 3)y = D(D-1)(D+1)(D-3)y = 0.$$

Then

$$y = \underbrace{c_1 e^{3x} + c_2 e^{-x}}_{y_c} + c_3 e^x + c_4$$

and $y_p = Ae^x + B$. Substituting y_p into the differential equation yields $-4Ae^x - 3B = 4e^x - 9$. Equating coefficients gives $A = -1$ and $B = 3$. The general solution is

$$y = c_1 e^{3x} + c_2 e^{-x} - e^x + 3.$$

46. Applying $D^2(D+2)$ to the differential equation we obtain

$$D^2(D+2)(D^2 + 6D + 8)y = D^2(D+2)^2(D+4)y = 0.$$

Then

$$y = \underbrace{c_1 e^{-2x} + c_2 e^{-4x}}_{y_c} + c_3 x e^{-2x} + c_4 x + c_5$$

and $y_p = Axe^{-2x} + Bx + C$. Substituting y_p into the differential equation yields $2Ae^{-2x} + 8Bx + (6B + 8C) = 3e^{-2x} + 2x$. Equating coefficients gives

$$2A = 3$$

$$8B = 2$$

$$6B + 8C = 0.$$

Then $A = 3/2$, $B = 1/4$, $C = -3/16$, and the general solution is

$$y = c_1 e^{-2x} + c_2 e^{-4x} + \frac{3}{2} x e^{-2x} + \frac{1}{4}x - \frac{3}{16}.$$

47. Applying $D^2 + 1$ to the differential equation we obtain

$$(D^2 + 1)(D^2 + 25)y = 0.$$

Then

$$y = \underbrace{c_1 \cos 5x + c_2 \sin 5x}_{y_c} + c_3 \cos x + c_4 \sin x$$

99

and $y_p = A \cos x + B \sin x$. Substituting y_p into the differential equation yields $24A \cos x + 24B \sin x = 6 \sin x$. Equating coefficients gives $A = 0$ and $B = 1/4$. The general solution is

$$y = c_1 \cos 5x + c_2 \sin 5x + \frac{1}{4} \sin x.$$

48. Applying $D(D^2 + 1)$ to the differential equation we obtain

$$D(D^2 + 1)(D^2 + 4)y = 0.$$

Then

$$y = \underbrace{c_1 \cos 2x + c_2 \sin 2x}_{y_c} + c_3 \cos x + c_4 \sin x + c_5$$

and $y_p = A \cos x + B \sin x + C$. Substituting y_p into the differential equation yields $3A \cos x + 3B \sin x + 4C = 4 \cos x + 3 \sin x - 8$. Equating coefficients gives $A = 4/3$, $B = 1$, and $C = -2$. The general solution is

$$y = c_1 \cos 2x + c_2 \sin 2x + \frac{4}{3} \cos x + \sin x - 2.$$

49. Applying $(D - 4)^2$ to the differential equation we obtain

$$(D - 4)^2(D^2 + 6D + 9)y = (D - 4)^2(D + 3)^2 y = 0.$$

Then

$$y = \underbrace{c_1 e^{-3x} + c_2 x e^{-3x}}_{y_c} + c_3 x e^{4x} + c_4 e^{4x}$$

and $y_p = A x e^{4x} + B e^{4x}$. Substituting y_p into the differential equation yields $49 A x e^{4x} + (14A + 49B)e^{4x} = -x e^{4x}$. Equating coefficients gives

$$49A = -1$$

$$14A + 49B = 0.$$

Then $A = -1/49$, $B = 2/343$, and the general solution is

$$y = c_1 e^{-3x} + c_2 x e^{-3x} - \frac{1}{49} x e^{4x} + \frac{2}{343} e^{4x}.$$

50. Applying $D^2(D - 1)^2$ to the differential equation we obtain

$$D^2(D - 1)^2(D^2 + 3D - 10)y = D^2(D - 1)^2(D - 2)(D + 5)y = 0.$$

Then

$$y = \underbrace{c_1 e^{2x} + c_2 e^{-5x}}_{y_c} + c_3 x e^x + c_4 e^x + c_5 x + c_6$$

and $y_p = Axe^x + Be^x + Cx + D$. Substituting y_p into the differential equation yields
$-6Axe^x + (5A - 6B)e^x - 10Cx + (3C - 10D) = xe^x + x$. Equating coefficients gives

$$-6A = 1$$

$$5A - 6B = 0$$

$$-10C = 1$$

$$3C - 10D = 0.$$

Then $A = -1/6$, $B = -5/36$, $C = -1/10$, $D = -3/100$, and the general solution is

$$y = c_1 e^{2x} + c_2 e^{-5x} - \frac{1}{6}xe^x - \frac{5}{36}e^x - \frac{1}{10}x - \frac{3}{100}.$$

51. Applying $D(D-1)^3$ to the differential equation we obtain

$$D(D-1)^3(D^2-1)y = D(D-1)^4(D+1)y = 0.$$

Then

$$y = \underbrace{c_1 e^x + c_2 e^{-x}}_{y_c} + c_3 x^3 e^x + c_4 x^2 e^x + c_5 x e^x + c_6$$

and $y_p = Ax^3 e^x + Bx^2 e^x + Cxe^x + D$. Substituting y_p into the differential equation yields
$6Ax^2 e^x + (6A + 4B)xe^x + (2B + 2C)e^x - D = x^2 e^x + 5$. Equating coefficients gives

$$6A = 1$$

$$6A + 4B = 0$$

$$2B + 2C = 0$$

$$-D = 5.$$

Then $A = 1/6$, $B = -1/4$, $C = 1/4$, $D = -5$, and the general solution is

$$y = c_1 e^x + c_2 e^{-x} + \frac{1}{6}x^3 e^x - \frac{1}{4}x^2 e^x + \frac{1}{4}xe^x - 5.$$

52. Applying $(D+1)^3$ to the differential equation we obtain

$$(D+1)^3(D^2 + 2D + 1)y = (D+1)^5 y = 0.$$

Then

$$y = \underbrace{c_1 e^{-x} + c_2 x e^{-x}}_{y_c} + c_3 x^4 e^{-x} + c_4 x^3 e^{-x} + c_5 x^2 e^{-x}$$

and $y_p = Ax^4 e^{-x} + Bx^3 e^{-x} + Cx^2 e^{-x}$. Substituting y_p into the differential equation yields $12Ax^2 e^{-x} + 6Bxe^{-x} + 2Ce^{-x} = x^2 e^{-x}$. Equating coefficients gives $A = 1/12$, $B = 0$, and $C = 0$. The general solution is

$$y = c_1 e^{-x} + c_2 x e^{-x} + \frac{1}{2} x^4 e^{-x}.$$

53. Applying $D^2 - 2D + 2$ to the differential equation we obtain

$$(D^2 - 2D + 2)(D^2 - 2D + 5)y = 0.$$

Then

$$y = \underbrace{e^x (c_1 \cos 2x + c_2 \sin 2x)}_{y_c} + e^x (c_3 \cos x + c_4 \sin x)$$

and $y_p = Ae^x \cos x + Be^x \sin x$. Substituting y_p into the differential equation yields we obtain $3Ae^x \cos x + 3Be^x \sin x = e^x \sin x$. Equating coefficients gives $A = 0$ and $B = 1/3$. The general solution is

$$y = e^x (c_1 \cos 2x + c_2 \sin 2x) + \frac{1}{3} e^x \sin x.$$

54. Applying $D^2 - 2D + 10$ to the differential equation we obtain

$$(D^2 - 2D + 10)\left(D^2 + D + \frac{1}{4}\right)y = (D^2 - 2D + 10)\left(D + \frac{1}{2}\right)^2 y = 0.$$

Then

$$y = \underbrace{c_1 e^{-x/2} + c_2 x e^{-x/2}}_{y_c} + c_3 e^x \cos 3x + c_4 e^x \sin 3x$$

and $y_p = Ae^x \cos 3x + Be^x \sin 3x$. Substituting y_p into the differential equation yields $(9B - 27A/4)e^x \cos 3x - (9A + 27B/4)e^x \sin 3x = -e^x \cos 3x + e^x \sin 3x$. Equating coefficients gives

$$-\frac{27}{4}A + 9B = -1$$

$$-9A - \frac{27}{4}B = 1.$$

Then $A = -4/225$, $B = -28/225$, and the general solution is

$$y = c_1 e^{-x/2} + c_2 x e^{-x/2} - \frac{4}{225} e^x \cos 3x - \frac{28}{225} e^x \sin 3x.$$

55. Applying $D^2 + 25$ to the differential equation we obtain

$$(D^2 + 25)(D^2 + 25) = (D^2 + 25)^2 = 0.$$

Then

$$y = \underbrace{c_1 \cos 5x + c_2 \sin 5x}_{y_c} + c_3 x \cos 5x + c_4 x \cos 5x$$

and $y_p = Ax \cos 5x + Bx \sin 5x$. Substituting y_p into the differential equation yields $10B \cos 5x -$ $10A \sin 5x = 20 \sin 5x$. Equating coefficients gives $A = -2$ and $B = 0$. The general solution is

$$y = c_1 \cos 5x + c_2 \sin 5x - 2x \cos 5x.$$

56. Applying $D^2 + 1$ to the differential equation we obtain

$$(D^2 + 1)(D^2 + 1) = (D^2 + 1)^2 = 0.$$

Then

$$y = \underbrace{c_1 \cos x + c_2 \sin x}_{y_c} + c_3 x \cos x + c_4 x \cos x$$

and $y_p = Ax \cos x + Bx \sin x$. Substituting y_p into the differential equation yields $2B \cos x -$ $2A \sin x = 4 \cos x - \sin x$. Equating coefficients gives $A = 1/2$ and $B = 2$. The general solution is

$$y = c_1 \cos x + c_2 \sin x + \frac{1}{2}x \cos x - 2x \sin x.$$

57. Applying $(D^2 + 1)^2$ to the differential equation we obtain

$$(D^2 + 1)^2(D^2 + D + 1) = 0.$$

Then

$$y = \underbrace{e^{-x/2}\left[c_1 \cos \frac{\sqrt{3}}{2}x + c_2 \sin \frac{\sqrt{3}}{2}x\right]}_{y_c} + c_3 \cos x + c_4 \sin x + c_5 x \cos x + c_6 x \sin x$$

and $y_p = A \cos x + B \sin x + Cx \cos x + Dx \sin x$. Substituting y_p into the differential equation yields

$$(B + C + 2D) \cos x + Dx \cos x + (-A - 2C + D) \sin x - Cx \sin x = x \sin x.$$

Equating coefficients gives

$$B + C + 2D = 0$$

$$D = 0$$

$$-A - 2C + D = 0$$

$$-C = 1.$$

Then $A = 2$, $B = 1$, $C = -1$, and $D = 0$, and the general solution is

$$y = e^{-x/2}\left[c_1 \cos \frac{\sqrt{3}}{2}x + c_2 \sin \frac{\sqrt{3}}{2}x\right] + 2 \cos x + \sin x - x \cos x.$$

58. Writing $\cos^2 x = \frac{1}{2}(1 + \cos 2x)$ and applying $D(D^2 + 4)$ to the differential equation we obtain

$$D(D^2 + 4)(D^2 + 4) = D(D^2 + 4)^2 = 0.$$

103

Then

$$y = \underbrace{c_1 \cos 2x + c_2 \sin 2x}_{y_c} + c_3 x \cos 2x + c_4 x \sin 2x + c_5$$

and $y_p = Ax \cos 2x + Bx \sin 2x + C$. Substituting y_p into the differential equation yields $-4A \sin 2x + 4B \cos 2x + 4C = \frac{1}{2} + \frac{1}{2} \cos 2x$. Equating coefficients gives $A = 0$, $B = 1/8$, and $C = 1/8$. The general solution is

$$y = c_1 \cos 2x + c_2 \sin 2x + \frac{1}{8} x \sin 2x + \frac{1}{8}.$$

59. Applying D^3 to the differential equation we obtain

$$D^3(D^3 + 8D^2) = D^5(D + 8) = 0.$$

Then

$$y = \underbrace{c_1 + c_2 x + c_3 e^{-8x}}_{y_c} + c_4 x^2 + c_5 x^3 + c_6 x^4$$

and $y_p = Ax^2 + Bx^3 + Cx^4$. Substituting y_p into the differential equation yields $16A + 6B + (48B + 24C)x + 96Cx^2 = 2 + 9x - 6x^2$. Equating coefficients gives

$$16A + 6B = 2$$

$$48B + 24C = 9$$

$$96C = -6.$$

Then $A = 11/256$, $B = 7/32$, and $C = -1/16$, and the general solution is

$$y = c_1 + c_2 x + c_3 e^{-8x} + \frac{11}{256}x^2 + \frac{7}{32}x^3 - \frac{1}{16}x^4.$$

60. Applying $D(D-1)^2(D+1)$ to the differential equation we obtain

$$D(D-1)^2(D+1)(D^3 - D^2 + D - 1) = D(D-1)^3(D+1)(D^2 + 1) = 0.$$

Then

$$y = \underbrace{c_1 e^x + c_2 \cos x + c_3 \sin x}_{y_c} + c_4 + c_5 e^{-x} + c_6 x e^x + c_7 x^2 e^x$$

and $y_p = A + Be^{-x} + Cxe^x + Dx^2 e^x$. Substituting y_p into the differential equation yields

$$4Dxe^x + (2C + 4D)e^x - 4Be^{-x} - A = xe^x - e^{-x} + 7.$$

Equating coefficients gives

$$4D = 1$$

$$2C + 4D = 0$$

$$-4B = -1$$

$$-A = 7.$$

Then $A = -7$, $B = 1/4$, $C = -1/2$, and $D = 1/4$, and the general solution is

$$y = c_1 e^x + c_2 \cos x + c_3 \sin x - 7 + \frac{1}{4} e^{-x} - \frac{1}{2} x e^x + \frac{1}{4} x^2 e^x.$$

61. Applying $D^2(D-1)$ to the differential equation we obtain

$$D^2(D-1)(D^3 - 3D^2 + 3D - 1) = D^2(D-1)^4 = 0.$$

Then

$$y = \underbrace{c_1 e^x + c_2 x e^x + c_3 x^2 e^x}_{y_c} + c_4 + c_5 x + c_6 x^3 e^x$$

and $y_p = A + Bx + Cx^3 e^x$. Substituting y_p into the differential equation yields $(-A + 3B) - Bx + 6Ce^x = 16 - x + e^x$. Equating coefficients gives

$$-A + 3B = 16$$

$$-B = -1$$

$$6C = 1.$$

Then $A = -13$, $B = 1$, and $C = 1/6$, and the general solution is

$$y = c_1 e^x + c_2 x e^x + c_3 x^2 e^x - 13 + x + \frac{1}{6} x^3 e^x.$$

62. Writing $(e^x + e^{-x})^2 = 2 + e^{2x} + e^{-2x}$ and applying $D(D-2)(D+2)$ to the differential equation we obtain

$$D(D-2)(D+2)(2D^3 - 3D^2 - 3D + 2) = D(D-2)^2(D+2)(D+1)(2D-1) = 0.$$

Then

$$y = \underbrace{c_1 e^{-x} + c_2 e^{2x} + c_3 e^{x/2}}_{y_c} + c_4 + c_5 x e^{2x} + c_6 e^{-2x}$$

and $y_p = A + Bx e^{2x} + Ce^{-2x}$. Substituting y_p into the differential equation yields $2A + 9Be^{2x} - 20Ce^{-2x} = 2 + e^{2x} + e^{-2x}$. Equating coefficients gives $A = 1$, $B = 1/9$, and $C = -1/20$, and the general solution is

$$y = c_1 e^{-x} + c_2 e^{2x} + c_3 e^{x/2} + 1 + \frac{1}{9} x e^{2x} - \frac{1}{20} e^{-2x}.$$

63. Applying $D(D-1)$ to the differential equation we obtain

$$D(D-1)(D^4 - 2D^3 + D^2) = D^3(D-1)^3 = 0.$$

Then

$$y = \underbrace{c_1 + c_2 x + c_3 e^x + c_4 x e^x}_{y_c} + c_5 x^2 + c_6 x^2 e^x$$

and $y_p = Ax^2 + Bx^2 e^x$. Substituting y_p into the differential equation yields $2A + 2Be^x = 1 + e^x$. Equating coefficients gives $A = 1/2$ and $B = 1/2$. The general solution is

$$y = c_1 + c_2 x + c_3 e^x + c_4 x e^x + \frac{1}{2}x^2 + \frac{1}{2}x^2 e^x.$$

64. Applying $D^3(D-2)$ to the differential equation we obtain

$$D^3(D-2)(D^4 - 4D^2) = D^5(D-2)^2(D+2) = 0.$$

Then

$$y = \underbrace{c_1 + c_2 x + c_3 e^{2x} + c_4 e^{-2x}}_{y_c} + c_5 x^2 + c_6 x^3 + c_7 x^4 + c_8 x e^{2x}$$

and $y_p = Ax^2 + Bx^3 + Cx^4 + Dxe^{2x}$. Substituting y_p into the differential equation yields $(-8A + 24C) - 24Bx - 48Cx^2 + 16De^{2x} = 5x^2 - e^{2x}$. Equating coefficients gives

$$-8A + 24C = 0$$

$$-24B = 0$$

$$-48C = 5$$

$$16D = -1.$$

Then $A = -5/16$, $B = 0$, $C = -5/48$, and $D = -1/16$, and the general solution is

$$y = c_1 + c_2 x + c_3 e^{2x} + c_4 e^{-2x} - \frac{5}{16}x^2 - \frac{5}{48}x^4 - \frac{1}{16}x e^{2x}.$$

65. The complementary function is $y_c = c_1 e^{8x} + c_2 e^{-8x}$. Using D to annihilate 16 we find $y_p = A$. Substituting y_p into the differential equation we obtain $-64A = 16$. Thus $A = -1/4$ and

$$y = c_1 e^{8x} + c_2 e^{-8x} - \frac{1}{4}$$

$$y' = 8c_1 e^{8x} - 8c_2 e^{-8x}.$$

The initial conditions imply

$$c_1 + c_2 = \frac{5}{4}$$

$$8c_1 - 8c_2 = 0.$$

106

Thus $c_1 = c_2 = 5/8$ and

$$y = \frac{5}{8}e^{8x} + \frac{5}{8}e^{-8x} - \frac{1}{4}.$$

66. The complementary function is $y_c = c_1 + c_2 e^{-x}$. Using D^2 to annihilate x we find $y_p = Ax + Bx^2$. Substituting y_p into the differential equation we obtain $(A + 2B) + 2Bx = x$. Thus $A = -1$ and $B = 1/2$, and

$$y = c_1 + c_2 e^{-x} - x + \frac{1}{2}x^2$$

$$y' = -c_2 e^{-x} - 1 + x.$$

The initial conditions imply

$$c_1 + c_2 = 1$$

$$-c_2 = 1.$$

Thus $c_1 = 2$ and $c_2 = -1$, and

$$y = 2 - e^{-x} - x + \frac{1}{2}x^2.$$

67. The complementary function is $y_c = c_1 + c_2 e^{5x}$. Using D^2 to annihilate $x - 2$ we find $y_p = Ax + Bx^2$. Substituting y_p into the differential equation we obtain $(-5A + 2B) - 10Bx = -2 + x$. Thus $A = 9/25$ and $B = -1/10$, and

$$y = c_1 + c_2 e^{5x} + \frac{9}{25}x - \frac{1}{10}x^2$$

$$y' = 5c_2 e^{5x} + \frac{9}{25} - \frac{1}{5}x.$$

The initial conditions imply

$$c_1 + c_2 = 0$$

$$5c_2 = \frac{41}{125}.$$

Thus $c_1 = -41/125$ and $c_2 = 41/125$, and

$$y = -\frac{41}{125} + \frac{41}{125}e^{5x} + \frac{9}{25}x - \frac{1}{10}x^2.$$

68. The complementary function is $y_c = c_1 e^x + c_2 e^{-6x}$. Using $D - 2$ to annihilate $10e^{2x}$ we find $y_p = Ae^{2x}$. Substituting y_p into the differential equation we obtain $8Ae^{2x} = 10e^{2x}$. Thus $A = 5/4$ and

$$y = c_1 e^x + c_2 e^{-6x} + \frac{5}{4}e^{2x}$$

$$y' = c_1 e^x - 6c_2 e^{-6x} + \frac{5}{2}e^{2x}.$$

The initial conditions imply

$$c_1 + c_2 = -\frac{1}{4}$$

$$c_1 - 6c_2 = -\frac{3}{2}.$$

Thus $c_1 = -3/7$ and $c_2 = 5/28$, and

$$y = -\frac{3}{7}e^x + \frac{5}{28}e^{-6x} + \frac{5}{4}e^{2x}$$

69. The complementary function is $y_c = c_1 \cos x + c_2 \sin x$. Using $(D^2 + 1)(D^2 + 4)$ to annihilate $8\cos 2x - 4\sin x$ we find $y_p = Ax\cos x + Bx\sin x + C\cos 2x + D\sin 2x$. Substituting y_p into the differential equation we obtain $2B\cos x - 3C\cos 2x - 2A\sin x - 3D\sin 2x = 8\cos 2x - 4\sin x$. Thus $A = 2$, $B = 0$, $C = -8/3$, and $D = 0$, and

$$y = c_1 \cos x + c_2 \sin x + 2x\cos x - \frac{8}{3}\cos 2x$$

$$y' = -c_1 \sin x + c_2 \cos x + 2\cos x - 2x\sin x + \frac{16}{3}\sin 2x.$$

The initial conditions imply

$$c_2 + \frac{8}{3} = -1$$

$$-c_1 - \pi = 0.$$

Thus $c_1 = -\pi$ and $c_2 = -11/3$, and

$$y = -\pi\cos x - \frac{11}{3}\sin x + 2x\cos x - \frac{8}{3}\cos 2x.$$

70. The complementary function is $y_c = c_1 + c_2 e^x + c_3 x e^x$. Using $D(D-1)^2$ to annihilate $xe^x + 5$ we find $y_p = Ax + Bx^2 e^x + Cx^3 e^x$. Substituting y_p into the differential equation we obtain $A + (2B + 6C)e^x + 6Cxe^x = xe^x + 5$. Thus $A = 5$, $B = -1/2$, and $C = 1/6$, and

$$y = c_1 + c_2 e^x + c_3 x e^x + 5x - \frac{1}{2}x^2 e^x + \frac{1}{6}x^3 e^x$$

$$y' = c_2 e^x + c_3(xe^x + e^x) + 5 - xe^x + \frac{1}{6}x^3 e^x$$

$$y'' = c_2 e^x + c_3(xe^x + 2e^x) - e^x - xe^x + \frac{1}{2}x^2 e^x + \frac{1}{6}x^3 e^x.$$

The initial conditions imply

$$c_1 + c_2 = 2$$

$$c_2 + c_3 + 5 = 2$$

$$c_2 + 2c_3 - 1 = -1.$$

Thus $c_1 = 8$, $c_2 = -6$, and $c_3 = 3$, and

$$y = 8 - 6e^x + 3xe^x + 5x - \frac{1}{2}x^2 e^x + \frac{1}{6}x^3 e^x.$$

71. The complementary function is $y_c = e^{2x}(c_1 \cos 2x + c_2 \sin 2x)$. Using D^4 to annihilate x^3 we find $y_p = A + Bx + Cx^2 + Dx^3$. Substituting y_p into the differential equation we obtain $(8A - 4B + 2C) + (8B - 8C + 6D)x + (8C - 12D)x^2 + 8Dx^3 = x^3$. Thus $A = 0$, $B = 3/32$, $C = 3/16$, and $D = 1/8$, and

$$y = e^{2x}(c_1 \cos 2x + c_2 \sin 2x) + \frac{3}{32}x + \frac{3}{16}x^2 + \frac{1}{8}x^3$$

$$y' = e^{2x}[c_1(2\cos 2x - 2\sin 2x) + c_2(2\cos 2x + 2\sin 2x)] + \frac{3}{32} + \frac{3}{8}x + \frac{3}{8}x^2.$$

The initial conditions imply

$$c_1 = 2$$

$$2c_1 + 2c_2 + \frac{3}{32} = 4.$$

Thus $c_1 = 2$, $c_2 = -3/64$, and

$$y = e^{2x}\left(2\cos 2x - \frac{3}{64}\sin 2x\right) + \frac{3}{32}x + \frac{3}{16}x^2 + \frac{1}{8}x^3.$$

72. The complementary function is $y_c = c_1 + c_2 x + c_3 x^2 + c_4 e^x$. Using $D^2(D-1)$ to annihilate $x + e^x$ we find $y_p = Ax^3 + Bx^4 + Cxe^x$. Substituting y_p into the differential equation we obtain $(-6A + 24B) - 24Bx + Ce^x = x + e^x$. Thus $A = -1/6$, $B = -1/24$, and $C = 1$, and

$$y = c_1 + c_2 x + c_3 x^2 + c_4 e^x - \frac{1}{6}x^3 - \frac{1}{24}x^4 + xe^x$$

$$y' = c_2 + 2c_3 x + c_4 e^x - \frac{1}{2}x^2 - \frac{1}{6}x^3 + e^x + xe^x$$

$$y'' = 2c_3 + c_4 e^x - x - \frac{1}{2}x^2 + 2e^x + xe^x.$$

$$y''' = c_4 e^x - 1 - x + 3e^x + xe^x$$

The initial conditions imply

$$c_1 + c_4 = 0$$

$$c_2 + c_4 + 1 = 0$$

$$2c_3 + c_4 + 2 = 0$$

$$2 + c_4 = 0.$$

Thus $c_1 = 2$, $c_2 = 1$, $c_3 = 0$, and $c_4 = -2$, and

$$y = 2 + x - 2e^x - \frac{1}{6}x^3 - \frac{1}{24}x^4 + xe^x.$$

73. To see in this case that the factors of L do not commute consider the operators $(xD - 1)(D + 4)$ and $(D + 4)(xD - 1)$. Applying the operators to the function x we find

$$(xD - 1)(D + 4)x = (xD^2 + 4xD - D - 4)x$$

$$= xD^2x + 4xDx - Dx - 4x$$

$$= x(0) + 4x(1) - 1 - 4x = -1$$

and

$$(D + 4)(xD - 1)x = (D + 4)(xDx - x)$$

$$= (D + 4)(x \cdot 1 - x) = 0.$$

Thus, the operators are not the same.

Exercises 4.6

The particular solution, $y_p = u_1 y_1 + u_2 y_2$, in the following problems can take on a variety of forms, especially where trigonometric functions are involved. The validity of a particular form can best be checked by substituting it back into the differential equation.

1. The auxiliary equation is $m^2 + 1 = 0$, so $y_c = c_1 \cos x + c_2 \sin x$ and

$$W = \begin{vmatrix} \cos x & \sin x \\ -\sin x & \cos x \end{vmatrix} = 1.$$

Identifying $f(x) = \sec x$ we obtain

$$u_1' = -\frac{\sin x \sec x}{1} = -\tan x$$

$$u_2' = \frac{\cos x \sec x}{1} = 1.$$

Then $u_1 = \ln|\cos x|$, $u_2 = x$, and

$$y = c_1 \cos x + c_2 \sin x + \cos x \ln|\cos x| + x \sin x$$

for $-\pi/2 < x < \pi/2$.

2. The auxiliary equation is $m^2 + 1 = 0$, so $y_c = c_1 \cos x + c_2 \sin x$ and

$$W = \begin{vmatrix} \cos x & \sin x \\ -\sin x & \cos x \end{vmatrix} = 1.$$

Identifying $f(x) = \tan x$ we obtain

$$u_1' = -\sin x \tan x = \frac{\cos^2 x - 1}{\cos x} = \cos x - \sec x$$

$$u_2' = \sin x.$$

Then $u_1 = \sin x - \ln|\sec x + \tan x|$, $u_2 = -\cos x$, and

$$y = c_1 \cos x + c_2 \sin x + \cos x \left(\sin x - \ln|\sec x + \tan x|\right) - \cos x \sin x$$

for $-\pi/2 < x < \pi/2$.

3. The auxiliary equation is $m^2 + 1 = 0$, so $y_c = c_1 \cos x + c_2 \sin x$ and

$$W = \begin{vmatrix} \cos x & \sin x \\ -\sin x & \cos x \end{vmatrix} = 1.$$

Identifying $f(x) = \sin x$ we obtain

$$u_1' = -\sin^2 x$$

$$u_2' = \cos x \sin x.$$

Then

$$u_1 = \frac{1}{4}\sin 2x - \frac{1}{2}x = \frac{1}{2}\sin x \cos x - \frac{1}{2}x$$

$$u_2 = -\frac{1}{2}\cos^2 x.$$

and

$$y = c_1 \cos x + c_2 \sin x + \frac{1}{2}\sin x \cos^2 x - \frac{1}{2}x \cos x - \frac{1}{2}\cos^2 x \sin x$$

$$= c_1 \cos x + c_2 \sin x - \frac{1}{2}x \cos x$$

for $-\infty < x < \infty$.

4. The auxiliary equation is $m^2 + 1 = 0$, so $y_c = c_1 \cos x + c_2 \sin x$ and

$$W = \begin{vmatrix} \cos x & \sin x \\ -\sin x & \cos x \end{vmatrix} = 1.$$

Identifying $f(x) = \sec x \tan x$ we obtain

$$u_1' = -\sin x(\sec x \tan x) = -\tan^2 x = 1 - \sec^2 x$$

$$u_2' = \cos x(\sec x \tan x) = \tan x.$$

Then $u_1 = x - \tan x$, $u_2 = -\ln|\cos x|$, and

$$y = c_1 \cos x + c_2 \sin x + x \cos x - \sin x - \sin x \ln|\cos x|$$

$$= c_1 \cos x + c_3 \sin x + x \cos x - \sin x \ln|\cos x|$$

for $-\pi/2 < x < \pi/2$.

5. The auxiliary equation is $m^2 + 1 = 0$, so $y_c = c_1 \cos x + c_2 \sin x$ and

$$W = \begin{vmatrix} \cos x & \sin x \\ -\sin x & \cos x \end{vmatrix} = 1.$$

Identifying $f(x) = \cos^2 x$ we obtain

$$u_1' = -\sin x \cos^2 x$$

$$u_2' = \cos^3 x = \cos x \left(1 - \sin^2 x\right).$$

Then $u_1 = \frac{1}{3}\cos^3 x$, $u_2 = \sin x - \frac{1}{3}\sin^3 x$, and

$$y = c_1 \cos x + c_2 \sin x + \frac{1}{3}\cos^4 x + \sin^2 x - \frac{1}{3}\sin^4 x$$

$$= c_1 \cos x + c_2 \sin x + \frac{1}{3}\left(\cos^2 x + \sin^2 x\right)\left(\cos^2 x - \sin^2 x\right) + \sin^2 x$$

$$= c_1 \cos x + c_2 \sin x + \frac{1}{3}\cos^2 x + \frac{2}{3}\sin^2 x$$

$$= c_1 \cos x + c_2 \sin x + \frac{1}{3} + \frac{1}{3}\sin^2 x$$

for $-\infty < x < \infty$.

6. The auxiliary equation is $m^2 + 1 = 0$, so $y_c = c_1 \cos x + c_2 \sin x$ and

$$W = \begin{vmatrix} \cos x & \sin x \\ -\sin x & \cos x \end{vmatrix} = 1.$$

Identifying $f(x) = \sec^2 x$ we obtain

$$u_1' = -\frac{\sin x}{\cos^2 x}$$

$$u_2' = \sec x.$$

Then

$$u_1 = -\frac{1}{\cos x} = -\sec x$$

$$u_2 = \ln|\sec x + \tan x|$$

and

$$y = c_1 \cos x + c_2 \sin x - \cos x \sec x + \sin x \ln|\sec x + \tan x|$$

$$= c_1 \cos x + c_2 \sin x - 1 + \sin x \ln|\sec x + \tan x|$$

for $-\pi/2 < x < \pi/2$.

7. The auxiliary equation is $m^2 - 1 = 0$, so $y_c = c_1 e^x + c_2 e^{-x}$ and

$$W = \begin{vmatrix} e^x & e^{-x} \\ e^x & -e^{-x} \end{vmatrix} = -2.$$

Identifying $f(x) = \cosh x = \frac{1}{2}(e^{-x} + e^x)$ we obtain

$$u_1' = \frac{1}{4}e^{2x} + \frac{1}{4}$$

$$u_2' = -\frac{1}{4} - \frac{1}{4}e^{2x}.$$

Then

$$u_1 = -\frac{1}{8}e^{-2x} + \frac{1}{4}x$$

$$u_2 = -\frac{1}{8}e^{2x} - \frac{1}{4}x$$

and

$$y = c_1 e^x + c_2 e^{-x} - \frac{1}{8}e^{-x} + \frac{1}{4}xe^x - \frac{1}{8}e^x - \frac{1}{4}xe^{-x}$$

$$= c_3 e^x + c_4 e^{-x} + \frac{1}{4}x(e^x - e^{-x})$$

$$= c_3 e^x + c_4 e^{-x} + \frac{1}{2}x \sinh x$$

for $-\infty < x < \infty$.

8. The auxiliary equation is $m^2 - 1 = 0$, so $y_c = c_1 e^x + c_2 e^{-x}$ and

$$W = \begin{vmatrix} e^x & e^{-x} \\ e^x & -e^{-x} \end{vmatrix} = -2.$$

Identifying $f(x) = \sinh 2x$ we obtain

$$u_1' = -\frac{1}{4}e^{-3x} + \frac{1}{4}e^x$$

$$u_2' = \frac{1}{4}e^{-x} - \frac{1}{4}e^{3x}.$$

Then

$$u_1 = \frac{1}{12}e^{-3x} + \frac{1}{4}e^x$$

$$u_2 = -\frac{1}{4}e^{-x} - \frac{1}{12}e^{3x}.$$

and

$$y = c_1 e^x + c_2 e^{-x} + \frac{1}{12}e^{-2x} + \frac{1}{4}e^{2x} - \frac{1}{4}e^{-2x} - \frac{1}{12}e^{2x}$$

$$= c_1 e^x + c_2 e^{-x} + \frac{1}{6}\left(e^{2x} - e^{-2x}\right)$$

$$= c_1 e^x + c_2 e^{-x} + \frac{1}{3}\sinh 2x$$

for $-\infty < x < \infty$.

9. The auxiliary equation is $m^2 - 4 = 0$, so $y_c = c_1 e^{2x} + c_2 e^{-2x}$ and

$$W = \begin{vmatrix} e^{2x} & e^{-2x} \\ 2e^{2x} & -2e^{-2x} \end{vmatrix} = -4.$$

Identifying $f(x) = e^{2x}/x$ we obtain $u_1' = 1/4x$ and $u_2' = -e^{4x}/4x$. Then

$$u_1 = \frac{1}{4}\ln|x|, \qquad u_2 = -\frac{1}{4}\int_{x_0}^x \frac{e^{4t}}{t}\,dt$$

and

$$y = c_1 e^{2x} + c_2 e^{-2x} + \frac{1}{4}\left(e^{2x}\ln|x| - e^{-2x}\int_{x_0}^x \frac{e^{4t}}{t}\,dt\right), \qquad x_0 > 0$$

for $x > 0$.

10. The auxiliary equation is $m^2 - 9 = 0$, so $y_c = c_1 e^{3x} + c_2 e^{-3x}$ and

$$W = \begin{vmatrix} e^{3x} & e^{-3x} \\ 3e^{3x} & -3e^{-3x} \end{vmatrix} = -6.$$

Identifying $f(x) = 9x/e^{3x}$ we obtain $u_1' = \frac{3}{2}xe^{-6x}$ and $u_2' = -\frac{3}{2}x$. Then

$$u_1 = -\frac{1}{24}e^{-6x} - \frac{1}{4}xe^{-6x}, \quad u_2 = -\frac{3}{4}x^2$$

and

$$y = c_1 e^{3x} + c_2 e^{-3x} - \frac{1}{24}e^{-3x} - \frac{1}{4}xe^{-3x} - \frac{3}{4}x^2 e^{-3x}$$

$$= c_1 e^{3x} + c_3 e^{-3x} - \frac{1}{4}xe^{-3x}(1 - 3x)$$

for $-\infty < x < \infty$.

11. The auxiliary equation is $m^2 + 3m + 2 = (m+1)(m+2) = 0$, so $y_c = c_1 e^{-x} + c_2 e^{-2x}$ and

$$W = \begin{vmatrix} e^{-x} & e^{-2x} \\ -e^{-x} & -2e^{-2x} \end{vmatrix} = -e^{-3x}.$$

114

Identifying $f(x) = 1/(1 + e^x)$ we obtain

$$u_1' = \frac{e^x}{1 + e^x}$$

$$u_2' = -\frac{e^{2x}}{1 + e^x} = \frac{e^x}{1 + e^x} - e^x.$$

Then $u_1 = \ln(1 + e^x)$, $u_2 = \ln(1 + e^x) - e^x$, and

$$y = c_1 e^{-x} + c_2 e^{-2x} + e^{-x}\ln(1 + e^x) + e^{-2x}\ln(1 + e^x) - e^{-x}$$

$$= c_3 e^{-x} + c_2 e^{-2x} + (1 + e^{-x})e^{-x}\ln(1 + e^x)$$

for $-\infty < x < \infty$.

12. The auxiliary equation is $m^2 - 3m + 2 = (m - 1)(m - 2) = 0$, so $y_c = c_1 e^x + c_2 e^{2x}$ and

$$W = \begin{vmatrix} e^x & e^{2x} \\ e^x & 2e^{2x} \end{vmatrix} = e^{3x}.$$

Identifying $f(x) = e^{3x}/(1 + e^x)$ we obtain

$$u_1' = -\frac{e^{2x}}{1 + e^x} = \frac{e^x}{1 + e^x} - e^x$$

$$u_2' = \frac{e^x}{1 + e^x}.$$

Then $u_1 = \ln(1 + e^x) - e^x$, $u_2 = \ln(1 + e^x)$, and

$$y = c_1 e^x + c_2 e^{2x} + e^x \ln(1 + e^x) - e^{2x} + e^{2x}\ln(1 + e^x)$$

$$= c_1 e^x + c_3 e^{2x} + (1 + e^x)e^x \ln(1 + e^x)$$

for $-\infty < x < \infty$.

13. The auxiliary equation is $m^2 + 3m + 2 = (m + 1)(m + 2) = 0$, so $y_c = c_1 e^{-x} + c_2 e^{-2x}$ and

$$W = \begin{vmatrix} e^{-x} & e^{-2x} \\ -e^{-x} & -2e^{-2x} \end{vmatrix} = -e^{-3x}.$$

Identifying $f(x) = \sin e^x$ we obtain

$$u_1' = \frac{e^{-2x} \sin e^x}{e^{-3x}} = e^x \sin e^x$$

$$u_2' = \frac{e^{-x} \sin e^x}{-e^{-3x}} = -e^{2x} \sin e^x.$$

Then $u_1 = -\cos e^x$, $u_2 = e^x \cos x - \sin e^x$, and

$$y = c_1 e^{-x} + c_2 e^{-2x} - e^{-x} \cos e^x + e^{-x} \cos e^x - e^{-2x} \sin e^x$$

$$= c_1 e^{-x} + c_2 e^{-2x} - e^{-2x} \sin e^x$$

for $-\infty < x < \infty$.

14. The auxiliary equation is $m^2 - 2m + 1 = (m-1)^2 = 0$, so $y_c = c_1 e^x + c_2 x e^x$ and

$$W = \begin{vmatrix} e^x & x e^x \\ e^x & x e^x + e^x \end{vmatrix} = e^{2x}.$$

Identifying $f(x) = e^x \tan^{-1} x$ we obtain

$$u_1' = -\frac{x e^x e^x \tan^{-1} x}{e^{2x}} = -x \tan^{-1} x$$

$$u_2' = \frac{e^x e^x \tan^{-1} x}{e^{2x}} = \tan^{-1} x.$$

Then

$$u_1 = -\frac{1+x^2}{2} \tan^{-1} x + \frac{x}{2}$$

$$u_2 = x \tan^{-1} x - \frac{1}{2} \ln\left(1+x^2\right)$$

and

$$y = c_1 e^x + c_2 x e^x + \left(-\frac{1+x^2}{2} \tan^{-1} x + \frac{x}{2}\right) e^x + \left(x \tan^{-1} x - \frac{1}{2} \ln\left(1+x^2\right)\right) x e^x$$

$$= c_1 e^x + c_3 x e^x + \frac{1}{2} e^x \left[\left(x^2 - 1\right) \tan^{-1} x - \ln\left(1+x^2\right)\right]$$

for $-\infty < x < \infty$.

15. The auxiliary equation is $m^2 - 2m + 1 = (m-1)^2 = 0$, so $y_c = c_1 e^x + c_2 x e^x$ and

$$W = \begin{vmatrix} e^x & x e^x \\ e^x & x e^x + e^x \end{vmatrix} = e^{2x}.$$

Identifying $f(x) = e^x / \left(1+x^2\right)$ we obtain

$$u_1' = -\frac{x e^x e^x}{e^{2x}\left(1+x^2\right)} = -\frac{x}{1+x^2}$$

$$u_2' = \frac{e^x e^x}{e^{2x}\left(1+x^2\right)} = \frac{1}{1+x^2}.$$

Then $u_1 = -\frac{1}{2} \ln\left(1+x^2\right)$, $u_2 = \tan^{-1} x$, and

$$y = c_1 e^x + c_2 x e^x - \frac{1}{2} e^x \ln\left(1+x^2\right) + x e^x \tan^{-1} x$$

for $-\infty < x < \infty$.

16. The auxiliary equation is $m^2 - 2m + 2 = [m - (1+i)][m - (1-i)] = 0$, so $y_c = c_1 c^x \sin x + c_2 e^x \cos x$ and

$$W = \begin{vmatrix} e^x \sin x & e^x \cos x \\ e^x \cos x + e^x \sin x & -e^x \sin x + e^x \cos x \end{vmatrix} = -e^{2x}.$$

Identifying $f(x) = e^x \sec x$ we obtain

$$u_1' = -\frac{e^x \cos x e^x \sec x}{-e^{2x}} = 1$$

$$u_2' = \frac{(e^x \sin x)(e^x \sec x)}{-e^{2x}} = -\tan x.$$

Then $u_1 = x$, $u_2 = \ln|\cos x|$, and

$$y = c_1 e^x \sin x + c_2 e^x \cos x + x e^x \sin x + e^x \cos x \ln|\cos x|$$

for $-\pi/2 < x < \pi/2$.

17. The auxiliary equation is $m^2 + 2m + 1 = (m+1)^2 = 0$, so $y_c = c_1 e^{-x} + c_2 x e^{-x}$ and

$$W = \begin{vmatrix} e^{-x} & x e^{-x} \\ -e^{-x} & -x e^{-x} + e^{-x} \end{vmatrix} = e^{-2x}.$$

Identifying $f(x) = e^{-x} \ln x$ we obtain

$$u_1' = -\frac{x e^{-x} e^{-x} \ln x}{e^{-2x}} = -x \ln x$$

$$u_2' = \frac{e^{-x} e^{-x} \ln x}{e^{-2x}} = \ln x.$$

Then

$$u_1 = -\frac{1}{2} x^2 \ln x + \frac{1}{4} x^2$$

$$u_2 = x \ln x - x$$

and

$$y = c_1 e^{-x} + c_2 x e^{-x} - \frac{1}{2} x^2 e^{-x} \ln x + \frac{1}{4} x^2 e^{-x} + x^2 e^{-x} \ln x - x^2 e^{-x}$$

$$= c_1 e^{-x} + c_2 x e^{-x} + \frac{1}{2} x^2 e^{-x} \ln x - \frac{3}{4} x^2 e^{-x}$$

for $x > 0$.

18. The auxiliary equation is $m^2 + 10m + 25 = (m+5)^2 = 0$, so $y_c = c_1 e^{-5x} + c_2 x e^{-5x}$ and

$$W = \begin{vmatrix} e^{-5x} & x e^{-5x} \\ -5e^{-5x} & -5x e^{-5x} + e^{-5x} \end{vmatrix} = e^{-10x}.$$

Identifying $f(x) = e^{-10x}/x^2$ we obtain

$$u_1' = -\frac{xe^{-5x}e^{-10x}}{x^2 e^{-10x}} = -\frac{e^{-5x}}{x}$$

$$u_2' = \frac{e^{-5x}e^{-10x}}{x^2 e^{-10x}} = \frac{e^{-5x}}{x^2}.$$

Then

$$u_1 = -\int_{x_0}^{x} \frac{e^{-5t}}{t}\, dt, \quad x_0 > 0$$

$$u_2 = \int_{x_0}^{x} \frac{e^{-5t}}{t^2}\, dt, \quad x_0 > 0$$

and

$$y = c_1 e^{-5x} + c_2 x e^{-5x} - e^{-5x}\int_{x_0}^{x} \frac{e^{-5t}}{t}\, dt + x e^{-5x}\int_{x_0}^{x} \frac{e^{-5t}}{t^2}\, dt$$

for $x > 0$.

19. The auxiliary equation is $3m^2 - 6m + 30 = 3[m - (1+3i)][m - (1-3i)] = 0$, so $y_c = c_1 e^x \cos 3x + c_2 e^x \sin 3x$ and

$$W = \begin{vmatrix} e^x \cos 3x & e^x \sin 3x \\ -3e^x \sin x + e^x \cos 3x & 3e^x \cos 3x + e^x \sin 3x \end{vmatrix} = 3e^{2x}.$$

Identifying $f(x) = \frac{1}{3}e^x \tan 3x$ we obtain

$$u_1' = -\frac{(e^x \sin 3x)(e^x \tan 3x)}{9e^{2x}} = -\frac{1}{9}\frac{\sin^2 3x}{\cos 3x} = \frac{1}{9}(\cos 3x - \sec 3x)$$

$$u_2' = \frac{(e^x \cos 3x)(e^x \tan 3x)}{9e^{2x}} = \frac{1}{9}\sin 3x.$$

Then

$$u_1 = \frac{1}{27}\sin 3x - \frac{1}{27}\ln|\sec 3x + \tan 3x|$$

$$u_2 = -\frac{1}{27}\cos 3x$$

and

$$y = c_1 e^x \cos 3x + c_2 e^x \sin 3x + \frac{1}{27}e^x \sin 3x \cos 3x$$

$$-\frac{1}{27}e^x \cos 3x \ln|\sec 3x + \tan 3x| - \frac{1}{27}e^x \sin 3x \cos 3x$$

$$= c_1 e^x \cos 3x + c_2 e^x \sin 3x - \frac{1}{27}e^x \cos 3x \ln|\sec 3x + \tan 3x|$$

for $-\pi/6 < x < \pi/6$.

20. The auxiliary equation is $4m^2 - 4m + 1 = (2m-1)^2 = 0$, so $y_c = c_1 e^{x/2} + c_2 x e^{x/2}$ and

$$W = \begin{vmatrix} e^{x/2} & x e^{x/2} \\ \frac{1}{2} e^{x/2} & \frac{1}{2} x e^{x/2} + e^{x/2} \end{vmatrix} = e^x.$$

Identifying $f(x) = \frac{1}{4} e^{x/2} \sqrt{1 - x^2}$ we obtain

$$u_1' = -\frac{x e^{x/2} e^{x/2} \sqrt{1 - x^2}}{4 e^x} = -\frac{1}{4} x \sqrt{1 - x^2}$$

$$u_2' = \frac{e^{x/2} e^{x/2} \sqrt{1 - x^2}}{4 e^x} = \frac{1}{4} \sqrt{1 - x^2}.$$

Then

$$u_1 = \frac{1}{12} \left(1 - x^2\right)^{3/2}$$

$$u_2 = \frac{x}{8} \sqrt{1 - x^2} + \frac{1}{8} \sin^{-1} x$$

and

$$y = c_1 e^{x/2} + c_2 x e^{x/2} + \frac{1}{12} e^{x/2} \left(1 - x^2\right)^{3/2} + \frac{1}{8} x^2 e^{x/2} \sqrt{1 - x^2} + \frac{1}{8} x e^{x/2} \sin^{-1} x$$

for $-1 \leq x \leq 1$.

21. The auxiliary equation is $m^3 + m = m(m^2 + 1) = 0$, so $y_c = c_1 + c_2 \cos x + c_3 \sin x$ and

$$W = \begin{vmatrix} 1 & \cos x & \sin x \\ 0 & -\sin x & \cos x \\ 0 & -\cos x & -\sin x \end{vmatrix} = 1.$$

Identifying $f(x) = \tan x$ we obtain

$$u_1' = W_1 = \begin{vmatrix} 0 & \cos x & \sin x \\ 0 & -\sin x & \cos x \\ \tan x & -\cos x & -\sin x \end{vmatrix} = \tan x$$

$$u_2' = W_2 = \begin{vmatrix} 1 & 0 & \sin x \\ 0 & 0 & \cos x \\ 0 & \tan x & -\sin x \end{vmatrix} = -\sin x$$

$$u_3' = W_3 = \begin{vmatrix} 1 & \cos x & 0 \\ 0 & -\sin x & 0 \\ 0 & -\cos x & \tan x \end{vmatrix} = -\sin x \tan x = \frac{\cos^2 x - 1}{\cos x} = \cos x - \sec x.$$

Then

$$u_1 = -\ln|\cos x|$$

$$u_2 = \cos x$$

$$u_3 = \sin x - \ln|\sec x + \tan x|$$

and

$$y = c_1 + c_2 \cos x + c_3 \sin x - \ln|\cos x| + \cos^2 x$$

$$+ \sin^2 x - \sin x \ln|\sec x + \tan x|$$

$$= c_4 + c_2 \cos x + c_3 \sin x - \ln|\cos x| - \sin x \ln|\sec x + \tan x|$$

for $-\infty < x < \infty$.

22. The auxiliary equation is $m^3 + 4m = m\left(m^2 + 4\right) = 0$, so $y_c = c_1 + c_2 \cos 2x + c_3 \sin 2x$ and

$$W = \begin{vmatrix} 1 & \cos 2x & \sin 2x \\ 0 & -2\sin 2x & 2\cos 2x \\ 0 & -4\cos 2x & -4\sin 2x \end{vmatrix} = 8.$$

Identifying $f(x) = \sec 2x$ we obtain

$$u_1' = \frac{1}{8}W_1 = \frac{1}{8}\begin{vmatrix} 0 & \cos 2x & \sin 2x \\ 0 & -2\sin 2x & 2\cos 2x \\ \sec 2x & -4\cos 2x & -4\sin 2x \end{vmatrix} = \frac{1}{4}\sec 2x$$

$$u_2' = \frac{1}{8}W_2 = \frac{1}{8}\begin{vmatrix} 1 & 0 & \sin 2x \\ 0 & 0 & 2\cos 2x \\ 0 & \sec 2x & -4\sin 2x \end{vmatrix} = -\frac{1}{4}$$

$$u_3' = \frac{1}{8}W_3 = \frac{1}{8}\begin{vmatrix} 1 & \cos 2x & 0 \\ 0 & -2\sin 2x & 0 \\ 0 & -4\cos 2x & \sec 2x \end{vmatrix} = -\frac{1}{4}\tan 2x.$$

Then

$$u_1 = \frac{1}{8}\ln|\sec 2x + \tan 2x|$$

$$u_2 = -\frac{1}{4}x$$

$$u_3 = \frac{1}{8}\ln|\cos 2x|$$

and

$$y = c_1 + c_2 \cos 2x + c_3 \sin 2x + \frac{1}{8}\ln|\sec 2x + \tan 2x| - \frac{1}{4}x \cos 2x + \frac{1}{8}\sin 2x \ln|\cos 2x|$$

120

for $-\pi/4 < x < \pi/4$.

23. The auxiliary equation is $m^3 - 2m^2 - m + 2 = (m-1)(m-2)(m+1) = 0$, so $y_c = c_1 e^x + c_2 e^{2x} + c_3 e^{-x}$ and

$$W = \begin{vmatrix} e^x & e^{2x} & e^{-x} \\ e^x & 2e^{2x} & -e^{-x} \\ e^x & 4e^{2x} & e^{-x} \end{vmatrix} = 6e^{2x}.$$

Identifying $f(x) = e^{3x}$ we obtain

$$u_1' = \frac{1}{6e^{2x}} W_1 = \frac{1}{6e^{2x}} \begin{vmatrix} 0 & e^{2x} & e^{-x} \\ 0 & 2e^{2x} & -e^{-x} \\ e^{3x} & 4e^{2x} & e^{-x} \end{vmatrix} = \frac{-3e^{4x}}{6e^{2x}} = -\frac{1}{2}e^{2x}$$

$$u_2' = \frac{1}{6e^{2x}} W_2 = \frac{1}{6e^{2x}} \begin{vmatrix} e^x & 0 & e^{-x} \\ e^x & 0 & -e^{-x} \\ e^x & e^{3x} & e^{-x} \end{vmatrix} = \frac{2e^{3x}}{6e^{2x}} = \frac{1}{3}e^x$$

$$u_3' = \frac{1}{6e^{2x}} W_1^3 = \frac{1}{6e^{2x}} \begin{vmatrix} e^x & e^{2x} & 0 \\ e^x & 2e^{2x} & 0 \\ e^x & 4e^{2x} & e^{3x} \end{vmatrix} = \frac{e^{6x}}{6e^{2x}} = \frac{1}{6}e^{4x}.$$

Then $u_1 = -\frac{1}{4}e^{2x}$, $u_2 = \frac{1}{3}e^x$, and $u_3 = \frac{1}{24}e^{4x}$, and

$$y = c_1 e^x + c_2 e^{2x} + c_3 e^{-x} - \frac{1}{4}e^{3x} + \frac{1}{3}e^{3x} + \frac{1}{24}e^{3x}$$

$$= c_1 e^x + c_2 e^{2x} + c_3 e^{-x} + \frac{1}{8}e^{3x}$$

for $-\infty < x < \infty$.

24. The auxiliary equation is $2m^3 - 6m^2 = 2m^2(m-3) = 0$, so $y_c = c_1 + c_2 x + c_3 e^{3x}$ and

$$W = \begin{vmatrix} 1 & x & e^{3x} \\ 0 & 1 & 3e^{3x} \\ 0 & 0 & 9e^{3x} \end{vmatrix} = 9e^{3x}.$$

Identifying $f(x) = x^2/2$ we obtain

$$u_1' = \frac{1}{9e^{3x}} W_1 = \frac{1}{9e^{3x}} \begin{vmatrix} 0 & x & e^{3x} \\ 0 & 1 & 3e^{3x} \\ x^2/2 & 0 & 9e^{3x} \end{vmatrix} = \frac{\frac{3}{2}x^3 e^{3x} - \frac{1}{2}x^2 e^{3x}}{9e^{3x}} = \frac{1}{6}x^3 - \frac{1}{18}x^2$$

$$u_2' = \frac{1}{9e^{3x}} W_2 = \frac{1}{9e^{3x}} \begin{vmatrix} 1 & 0 & e^{3x} \\ 0 & 0 & 3e^{3x} \\ 0 & x^2/2 & 9e^{3x} \end{vmatrix} = \frac{-\frac{3}{2}x^2 e^{3x}}{9e^{3x}} = -\frac{1}{6}x^2$$

$$u_3' = \frac{1}{9e^{3x}} W_3 = \frac{1}{9e^{3x}} \begin{vmatrix} 1 & x & 0 \\ 0 & 1 & 0 \\ 0 & 0 & x^2/2 \end{vmatrix} = \frac{\frac{1}{2}x^2}{9e^{3x}} = \frac{1}{18}x^2 e^{-3x}.$$

Then

$$u_1 = \frac{1}{24}x^4 - \frac{1}{54}x^3$$

$$u_2 = -\frac{1}{18}x^3$$

$$u_3 = -\frac{1}{54}x^2 e^{-3x} - \frac{1}{81}x e^{-3x} - \frac{1}{243}e^{-3x}$$

and

$$y = c_1 + c_2 x + c_3 e^{3x} + \frac{1}{24}x^4 - \frac{1}{54}x^3 - \frac{1}{18}x^4 - \frac{1}{54}x^2 - \frac{1}{81}x - \frac{1}{243}$$

$$= c_4 + c_5 x + c_3 e^{3x} - \frac{1}{72}x^4 - \frac{1}{54}x^3 - \frac{1}{54}x^2$$

for $-\infty < x < \infty$.

25. The auxiliary equation is $4m^2 - 1 = (2m - 1)(2m + 1) = 0$, so $y_c = c_1 e^{x/2} + c_2 e^{-x/2}$ and

$$W = \begin{vmatrix} e^{x/2} & e^{-x/2} \\ \frac{1}{2}e^{x/2} & -\frac{1}{2}e^{-x/2} \end{vmatrix} = -1.$$

Identifying $f(x) = xe^{x/2}/4$ we obtain $u_1' = x/4$ and $u_2' = -xe^x/4$. Then $u_1 = x^2/8$ and $u_2 = -xe^x/4 + e^x/4$. Thus

$$y = c_1 e^{x/2} + c_2 e^{-x/2} + \frac{1}{8}x^2 e^{x/2} - \frac{1}{4}x e^{x/2} + \frac{1}{4}e^{x/2}$$

$$= c_3 e^{x/2} + c_2 e^{-x/2} + \frac{1}{8}x^2 e^{x/2} - \frac{1}{4}x e^{x/2}$$

and

$$y' = \frac{1}{2}c_3 e^{x/2} - \frac{1}{2}c_2 e^{-x/2} + \frac{1}{16}x^2 e^{x/2} + \frac{1}{8}x e^{x/2} - \frac{1}{4}e^{x/2}.$$

The initial conditions imply

$$c_3 + c_2 = 1$$

$$\frac{1}{2}c_3 - \frac{1}{2}c_2 - \frac{1}{4} = 0.$$

Thus $c_3 = 3/4$ and $c_2 = 1/4$, and

$$y = \frac{3}{4}e^{x/2} + \frac{1}{4}e^{-x/2} + \frac{1}{8}x^2e^{x/2} - \frac{1}{4}xe^{x/2}.$$

26. The auxiliary equation is $2m^2 + m - 1 = (2m - 1)(m + 1) = 0$, so $y_c = c_1e^{x/2} + c_2e^{-x}$ and

$$W = \begin{vmatrix} e^{x/2} & e^{-x} \\ \frac{1}{2}e^{x/2} & -e^{-x} \end{vmatrix} = -\frac{3}{2}e^{-x/2}.$$

Identifying $f(x) = (x + 1)/2$ we obtain

$$u_1' = \frac{1}{3}e^{-x/2}(x + 1)$$

$$u_2' = -\frac{1}{3}e^x(x + 1).$$

Then

$$u_1 = -e^{-x/2}\left(\frac{2}{3}x - 2\right)$$

$$u_2 = -\frac{1}{3}xe^x.$$

Thus

$$y = c_1e^{x/2} + c_2e^{-x} - x - 2$$

and

$$y' = \frac{1}{2}c_1e^{x/2} - c_2e^{-x} - 1.$$

The initial conditions imply

$$c_1 - c_2 - 2 = 1$$

$$\frac{1}{2}c_1 - c_2 - 1 = 0.$$

Thus $c_1 = 8/3$ and $c_2 = 1/3$, and

$$y = \frac{8}{3}e^{x/2} + \frac{1}{3}e^{-x} - x - 2.$$

27. The auxiliary equation is $m^2 + 2m - 8 = (m - 2)(m + 4) = 0$, so $y_c = c_1e^{2x} + c_2e^{-4x}$ and

$$W = \begin{vmatrix} e^{2x} & e^{-4x} \\ 2e^{2x} & -4e^{-4x} \end{vmatrix} = -6e^{-2x}.$$

Identifying $f(x) = 2e^{-2x} - e^{-x}$ we obtain

$$u_1' = \frac{1}{3}e^{-4x} - \frac{1}{6}e^{-3x}$$

$$u_2' = -\frac{1}{6}e^{3x} - \frac{1}{3}e^{2x}.$$

Then

$$u_1 = -\frac{1}{12}e^{-4x} + \frac{1}{18}e^{-3x}$$

$$u_2 = \frac{1}{18}e^{3x} - \frac{1}{6}e^{2x}.$$

Thus

$$y = c_1 e^{2x} + c_2 e^{-4x} - \frac{1}{12}e^{-2x} + \frac{1}{18}e^{-x} + \frac{1}{18}e^{-x} - \frac{1}{6}e^{-2x}$$

$$= c_1 e^{2x} + c_2 e^{-4x} - \frac{1}{4}e^{-2x} + \frac{1}{9}e^{-x}$$

and

$$y' = 2c_1 e^{2x} - 4c_2 e^{-4x} + \frac{1}{2}e^{-2x} - \frac{1}{9}e^{-x}.$$

The initial conditions imply

$$c_1 + c_2 - \frac{5}{36} = 1$$

$$2c_1 - 4c_2 + \frac{7}{18} = 0.$$

Thus $c_1 = 25/36$ and $c_2 = 4/9$, and

$$y = \frac{25}{36}e^{2x} + \frac{4}{9}e^{-4x} - \frac{1}{4}e^{-2x} + \frac{1}{9}e^{-x}.$$

28. The auxiliary equation is $m^2 - 4m + 4 = (m-2)^2 = 0$, so $y_c = c_1 e^{2x} + c_2 x e^{2x}$ and

$$W = \begin{vmatrix} e^{2x} & xe^{2x} \\ 2e^{2x} & 2xe^{2x} + e^{2x} \end{vmatrix} = e^{4x}.$$

Identifying $f(x) = \left(12x^2 - 6x\right) e^{2x}$ we obtain

$$u_1' = 6x^2 - 12x^3$$

$$u_2' = 12x^2 - 6x.$$

Then

$$u_1 = 2x^3 - 3x^4$$

$$u_2 = 4x^3 - 3x^2.$$

Thus

$$y = c_1 e^{2x} + c_2 x e^{2x} + \left(2x^3 - 3x^4\right)e^{2x} + \left(4x^3 - 3x^2\right)xe^{2x}$$

$$= c_1 e^{2x} + c_2 x e^{2x} + e^{2x}\left(x^4 - x^3\right)$$

and

$$y' = 2c_1 e^{2x} + c_2\left(2xe^{2x} + e^{2x}\right) + e^{2x}\left(4x^3 - 3x^2\right) + 2e^{2x}\left(x^4 - x^3\right).$$

The initial conditions imply

$$c_1 \qquad = 1$$

$$2c_1 + c_2 = 0.$$

Thus $c_1 = 1$ and $c_2 = -2$, and

$$y = e^{2x} - 2xe^{2x} + e^{2x}\left(x^4 - x^3\right)$$

$$= e^{2x}\left(x^4 - x^3 - 2x + 1\right).$$

29. Write the equation in the form

$$y'' + \frac{1}{x}y' + \left(1 - \frac{1}{4x^2}\right)y = x^{-1/2}$$

and identify $f(x) = x^{-1/2}$. From $y_1 = x^{-1/2}\cos x$ and $y_2 = x^{-1/2}\sin x$ we compute

$$W(y_1, y_2) = \begin{vmatrix} x^{-1/2}\cos x & x^{-1/2}\sin x \\ -x^{-1/2}\sin x - \frac{1}{2}x^{-3/2}\cos x & x^{-1/2}\cos x - \frac{1}{2}x^{-3/2}\sin x \end{vmatrix} = \frac{1}{x}.$$

Now

$$u_1' = \sin x \quad \text{so} \quad u_1 = \cos x,$$

and

$$u_2' = \cos x \quad \text{so} \quad u_2 = \sin x.$$

Thus

$$y = c_1 x^{-1/2}\cos x + c_2 x^{-1/2}\sin x + x^{-1/2}\cos^2 x + x^{-1/2}\sin^2 x$$

$$= c_1 x^{-1/2}\cos x + c_2 x^{-1/2}\sin x + x^{-1/2}.$$

30. Write the equation in the form

$$y'' + \frac{1}{x}y' + \frac{1}{x^2}y = \frac{\sec(\ln x)}{x^2}$$

and identify $f(x) = \sec(\ln x)/x^2$. From $y_1 = \cos(\ln x)$ and $y_2 = \sin(\ln x)$ we compute

$$W = \begin{vmatrix} \cos(\ln x) & \sin(\ln x) \\ -\dfrac{\sin(\ln x)}{x} & \dfrac{\cos(\ln x)}{x} \end{vmatrix} = \frac{1}{x}.$$

Now

$$u_1' = -\frac{\tan(\ln x)}{x} \quad \text{so} \quad u_1 = \ln|\cos(\ln x)|,$$

and

$$u_2' = \frac{1}{x} \quad \text{so} \quad u_2 = \ln x.$$

Thus, a particular solution is

$$y_p = \cos(\ln x) \ln|\cos(\ln x)| + (\ln x)\sin(\ln x).$$

31. The general solution is

$$y = c_1 \cos(\ln x) + c_2 \sin(\ln x) + \cos(\ln x)\ln|\cos(\ln x)| + (\ln x)\sin(\ln x)$$

for $-\pi/2 < \ln x < \pi/2$ or $e^{-\pi/2} < x < e^{\pi/2}$. The bounds on $\ln x$ are due to the presence of $\sec(\ln x)$ in the differential equation .

32. We have $y_c = c_1 + c_2 e^{-x}$ and we assume $y_{p_1} = Ax^3 + Bx^2 + Cx$. Substituting into the differential equation $y'' + y' = 4x^2 - 3$ we obtain

$$3Ax^2 + (6A + 2B)x + (2B + C) = 4x^2 - 3.$$

Equating coefficients we find $A = \frac{4}{3}$, $B = -4$, $C = 5$. A particular solution is $y_{p_1} = \frac{4}{3}x^3 - 4x^2 + 5x$. Next we use variation of parameters to find a particular solution y_{p_2} of $y'' + y' = e^x/x$. With $y_1 = 1$ and $y_2 = e^{-x}$ we have

$$W = \begin{vmatrix} 1 & e^{-x} \\ 0 & -e^{-x} \end{vmatrix} = -e^{-x}.$$

Identifying $f(x) = e^x/x$ we obtain

$$u_1' = -\frac{1/x}{-e^{-x}} = \frac{e^x}{x}$$

$$u_2' = \frac{e^x/x}{-e^{-x}} = -\frac{e^{2x}}{x}.$$

Thus $y_{p_2} = \int_1^x e^t \, dt/t - e^{-x}\int_1^x e^{2t} \, dt/t$ and the general solution of the differential equation is

$$y = c_1 + c_2 e^{-x} + \frac{4}{3}x^3 - 4x^2 + 5x + \int_1^x \frac{e^t}{t} \, dt - e^{-x} \int_0^x \frac{e^{2t}}{t} \, dt.$$

Exercises 4.7

1. The auxiliary equation is $m^2 - m - 2 = (m+1)(m-2) = 0$ so that $y = c_1 x^{-1} + c_2 x^2$.

2. The auxiliary equation is $4m^2 - 4m + 1 = (2m-1)^2 = 0$ so that $y = c_1 x^{1/2} + c_2 x^{1/2} \ln x$.

3. The auxiliary equation is $m^2 = 0$ so that $y = c_1 + c_2 \ln x$.

4. The auxiliary equation is $m^2 - 2m = m(m-2) = 0$ so that $y = c_1 + c_2 x^2$.

5. The auxiliary equation is $m^2 + 4 = 0$ so that $y = c_1 \cos(2 \ln x) + c_2 \sin(2 \ln x)$.

6. The auxiliary equation is $m^2 + 4m + 3 = (m+1)(m+3) = 0$ so that $y = c_1 x^{-1} + c_2 x^{-3}$.

7. The auxiliary equation is $m^2 - 4m - 2 = 0$ so that $y = c_1 x^{2-\sqrt{6}} + c_2 x^{2+\sqrt{6}}$.

8. The auxiliary equation is $m^2 + 2m - 4 = 0$ so that $y = c_1 x^{-1+\sqrt{5}} + c_2 x^{-1-\sqrt{5}}$.

9. The auxiliary equation is $25m^2 + 1 = 0$ so that $y = c_1 \cos\left(\frac{1}{5} \ln x\right) + c_2 \sin\left(\frac{1}{5} \ln x\right)$.

10. The auxiliary equation is $4m^2 - 1 = (2m-1)(2m+1) = 0$ so that $y = c_1 x^{1/2} + c_2 x^{-1/2}$.

11. The auxiliary equation is $m^2 + 4m + 4 = (m+2)^2 = 0$ so that $y = c_1 x^{-2} + c_2 x^{-2} \ln x$.

12. The auxiliary equation is $m^2 + 7m + 6 = (m+1)(m+6) = 0$ so that $y = c_1 x^{-1} + c_2 x^{-6}$.

13. The auxiliary equation is $m^2 - 2m + 2 = 0$ so that $y = x\left[c_1 \cos(\ln x) + c_2 \sin(\ln x)\right]$.

14. The auxiliary equation is $m^2 - 8m + 41 = 0$ so that $y = x^4\left[c_1 \cos(5 \ln x) + c_2 \sin(5 \ln x)\right]$.

15. The auxiliary equation is $3m^2 + 3m + 1 = 0$ so that $y = x^{-1/2}\left[c_1 \cos\left(\frac{\sqrt{3}}{6} \ln x\right) + c_2 \sin\left(\frac{\sqrt{3}}{6} \ln x\right)\right]$.

16. The auxiliary equation is $2m^2 - m + 1 = 0$ so that $y = x^{1/4}\left[c_1 \cos\left(\frac{\sqrt{7}}{4} \ln x\right) + c_2 \sin\left(\frac{\sqrt{7}}{4} \ln x\right)\right]$.

17. Assuming that $y = x^m$ and substituting into the differential equation we obtain

$$m(m-1)(m-2) - 6 = m^3 - 3m^2 + 2m - 6 = (m-3)(m^2+2) = 0.$$

Thus

$$y = c_1 x^3 + c_2 \cos\left(\sqrt{2} \ln x\right) + c_3 \sin\left(\sqrt{2} \ln x\right).$$

18. Assuming that $y = x^m$ and substituting into the differential equation we obtain

$$m(m-1)(m-2) + m - 1 = m^3 - 3m^2 + 3m - 1 = (m-1)^3 = 0.$$

Thus

$$y = c_1 x + c_2 x \ln x + c_3 x(\ln x)^2.$$

19. Assuming that $y = x^m$ and substituting into the differential equation we obtain

$$m(m-1)(m-2) - 2m(m-1) - 2m + 8 = m^3 - 5m^2 + 2m + 8 = (m+1)(m-2)(m-4) = 0.$$

127

Thus

$$y = c_1 x^{-1} + c_2 x^2 + c_3 x^4.$$

20. Assuming that $y = x^m$ and substituting into the differential equation we obtain

$$m(m-1)(m-2) - 2m(m-1) + 4m - 4 = m^3 - 5m^2 + 8m - 4 = (m-1)(m-2)^2 = 0.$$

Thus

$$y = c_1 x + c_2 x^2 + c_3 x^2 \ln x.$$

21. Assuming that $y = x^m$ and substituting into the differential equation we obtain

$$m(m-1)(m-2)(m-3) + 6m(m-1)(m-2) = m^4 - 7m^2 + 6m = m(m-1)(m-2)(m+3) = 0.$$

Thus

$$y = c_1 + c_2 x + c_3 x^2 + c_4 x^{-3}.$$

22. Assuming that $y = x^m$ and substituting into the differential equation we obtain

$$m(m-1)(m-2)(m-3) + 6m(m-1)(m-2) + 9m(m-1) + 3m + 1 = m^4 + 2m^2 + 1 = (m^2+1)^2 = 0.$$

Thus

$$y = c_1 \cos(\ln x) + c_2 \sin(\ln x) + c_3 \ln x \cos(\ln x) + c_4 \ln x \sin(\ln x).$$

23. The auxiliary equation is $m^2 + 2m = m(m+2) = 0$, so that

$$y = c_1 + c_2 x^{-2} \quad \text{and} \quad y' = -2c_2 x^{-3}.$$

The initial conditions imply

$$c_1 + c_2 = 0$$

$$-2c_2 = 4.$$

Thus, $c_1 = 2$, $c_2 = -2$, and $y = 2 - 2x^{-2}$.

24. The auxiliary equation is $m^2 - 6m + 8 = (m-2)(m-4) = 0$, so that

$$y = c_1 x^2 + c_2 x^4 \quad \text{and} \quad y' = 2c_1 x + 4c_2 x^3.$$

The initial conditions imply

$$4c_1 + 16c_2 = 32$$

$$4c_1 + 32c_2 = 0.$$

Thus, $c_1 = 16$, $c_2 = -2$, and $y = 16x^2 - 2x^4$.

25. The auxiliary equation is $m^2 + 1 = 0$, so that

$$y = c_1 \cos(\ln x) + c_2 \sin(\ln x) \quad \text{and} \quad y' = -c_1 \frac{1}{x} \sin(\ln x) + c_2 \frac{1}{x} \cos(\ln x).$$

The initial conditions imply $c_1 = 1$ and $c_2 = 2$. Thus $y = \cos(\ln x) + 2\sin(\ln x)$.

26. The auxiliary equation is $m^2 - 4m + 4 = (m-2)^2 = 0$, so that

$$y = c_1 x^2 + c_2 x^2 \ln x \quad \text{and} \quad y' = 2c_1 x + c_2(x + 2x \ln x).$$

The initial conditions imply $c_1 = 5$ and $c_2 + 10 = 3$. Thus $y = 5x^2 - 7x^2 \ln x$.

In the next two problems we use the substitution $t = -x$ since the initial conditions are on the interval $(-\infty, 0)$. In this case

$$\frac{dy}{dt} = \frac{dy}{dx}\frac{dx}{dt} = -\frac{dy}{dx}$$

and

$$\frac{d^2 y}{dt^2} = \frac{d}{dt}\left(\frac{dy}{dt}\right) = \frac{d}{dt}\left(-\frac{dy}{dx}\right) = -\frac{d}{dt}(y') = -\frac{dy'}{dx}\frac{dx}{dt} = -\frac{d^2 y}{dx^2}\frac{dx}{dt} = \frac{d^2 y}{dx^2}.$$

27. The differential equation and initial conditions become

$$4t^2 \frac{d^2 y}{dt^2} + y = 0; \quad y(t)\Big|_{t=1} = 2, \quad y'(t)\Big|_{t=1} = -4.$$

The auxiliary equation is $4m^2 - 4m + 1 = (2m-1)^2 = 0$, so that

$$y = c_1 t^{1/2} + c_2 t^{1/2} \ln t \quad \text{and} \quad y' = \frac{1}{2}c_1 t^{-1/2} + c_2\left(t^{-1/2} + \frac{1}{2}t^{-1/2}\ln t\right).$$

The initial conditions imply $c_1 = 2$ and $1 + c_2 = -4$. Thus

$$y = 2t^{1/2} - 5t^{1/2}\ln t = 2(-x)^{1/2} - 5(-x)^{1/2}\ln(-x), \quad x < 0.$$

28. The differential equation and initial conditions become

$$t^2 \frac{d^2 y}{dt^2} - 4t \frac{dy}{dt} + 6y = 0; \quad y(t)\Big|_{t=2} = 8, \quad y'(t)\Big|_{t=2} = 0.$$

The auxiliary equation is $m^2 - 5m + 6 = (m-2)(m-3) = 0$, so that

$$y = c_1 t^2 + c_2 t^3 \quad \text{and} \quad y' = 2c_1 t + 3c_2 t^2.$$

The initial conditions imply

$$4c_1 + 8c_2 = 8$$

$$4c_1 + 12c_2 = 0$$

from which we find $c_1 = 6$ and $c_2 = -2$. Thus

$$y = 6t^2 - 2t^3 = 6x^2 + 2x^3, \quad x < 0.$$

29. The auxiliary equation is $m^2 = 0$ so that $y_c = c_1 + c_2 \ln x$ and

$$W(1, \ln x) = \begin{vmatrix} 1 & \ln x \\ 0 & 1/x \end{vmatrix} = \frac{1}{x}.$$

Identifying $f(x) = 1$ we obtain $u_1' = -x \ln x$ and $u_2' = x$. Then $u_1 = \frac{1}{4}x^2 - \frac{1}{2}x^2 \ln x$, $u_2 = \frac{1}{2}x^2$, and

$$y = c_1 + c_2 \ln x + \frac{1}{4}x^2 - \frac{1}{2}x^2 \ln x + \frac{1}{2}x^2 \ln x = c_1 + c_2 \ln x + \frac{1}{4}x^2.$$

30. The auxiliary equation is $m^2 - 5m = m(m-5) = 0$ so that $y_c = c_1 + c_2 x^5$ and

$$W(1, x^5) = \begin{vmatrix} 1 & x^5 \\ 0 & 5x^4 \end{vmatrix} = 5x^4.$$

Identifying $f(x) = x^3$ we obtain $u_1' = -\frac{1}{5}x^4$ and $u_2' = 1/5x$. Then $u_1 = -\frac{1}{25}x^5$, $u_2 = \frac{1}{5}\ln x$, and

$$y = c_1 + c_2 x^5 - \frac{1}{25}x^5 + \frac{1}{5}x^5 \ln x = c_1 + c_3 x^5 + \frac{1}{5}x^5 \ln x.$$

31. The auxiliary equation is $2m^2 + 3m + 1 = (2m+1)(m+1) = 0$ so that $y_c = c_1 x^{-1} + c_2 x^{-1/2}$ and

$$W(x^{-1}, x^{-1/2}) = \begin{vmatrix} x^{-1} & x^{-1/2} \\ -x^{-2} & -\frac{1}{2}x^{-3/2} \end{vmatrix} = \frac{1}{2}x^{-5/2}.$$

Identifying $f(x) = \frac{1}{2} - \frac{1}{2x}$ we obtain $u_1' = x - x^2$ and $u_2' = x^{3/2} - x^{1/2}$. Then $u_1 = \frac{1}{2}x^2 - \frac{1}{3}x^3$,
$u_2 = \frac{2}{5}x^{5/2} - \frac{2}{3}x^{3/2}$, and

$$y = c_1 x^{-1} + c_2 x^{-1/2} + \frac{1}{2}x - \frac{1}{3}x^2 + \frac{2}{5}x^2 - \frac{2}{3}x = c_1 x^{-1} + c_2 x^{-1/2} - \frac{1}{6}x + \frac{1}{15}x^2.$$

32. The auxiliary equation is $m^2 - 3m + 2 = (m-1)(m-2) = 0$ so that $y_c = c_1 x + c_2 x^2$ and

$$W(x, x^2) = \begin{vmatrix} x & x^2 \\ 1 & 2x \end{vmatrix} = x^2.$$

Identifying $f(x) = x^2 e^x$ we obtain $u_1' = -x^2 e^x$ and $u_2' = xe^x$. Then $u_1 = -x^2 e^x + 2xe^x - 2e^x$,
$u_2 = xe^x - e^x$, and

$$y = c_1 x + c_2 x^2 - x^3 e^x + 2x^2 e^x - 2xe^x + x^3 e^x - x^2 e^x$$

$$= c_1 x + c_2 x^2 + x^2 e^x - 2xe^x.$$

33. The auxiliary equation is $m^2 - 2m + 1 = (m-1)^2 = 0$ so that $y_c = c_1 x + c_2 x \ln x$ and

$$W(x, x \ln x) = \begin{vmatrix} x & x \ln x \\ 1 & 1 + \ln x \end{vmatrix} = x.$$

Identifying $f(x) = 2/x$ we obtain $u_1' = -2\ln x/x$ and $u_2' = 2/x$. Then $u_1 = -(\ln x)^2$, $u_2 = 2\ln x$, and

$$y = c_1 x + c_2 x \ln x - x(\ln x)^2 + 2x(\ln x)^2$$

$$= c_1 x + c_2 x \ln x + x(\ln x)^2.$$

34. The auxiliary equation is $m^2 - 3m + 2 = (m-1)(m-2) = 0$ so that $y_c = c_1 x + c_2 x^2$ and

$$W(x, x^2) = \begin{vmatrix} x & x^2 \\ 1 & 2x \end{vmatrix} = x^2.$$

Identifying $f(x) = x \ln x$ we obtain $u_1' = -x \ln x$ and $u_2' = \ln x$. Then $u_1 = \frac{1}{4}x^2 - \frac{1}{2}x^2 \ln x$, $u_2 = x \ln x - x$, and

$$y = c_1 x + c_2 x^2 + \frac{1}{4}x^3 - \frac{1}{2}x^3 \ln x + x^3 \ln x - x^3 = c_1 x + c_2 x^2 - \frac{3}{4}x^3 + \frac{1}{2}x^3 \ln x.$$

In Problems 35-40 we use the following results: When $x = e^t$ or $t = \ln x$, then

$$\frac{dy}{dx} = \frac{1}{x}\frac{dy}{dt} \quad \text{and} \quad \frac{d^2 y}{dx^2} = \frac{1}{x^2}\left[\frac{d^2 y}{dt^2} - \frac{dy}{dt}\right].$$

35. Substituting into the differential equation we obtain

$$\frac{d^2 y}{dt^2} + 9\frac{dy}{dt} + 8y = e^{2t}.$$

The auxiliary equation is $m^2 + 9 + 8 = (m+1)(m+8) = 0$ so that $y_c = c_1 e^{-t} + c_2 e^{-8t}$. Using undetermined coefficients we try $y_p = Ae^{2t}$. This leads to $30Ae^{2t} = e^{2t}$, so that $A = 1/30$ and

$$y = c_1 e^{-t} + c_2 e^{-8t} + \frac{1}{30}e^{2t} = c_1 x^{-1} + c_2 x^{-8} + \frac{1}{30}x^2.$$

36. Substituting into the differential equation we obtain

$$\frac{d^2 y}{dt^2} - 5\frac{dy}{dt} + 6y = 2t.$$

The auxiliary equation is $m^2 - 5m + 6 = (m-2)(m-3) = 0$ so that $y_c = c_1 e^{2t} + c_2 e^{3t}$. Using undetermined coefficients we try $y_p = At + B$. This leads to $(-5A + 6B) + 6At = 2t$, so that $A = 1/3$, $B = 5/18$, and

$$y = c_1 e^{2t} + c_2 e^{3t} + \frac{1}{3}t + \frac{5}{18} = c_1 x^2 + c_2 x^3 + \frac{1}{3}\ln x + \frac{5}{18}.$$

37. Substituting into the differential equation we obtain

$$\frac{d^2 y}{dt^2} - 4\frac{dy}{dt} + 13y = 4 + 3e^t.$$

The auxiliary equation is $m^2 - 4m + 13 = 0$ so that $y_c = e^{2t}(c_1 \cos 3t + c_2 \sin 3t)$. Using undetermined coefficients we try $y_p = A + Be^t$. This leads to $13A + 10Be^t = 4 + 3e^t$, so that $A = 4/13$, $B = 3/10$, and

$$y = e^{2t}(c_1 \cos 3t + c_2 \sin 3t) + \frac{4}{13} + \frac{3}{10}e^t$$

$$= x^2[c_1 \cos(3 \ln x) + c_2 \sin(3 \ln x)] + \frac{4}{13} + \frac{3}{10}x.$$

38. Substituting into the differential equation we obtain

$$2\frac{d^2y}{dt^2} - 5\frac{dy}{dt} - 3y = 1 + 2e^t + e^{2t}.$$

The auxiliary equation is $2m^2 - 5m - 3 = (2m+1)(m-3) = 0$ so that $y_c = c_1 e^{-t/2} + c_2 e^{3t}$. Using undetermined coefficients we try $y_p = A + Be^t + Ce^{2t}$. This leads to $-3A - 6Be^t - 5Ce^{2t} = 1 + 2e^t + e^{2t}$, so that $A = -1/3$, $B = -1/3$, $C = -1/5$, and

$$y = c_1 e^{-t/2} + c_2 e^{3t} - \frac{1}{3}e^t - \frac{1}{5}e^{2t} = c_1 x^{-1/2} + c_2 x^3 - \frac{1}{3}x - \frac{1}{5}x^2.$$

39. Substituting into the differential equation we obtain

$$\frac{d^2y}{dt^2} + 8\frac{dy}{dt} - 20y = 5e^{-3t}.$$

The auxiliary equation is $m^2 + 8m - 20 = (m+10)(m-2) = 0$ so that $y_c = c_1 e^{-10t} + c_2 e^{2t}$. Using undetermined coefficients we try $y_p = Ae^{-3t}$. This leads to $-35Ae^{-3t} = 5e^{-3t}$, so that $A = -1/7$ and

$$y = c_1 e^{-10t} + c_2 e^{2t} - \frac{1}{7}e^{-3t} = c_1 x^{-10} + c_2 x^2 - \frac{1}{7}x^{-3}.$$

40. From

$$\frac{d^2y}{dx^2} = \frac{1}{x^2}\left(\frac{d^2y}{dt^2} - \frac{dy}{dt}\right).$$

it follows that

$$\frac{d^3y}{dx^3} = \frac{1}{x^2}\frac{d}{dx}\left(\frac{d^2y}{dt^2} - \frac{dy}{dt}\right) - \frac{2}{x^3}\left(\frac{d^2y}{dt^2} - \frac{dy}{dt}\right)$$

$$= \frac{1}{x^2}\frac{d}{dx}\left(\frac{d^2y}{dt^2}\right) - \frac{1}{x^2}\frac{d}{dx}\left(\frac{dy}{dt}\right) - \frac{2}{x^3}\frac{d^2y}{dt^2} + \frac{2}{x^3}\frac{dy}{dt}$$

$$= \frac{1}{x^2}\frac{d^3y}{dt^3}\left(\frac{1}{x}\right) - \frac{1}{x^2}\frac{d^2y}{dt^2}\left(\frac{1}{x}\right) - \frac{2}{x^3}\frac{d^2y}{dt^2} + \frac{2}{x^3}\frac{dy}{dt}$$

$$= \frac{1}{x^3}\left(\frac{d^3y}{dt^3} - 3\frac{d^2y}{dt^2} + 2\frac{dy}{dt}\right).$$

Substituting into the differential equation we obtain

$$\frac{d^3y}{dt^3} - 3\frac{d^2y}{dt^2} + 2\frac{dy}{dt} - 3\left(\frac{d^2y}{dt^2} - \frac{dy}{dt}\right) + 6\frac{dy}{dt} - 6y = 3 + 3t$$

or

$$\frac{d^3y}{dt^3} - 6\frac{d^2y}{dt^2} + 11\frac{dy}{dt} - 6y = 3 + 3t.$$

The auxiliary equation is $m^3 - 6m^2 + 11m - 6 = (m-1)(m-2)(m-3) = 0$ so that $y_c = c_1 e^t + c_2 e^{2t} + c_3 e^{3t}$. Using undetermined coefficients we try $y_p = A + Bt$. This leads to $(11B - 6A) - 6Bt = 3 + 3t$, so that $A = -17/12$, $B = -1/2$, and

$$y = c_1 e^t + c_2 e^{2t} + c_3 e^{3t} - \frac{17}{12} - \frac{1}{2}t = c_1 x + c_2 x^2 + c_3 x^3 - \frac{17}{12} - \frac{1}{2}\ln x.$$

41. When m_1 and m_2 are distinct real roots, the corresponding solutions have the form $y = c_1 x^{m_1}$ and $y = c_2 x^{m_2}$. Graphs of these solutions are shown below. We see that as $x \to 0^+$ the solutions are 0 when $m > 0$, are bounded when $m = 0$, and are unbounded when $m < 0$.

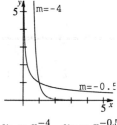
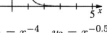

$y_1 = x^{-4}, \quad y_2 = x^{-0.5}$

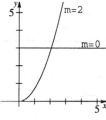

$y_1 = 3, \quad y_2 = 0.7x^2$

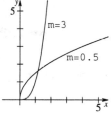

$y_1 = x^3, \quad y_2 = \frac{3}{2}x^{0.5}$

When $m_1 = m_2$ are real equal roots the corresponding solutions have the forms $y = c_1 x^{m_1}$ and $y = c_2 x^{m_1} \ln x$. Graphs are shown below. We see that the solutions are 0 at $x = 0$ when $m > 0$ and unbounded when $m < 0$. When $m = 0$ the solutions are $y = c_1$ and $y = c_2 \ln x$. The first of these is bounded at $x = 0$ and the second is unbounded as $x \to 0^+$.

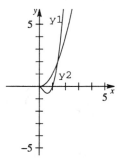

$y_1 = x^2, \quad y_2 = x^2 \ln x$

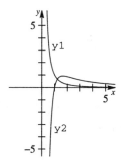

$y_1 = x^{-2}, \quad y_2 = 5x^{-2} \ln x$

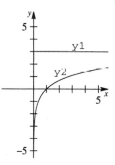

$y_1 = 3, \quad y_2 = \ln 2$

When m_1 and m_2 are conjugate complex roots $\alpha \pm \beta i$ the corresponding solutions have the forms $y_1 = c_1 x^\alpha \cos(\beta \ln x)$ and $y_2 = c_2 x^\alpha \sin(\beta \ln x)$. Graphs are shown below. We see that the solutions appear to be bounded as $x \to 0^+$ for $\alpha > 0$ and unbounded when $\alpha < 0$. When $\alpha = 0$ the solutions are bounded as $x \to 0^+$ but oscillate more and more rapidly near 0.

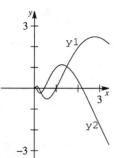
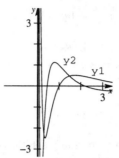
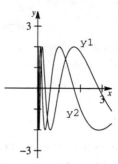

$y_1 = x \sin(2 \ln x)$

$y_2 = x \cos(2 \ln x)$

$y_1 = x^{-1} \sin(2 \ln x)$

$y_2 = x^{-1} \cos 2(\ln x)$

$y_1 = 2 \sin(3 \ln x)$

$y_2 = 2 \cos(3 \ln x)$

Exercises 4.8

1. From $Dx = 2x - y$ and $Dy = x$ we obtain $y = 2x - Dx$, $Dy = 2Dx - D^2 x$, and $(D^2 - 2D + 1)x = 0$. Then

$$x = c_1 e^t + c_2 t e^t \quad \text{and} \quad y = (c_1 - c_2)e^t + c_2 t e^t.$$

2. From $Dx = 4x + 7y$ and $Dy = x - 2y$ we obtain $y = \frac{1}{7}Dx - \frac{4}{7}x$, $Dy = \frac{1}{7}D^2 x - \frac{4}{7}Dx$, and $(D^2 - 2D - 15)x = 0$. Then

$$x = c_1 e^{5t} + c_2 e^{-3t} \quad \text{and} \quad y = \frac{1}{7}c_1 e^{5t} - c_2 e^{-3t}.$$

3. From $Dx = -y + t$ and $Dy = x - t$ we obtain $y = t - Dx$, $Dy = 1 - D^2 x$, and $(D^2 + 1)x = 1 + t$. Then

$$x = c_1 \cos t + c_2 \sin t + 1 + t$$

and

$$y = c_1 \sin t - c_2 \cos t + t - 1.$$

4. From $Dx - 4y = 1$ and $x + Dy = 2$ we obtain $y = \frac{1}{4}Dx - \frac{1}{4}$, $Dy = \frac{1}{4}D^2 x$, and $(D^2 + 1)x = 2$. Then

$$x = c_1 \cos t + c_2 \sin t + 2$$

and

$$y = \frac{1}{4}c_2 \cos t - \frac{1}{4}c_1 \sin t - \frac{1}{4}c_1 \sin t - \frac{1}{4}.$$

5. From $(D^2+5)x - 2y = 0$ and $-2x + (D^2+2)y = 0$ we obtain $y = \frac{1}{2}(D^2+5)x$, $D^2y = \frac{1}{2}(D^4+5D^2)x$, and $(D^2+1)(D^2+6)x = 0$. Then

$$x = c_1 \cos t + c_2 \sin t + c_3 \cos \sqrt{6}\,t + c_4 \sin \sqrt{6}\,t$$

and

$$y = 2c_1 \cos t + 2c_2 \sin t - \frac{1}{2}c_3 \cos \sqrt{6}\,t - \frac{1}{2}c_4 \sin \sqrt{6}\,t.$$

6. From $(D+1)x + (D-1)y = 2$ and $3x + (D+2)y = -1$ we obtain $x = -\frac{1}{3} - \frac{1}{3}(D+2)y$, $Dx = -\frac{1}{3}(D^2 + 2D)y$, and $(D^2+5)y = -7$. Then

$$y = c_1 \cos \sqrt{5}\,t + c_2 \sin \sqrt{5}\,t - \frac{7}{5}$$

and

$$x = \left(-\frac{2}{3}c_1 - \frac{\sqrt{5}}{3}c_2\right) \cos \sqrt{5}\,t + \left(\frac{\sqrt{5}}{3}c_1 - \frac{2}{3}c_2\right) \sin \sqrt{5}\,t + \frac{3}{5}.$$

7. From $D^2x = 4y + e^t$ and $D^2y = 4x - e^t$ we obtain $y = \frac{1}{4}D^2x - \frac{1}{4}e^t$, $D^2y = \frac{1}{4}D^4x - \frac{1}{4}e^t$, and $(D^2+4)(D-2)(D+2)x = -3e^t$. Then

$$x = c_1 \cos 2t + c_2 \sin 2t + c_3 e^{2t} + c_4 e^{-2t} + \frac{1}{5}e^t$$

and

$$y = -c_1 \cos 2t - c_2 \sin 2t + c_3 e^{2t} + c_4 e^{-2t} - \frac{1}{5}e^t.$$

8. From $(D^2+5)x + Dy = 0$ and $(D+1)x + (D-4)y = 0$ we obtain $(D-5)(D^2+4)x = 0$ and $(D-5)(D^2+4)y = 0$. Then

$$x = c_1 e^{5t} + c_2 \cos 2t + c_3 \sin 2t$$

and

$$y = c_4 e^{5t} + c_5 \cos 2t + c_6 \sin 2t.$$

Substituting into $(D+1)x + (D-4)y = 0$ gives

$$(6c_1 + c_4)e^{5t} + (c_2 + 2c_3 - 4c_5 + 2c_6) \cos 2t + (-2c_2 + c_3 - 2c_5 - 4c_6) \sin 2t = 0$$

so that $c_4 = -6c_1$, $c_5 = \frac{1}{2}c_3$, $c_6 = -\frac{1}{2}c_2$, and

$$y = -6c_1 e^{5t} + \frac{1}{2}c_3 \cos 2t - \frac{1}{2}c_2 \sin 2t.$$

9. From $Dx + D^2y = e^{3t}$ and $(D+1)x + (D-1)y = 4e^{3t}$ we obtain $D(D^2+1)x = 34e^{3t}$ and $D(D^2+1)y = -8e^{3t}$. Then

$$y = c_1 + c_2 \sin t + c_3 \cos t - \frac{4}{15}e^{3t}$$

and

$$x = c_4 + c_5 \sin t + c_6 \cos t + \frac{17}{15}e^{3t}.$$

Substituting into $(D+1)x + (D-1)y = 4e^{3t}$ gives

$$(c_4 - c_1) + (c_5 - c_6 - c_3 - c_2)\sin t + (c_6 + c_5 + c_2 - c_3)\cos t = 0$$

so that $c_4 = c_1$, $c_5 = c_3$, $c_6 = -c_2$, and

$$x = c_1 - c_2 \cos t + c_3 \sin t + \frac{17}{15}e^{3t}.$$

10. From $D^2 x - Dy = t$ and $(D+3)x + (D+3)y = 2$ we obtain $D(D+1)(D+3)x = 1 + 3t$ and $D(D+1)(D+3)y = -1 - 3t$. Then

$$x = c_1 + c_2 e^{-t} + c_3 e^{-3t} - t + \frac{1}{2}t^2$$

and

$$y = c_4 + c_5 e^{-t} + c_6 e^{-3t} + t - \frac{1}{2}t^2.$$

Substituting into $(D+3)x + (D+3)y = 2$ and $D^2 x - Dy = t$ gives

$$3(c_1 + c_4) + 2(c_2 + c_5)e^{-t} = 2$$

and

$$(c_2 + c_5)e^{-t} + 3(3c_3 + c_6)e^{-3t} = 0$$

so that $c_4 = -c_1$, $c_5 = -c_2$, $c_6 = -3c_3$, and

$$y = -c_1 - c_2 e^{-t} - 3c_3 e^{-3t} + t - \frac{1}{2}t^2.$$

11. From $(D^2 - 1)x - y = 0$ and $(D-1)x + Dy = 0$ we obtain $y = (D^2 - 1)x$, $Dy = (D^3 - D)x$, and $(D-1)(D^2 + D + 1)x = 0$. Then

$$x = c_1 e^t + e^{-t/2}\left[c_2 \cos \frac{\sqrt{3}}{2}t + c_3 \sin \frac{\sqrt{3}}{2}t\right]$$

and

$$y = \left(-\frac{3}{2}c_2 - \frac{\sqrt{3}}{2}c_3\right)e^{-t/2}\cos \frac{\sqrt{3}}{2}t + \left(\frac{\sqrt{3}}{2}c_2 - \frac{3}{2}c_3\right)e^{-t/2}\sin \frac{\sqrt{3}}{2}t.$$

12. From $(2D^2 - D - 1)x - (2D+1)y = 1$ and $(D-1)x + Dy = -1$ we obtain $(2D+1)(D-1)(D+1)x = -1$ and $(2D+1)(D+1)y = -2$. Then

$$x = c_1 e^{-t/2} + c_2 e^{-t} + c_3 e^t + 1$$

and

$$y = c_4 e^{-t/2} + c_5 e^{-t} - 2.$$

Substituting into $(D-1)x + Dy = -1$ gives

$$\left(-\frac{3}{2}c_1 - \frac{1}{2}c_4\right)e^{-t/2} + (-2c_2 - c_5)e^{-t} = 0$$

so that $c_4 = -3c_1$, $c_5 = -2c_2$, and

$$y = -3c_1e^{-t/2} - 2c_2e^{-t} - 2.$$

13. From $(2D-5)x+Dy = e^t$ and $(D-1)x+Dy = 5e^t$ we obtain $Dy = (5-2D)x+e^t$ and $(4-D)x = 4e^t$. Then

$$x = c_1e^{4t} + \frac{4}{3}e^t$$

and $Dy = -3c_1e^{4t} + 5e^t$ so that

$$y = -\frac{3}{4}c_1e^{4t} + c_2 + 5e^t.$$

14. From $Dx+Dy = e^t$ and $(-D^2+D+1)x+y = 0$ we obtain $y = (D^2-D-1)x$, $Dy = (D^3-D^2-D)x$, and $D^2(D-1)x = e^t$. Then

$$x = c_1 + c_2t + c_3e^t + te^t$$

and

$$y = -c_1 - c_2 - c_2t - c_3e^t - te^t + e^t.$$

15. From $(D-1)x + (D^2+1)y = 1$ and $(D^2-1)x + (D+1)y = 2$ we obtain $D^2(D-1)(D+1)x = 1$ and $D^2(D-1)(D+1)y = 1$. Then

$$x = c_1 + c_2t + c_3e^t + c_4e^{-t} - \frac{1}{2}t^2$$

and

$$y = c_5 + c_6t + c_7e^t + c_8e^{-t} - \frac{1}{2}t^2.$$

Substituting into $(D-1)x + (D^2+1)y = 1$ gives

$$(c_2 - c_1 - 1 + c_5) + (c_6 - c_2 - 1)t + (2c_8 - 2c_4)e^{-t} + (2c_7)e^t = 1$$

so that $c_6 = c_2 + 1$, $c_8 = c_4$, $c_7 = 0$, $c_5 = c_1 - c_2 + 2$, and

$$y = (c_1 - c_2 + 2) + (c_2 + 1)t + c_4e^{-t} - \frac{1}{2}t^2.$$

16. From $D^2x - 2(D^2 + D)y = \sin t$ and $x + Dy = 0$ we obtain $x = -Dy$, $D^2x = -D^3y$, and $D(D^2 + 2D + 2)y = -\sin t$. Then

$$y = c_1 + c_2e^{-t}\cos t + c_3e^{-t}\sin t + \frac{1}{5}\cos t + \frac{2}{5}\sin t$$

and

$$x = (c_2 + c_3)e^{-t}\sin t + (c_2 - c_3)e^{-t}\cos t + \frac{1}{5}\sin t - \frac{2}{5}\cos t.$$

137

17. From $Dx = y$, $Dy = z$. and $Dz = x$ we obtain $x = D^2y = D^3x$ so that $(D-1)(D^2+D+1)x = 0$,

$$x = c_1e^t + e^{-t/2}\left[c_2 \sin\frac{\sqrt{3}}{2}t + c_3 \cos\frac{\sqrt{3}}{2}t\right],$$

$$y = c_1e^t + \left(-\frac{1}{2}c_2 - \frac{\sqrt{3}}{2}c_3\right)e^{-t/2}\sin\frac{\sqrt{3}}{2}t + \left(\frac{\sqrt{3}}{2}c_2 - \frac{1}{2}c_3\right)e^{-t/2}\cos\frac{\sqrt{3}}{2}t,$$

and

$$z = c_1e^t + \left(-\frac{1}{2}c_2 + \frac{\sqrt{3}}{2}c_3\right)e^{-t/2}\sin\frac{\sqrt{3}}{2}t + \left(-\frac{\sqrt{3}}{2}c_2 - \frac{1}{2}c_3\right)e^{-t/2}\cos\frac{\sqrt{3}}{2}t.$$

18. From $Dx + z = e^t$, $(D-1)x + Dy + Dz = 0$, and $x + 2y + Dz = e^t$ we obtain $z = -Dx + e^t$, $Dz = -D^2x + e^t$, and the system $(-D^2 + D - 1)x + Dy = -e^t$ and $(-D^2 + 1)x + 2y = 0$. Then $y = \frac{1}{2}(D^2 - 1)x$, $Dy = \frac{1}{2}D(D^2 - 1)x$, and $(D-2)(D^2+1)x = -2e^t$ so that

$$x = c_1e^{2t} + c_2\cos t + c_3\sin t + e^t,$$

$$y = \frac{3}{2}c_1e^{2t} - c_2\cos t - c_3\sin t,$$

and

$$z = -2c_1e^{2t} - c_3\cos t + c_2\sin t.$$

19. From $Dx - 6y = 0$, $x - Dy + z = 0$, and $x + y - Dz = 0$ we obtain

$$\begin{vmatrix} D & -6 & 0 \\ 1 & -D & 1 \\ 1 & 1 & -D \end{vmatrix} x = \begin{vmatrix} 0 & -6 & 0 \\ 0 & -D & 1 \\ 0 & 1 & -D \end{vmatrix}$$

so that $(D+1)(D-3)(D+2)x = 0$. Then

$$x = c_1e^{-t} + c_2e^{3t} + c_3e^{-2t},$$

$$y = -\frac{1}{6}c_1e^{-t} + \frac{1}{2}c_2e^{3t} - \frac{1}{3}c_3e^{-2t},$$

and

$$z = -\frac{5}{6}c_1e^{-t} + \frac{1}{2}c_2e^{3t} - \frac{1}{3}c_3e^{-2t}.$$

20. From $(D+1)x - z = 0$, $(D+1)y - z = 0$, and $x - y + Dz = 0$ we obtain

$$\begin{vmatrix} D+1 & 0 & -1 \\ 0 & D+1 & -1 \\ 1 & -1 & D \end{vmatrix} x = \begin{vmatrix} 0 & 0 & -1 \\ 0 & D+1 & -1 \\ 0 & -1 & D \end{vmatrix}$$

so that $D(D+1)^2x = 0$. Then

$$x = c_1 + c_2e^{-t} + c_3te^{-t},$$

$$y = c_1 + (c_2 - c_3)e^{-t} + c_3 t e^{-t},$$

and

$$z = c_1 + c_3 e^{-t}.$$

21. From $2Dx + (D-1)y = t$ and $Dx + Dy = t^2$ we obtain $(D+1)y = 2t^2 - t$. Then

$$y = c_1 e^{-t} + 2t^2 - 5t + 5$$

and $Dx = c_1 e^{-t} + t^2 - 4t + 5$ so that

$$x = -c_1 e^{-t} + c_2 + \frac{1}{3}t^3 - 2t^2 + 5t.$$

22. From $Dx - 2Dy = t^2$ and $(D+1)x - 2(D+1)y = 1$ we obtain $0 = 2t + t^2$ so that the system has no solution.

23. From $(D+5)x + y = 0$ and $4x - (D+1)y = 0$ we obtain $y = -(D+5)x$ so that $Dy = -(D^2+5D)x$. Then $4x + (D^2 + 5D)x + (D+5)x = 0$ and $(D+3)^2 x = 0$. Thus

$$x = c_1 e^{-3t} + c_2 t e^{-3t}$$

and

$$y = -(2c_1 + c_2)e^{-3t} - 2c_2 t e^{-3t}.$$

Using $x(1) = 0$ and $y(1) = 1$ we obtain

$$c_1 e^{-3} + c_2 e^{-3} = 0$$

$$-(2c_1 + c_2)e^{-3} - 2c_2 e^{-3} = 1$$

or

$$c_1 + c_2 = 0$$

$$2c_1 + 3c_2 = -e^3.$$

Thus $c_1 = e^3$ and $c_2 = -e^3$. The solution of the initial value problem is

$$x = e^{-3t+3} - te^{-3t+3}$$

$$y = -e^{-3t+3} + 2te^{-3t+3}.$$

24. From $Dx - y = -1$ and $3x + (D-2)y = 0$ we obtain $x = -\frac{1}{3}(D-2)y$ so that $Dx = -\frac{1}{3}(D^2 - 2D)y$. Then $-\frac{1}{3}(D^2 - 2D)y = y - 1$ and $(D^2 - 2D + 3)y = 3$. Thus

$$y = e^t \left(c_1 \cos \sqrt{2}\,t + c_2 \sin \sqrt{2}\,t \right) + 1$$

and

$$x = \frac{1}{3}e^t \left[\left(c_1 - \sqrt{2}\,c_2 \right) \cos \sqrt{2}\,t + \left(\sqrt{2}\,c_1 + c_2 \right) \sin \sqrt{2}\,t \right] + \frac{2}{3}.$$

139

Using $x(0) = y(0) = 0$ we obtain

$$c_1 + 1 = 0$$

$$\frac{1}{3}\left(c_1 - \sqrt{2}\,c_2\right) + \frac{2}{3} = 0.$$

Thus $c_1 = -1$ and $c_2 = \sqrt{2}/2$. The solution of the initial value problem is

$$x = e^t\left(-\frac{2}{3}\cos\sqrt{2}\,t - \frac{\sqrt{2}}{6}\sin\sqrt{2}\,t\right) + \frac{2}{3}$$

$$y = e^t\left(-\cos\sqrt{2}\,t + \frac{\sqrt{2}}{2}\sin\sqrt{2}\,t\right) + 1.$$

25. Equating Newton's law with the net forces in the x- and y-directions gives $m\dfrac{d^2x}{dt^2} = 0$ and

$m\dfrac{d^2y}{dt^2} = -mg$, respectively. From $mD^2x = 0$ we obtain $x(t) = c_1 t + c_2$, and from $mD^2y = -mg$

or $D^2y = -g$ we obtain $y(t) = -\dfrac{1}{2}gt^2 + c_3 t + c_4$.

26. From Newton's second law in the x-direction we have

$$m\frac{d^2x}{dt^2} = -k\cos\theta = -k\frac{1}{v}\frac{dx}{dt} = -|c|\frac{dx}{dt}\,.$$

In the y-direction we have

$$m\frac{d^2y}{dt^2} = -mg - k\sin\theta = -mg - k\frac{1}{v}\frac{dy}{dt} = -mg - |c|\frac{dy}{dt}\,.$$

From $mD^2x + |c|Dx = 0$ we have $D(mD + |c|)x = 0$ so that $(mD + |c|)x = c_1$. This is a first-order linear equation. An integrating factor is $e^{\int |c|dt/m} e^{|c|t/m}$ so that

$$\frac{d}{dt}[e^{|c|t/m}x] = c_1 e^{|c|t/m}$$

and $e^{|c|t}x = (c_1 m/|c|)e^{|c|t/m} + c_2$. The general solution of this equation is $x(t) = c_3 + c_2 e^{|c|t/m}$.
From $(mD^2 + |c|D)y = -mg$ we have $D(mD + |c|)y = -mg$ so that $(mD + |c|)y = -mgt + c_1$.
This is a first-order linear equation with integrating factor $e^{|c|t/m}$. Thus

$$\frac{d}{dt}[e^{|c|t/m}y] = (-mgt + c_1)e^{|c|t/m}$$

$$e^{|c|t/m}y = -\frac{m^2g}{|c|}te^{|c|t/m} + \frac{m^3g}{c^2}e^{|c|t/m} + \frac{c_1 m}{|c|}e^{|c|t/m} + c_2$$

and

$$y(t) = -\frac{m^2g}{|c|}t + \frac{m^3g}{c^2} + c_3 + c_2 e^{-|c|t/m}.$$

Exercises 4.9

1. We have $y_1' = y_1'' = e^x$, so
$$(y_1'')^2 = (e^x)^2 = e^{2x} = y_1^2.$$

Also, $y_2' = -\sin x$ and $y_2'' = -\cos x$, so
$$(y_2'')^2 = (-\cos x)^2 = \cos^2 x = y_2^2.$$

However, if $y = c_1 y_1 + c_2 y_2$, we have $(y'')^2 = (c_1 e^x - c_2 \cos x)^2$ and $y^2 = (c_1 e^x + c_2 \cos x)^2$. Thus $(y'')^2 \neq y^2$.

2. We have $y_1' = y_1'' = 0$, so
$$y_1 y_1'' = 1 \cdot 0 = 0 = \frac{1}{2}(0)^2 = \frac{1}{2}(y_1')^2.$$

Also, $y_2' = 2x$ and $y_2'' = 2$, so
$$y_2 y_2'' = x^2(2) = 2x^2 = \frac{1}{2}(2x)^2 = \frac{1}{2}(y_2')^2.$$

However, if $y = c_1 y_1 + c_2 y_2$, we have $yy'' = (c_1 \cdot 1 + c_2 x^2)(c_1 \cdot 0 + 2c_2) = 2c_2(c_1 + c_2 x^2)$ and $\frac{1}{2}(y')^2 = \frac{1}{2}[c_1 \cdot 0 + c_2(2x)]^2 = 2c_2^2 x^2$. Thus $yy'' \neq \frac{1}{2}(y')^2$.

3. Let $u = y'$ so that $u' = y''$. The equation becomes $u' = -u - 1$ which is separable. Thus
$$\frac{du}{u^2 + 1} = -dx \implies \tan^{-1} u = -x + c_1 \implies y' = \tan(c_1 - x) \implies y = \ln|\cos(c_1 - x)| + c_2.$$

4. Let $u = y'$ so that $u' = y''$. The equation becomes $u' = 1 + u^2$. Separating variables we obtain
$$\frac{du}{1 + u^2} = dx \implies \tan^{-1} u = x + c_1 \implies u = \tan(x + c_1) \implies y = -\ln|\cos(x + c_1)| + c_2.$$

5. Let $u = y'$ so that $u' = y''$. The equation becomes $x^2 u' + u^2 = 0$. Separating variables we obtain
$$\frac{du}{u^2} = -\frac{dx}{x^2} \implies -\frac{1}{u} = \frac{1}{x} + c_1 = \frac{c_1 x + 1}{x} \implies u = -\frac{1}{c_1}\left(\frac{x}{x + 1/c_1}\right) = \frac{1}{c_1}\left(\frac{1}{c_1 x + 1} - 1\right)$$
$$\implies y = \frac{1}{c_1^2}\ln|c_1 x + 1| - \frac{1}{c_1}x + c_2.$$

6. Let $u = y'$ so that $y'' = u\dfrac{du}{dy}$. The equation becomes $(y + 1)u\dfrac{du}{dy} = u^2$. Separating variables we obtain
$$\frac{du}{u} = \frac{dy}{y + 1} \implies \ln|u| = \ln|y + 1| + \ln c_1 \implies u = c_1(y + 1)$$
$$\implies \frac{dy}{dx} = c_1(y + 1) \implies \frac{dy}{y + 1} = c_1\,dx$$
$$\implies \ln|y + 1| = c_1 x + c_2 \implies y + 1 = c_3 e^{c_1 x}.$$

141

7. Let $u = y'$ so that $y'' = u\dfrac{du}{dy}$. The equation becomes $u\dfrac{du}{dy} + 2yu^3 = 0$. Separating variables we obtain

$$\frac{du}{u^2} + 2y\,dy = 0 \implies -\frac{1}{u} + y^2 = c \implies u = \frac{1}{y^2 + c_1} \implies y' = \frac{1}{y^2 + c_1}$$

$$\implies \left(y^2 + c_1\right) dy = dx \implies \frac{1}{3}y^3 + c_1 y = x + c_2.$$

8. Let $u = y'$ so that $y'' = u\dfrac{du}{dy}$. The equation becomes $y^2 u\dfrac{du}{dy} = u$. Separating variables we obtain

$$du = \frac{dy}{y^2} \implies u = -\frac{1}{y} + c_1 \implies y' = \frac{c_1 y - 1}{y} \implies \frac{y}{c_1 y - 1}\,dy = dx$$

$$\implies \frac{1}{c_1}\left(1 + \frac{1}{c_1 y - 1}\right) dy = dx \text{ (for } c_1 \neq 0) \implies \frac{1}{c_1}y + \frac{1}{c_1^2}\ln|y - 1| = x + c_2.$$

If $c_1 = 0$, then $y\,dy = -dx$ and another solution is $\dfrac{1}{2}y^2 = -x + c_2$.

9. Let $u = y'$ so that $y'' = u\dfrac{du}{dy}$. The equation becomes $u\dfrac{du}{dy} + yu = 0$. Separating variables we obtain

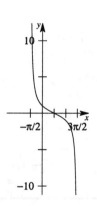

$$du = -y\,dy \implies u = -\frac{1}{2}y^2 + c_1 \implies y' = -\frac{1}{2}y^2 + c_1.$$

When $x = 0$, $y = 1$ and $y' = -1$ so $-1 = -\frac{1}{2} + c_1$ and $c_1 = -\frac{1}{2}$. Then

$$\frac{dy}{dx} = -\frac{1}{2}y^2 - \frac{1}{2} \implies \frac{dy}{y^2 + 1} = -\frac{1}{2}\,dx \implies \tan^{-1}y = -\frac{1}{2}x + c_2$$

$$\implies y = \tan\left(-\frac{1}{2}x + c_2\right).$$

When $x = 0$, $y = 1$ so $1 = \tan c_2$ and $c_2 = \pi/4$. The solution of the initial-value problem is

$$y = \tan\left(\frac{\pi}{4} - \frac{1}{2}x\right), \qquad -\frac{\pi}{2} < x < \frac{3\pi}{2}.$$

10. Let $u = y'$ so that $u' = y''$. The equation becomes $(u')^2 + u^2 = 1$ which results in $u' = \pm\sqrt{1 - u^2}$. To solve $u' = \sqrt{1 - u^2}$ we separate variables:

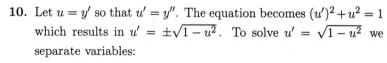

$$\frac{du}{\sqrt{1 - u^2}} = dx \implies \sin^{-1}u = x + c_1 \implies u = \sin(x + c_1)$$

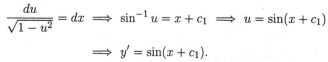

$$\implies y' = \sin(x + c_1).$$

When $x = \dfrac{\pi}{2}$, $y' = \dfrac{\sqrt{3}}{2}$, so $\dfrac{\sqrt{3}}{2} = \sin\left(\dfrac{\pi}{2} + c_1\right)$ and $c_1 = -\dfrac{\pi}{6}$.

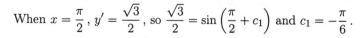

Thus

$$y' = \sin\left(x - \frac{\pi}{6}\right) \implies y = -\cos\left(x - \frac{\pi}{6}\right) + c_2.$$

When $x = \frac{\pi}{2}$, $y = \frac{1}{2}$, so $\frac{1}{2} = -\cos\left(\frac{\pi}{2} - \frac{\pi}{6}\right) + c_2 = -\frac{1}{2} + c_2$ and $c_2 = 1$. The solution of the initial-value problem is $y = 1 - \cos\left(x - \frac{\pi}{6}\right)$.

To solve $u' = -\sqrt{1 - u^2}$ we separate variables:

$$\frac{du}{\sqrt{1 - u^2}} = -dx \implies \cos^{-1} u = x + c_1$$

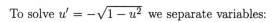

$$\implies u = \cos(x + c_1) \implies y' = \cos(x + c_1).$$

When $x = \frac{\pi}{2}$, $y' = \frac{\sqrt{3}}{2}$, so $\frac{\sqrt{3}}{2} = \cos\left(\frac{\pi}{2} + c_1\right)$ and $c_1 = -\frac{\pi}{3}$. Thus

$$y' = \cos\left(x - \frac{\pi}{3}\right) \implies y = \sin\left(x - \frac{\pi}{3}\right) + c_2.$$

When $x = \frac{\pi}{2}$, $y = \frac{1}{2}$, so $\frac{1}{2} = \sin\left(\frac{\pi}{2} - \frac{\pi}{3}\right) + c_2 = \frac{1}{2} + c_2$ and $c_2 = 0$. The solution of the initial-value problem is $y = \sin\left(x - \frac{\pi}{3}\right)$.

11. Let $u = y'$ so that $u' = y''$. The equation becomes $u' - \frac{1}{x}u = \frac{1}{x}u^3$, which is Bernoulli. Using $w = u^{-2}$ we obtain $\frac{dw}{dx} + \frac{2}{x}w = -\frac{2}{x}$. An integrating factor is x^2, so

$$\frac{d}{dx}[x^2 w] = -2x \implies x^2 w = -x^2 + c_1 \implies w = -1 + \frac{c_1}{x^2}$$

$$\implies u^{-2} = -1 + \frac{c_1}{x^2} \implies u = \frac{x}{\sqrt{c_1 - x^2}}$$

$$\implies \frac{dy}{dx} = \frac{x}{\sqrt{c_1 - x^2}} \implies y = -\sqrt{c_1 - x^2} + c_2$$

$$\implies c_1 - x^2 = (c_2 - y)^2 \implies x^2 + (c_2 - y)^2 = c_1.$$

12. Let $u = y'$ so that $u' = y''$. The equation becomes $u' - \frac{1}{x}u = u^2$, which is Bernoulli. Using the substitution $w = u^{-1}$ we obtain $\frac{dw}{dx} + \frac{1}{x}w = -1$. An integrating factor is x, so

$$\frac{d}{dx}[xw] = -x \implies w = -\frac{1}{2}x + \frac{1}{x}c \implies \frac{1}{u} = \frac{c_1 - x^2}{2x} \implies u = \frac{2x}{c_1 - x^2} \implies y = -\ln\left|c_1 - x^2\right| + c_2.$$

13. We look for a solution of the form

$$y(x) = y(0) + y'(0) + \frac{1}{2}y''(0) + \frac{1}{3!}y'''(0) + \frac{1}{4!}y^{(4)}(x) + \frac{1}{5!}y^{(5)}(x).$$

From $y''(x) = x + y^2$ we compute

$$y'''(x) = 1 + 2yy'$$

$$y^{(4)}(x) = 2yy'' + 2(y')^2$$

$$y^{(5)}(x) = 2yy''' + 6y'y''.$$

Using $y(0) = 1$ and $y'(0) = 1$ we find

$$y''(0) = 1, \quad y'''(0) = 3, \quad y^{(4)}(0) = 4, \quad y^{(5)}(0) = 12.$$

An approximate solution is

$$y(x) = 1 + x + \frac{1}{2}x^2 + \frac{1}{2}x^3 + \frac{1}{6}x^4 + \frac{1}{10}x^5.$$

14. We look for a solution of the form

$$y(x) = y(0) + y'(0) + \frac{1}{2}y''(0) + \frac{1}{3!}y'''(0) + \frac{1}{4!}y^{(4)}(x) + \frac{1}{5!}y^{(5)}(x).$$

From $y''(x) = 1 - y^2$ we compute

$$y'''(x) = -2yy'$$

$$y^{(4)}(x) = -2yy'' - 2(y')^2$$

$$y^{(5)}(x) = -2yy''' - 6y'y''.$$

Using $y(0) = 2$ and $y'(0) = 3$ we find

$$y''(0) = -3, \quad y'''(0) = -12, \quad y^{(4)}(0) = -6, \quad y^{(5)}(0) = 102.$$

An approximate solution is

$$y(x) = 2 + 3x - \frac{3}{2}x^2 - 2x^3 - \frac{1}{4}x^4 + \frac{17}{20}x^5.$$

15. We look for a solution of the form

$$y(x) = y(0) + y'(0) + \frac{1}{2}y''(0) + \frac{1}{3!}y'''(0) + \frac{1}{4!}y^{(4)}(x) + \frac{1}{5!}y^{(5)}(x).$$

From $y''(x) = x^2 + y^2 - 2y'$ we compute

$$y'''(x) = 2x + 2yy' - 2y''$$

$$y^{(4)}(x) = 2 + 2(y')^2 + 2yy'' - 2y'''$$

$$y^{(5)}(x) = 6y'y'' + 2yy''' - 2y^{(4)}.$$

Using $y(0) = 1$ and $y'(0) = 1$ we find
$$y''(0) = -1, \quad y'''(0) = 4, \quad y^{(4)}(0) = -6, \quad y^{(5)}(0) = 14.$$

An approximate solution is
$$y(x) = 1 + x - \frac{1}{2}x^2 + \frac{2}{3}x^3 - \frac{1}{4}x^4 + \frac{7}{60}x^5.$$

16. We look for a solution of the form
$$y(x) = y(0) + y'(0) + \frac{1}{2}y''(0) + \frac{1}{3!}y'''(0) + \frac{1}{4!}y^{(4)}(x) + \frac{1}{5!}y^{(5)}(x).$$

From $y''(x) = e^y$ we compute
$$y'''(x) = e^y y'$$
$$y^{(4)}(x) = e^y(y')^2 + e^y y''$$
$$y^{(5)}(x) = e^y(y')^3 + 3e^y y' y'' + e^y y'''.$$

Using $y(0) = 0$ and $y'(0) = -1$ we find
$$y''(0) = 1, \quad y'''(0) = -1, \quad y^{(4)}(0) = 2, \quad y^{(5)}(0) = -5.$$

An approximate solution is
$$y(x) = -x + \frac{1}{2}x^2 - \frac{1}{6}x^3 + \frac{1}{12}x^4 + \frac{1}{24}x^5.$$

17. We need to solve $\left[1 + (y')^2\right]^{3/2} = y''$. Let $u = y'$ so that $u' = y''$. The equation becomes $\left(1 + u^2\right)^{3/2} = u'$ or $\left(1 + u^2\right)^{3/2} = \dfrac{du}{dx}$. Separating variables and using the substitution $u = \tan\theta$ we have

$$\frac{du}{(1+u^2)^{3/2}} = dx \implies \int \frac{\sec^2\theta}{\left(1+\tan^2\theta\right)^{3/2}}\,d\theta = x \implies \int \frac{\sec^2\theta}{\sec^3\theta}\,d\theta = x$$

$$\implies \int \cos\theta\,d\theta = x \implies \sin\theta = x \implies \frac{u}{\sqrt{1+u^2}} = x$$

$$\implies \frac{y'}{\sqrt{1+(y')^2}} = x \implies (y')^2 = x^2\left[1 + (y')^2\right] = \frac{x^2}{1-x^2}$$

$$\implies y' = \frac{x}{\sqrt{1-x^2}} \quad (\text{for } x > 0) \implies y = -\sqrt{1 - x^2}.$$

18. Let $u = \dfrac{dx}{dt}$ so that $\dfrac{d^2x}{dt^2} = u\dfrac{du}{dx}$. The equation becomes $u\dfrac{du}{dx} = \dfrac{-k^2}{x^2}$. Separating variables we obtain

$$u\,du = -\frac{k^2}{x^2}\,dx \implies \frac{1}{2}u^2 = \frac{k^2}{x} + c \implies \frac{1}{2}v^2 = \frac{k^2}{x} + c.$$

When $t = 0$, $x = x_0$ and $v = 0$ so $0 = \dfrac{k^2}{x_0} + c$ and $c = -\dfrac{k^2}{x_0}$. Then

$$\frac{1}{2}v^2 = k^2\left(\frac{1}{x} - \frac{1}{x_0}\right) \quad \text{and} \quad \frac{dx}{dt} = -k\sqrt{2}\sqrt{\frac{x_0 - x}{xx_0}}.$$

Separating variables we have

$$-\sqrt{\frac{xx_0}{x_0 - x}}\, dx = k\sqrt{2}\, dt \implies t = -\frac{1}{k}\sqrt{\frac{x_0}{2}}\int \sqrt{\frac{x}{x_0 - x}}\, dx.$$

Using *Mathematica* to integrate we obtain

$$t = -\frac{1}{k}\sqrt{\frac{x_0}{2}}\left[-\sqrt{x(x_0 - x)} - \frac{x_0}{2}\tan^{-1}\frac{(x_0 - 2x)}{2x}\sqrt{\frac{x}{x_0 - x}}\right]$$

$$= \frac{1}{k}\sqrt{\frac{x_0}{2}}\left[\sqrt{x(x_0 - x)} + \frac{x_0}{2}\tan^{-1}\frac{x_0 - 2x}{2\sqrt{x(x_0 - x)}}\right].$$

19.

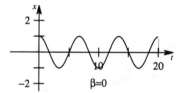

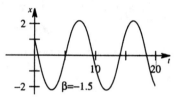

For $\dfrac{d^2x}{dt^2} + \sin x = 0$ the motion appears to be periodic with amplitude 1 when $\beta = 0$. The amplitude and period are larger for larger magnitudes of β.

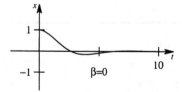

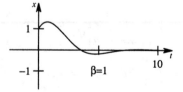

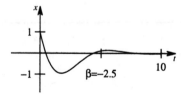

For $\dfrac{d^2x}{dt^2} + \dfrac{dx}{dt} + \sin x = 0$ the motion appears to be periodic with decreasing amplitude. The dx/dt term could be said to have a damping effect.

20. From $(y'')^2 - y^2 = 0$ we have $y'' = \pm y$, which can be treated as two linear equations. Since linear combinations of solutions of linear homogeneous differential equations are also solutions, we see that $y = c_1 e^x + c_2 e^{-x}$ and $y = c_3\cos x + c_4\sin x$ must satisfy the differential equation.

1. $y = 0$

2. False; see Problem 31, Section 4.1.

3. False; consider $f_1(x) = 0$ and $f_2(x) = x$. These are linearly dependent even though x is not a multiple of 0. The statement would be true if it read "Two functions $f_1(x)$ and $f_2(x)$ are linearly independent on an interval if *neither* is a constant multiple of the other."

4. dependent; $-3x^2 + (-2)\left(1 - x^2\right) + \left(2 + x^2\right) = 0$

5. $(-\infty, \infty)$; $(0, \infty)$ or $(-\infty, 0)$

6. True

7. False; consider the solution $y = 1/x$ of $y'' = 2y^3$.

8. $x = 2$

9. $A + Bxe^x$

10. $(D - 2)^2 \left(D^2 - 4D + 5\right)$

11. Identifying $P(x) = 0$ we have

$$y_2 = \cos 2x \int \frac{e^{-\int 0\,dx}}{\cos^2 2x}\,dx = \cos 2x \int \sec^2 2x\,dx$$

$$= \cos 2x \left(\frac{1}{2}\tan 2x\right)$$

$$= \frac{1}{2}\sin 2x.$$

12. Identifying $P(x) = -2 - 2/x$ we have $\int P\,dx = -2x - 2\ln x$ and

$$y_2 = e^x \int \frac{e^{2x + \ln x^2}}{e^{2x}}\,dx = e^x \int x^2\,dx = \frac{1}{3}x^3 e^x.$$

13. From $m^2 - 2m - 2 = 0$ we obtain $m = 1 \pm \sqrt{3}$ so that

$$y = c_1 e^{(1+\sqrt{3})x} + c_2 e^{(1-\sqrt{3})x}.$$

14. From $2m^2 + 2m + 3 = 0$ we obtain $m = -1/2 \pm \sqrt{5}/2$ so that

$$y = e^{-x/2}\left(c_1 \cos \frac{\sqrt{5}}{2}x + c_2 \sin \frac{\sqrt{5}}{2}x\right).$$

15. From $m^3 + 10m^2 + 25m = 0$ we obtain $m = 0$, $m = -5$, and $m = -5$ so that

$$y = c_1 + c_2 e^{-5x} + c_3 x e^{-5x}.$$

16. From $2m^3 + 9m^2 + 12m + 5 = 0$ we obtain $m = -1$, $m = -1$, and $m = -5/2$ so that

$$y = c_1 e^{-5x/2} + c_2 e^{-x} + c_3 x e^{-x}.$$

17. From $3m^3 + 10m^2 + 15m + 4 = 0$ we obtain $m = -1/3$ and $m = -3/2 \pm \sqrt{7}/2$ so that

$$y = c_1 e^{-x/3} + e^{-3x/2}\left(c_2 \cos\frac{\sqrt{7}}{2}x + c_3 \sin\frac{\sqrt{7}}{2}x\right).$$

18. From $2m^4 + 3m^3 + 2m^2 + 6m - 4 = 0$ we obtain $m = 1/2$, $m = -2$, and $m = \pm\sqrt{2}\,i$ so that

$$y = c_1 e^{x/2} + c_2 e^{-2x} + c_3 \cos\sqrt{2}\,x + c_4 \sin\sqrt{2}\,x.$$

19. The auxiliary equation is $6m^2 - m - 1 = 0$ so that

$$y = c_1 x^{1/2} + c_2 x^{-1/3}.$$

20. The auxiliary equation is $2m^3 + 13m^2 + 24m + 9 = (m+3)^2(m+1/2) = 0$ so that

$$y = c_1 x^{-3} + c_2 x^{-3}\ln x + \frac{1}{4}x^3.$$

21. Applying D^4 to the differential equation we obtain $D^4(D^2 - 3D + 5) = 0$. Then

$$y = \underbrace{e^{3x/2}\left(c_1 \cos\frac{\sqrt{11}}{2}x + c_2 \sin\frac{\sqrt{11}}{2}x\right)}_{y_c} + c_3 + c_4 x + c_5 x^2 + c_6 x^3$$

and $y_p = A + Bx + Cx^2 + Dx^3$. Substituting y_p into the differential equation yields

$$(5A - 3B + 2C) + (5B - 6C + 6D)x + (5C - 9D)x^2 + 5Dx^3 = -2x + 4x^3.$$

Equating coefficients gives $A = -222/625$, $B = 46/125$, $C = 36/25$, and $D = 4/5$. The general solution is

$$y = e^{3x/2}\left(c_1 \cos\frac{\sqrt{11}}{2}x + c_2 \sin\frac{\sqrt{11}}{2}x\right) - \frac{222}{625} + \frac{46}{125}x + \frac{36}{25}x^2 + \frac{4}{5}x^3.$$

22. Applying $(D-1)^3$ to the differential equation we obtain $(D-1)^3(D-2D+1) = (D-1)^5 = 0$. Then

$$y = \underbrace{c_1 e^x + c_2 x e^x}_{y_c} + c_3 x^2 e^x + c_4 x^3 e^x + c_5 x^4 e^x$$

and $y_p = Ax^2 e^x + Bx^3 e^x + Cx^4 e^x$. Substituting y_p into the differential equation yields

$$12Cx^2 e^x + 6Bx e^x + 2A e^x = x^2 e^x.$$

Equating coefficients gives $A = 0$, $B = 0$, and $C = 1/12$. The general solution is

$$y = c_1 e^x + c_2 x e^x + \frac{1}{12}x^4 e^x.$$

23. Applying $D(D^2 + 1)$ to the differential equation we obtain

$$D(D^2 + 1)(D^3 - 5D^2 + 6D) = D^2(D^2 + 1)(D - 2)(D - 3) = 0.$$

Then

$$y = \underbrace{c_1 + c_2 e^{2x} + c_3 e^{3x}}_{y_c} + c_4 x + c_5 \cos x + c_6 \sin x$$

and $y_p = Ax + B \cos x + C \sin x$. Substituting y_p into the differential equation yields

$$6A + (5B + 5C) \cos x + (-5B + 5C) \sin x = 8 + 2 \sin x.$$

Equating coefficients gives $A = 4/3$, $B = -1/5$, and $C = 1/5$. The general solution is

$$y = c_1 + c_2 e^{2x} + c_3 e^{3x} + \frac{4}{3}x - \frac{1}{5}\cos x + \frac{1}{5}\sin x.$$

24. Applying D to the differential equation we obtain $D(D^3 - D^2) = D^3(D - 1) = 0$. Then

$$y = \underbrace{c_1 + c_2 x + c_3 e^x}_{y_c} + c_4 x^2$$

and $y_p = Ax^2$. Substituting y_p into the differential equation yields $-2A = 6$. Equating coefficients gives $A = -3$. The general solution is

$$y = c_1 + c_2 x + c_3 e^x - 3x^2.$$

25. The auxiliary equation is $m^2 - 2m + 2 = 0$ so that $m = 1 \pm i$ and $y = e^x(c_1 \cos x + c_2 \sin x)$. Setting $y(\pi/2) = 0$ and $y(\pi) = -1$ we obtain $c_1 = e^{-\pi}$ and $c_2 = 0$. Thus, $y = e^{x-\pi} \cos x$.

26. The auxiliary equation is $m^2 - 1 = (m - 1)(m + 1) = 0$ so that $m = \pm 1$ and $y = c_1 e^x + c_2 e^{-x}$. Assuming $y_p = Ax + B + C \sin x$ and substituting into the differential equation we find $A = -1$, $B = 0$, and $C = -\frac{1}{2}$. Thus $y_p = -x - \frac{1}{2}\sin x$ and

$$y = c_1 e^x + c_2 e^{-x} - x - \frac{1}{2}\sin x.$$

Setting $y(0) = 2$ and $y'(0) = 3$ we obtain

$$c_1 + c_2 = 2$$

$$c_1 - c_2 - \frac{3}{2} = 3.$$

Solving this system we find $c_1 = \frac{13}{4}$ and $c_2 = -\frac{5}{4}$. The solution of the initial-value problem is

$$y = \frac{13}{4}e^x - \frac{5}{4}e^{-x} - x - \frac{1}{2}\sin x.$$

27. Let $u = y'$ so that $u' = y''$. The equation becomes $u\dfrac{du}{dx} = 4x$. Separating variables we obtain

$$u\, du = 4x\, dx \implies \frac{1}{2}u^2 = 2x^2 + c_1 \implies u^2 = 4x^2 + c_2.$$

When $x = 1$, $y' = u = 2$, so $4 = 4 + c_2$ and $c_2 = 0$. Then

$$u^2 = 4x^2 \implies \frac{dy}{dx} = 2x \quad \text{or} \quad \frac{dy}{dx} = -2x$$

$$\implies y = x^2 + c_3 \quad \text{or} \quad y = -x^2 + c_4.$$

When $x = 1$, $y = 5$, so $5 = 1 + c_3$ and $5 = -1 + c_4$. Thus $c_3 = 4$ and $c_4 = 6$. We have $y = x^2 + 4$ and $y = -x^2 + 6$. Note however that when $y = -x^2 + 6$, $y' = -2x$ and $y'(1) = -2 \neq 2$. Thus, the solution of the initial-value problem is $y = x^2 + 4$.

28. Let $u = y'$ so that $y'' = u \dfrac{du}{dy}$. The equation becomes $2u \dfrac{du}{dy} = 3y^2$. Separating variables we obtain

$$2u \, dy = 3y^2 \, dy \implies u^2 = y^3 + c_1.$$

When $x = 0$, $y = 1$ and $y' = u = 1$ so $1 = 1 + c_1$ and $c_1 = 0$. Then

$$u^2 = y^3 \implies \left(\frac{dy}{dx}\right)^2 = y^3 \implies \frac{dy}{dx} = y^{3/2} \implies y^{-3/2} \, dy = dx$$

$$\implies -2y^{-1/2} = x + c_2 \implies y = \frac{4}{(x + c_2)^2}.$$

When $x = 0$, $y = 1$, so $1 = \dfrac{4}{c_2^2} \implies c_2 = \pm 2$. Thus, $y = \dfrac{4}{(x + 2)^2}$ and $y = \dfrac{4}{(x - 2)^2}$. Note however that when $y = \dfrac{4}{(x + 2)^2}$, $y' = -\dfrac{8}{(x + 2)^3}$ and $y'(0) = -1 \neq 1$. Thus, the solution of the initial-value problem is $y = \dfrac{4}{(x - 2)^2}$.

29. The auxiliary equation is $m^2 - 2m + 2 = [m - (1 + i)][m - (1 - i)] = 0$, so $y_c = c_1 e^x \sin x + c_2 e^x \cos x$ and

$$W = \begin{vmatrix} e^x \sin x & e^x \cos x \\ e^x \cos x + e^x \sin x & -e^x \sin x + e^x \cos x \end{vmatrix} = -e^{2x}.$$

Identifying $f(x) = e^x \tan x$ we obtain

$$u_1' = -\frac{(e^x \cos x)(e^x \tan x)}{-e^{2x}} = \sin x$$

$$u_2' = \frac{(e^x \sin x)(e^x \tan x)}{-e^{2x}} = -\frac{\sin^2 x}{\cos x} = \cos x - \sec x.$$

Then $u_1 = -\cos x$, $u_2 = \sin x - \ln|\sec x + \tan x|$, and

$$y = c_1 e^x \sin x + c_2 e^x \cos x - e^x \sin x \cos x + e^x \sin x \cos x - e^x \cos x \ln|\sec x + \tan x|$$

$$= c_1 e^x \sin x + c_2 e^x \cos x - e^x \cos x \ln|\sec x + \tan x|.$$

30. The auxiliary equation is $m^2 - 1 = 0$, so $y_c = c_1 e^x + c_2 e^{-x}$ and

$$W = \begin{vmatrix} e^x & e^{-x} \\ e^x & -e^{-x} \end{vmatrix} = -2.$$

Identifying $f(x) = 2e^x/(e^x + e^{-x})$ we obtain

$$u_1' = \frac{1}{e^x + e^{-x}} = \frac{e^x}{1 + e^{2x}}$$

$$u_2' = -\frac{e^{2x}}{e^x + e^{-x}} = -\frac{e^{3x}}{1 + e^{2x}} = -e^x + \frac{e^x}{1 + e^{2x}}.$$

Then $u_1 = \tan^{-1} e^x$, $u_2 = -e^x + \tan^{-1} e^x$, and

$$y = c_1 e^x + c_2 e^{-x} + e^x \tan^{-1} e^x - 1 + e^{-x} \tan^{-1} e^x.$$

31. The auxiliary equation is $m^2 - 5m + 6 = (m-2)(m-3) = 0$ and a particular solution is $y_p = x^4 - x^2 \ln x$ so that

$$y = c_1 x^2 + c_2 x^3 + x^4 - x^2 \ln x.$$

32. The auxiliary equation is $m^2 - 2m + 1 = (m-1)^2 = 0$ and a particular solution is $y_p = \frac{1}{4}x^3$ so that

$$y = c_1 x + c_2 x \ln x + \frac{1}{4}x^3.$$

33. The auxiliary equation is $2m^3 - 13m^2 + 24m - 9 = (2m-1)(m-3)^2 = 0$ so that

$$y_c = c_1 e^{x/2} + c_2 e^{3x} + c_3 x e^{3x}.$$

A particular solution is $y_p = -4$ and the general solution is

$$y = c_1 e^{x/2} + c_2 e^{3x} + c_3 x e^{3x} - 4.$$

Setting $y(0) = -4$, $y'(0) = 0$, and $y''(0) = \frac{5}{2}$ we obtain

$$c_1 + c_2 - 4 = -4$$

$$\frac{1}{2}c_1 + 3c_2 + c_3 = 0$$

$$\frac{1}{4}c_1 + 9c_2 + 6c_3 = \frac{5}{2}.$$

Solving this system we find $c_1 = \frac{2}{5}$, $c_2 = -\frac{2}{5}$, and $c_3 = 1$. Thus

$$y = \frac{2}{5}e^{x/2} - \frac{2}{5}e^{3x} + x e^{3x} - 4.$$

34. The auxiliary equation is $m^2 + 1 = 0$, so $y_c = c_1 \cos x + c_2 \sin x$ and

$$W = \begin{vmatrix} \cos x & \sin x \\ -\sin x & \cos x \end{vmatrix} = 1.$$

Identifying $f(x) = \sec^3 x$ we obtain

$$u_1' = -\sin x \sec^3 x = -\frac{\sin x}{\cos^3 x}$$

$$u_2' = \cos x \sec^3 x = \sec^2 x.$$

Then

$$u_1 = -\frac{1}{2}\frac{1}{\cos^2 x} = -\frac{1}{2}\sec^2 x$$

$$u_2 = \tan x.$$

Thus

$$y = c_1 \cos x + c_2 \sin x - \frac{1}{2}\cos x \sec^2 x + \sin x \tan x$$

$$= c_1 \cos x + c_2 \sin x - \frac{1}{2}\sec x + \frac{1 - \cos^2 x}{\cos x}$$

$$= c_3 \cos x + c_2 \sin x + \frac{1}{2}\sec x.$$

and

$$y_p' = -c_3 \sin x + c_2 \cos x + \frac{1}{2}\sec x \tan x.$$

The initial conditions imply

$$c_3 + \frac{1}{2} = 1$$

$$c_2 = \frac{1}{2}.$$

Thus $c_3 = c_2 = 1/2$ and

$$y = \frac{1}{2}\cos x + \frac{1}{2}\sin x + \frac{1}{2}\sec x.$$

35. From $(D-2)x + (D-2)y = 1$ and $Dx + (2D-1)y = 3$ we obtain $(D-1)(D-2)y = -6$ and $Dx = 3 - (2D-1)y$. Then

$$y = c_1 e^{2t} + c_2 e^t - 3 \quad \text{and} \quad x = -c_2 e^t - \frac{3}{2}c_1 e^{2t} + c_3.$$

Substituting into $(D-2)x + (D-2)y = 1$ gives $c_3 = 5/2$ so that

$$x = -c_2 e^t - \frac{3}{2}c_1 e^{2t} + \frac{5}{2}.$$

36. From $(D-2)x - y = t - 2$ and $-3x + (D-4)y = -4t$ we obtain $(D-1)(D-5)x = 9 - 8t$. Then

$$x = c_1 e^t + c_2 e^{5t} - \frac{8}{5}t - \frac{3}{25}$$

and

$$y = (D-2)x - t + 2 = -c_1 e^t + 3c_2 e^{5t} + \frac{16}{25} + \frac{11}{25}t.$$

37. From $(D-2)x - y = -e^t$ and $-3x + (D-4)y = -7e^t$ we obtain $(D-1)(D-5)x = -4e^t$ so that

$$x = c_1 e^t + c_2 e^{5t} + te^t.$$

Then

$$y = (D-2)x + e^t = -c_1 e^t + 3c_2 e^{5t} - te^t + 2e^t.$$

38. From $(D+2)x + (D+1)y = \sin 2t$ and $5x + (D+3)y = \cos 2t$ we obtain $(D^2+5)y = 2\cos 2t - 7\sin 2t$. Then

$$y = c_1 \cos t + c_2 \sin t - \frac{2}{3}\cos 2t + \frac{7}{3}\sin 2t$$

and

$$x = -\frac{1}{5}(D+3)y + \frac{1}{5}\cos 2t$$

$$= \left(\frac{1}{5}c_1 - \frac{3}{5}c_2\right)\sin t + \left(-\frac{1}{5}c_2 - \frac{3}{5}c_1\right)\cos t - \frac{5}{3}\sin 2t - \frac{1}{3}\cos 2t.$$

5 Modeling with Higher-Order Differential Equations

_____ **Exercises 5.1** _____

1. From $\frac{1}{8}x' + 16x = 0$ we obtain

$$x = c_1 \cos 8\sqrt{2}\,t + c_2 \sin 8\sqrt{2}\,t$$

so that the period of motion is $2\pi/8\sqrt{2} = \sqrt{2}\,\pi/8$ seconds.

2. From $20x'' + kx = 0$ we obtain

$$x = c_1 \cos \frac{1}{2}\sqrt{\frac{k}{5}}\,t + c_2 \sin \frac{1}{2}\sqrt{\frac{k}{5}}\,t$$

so that the frequency $2/\pi = \frac{1}{4}\sqrt{k/5}\,\pi$ and $k = 320$ N/m. If $80x'' + 320x = 0$ then $x = c_1 \cos 2t + c_2 \sin 2t$ so that the frequency is $2/2\pi = 1/\pi$ vibrations/second.

3. From $\frac{3}{4}x'' + 72x = 0$, $x(0) = -1/4$, and $x'(0) = 0$ we obtain $x = -\frac{1}{4}\cos 4\sqrt{6}\,t$.

4. From $\frac{3}{4}x'' + 72x = 0$, $x(0) = 0$, and $x'(0) = 2$ we obtain $x = \frac{\sqrt{6}}{12}\sin 4\sqrt{6}\,t$.

5. From $\frac{5}{8}x'' + 40x = 0$, $x(0) = 1/2$, and $x'(0) = 0$ we obtain $x = \frac{1}{2}\cos 8t$.

    (a) $x(\pi/12) = -1/4$, $x(\pi/8) = -1/2$, $x(\pi/6) = -1/4$, $x(\pi/8) = 1/2$, $x(9\pi/32) = \sqrt{2}/4$.
    (b) $x' = -4\sin 8t$ so that $x'(3\pi/16) = 4$ ft/s directed downward.
    (c) If $x = \frac{1}{2}\cos 8t = 0$ then $t = (2n+1)\pi/16$ for $n = 0, 1, 2, \ldots$.

6. From $50x'' + 200x = 0$, $x(0) = 0$, and $x'(0) = -10$ we obtain $x = -5\sin 2t$ and $x' = -10\cos 2t$.

7. From $20x'' + 20x = 0$, $x(0) = 0$, and $x'(0) = -10$ we obtain $x = -10\sin t$ and $x' = -10\cos t$.

    (a) The 20 kg mass has the larger amplitude.
    (b) 20 kg: $x'(\pi/4) = -5\sqrt{2}$ m/s, $x'(\pi/2) = 0$ m/s;   50 kg: $x'(\pi/4) = 0$ m/s, $x'(\pi/2) = 10$ m/s
    (c) If $-5\sin 2t = -10\sin t$ then $2\sin t(\cos t - 1) = 0$ so that $t = n\pi$ for $n = 0, 1, 2, \ldots$, placing both masses at the equilibrium position. The 50 kg mass is moving upward; the 20 kg mass is moving upward when n is even and downward when n is odd.

8. From $x'' + 16x = 0$, $x(0) = -1$, and $x'(0) = -2$ we obtain

$$x = -\cos 4t - \frac{1}{2}\sin 4t = \frac{\sqrt{5}}{2}\cos(4t - 3.6).$$

154

The period is $\pi/2$ seconds and the amplitude is $\sqrt{5}/2$ feet. In 4π seconds it will make 8 complete vibrations.

9. From $\frac{1}{4}x'' + x = 0$, $x(0) = 1/2$, and $x'(0) = 3/2$ we obtain

$$x = \frac{1}{2}\cos 2t + \frac{3}{4}\sin 2t = \frac{\sqrt{13}}{4}\sin(2t + 0.588).$$

10. From $1.6x'' + 40x = 0$, $x(0) = -1/3$, and $x'(0) = 5/4$ we obtain

$$x = -\frac{1}{3}\cos 5t + \frac{1}{4}\sin 5t = \frac{5}{12}\sin(5t + 0.927).$$

If $x = 5/24$ then $t = \frac{1}{5}\left(\frac{\pi}{6} + 0.927 + 2n\pi\right)$ and $t = \frac{1}{5}\left(\frac{5\pi}{6} + 0.927 + 2n\pi\right)$ for $n = 0, 1, 2, \ldots$.

11. From $2x'' + 200x = 0$, $x(0) = -2/3$, and $x'(0) = 5$ we obtain

(a) $x = -\frac{2}{3}\cos 10t + \frac{1}{2}\sin 10t = \frac{5}{6}\sin(10t - 0.927)$.

(b) The amplitude is $5/6$ ft and the period is $2\pi/10 = \pi/5$

(c) $3\pi = \pi k/5$ and $k = 15$ cycles.

(d) If $x = 0$ and the weight is moving downward for the second time, then $10t - 0.927 = 2\pi$ or $t = 0.721$ s.

(e) If $x' = \frac{25}{3}\cos(10t - 0.927) = 0$ then $10t - 0.927 = \pi/2 + n\pi$ or $t = (2n + 1)\pi/20 + 0.0927$ for $n = 0, 1, 2, \ldots$.

(f) $x(3) = -0.597$ ft

(g) $x'(3) = -5.814$ ft/s

(h) $x''(3) = 59.702$ ft/s^2

(i) If $x = 0$ then $t = \frac{1}{10}(0.927 + n\pi)$ for $n = 0, 1, 2, \ldots$ and $x'(t) = \pm\frac{25}{3}$ ft/s.

(j) If $x = 5/12$ then $t = \frac{1}{10}(\pi/6 + 0.927 + 2n\pi)$ and $t = \frac{1}{10}(5\pi/6 + 0.927 + 2n\pi)$ for $n = 0, 1, 2, \ldots$.

(k) If $x = 5/12$ and $x' < 0$ then $t = \frac{1}{10}(5\pi/6 + 0.927 + 2n\pi)$ for $n = 0, 1, 2, \ldots$.

12. From $x' + 9x = 0$, $x(0) = -1$, and $x'(0) = -\sqrt{3}$ we obtain

$$x = -\cos 3t - \frac{\sqrt{3}}{3}\sin 3t = \frac{2}{\sqrt{3}}\sin\left(37 + \frac{4\pi}{3}\right)$$

and $x' = 2\sqrt{3}\cos(3t + 4\pi/3)$. If $x' = 3$ then $t = -7\pi/18 + 2n\pi/3$ and $t = -\pi/2 + 2n\pi/3$ for $n = 1$, $2, 3, \ldots$.

13. From $k_1 = 40$ and $k_2 = 120$ we compute the effective spring constant $k = 4(40)(120)/160 = 120$. Now, $m = 20/32$ so $k/m = 120(32)/20 = 192$ and $x'' + 192x = 0$. Using $x(0) = 0$ and $x'(0) = 2$ we obtain $x(t) = \frac{\sqrt{3}}{12}\sin 8\sqrt{3}\,t$.

14. Let m denote the mass in slugs of the first weight. Let k_1 and k_2 be the spring constants and $k = 4k_1k_2/(k_1 + k_2)$ the effective spring constant of the system. Now, the numerical value of the first weight is $W = mg = 32m$, so

$$32m = k_1\left(\frac{1}{3}\right) \quad \text{and} \quad 32m = k_2\left(\frac{1}{2}\right).$$

From these equations we find $2k_1 = 3k_2$. The given period of the combined system is $2\pi/w = \pi/15$, so $w = 30$. Since the mas of an 8-pound weight is $1/4$ slug, we have from $w^2 = k/m$

$$30^2 = \frac{k}{1/4} = 4k \quad \text{or} \quad k = 225.$$

We now have the system of equations

$$\frac{4k_1 k_2}{k_1 + k_2} = 225$$

$$2k_1 = 3k_2.$$

Solving the second equation for k_1 and substituting in the first equation, we obtain

$$\frac{4(3k_2/2)k_2}{3k_2/2 + k_2} = \frac{12k_2^2}{5k_2} = \frac{12k_2}{5} = 225.$$

Thus, $k_2 = 375/4$ and $k_1 = 1125/8$. Finally, the value of the first weight is

$$W = 32m = \frac{k_1}{3} = \frac{1125/8}{3} = \frac{375}{8} \approx 46.88 \text{ lb.}$$

15. For large values of t the differential equation is approximated by $x'' = 0$. The solution of this equation is the linear function $x = c_1 t + c_2$. Thus, for large time, the restoring force will have decayed to the point where the spring is incapable of returning the mass, and the spring will simply keep on stretching.

16. As t becomes larger the spring constant increases; that is, the spring is stiffening. It would seem that the oscillations would become periodic and the spring would oscillate more rapidly. It is likely that the amplitudes of the oscillations would decrease as t increases.

17. (a) above   (b) heading upward

18. (a) below   (b) from rest

19. (a) below   (b) heading upward

20. (a) above   (b) heading downward

21. From $\frac{1}{8}x'' + x' + 2x = 0$, $x(0) = -1$, and $x'(0) = 8$ we obtain $x = 4te^{-4t} - e^{-4t}$ and $x' = 8e^{-4t} - 16te^{-4t}$. If $x = 0$ then $t = 1/4$ second. If $x' = 0$ then $t = 1/2$ second and the extreme displacement is $x = e^{-2}$ feet.

22. From $\frac{1}{4}x'' + \sqrt{2}\,x' + 2x = 0$, $x(0) = 0$, and $x'(0) = 5$ we obtain $x = 5te^{-2\sqrt{2}t}$ and

$x' = 5e^{-2\sqrt{2}t}\left(1 - 2\sqrt{2}t\right)$. If $x' = 0$ then $t = \sqrt{2}/4$ second and the extreme displacement is $x = 5\sqrt{2}\,e^{-1}/4$ feet.

23. (a) From $x'' + 10x' + 16x = 0$, $x(0) = 1$, and $x'(0) = 0$ we obtain $x = \frac{4}{3}e^{-2t} - \frac{1}{3}e^{-8t}$.

   (b) From $x'' + x' + 16x = 0$, $x(0) = 1$, and $x'(0) = -12$ then $x = -\frac{2}{3}e^{-2t} + \frac{5}{3}e^{-8t}$.

24. (a) $x = \frac{1}{3}e^{-8t}\left(4e^{6t} - 1\right)$ is never zero; the extreme displacement is $x(0) = 1$ meter.

   (b) $x = \frac{1}{3}e^{-8t}\left(5 - 2e^{6t}\right) = 0$ when $t = \frac{1}{6}\ln\frac{5}{2} \approx 0.153$ second; if $x' = \frac{4}{3}e^{-8t}\left(e^{6t} - 10\right) = 0$ then $t = \frac{1}{6}\ln 10 \approx 0.384$ second and the extreme displacement is $x = -0.232$ meter.

25. (a) From $0.1x'' + 0.4x' + 2x = 0$, $x(0) = -1$, and $x'(0) = 0$ we obtain $= e^{-2t}\left[-\cos 4t - \frac{1}{2}\sin 4t\right]$.

   (b) $x = e^{-2t}\frac{\sqrt{5}}{2}\left[-\frac{2}{\sqrt{5}}\cos 4t - \frac{1}{\sqrt{5}}\sin 4t\right] = \frac{\sqrt{5}}{2}e^{-2t}\sin(4t + 4.25)$.

   (c) If $x = 0$ then $4t + 4.25 = 2\pi$, 3π, 4π, ... so that the first time heading upward is $t = 1.294$ seconds.

26. (a) From $\frac{1}{4}x'' + x' + 5x = 0$, $x(0) = 1/2$, and $x'(0) = 1$ we obtain $x = e^{-2t}\left(\frac{1}{2}\cos 4t + \frac{1}{2}\sin 4t\right)$.

   (b) $x = e^{-2t}\frac{1}{\sqrt{2}}\left(\frac{\sqrt{2}}{2}\cos 4t + \frac{\sqrt{2}}{2}\sin 4t\right) = \frac{1}{\sqrt{2}}e^{-2t}\sin\left(4t + \frac{\pi}{4}\right)$.

   (c) If $x = 0$ then $4t + \pi/4 = \pi$, 2π, 3π, ... so that the times heading downward are $t = (7+8n)\pi/16$ for $n = 0, 1, 2, \ldots$.

   (d)

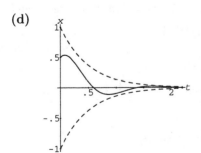

27. From $\frac{5}{16}x'' + \beta x' + 5x = 0$ we find that the roots of the auxiliary equation are $m = -\frac{8}{5}\beta \pm \frac{4}{5}\sqrt{4\beta^2 - 25}$.

   (a) If $4\beta^2 - 25 > 0$ then $\beta > 5/2$.
   (b) If $4\beta^2 - 25 = 0$ then $\beta = 5/2$.
   (c) If $4\beta^2 - 25 < 0$ then $0 < \beta < 5/2$.

28. From $0.75x'' + \beta x' + 6x = 0$ and $\beta > 3\sqrt{2}$ we find that the roots of the auxiliary equation are $m = -\frac{2\beta}{3} \pm \frac{2}{3}\sqrt{\beta^2 - 18}$ and

$$x = e^{-2\beta t/3}\left[c_1 \cosh\frac{2}{3}\sqrt{\beta^2 - 18}\,t + c_2 \sinh\frac{2}{3}\sqrt{\beta^2 - 18}\,t\right].$$

If $x(0) = 0$ and $x'(0) = -2$ then $c_1 = 0$ and $c_2 = -3/\sqrt{\beta^2 - 18}$.

29. If $\frac{1}{2}x'' + \frac{1}{2}x' + 6x = 10\cos 3t$, $x(0) = -2$, and $x'(0) = 0$ then

$$x_c = e^{-t/2}\left(c_1 \cos\frac{\sqrt{47}}{2}t + c_2 \sin\frac{\sqrt{47}}{2}t\right)$$

and $x_p = \frac{10}{3}(\cos 3t + \sin 3t)$ so that the equation of motion is

$$x = e^{-t/2}\left(-\frac{4}{3}\cos\frac{\sqrt{47}}{2}t - \frac{64}{3\sqrt{47}}\sin\frac{\sqrt{47}}{2}t\right) + \frac{10}{3}(\cos 3t + \sin 3t).$$

30. (a) If $x'' + 2x' + 5x = 12\cos 2t + 3\sin 2t$, $x(0) = -1$, and $x'(0) = 5$ then $x_c = e^{-t}(c_1 \cos 2t + c_2 \sin 2t)$ and $x_p = 3\sin 2t$ so that the equation of motion is

$$x = e^{-t}\cos 2t + 3\sin 2t.$$

(b)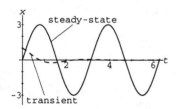

(c)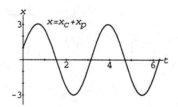

31. From $x'' + 8x' + 16x = 8\sin 4t$, $x(0) = 0$, and $x'(0) = 0$ we obtain $x_c = c_1 e^{-4t} + c_2 t e^{-4t}$ and $x_p = -\frac{1}{4}\cos 4t$ so that the equation of motion is

$$x = \frac{1}{4}e^{-4t} + te^{-4t} - \frac{1}{4}\cos 4t.$$

32. From $x'' + 8x' + 16x = e^{-t}\sin 4t$, $x(0) = 0$, and $x'(0) = 0$ we obtain $x_c = c_1 e^{-4t} + c_2 t e^{-4t}$ and $x_p = -\frac{24}{625}e^{-t}\cos 4t - \frac{7}{625}e^{-t}\sin 4t$ so that

$$x = \frac{1}{625}e^{-4t}(24 + 100t) - \frac{1}{625}e^{-t}(24\cos 4t + 7\sin 4t).$$

As $t \to \infty$ the displacement $x \to 0$.

33. From $2x'' + 32x = 68e^{-2t}\cos 4t$, $x(0) = 0$, and $x'(0) = 0$ we obtain $x_c = c_1 \cos 4t + c_2 \sin 4t$ and $x_p = \frac{1}{2}e^{-2t}\cos 4t - 2e^{-2t}\sin 4t$ so that

$$x = -\frac{1}{2}\cos 4t + \frac{9}{4}\sin 4t + \frac{1}{2}e^{-2t}\cos 4t - 2e^{-2t}\sin 4t.$$

34. Since $x = \frac{\sqrt{85}}{4} \sin(4t - 0.219) - \frac{\sqrt{17}}{2} e^{-2t} \sin(4t - 2.897)$, the amplitude approaches $\sqrt{85}/4$ as $t \to \infty$.

35. (a) By Hooke's law the external force is $F(t) = kh(t)$ so that $mx'' + \beta x' + kx = kh(t)$.

(b) From $\frac{1}{2}x'' + 2x' + 4x = 20 \cos t$, $x(0) = 0$, and $x'(0) = 0$ we obtain $x_c = e^{-2t}(c_1 \cos 2t + c_2 \sin 2t)$ and $x_p = \frac{56}{13} \cos t + \frac{32}{13} \sin t$ so that

$$x = e^{-2t}\left(-\frac{56}{13} \cos 2t - \frac{72}{13} \sin 2t\right) + \frac{56}{13} \cos t + \frac{32}{13} \sin t.$$

36. (a) From $100x'' + 1600x = 1600 \sin 8t$, $x(0) = 0$, and $x'(0) = 0$ we obtain $x_c = c_1 \cos 4t + c_2 \sin 4t$ and $x_p = -\frac{1}{3} \sin 8t$ so that

$$x = \frac{2}{3} \sin 4t - \frac{1}{3} \sin 8t.$$

(b) If $x = \frac{1}{3} \sin 4t(2 - 2 \cos 4t) = 0$ then $t = n\pi/4$ for $n = 0, 1, 2, \ldots$.

(c) If $x' = \frac{8}{3} \cos 4t - \frac{8}{3} \cos 8t = \frac{8}{3}(1 - \cos 4t)(1 + 2 \cos 4t) = 0$ then $t = \pi/3 + n\pi/2$ and $t = \pi/6 + n\pi/2$ for $n = 0, 1, 2, \ldots$ at the extreme values. *Note:* There are many other values of t for which $x' = 0$.

(d) $x(\pi/6 + n\pi/2) = \sqrt{3}/2$ cm. and $x(\pi/3 + n\pi/2) = -\sqrt{3}/2$ cm.

(e)

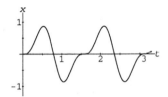

37. From $x'' + 4x = -5 \sin 2t + 3 \cos 2t$, $x(0) = -1$, and $x'(0) = 1$ we obtain $x_c = c_1 \cos 2t + c_2 \sin 2t$, $x_p = \frac{3}{4}t \sin 2t + \frac{5}{4}t \cos 2t$, and

$$x = -\cos 2t - \frac{1}{8} \sin 2t + \frac{3}{4}t \sin 2t + \frac{5}{4}t \cos 2t.$$

38. From $x'' + 9x = 5 \sin 3t$, $x(0) = 2$, and $x'(0) = 0$ we obtain $x_c = c_1 \cos 3t + c_2 \sin 3t$, $x_p = -\frac{5}{6}t \cos 3t$, and

$$x = 2 \cos 3t + \frac{5}{18} \sin 3t - \frac{5}{6}t \cos 3t.$$

39. (a) From $x'' + \omega^2 x = F_0 \cos \gamma t$, $x(0) = 0$, and $x'(0) = 0$ we obtain $x_c = c_1 \cos \omega t + c_2 \sin \omega t$ and $x_p = (F_0 \cos \gamma t)/(\omega^2 - \gamma^2)$ so that

$$x = -\frac{F_0}{\omega^2 - \gamma^2} \cos \omega t + \frac{F_0}{\omega^2 - \gamma^2} \cos \gamma t.$$

(b) $\lim\limits_{\gamma \to \omega} \dfrac{F_0}{\omega^2 - \gamma^2}(\cos \gamma t - \cos \omega t) = \lim\limits_{\gamma \to \omega} \dfrac{-F_0 t \sin \gamma t}{-2\gamma} = \dfrac{F_0}{2\omega}t \sin \omega t.$

40. From $x'' + \omega^2 x = F_0 \cos \omega t$, $x(0) = 0$, and $x'(0) = 0$ we obtain $x_c = c_1 \cos \omega t + c_2 \sin \omega t$ and $x_p = (F_0 t/2\omega)\sin \omega t$ so that $x = (F_0 t/2\omega)\sin \omega t$ and $\lim\limits_{\gamma \to \omega} \dfrac{F_0}{2\omega}t \sin \omega t = \dfrac{F_0}{2\omega}t \sin \omega t.$

41. (a) From $\cos(u - v) = \cos u \cos v + \sin u \sin v$ and $\cos(u + v) = \cos u \cos v - \sin u \sin v$ we obtain $\sin u \sin v = \frac{1}{2}[\cos(u - v) - \cos(u + v)]$. Letting $u = \frac{1}{2}(\gamma - \omega)t$ and $v = \frac{1}{2}(\gamma + \omega)t$, the result follows.

(b) If $\epsilon = \frac{1}{2}(\gamma - \omega)$ then $\gamma \approx \omega$ so that $x = (F_0/2\epsilon\gamma)\sin \epsilon t \sin \gamma t.$

42. See the article "Distinguished Oscillations of a Forced Harmonic Oscillator" by T.G. Procter in *The College Mathematics Journal*, March, 1995. In this article the author illustrates that for $F_0 = 1$, $\lambda = 0.01$, $\gamma = 22/9$, and $\omega = 2$ the system exhibits beats oscillations on the interval $[0, 9\pi]$, but that this phenomenon is transient as $t \to \infty$.

43. (a) The general solution of the homogeneous equation is

$$x_c(t) = c_1 e^{-\lambda t}\cos(\sqrt{\omega^2 - \lambda^2}\,t) + c_2 e^{-\lambda t}\sin(\sqrt{\omega^2 - \lambda^2}\,t)$$

$$= Ae^{-\lambda t}\sin[\sqrt{\omega^2 - \lambda^2}\,t + \phi],$$

where $A = \sqrt{c_1^2 + c_2^2}$, $\sin \phi = c_1/A$, and $\cos \phi = c_2/A$. Now

$$x_p(t) = \frac{F_0(\omega^2 - \gamma^2)}{(\omega^2 - \gamma^2)^2 + 4\lambda^2\gamma^2}\sin \gamma t + \frac{F_0(-2\lambda\gamma)}{(\omega^2 - \gamma^2)^2 + 4\lambda^2\gamma^2}\cos \gamma t$$

$$= A\sin(\gamma t + \theta),$$

where

$$\sin \theta = \frac{\dfrac{F_0(-2\lambda\gamma)}{(\omega^2 - \gamma^2)^2 + 4\lambda^2\gamma^2}}{\dfrac{F_0}{\sqrt{\omega^2 - \gamma^2 + 4\lambda^2\gamma^2}}} = \frac{-2\lambda\gamma}{\sqrt{(\omega^2 - \gamma^2)^2 + 4\lambda^2\gamma^2}}$$

and

$$\cos\theta = \frac{\dfrac{F_0(\omega^2 - \gamma^2)}{(\omega^2 - \gamma^2)^2 + 4\lambda^2\gamma^2}}{\dfrac{F_0}{\sqrt{(\omega^2 - \gamma^2)^2 + 4\lambda^2\gamma^2}}} = \frac{\omega^2 - \gamma^2}{\sqrt{(\omega^2 - \gamma^2)^2 + 4\lambda^2\gamma^2}}.$$

(b) If $g'(\gamma) = 0$ then $\gamma\left(\gamma^2 + 2\lambda^2 - \omega^2\right) = 0$ so that $\gamma = 0$ or $\gamma = \sqrt{\omega^2 - 2\lambda^2}$. The first derivative test shows that g has a maximum value at $\gamma = \sqrt{\omega^2 - 2\lambda^2}$. The maximum value of g is

$$g\left(\sqrt{\omega^2 - 2\lambda^2}\right) = F_0 / 2\lambda\sqrt{\omega^2 - \lambda^2}.$$

(c) We identify $\omega^2 = k/m = 4$, $\lambda = \beta/2$, and $\gamma_1 = \sqrt{\omega^2 - 2\lambda^2} = \sqrt{4 - \beta^2/2}$. As $\beta \to 0$, $\gamma_1 \to 2$ and the resonance curve grows without bound at $\gamma_1 = 2$. That is, the system approaches pure resonance.

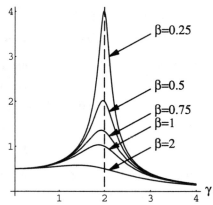

β	$\gamma 1$	g
2.00	1.41	0.58
1.00	1.87	1.03
0.75	1.93	1.36
0.50	1.97	2.02
0.25	1.99	4.01

44. (a) For $n = 2$, $\sin^2\gamma t = \frac{1}{2}(1 - \cos 2\gamma t)$. The system is in pure resonance when $2\gamma_1/2\pi = \omega/2\pi$, or when $\gamma_1 = \omega/2$.

(b) Note that

$$\sin^3 \gamma t = \sin \gamma t \sin^2 \gamma t = \frac{1}{2}[\sin \gamma t - \sin \gamma t \cos 2\gamma t]$$

Now

$$\sin(A + B) + \sin(A - B) = 2\sin A \cos B$$

so

$$\sin \gamma t \cos 2\gamma t = \frac{1}{2}[\sin 3\gamma t - \sin \gamma t]$$

and

$$\sin^3 \gamma t = \frac{3}{4}\sin \gamma t - \frac{1}{4}\sin 3\gamma t.$$

161

Thus

$$x'' + \omega^2 x = \frac{3}{4}\sin\gamma t - \frac{1}{4}\sin 3\gamma t.$$

The frequency of free vibration is $\omega/2\pi$. Thus, when $\gamma_1/2\pi = \omega/2\pi$ or $\gamma_1 = \omega$, and when $3\gamma_2/2\pi = \omega/2\pi$ or $3\gamma_2 = \omega$ or $\gamma_3 = \omega/3$, the system will be in pure resonance.

(c)

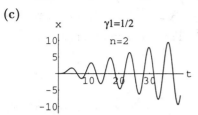 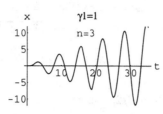 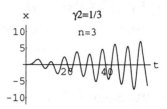

45. Solving $\frac{1}{20}q'' + 2q' + 100q = 0$ we obtain $q(t) = e^{-20t}(c_1\cos 40t + c_2\sin 40t)$. The initial conditions $q(0) = 5$ and $q'(0) = 0$ imply $c_1 = 5$ and $c_2 = 5/2$. Thus

$$q(t) = e^{-20t}\left(5\cos 40t + \frac{5}{2}\sin 40t\right) \approx \sqrt{25 + 25/4}\, e^{-20t}\sin(40t + 1.1071)$$

and $q(0.01) \approx 4.5676$ coulombs. The charge is zero for the first time when $40t + 0.4636 = \pi$ or $t \approx 0.0509$ second.

46. Solving $\frac{1}{4}q'' + 20q' + 300q = 0$ we obtain $q(t) = c_1 e^{-20t} + c_2 e^{-60t}$. The initial conditions $q(0) = 4$ and $q'(0) = 0$ imply $c_1 = 6$ and $c_2 = -2$. Thus

$$q(t) = 6e^{-20t} - 2e^{-60t}.$$

Setting $q = 0$ we find $e^{40t} = 1/3$ which implies $t < 0$. Therefore the charge is never 0.

47. Solving $\frac{5}{3}q'' + 10q' + 30q = 300$ we obtain $q(t) = e^{-3t}(c_1\cos 3t + c_2\sin 3t) + 10$. The initial conditions $q(0) = q'(0) = 0$ imply $c_1 = c_2 = -10$. Thus

$$q(t) = 10 - 10e^{-3t}(\cos 3t + \sin 3t) \quad \text{and} \quad i(t) = 60e^{3t}\sin 3t.$$

Solving $i(t) = 0$ we see that the maximum charge occurs when $t = \pi/3$ and $q(\pi/3) \approx 10.432$ coulombs.

48. Solving $q'' + 100q' + 2500q = 30$ we obtain $q(t) = c_1 e^{-50t} + c_2 t e^{-50t} + 0.012$. The initial conditions $q(0) = 0$ and $q'(0) = 2$ imply $c_1 = -0.012$ and $c_2 = 1.4$. Thus

$$q(t) = -0.012e^{-50t} + 1.4te^{-50t} + 0.012 \quad \text{and} \quad i(t) = 2e^{-50t} - 70te^{-50t}.$$

Solving $i(t) = 0$ we see that the maximum charge occurs when $t = 1/35$ and $q(1/35) \approx 0.01871$.

49. Solving $q'' + 2q' + 4q = 0$ we obtain $y_c = e^{-t}\left(\cos\sqrt{3}\,t + \sin\sqrt{3}\,t\right)$. The steady-state charge has the form $y_p = A\cos t + B\sin t$. Substituting into the differential equation we find

$$(3A + 2B)\cos t + (3B - 2A)\sin t = 50\cos t.$$

162

Thus, $A = 150/13$ and $B = 100/13$. The steady-state charge is

$$q_p(t) = \frac{150}{13}\cos t + \frac{100}{13}\sin t$$

and the steady-state current is

$$i_p(t) = -\frac{150}{13}\sin t + \frac{100}{13}\cos t.$$

50. From

$$i_p(t) = \frac{E_0}{Z}\left(\frac{R}{Z}\sin\gamma t - \frac{X}{Z}\cos\gamma t\right)$$

and $Z = \sqrt{X^2 + R^2}$ we see that the amplitude of $i_p(t)$ is

$$A = \sqrt{\frac{E_0^2 R^2}{Z^4} + \frac{E_0^2 X^2}{Z^4}} = \frac{E_0}{Z^2}\sqrt{R^2 + X^2} = \frac{E_0}{Z}.$$

51. The differential equation is $\frac{1}{2}q'' + 20q' + 1000q = 100\sin t$. To use Example 11 in the text we identify $E_0 = 100$ and $\gamma = 60$. Then

$$X = L\gamma - \frac{1}{c\gamma} = \frac{1}{2}(60) - \frac{1}{0.001(60)} \approx 13.3333,$$

$$Z = \sqrt{X^2 + R^2} = \sqrt{X^2 + 400} \approx 24.0370,$$

and

$$\frac{E_0}{Z} = \frac{100}{Z} \approx 4.1603.$$

From Problem 50, then

$$i_p(t) \approx 4.1603(60t + \phi)$$

where $\sin\phi = -X/Z$ and $\cos\phi = R/Z$. Thus $\tan\phi = -X/R \approx -0.6667$ and ϕ is a fourth quadrant angle. Now $\phi \approx -0.5880$ and

$$i_p(t) \approx 4.1603(60t - 0.5880).$$

52. Solving $\frac{1}{2}q'' + 20q' + 1000q = 0$ we obtain $q_c(t) = (c_1\cos 40t + c_2\sin 40t)$. The steady-state charge has the form $q_p(t) = A\sin 60t + B\cos 60t + C\sin 40t + D\cos 40t$. Substituting into the differential equation we find

$$(-1600A - 2400B)\sin 60t + (2400A - 1600B)\cos 60t$$

$$+ (400C - 1600D)\sin 40t + (1600C + 400D)\cos 40t$$

$$= 200\sin 60t + 400\cos 40t.$$

163

Equating coefficients we obtain $A = -1/26$, $B = -3/52$, $C = 4/17$, and $D = 1/17$. The steady-state charge is

$$q_p(t) = -\frac{1}{26}\sin 60t - \frac{3}{52}\cos 60t + \frac{4}{17}\sin 40t + \frac{1}{17}\cos 40t$$

and the steady-state current is

$$i_p(t) = -\frac{30}{13}\cos 60t + \frac{45}{13}\sin 60t + \frac{160}{17}\cos 40t - \frac{40}{17}\sin 40t.$$

53. Solving $\frac{1}{2}q'' + 10q' + 100q = 150$ we obtain $q(t) = e^{-10t}(c_1\cos 10t + c_2\sin 10t) + 3/2$. The initial conditions $q(0) = 1$ and $q'(0) = 0$ imply $c_1 = c_2 = -1/2$. Thus

$$q(t) = -\frac{1}{2}e^{-10t}(\cos 10t + \sin 10t) + \frac{3}{2}.$$

As $t \to \infty$, $q(t) \to 3/2$.

54. By Problem 50 the amplitude of the steady-state current is E_0/Z, where $Z = \sqrt{X^2 + R^2}$ and $X = L\gamma - 1/C\gamma$. Since E_0 is constant the amplitude will be a maximum when Z is a minimum. Since R is constant, Z will be a minimum when $X = 0$. Solving $L\gamma - 1/C\gamma = 0$ for γ we obtain $\gamma = 1/\sqrt{LC}$. The maximum amplitude will be E_0/R.

55. By Problem 50 the amplitude of the steady-state current is E_0/Z, where $Z = \sqrt{X^2 + R^2}$ and $X = L\gamma - 1/C\gamma$. Since E_0 is constant the amplitude will be a maximum when Z is a minimum. Since R is constant, Z will be a minimum when $X = 0$. Solving $L\gamma - 1/C\gamma = 0$ for C we obtain $C = 1/L\gamma^2$.

56. Solving $0.1q'' + 10q = 100\sin\gamma t$ we obtain $q(t) = c_1\cos 10t + c_2\sin 10t + q_p(t)$ where $q_p(t) = A\sin\gamma t + B\cos\gamma t$. Substituting $q_p(t)$ into the differential equation we find

$$(100 - \gamma^2)A\sin\gamma t + (100 - \gamma^2)B\cos\gamma t = 100\sin\gamma t.$$

Equating coefficients we obtain $A = 100/(100 - \gamma^2)$ and $B = 0$. Thus, $q_p(t) = \dfrac{100}{100 - \gamma^2}\sin\gamma t$. The initial conditions $q(0) = q'(0) = 0$ imply $c_1 = 0$ and $c_2 = -10\gamma/(100 - \gamma^2)$. The charge is

$$q(t) = \frac{10}{100 - \gamma^2}(10\sin\gamma t - \gamma\sin 10t)$$

and the current is

$$i(t) = \frac{100\gamma}{100 - \gamma^2}(\cos\gamma t - \cos 10t).$$

57. In an L-C series circuit there is no resistor, so the differential equation is

$$L\frac{d^2q}{dt^2} + \frac{1}{C}q = E(t).$$

164

Then $q(t) = c_1 \cos\left(t/\sqrt{LC}\right) + c_2 \sin\left(t/\sqrt{LC}\right) + q_p(t)$ where $q_p(t) = A \sin \gamma t + B \cos \gamma t$. Substituting $q_p(t)$ into the differential equation we find

$$\left(\frac{1}{C} - L\gamma^2\right) A \sin \gamma t + \left(\frac{1}{C} - L\gamma^2\right) B \cos \gamma t = E_0 \cos \gamma t.$$

Equating coefficients we obtain $A = 0$ and $B = E_0 C/(1 - LC\gamma^2)$. Thus, the charge is

$$q(t) = c_1 \cos \frac{1}{\sqrt{LC}} t + c_2 \sin \frac{1}{\sqrt{LC}} t + \frac{E_0 C}{1 - LC\gamma^2} \cos \gamma t.$$

The initial conditions $q(0) = q_0$ and $q'(0) = i_0$ imply $c_1 = q_o - E_0 C/(1 - LC\gamma^2)$ and $c_2 = i_0 \sqrt{LC}$. The current is

$$i(t) = -\frac{c_1}{\sqrt{LC}} \sin \frac{1}{\sqrt{LC}} t + \frac{c_2}{\sqrt{LC}} \cos \frac{1}{\sqrt{LC}} t - \frac{E_0 C\gamma}{1 - LC\gamma^2} \sin \gamma t$$

$$= i_0 \cos \frac{1}{\sqrt{LC}} t - \frac{1}{\sqrt{LC}} \left(q_0 - \frac{E_0 C}{1 - LC\gamma^2}\right) \sin \frac{1}{\sqrt{LC}} t - \frac{E_0 C\gamma}{1 - LC\gamma^2} \sin \gamma t.$$

58. When the circuit is in resonance the form of $q_p(t)$ is $q_p(t) = At \cos kt + Bt \sin kt$ where $k = 1/\sqrt{LC}$. Substituting $q_p(t)$ into the differential equation we find

$$q_p'' + k^2 q = -2kA \sin kt + 2kB \cos kt = \frac{E_0}{L} \cos kt.$$

Equating coefficients we obtain $A = 0$ and $B = E_0/2kL$. The charge is

$$q(t) = c_1 \cos kt + c_2 \sin kt + \frac{E_0}{2kL} t \sin kt.$$

The initial conditions $q(0) = q_0$ and $q'(0) = i_0$ imply $c_1 = q_0$ and $c_2 = i_0/k$. The current is

$$i(t) = -c_1 k \sin kt + c_2 k \cos kt + \frac{E_0}{2kL}(kt \cos kt + \sin kt)$$

$$= \left(\frac{E_0}{2kL} - q_0 k\right) \sin kt + i_0 \cos kt + \frac{E_0}{2L} t \cos kt.$$

1. (a) The general solution is

$$y(x) = c_1 + c_2 x + c_3 x^2 + c_4 x^3 + \frac{w_0}{24EI} x^4.$$

The boundary conditions are $y(0) = 0$, $y'(0) = 0$, $y''(L) = 0$, $y'''(L) = 0$. The first two conditions give $c_1 = 0$ and $c_2 = 0$. The conditions at $x = L$ give the system

$$2c_3 + 6c_4 L + \frac{w_0}{2EI} L^2 = 0$$

$$6c_4 + \frac{w_0}{EI} L = 0.$$

Solving, we obtain $c_3 = w_0 L^2/4EI$ and $c_4 = -w_0 L/6EI$. The deflection is

$$y(x) = \frac{w_0}{24EI}(6L^2 x^2 - 4Lx^3 + x^4).$$

(b)

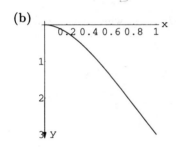

2. (a) The general solution is

$$y(x) = c_1 + c_2 x + c_3 x^2 + c_4 x^3 + \frac{w_0}{24EI} x^4.$$

The boundary conditions are $y(0) = 0$, $y''(0) = 0$, $y(L) = 0$, $y''(L) = 0$. The first two conditions give $c_1 = 0$ and $c_3 = 0$. The conditions at $x = L$ give the system

$$c_2 L + c_4 L^3 + \frac{w_0}{24EI} L^4 = 0$$

$$6c_4 L + \frac{w_0}{2EI} L^2 = 0.$$

Solving, we obtain $c_2 = w_0 L^3/24EI$ and $c_4 = -w_0 L/12EI$. The deflection is

$$y(x) = \frac{w_0}{24EI}(L^3 x - 2Lx^3 + x^4).$$

(b)

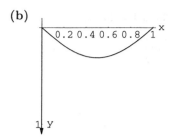

3. (a) The general solution is

$$y(x) = c_1 + c_2x + c_3x^2 + c_4x^3 + \frac{w_0}{24EI}x^4.$$

The boundary conditions are $y(0) = 0$, $y'(0) = 0$, $y(L) = 0$, $y''(L) = 0$. The first two conditions give $c_1 = 0$ and $c_2 = 0$. The conditions at $x = L$ give the system

$$c_3L^2 + c_4L^3 + \frac{w_0}{24EI}L^4 = 0$$

$$2c_3 + 6c_4L + \frac{w_0}{2EI}L^2 = 0.$$

Solving, we obtain $c_3 = w_0L^2/16EI$ and $c_4 = -5w_0L/48EI$. The deflection is

$$y(x) = \frac{w_0}{48EI}(3L^2x^2 - 5Lx^3 + 2x^4).$$

(b)

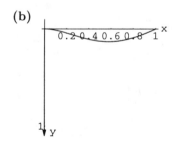

4. (a) The general solution is

$$y(x) = c_1 + c_2x + c_3x^2 + c_4x^3 + \frac{w_0L^4}{EI\pi^4}\sin\frac{\pi}{L}x.$$

The boundary conditions are $y(0) = 0$, $y'(0) = 0$, $y(L) = 0$, $y''(L) = 0$. The first two conditions give $c_1 = 0$ and $c_2 = -w_0L^3/EI\pi^3$. The conditions at $x = L$ give the system

$$c_3L^2 + c_4L^3 + \frac{w_0}{EI\pi^3}L^4 = 0$$

$$2c_3 + 6c_4L = 0.$$

Solving, we obtain $c_3 = 3w_0 L^2/2EI\pi^3$ and $c_4 = -w_0 L/2EI\pi^3$. The deflection is

$$y(x) = \frac{w_0 L}{2EI\pi^3}\left(-2L^2 x + 3Lx^2 - x^3 + \frac{2L^3}{\pi}\sin\frac{\pi}{L}x\right).$$

(b)

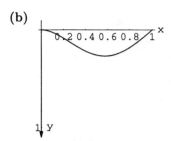

5. (a) $y_{max} = y(L) = \dfrac{w_0 L^4}{8EI}$

(b) Replacing both L and x by $L/2$ in $y(x)$ we obtain $w_0 L^4/128EI$, which is $1/16$ of the maximum deflection when the length of the beam is L.

6. (a) $y_{max} = y(L/2) = \dfrac{5w_0 L^4}{384EI}$

(b) The maximum deflection of the beam in Example 1 is $y(L/2) = (w_0/24EI)L^4/16 = w_0 L^4/384EI$, which is $1/5$ of the maximum displacement of the beam in Problem 2.

7. The general solution of the differential equation is

$$y = c_1 \cosh\sqrt{\frac{P}{EI}}\,x + c_2 \sinh\sqrt{\frac{P}{EI}}\,x + \frac{w_0}{2P}x^2 + \frac{w_0 EI}{P^2}.$$

Setting $y(0) = 0$ we obtain $c_1 = -w_0 EI/P^2$, so that

$$y = -\frac{w_0 EI}{P^2}\cosh\sqrt{\frac{P}{EI}}\,x + c_2 \sinh\sqrt{\frac{P}{EI}}\,x + \frac{w_0}{2P}x^2 + \frac{w_0 EI}{P^2}.$$

Setting $y'(L) = 0$ we find

$$c_2 = \left(\sqrt{\frac{P}{EI}}\,\frac{w_0 EI}{P^2}\sinh\sqrt{\frac{P}{EI}}\,L - \frac{w_0 L}{P}\right)\Big/\sqrt{\frac{P}{EI}}\cosh\sqrt{\frac{P}{EI}}\,L.$$

8. The general solution of the differential equation is

$$y = c_1 \cos\sqrt{\frac{P}{EI}}\,x + c_2 \sin\sqrt{\frac{P}{EI}}\,x + \frac{w_0}{2P}x^2 + \frac{w_0 EI}{P^2}.$$

Setting $y(0) = 0$ we obtain $c_1 = -w_0 EI/P^2$, so that

$$y = -\frac{w_0 EI}{P^2}\cos\sqrt{\frac{P}{EI}}\,x + c_2 \sin\sqrt{\frac{P}{EI}}\,x + \frac{w_0}{2P}x^2 + \frac{w_0 EI}{P^2}.$$

Setting $y'(L) = 0$ we find

$$c_2 = \left(-\sqrt{\frac{P}{EI}}\,\frac{w_0 EI}{P^2}\sin\sqrt{\frac{P}{EI}}\,L - \frac{w_0 L}{P}\right) \Big/ \sqrt{\frac{P}{EI}}\cos\sqrt{\frac{P}{EI}}\,L.$$

9. For $\lambda \le 0$ the only solution of the boundary-value problem is $y = 0$. For $\lambda > 0$ we have

$$y = c_1\cos\sqrt{\lambda}\,x + c_2\sin\sqrt{\lambda}\,x.$$

Now $y(0) = 0$ implies $c_1 = 0$, so

$$y(\pi) = c_2\sin\sqrt{\lambda}\,\pi = 0$$

gives

$$\sqrt{\lambda}\,\pi = n\pi \quad\text{or}\quad \lambda = n^2,\ n = 1,\ 2,\ 3,\dots.$$

The eigenvalues n^2 correspond to the eigenfunctions $\sin nx$ for $n = 1,\ 2,\ 3,\ \dots$.

10. For $\lambda \le 0$ the only solution of the boundary-value problem is $y = 0$. For $\lambda > 0$ we have

$$y = c_1\cos\sqrt{\lambda}\,x + c_2\sin\sqrt{\lambda}\,x.$$

Now $y(0) = 0$ implies $c_1 = 0$, so

$$y\left(\frac{\pi}{4}\right) = c_2\sin\sqrt{\lambda}\,\frac{\pi}{4} = 0$$

gives

$$\sqrt{\lambda}\,\frac{\pi}{4} = n\pi \quad\text{or}\quad \lambda = 16n^2,\ n = 1,\ 2,\ 3,\dots.$$

The eigenvalues $16n^2$ correspond to the eigenfunctions $\sin 4nx$ for $n = 1,\ 2,\ 3,\ \dots$.

11. For $\lambda \le 0$ the only solution of the boundary-value problem is $y = 0$. For $\lambda > 0$ we have

$$y = c_1\cos\sqrt{\lambda}\,x + c_2\sin\sqrt{\lambda}\,x.$$

Now

$$y'(x) = -c_1\sqrt{\lambda}\,\sin\sqrt{\lambda}\,x + c_2\sqrt{\lambda}\cos\sqrt{\lambda}\,x$$

and $y'(0) = 0$ implies $c_2 = 0$, so

$$y(L) = c_1\cos\sqrt{\lambda}\,L = 0$$

gives

$$\sqrt{\lambda}\,L = \frac{(2n-1)\pi}{2} \quad\text{or}\quad \lambda = \frac{(2n-1)^2\pi^2}{4L^2},\ n = 1,\ 2,\ 3,\dots.$$

The eigenvalues $(2n-1)^2\pi^2/4L^2$ correspond to the eigenfunctions $\cos\dfrac{(2n-1)\pi}{2L}x$ for $n = 1,\ 2,\ 3,\ \dots$.

12. For $\lambda \le 0$ the only solution of the boundary-value problem is $y = 0$. For $\lambda > 0$ we have

$$y = c_1\cos\sqrt{\lambda}\,x + c_2\sin\sqrt{\lambda}\,x.$$

Now $y(0) = 0$ implies $c_1 = 0$, so

$$y'\left(\frac{\pi}{2}\right) = c_2\sqrt{\lambda}\cos\sqrt{\lambda}\,\frac{\pi}{2} = 0$$

gives

$$\sqrt{\lambda}\,\frac{\pi}{2} = \frac{(2n-1)\pi}{2} \quad \text{or} \quad \lambda = (2n-1)^2, \ n = 1, 2, 3, \dots.$$

The eigenvalues $(2n-1)^2$ correspond to the eigenfunctions $\sin(2n-1)x$.

13. For $\lambda < 0$ the only solution of the boundary-value problem is $y = 0$. For $\lambda = 0$ we have $y = c_1 x + c_2$. Now $y' = c_1$ and $y'(0) = 0$ implies $c_1 = 0$. Then $y = c_2$ and $y'(\pi) = 0$. Thus, $\lambda = 0$ is an eigenvalue with corresponding eigenfunction $y = 1$.

For $\lambda > 0$ we have

$$y = c_1 \cos\sqrt{\lambda}\,x + c_2 \sin\sqrt{\lambda}\,x.$$

Now

$$y'(x) = -c_1\sqrt{\lambda}\sin\sqrt{\lambda}\,x + c_2\sqrt{\lambda}\cos\sqrt{\lambda}\,x$$

and $y'(0) = 0$ implies $c_2 = 0$, so

$$y'(\pi) = -c_1\sqrt{\lambda}\sin\sqrt{\lambda}\,\pi = 0$$

gives

$$\sqrt{\lambda}\,\pi = n\pi \quad \text{or} \quad \lambda = n^2, \ n = 1, 2, 3, \dots.$$

The eigenvalues n^2 correspond to the eigenfunctions $\cos nx$ for $n = 0, 1, 2, \dots$.

14. For $\lambda \leq 0$ the only solution of the boundary-value problem is $y = 0$. For $\lambda > 0$ we have

$$y = c_1 \cos\sqrt{\lambda}\,x + c_2 \sin\sqrt{\lambda}\,x.$$

Now $y(-\pi) = y(\pi) = 0$ implies

$$c_1 \cos\sqrt{\lambda}\,\pi - c_2 \sin\sqrt{\lambda}\,\pi = 0$$

$$c_1 \cos\sqrt{\lambda}\,\pi + c_2 \sin\sqrt{\lambda}\,\pi = 0.$$

(1)

This homogeneous system will have a nontrivial solution when

$$\begin{vmatrix} \cos\sqrt{\lambda}\,\pi & -\sin\sqrt{\lambda}\,\pi \\ \cos\sqrt{\lambda}\,\pi & \sin\sqrt{\lambda}\,\pi \end{vmatrix} = 2\sin\sqrt{\lambda}\,\pi\cos\sqrt{\lambda}\,\pi = \sin 2\sqrt{\lambda}\,\pi = 0.$$

Then

$$2\sqrt{\lambda}\,\pi = n\pi \quad \text{or} \quad \lambda = \frac{n^2}{4}; \quad n = 1, 2, 3, \dots.$$

When $n = 2k - 1$ is odd, the eigenvalues are $(2k-1)^2/4$. Since $\cos(2k-1)\pi/2 = 0$ and $\sin(2k-1)\pi/2 \neq 0$, we see from either equation in (1) that $c_2 = 0$. Thus, the eigenfunctions corresponding to the eigenvalues $(2k-1)^2/4$ are $y = \cos(2k-1)x/2$ for $k = 1, 2, 3, \dots$. Similarly,

170

when $n = 2k$ is even, the eigenvalues are k^2 with corresponding eigenfunctions $y = \sin kx$ for $k = 1, 2, 3, \ldots$.

15. The auxiliary equation has solutions

$$m = \frac{1}{2}\left(-2 \pm \sqrt{4 - 4(\lambda + 1)}\right) = -1 \pm \sqrt{-\lambda}.$$

For $\lambda < 0$ we have

$$y = e^{-x}\left(c_1 \cosh \sqrt{-\lambda}\, x + c_2 \sinh \sqrt{-\lambda}\, x\right).$$

The boundary conditions imply

$$y(0) = c_1 = 0$$

$$y(5) = c_2 e^{-5} \sinh 5\sqrt{-\lambda} = 0$$

so $c_1 = c_2 = 0$ and the only solution of the boundary-value problem is $y = 0$.

For $\lambda = 0$ we have

$$y = c_1 e^{-x} + c_2 x e^{-x}$$

and the only solution of the boundary-value problem is $y = 0$.

For $\lambda > 0$ we have

$$y = e^{-x}\left(c_1 \cos \sqrt{\lambda}\, x + c_2 \sin \sqrt{\lambda}\, x\right).$$

Now $y(0) = 0$ implies $c_1 = 0$, so

$$y(5) = c_2 e^{-5} \sin 5\sqrt{\lambda} = 0$$

gives

$$5\sqrt{\lambda} = n\pi \quad \text{or} \quad \lambda = \frac{n^2 \pi^2}{25}, \quad n = 1, 2, 3, \ldots .$$

The eigenvalues $n^2 \pi^2 / 25$ correspond to the eigenfunctions $e^{-x} \sin \dfrac{n\pi}{5} x$ for $n = 1, 2, 3, \ldots$.

16. For $\lambda < -1$ the only solution of the boundary-value problem is $y = 0$. For $\lambda = -1$ we have $y = c_1 x + c_2$. Now $y' = c_1$ and $y'(0) = 0$ implies $c_1 = 0$. Then $y = c_2$ and $y'(1) = 0$. Thus, $\lambda = -1$ is an eigenvalue with corresponding eigenfunction $y = 1$.

For $\lambda > -1$ we have

$$y = c_1 \cos \sqrt{\lambda + 1}\, x + c_2 \sin \sqrt{\lambda + 1}\, x.$$

Now

$$y' = -c_1 \sqrt{\lambda + 1} \sin \sqrt{\lambda + 1}\, x + c_2 \sqrt{\lambda + 1} \cos \sqrt{\lambda + 1}\, x$$

and $y'(0) = 0$ implies $c_2 = 0$, so

$$y'(1) = -c_1 \sqrt{\lambda + 1} \sin \sqrt{\lambda + 1} = 0$$

gives

$$\sqrt{\lambda + 1} = n\pi \quad \text{or} \quad \lambda = n^2\pi^2 - 1, \ n = 1, 2, 3, \ldots.$$

The eigenvalues $n^2\pi^2 - 1$ correspond to the eigenfunctions $\cos n\pi x$ for $n = 0, 1, 2, \ldots$.

17. For $\lambda = 0$ the only solution of the boundary-value problem is $y = 0$. For $\lambda \neq 0$ we have

$$y = c_1 \cos \lambda x + c_2 \sin \lambda x.$$

Now $y(0) = 0$ implies $c_1 = 0$, so

$$y(L) = c_2 \sin \lambda L = 0$$

gives

$$\lambda L = n\pi \quad \text{or} \quad \lambda = \frac{n\pi}{L}, \ n = 1, 2, 3, \ldots.$$

The eigenvalues $n\pi/L$ correspond to the eigenfunctions $\sin \frac{n\pi}{L} x$ for $n = 1, 2, 3, \ldots$.

18. For $\lambda = 0$ the only solution of the boundary-value problem is $y = 0$. For $\lambda \neq 0$ we have

$$y = c_1 \cos \lambda x + c_2 \sin \lambda x.$$

Now $y(0) = 0$ implies $c_1 = 0$, so

$$y'(3\pi) = c_2 \lambda \cos 3\pi\lambda = 0$$

gives

$$3\pi\lambda = \frac{(2n-1)\pi}{2} \quad \text{or} \quad \lambda = \frac{2n-1}{6}, \ n = 1, 2, 3, \ldots.$$

The eigenvalues $(2n-1)/6$ correspond to the eigenfunctions $\sin \frac{2n-1}{6} x$ for $n = 1, 2, 3, \ldots$.

19. For $\lambda > 0$ a general solution of the given differential equation is

$$y = c_1 \cos(\sqrt{\lambda} \ln x) + c_2 \sin(\sqrt{\lambda} \ln x).$$

Since $\ln 1 = 0$, the boundary condition $y(1) = 0$ implies $c_1 = 0$. Therefore

$$y = c_2 \sin(\sqrt{\lambda} \ln x).$$

Using $\ln e^\pi = \pi$ we find that $y(e^\pi) = 0$ implies

$$c_2 \sin \sqrt{\lambda} \pi = 0$$

or $\sqrt{\lambda} \pi = n\pi$, $n = 1, 2, 3, \ldots$. The eigenvalues and eigenfunctions are, in turn,

$$\lambda = n^2, \quad n = 1, 2, 3, \ldots \quad \text{and} \quad y = \sin(n \ln x).$$

For $\lambda \leq 0$ the only solution of the boundary-value problem is $y = 0$.

To obtain the self-adjoint form we note that the integrating factor is $(1/x^2)e^{\int dx/x} = 1/x$. That is, the self-adjoint form is

$$\frac{d}{dx}[xy'] + \frac{\lambda}{x} y = 0.$$

Identifying the weight function $p(x) = 1/x$ we can then write the orthogonality relation

$$\int_1^{e^\pi} \frac{1}{x} \sin(n \ln x) \sin(m \ln x) \, dx = 0, \quad m \neq n.$$

20. For $\lambda = 0$ the general solution is $y = c_1 + c_2 \ln x$. Now $y' = c_2/x$, so $y'(e^{-1}) = c_2 e = 0$ implies $c_2 = 0$. Then $y = c_1$ and $y(1) = 0$ gives $c_1 = 0$. Thus $y(x) = 0$.

For $\lambda < 0$, $y = c_1 x^{-\sqrt{-\lambda}} + c_2 x^{\sqrt{-\lambda}}$. The initial conditions give $c_2 = c_1 e^{2\sqrt{-\lambda}}$ and $c_1 = 0$, so that $c_2 = 0$ and $y(x) = 0$.

For $\lambda > 0$, $y = c_1 \cos(\sqrt{\lambda} \ln x) + c_2 \sin(\sqrt{\lambda} \ln x)$. From $y(1) = 0$ we obtain $c_1 = 0$ and $y = c_2 \sin(\sqrt{\lambda} \ln x)$. Now $y' = c_2(\sqrt{\lambda}/x) \cos(\sqrt{\lambda} \ln x)$, so $y'(e^{-1}) = c_2 e \sqrt{\lambda} \cos \sqrt{\lambda} = 0$ implies $\cos \sqrt{\lambda} = 0$ or $\lambda = (2n-1)^2 \pi^2/4$ for $n = 1, 2, 3, \ldots$. The corresponding eigenfunctions are

$$y = \sin\left(\frac{2n-1}{2}\pi \ln x\right).$$

21. For $\lambda = 0$ the general solution is $y = c_1 + c_2 \ln x$. Now $y' = c_2/x$, so $y'(1) = c_2 = 0$ and $y = c_1$. Since $y'(e^2) = 0$ for any c_1 we see that $y(x) = 1$ is an eigenfunction corresponding to the eigenvalue $\lambda = 0$.

For $\lambda < 0$, $y = c_1 x^{-\sqrt{-\lambda}} + c_2 x^{\sqrt{-\lambda}}$. The initial conditions imply $c_1 = c_2 = 0$, so $y(x) = 0$.

For $\lambda > 0$, $y = c_1 \cos(\sqrt{\lambda} \ln x) + c_2 \sin(\sqrt{\lambda} \ln x)$. Now

$$y' = -c_1 \frac{\sqrt{\lambda}}{x} \sin(\sqrt{\lambda} \ln x) + c_2 \frac{\sqrt{\lambda}}{x} \cos(\sqrt{\lambda} \ln x),$$

and $y'(1) = c_2 \sqrt{\lambda} = 0$ implies $c_2 = 0$. Finally, $y'(e^2) = -(c_1 \sqrt{\lambda}/e^2) \sin(2\sqrt{\lambda}) = 0$ implies $\lambda = n^2 \pi^2/4$ for $n = 1, 2, 3, \ldots$. The corresponding eigenfunctions are

$$y = \cos\left(\frac{n\pi}{2} \ln x\right).$$

22. For $\lambda > 1/4$ a general solution of the given differential equation is

$$y = c_1 x^{-1/2} \cos\left(\frac{\sqrt{4\lambda - 1}}{2} \ln x\right) + c_2 x^{-1/2} \sin\left(\frac{\sqrt{4\lambda - 1}}{2} \ln x\right).$$

Since $\ln 1 = 0$, the boundary condition $y(1) = 0$ implies $c_1 = 0$. Therefore

$$y = c_2 x^{-1/2} \sin\left(\frac{\sqrt{4\lambda - 1}}{2} \ln x\right).$$

Using $\ln e^2 = 2$ we find that $y(e^2) = 0$ implies

$$c_2 e^{-1} \sin\left(\sqrt{4\lambda - 1}\right) = 0$$

or $\sqrt{4\lambda - 1} = n\pi$, $n = 1, 2, 3, \ldots$. The eigenvalues and eigenfunctions are, in turn,

$$\lambda = \left(n^2 \pi^2 + 1\right)/4, \quad n = 1, 2, 3, \ldots \quad \text{and} \quad y = x^{-1/2} \sin\left(\frac{n\pi}{2} \ln x\right).$$

For $\lambda < 0$ the only solution of the boundary-value problem is $y = 0$.

For $\lambda = 1/4$ a general solution of the differential equation is

$$y = c_1 x^{-1/2} + c_2 x^{-1/2} \ln x.$$

From $y(1) = 0$ we obtain $c_1 = 0$, so $y = c_2 x^{-1/2} \ln x$. From $y(e^2) = 0$ we obtain $2c_2 e^{-1} = 0$ or $c_2 = 0$. Thus, there are no eigenvalues and eigenfunctions in this case.

To obtain the self-adjoint form we note that the integrating factor is $(1/x^2) e^{\int (2/x)\, dx} = (1/x^2) \cdot x^2 = 1$. That is, the self-adjoint form is

$$\frac{d}{dx}\left[x^2 y'\right] + \lambda y = 0.$$

Identifying the weight function $p(x) = 1$ we can then write the orthogonality relation

$$\int_1^{e^2} 1 \cdot x^{-1/2} \sin\left(\frac{m\pi}{2} \ln x\right) x^{-1/2} \sin\left(\frac{n\pi}{2} \ln x\right) dx = 0, \quad m \neq n,$$

or

$$\int_1^{e^2} x^{-1} \sin\left(\frac{m\pi}{2} \ln x\right) \sin\left(\frac{n\pi}{2} \ln x\right) dx = 0, \quad m \neq n.$$

23. For $\lambda > 0$ the general solution is $y = c_1 \cos\sqrt{\lambda}\, x + c_2 \sin\sqrt{\lambda}\, x$. Setting $y(0) = 0$ we find $c_1 = 0$, so that $y = c_2 \sin\sqrt{\lambda}\, x$. The boundary condition $y(1) + y'(1) = 0$ implies

$$c_2 \sin\sqrt{\lambda} + c_2\sqrt{\lambda} \cos\sqrt{\lambda} = 0.$$

Taking $c_2 \neq 0$, this equation is equivalent to $\tan\sqrt{\lambda} = -\sqrt{\lambda}$. Thus, the eigenvalues are $\lambda_n = x_n^2$, $n = 1, 2, 3, \ldots$, where the x_n are the consecutive positive roots of $\tan\sqrt{\lambda} = -\sqrt{\lambda}$.

24. (a) Since $\lambda_n = x_n^2$, there are no new eigenvalues when $x_n < 0$. For $\lambda = 0$, the differential equation $y'' = 0$ has general solution $y = c_1 x + c_2$. The boundary conditions imply $c_1 = c_2 = 0$, so $y = 0$.

(b) $\lambda_1 = 4.1159$, $\lambda_2 = 24.1393$, $\lambda_3 = 63.6591$, $\lambda_4 = 122.8892$.

25. The general solution is

$$y = c_1 \cos\sqrt{\frac{\rho\omega^2}{T}}\, x + c_2 \sin\sqrt{\frac{\rho\omega^2}{T}}\, x.$$

From $y(0) = 0$ we obtain $c_1 = 0$. Setting $y(L) = 0$ we find $\sqrt{\rho\omega^2/T}\, L = n\pi$, $n = 1, 2, 3, \ldots$. Thus, critical speeds are $\omega_n = n\pi\sqrt{T}/L\sqrt{\rho}$, $n = 1, 2, 3, \ldots$. The corresponding deflection curves are

$$y(x) = c_2 \sin\frac{n\pi}{L}\, x, \quad n = 1, 2, 3, \ldots,$$

where $c_2 \neq 0$.

26. (a) When $T(x) = x^2$ the given differential equation is the Cauchy-Euler equation

$$x^2 y'' + 2x y' + \rho\omega^2 y = 0.$$

The solutions of the auxiliary equation

$$m(m-1) + 2m + \rho\omega^2 = m^2 + m + \rho\omega^2 = 0$$

are

$$m_1 = -\frac{1}{2} - \frac{1}{2}\sqrt{4\rho\omega^2 - 1}\,i, \quad m_2 = -\frac{1}{2} + \frac{1}{2}\sqrt{4\rho\omega^2 - 1}\,i$$

when $\rho\omega^2 > 0.25$. Thus

$$y = c_1 x^{-1/2}\cos(\lambda \ln x) + c_2 x^{-1/2}\sin(\lambda \ln x)$$

where $\lambda = \sqrt{4\rho\omega^2 - 1}/2$. Applying $y(1) = 0$ gives $c_1 = 0$ and consequently

$$y = c_2 x^{-1/2}\sin(\lambda \ln x).$$

The condition $y(e) = 0$ requires $c_2 e^{-1/2}\sin\lambda = 0$. We obtain a nontrivial solution when $\lambda_n = n\pi$, $n = 1, 2, 3, \ldots$. But

$$\lambda_n = \sqrt{4\rho\omega_n^2 - 1}/2 = n\pi.$$

Solving for ω_n gives

$$\omega_n = \frac{1}{2}\sqrt{(4n^2\pi^2 + 1)/\rho}.$$

The corresponding solutions are

$$y_n(x) = c_2 x^{-1/2}\sin(n\pi \ln x).$$

(b)

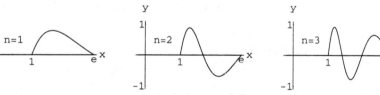

27. The auxiliary equation is $m^2 + m = m(m+1) = 0$ so that $u(r) = c_1 r^{-1} + c_2$. The boundary conditions $u(a) = u_0$ and $u(b) = u_1$ yield the system $c_1 a^{-1} + c_2 = u_0$, $c_1 b^{-1} + c_2 = u_1$. Solving gives

$$c_1 = \left(\frac{u_0 - u_1}{b - a}\right)ab \quad \text{and} \quad c_2 = \frac{u_1 b - u_0 a}{b - a}.$$

Thus

$$u(r) = \left(\frac{u_0 - u_1}{b - a}\right)\frac{ab}{r} + \frac{u_1 b - u_0 a}{b - a}.$$

28. The auxiliary equation is $m^2 = 0$ so that $u(r) = c_1 + c_2 \ln r$. The boundary conditions $u(a) = u_0$ and $u(b) = u_1$ yield the system $c_1 + c_2 \ln a = u_0$, $c_1 + c_2 \ln b = u_1$. Solving gives

$$c_1 = \frac{u_1 \ln a - u_0 \ln b}{\ln(a/b)} \quad \text{and} \quad c_2 = \frac{u_0 - u_1}{\ln(a/b)}.$$

175

Thus $$u(r) = \frac{u_1 \ln a - u_0 \ln b}{\ln(a/b)} + \frac{u_0 - u_1}{\ln(a/b)} \ln r = \frac{u_0 \ln(r/b) - u_1 \ln(r/a)}{\ln(a/b)}.$$

29. **(a)** The general solution of the differential equation is $y = c_1 \cos 4x + c_2 \sin 4x$. From $y_0 = y(0) = c_1$ we see that $y = y_0 \cos 4x + c_2 \sin 4x$. From $y_1 = y(\pi/2) = y_0$ we see that any solution must satisfy $y_0 = y_1$. We also see that when $y_0 = y_1$, $y = y_0 \cos 4x + c_2 \sin 4x$ is a solution of the boundary-value problem for any choice of c_2. Thus, the boundary-value problem does not have a unique solution for any choice of y_0 and y_1.

   (b) Whenever $y_0 = y_1$ there are infinitely many solutions.

   (c) When $y_0 \neq y_1$ there will be no solutions.

   (d) The boundary-value problem will have the trivial solution when $y_0 = y_1 = 0$. This solution will not be unique.

30. **(a)** The general solution of the differential equation is $y = c_1 \cos 4x + c_2 \sin 4x$. From $1 = y(0) = c_1$ we see that $y = \cos 4x + c_2 \sin 4x$. From $1 = y(L) = \cos 4L + c_2 \sin 4L$ we see that $c_2 = (1 - \cos 4L)/\sin 4L$. Thus,

$$y = \cos 4x + \left(\frac{1 - \cos 4L}{\sin 4L}\right) \sin 4x$$

   will be a unique solution when $\sin 4L \neq 0$; that is, when $L \neq k\pi/4$ where $k = 1, 2, 3, \ldots$.

   (b) There will be infinitely many solutions when $\sin 4L = 0$ and $1 - \cos 4L = 0$; that is, when $L = k\pi/2$ where $k = 1, 2, 3, \ldots$.

   (c) There will be no solution when $\sin 4L \neq 0$ and $1 - \cos 4L \neq 0$; that is, when $L = k\pi/4$ where $k = 1, 3, 5, \ldots$.

   (d) There can be no trivial solution since it would fail to satisfy the boundary conditions.

Exercises 5.3

1. The period corresponding to $x(0) = 1$, $x'(0) = 1$ is approximately 5.6. The period corresponding to $x(0) = 1/2$, $x'(0) = -1$ is approximately 6.2.

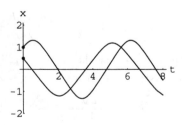

2. The solutions are not periodic.

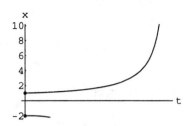

3. The period corresponding to $x(0) = 1$, $x'(0) = 1$ is approximately 5.8. The second initial-value problem does not have a periodic solution.

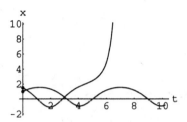

4. Both solutions have periods of approximately 6.3.

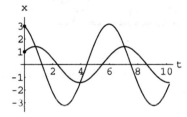

5. From the graph we see that $|x_1| \approx 1.2$.

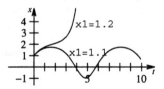

6. From the graphs we see that the interval is approximately $(-0.8, 1.1)$.

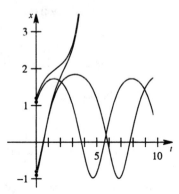

7. Since

$$xe^{0.01x} = x\left[1 + 0.01x + \frac{1}{2!}(0.01x)^2 + \cdots\right] \approx x$$

for small values of x, a linearization is

$$\frac{d^2x}{dt^2} + x = 0.$$

8.

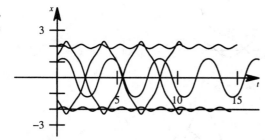

For $x(0) = 1$ and $x'(0) = 1$ the oscillations are symmetric about the line $x = 0$ with amplitude slightly greater than 1.

For $x(0) = -2$ and $x'(0) = 0.5$ the oscillations are symmetric about the line $x = -2$ with small amplitude.

For $x(0) = \sqrt{2}$ and $x'(0) = 1$ the oscillations are symmetric about the line $x = 0$ with amplitude a little greater than 2.

For $x(0) = 2$ and $x'(0) = 0.5$ the oscillations are symmetric about the line $x = 2$ with small amplitude.

For $x(0) = -2$ and $x'(0) = 0$ there is no oscillation; the solution is constant.

For $x(0) = -\sqrt{2}$ and $x'(0) = -1$ the oscillations are symmetric about the line $x = 0$ with amplitude a little greater than 2.

9. This is a damped hard spring, so all solutions should be oscillatory with $x \to 0$ as $t \to \infty$.

10. This is a damped soft spring, so we expect no oscillatory solutions.

11.

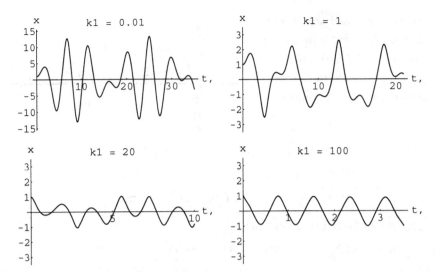

When k_1 is very small the effect of the nonlinearity is greatly diminished, and the system is close to pure resonance.

12. (a)

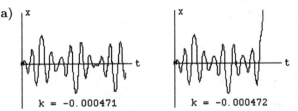

The system appears to be oscillatory for $-0.000471 \leq k_1 < 0$ and nonoscillatory for

$k_1 \leq -0.000472$.

(b)

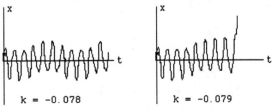

The system appears to be oscillatory for $-0.077 \leq k_1 < 0$ and nonoscillatory for $k_1 \leq 0.078$.

179

13. For $\lambda^2 - \omega^2 > 0$ we choose $\lambda = 2$ and $\omega = 1$ with $x(0) = 1$ and $x'(0) = 2$. For $\lambda^2 - \omega^2 < 0$ we choose $\lambda = 1/3$ and $\omega = 1$ with $x(0) = -2$ and $x'(0) = 4$. In both cases the motion corresponds to the over-damped and underdamped cases for spring/mass systems.

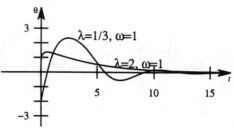

14. (a) As $y \to \infty$ we assume that $v \to 0^+$. Then $v_0^2 = 2gR$ and $v_0 = \sqrt{2gR}$.

(b) Using $g = 32$ ft/s and $R = 4000(5280)$ ft we find

$$v_0 = \sqrt{2(32)(4000)(5280)} \approx 36765.2 \text{ ft/s} \approx 25067 \text{ mi/hr}.$$

(c) $v_0 = \sqrt{2(0.165)(32)(1080)} \approx 7760 \text{ ft/s} \approx 5291 \text{ mi/hr}$

15. (a) Let (x, y) be the coordinates of S_2 on the curve C. The slope at (x, y) is then

$$dy/dx = (v_1 t - y)/(0 - x) = (y - v_1 t)/x \quad \text{or} \quad xy' - y = -v_1 t.$$

(b) Differentiating with respect to x gives

$$xy'' + y' - y' = -v_1 \frac{dt}{dx}$$

$$xy'' = -v_1 \frac{dt}{ds} \frac{ds}{dx}$$

$$xy'' = -v_1 \frac{1}{v_2} \left(-\sqrt{1 + (y')^2}\right)$$

$$xy'' = r\sqrt{1 + (y')^2}.$$

Letting $u = y'$ and separating variables, we obtain

$$x \frac{du}{dx} = r\sqrt{1 + u^2}$$

$$\frac{du}{\sqrt{1 + u^2}} = \frac{r}{x} dx$$

$$\sinh^{-1} u = r \ln x + \ln c = \ln(cx^r)$$

$$u = \sinh(\ln cx^r)$$

$$\frac{dy}{dx} = \frac{1}{2}\left(cx^r - \frac{1}{cx^r}\right).$$

At $t = 0$, $dy/dx = 0$ and $x = a$, so $0 = ca^r - 1/ca^r$. Thus $c = 1/a^r$ and

$$\frac{dy}{dx} = \frac{1}{2}\left[\left(\frac{x}{a}\right)^r - \left(\frac{a}{x}\right)^r\right] = \frac{1}{2}\left[\left(\frac{x}{a}\right)^r - \left(\frac{x}{a}\right)^{-r}\right].$$

If $r > 1$ or $r < 1$, integrating gives

$$y = \frac{a}{2}\left[\frac{1}{1+r}\left(\frac{x}{a}\right)^{1+r} - \frac{1}{1-r}\left(\frac{x}{a}\right)^{1-r}\right] + c_1.$$

When $t = 0$, $y = 0$ and $x = a$, so $0 = (a/2)[1/(1+r) - 1/(1-r)] + c_1$. Thus $c_1 = ar/(1 - r^2)$ and

$$y = \frac{a}{2}\left[\frac{1}{1+r}\left(\frac{x}{a}\right)^{1+r} - \frac{1}{1-r}\left(\frac{x}{a}\right)^{1-r}\right] + \frac{ar}{1-r^2}.$$

(c) If $r > 1$, $v_1 > v_2$ and $y \to \infty$ as $x \to 0^+$. In other words, S_2 always lags behind S_1. If $r < 1$, $v_1 < v_2$ and $y = ar/(1-r^2)$ when $x = 0$. In other words, when the submarine's speed is greater than the ship's, their paths will intersect at the point $(0, ar/(1 - r^2))$.

If $r = 1$, integration gives

$$y = \frac{1}{2}\left[\frac{x^2}{2a} - \frac{1}{a}\ln x\right] + c_2.$$

When $t = 0$, $y = 0$ and $x = a$, so $0 = (1/2)[a/2 - (1/a)\ln a] + c_2$. Thus $c_2 = -(1/2)[a/2 - (1/a)\ln a]$ and

$$y = \frac{1}{2}\left[\frac{x^2}{2a} - \frac{1}{a}\ln x\right] - \frac{1}{2}\left[\frac{a}{2} - \frac{1}{a}\ln a\right] = \frac{1}{2}\left[\frac{1}{2a}(x^2 - a^2) + \frac{1}{a}\ln\frac{a}{x}\right].$$

Since $y \to \infty$ as $x \to 0^+$, S_2 will never catch up with S_1.

16. (a) Let (r, θ) denote the polar coordinates of the destroyer S_1. When S_1 travels the 6 miles from $(9, 0)$ to $(3, 0)$ it stands to reason, since S_2 travels half as fast as S_1, that the polar coordinates of S_2 are $(3, \theta_2)$, where θ_2 is unknown. In other words, the distances of the ships from $(0, 0)$ are the same and $r(t) = 15t$ then gives the radial distance of both ships. This is necessary if S_1 is to intercept S_2.

(b) The differential of arc length in polar coordinates is $(ds)^2 = (r\, d\theta)^2 + (dr)^2$, so that

$$\left(\frac{ds}{dt}\right)^2 = r^2\left(\frac{d\theta}{dt}\right)^2 + \left(\frac{dr}{dt}\right)^2.$$

Using $ds/dt = 30$ and $dr/dt = 15$ then gives

$$900 = 225t^2\left(\frac{d\theta}{dt}\right)^2 + 225$$

$$675 = 225t^2\left(\frac{d\theta}{dt}\right)^2$$

$$\frac{d\theta}{dt} = \frac{\sqrt{3}}{t}$$

$$\theta(t) = \sqrt{3}\,\ln t + c = \sqrt{3}\,\ln\frac{r}{15} + c.$$

181

When $r = 3$, $\theta = 0$, so $c = -\sqrt{3}\ln(1/5)$ and

$$\theta(t) = \sqrt{3}\left(\ln\frac{r}{15} - \ln\frac{1}{5}\right) = \sqrt{3}\ln\frac{r}{3}.$$

Thus $r = 3e^{\theta/\sqrt{3}}$, whose graph is a logarithmic spiral.

(c) The time for S_1 to go from $(9,0)$ to $(3,0) = \frac{1}{5}$ hour. Now S_1 must intercept the path of S_2 for some angle β, where $0 < \beta < 2\pi$. At the time of interception t_2 we have $15t_2 = 3e^{\beta/\sqrt{3}}$ or $t = e^{\beta/\sqrt{3}}/5$. The total time is then

$$t = \frac{1}{5} + \frac{1}{5}e^{\beta/\sqrt{3}} < \frac{1}{5}(1 + e^{2\pi/\sqrt{3}}).$$

17. (a) Write the differential equation as

$$\frac{d^2\theta}{dt^2} + w\sin\theta = 0$$

where $w = g/l$. To test for differences between the earth and moon we use the fact that g, and hence w, is greater on the earth than on the moon. With initial conditions $\theta(0) = 1$ and $\theta'(0) = 2$ we compare solutions obtained using

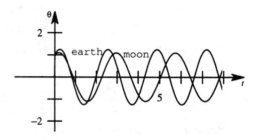

$w = 9$ and $w = 6$. We see that the pendulum oscillates faster on the earth than on the moon.

(b) The amplitude is greater on the earth than on the moon.

(c) The linear model is

$$\frac{d^2\theta}{dt^2} + w\theta = 0,$$

Where $w = g/l$. When $w = 9$ the general solution is $\theta = c_1\cos 3t + c_2\sin 3t$. The initial conditions $\theta(0) = 1$ and $\theta'(0) = 2$ lead to $\theta = \cos 3t + \frac{2}{3}\sin 3t$. When $w = 1$ the general solution is $\theta = c_1\cos t + c_2\sin t$. The initial

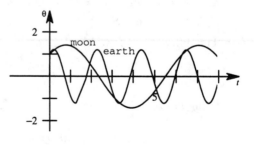

conditions $\theta(0) = 1$ and $\theta'(0) = 2$ lead to $\theta = \cos t + \sin t$. As in the nonlinear case, the pendulum oscillates faster on the earth than on the moon, but has greater amplitude on the moon.

18. (a) The general solution of

$$\frac{d^2\theta}{dt^2} + \theta = 0$$

is $\theta(t) = c_1 \cos t + c_2 \sin t$. From $\theta(0) = \pi/12$ and $\theta'(0) = -1/3$ we find $\theta(t) = (\pi/12) \cos t - (1/3) \sin t$. Setting $\theta(t) = 0$ we have $\tan t = \pi/4$ which implies $t_1 = \tan^{-1}(\pi/4) \approx 0.66577$.

(b) We set $\theta(t) = \theta(0) + \theta'(0)t + \frac{1}{2}\theta''(0)t + \frac{1}{6}\theta'''(0)t + \cdots$ and use $\theta''(t) = -\sin\theta(t)$ together with $\theta(0) = \pi/12$ and $\theta'(0) = -1/3$. Then $\theta''(0) = -\sin(\pi/12) = -\sqrt{2}(\sqrt{3}-1)/4$ and $\theta'''(0) = -\cos\theta(0) \cdot \theta'(0) = -\cos(\pi/12)(-1/3) = \sqrt{2}(\sqrt{3}+1)/12$. Thus

$$\theta(t) = \frac{\pi}{12} - \frac{1}{3}t - \frac{\sqrt{2}(\sqrt{3}-1)}{8}t^2 + \frac{\sqrt{2}(\sqrt{3}+1)}{72}t^3 + \cdots.$$

(c) Setting $\pi/12 - t/3 = 0$ we obtain $t_1 = \pi/4 \approx 0.785398$.

(d) Setting

$$\frac{\pi}{12} - \frac{1}{3}t - \frac{\sqrt{2}(\sqrt{3}-1)}{8}t^2 = 0$$

and using the positive root we obtain $t_1 \approx 0.63088$.

(e) Setting

$$\frac{\pi}{12} - \frac{1}{3}t - \frac{\sqrt{2}(\sqrt{3}-1)}{8}t^2 + \frac{\sqrt{2}(\sqrt{3}+1)}{72}t^3 = 0$$

we find $t_1 \approx 0.661973$ to be the first positive root.

19. (a) From the output we see that $y(t)$ is an interpolating function on the interval $0 \le t \le 5$, whose graph is shown. The positive root of $y(t) = 0$ near $t = 1$ is $t_1 = 0.666404$.

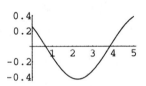

(b) To find the next two positive roots we change the interval used in **NDSolve** and **Plot** from $\{t,0,5\}$ to $\{t,0,10\}$. We see from the graph that the second and third positive roots are near 4 and 7, respectively. Replacing $\{t,1\}$ in **FindRoot** with $\{t,4\}$ and then $\{t,7\}$ we obtain $t_2 = 3.84411$ and $t_3 = 7.0218$.

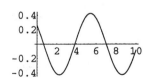

——— Chapter 5 Review Exercises ———

1. 8 ft., since $k = 4$.

2. $2\pi/5$, since $\frac{1}{4}x'' + 6.25x = 0$.

3. 5/4 m., since $x = -\cos 4t + \frac{3}{4}\sin 4t$.

4. True

5. False; since an external force may exist.

6. False

7. overdamped

8. $-\pi/4$

9. $9/2$, since $x = c_1 \cos \sqrt{2k}\, t + c_2 \sin \sqrt{2k}\, t$.

10. (a) Solving $\frac{3}{8}x'' + 6x = 0$ subject to $x(0) = 1$ and $x'(0) = -4$ we obtain

$$x = \cos 4t - \sin 4t = \sqrt{2} \sin (4t + 3\pi/4).$$

(b) The amplitude is $\sqrt{2}$, period is $\pi/2$, and frequency is $2/\pi$.

(c) If $x = 1$ then $t = n\pi/2$ and $t = -\pi/8 + n\pi/2$ for $n = 1,\, 2,\, 3,\, \ldots$.

(d) If $x = 0$ then $t = \pi/16 + n\pi/4$ for $n = 0,\, 1,\, 2,\, \ldots$. The motion is upward for n even and downward for n odd.

(e) $x'(3\pi/16) = 0$

(f) If $x' = 0$ then $4t + 3\pi/4 = \pi/2 + n\pi$ or $t = 3\pi/16 + n\pi$.

11. From $\frac{1}{4}x'' + \frac{3}{2}x' + 2x = 0$, $x(0) = 1/3$, and $x'(0) = 0$ we obtain $x = \frac{2}{3}e^{-2t} - \frac{1}{3}e^{-4t}$.

12. From $x'' + \beta x' + 64x = 0$ we see that oscillatory motion results if $\beta^2 - 256 < 0$ or $0 \leq |\beta| < 16$.

13. From $mx'' + 4x' + 2x = 0$ we see that non-oscillatory motion results if $16 - 8m \geq 0$ or $0 < m \leq 2$.

14. From $\frac{1}{4}x'' + x' + x = 0$, $x(0) = 4$, and $x'(0) = 2$ we obtain $x = 4e^{-2t} + 10te^{-2t}$. If $x'(t) = 0$, then $t = 1/10$, so that the maximum displacement is $x = 5e^{-0.2} \approx 4.094$.

15. Writing $\frac{1}{8}x'' + \frac{8}{3}x = \cos \gamma t + \sin \gamma t$ in the form $x'' + \frac{64}{3}x = 8\cos \gamma t + 8\sin \gamma t$ we identify $\lambda = 0$ and $\omega^2 = 64/3$. The system is in a state of pure resonance when $\gamma = \omega = \sqrt{64/3} = 8/\sqrt{3}$.

16. Clearly $x_p = A/\omega^2$ suffices.

17. From $\frac{1}{8}x'' + x' + 3x = e^{-t}$, $x(0) = 2$, and $x'(0) = 0$ we obtain $x_c = e^{-4t}\left(c_1 \cos 2\sqrt{2}\, t + c_2 \sin 2\sqrt{2}\, t\right)$, $x_p = \frac{8}{17}e^{-t}$, and

$$x = e^{-4t}\left(\frac{26}{17}\cos 2\sqrt{2}\, t + \frac{28\sqrt{2}}{17}\sin 2\sqrt{2}\, t\right) + \frac{8}{17}e^{-t}.$$

18. (a) Let k be the effective spring constant and x_1 and x_2 the elongation of springs k_1 and k_2. The

184

restoring forces satisfy $k_1 x_1 = k_2 x_2$ so $x_2 = (k_1/k_2)x_1$. From $k(x_1 + x_2) = k_1 x_1$ we have

$$k\left(x_1 + \frac{k_1}{k_2}x_2\right) = k_1 x_1$$

$$k\left(\frac{k_2 + k_1}{k_2}\right) = k_1$$

$$k = \frac{k_1 k_2}{k_1 + k_2}$$

$$\frac{1}{k} = \frac{1}{k_1} + \frac{1}{k_2}.$$

(b) From $k_1 = 2W$ and $k_2 = 4W$ we find $1/k = 1/2W + 1/4W = 3/4W$. Then $k = 4W/3 = 4mg/3$. The differential equation $mx'' + kx = 0$ then becomes $x'' + (4g/3)x = 0$. The solution is

$$x(t) = c_1 \cos 2\sqrt{\frac{g}{3}}\,t + c_2 \sin 2\sqrt{\frac{g}{3}}\,t.$$

The initial conditions $x(0) = 1$ and $x'(0) = 2/3$ imply $c_1 = 1$ and $c_2 = 1/\sqrt{3g}$.

(c) To compute the maximum speed of the weight we compute

$$x'(t) = 2\sqrt{\frac{g}{3}}\sin 2\sqrt{\frac{g}{3}}\,t + \frac{2}{3}\cos 2\sqrt{\frac{g}{3}}\,t \quad \text{and} \quad |x'(t)| = \sqrt{4\frac{g}{3} + \frac{4}{9}} = \frac{2}{3}\sqrt{3g + 1}.$$

19. From $q'' + 10^4 q = 100\sin 50t$, $q(0) = 0$, and $q'(t) = 0$ we obtain $q_c = c_1 \cos 100t + c_2 \sin 100t$, $q_p = \frac{1}{75}\sin 50t$, and

(a) $q = -\frac{1}{150}\sin 100t + \frac{1}{75}\sin 50t$,

(b) $i = -\frac{2}{3}\cos 100t + \frac{2}{3}\cos 50t$, and

(c) $q = 0$ when $\sin 50t(1 - \cos 50t) = 0$ or $t = n\pi/50$ for $n = 0, 1, 2, \ldots$.

20. Differentiate $L\dfrac{d^2 q}{dt^2} + R\dfrac{dq}{dt} + \dfrac{1}{C}q = E(t)$ and use $q'(t) = i(t)$ to obtain the desired result.

21. For $\lambda > 0$ the general solution is $y = c_1 \cos\sqrt{\lambda}\,x + c_2 \sin\sqrt{\lambda}\,x$. Now $y(0) = c_1$ and $y(2\pi) = c_1 \cos 2\pi\sqrt{\lambda} + c_2 \sin 2\pi\sqrt{\lambda}$, so the condition $y(0) = y(2\pi)$ implies

$$c_1 = c_1 \cos 2\pi\sqrt{\lambda} + c_2 \sin 2\pi\sqrt{\lambda}$$

which is true when $\sqrt{\lambda} = n$ or $\lambda = n^2$ for $n = 1, 2, 3, \ldots$. Since $y' = -\sqrt{\lambda}\,c_1 \sin\sqrt{\lambda}\,x + \sqrt{\lambda}\,c_2 \cos\sqrt{\lambda}\,x = -nc_1 \sin nx + nc_2 \cos nx$, we see that $y'(0) = nc_2 = y'(2\pi)$ for $n = 1, 2, 3, \ldots$. Thus, the eigenvalues are n^2 for $n = 1, 2, 3, \ldots$, with corresponding eigenfunctions $\cos nx$ and $\sin x$. When $\lambda = 0$, the general solution is $y = c_1 x + c_2$ and the corresponding eigenfunction is $y = 1$. For $\lambda < 0$ the general solution is $y = c_1 \cosh\sqrt{-\lambda}\,x + c_2 \sinh\sqrt{-\lambda}\,x$. In this case $y(0) = c_1$

and $y(2\pi) = c_1 \cosh 2\pi\sqrt{-\lambda} + c_2 \sinh 2\pi\sqrt{-\lambda}$, so $y(0) = y(2\pi)$ can only be valid for $\lambda = 0$. Thus, there are no eigenvalues corresponding to $\lambda < 0$.

6 Series Solutions of Linear Equations

_____ **Exercises 6.1** _____

1. $\lim\limits_{n\to\infty}\left|\dfrac{a_{n+1}}{a_n}\right| = \lim\limits_{n\to\infty}\left|\dfrac{x^{n+1}/(n+1)}{x^n/n}\right| = \lim\limits_{n\to\infty}\dfrac{n}{n+1}|x| = |x|$

The series is absolutely convergent on $(-1,1)$. At $x = -1$, the series $\sum\limits_{n=1}^{\infty}\dfrac{1}{n}$ is the harmonic series

which diverges. At $x = 1$, the series $\sum\limits_{n=1}^{\infty}\dfrac{(-1)^n}{n}$ converges by the alternating series test. Thus, the

given series converges on $(-1,1]$.

2. $\lim\limits_{n\to\infty}\left|\dfrac{a_{n+1}}{a_n}\right| = \lim\limits_{n\to\infty}\left|\dfrac{x^{n+1}/(n+1)^2}{x^n/n^2}\right| = \lim\limits_{n\to\infty}\left(\dfrac{n}{n+1}\right)^2|x| = |x|$

The series is absolutely convergent on $(-1,1)$. At $x = -1$, the series $\sum\limits_{n=1}^{\infty}\dfrac{(-1)^n}{n^2}$ converges by the

alternating series test. At $x = 1$, the series $\sum\limits_{n=1}^{\infty}\dfrac{1}{n^2}$ is a convergent p-series. Thus, the given series

converges on $[-1,1]$.

3. $\lim\limits_{n\to\infty}\left|\dfrac{a_{n+1}}{a_n}\right| = \lim\limits_{n\to\infty}\left|\dfrac{2^{n+1}x^{n+1}/(n+1)}{2^n x^n/n}\right| = \lim\limits_{n\to\infty}\dfrac{2n}{n+1}|x| = 2|x|$

The series is absolutely convergent for $2|x| < 1$ or $|x| < 1/2$. At $x = -1/2$, the series $\sum\limits_{k=1}^{\infty}\dfrac{(-1)^k}{k}$

converges by the alternating series test. At $x = 1/2$, the series $\sum\limits_{k=1}^{\infty}\dfrac{1}{k}$ is the harmonic series which

diverges. Thus, the given series converges on $[-1/2, 1/2)$.

4. $\lim\limits_{n\to\infty}\left|\dfrac{a_{n+1}}{a_n}\right| = \lim\limits_{n\to\infty}\left|\dfrac{5^{n+1}x^{n+1}/(n+1)!}{5^n x^n/n!}\right| = \lim\limits_{n\to\infty}\dfrac{5}{n+1}|x| = 0$

The series is absolutely convergent on $(-\infty, \infty)$.

5. $\lim\limits_{n\to\infty}\left|\dfrac{a_{n+1}}{a_n}\right| = \lim\limits_{n\to\infty}\left|\dfrac{(x-3)^{n+1}/(n+1)^3}{(x-3)^n/n^3}\right| = \lim\limits_{n\to\infty}\left(\dfrac{n}{n+1}\right)^3|x-3| = |x-3|$

The series is absolutely convergent for $|x - 3| < 1$ or on $(2, 4)$. At $x = 2$, the series $\sum\limits_{n=1}^{\infty}\dfrac{(-1)^n}{n^3}$

converges by the alternating series test. At $x = 4$, the series $\sum\limits_{n=1}^{\infty} \dfrac{1}{n^3}$ is a convergent p-series. Thus, the given series converges on $[2, 4]$.

6. $\lim\limits_{n\to\infty} \left| \dfrac{a_{n+1}}{a_n} \right| = \lim\limits_{n\to\infty} \left| \dfrac{(x+7)^{n+1}/\sqrt{n+1}}{(x+7)^n \sqrt{n}} \right| = \lim\limits_{n\to\infty} \sqrt{\dfrac{n}{n+1}}\, |x+7| = |x+7|$

The series is absolutely convergent for $|x + 7| < 1$ or on $(-8, 6)$. At $x = -8$, the series $\sum\limits_{n=1}^{\infty} \dfrac{(-1)^n}{\sqrt{n}}$ converges by the alternating series test. At $x = -6$, the series $\sum\limits_{n=1}^{\infty} \dfrac{1}{\sqrt{n}}$ is a divergent p-series. Thus, the given series converges on $[-8, -6)$.

7. $\lim\limits_{n\to\infty} \left| \dfrac{a_{n+1}}{a_n} \right| = \lim\limits_{n\to\infty} \left| \dfrac{(x-5)^{n+1}/10^{n+1}}{(x-5)^n/10^n} \right| = \lim\limits_{n\to\infty} \dfrac{1}{10}|x-5| = \dfrac{1}{10}|x-5|$

The series is absolutely convergent for $\dfrac{1}{10}|x - 5| < 1$, $|x - 5| < 10$, or on $(-5, 15)$. At $x = -5$, the series $\sum\limits_{k=1}^{\infty} \dfrac{(-1)^k(-10)^k}{10^k} = \sum\limits_{k=1}^{\infty} 1$ diverges by the n-th term test. At $x = 15$, the series $\sum\limits_{k=1}^{\infty} \dfrac{(-1)^k 10^k}{10^k} = \sum\limits_{k=1}^{\infty} (-1)^k$ diverges by the n-th term test. Thus, the series converges on $(-5, 15)$.

8. $\lim\limits_{n\to\infty} \left| \dfrac{a_{n+1}}{a_n} \right| = \lim\limits_{n\to\infty} \left| \dfrac{(n+1)(x-4)^{n+1}/(n+3)^2}{n(x-4)^n/(n+2)^2} \right| = \lim\limits_{n\to\infty} \dfrac{(n+1)(n+2)^2}{x(n+3)^2}|x-4| = |x-4|$

The series is absolutely convergent for $|x - 4| < 1$ or on $(3, 5)$. At $x = 3$, the series $\sum\limits_{k=1}^{\infty} \dfrac{(-1)^k k}{(k+2)^2}$ converges by the alternating series test. At $x = 5$, the series $\sum\limits_{k=1}^{\infty} \dfrac{k}{(k+2)^2}$ diverges by the limit comparison test with $\sum\limits_{k=1}^{\infty} \dfrac{1}{k}$. Thus, the series converges on $[3, 5)$.

9. $\lim\limits_{n\to\infty} \left| \dfrac{a_{n+1}}{a_n} \right| = \lim\limits_{n\to\infty} \left| \dfrac{(n+1)!2^{n+1}x^{n+1}}{n!2^n x^n} \right| = \lim\limits_{n\to\infty} 2(n+1)|x| = \infty, \; x \neq 0$

The series converges only at $x = 0$.

10. $\displaystyle\lim_{n\to\infty}\left|\frac{a_{n+1}}{a_n}\right| = \lim_{n\to\infty}\left|\frac{nx^{n+1}/(n+1)^{2(n+1)}}{(n-1)x^n/n^{2n}}\right| = \lim_{n\to\infty}\frac{nn^{2n}}{(n-1)(n+1)^{2n+2}}|x|$

$\displaystyle = \lim_{n\to\infty}\frac{n}{(n-1)(n+1)^2}\left(\frac{n}{n+1}\right)^{2n}|x| = \lim_{n\to\infty}\frac{n}{(n-1)(n+1)^2}\cdot\frac{1}{\left[\left(\frac{n+1}{n}\right)^n\right]^2}|x|$

$\displaystyle = \lim_{n\to\infty}\frac{n}{(n-1)(n+1)^2}\cdot\frac{1}{[(1+1/n)^n]^2}|x| = 0\cdot\frac{1}{e^2}|x| = 0$

The series is convergent on $(-\infty, \infty)$.

11. $e^x\sin x = \left(1 + x + \dfrac{x^2}{2} + \dfrac{x^3}{6} + \dfrac{x^4}{24} + \cdots\right)\left(x - \dfrac{x^3}{6} + \dfrac{x^5}{120} - \cdots\right) = x + x^2 + \dfrac{x^3}{3} - \dfrac{x^5}{30} - \cdots$

12. $e^{-x}\cos x = \left(1 - x + \dfrac{x^2}{2} - \dfrac{x^3}{6} + \dfrac{x^4}{24} - \cdots\right)\left(1 - \dfrac{x^2}{2} + \dfrac{x^4}{24} - \cdots\right) = 1 - x + \dfrac{x^3}{3} - \dfrac{x^4}{6} + \cdots$

13. $\sin x\cos x = \left(x - \dfrac{x^3}{6} + \dfrac{x^5}{120} - \dfrac{x^7}{5040} + \cdots\right)\left(1 - \dfrac{x^2}{2} + \dfrac{x^4}{24} - \dfrac{x^6}{720} + \cdots\right) = x - \dfrac{2x^3}{3} + \dfrac{2x^5}{15} - \dfrac{4x^7}{315} + \cdots$

14. $e^x\ln(1-x) = \left(1 + x + \dfrac{x^2}{2} + \dfrac{x^3}{6} + \cdots\right)\left(-x - \dfrac{x^2}{2} - \dfrac{x^3}{3} - \dfrac{x^4}{4} - \cdots\right) = -x - \dfrac{3x^2}{2} - \dfrac{4x^3}{3} - x^4 - \cdots$

15. Separating variables we obtain

$$\frac{dy}{y} = -dx \implies \ln|y| = -x + c \implies y = c_1 e^{-x}.$$

Substituting $y = \sum_{n=0}^{\infty} c_n x^n$ into the differential equation leads to

$$y' + y = \underbrace{\sum_{n=1}^{\infty} nc_n x^{n-1}}_{k=n-1} + \underbrace{\sum_{n=0}^{\infty} c_n x^n}_{k=n} = \sum_{k=0}^{\infty}(k+1)c_{k+1}x^k + \sum_{k=0}^{\infty} c_k x^k = \sum_{k=0}^{\infty}[(k+1)c_{k+1} + c_k]x^k = 0.$$

Thus

$$(k+1)c_{k+1} + c_k = 0$$

and

$$c_{k+1} = -\frac{1}{k+1}c_k, \quad k = 0, 1, 2, \ldots.$$

Iterating we find

$$c_1 = -c_0$$

$$c_2 = -\frac{1}{2}c_1 = \frac{1}{2}c_0$$

$$c_3 = -\frac{1}{3}c_2 = -\frac{1}{6}c_0$$

$$c_4 = -\frac{1}{4}c_3 = \frac{1}{24}c_0$$

and so on. Therefore

$$y = c_0 - c_0 x + \frac{1}{2}c_0 x^2 - \frac{1}{6}c_0 x^3 + \frac{1}{24}c_0 x^4 - \cdots = c_0\left[1 - x + \frac{1}{2}x^2 - \frac{1}{6}x^3 + \frac{1}{24}x^4 - \cdots\right]$$

$$= c_0 \sum_{n=0}^{\infty} \frac{1}{n!}(-x)^n = c_0 e^{-x}.$$

16. Separating variables we obtain

$$\frac{dy}{y} = 2\,dx \implies \ln|y| = 2x + c \implies y = c_1 e^{2x}.$$

Substituting $y = \sum_{n=0}^{\infty} c_n x^n$ into the differential equation leads to

$$y' - 2y = \sum_{n=1}^{\infty} n c_n x^{n-1} - 2\sum_{n=0}^{\infty} c_n x^n = \sum_{k=0}^{\infty}(k+1)c_{k+1}x^k - 2\sum_{k=0}^{\infty} c_k x^k$$

$$\underbrace{}_{k=n-1} \qquad \underbrace{}_{k=n}$$

$$= \sum_{k=0}^{\infty}[(k+1)c_{k+1} - 2c_k]x^k = 0.$$

Thus

$$(k+1)c_{k+1} - 2c_k = 0$$

and

$$c_{k+1} = \frac{2}{k+1}c_k, \quad k = 0, 1, 2, \ldots.$$

Iterating we find

$$c_1 = 2c_0$$

$$c_2 = \frac{2}{2}c_1 = \frac{4}{2}c_0$$

$$c_3 = \frac{2}{3}c_2 = \frac{8}{6}c_0$$

$$c_4 = \frac{2}{4}c_3 = \frac{16}{24}c_0$$

and so on. Therefore

$$y = c_0 + 2c_0 x + \frac{4}{2}c_0 x^2 + \frac{8}{6}c_0 x^3 + \frac{16}{24}c_0 x^4 + \cdots$$

$$= c_0\left[1 + 2x + \frac{1}{2}(2x)^2 + \frac{1}{6}(2x)^3 + \frac{1}{24}(2x)^4 + \cdots\right] = c_0 \sum_{n=0}^{\infty} \frac{1}{n!}(2x)^n = c_0 e^{2x}.$$

17. Separating variables we obtain

$$\frac{dy}{y} = x^2 dx \implies \ln|y| = \frac{1}{3}x^3 + c \implies y = c_1 e^{x^3/3}.$$

Substituting $y = \sum_{n=0}^{\infty} c_n x^n$ into the differential equation leads to

$$y' - x^2 y = \underbrace{\sum_{n=1}^{\infty} n c_n x^{n-1}}_{k=n-3} - \underbrace{\sum_{n=0}^{\infty} c_n x^{n+2}}_{k=n} = \sum_{k=-2}^{\infty} (k+3) c_{k+3} x^{k+2} - \sum_{k=0}^{\infty} c_k x^{k+2}$$

$$= c_1 + 2c_2 x + \sum_{k=0}^{\infty} [(k+3)c_{k+3} - c_k] x^{k+2} = 0.$$

Thus
$$c_1 = c_2 = 0,$$

$$(k+3)c_{k+3} - c_k = 0,$$

and
$$c_{k+3} = \frac{1}{k+3} c_k, \quad k = 0, 1, 2, \dots.$$

Iterating we find

$$c_3 = \frac{1}{3} c_0$$

$$c_4 = c_5 = 0$$

$$c_6 = \frac{1}{6} c_3 = \frac{1}{2} \cdot \frac{1}{3^2} c_0$$

$$c_7 = c_8 = 0$$

$$c_9 = \frac{1}{9} c_6 = \frac{1}{2 \cdot 3} \cdot \frac{1}{3^3} c_0$$

and so on. Therefore

$$y = c_0 + \frac{1}{3} c_0 x^3 + \frac{1}{2} \cdot \frac{1}{3^2} c_0 x^6 + \frac{1}{2 \cdot 3} \cdot \frac{1}{3^3} c_0 x^9 + \cdots$$

$$= c_0 \left[1 + \frac{x^3}{3} + \frac{1}{2} \left(\frac{x^3}{3} \right)^2 + \frac{1}{2 \cdot 3} \left(\frac{x^3}{3} \right)^3 + \cdots \right] = c_0 \sum_{n=0}^{\infty} \frac{1}{n!} \left(\frac{x^3}{3} \right)^n = c_0 e^{x^3/3}.$$

18. Separating variables we obtain

$$\frac{dy}{y} = -x^3 dx \implies \ln|y| = -\frac{1}{4} x^4 + c \implies y = c_1 e^{-x^4/4}.$$

Substituting $y = \sum_{n=0}^{\infty} c_n x^n$ into the differential equation leads to

$$y' + x^3 y = \underbrace{\sum_{n=1}^{\infty} n c_n x^{n-1}}_{k=n-4} + \underbrace{\sum_{n=0}^{\infty} c_n x^{n+3}}_{k=n} = \sum_{k=-3}^{\infty} (k+4) c_{k+4} x^{k+3} - \sum_{k=0}^{\infty} c_k x^{k+3}$$

$$= c_1 + 2c_2 x + 3c_3 x^2 + \sum_{k=0}^{\infty} [(k+4)c_{k+4} + c_k] x^{k+2} = 0.$$

Thus

$$c_1 = c_2 = c_3 = 0,$$

$$(k+4)c_{k+4} + c_k = 0,$$

and

$$c_{k+4} = -\frac{1}{k+4}c_k, \quad k = 0, 1, 2, \ldots.$$

Iterating we find

$$c_4 = -\frac{1}{4}c_0$$

$$c_5 = c_6 = c_7 = 0$$

$$c_8 = -\frac{1}{8}c_4 = \frac{1}{2} \cdot \frac{1}{4^2}c_0$$

$$c_9 = c_{10} = c_{11} = 0$$

$$c_{12} = -\frac{1}{12}c_8 = -\frac{1}{2 \cdot 3} \cdot \frac{1}{4^3}c_0$$

and so on. Therefore

$$y = c_0 - \frac{1}{4}c_0x^4 + \frac{1}{2} \cdot \frac{1}{4^2}c_0x^8 - \frac{1}{2 \cdot 3} \cdot \frac{1}{4^3}c_0x^{12} + \cdots$$

$$= c_0\left[1 - \frac{x^4}{4} + \frac{1}{2}\left(\frac{x^4}{4}\right)^2 - \frac{1}{2 \cdot 3}\left(\frac{x^4}{4}\right)^3 + \cdots\right] = c_0\sum_{n=0}^{\infty}\frac{1}{n!}\left(\frac{-x^4}{4}\right)^n = c_0e^{-x^4/4}.$$

19. Separating variables we obtain

$$\frac{dy}{y} = \frac{dx}{1-x} \implies \ln|y| = -\ln|1-x| + c \implies y = \frac{c_1}{1-x}.$$

Substituting $y = \sum_{n=0}^{\infty} c_n x^n$ into the differential equation leads to

$$(1-x)y' - y = \underbrace{\sum_{n=1}^{\infty} nc_nx^{n-1}}_{k=n-1} - \underbrace{\sum_{n=1}^{\infty} nc_nx^n}_{k=n} - \underbrace{\sum_{n=0}^{\infty} c_nx^n}_{k=n}$$

$$= \sum_{k=0}^{\infty}(k+1)c_{k+1}x^k - \sum_{k=1}^{\infty} kc_kx^k - \sum_{k=0}^{\infty} c_kx^k$$

$$= c_1 - c_0 + \sum_{k=1}^{\infty}[(k+1)c_{k+1} - (k+1)c_k]x^k = 0.$$

Thus

$$c_1 - c_0 = 0,$$

$$(k+1)c_{k+1} - (k+1)c_k = 0,$$

and

$$c_1 = c_0$$

$$c_{k+1} = c_k, \quad k = 1, 2, 3, \dots .$$

Iterating we find

$$c_1 = c_0,$$

$$c_2 = c_1 = c_0$$

$$c_3 = c_2 = c_0$$

and so on. Therefore

$$y = c_0 + c_0 x + c_0 x^2 + c_0 x^3 + \cdots = c_0 \left[1 + x + x^2 + x^3 + \cdots\right] = c_0 \sum_{n=0}^{\infty} x^n = \frac{c_0}{1-x}.$$

20. Separating variables we obtain

$$\frac{dy}{y} = \frac{2}{1+x} dx \implies \ln|y| = 2\ln|1+x| + c \implies y = c_1(1+x)^2.$$

Substituting $y = \sum_{n=0}^{\infty} c_n x^n$ into the differential equation leads to

$$(1+x)y' - 2y = \underbrace{\sum_{n=1}^{\infty} n c_n x^{n-1}}_{k=n-1} + \underbrace{\sum_{n=1}^{\infty} n c_n x^n}_{k=n} - 2\underbrace{\sum_{n=0}^{\infty} c_n x^n}_{k=n}$$

$$= \sum_{k=0}^{\infty} (k+1)c_{k+1} x^k + \sum_{k=1}^{\infty} k c_k x^k - 2\sum_{k=0}^{\infty} c_k x^k$$

$$= c_1 - 2c_0 + \sum_{k=1}^{\infty} [(k+1)c_{k+1} + (k-2)c_k]x^k = 0.$$

Thus

$$c_1 - 2c_0 = 0,$$

$$(k+1)c_{k+1} + (k-2)c_k = 0,$$

and

$$c_1 = 2c_0$$

$$c_{k+1} = -\frac{k-2}{k+1} c_k, \quad k = 1, 2, 3, \dots .$$

Iterating we find

$$c_1 = 2c_0$$

$$c_2 = \frac{1}{2}c_1 = c_0$$

$$c_3 = 0c_2 = 0$$

$$c_4 = 0$$

and so on. Therefore

$$y = c_0 + 2c_0x + c_0x^2 = c_0\left(1 + 2x + x^2\right) = c_0(1+x)^2.$$

21. The auxiliary equation is $m^2 + 1 = 0$, so $y = c_1 \cos x + c_2 \sin x$. Substituting $y = \sum_{n=0}^{\infty} c_n x^n$ into the differential equation leads to

$$y'' + y = \underbrace{\sum_{n=2}^{\infty} n(n-1)c_n x^{n-2}}_{k=n-2} + \underbrace{\sum_{n=0}^{\infty} c_n x^n}_{k=n} = \sum_{k=0}^{\infty}(k+2)(k+1)c_{k+2}x^k + \sum_{k=0}^{\infty} c_k x^k$$

$$= \sum_{k=0}^{\infty}[(k+2)(k+1)c_{k+2} + c_k]x^k = 0.$$

Thus

$$(k+2)(k+1)c_{k+2} + c_k = 0$$

and

$$c_{k+2} = -\frac{1}{(k+2)(k+1)}c_k, \quad k = 0,1,2,\ldots.$$

Iterating we find

$$c_2 = -\frac{1}{2}c_0$$

$$c_3 = -\frac{1}{3\cdot 2}c_1$$

$$c_4 = -\frac{1}{4\cdot 3}c_2 = \frac{1}{4\cdot 3\cdot 2}c_0$$

$$c_5 = -\frac{1}{5\cdot 4}c_3 = \frac{1}{5\cdot 4\cdot 3\cdot 2}c_1$$

$$c_6 = -\frac{1}{6\cdot 5}c_4 = -\frac{1}{6!}c_0$$

$$c_7 = -\frac{1}{7\cdot 6}c_5 = -\frac{1}{7!}c_1$$

and so on. Therefore

$$y = c_0 + c_1 x - \frac{1}{2}c_0 x^2 - \frac{1}{3!}c_1 x^3 + \frac{1}{4!}c_0 x^4 + \frac{1}{5!}c_1 x^5 - \cdots$$

$$= c_0\left[1 - \frac{1}{2}x^2 + \frac{1}{4!}x^4 - \cdots\right] + c_1\left[1 - \frac{1}{3!}x^3 + \frac{1}{5!}x^5 - \cdots\right]$$

$$= c_0 \sum_{n=0}^{\infty} \frac{(-1)^n x^{2n}}{(2n)!} + c_1 \sum_{n=0}^{\infty} \frac{(-1)^n x^{2n+1}}{(2n+1)!} = c_0 \cos x + c_1 \sin x.$$

22. The auxiliary equation is $m^2 - 1 = 0$, so $y = c_1 e^x + c_2 e^{-x}$. Substituting $y = \sum_{n=0}^{\infty} c_n x^n$ into the differential equation leads to

$$y'' - y = \underbrace{\sum_{n=2}^{\infty} n(n-1)c_n x^{n-2}}_{k=n-2} - \underbrace{\sum_{n=0}^{\infty} c_n x^n}_{k=n} = \sum_{k=0}^{\infty} (k+2)(k+1)c_{k+2}x^k - \sum_{k=0}^{\infty} c_k x^k$$

$$= \sum_{k=0}^{\infty} [(k+2)(k+1)c_{k+2} - c_k]x^k = 0.$$

Thus
$$(k+2)(k+1)c_{k+2} - c_k = 0, \quad k = 0,\ 1,\ 2,\ \ldots$$

and
$$c_{k+2} = \frac{c_k}{(k+2)(k+1)}, \quad k = 0,\ 1,\ 2,\ \ldots.$$

Iterating, we find

$$c_2 = \frac{c_0}{2\cdot 1} = \frac{1}{2}c_0$$

$$c_3 = \frac{c_1}{3\cdot 2} = \frac{1}{3\cdot 2}c_1$$

$$c_4 = \frac{c_2}{4\cdot 3} = \frac{1}{4\cdot 3\cdot 2}c_0$$

$$c_5 = \frac{c_3}{5\cdot 4} = \frac{1}{5\cdot 4\cdot 3\cdot 2}c_1$$

$$c_6 = \frac{c_4}{6\cdot 5} = \frac{1}{6\cdot 5\cdot 4\cdot 3\cdot 2}c_0$$

$$c_7 = \frac{c_5}{7\cdot 6} = \frac{1}{7\cdot 6\cdot 5\cdot 4\cdot 3\cdot 2}c_1.$$

and so on. Therefore,

$$y = c_0 + c_1 x + c_2 x^2 + c_3 x^3 + \ldots$$

$$= c_0 + c_1 x + \frac{1}{2} c_0 x^2 + \frac{1}{3 \cdot 2} c_1 x^3 + \frac{1}{4 \cdot 3 \cdot 2} c_0 x^4 + \frac{1}{5 \cdot 4 \cdot 3 \cdot 2} c_1 x^5 + \ldots$$

$$= c_0 \left[1 + \frac{1}{2!} x^2 + \frac{1}{4!} x^4 + \ldots \right] + c_1 \left[x + \frac{1}{3!} x^3 + \frac{1}{5!} x^5 + \ldots \right]$$

$$= c_0 \sum_{n=0}^{\infty} \frac{1}{(2n)!} x^{2n} + c_1 \sum_{n=0}^{\infty} \frac{1}{(2n+1)!} x^{2n+1} = c_0 \cosh x + c_1 \sinh x$$

$$= c_0 \frac{e^x + e^{-x}}{2} + c_1 \frac{e^x - e^{-x}}{2} = \left(\frac{c_0 + c_1}{2} \right) e^x + \left(\frac{c_0 - c_1}{2} \right) e^{-x} = C_0 e^x + C_1 e^{-x}.$$

23. The auxiliary equation is $m^2 - m = m(m - 1) = 0$, so $y = c_1 + c_2 e^x$. Substituting $y = \sum_{n=0}^{\infty} c_n x^n$ into the differential equation leads to

$$y'' - y' = \underbrace{\sum_{n=2}^{\infty} n(n-1) c_n x^{n-2}}_{k=n-2} - \underbrace{\sum_{n=1}^{\infty} n c_n x^{n-1}}_{k=n-1}$$

$$= \sum_{k=0}^{\infty} (k+2)(k+1) c_{k+2} x^k - \sum_{k=0}^{\infty} (k+1) c_{k+1} x^k$$

$$= \sum_{k=0}^{\infty} [(k+2)(k+1) c_{k+2} - (k+1) c_{k+1}] x^k = 0.$$

Thus

$$(k+2)(k+1) c_{k+2} - (k+1) c_{k+1} = 0$$

and

$$c_{k+2} = \frac{1}{(k+2)} c_{k+1}, \quad k = 0, 1, 2, \ldots .$$

Iterating we find

$$c_2 = \frac{1}{2} c_1$$

$$c_3 = \frac{1}{3} c_2 = \frac{1}{3 \cdot 2} c_1$$

$$c_4 = \frac{1}{4} c_3 = \frac{1}{4!} c_1$$

and so on. Therefore

$$y = c_0 + c_1 x + \frac{1}{2}c_1 x^2 + \frac{1}{3!}c_1 x^3 + \frac{1}{4!}c_1 x^4 + \cdots$$

$$\boxed{c_0 = C_0 + c_1}$$

$$= C_0 + c_1 \left[1 + x + \frac{1}{2}x^2 + \frac{1}{3!}x^3 + \frac{1}{4!}x^4 + \cdots \right] = C_0 + c_1 \sum_{n=0}^{\infty} \frac{1}{n!}x^n = C_0 + c_1 e^x.$$

24. The auxiliary equation is $2m^2 + m = m(2m + 1) = 0$, so $y = c_1 + c_2 e^{-x/2}$. Substituting $y = \sum_{n=0}^{\infty} c_n x^n$ into the differential equation leads to

$$2y'' + y' = 2 \underbrace{\sum_{n=2}^{\infty} n(n-1)c_n x^{n-2}}_{k=n-2} + \underbrace{\sum_{n=1}^{\infty} nc_n x^{n-1}}_{k=n-1}$$

$$= 2 \sum_{k=0}^{\infty} (k+2)(k+1)c_{k+2}x^k + \sum_{k=0}^{\infty} (k+1)c_{k+1}x^k$$

$$= \sum_{k=0}^{\infty} [2(k+2)(k+1)c_{k+2} + (k+1)c_{k+1}]x^k = 0.$$

Thus

$$2(k+2)(k+1)c_{k+2} + (k+1)c_{k+1} = 0$$

and

$$c_{k+2} = -\frac{1}{2(k+2)} c_{k+1}, \quad k = 0, 1, 2, \ldots .$$

Iterating we find

$$c_2 = -\frac{1}{2}\frac{1}{2}c_1$$

$$c_3 = -\frac{1}{2}\frac{1}{3}c_2 = \frac{1}{2^2}\frac{1}{3 \cdot 2}c_1$$

$$c_4 = -\frac{1}{2}\frac{1}{4}c_3 = \frac{1}{2^3}\frac{1}{4!}c_1$$

and so on. Therefore

$$y = c_0 + c_1 x - \frac{1}{2}\frac{1}{2}c_1 x^2 + \frac{1}{2^2 3!}c_1 x^3 - \frac{1}{2^3 4!}c_1 x^4 + \cdots$$

$$\boxed{\begin{array}{l} c_0 = C_0 - 2c_1 \\ c_1 = -\frac{1}{2}C_1 \end{array}}$$

197

$$= C_0 + \left[C_1 - \frac{1}{2}C_1 x + \frac{1}{2}\frac{1}{2}\frac{1}{2}C_1 x^2 - \frac{1}{2^2 2 \cdot 3!}\frac{1}{2}C_1 x^3 + \cdots \right]$$

$$= C_0 + C_1 \left[1 - \frac{x}{2} + \frac{1}{2}\left(\frac{x}{2}\right)^2 - \frac{1}{3!}\left(\frac{x}{3}\right)^3 + \cdots \right]$$

$$= C_0 + C_1 \sum_{n=0}^{\infty} \frac{(-1)^n}{n!}\left(\frac{x}{2}\right)^n = C_0 + C_1 \sum_{n=0}^{\infty} \frac{1}{n!}\left(-\frac{x}{n}\right)^n = C_0 + C_1 e^{-x/2}.$$

25. Substituting $y = J_0(x)$ into the differential equation gives

$$xy'' + y' + xy = \underbrace{\sum_{n=1}^{\infty} \frac{2n(2n-1)(-1)^n}{2^{2n}(n!)^2} x^{2n-1}}_{k=n} + \underbrace{\sum_{n=1}^{\infty} \frac{2n(-1)^n}{2^{2n}(n!)^2} x^{2n-1}}_{k=n}$$

$$+ \underbrace{\sum_{n=0}^{\infty} \frac{(-1)^n}{2^{2n}(n!)^2} x^{2n+1}}_{k=n+1}$$

$$= \sum_{k=1}^{\infty} \frac{2k(2k-1)(-1)^k}{2^{2k}(k!)^2} x^{2k-1} + \sum_{k=1}^{\infty} \frac{2k(-1)^k}{2^{2k}(k!)^2} x^{2k-1}$$

$$+ \sum_{k=1}^{\infty} \frac{(-1)^{k-1}}{2^{2k-2}[(k-1)!]^2} x^{2k-1}$$

$$= \sum_{k=1}^{\infty} \frac{(-1)^k}{2^{2k}(k!)^2} [2k(2k-1) + 2k - 2^2 k^2] x^{2k-1}$$

$$= \sum_{k=1}^{\infty} \frac{(-1)^k}{2^{2k}(k!)^2} (4k^2 - 2k + 2k - 4k^2) x^{2k-1} = 0.$$

26. The series converges on an interval centered at $x_0 = 4$. Since it converges at -2, it must converge on $[-2, 10)$. Since it diverges at 13 it must diverge for $|x - 4| > 9$, or on $(-\infty, -5)$ and $(13, \infty)$. Thus, the series might converge at 10, does converge at 7, does not converge at -7, and might converge at 11.

Exercises 6.2

1. Substituting $y = \sum_{n=0}^{\infty} c_n x^n$ into the differential equation we have

$$y'' - xy = \underbrace{\sum_{n=2}^{\infty} n(n-1)c_n x^{n-2}}_{k=n-2} - \underbrace{\sum_{n=0}^{\infty} c_n x^{n+1}}_{k=n+1} = \sum_{k=0}^{\infty} (k+2)(k+1)c_{k+2}x^k - \sum_{k=1}^{\infty} c_{k-1}x^k$$

$$= 2c_2 + \sum_{k=1}^{\infty} [(k+2)(k+1)c_{k+2} - c_{k-1}]x^k = 0.$$

Thus
$$c_2 = 0$$

$$(k+2)(k+1)c_{k+2} - c_{k-1} = 0$$

and
$$c_{k+2} = \frac{1}{(k+2)(k+1)} c_{k-1}, \quad k = 1, 2, 3, \ldots .$$

Choosing $c_0 = 1$ and $c_1 = 0$ we find

$$c_3 = \frac{1}{6}$$

$$c_4 = c_5 = 0$$

$$c_6 = \frac{1}{180}$$

and so on. For $c_0 = 0$ and $c_1 = 1$ we obtain

$$c_3 = 0$$

$$c_4 = \frac{1}{12}$$

$$c_5 = c_6 = 0$$

$$c_7 = \frac{1}{504}$$

and so on. Thus, two solutions are

$$y_1 = 1 + \frac{1}{6}x^3 + \frac{1}{180}x^6 + \cdots \quad \text{and} \quad y_2 = x + \frac{1}{12}x^4 + \frac{1}{504}x^7 + \cdots .$$

2. Substituting $y = \sum_{n=0}^{\infty} c_n x^n$ into the differential equation we have

$$y'' + x^2 y = \underbrace{\sum_{n=2}^{\infty} n(n-1)c_n x^{n-2}}_{k=n-2} + \underbrace{\sum_{n=0}^{\infty} c_n x^{n+2}}_{k=n+2} = \sum_{k=0}^{\infty} (k+2)(k+1)c_{k+2}x^k + \sum_{k=2}^{\infty} c_{k-2}x^k$$

$$= 2c_2 + 6c_3 x + \sum_{k=2}^{\infty} [(k+2)(k+1)c_{k+2} + c_{k-2}]x^k = 0.$$

Thus
$$c_2 = c_3 = 0$$

$$(k+2)(k+1)c_{k+2} + c_{k-2} = 0$$

and
$$c_{k+2} = -\frac{1}{(k+2)(k+1)}c_{k-2}, \quad k = 2, 3, 4, \ldots.$$

Choosing $c_0 = 1$ and $c_1 = 0$ we find

$$c_4 = -\frac{1}{12}$$

$$c_5 = c_6 = c_7 = 0$$

$$c_8 = \frac{1}{672}$$

and so on. For $c_0 = 0$ and $c_1 = 1$ we obtain

$$c_4 = 0$$

$$c_5 = -\frac{1}{20}$$

$$c_6 = c_7 = c_8 = 0$$

$$c_9 = \frac{1}{1440}$$

and so on. Thus, two solutions are

$$y_1 = 1 - \frac{1}{12}x^4 + \frac{1}{672}x^8 - \cdots \quad \text{and} \quad y_2 = x - \frac{1}{20}x^5 + \frac{1}{1440}x^9 - \cdots.$$

3. Substituting $y = \sum_{n=0}^{\infty} c_n x^n$ into the differential equation we have

$$y'' - 2xy' + y = \underbrace{\sum_{n=2}^{\infty} n(n-1)c_n x^{n-2}}_{k=n-2} - 2\underbrace{\sum_{n=1}^{\infty} nc_n x^n}_{k=n} + \underbrace{\sum_{n=0}^{\infty} c_n x^n}_{k=n}$$

$$= \sum_{k=0}^{\infty}(k+2)(k+1)c_{k+2}x^k - 2\sum_{k=1}^{\infty} kc_k x^k + \sum_{k=0}^{\infty} c_k x^k$$

$$= 2c_2 + c_0 + \sum_{k=1}^{\infty}[(k+2)(k+1)c_{k+2} - (2k-1)c_k]x^k = 0.$$

Thus
$$2c_2 + c_0 = 0$$

$$(k+2)(k+1)c_{k+2} - (2k-1)c_k = 0$$

200

and
$$c_2 = -\frac{1}{2}c_0$$

$$c_{k+2} = \frac{2k-1}{(k+2)(k+1)}\, c_k, \quad k = 1, 2, 3, \ldots .$$

Choosing $c_0 = 1$ and $c_1 = 0$ we find

$$c_2 = -\frac{1}{2}$$

$$c_3 = c_5 = c_7 = \cdots = 0$$

$$c_4 = -\frac{1}{8}$$

$$c_6 = -\frac{7}{336}$$

and so on. For $c_0 = 0$ and $c_1 = 1$ we obtain

$$c_2 = c_4 = c_6 = \cdots = 0$$

$$c_3 = \frac{1}{6}$$

$$c_5 = \frac{1}{24}$$

$$c_7 = \frac{1}{112}$$

and so on. Thus, two solutions are

$$y_1 = 1 - \frac{1}{2}x^2 - \frac{1}{8}x^4 - \frac{7}{336}x^6 - \cdots \quad \text{and} \quad y_2 = x + \frac{1}{6}x^3 + \frac{1}{24}x^5 + \frac{1}{112}x^7 + \cdots .$$

4. Substituting $y = \sum_{n=0}^{\infty} c_n x^n$ into the differential equation we have

$$y'' - xy' + 2y = \underbrace{\sum_{n=2}^{\infty} n(n-1)c_n x^{n-2}}_{k=n-2} - \underbrace{\sum_{n=1}^{\infty} nc_n x^n}_{k=n} + 2\underbrace{\sum_{n=0}^{\infty} c_n x^n}_{k=n}$$

$$= \sum_{k=0}^{\infty} (k+2)(k+1)c_{k+2}x^k - \sum_{k=1}^{\infty} kc_k x^k + 2\sum_{k=0}^{\infty} c_k x^k$$

$$= 2c_2 + 2c_0 + \sum_{k=1}^{\infty} [(k+2)(k+1)c_{k+2} - (k-2)c_k]x^k = 0.$$

Thus
$$2c_2 + 2c_0 = 0$$

$$(k+2)(k+1)c_{k+2} - (k-2)c_k = 0$$

201

and
$$c_2 = -c_0$$

$$c_{k+2} = \frac{k-2}{(k+2)(k+1)} c_k, \quad k = 1, 2, 3, \ldots.$$

Choosing $c_0 = 1$ and $c_1 = 0$ we find

$$c_2 = -1$$

$$c_3 = c_5 = c_7 = \cdots = 0$$

$$c_4 = 0$$

$$c_6 = c_8 = c_{10} = \cdots = 0.$$

For $c_0 = 0$ and $c_1 = 1$ we obtain

$$c_2 = c_4 = c_6 = \cdots = 0$$

$$c_3 = -\frac{1}{6}$$

$$c_5 = -\frac{1}{120}$$

and so on. Thus, two solutions are

$$y_1 = 1 - x^2 \quad \text{and} \quad y_2 = x - \frac{1}{6}x^3 - \frac{1}{120}x^5 - \cdots.$$

5. Substituting $y = \sum_{n=0}^{\infty} c_n x^n$ into the differential equation we have

$$y'' + x^2 y' + xy = \underbrace{\sum_{n=2}^{\infty} n(n-1)c_n x^{n-2}}_{k=n-2} + \underbrace{\sum_{n=1}^{\infty} nc_n x^{n+1}}_{k=n+1} + \underbrace{\sum_{n=0}^{\infty} c_n x^{n+1}}_{k=n+1}$$

$$= \sum_{k=0}^{\infty} (k+2)(k+1)c_{k+2} x^k + \sum_{k=2}^{\infty} (k-1)c_{k-1} x^k + \sum_{k=1}^{\infty} c_{k-1} x^k$$

$$= 2c_2 + (6c_3 + c_0)x + \sum_{k=2}^{\infty} [(k+2)(k+1)c_{k+2} + kc_{k-1}]x^k = 0.$$

Thus

$$c_2 = 0$$

$$6c_3 + c_0 = 0$$

$$(k+2)(k+1)c_{k+2} + kc_{k-1} = 0$$

202

and
$$c_3 = -\frac{1}{6}c_0$$

$$c_{k+2} = -\frac{k}{(k+2)(k+1)}c_{k-1}, \quad k = 2, 3, 4, \dots .$$

Choosing $c_0 = 1$ and $c_1 = 0$ we find
$$c_3 = -\frac{1}{6}$$
$$c_4 = c_5 = 0$$
$$c_6 = \frac{1}{45}$$

and so on. For $c_0 = 0$ and $c_1 = 1$ we obtain
$$c_3 = 0$$
$$c_4 = -\frac{1}{6}$$
$$c_5 = c_6 = 0$$
$$c_7 = \frac{5}{252}$$

and so on. Thus, two solutions are
$$y_1 = 1 - \frac{1}{6}x^3 + \frac{1}{45}x^6 - \cdots \quad \text{and} \quad y_2 = x - \frac{1}{6}x^4 + \frac{5}{232}x^7 - \cdots .$$

6. Substituting $y = \sum_{n=0}^{\infty} c_n x^n$ into the differential equation we have

$$y'' + 2xy' + 2y = \underbrace{\sum_{n=2}^{\infty} n(n-1)c_n x^{n-2}}_{k=n-2} + 2\underbrace{\sum_{n=1}^{\infty} nc_n x^n}_{k=n} + 2\underbrace{\sum_{n=0}^{\infty} c_n x^n}_{k=n}$$

$$= \sum_{k=0}^{\infty}(k+2)(k+1)c_{k+2}x^k + 2\sum_{k=1}^{\infty} kc_k x^k + 2\sum_{k=0}^{\infty} c_k x^k$$

$$= 2c_2 + 2c_0 + \sum_{k=1}^{\infty}[(k+2)(k+1)c_{k+2} + 2(k+1)c_k]x^k = 0.$$

Thus
$$2c_2 + 2c_0 = 0$$
$$(k+2)(k+1)c_{k+2} + 2(k+1)c_k = 0$$

and
$$c_2 = -c_0$$

$$c_{k+2} = -\frac{2}{k+2}\, c_k, \quad k = 1, 2, 3, \ldots .$$

Choosing $c_0 = 1$ and $c_1 = 0$ we find

$$c_2 = -1$$

$$c_3 = c_5 = c_7 = \cdots = 0$$

$$c_4 = \frac{1}{2}$$

$$c_6 = -\frac{1}{6}$$

and so on. For $c_0 = 0$ and $c_1 = 1$ we obtain

$$c_2 = c_4 = c_6 = \cdots = 0$$

$$c_3 = -\frac{2}{3}$$

$$c_5 = \frac{4}{15}$$

$$c_7 = -\frac{8}{105}$$

and so on. Thus, two solutions are

$$y_1 = 1 - x^2 + \frac{1}{2}x^4 - \frac{1}{6}x^6 + \cdots \quad \text{and} \quad y_2 = x - \frac{2}{3}x^3 + \frac{4}{15}x^5 - \frac{8}{105}x^7 + \cdots .$$

7. Substituting $y = \sum_{n=0}^{\infty} c_n x^n$ into the differential equation we have

$$(x-1)y'' + y' = \underbrace{\sum_{n=2}^{\infty} n(n-1)c_n x^{n-1}}_{k=n-1} - \underbrace{\sum_{n=2}^{\infty} n(n-1)c_n x^{n-2}}_{k=n-2} + \underbrace{\sum_{n=1}^{\infty} nc_n x^{n-1}}_{k=n-1}$$

$$= \sum_{k=1}^{\infty} (k+1)k c_{k+1} x^k - \sum_{k=0}^{\infty} (k+2)(k+1)c_{k+2} x^k + \sum_{k=0}^{\infty} (k+1)c_{k+1} x^k$$

$$= -2c_2 + c_1 + \sum_{k=1}^{\infty} [(k+1)k c_{k+1} - (k+2)(k+1)c_{k+2} + (k+1)c_{k+1}] x^k = 0.$$

Thus
$$-2c_2 + c_1 = 0$$

$$(k+1)^2 c_{k+1} - (k+2)(k+1)c_{k+2} = 0$$

and
$$c_2 - \frac{1}{2}c_1$$

$$c_{k+2} = \frac{k+1}{k+2}c_{k+1}, \quad k = 1, 2, 3, \ldots.$$

Choosing $c_0 = 1$ and $c_1 = 0$ we find $c_2 = c_3 = c_4 = \cdots = 0$. For $c_0 = 0$ and $c_1 = 1$ we obtain

$$c_2 = \frac{1}{2}, \qquad c_3 = \frac{1}{3}, \qquad c_4 = \frac{1}{4},$$

and so on. Thus, two solutions are

$$y_1 = 1 \quad \text{and} \quad y_2 = x + \frac{1}{2}x^2 + \frac{1}{3}x^3 + \frac{1}{4}x^4 + \cdots.$$

8. Substituting $y = \sum_{n=0}^{\infty} c_n x^n$ into the differential equation we have

$$(x+2)y'' + xy' - y = \underbrace{\sum_{n=2}^{\infty} n(n-1)c_n x^{n-1}}_{k=n-1} + \underbrace{\sum_{n=2}^{\infty} 2n(n-1)c_n x^{n-2}}_{k=n-2} + \underbrace{\sum_{n=1}^{\infty} nc_n x^n}_{k=n} - \underbrace{\sum_{n=0}^{\infty} c_n x^n}_{k=n}$$

$$= \sum_{k=1}^{\infty} (k+1)kc_{k+1}x^k + \sum_{k=0}^{\infty} 2(k+2)(k+1)c_{k+2}x^k + \sum_{k=1}^{\infty} kc_k x^k - \sum_{k=0}^{\infty} c_k x^k$$

$$= 4c_2 - c_0 + \sum_{k=1}^{\infty} [(k+1)kc_{k+1} + 2(k+2)(k+1)c_{k+2} + (k-1)c_k]x^k = 0.$$

Thus
$$4c_2 - c_0 = 0$$

$$(k+1)kc_{k+1} + 2(k+2)(k+1)c_{k+2} + (k-1)c_k = 0, \quad k = 1, 2, 3, \ldots$$

and
$$c_2 = \frac{1}{4}c_0$$

$$c_{k+2} = -\frac{(k+1)kc_{k+1} + (k-1)c_k}{2(k+2)(k+1)}, \quad k = 1, 2, 3, \ldots.$$

Choosing $c_0 = 1$ and $c_1 = 0$ we find

$$c_1 = 0, \qquad c_2 = \frac{1}{4}, \qquad c_3 = -\frac{1}{24}, \qquad c_4 = 0, \qquad c_5 = \frac{1}{480}$$

and so on. For $c_0 = 0$ and $c_1 = 1$ we obtain

$$c_2 = 0$$

$$c_3 = 0$$

$$c_4 = c_5 = c_6 = \cdots = 0.$$

Thus, two solutions are

$$y_1 = c_0\left[1 + \frac{1}{4}x^2 - \frac{1}{24}x^3 + \frac{1}{480}x^5 + \cdots\right] \quad \text{and} \quad y_2 = c_1 x.$$

9. Substituting $y = \sum_{n=0}^{\infty} c_n x^n$ into the differential equation we have

$$\left(x^2 - 1\right)y'' + 4xy' + 2y = \underbrace{\sum_{n=2}^{\infty} n(n-1)c_n x^n}_{k=n} - \underbrace{\sum_{n=2}^{\infty} n(n-1)c_n x^{n-2}}_{k=n-2} + 4\underbrace{\sum_{n=1}^{\infty} nc_n x^n}_{k=n} + 2\underbrace{\sum_{n=0}^{\infty} c_n x^n}_{k=n}$$

$$= \sum_{k=2}^{\infty} k(k-1)c_k x^k - \sum_{k=0}^{\infty}(k+2)(k+1)c_{k+2}x^k + 4\sum_{k=1}^{\infty} kc_k x^k + 2\sum_{k=0}^{\infty} c_k x^k$$

$$= -2c_2 + 2c_0 + (-6c_3 + 6c_1)x + \sum_{k=2}^{\infty}\left[\left(k^2 - k + 4k + 2\right)c_k - (k+2)(k+1)c_{k+2}\right]x^k = 0.$$

Thus

$$-2c_2 + 2c_0 = 0$$

$$-6c_3 + 6c_1 = 0$$

$$\left(k^2 + 3k + 2\right)c_k - (k+2)(k+1)c_{k+2} = 0$$

and

$$c_2 = c_0$$

$$c_3 = c_1$$

$$c_{k+2} = c_k, \quad k = 2, 3, 4, \ldots .$$

Choosing $c_0 = 1$ and $c_1 = 0$ we find

$$c_2 = 1$$

$$c_3 = c_5 = c_7 = \cdots = 0$$

$$c_4 = c_6 = c_8 = \cdots = 1.$$

For $c_0 = 0$ and $c_1 = 1$ we obtain

$$c_2 = c_4 = c_6 = \cdots = 0$$

$$c_3 = c_5 = c_7 = \cdots = 1.$$

Thus, two solutions are

$$y_1 = 1 + x^2 + x^4 + \cdots \quad \text{and} \quad y_2 = x + x^3 + x^5 + \cdots .$$

10. Substituting $y = \sum_{n=0}^{\infty} c_n x^n$ into the differential equation we have

$$\left(x^2 + 1\right) y'' - 6y = \underbrace{\sum_{n=2}^{\infty} n(n-1)c_n x^n}_{k=n} + \underbrace{\sum_{n=2}^{\infty} n(n-1)c_n x^{n-2}}_{k=n-2} - \underbrace{6 \sum_{n=0}^{\infty} c_n x^n}_{k=n}$$

$$= \sum_{k=2}^{\infty} k(k-1)c_k x^k + \sum_{k=0}^{\infty} (k+2)(k+1)c_{k+2}x^k - 6 \sum_{k=0}^{\infty} c_k x^k$$

$$= 2c_2 - 6c_0 + (6c_3 - 6c_1)x + \sum_{k=2}^{\infty} \left[\left(k^2 - k - 6\right)c_k + (k+2)(k+1)c_{k+2}\right] x^k = 0.$$

Thus
$$2c_2 - 6c_0 = 0$$

$$6c_3 - 6c_1 = 0$$

$$(k-3)(k+2)c_k + (k+2)(k+1)c_{k+2} = 0$$

and
$$c_2 = 3c_0$$

$$c_3 = c_1$$

$$c_{k+2} = -\frac{k-3}{k+1} c_k, \quad k = 2, 3, 4, \ldots.$$

Choosing $c_0 = 1$ and $c_1 = 0$ we find

$$c_2 = 3$$

$$c_3 = c_5 = c_7 = \cdots = 0$$

$$c_4 = 1$$

$$c_6 = -\frac{1}{5}$$

and so on. For $c_0 = 0$ and $c_1 = 1$ we obtain

$$c_2 = c_4 = c_6 = \cdots = 0$$

$$c_3 = 1$$

$$c_5 = c_7 = c_9 = \cdots = 0.$$

Thus, two solutions are

$$y_1 = 1 + 3x^2 + x^4 - \frac{1}{5}x^6 + \cdots \quad \text{and} \quad y_2 = x + x^3.$$

207

11. Substituting $y = \sum_{n=0}^{\infty} c_n x^n$ into the differential equation we have

$$\left(x^2 + 2\right)y'' + 3xy' - y = \underbrace{\sum_{n=2}^{\infty} n(n-1)c_n x^n}_{k=n} + \underbrace{2\sum_{n=2}^{\infty} n(n-1)c_n x^{n-2}}_{k=n-2} + \underbrace{3\sum_{n=1}^{\infty} nc_n x^n}_{k=n} - \underbrace{\sum_{n=0}^{\infty} c_n x^n}_{k=n}$$

$$= \sum_{k=2}^{\infty} k(k-1)c_k x^k + 2\sum_{k=0}^{\infty} (k+2)(k+1)c_{k+2} x^k + 3\sum_{k=1}^{\infty} kc_k x^k - \sum_{k=0}^{\infty} c_k x^k$$

$$= (4c_2 - c_0) + (12c_3 + 2c_1)x + \sum_{k=2}^{\infty} \left[2(k+2)(k+1)c_{k+2} + \left(k^2 + 2k - 1\right)c_k\right]x^k = 0.$$

Thus
$$4c_2 - c_0 = 0$$

$$12c_3 + 2c_1 = 0$$

$$2(k+2)(k+1)c_{k+2} + \left(k^2 + 2k - 1\right)c_k = 0$$

and
$$c_2 = \frac{1}{4}c_0$$

$$c_3 = -\frac{1}{6}c_1$$

$$c_{k+2} = -\frac{k^2 + 2k - 1}{2(k+2)(k+1)}c_k, \quad k = 2, 3, 4, \dots .$$

Choosing $c_0 = 1$ and $c_1 = 0$ we find

$$c_2 = \frac{1}{4}$$

$$c_3 = c_5 = c_7 = \cdots = 0$$

$$c_4 = -\frac{7}{96}$$

and so on. For $c_0 = 0$ and $c_1 = 1$ we obtain

$$c_2 = c_4 = c_6 = \cdots = 0$$

$$c_3 = -\frac{1}{6}$$

$$c_5 = \frac{7}{120}$$

208

and so on. Thus, two solutions are

$$y_1 = 1 + \frac{1}{4}x^2 - \frac{7}{96}x^4 + \cdots \quad \text{and} \quad y_2 = x - \frac{1}{6}x^3 + \frac{7}{120}x^5 - \cdots .$$

12. Substituting $y = \sum_{n=0}^{\infty} c_n x^n$ into the differential equation we have

$$\left(x^2 - 1\right)y'' + xy' - y = \underbrace{\sum_{n=2}^{\infty} n(n-1)c_n x^n}_{k=n} - \underbrace{\sum_{n=2}^{\infty} n(n-1)c_n x^{n-2}}_{k=n-2} + \underbrace{\sum_{n=1}^{\infty} nc_n x^n}_{k=n} - \underbrace{\sum_{n=0}^{\infty} c_n x^n}_{k=n}$$

$$= \sum_{k=2}^{\infty} k(k-1)c_k x^k - \sum_{k=0}^{\infty} (k+2)(k+1)c_{k+2}x^k + \sum_{k=1}^{\infty} kc_k x^k - \sum_{k=0}^{\infty} c_k x^k$$

$$= (-c_2 - c_0) - 6c_3 x + \sum_{k=2}^{\infty} \left[-(k+2)(k+1)c_{k+2} + \left(k^2 - 1\right)c_k\right]x^k = 0.$$

Thus

$$-2c_2 - c_0 = 0$$

$$-6c_3 = 0$$

$$-(k+2)(k+1)c_{k+2} + (k-1)(k+1)c_k = 0$$

and

$$c_2 = -\frac{1}{2}c_0$$

$$c_3 = 0$$

$$c_{k+2} = \frac{k-1}{k+2}c_k, \quad k = 2, 3, 4, \ldots .$$

Choosing $c_0 = 1$ and $c_1 = 0$ we find

$$c_2 = -\frac{1}{2}$$

$$c_3 = c_5 = c_7 = \cdots = 0$$

$$c_4 = -\frac{1}{8}$$

and so on. For $c_0 = 0$ and $c_1 = 1$ we obtain

$$c_2 = c_4 = c_6 = \cdots = 0$$

$$c_3 = c_5 = c_7 = \cdots = 0.$$

Thus, two solutions are

$$y_1 = 1 - \frac{1}{2}x^2 - \frac{1}{8}x^4 - \cdots \quad \text{and} \quad y_2 = x.$$

13. Substituting $y = \sum_{n=0}^{\infty} c_n x^n$ into the differential equation we have

$$y'' - (x+1)y' - y = \underbrace{\sum_{n=2}^{\infty} n(n-1)c_n x^{n-2}}_{k=n-2} - \underbrace{\sum_{n=1}^{\infty} nc_n x^n}_{k=n} - \underbrace{\sum_{n=1}^{\infty} nc_n x^{n-1}}_{k=n-1} - \underbrace{\sum_{n=0}^{\infty} c_n x^n}_{k=n}$$

$$= \sum_{k=0}^{\infty} (k+2)(k+1)c_{k+2} x^k - \sum_{k=1}^{\infty} kc_k x^k - \sum_{k=0}^{\infty} (k+1)c_{k+1} x^k - \sum_{k=0}^{\infty} c_k x^k$$

$$= 2c_2 - c_1 - c_0 + \sum_{k=1}^{\infty} [(k+2)(k+1)c_{k+2} - (k+1)c_{k+1} - (k+1)c_k]x^k = 0.$$

Thus

$$2c_2 - c_1 - c_0 = 0$$

$$(k+2)(k+1)c_{k+2} - (k-1)(c_{k+1} + c_k) = 0$$

and

$$c_2 = \frac{c_1 + c_0}{2}$$

$$c_{k+2} = \frac{c_{k+1} + c_k}{k+2} c_k, \quad k = 2, 3, 4, \ldots.$$

Choosing $c_0 = 1$ and $c_1 = 0$ we find

$$c_2 = \frac{1}{2}, \qquad c_3 = \frac{1}{6}, \qquad c_4 = \frac{1}{6}$$

and so on. For $c_0 = 0$ and $c_1 = 1$ we obtain

$$c_2 = \frac{1}{2}, \qquad c_3 = \frac{1}{2}, \qquad c_4 = \frac{1}{4}$$

and so on. Thus, two solutions are

$$y_1 = 1 + \frac{1}{2}x^2 + \frac{1}{6}x^3 + \frac{1}{6}x^4 + \cdots \quad \text{and} \quad y_2 = x + \frac{1}{2}x^2 + \frac{1}{2}x^3 + \frac{1}{4}x^4 + \cdots.$$

14. Substituting $y = \sum_{n=0}^{\infty} c_n x^n$ into the differential equation we have

$$y'' - xy' - (x+2)y = \underbrace{\sum_{n=2}^{\infty} n(n-1)c_n x^{n-2}}_{k=n-2} - \underbrace{\sum_{n=1}^{\infty} nc_n x^n}_{k=n} - \underbrace{\sum_{n=0}^{\infty} c_n x^{n+1}}_{k=n+1} - \underbrace{\sum_{n=0}^{\infty} 2c_n x^n}_{k=n}$$

$$= \sum_{k=0}^{\infty} (k+2)(k+1)c_{k+2} x^k - \sum_{k=1}^{\infty} kc_k x^k - \sum_{k=1}^{\infty} c_{k-1} x^k - \sum_{k=0}^{\infty} 2c_k x^k$$

$$= 2c_2 - 2c_0 + \sum_{k=1}^{\infty} [(k+2)(k+1)c_{k+2} - (k+2)c_k - c_{k-1}]x^k = 0.$$

Thus
$$2c_2 - 2c_0 = 0$$

$$(k+2)(k+1)c_{k+2} - (k+2)c_k - c_{k-1} = 0, \qquad k = 1, 2, 3, \ldots,$$

and
$$c_2 = c_0$$

$$c_{k+2} = \frac{c_k}{k+1} + \frac{c_{k-1}}{(k+2)(k+1)}, \qquad k = 1, 2, 3, \ldots.$$

Choosing $c_0 = 1$ and $c_1 = 0$ we find

$$c_1 = 0, \qquad c_2 = 1, \qquad c_3 = \frac{1}{6}, \qquad c_4 = \frac{1}{3}, \qquad c_5 = \frac{11}{5!}$$

and so on. For $c_0 = 0$ and $c_1 = 1$ we obtain

$$c_2 = c_0 = 0, \qquad c_3 = \frac{1}{2}, \qquad c_4 = \frac{1}{12}, \qquad c_5 = \frac{1}{8}$$

and so on. Thus, two solutions are

$$y_1 = 1 + x^2 + \frac{1}{6}x^3 + \frac{1}{3}x^4 + \frac{11}{5!}x^5 + \cdots \quad \text{and} \quad y_2 = x + \frac{1}{2}x^3 + \frac{1}{12}x^4 + \frac{1}{8}x^5 + \cdots.$$

15. Substituting $y = \sum_{n=0}^{\infty} c_n x^n$ into the differential equation we have

$$(x-1)y'' - xy' + y = \underbrace{\sum_{n=2}^{\infty} n(n-1)c_n x^{n-1}}_{k=n-1} - \underbrace{\sum_{n=2}^{\infty} n(n-1)c_n x^{n-2}}_{k=n-2} - \underbrace{\sum_{n=1}^{\infty} nc_n x^n}_{k=n} + \underbrace{\sum_{n=0}^{\infty} c_n x^n}_{k=n}$$

$$= \sum_{k=1}^{\infty} (k+1)kc_{k+1} x^k - \sum_{k=0}^{\infty} (k+2)(k+1)c_{k+2} x^k - \sum_{k=1}^{\infty} kc_k x^k + \sum_{k=0}^{\infty} c_k x^k$$

$$= -2c_2 + c_0 + \sum_{k=1}^{\infty} [-(k+2)(k+1)c_{k+2} + (k+1)kc_{k+1} - (k-1)c_k]x^k = 0.$$

211

Thus
$$-2c_2 + c_0 = 0$$

$$-(k+2)(k+1)c_{k+2} + (k-1)kc_{k+1} - (k-1)c_k = 0$$

and
$$c_2 = \frac{1}{2}c_0$$

$$c_{k+2} = \frac{kc_{k+1}}{k+2} - \frac{(k-1)c_k}{(k+2)(k+1)}, \quad k = 1, 2, 3, \ldots.$$

Choosing $c_0 = 1$ and $c_1 = 0$ we find

$$c_2 = \frac{1}{2}, \qquad c_3 = \frac{1}{6}, \qquad c_4 = 0$$

and so on. For $c_0 = 0$ and $c_1 = 1$ we obtain $c_2 = c_3 = c_4 = \cdots = 0$. Thus,

$$y = C_1\left(1 + \frac{1}{2}x^2 + \frac{1}{6}x^3 + \cdots\right) + C_2 x$$

and

$$y' = C_1\left(x + \frac{1}{2}x^2 + \cdots\right) + C_2.$$

The initial conditions imply $C_1 = -2$ and $C_2 = 6$, so

$$y = -2\left(1 + \frac{1}{2}x^2 + \frac{1}{6}x^3 + \cdots\right) + 6x = 8x - 2e^x.$$

16. Substituting $y = \sum_{n=0}^{\infty} c_n x^n$ into the differential equation we have

$$(x+1)y'' - (2-x)y' + y$$

$$= \underbrace{\sum_{n=2}^{\infty} n(n-1)c_n x^{n-1}}_{k=n-1} + \underbrace{\sum_{n=2}^{\infty} n(n-1)c_n x^{n-2}}_{k=n-2} - 2\underbrace{\sum_{n=1}^{\infty} nc_n x^{n-1}}_{k=n-1} + \underbrace{\sum_{n=1}^{\infty} nc_n x^n}_{k=n} + \underbrace{\sum_{n=0}^{\infty} c_n x^n}_{k=n}$$

$$= \sum_{k=1}^{\infty}(k+1)kc_{k+1}x^k + \sum_{k=0}^{\infty}(k+2)(k+1)c_{k+2}x^k - 2\sum_{k=0}^{\infty}(k+1)c_{k+1}x^k + \sum_{k=1}^{\infty} kc_k x^k + \sum_{k=0}^{\infty} c_k x^k$$

$$= 2c_2 - 2c_1 + c_0 + \sum_{k=1}^{\infty}[(k+2)(k+1)c_{k+2} - (k+1)c_{k+1} + (k+1)c_k]x^k = 0.$$

Thus
$$2c_2 - 2c_1 + c_0 = 0$$

$$(k+2)(k+1)c_{k+2} - (k+1)c_{k+1} + (k+1)c_k = 0$$

and
$$c_2 = c_1 - \frac{1}{2}c_0$$

$$c_{k+2} = \frac{1}{k+2}c_{k+1} - \frac{1}{k+2}c_k, \quad k = 1, 2, 3, \ldots .$$

Choosing $c_0 = 1$ and $c_1 = 0$ we find
$$c_2 = -\frac{1}{2}, \qquad c_3 = -\frac{1}{6}, \qquad c_4 = \frac{1}{12}$$

and so on. For $c_0 = 0$ and $c_1 = 1$ we obtain
$$c_2 = 1, \qquad c_3 = 0, \qquad c_4 = -\frac{1}{4}$$

and so on. Thus,
$$y = C_1 \left(1 - \frac{1}{2}x^2 - \frac{1}{6}x^3 + \frac{1}{12}x^4 + \cdots\right) + C_2 \left(x + x^2 - \frac{1}{4}x^4 + \cdots\right)$$

and
$$y' = C_1 \left(-x - \frac{1}{2}x^2 + \frac{1}{3}x^3 + \cdots\right) + C_2 \left(1 + 2x - x^3 + \cdots\right).$$

The initial conditions imply $C_1 = 2$ and $C_2 = -1$, so
$$y = 2\left(1 - \frac{1}{2}x^2 - \frac{1}{6}x^3 + \frac{1}{12}x^4 + \cdots\right) - \left(x + x^2 - \frac{1}{4}x^4 + \cdots\right)$$

$$= 2 - x - 2x^2 - \frac{1}{3}x^3 + \frac{5}{12}x^4 + \cdots .$$

17. Substituting $y = \sum_{n=0}^{\infty} c_n x^n$ into the differential equation we have

$$y'' - 2xy' + 8y = \underbrace{\sum_{n=2}^{\infty} n(n-1)c_n x^{n-2}}_{k=n-2} - 2\underbrace{\sum_{n=1}^{\infty} nc_n x^n}_{k=n} + 8\underbrace{\sum_{n=0}^{\infty} c_n x^n}_{k=n}$$

$$= \sum_{k=0}^{\infty} (k+2)(k+1)c_{k+2}x^k - 2\sum_{k=1}^{\infty} kc_k x^k + 8\sum_{k=0}^{\infty} c_k x^k$$

$$= 2c_2 + 8c_0 + \sum_{k=1}^{\infty} [(k+2)(k+1)c_{k+2} + (8-2k)c_k]x^k = 0.$$

Thus
$$2c_2 + 8c_0 = 0$$

$$(k+2)(k+1)c_{k+2} + (8-2k)c_k = 0$$

and
$$c_2 = -4c_0$$

$$c_{k+2} = \frac{2k-8}{(k+2)(k+1)}c_k, \quad k = 1, 2, 3, \ldots .$$

213

Choosing $c_0 = 1$ and $c_1 = 0$ we find

$$c_2 = -4$$

$$c_3 = c_5 = c_7 = \cdots = 0$$

$$c_4 = \frac{4}{3}$$

$$c_6 = c_8 = c_{10} = \cdots = 0.$$

For $c_0 = 0$ and $c_1 = 1$ we obtain

$$c_2 = c_4 = c_6 = \cdots = 0$$

$$c_3 = -1$$

$$c_5 = \frac{1}{10}$$

and so on. Thus,

$$y = C_1\left(1 - 4x^2 + \frac{4}{3}x^4\right) + C_2\left(x - x^3 + \frac{1}{10}x^5 + \cdots\right)$$

and

$$y' = C_1\left(-8x + \frac{16}{3}x^3\right) + C_2\left(1 - 3x^2 + \frac{1}{2}x^4 + \cdots\right).$$

The initial conditions imply $C_1 = 3$ and $C_2 = 0$, so

$$y = 3\left(1 - 4x^2 + \frac{4}{3}x^4\right) = 3 - 12x^2 + 4x^4.$$

18. Substituting $y = \sum_{n=0}^{\infty} c_n x^n$ into the differential equation we have

$$(x^2 + 1)y'' + 2xy' = \underbrace{\sum_{n=2}^{\infty} n(n-1)c_n x^n}_{k=n} + \underbrace{\sum_{n=2}^{\infty} n(n-1)c_n x^{n-2}}_{k=n-2} + \underbrace{\sum_{n=1}^{\infty} 2nc_n x^n}_{k=n}$$

$$= \sum_{k=2}^{\infty} k(k-1)c_k x^k + \sum_{k=0}^{\infty} (k+2)(k+1)c_{k+2}x^k + \sum_{k=1}^{\infty} 2kc_k x^k$$

$$= 2c_2 + (6c_3 + 2c_1)x + \sum_{k=2}^{\infty}[k(k+1)c_k + (k+2)(k+1)c_{k+2}]x^k = 0.$$

Thus

$$2c_2 = 0$$

$$6c_3 + 2c_1 = 0$$

$$k(k+1)c_k + (k+2)(k+1)c_{k+2} = 0$$

and
$$c_2 = 0$$

$$c_3 = -\frac{1}{3}c_1$$

$$c_{k+2} = -\frac{k}{k+2}c_k, \quad k = 2, 3, 4, \ldots.$$

Choosing $c_0 = 1$ and $c_1 = 0$ we find $c_3 = c_4 = c_5 = \cdots = 0$. For $c_0 = 0$ and $c_1 = 1$ we obtain

$$c_3 = -\frac{1}{3}$$

$$c_4 = c_6 = c_8 = \cdots = 0$$

$$c_5 = -\frac{1}{5}$$

$$c_7 = \frac{1}{7}$$

and so on. Thus

$$y = c_0 + c_1 \left(x - \frac{1}{3}x^3 + \frac{1}{5}x^5 - \frac{1}{7}x^7 + \cdots \right)$$

and

$$y' = c_1 \left(1 - x^2 + x^4 - x^6 + \cdots \right).$$

The initial conditions imply $c_0 = 0$ and $c_1 = 1$, so

$$y = x - \frac{1}{3}x^3 + \frac{1}{5}x^5 - \frac{1}{7}x^7 + \cdots.$$

19. Substituting $y = \sum_{n=0}^{\infty} c_n x^n$ into the differential equation we have

$$y'' + (\sin x)y = \sum_{n=2}^{\infty} n(n-1)c_n x^{n-2} + \left(x - \frac{1}{6}x^3 + \frac{1}{120}x^5 - \cdots \right)\left(c_0 + c_1 x + c_2 x^2 + \cdots \right)$$

$$= \left[2c_2 + 6c_3 x + 12c_4 x^2 + 20c_5 x^3 + \cdots \right] + \left[c_0 x + c_1 x^2 + \left(c_2 - \frac{1}{6}c_0 \right)x^3 + \cdots \right]$$

$$= 2c_2 + (6c_3 + c_0)x + (12c_4 + c_1)x^2 + \left(20c_5 + c_2 - \frac{1}{6}c_0 \right)x^3 + \cdots = 0.$$

Thus
$$2c_2 = 0$$

$$6c_3 + c_0 = 0$$

$$12c_4 + c_1 = 0$$

$$20c_5 + c_2 - \frac{1}{6}c_0 = 0$$

215

and
$$c_2 = 0$$

$$c_3 = -\frac{1}{6}c_0$$

$$c_4 = -\frac{1}{12}c_1$$

$$c_5 = -\frac{1}{20}c_2 + \frac{1}{120}c_0.$$

Choosing $c_0 = 1$ and $c_1 = 0$ we find

$$c_2 = 0, \qquad c_3 = -\frac{1}{6}, \qquad c_4 = 0, \qquad c_5 = \frac{1}{120}$$

and so on. For $c_0 = 0$ and $c_1 = 1$ we obtain

$$c_2 = 0, \qquad c_3 = 0, \qquad c_4 = -\frac{1}{12}, \qquad c_5 = 0$$

and so on. Thus, two solutions are

$$y_1 = 1 - \frac{1}{6}x^3 + \frac{1}{120}x^5 + \cdots \quad \text{and} \quad y_2 = x - \frac{1}{12}x^4 + \cdots.$$

20. Substituting $y = \sum_{n=0}^{\infty} c_n x^n$ into the differential equation we have

$$y'' + \frac{\sin x}{x}y = \sum_{n=2}^{\infty} n(n-1)c_n x^{n-2} + \left(1 - \frac{1}{6}x^2 + \frac{1}{120}x^4 - \cdots\right)\left(c_0 + c_1 x + c_2 x^2 + c_3 x^3 + \cdots\right)$$

$$= \left[2c_2 + 6c_3 x + 12c_4 x^2 + 20c_5 x^3 + \cdots\right]$$

$$+ \left[c_0 + c_1 x + \left(c_2 - \frac{1}{6}c_0\right)x^2 + \left(c_3 - \frac{1}{6}c_1\right)x^3 + \cdots\right]$$

$$= (2c_2 + c_0) + (6c_3 + c_1)x + \left(12c_4 + c_2 - \frac{1}{6}c_0\right)x^2 + \cdots = 0.$$

Thus
$$2c_2 + c_0 = 0$$

$$6c_3 + c_1 = 0$$

$$12c_4 + c_2 - \frac{1}{6}c_0 = 0$$

and
$$c_2 = -\frac{1}{2}$$

$$c_3 = -\frac{1}{6}c_1$$

$$c_4 = -\frac{1}{12}c_2 + \frac{1}{72}c_0.$$

Choosing $c_0 = 1$ and $c_1 = 0$ we find

$$c_2 = -\frac{1}{2}, \qquad c_3 = 0, \qquad c_4 = \frac{1}{18}$$

and so on. For $c_0 = 0$ and $c_1 = 1$ we obtain

$$c_2 = 0, \qquad c_3 = -\frac{1}{6}, \qquad c_4 = 0$$

and so on. Thus, two solutions are

$$y_1 = 1 - \frac{1}{2}x^2 + \frac{1}{18}x^4 - \cdots \quad \text{and} \quad y_2 = x - \frac{1}{6}x^3 + \cdots .$$

21. Substituting $y = \sum_{n=0}^{\infty} c_n x^n$ into the differential equation we have

$$y'' + e^{-x}y = \sum_{n=2}^{\infty} n(n-1)c_n x^{n-2}$$

$$+ \left(1 - x + \frac{1}{2}x^2 - \frac{1}{6}x^3 + \frac{1}{24}x^4 - \cdots\right)\left(c_0 + c_1 x + c_2 x^2 + c_3 x^3 + \cdots\right)$$

$$= \left[2c_2 + 6c_3 x + 12c_4 x^2 + 20c_5 x^3 + \cdots\right] + \left[c_0 + (c_1 - c_0)x + \left(c_2 - c_1 + \frac{1}{2}c_0\right)x^2 + \cdots\right]$$

$$= (2c_2 + c_0) + (6c_3 + c_1 - c_0)x + \left(12c_4 + c_2 - c_1 + \frac{1}{2}c_0\right)x^2 + \cdots = 0.$$

Thus
$$2c_2 + c_0 = 0$$

$$6c_3 + c_1 - c_0 = 0$$

$$12c_4 + c_2 - c_1 + \frac{1}{2}c_0 = 0$$

and
$$c_2 = -\frac{1}{2}c_0$$

$$c_3 = -\frac{1}{6}c_1 + \frac{1}{6}c_0$$

$$c_4 = -\frac{1}{12}c_2 + \frac{1}{12}c_1 - \frac{1}{24}c_0.$$

Choosing $c_0 = 1$ and $c_1 = 0$ we find

$$c_2 = -\frac{1}{2}, \qquad c_3 = \frac{1}{6}, \qquad c_4 = 0$$

and so on. For $c_0 = 0$ and $c_1 = 1$ we obtain

$$c_2 = 0, \qquad c_3 = -\frac{1}{6}, \qquad c_4 = \frac{1}{12}.$$

217

Thus, two solutions are

$$y_1 = 1 - \frac{1}{2}x^2 + \frac{1}{6}x^3 + \cdots \quad \text{and} \quad y_2 = x - \frac{1}{6}x^3 + \frac{1}{12}x^4 + \cdots .$$

22. Substituting $y = \sum_{n=0}^{\infty} c_n x^n$ into the differential equation we have

$$y'' + e^x y' - y = \sum_{n=2}^{\infty} n(n-1)c_n x^{n-2}$$

$$+ \left(1 + x + \frac{1}{2}x^2 + \frac{1}{6}x^3 + \cdots\right)\left(c_1 + 2c_2 x + 3c_3 x^2 + 4c_4 x^3 + \cdots\right) - \sum_{n=0}^{\infty} c_n x^n$$

$$= \left[2c_2 + 6c_3 x + 12c_4 x^2 + 20c_5 x^3 + \cdots\right]$$

$$+ \left[c_1 + (2c_2 + c_1)x + \left(3c_3 + 2c_2 + \frac{1}{2}c_1\right)x^2 + \cdots\right] - [c_0 + c_1 x + c_2 x^2 + \cdots]$$

$$= (2c_2 + c_1 - c_0) + (6c_3 + 2c_2)x + \left(12c_4 + 3c_3 + c_2 + \frac{1}{2}c_1\right)x^2 + \cdots = 0.$$

Thus
$$2c_2 + c_1 - c_0 = 0$$

$$6c_3 + 2c_2 = 0$$

$$12c_4 + 3c_3 + c_2 + \frac{1}{2}c_1 = 0$$

and
$$c_2 = \frac{1}{2}c_0 - \frac{1}{2}c_1$$

$$c_3 = -\frac{1}{3}c_2$$

$$c_4 = -\frac{1}{4}c_3 + \frac{1}{12}c_2 - \frac{1}{24}c_1.$$

Choosing $c_0 = 1$ and $c_1 = 0$ we find

$$c_2 = \frac{1}{2}, \qquad c_3 = -\frac{1}{6}, \qquad c_4 = 0$$

and so on. For $c_0 = 0$ and $c_1 = 1$ we obtain

$$c_2 = -\frac{1}{2}, \qquad c_3 = \frac{1}{6}, \qquad c_4 = -\frac{1}{24}$$

and so on. Thus, two solutions are

$$y_1 = 1 + \frac{1}{2}x^2 - \frac{1}{6}x^3 + \cdots \quad \text{and} \quad y_2 = x - \frac{1}{2}x^2 + \frac{1}{6}x^3 - \frac{1}{24}x^4 + \cdots .$$

23. Substituting $y = \sum_{n=0}^{\infty} c_n x^n$ into the differential equation leads to

$$y'' - xy = \underbrace{\sum_{n=2}^{\infty} n(n-1)c_n x^{n-2}}_{k=n-2} - \underbrace{\sum_{n=0}^{\infty} c_n x^{n+1}}_{k=n+1} = \sum_{k=0}^{\infty} (k+2)(k+1)c_{k+2}x^k - \sum_{k=1}^{\infty} c_{k-1}x^k$$

$$= 2c_2 + \sum_{k=1}^{\infty} [(k+2)(k+1)c_{k+2} - c_{k-1}]x^k = 1.$$

Thus
$$2c_2 = 1$$

$$(k+2)(k+1)c_{k+2} - c_{k-1} = 0$$

and
$$c_2 = \frac{1}{2}$$

$$c_{k+2} = \frac{c_{k-1}}{(k+2)(k+1)}, \quad k = 1, 2, 3, \dots.$$

Let c_0 and c_1 be arbitrary and iterate to find

$$c_2 = \frac{1}{2}$$

$$c_3 = \frac{1}{6}c_0$$

$$c_4 = \frac{1}{12}c_1$$

$$c_5 = \frac{1}{20}c_2 = \frac{1}{40}$$

and so on. The solution is

$$y = c_0 + c_1 x + \frac{1}{2}x^2 + \frac{1}{6}c_0 x^3 + \frac{1}{12}c_1 x^4 + \frac{1}{40}c_5 + \cdots$$

$$= c_0 \left(1 + \frac{1}{6}x^3 + \cdots\right) + c_1 \left(x + \frac{1}{12}x^4 + \cdots\right) + \frac{1}{2}x^2 + \frac{1}{40}x^5 + \cdots.$$

24. Substituting $y = \sum_{n=0}^{\infty} c_n x^n$ into the differential equation leads to

$$y'' - 4xy' - 4y = \underbrace{\sum_{n=2}^{\infty} n(n-1)c_n x^{n-2}}_{k=n-2} - \underbrace{\sum_{n=1}^{\infty} 4nc_n x^n}_{k=n} - \underbrace{\sum_{n=0}^{\infty} 4c_n x^n}_{k=n}$$

$$= \sum_{k=0}^{\infty} (k+2)(k+1)c_{k+2}x^k - \sum_{k=1}^{\infty} 4kc_k x^k - \sum_{k=0}^{\infty} 4c_k x^k$$

$$= 2c_2 - 4c_0 + \sum_{k=1}^{\infty} [(k+2)(k+1)c_{k+2} - 4(k+1)c_k]x^k$$

$$= e^x = 1 + \sum_{k=1}^{\infty} \frac{1}{k!}x^k.$$

Thus
$$2c_2 - 4c_0 = 1$$

$$(k+2)(k+1)c_{k+2} - 4(k+1)c_k = \frac{1}{k!}$$

and
$$c_2 = \frac{1}{2} + 2c_0$$

$$c_{k+2} = \frac{1}{(k+2)!} + \frac{4}{k+2}c_k, \qquad k = 1, 2, 3, \ldots.$$

Let c_0 and c_1 be arbitrary and iterate to find

$$c_2 = \frac{1}{2} + 2c_0$$

$$c_3 = \frac{1}{3!} + \frac{4}{3}c_1 = \frac{1}{3!} + \frac{4}{3}c_1$$

$$c_4 = \frac{1}{4!} + \frac{4}{4}c_2 = \frac{1}{4!} + \frac{1}{2} + 2c_0 = \frac{13}{4!} + 2c_0$$

$$c_5 = \frac{1}{5!} + \frac{4}{5}c_3 = \frac{1}{5!} + \frac{4}{5 \cdot 3!} + \frac{16}{15}c_1 = \frac{17}{5!} + \frac{16}{15}c_1$$

$$c_6 = \frac{1}{6!} + \frac{4}{6}c_4 = \frac{1}{6!} + \frac{4 \cdot 13}{6 \cdot 4!} + \frac{8}{6}c_0 = \frac{261}{6!} + \frac{4}{3}c_0$$

$$c_7 = \frac{1}{7!} + \frac{4}{7}c_5 = \frac{1}{7!} + \frac{4 \cdot 17}{7 \cdot 5!} + \frac{64}{105}c_1 = \frac{409}{7!} + \frac{64}{105}c_1$$

and so on. The solution is

$$y = c_0 + c_1 x + \left(\frac{1}{2} + 2c_0\right)x^2 + \left(\frac{1}{3!} + \frac{4}{3}c_1\right)x^3 - \left(\frac{13}{4!} + 2c_0\right)x^4 + \left(\frac{17}{5!} + \frac{16}{15}c_1\right)x^5$$

$$+ \left(\frac{261}{6!} + \frac{4}{3}c_0\right)x^6 + \left(\frac{409}{7!} + \frac{64}{105}c_1\right)x^7 + \cdots$$

$$= c_0\left[1 + 2x^2 + 2x^4 + \frac{4}{3}x^6 + \cdots\right] + c_1\left[x + \frac{4}{3}x^3 + \frac{16}{15}x^5 + \frac{64}{105}x^7 + \cdots\right]$$

$$+ \frac{1}{2}x^2 + \frac{1}{3!}x^3 + \frac{13}{4!}x^4 + \frac{17}{5!}x^5 + \frac{261}{6!}x^6 + \frac{409}{7!}x^7 + \cdots.$$

1. Irregular singular point: $x = 0$

2. Regular singular points: $x = 0, -3$

3. Irregular singular point: $x = 3$; regular singular point: $x = -3$

4. Irregular singular point: $x = 1$; regular singular point: $x = 0$

5. Regular singular points: $x = 0, \pm 2i$

6. Irregular singular point: $x = 5$; regular singular point: $x = 0$

7. Regular singular points: $x = -3, 2$

8. Regular singular points: $x = 0, \pm i$

9. Irregular singular point: $x = 0$; regular singular points: $x = 2, \pm 5$

10. Irregular singular point: $x = -1$; regular singular points: $x = 0, 3$

11. Substituting $y = \sum_{n=0}^{\infty} c_n x^{n+r}$ into the differential equation and collecting terms, we obtain

$$2xy'' - y' + 2y = \left(2r^2 - 3r\right) c_0 x^{r-1} + \sum_{k=1}^{\infty} [2(k+r-1)(k+r)c_k - (k+r)c_k + 2c_{k-1}]x^{k+r-1} = 0,$$

which implies
$$2r^2 - 3r = r(2r - 3) = 0$$

and
$$(k+r)(2k + 2r - 3)c_k + 2c_{k-1} = 0.$$

The indicial roots are $r = 0$ and $r = 3/2$. For $r = 0$ the recurrence relation is

$$c_k = -\frac{2c_{k-1}}{k(2k-3)}, \quad k = 1, 2, 3, \ldots,$$

and
$$c_1 = 2c_0, \qquad c_2 = -2c_0, \qquad c_3 = \frac{4}{9}c_0.$$

For $r = 3/2$ the recurrence relation is

$$c_k = -\frac{2c_{k-1}}{(2k+3)k}, \quad k = 1, 2, 3, \ldots,$$

and
$$c_1 = -\frac{2}{5}c_0, \qquad c_2 = \frac{2}{35}c_0, \qquad c_3 = -\frac{4}{945}c_0.$$

The general solution on $(0, \infty)$ is

$$y = C_1 \left(1 + 2x - 2x^2 + \frac{4}{9}x^3 + \cdots \right) + C_2 x^{3/2} \left(1 - \frac{2}{5}x + \frac{2}{35}x^2 - \frac{4}{945}x^3 + \cdots \right).$$

Exercises 6.3

12. Substituting $y = \sum_{n=0}^{\infty} c_n x^{n+r}$ into the differential equation and collecting terms, we obtain

$$2xy'' + 5y' + xy = \left(2r^2 + 3r\right)c_0 x^{r-1} + \left(2r^2 + 7r + 5\right)c_1 x^r$$

$$+ \sum_{k=2}^{\infty} [2(k+r)(k+r-1)c_k + 5(k+r)c_k + c_{k-2}]x^{k+r-1}$$

$$= 0,$$

which implies

$$2r^2 + 3r = r(2r+3) = 0,$$

$$\left(2r^2 + 7r + 5\right)c_1 = 0,$$

and
$$(k+r)(2k+2r+3)c_k + c_{k-2} = 0.$$

The indicial roots are $r = -3/2$ and $r = 0$, so $c_1 = 0$. For $r = -3/2$ the recurrence relation is

$$c_k = -\frac{c_{k-2}}{(2k-3)k}, \quad k = 2, 3, 4, \ldots,$$

and
$$c_2 = -\frac{1}{2}c_0, \qquad c_3 = 0, \qquad c_4 = \frac{1}{40}c_0.$$

For $r = 0$ the recurrence relation is

$$c_k = -\frac{c_{k-2}}{k(2k+3)}, \quad k = 2, 3, 4, \ldots,$$

and
$$c_2 = -\frac{1}{14}c_0, \qquad c_3 = 0, \qquad c_4 = \frac{1}{616}c_0.$$

The general solution on $(0, \infty)$ is

$$y = C_1 x^{-3/2}\left(1 - \frac{1}{2}x^2 + \frac{1}{40}x^4 + \cdots\right) + C_2\left(1 - \frac{1}{14}x^2 + \frac{1}{616}x^4 + \cdots\right).$$

13. Substituting $y = \sum_{n=0}^{\infty} c_n x^{n+r}$ into the differential equation and collecting terms, we obtain

$$4xy'' + \frac{1}{2}y' + y = \left(4r^2 - \frac{7}{2}r\right)c_0 x^{r-1} + \sum_{k=1}^{\infty}\left[4(k+r)(k+r-1)c_k + \frac{1}{2}(k+r)c_k + c_{k-1}\right]x^{k+r-1}$$

$$= 0,$$

which implies
$$4r^2 - \frac{7}{2}r = r\left(4r - \frac{7}{2}\right) = 0$$

and
$$\frac{1}{2}(k+r)(8k+8r-7)c_k + c_{k-1} = 0.$$

The indicial roots are $r = 0$ and $r = 7/8$. For $r = 0$ the recurrence relation is

$$c_k = -\frac{2c_{k-1}}{k(8k-7)}, \quad k = 1, 2, 3, \ldots,$$

and
$$c_1 = -2c_0, \quad c_2 = \frac{2}{9}c_0, \quad c_3 = -\frac{4}{459}c_0.$$

For $r = 7/8$ the recurrence relation is

$$c_k = -\frac{2c_{k-1}}{(8k+7)k}, \quad k = 1, 2, 3, \ldots,$$

and
$$c_1 = -\frac{2}{15}c_0, \quad c_2 = \frac{2}{345}c_0, \quad c_3 = -\frac{4}{32{,}085}c_0.$$

The general solution on $(0, \infty)$ is

$$y = C_1\left(1 - 2x + \frac{2}{9}x^2 - \frac{4}{459}x^3 + \cdots\right) + C_2 x^{7/8}\left(1 - \frac{2}{15}x + \frac{2}{345}x^2 - \frac{4}{32{,}085}x^3 + \cdots\right).$$

14. Substituting $y = \sum_{n=0}^{\infty} c_n x^{n+r}$ into the differential equation and collecting terms, we obtain

$$2x^2 y'' - xy' + \left(x^2 + 1\right)y = \left(2r^2 - 3r + 1\right)c_0 x^r + \left(2r^2 + r\right)c_1 x^{r+1}$$

$$+ \sum_{k=2}^{\infty}[2(k+r)(k+r-1)c_k - (k+r)c_k + c_k + c_{k-2}]x^{k+r}$$

$$= 0,$$

which implies

$$2r^2 - 3r + 1 = (2r-1)(r-1) = 0,$$
$$\left(2r^2 + r\right)c_1 = 0,$$

and
$$[(k+r)(2k+2r-3) + 1]c_k + c_{k-2} = 0.$$

The indicial roots are $r = 1/2$ and $r = 1$, so $c_1 = 0$. For $r = 1/2$ the recurrence relation is

$$c_k = -\frac{c_{k-2}}{k(2k-1)}, \quad k = 2, 3, 4, \ldots,$$

and
$$c_2 = -\frac{1}{6}c_0, \quad c_3 = 0, \quad c_4 = \frac{1}{168}c_0.$$

For $r = 1$ the recurrence relation is

$$c_k = -\frac{c_{k-2}}{k(2k+1)}, \quad k = 2, 3, 4, \ldots,$$

and
$$c_2 = -\frac{1}{10}c_0, \qquad c_3 = 0, \qquad c_4 = \frac{1}{360}c_0.$$

The general solution on $(0, \infty)$ is

$$y = C_1 x^{1/2}\left(1 - \frac{1}{6}x^2 + \frac{1}{168}x^4 + \cdots\right) + C_2 x\left(1 - \frac{1}{10}x^2 + \frac{1}{360}x^4 + \cdots\right).$$

15. Substituting $y = \sum_{n=0}^{\infty} c_n x^{n+r}$ into the differential equation and collecting terms, we obtain

$$3xy'' + (2 - x)y' - y = \left(3r^2 - r\right)c_0 x^{r-1}$$

$$+ \sum_{k=1}^{\infty} [3(k + r - 1)(k + r)c_k + 2(k + r)c_k - (k + r)c_{k-1}]x^{k+r-1}$$

$$= 0,$$

which implies
$$3r^2 - r = r(3r - 1) = 0$$

and
$$(k + r)(3k + 3r - 1)c_k - (k + r)c_{k-1} = 0.$$

The indicial roots are $r = 0$ and $r = 1/3$. For $r = 0$ the recurrence relation is

$$c_k = \frac{c_{k-1}}{(3k - 1)}, \qquad k = 1, 2, 3, \ldots,$$

and
$$c_1 = \frac{1}{2}c_0, \qquad c_2 = \frac{1}{10}c_0, \qquad c_3 = \frac{1}{80}c_0.$$

For $r = 1/3$ the recurrence relation is

$$c_k = \frac{c_{k-1}}{3k}, \qquad k = 1, 2, 3, \ldots,$$

and
$$c_1 = \frac{1}{3}c_0, \qquad c_2 = \frac{1}{18}c_0, \qquad c_3 = \frac{1}{162}c_0.$$

The general solution on $(0, \infty)$ is

$$y = C_1\left(1 + \frac{1}{2}x + \frac{1}{10}x^2 + \frac{1}{80}x^3 + \cdots\right) + C_2 x^{1/3}\left(1 + \frac{1}{3}x + \frac{1}{18}x^2 + \frac{1}{162}x^3 + \cdots\right).$$

16. Substituting $y = \sum_{n=0}^{\infty} c_n x^{n+r}$ into the differential equation and collecting terms, we obtain

$$x^2 y'' - \left(x - \frac{2}{9}\right)y = \left(r^2 - r + \frac{2}{9}\right)c_0 x^r + \sum_{k=1}^{\infty}\left[(k + r)(k + r - 1)c_k + \frac{2}{9}c_k - c_{k-1}\right]x^{k+r}$$

$$= 0,$$

which implies
$$r^2 - r + \frac{2}{9} = \left(r - \frac{2}{3}\right)\left(r - \frac{1}{3}\right) = 0$$

224

and
$$\left[(k+r)(k+r-1)+\frac{2}{9}\right]c_k - c_{k-1} = 0.$$

The indicial roots are $r = 2/3$ and $r = 1/3$. For $r = 2/3$ the recurrence relation is

$$c_k = \frac{3c_{k-1}}{3k^2 + k}, \quad k = 1, 2, 3, \ldots,$$

and
$$c_1 = \frac{3}{4}c_0, \quad c_2 = \frac{9}{56}c_0, \quad c_3 = \frac{9}{560}c_0.$$

For $r = 1/3$ the recurrence relation is

$$c_k = \frac{3c_{k-1}}{3k^2 - k}, \quad k = 1, 2, 3, \ldots,$$

and
$$c_1 = \frac{3}{2}c_0, \quad c_2 = \frac{9}{20}c_0, \quad c_3 = \frac{9}{160}c_0.$$

The general solution on $(0, \infty)$ is

$$y = C_1 x^{2/3}\left(1 + \frac{3}{4}x + \frac{9}{56}x^2 + \frac{9}{560}x^3 + \cdots\right) + C_2 x^{1/3}\left(1 + \frac{3}{2}x + \frac{9}{20}x^2 + \frac{9}{160}x^3 + \cdots\right).$$

17. Substituting $y = \sum_{n=0}^{\infty} c_n x^{n+r}$ into the differential equation and collecting terms, we obtain

$$2xy'' - (3+2x)y' + y = \left(2r^2 - 5r\right)c_0 x^{r-1} + \sum_{k=1}^{\infty}[2(k+r)(k+r-1)c_k$$

$$- 3(k+r)c_k - 2(k+r-1)c_{k-1} + c_{k-1}]x^{k+r-1}$$

$$= 0,$$

which implies
$$2r^2 - 5r = r(2r-5) = 0$$

and
$$(k+r)(2k+2r-5)c_k - (2k+2r-3)c_{k-1} = 0.$$

The indicial roots are $r = 0$ and $r = 5/2$. For $r = 0$ the recurrence relation is

$$c_k = \frac{(2k-3)c_{k-1}}{k(2k-5)}, \quad k = 1, 2, 3, \ldots,$$

and
$$c_1 = \frac{1}{3}c_0, \quad c_2 = -\frac{1}{6}c_0, \quad c_3 = -\frac{1}{6}c_0.$$

For $r = 5/2$ the recurrence relation is

$$c_k = \frac{2(k+1)c_{k-1}}{k(2k+5)}, \quad k = 1, 2, 3, \ldots,$$

and
$$c_1 = \frac{4}{7}c_0, \quad c_2 = \frac{4}{21}c_0, \quad c_3 = \frac{32}{693}c_0.$$

The general solution on $(0, \infty)$ is

$$y = C_1 \left(1 + \frac{1}{3}x - \frac{1}{6}x^2 - \frac{1}{6}x^3 + \cdots\right) + C_2 x^{5/2}\left(1 + \frac{4}{7}x + \frac{4}{21}x^2 + \frac{32}{693}x^3 + \cdots\right).$$

18. Substituting $y = \sum_{n=0}^{\infty} c_n x^{n+r}$ into the differential equation and collecting terms, we obtain

$$2xy'' + xy' + \left(x^2 - \frac{4}{9}\right)y = \left(r^2 - \frac{4}{9}\right)c_0 x^r + \left(r^2 + 2r + \frac{5}{9}\right)c_1 x^{r+1}$$

$$+ \sum_{k=2}^{\infty} \left[(k+r)(k+r-1)c_k + (k+r)c_k - \frac{4}{9}c_k + c_{k-2}\right]x^{k+r}$$

$$= 0,$$

which implies

$$r^2 - \frac{4}{9} = \left(r + \frac{2}{3}\right)\left(r - \frac{2}{3}\right) = 0,$$

$$\left(r^2 + 2r + \frac{5}{9}\right)c_1 = 0,$$

and

$$\left[(k+r)^2 - \frac{4}{9}\right]c_k + c_{k-2} = 0.$$

The indicial roots are $r = -2/3$ and $r = 2/3$, so $c_1 = 0$. For $r = -2/3$ the recurrence relation is

$$c_k = -\frac{9c_{k-2}}{3k(3k-4)}, \quad k = 2, 3, 4, \ldots,$$

and

$$c_2 = -\frac{3}{4}c_0, \qquad c_3 = 0, \qquad c_4 = \frac{9}{128}c_0.$$

For $r = 2/3$ the recurrence relation is

$$c_k = -\frac{9c_{k-2}}{3k(3k+4)}, \quad k = 2, 3, 4, \ldots,$$

and

$$c_2 = -\frac{3}{20}c_0, \qquad c_3 = 0, \qquad c_4 = \frac{9}{1,280}c_0.$$

The general solution on $(0, \infty)$ is

$$y = C_1 x^{-2/3}\left(1 - \frac{3}{4}x^2 + \frac{9}{128}x^4 + \cdots\right) + C_2 x^{2/3}\left(1 - \frac{3}{20}x^2 + \frac{9}{1,280}x^4 + \cdots\right).$$

19. Substituting $y = \sum_{n=0}^{\infty} c_n x^{n+r}$ into the differential equation and collecting terms, we obtain

$$9x^2 y'' + 9x^2 y' + 2y = \left(9r^2 - 9r + 2\right)c_0 x^r$$

$$+ \sum_{k=1}^{\infty} [9(k+r)(k+r-1)c_k + 2c_k + 9(k+r-1)c_{k-1}]x^{k+r}$$

$$= 0,$$

which implies

$$9r^2 - 9r + 2 = (3r - 1)(3r - 2) = 0$$

and

$$[9(k + r)(k + r - 1) + 2]c_k + 9(k + r - 1)c_{k-1} = 0.$$

The indicial roots are $r = 1/3$ and $r = 2/3$. For $r = 1/3$ the recurrence relation is

$$c_k = -\frac{(3k - 2)c_{k-1}}{k(3k - 1)}, \quad k = 1, 2, 3, \ldots,$$

and

$$c_1 = -\frac{1}{2}c_0, \qquad c_2 = \frac{1}{5}c_0, \qquad c_3 = -\frac{7}{120}c_0.$$

For $r = 2/3$ the recurrence relation is

$$c_k = -\frac{(3k - 1)c_{k-1}}{k(3k + 1)}, \quad k = 1, 2, 3, \ldots,$$

and

$$c_1 = -\frac{1}{2}c_0, \qquad c_2 = \frac{5}{28}c_0, \qquad c_3 = -\frac{1}{21}c_0.$$

The general solution on $(0, \infty)$ is

$$y = C_1 x^{1/3}\left(1 - \frac{1}{2}x + \frac{1}{5}x^2 - \frac{7}{120}x^3 + \cdots\right) + C_2 x^{2/3}\left(1 - \frac{1}{2}x + \frac{5}{28}x^2 - \frac{1}{21}x^3 + \cdots\right).$$

20. Substituting $y = \sum_{n=0}^{\infty} c_n x^{n+r}$ into the differential equation and collecting terms, we obtain

$$2x^2 y'' + 3xy' + (2x - 1)y = \left(2r^2 + r - 1\right)c_0 x^r$$

$$+ \sum_{k=1}^{\infty} [2(k + r)(k + r - 1)c_k + 3(k + r)c_k - c_k + 2c_{k-1}]x^{k+r}$$

$$= 0,$$

which implies

$$2r^2 + r - 1 = (2r - 1)(r + 1) = 0$$

and

$$[(k + r)(2k + 2r + 1) - 1]c_k + 2c_{k-1} = 0.$$

The indicial roots are $r = -1$ and $r = 1/2$. For $r = -1$ the recurrence relation is

$$c_k = -\frac{2c_{k-1}}{k(2k - 3)}, \quad k = 1, 2, 3, \ldots,$$

and

$$c_1 = 2c_0, \qquad c_2 = -2c_0, \qquad c_3 = \frac{4}{9}c_0.$$

For $r = 1/2$ the recurrence relation is

$$c_k = -\frac{2c_{k-1}}{k(2k + 3)}, \quad k = 1, 2, 3, \ldots,$$

227

and
$$c_1 = -\frac{2}{5}c_0, \qquad c_2 = \frac{2}{35}c_0, \qquad c_3 = -\frac{4}{945}c_0.$$

The general solution on $(0, \infty)$ is

$$y = C_1 x^{-1}\left(1 + 2x - 2x^2 + \frac{4}{9}x^3 + \cdots\right) + C_2 x^{1/2}\left(1 - \frac{2}{5}x + \frac{2}{35}x^2 - \frac{4}{945}x^3 + \cdots\right).$$

21. Substituting $y = \sum_{n=0}^{\infty} c_n x^{n+r}$ into the differential equation and collecting terms, we obtain

$$2x^2 y'' - x(x-1)y' - y = \left(2r^2 - r - 1\right)c_0 x^r$$

$$+ \sum_{k=1}^{\infty} [2(k+r)(k+r-1)c_k + (k+r)c_k - c_k - (k+r-1)c_{k-1}]x^{k+r}$$

$$= 0,$$

which implies
$$2r^2 - r - 1 = (2r+1)(r-1) = 0$$

and
$$[(k+r)(2k+2r-1) - 1]c_k - (k+r-1)2c_{k-1} = 0.$$

The indicial roots are $r = -1/2$ and $r = 1$. For $r = -1/2$ the recurrence relation is

$$c_k = \frac{c_{k-1}}{2k}, \qquad k = 1, 2, 3, \ldots,$$

and
$$c_1 = \frac{1}{2}c_0, \qquad c_2 = \frac{1}{8}c_0, \qquad c_3 = \frac{1}{48}c_0.$$

For $r = 1$ the recurrence relation is

$$c_k = \frac{c_{k-1}}{2k+3}, \qquad k = 1, 2, 3, \ldots,$$

and
$$c_1 = \frac{1}{5}c_0, \qquad c_2 = \frac{1}{35}c_0, \qquad c_3 = \frac{1}{315}c_0.$$

The general solution on $(0, \infty)$ is

$$y = C_1 x^{-1/2}\left(1 + \frac{1}{2}x + \frac{1}{8}x^2 + \frac{1}{48}x^3 + \cdots\right) + C_2 x\left(1 + \frac{1}{5}x + \frac{1}{35}x^2 + \frac{1}{315}x^3 + \cdots\right).$$

22. Substituting $y = \sum_{n=0}^{\infty} c_n x^{n+r}$ into the differential equation and collecting terms, we obtain

$$x(x-2)y'' + y' - 2y = \left(-2r^2 + 3r\right)c_0 x^{r-1}$$

$$+ \sum_{k=0}^{\infty} [(k+r)(k+r-1)c_k - 2c_k - (k+r+1)(2k+2r-1)c_{k+1}]x^{k+r}$$

$$= 0,$$

which implies
$$-2r^2 + 3r = -r(2r - 3) = 0$$

and
$$[(k + r)(k + r - 1) - 2]c_k - (k + r + 1)(2k + 2r - 1)c_{k+1} = 0.$$

The indicial roots are $r = 3/2$ and $r = 0$. For $r = 3/2$ the recurrence relation is

$$c_{k+1} = \frac{2k - 1}{4(k + 1)} c_k, \quad k = 0, 1, 2, \ldots,$$

and
$$c_1 = -\frac{1}{4}c_0, \qquad c_2 = -\frac{1}{32}c_0, \qquad c_3 = -\frac{1}{128}c_0.$$

For $r = 0$ the recurrence relation is

$$c_{k+1} = \frac{k - 2}{2k - 1} c_k, \quad k = 0, 1, 2, \ldots,$$

and
$$c_1 = 2c_0, \qquad c_2 = -2c_0, \qquad c_3 = 0.$$

The general solution on $(0, \infty)$ is

$$y = C_1 x^{3/2} \left(1 - \frac{1}{4}x - \frac{1}{32}x^2 - \frac{1}{128}x^3 - \cdots\right) + C_2 \left(1 + 2x - 2x^2\right).$$

23. Substituting $y = \sum_{n=0}^{\infty} c_n x^{n+r}$ into the differential equation and collecting terms, we obtain
$$xy'' + 2y' - xy = \left(r^2 + r\right) c_0 x^{r-1} + \left(r^2 + 3r + 2\right) c_1 x^r$$

$$+ \sum_{k=2}^{\infty} [(k + r)(k + r - 1)c_k + 2(k + r)c_k - c_{k-2}]x^{k+r-1}$$

$$= 0,$$

which implies
$$r^2 + r = r(r + 1) = 0,$$
$$\left(r^2 + 3r + 2\right) c_1 = 0,$$

and
$$(k + r)(k + r + 1)c_k - c_{k-2} = 0.$$

The indicial roots are $r_1 = 0$ and $r_2 = -1$, so $c_1 = 0$. For $r_1 = 0$ the recurrence relation is
$$c_k = \frac{c_{k-2}}{k(k + 1)}, \quad k = 2, 3, 4, \ldots,$$

and
$$c_2 - \frac{1}{3!}c_0$$

$$c_3 = c_5 = c_7 = \cdots = 0$$

$$c_4 = \frac{1}{5!}c_0$$

$$c_{2n} = \frac{1}{(2n + 1)!}c_0.$$

For $r_2 = -1$ the recurrence relation is

$$c_k = \frac{c_{k-2}}{k(k-1)}, \quad k = 2, 3, 4, \ldots,$$

and

$$c_2 = \frac{1}{2!}c_0$$

$$c_3 = c_5 = c_7 = \cdots = 0$$

$$c_4 = \frac{1}{4!}c_0$$

$$c_{2n} = \frac{1}{(2n)!}c_0.$$

The general solution on $(0, \infty)$ is

$$y = C_1 \sum_{n=0}^{\infty} \frac{1}{(2n+1)!}x^{2n} + C_2 x^{-1} \sum_{n=0}^{\infty} \frac{1}{(2n)!}x^{2n}$$

$$= \frac{1}{x}\left[C_1 \sum_{n=0}^{\infty} \frac{1}{(2n+1)!}x^{2n+1} + C_2 \sum_{n=0}^{\infty} \frac{1}{(2n)!}x^{2n}\right]$$

$$= \frac{1}{x}[C_1 \sinh x + C_2 \cosh x].$$

24. Substituting $y = \sum_{n=0}^{\infty} c_n x^{n+r}$ into the differential equation and collecting terms, we obtain

$$x^2 y'' + xy' + \left(x^2 - \frac{1}{4}\right)y = \left(r^2 - \frac{1}{4}\right)c_0 x^r + \left(r^2 + 2r + \frac{3}{4}\right)c_1 x^{r+1}$$

$$+ \sum_{k=2}^{\infty}\left[(k+r)(k+r-1)c_k + (k+r)c_k - \frac{1}{4}c_k + c_{k-2}\right]x^{k+r}$$

$$= 0,$$

which implies

$$r^2 - \frac{1}{4} = \left(r - \frac{1}{2}\right)\left(r + \frac{1}{2}\right) = 0,$$

$$\left(r^2 + 2r + \frac{3}{4}\right)c_1 = 0,$$

and

$$\left[(k+r)^2 - \frac{1}{4}\right]c_k + c_{k-2} = 0.$$

The indicial roots are $r_1 = 1/2$ and $r_2 = -1/2$, so $c_1 = 0$. For $r_1 = 1/2$ the recurrence relation is

$$c_k = -\frac{c_{k-2}}{k(k+1)}, \quad k = 2, 3, 4, \ldots,$$

and
$$c_2 = -\frac{1}{3!}c_0$$

$$c_3 = c_5 = c_7 = \cdots = 0$$

$$c_4 = \frac{1}{5!}c_0$$

$$c_{2n} = \frac{(-1)^n}{(2n+1)!}c_0.$$

For $r_2 = -1/2$ the recurrence relation is

$$c_k = -\frac{c_{k-2}}{k(k-1)}, \quad k = 2, 3, 4, \ldots,$$

and
$$c_2 = -\frac{1}{2!}c_0$$

$$c_3 = c_5 = c_7 = \cdots = 0$$

$$c_4 = \frac{1}{4!}c_0$$

$$c_{2n} = \frac{(-1)^n}{(2n)!}c_0.$$

The general solution on $(0, \infty)$ is

$$y = C_1 x^{1/2} \sum_{n=0}^{\infty} \frac{(-1)^n}{(2n+1)!} x^{2n} + C_2 x^{-1/2} \sum_{n=0}^{\infty} \frac{(-1)^n}{(2n)!} x^{2n}$$

$$= C_1 x^{-1/2} \sum_{n=0}^{\infty} \frac{(-1)^n}{(2n+1)!} x^{2n+1} + C_2 x^{-1/2} \sum_{n=0}^{\infty} \frac{(-1)^n}{(2n)!} x^{2n}$$

$$= x^{-1/2}[C_1 \sin x + C_2 \cos x].$$

25. Substituting $y = \sum_{n=0}^{\infty} c_n x^{n+r}$ into the differential equation and collecting terms, we obtain

$$x(x-1)y'' + 3y' - 2y$$

$$= \left(4r - r^2\right) c_0 x^{r-1} + \sum_{k=1}^{\infty} [(k+r-1)(k+r-12)c_{k-1} - (k+r)(k+r-1)c_k$$

$$+ 3(k+r)c_k - 2c_{k-1}]x^{k+r-1}$$

$$= 0,$$

which implies
$$4r - r^2 = r(4-r) = 0$$

and
$$-(k+r)(k+r-4)c_k + [(k+r-1)(k+r-2) - 2]c_{k-1} = 0.$$

The indicial roots are $r_1 = 4$ and $r_2 = 0$. For $r_2 = 0$ the recurrence relation is

$$-k(k-4)c_k + k(k-3)c_{k-1} = 0, \quad k = 1, 2, 3, \ldots,$$

or

$$-(k-4)c_k + (k-3)c_{k-1} = 0, \quad k = 1, 2, 3, \ldots.$$

Then

$$3c_1 - 2c_0 = 0$$

$$2c_2 - c_1 = 0$$

$$c_3 + 0c_2 = 0 \quad \Rightarrow \quad c_3 = 0$$

$$0c_4 + c_3 = 0 \quad \Rightarrow \quad c_4 \text{ is arbitrary}$$

and

$$c_k = \frac{(k-3)c_{k-1}}{c-4}, \quad k = 5, 6, 7, \ldots.$$

Taking $c_0 \neq 0$ and $c_4 = 0$ we obtain

$$c_1 = \frac{2}{3}c_0$$

$$c_2 = \frac{1}{3}c_0$$

$$c_3 = c_4 = c_5 = \cdots = 0.$$

Taking $c_0 = 0$ and $c_4 \neq 0$ we obtain

$$c_1 = c_2 = c_3 = 0$$

$$c_5 = 2c_4$$

$$c_6 = 3c_4$$

$$c_7 = 4c_4.$$

The general solution on $(0, \infty)$ is

$$y = C_1 \left(1 + \frac{2}{3}x + \frac{1}{3}x^2\right) + C_2 \left(x^4 + 2x^5 + 3x^6 + 4x^7 + \cdots\right)$$

$$= C_1 \left(1 + \frac{2}{3}x + \frac{1}{3}x^2\right) + C_2 \sum_{n=1}^{\infty} nx^{n+3}.$$

26. Substituting $y = \sum_{n=0}^{\infty} c_n x^{n+r}$ into the differential equation and collecting terms, we obtain

$$y'' + \frac{3}{x}y' - 2y = \left(r^2 + 2r\right) c_0 x^{r-2} + \left(r^2 + 4r + 3\right) c_1 x^{r-1}$$

$$+ \sum_{k=2}^{\infty} [(k+r)(k+r-1)c_k + 3(k+r)c_k - 2c_{k-2}]x^{k+r-2}$$

$$= 0,$$

which implies

$$r^2 + 2r = r(r+2) = 0$$

$$\left(r^2 + 4r + 3\right)c_1 = 0$$

$$(k+r)(k+r+2)c_k - 2c_{k-2} = 0.$$

The indicial roots are $r_1 = 0$ and $r_2 = -2$, so $c_1 = 0$. For $r_1 = 0$ the recurrence relation is

$$c_k = \frac{2c_{k-2}}{k(k+2)}, \quad k = 2, 3, 4, \ldots,$$

and

$$c_2 = \frac{1}{4}c_0$$

$$c_3 = c_5 = c_7 = \cdots = 0$$

$$c_4 = \frac{1}{48}c_0$$

$$c_6 = \frac{1}{1,152}c_0.$$

The result is

$$y_1 = c_0 \left(1 + \frac{1}{4}x^2 + \frac{1}{48}x^4 + \frac{1}{1,152}c_6 + \cdots\right).$$

A second solution is

$$y_2 = y_1 \int \frac{e^{-\int (3/x)dx}}{y_1^2} \, dx = y_1 \int \frac{dx}{x^3 \left(1 + \frac{1}{4}x^2 + \frac{1}{48}x^4 + \cdots\right)^2}$$

$$= y_1 \int \frac{dx}{x^3 \left(1 + \frac{1}{2}x^2 + \frac{5}{48}x^4 + \frac{7}{576}x^6 + \cdots\right)} = y_1 \int \frac{1}{x^3}\left(1 - \frac{1}{2}x^2 + \frac{7}{48}x^4 + \frac{19}{576}x^6 + \cdots\right)$$

$$= y_1 \int \left(\frac{1}{x^3} - \frac{1}{2x} + \frac{7}{48}x - \frac{19}{576}x^3 + \cdots\right) = y_1 \left[-\frac{1}{2x^2} - \frac{1}{2}\ln x + \frac{7}{96}x^2 - \frac{19}{2,304}x^4 + \cdots\right]$$

$$= -\frac{1}{2}y_1 \ln x + y\left[-\frac{1}{2x^2} + \frac{7}{96}x^2 - \frac{19}{2,304}x^4 + \cdots\right].$$

The general solution on $(0, \infty)$ is

$$y = C_1 y_1(x) + C_2 y_2(x).$$

27. Substituting $y = \sum_{n=0}^{\infty} c_n x^{n+r}$ into the differential equation and collecting terms, we obtain

$$xy'' + (1-x)y' - y = r^2 c_0 x^{r-1} + \sum_{k=0}^{\infty} [(k+r)(k+r-1)c_k + (k+r)c_k - (k+r)c_{k-1}]x^{k+r-1} = 0,$$

which implies $r^2 = 0$ and

$$(k+r)^2 c_k - (k+r)c_{k-1} = 0.$$

The indicial roots are $r_1 = r_2 = 0$ and the recurrence relation is

$$c_k = \frac{c_{k-1}}{k}, \quad k = 1, 2, 3, \ldots.$$

One solution is

$$y_1 = c_0 \left(1 + x + \frac{1}{2}x^2 + \frac{1}{3!}x^3 + \cdots\right) = c_0 e^x.$$

A second solution is

$$y_2 = y_1 \int \frac{e^{-\int (1/x - 1)dx}}{e^{2x}}\, dx = e^x \int \frac{e^x/x}{e^{2x}}\, dx = e^x \int \frac{1}{x} e^{-x} dx$$

$$= e^x \int \frac{1}{x}\left(1 - x + \frac{1}{2}x^2 - \frac{1}{3!}x^3 + \cdots\right) dx = e^x \int \left(\frac{1}{x} - 1 + \frac{1}{2}x - \frac{1}{3!}x^2 + \cdots\right) dx$$

$$= e^x \left[\ln x - x + \frac{1}{2 \cdot 2}x^2 - \frac{1}{3 \cdot 3!}x^3 + \cdots\right] = e^x \ln x - e^x \sum_{n=1}^{\infty} \frac{(-1)^{n+1}}{n \cdot n!} x^n.$$

The general solution on $(0, \infty)$ is

$$y = C_1 e^x + C_2 e^x \left(\ln x - \sum_{n=1}^{\infty} \frac{(-1)^{n+1}}{n \cdot n!} x^n\right).$$

28. Substituting $y = \sum_{n=0}^{\infty} c_n x^{n+r}$ into the differential equation and collecting terms, we obtain

$$xy'' + y = \left(r^2 - r\right) c_0 x^{r-1} + \sum_{k=0}^{\infty} [(k+r+1)(k+r)c_{k+1} + c_k]x^{k+r} = 0,$$

which implies

$$r^2 - r = r(r-1) = 0$$

$$(k+r+1)(k+r)c_{k+1} + c_k = 0.$$

The indicial roots are $r_1 = 1$ and $r_2 = 0$. For $r_1 = 1$ the recurrence relation is

$$c_{k+1} = \frac{-c_k}{(k+2)(k+1)}, \quad k = 0, 1, 2, \ldots,$$

and
$$c_1 = -\frac{1}{2}c_0$$

$$c_2 = \frac{1}{3! \cdot 2}c_0$$

$$c_3 = -\frac{1}{4!3!}c_0$$

$$c_4 = \frac{1}{5!4!}c_0.$$

The result is

$$y_1 = c_0\left(x - \frac{1}{2}x^2 + \frac{1}{3!2}x^3 - \frac{1}{4!3!}x^4 + \frac{1}{5!4!}x^5 - \cdots\right).$$

A second solution is

$$y_2 = y_1 \int \frac{e^{\int 0\, dx}}{y_1^2}\, dx = y_1 \int \frac{1}{y_1^2}\, dx = y_1 \int \frac{dx}{\left(x - \frac{1}{2}x^2 + \frac{1}{3!2}x^3 - \frac{1}{4!3!}x^4 + \cdots\right)^2}$$

$$= y_1 \int \frac{dx}{x^2 - x^3 + \frac{5}{3!2}x^4 - \frac{14}{4!3!}x^5 + \cdots} = y_1 \int \frac{dx}{x^2\left(1 - x + \frac{5}{3!2}x^2 - \frac{14}{4!3!}x^3 + \cdots\right)}$$

$$= y_1 \int \frac{1}{x^2}\left(1 + x + \frac{7}{3!2}x^2 + \frac{38}{4!3!}x^3 + \cdots\right)dx = y_1 \int \left(\frac{1}{x^2} + \frac{1}{x} + \frac{7}{3!2} + \frac{38}{4!3!}x + \cdots\right)dx$$

$$= y_1\left(-\frac{1}{x} + \ln x + \frac{7}{3!2}x + \frac{19}{4!3!}x^2 + \cdots\right).$$

The general solution is

$$y_1(x) = C_1 y_1 + C_2 y_1\left(-\frac{1}{x} + \ln x + \frac{7}{3!2}x + \frac{19}{4!3!}x^2 + \cdots\right).$$

29. Substituting $y = \sum_{n=0}^{\infty} c_n x^{n+r}$ into the differential equation and collecting terms, we obtain

$$xy'' + y' + y = r^2 c_0 x^{r-1} + \sum_{k=1}^{\infty}[(k+r)(k+r-1)c_k + (k+r)c_k + c_{k-1}]x^{k+r-1} = 0$$

which implies $r^2 = 0$ and

$$(k+r)^2 c_k + c_{k-1} = 0.$$

The indicial roots are $r_1 = r_2 = 0$ and the recurrence relation is

$$c_k = -\frac{c_{k-1}}{k^2}, \quad k = 1, 2, 3, \ldots.$$

One solution is

$$y_1 = c_0\left(1 - x + \frac{1}{2^2}x^2 - \frac{1}{(3!)^2}x^3 + \frac{1}{(4!)^2}x^4 - \cdots\right) = c_0 \sum_{n=0}^{\infty} \frac{(-1)^n}{(n!)^2}x^n.$$

235

A second solution is

$$y_2 = y_1 \int \frac{e^{-\int (1/x)dx}}{y_1^2}\, dx = y_1 \int \frac{dx}{x\left(1 - x + \frac{1}{4}x^2 - \frac{1}{36}x^3 + \cdots\right)^2}$$

$$= y_1 \int \frac{dx}{x\left(1 - 2x + \frac{3}{2}x^2 - \frac{5}{9}x^3 + \frac{35}{288}x^4 - \cdots\right)}$$

$$= y_1 \int \frac{1}{x}\left(1 + 2x + \frac{5}{2}x^2 + \frac{23}{9}x^3 + \frac{677}{288}x^4 + \cdots\right) dx$$

$$= y_1 \int \left(\frac{1}{x} + 2 + \frac{5}{2}x + \frac{23}{9}x^2 + \frac{677}{288}x^3 + \cdots\right) dx$$

$$= y_1 \left[\ln x + 2x + \frac{5}{4}x^2 + \frac{23}{27}x^3 + \frac{677}{1{,}152}x^4 + \cdots\right]$$

$$= y_1 \ln x + y_1 \left(2x + \frac{5}{4}x^2 + \frac{23}{27}x^3 + \frac{677}{1{,}152}x^4 + \cdots\right).$$

The general solution on $(0, \infty)$ is

$$y = C_1 y_1(x) + C_2 y_2(x).$$

30. Substituting $y = \sum_{n=0}^{\infty} c_n x^{n+r}$ into the differential equation and collecting terms, we obtain

$$xy'' - xy' + y = \left(r^2 - r\right)c_0 x^{r-1} + \sum_{k=0}^{\infty}[(k+r+1)(k+r)c_{k+1} - (k+r)c_k + c_k]x^{k+r} = 0$$

which implies

$$r^2 - r = r(r - 1) = 0$$

and

$$(k+r+1)(k+r)c_{k+1} - (k+r-1)c_k = 0.$$

The indicial roots are $r_1 = 1$ and $r_2 = 0$. For $r_1 = 1$ the recurrence relation is

$$c_{k+1} = \frac{kc_k}{(k+2)(k+1)}, \quad k = 0, 1, 2, \ldots,$$

and one solution is $y_1 = c_0 x$. A second solution is

$$y_2 = x \int \frac{e^{-\int - dx}}{x^2}\, dx = x \int \frac{e^x}{x^2}\, dx = x \int \frac{1}{x^2}\left(1 + x + \frac{1}{2}x^2 + \frac{1}{3!}x^3 + \cdots\right) dx$$

$$= x \int \left(\frac{1}{x^2} + \frac{1}{x} + \frac{1}{2} + \frac{1}{3!}x + \frac{1}{4!}x^2 + \cdots\right) dx = x\left[-\frac{1}{x} + \ln x + \frac{1}{2}x + \frac{1}{12}x^2 + \frac{1}{72}x^3 + \cdots\right]$$

$$= x \ln x - 1 + \frac{1}{2}x^2 + \frac{1}{12}x^3 + \frac{1}{72}x^4 + \cdots.$$

The general solution on $(0, \infty)$ is

$$y = C_1 x + C_2 y_2(x).$$

Exercises 6.4

1. Since $\nu^2 = 1/9$ the general solution is $y = c_1 J_{1/3}(x) + c_2 J_{-1/3}(x)$.

2. Since $\nu^2 = 1$ the general solution is $y = c_1 J_1(x) + c_2 Y_1(x)$.

3. Since $\nu^2 = 25/4$ the general solution is $y = c_1 J_{5/2}(x) + c_2 J_{-5/2}(x)$.

4. Since $\nu^2 = 1/16$ the general solution is $y = c_1 J_{1/4}(x) + c_2 J_{-1/4}(x)$.

5. Since $\nu^2 = 0$ the general solution is $y = c_1 J_0(x) + c_2 Y_0(x)$.

6. Since $\nu^2 = 4$ the general solution is $y = c_1 J_2(x) + c_2 Y_2(x)$.

7. Since $\nu^2 = 2$ the general solution is $y = c_1 J_2(3x) + c_2 Y_2(3x)$.

8. Since $\nu^2 = 1/4$ the general solution is $y = c_1 J_{1/2}(6x) + c_2 J_{-1/2}(6x)$.

9. If $y = x^{-1/2} v(x)$ then

$$y' = x^{-1/2} v'(x) - \frac{1}{2} x^{-3/2} v(x),$$

$$y'' = x^{-1/2} v''(x) - x^{-3/2} v'(x) + \frac{3}{4} x^{-5/2} v(x),$$

and

$$x^2 y'' + 2xy' + \lambda^2 x^2 y = x^{3/2} v'' + x^{1/2} v' + \left(\lambda^2 x^{3/2} - \frac{1}{4} x^{-1/2} \right) v.$$

Multiplying by $x^{1/2}$ we obtain

$$x^2 v'' + xv' + \left(\lambda^2 x^2 - \frac{1}{4} \right) v = 0,$$

whose solution is $v = c_1 J_{1/2}(\lambda x) + c_2 J_{-1/2}(\lambda x)$. Then $y = c_1 x^{-1/2} J_{1/2}(\lambda x) + c_2 x^{-1/2} J_{-1/2}(\lambda x)$.

10. From $y = x^n J_n(x)$ we find

$$y' = x^n J_n' + nx^{n-1} J_n \qquad \text{and} \qquad y'' = x^n J_n'' + 2nx^{n-1} J_n' + n(n-1)x^{n-2} J_n.$$

237

Substituting into the differential equation, we have

$$x^{n+1}J_n'' + 2nx^n J_n' + n(n-1)x^{n-1}J_n + (1-2n)(x^n J_n' + nx^{n-1}J_n) + x^{n+1}J_n$$

$$= x^{n+1}J_n'' + (2n+1-2n)x^n J_n' + (n^2 - n + n - 2n^2)x^{n-1}J_n + x^{n+1}J_n$$

$$= x^{n+1}[x^2 J_n'' + x J_n' - n^2 J_n + x^2 J_n]$$

$$= x^{n+1}[x^2 J_n'' + x J_n' + (x^2 - n^2)J_n]$$

$$= x^{n-1} \cdot 0 \qquad \text{(since } J_n \text{ is a solution of Bessel's equation)}$$

$$= 0.$$

Therefore, $x^n J_n$ is a solution of the original equation.

11. From $y = x^{-n}J_n$ we find

$$y' = x^{-n}J_n' - nx^{-n-1}J_n \quad \text{and} \quad y'' = x^{-n}J_n'' - 2nx^{-n-1}J_n' + n(n+1)x^{-n-2}J_n.$$

Substituting into the differential equation, we have

$$xy'' + (1+2n)y' + xy = x^{-n-1}\left[x^2 J_n'' + x J_n' + \left(x^2 - n^2\right)J_n\right]$$

$$= x^{-n-1} \cdot 0 \qquad \text{(since } J_n \text{ is a solution of Bessel's equation)}$$

$$= 0.$$

Therefore, $x^{-n}J_n$ is a solution of the original equation.

12. From $y = \sqrt{x}\, J_\nu(\lambda x)$ we find

$$y' = \lambda\sqrt{x}\, J_\nu'(\lambda x) + \frac{1}{2}x^{-1/2}J_\nu(\lambda x)$$

and

$$y'' = \lambda^2\sqrt{x}\, J_\nu''(\lambda x) + \lambda x^{-1/2}J_\nu'(\lambda x) - \frac{1}{4}x^{-3/2}J_\nu(\lambda x).$$

Substituting into the differential equation, we have

$$x^2 y'' + \left(\lambda^2 x^2 - \nu^2 + \frac{1}{4}\right)y = \sqrt{x}\left[\lambda^2 x^2 J_\nu''(\lambda x) + \lambda x J_\nu'(\lambda x) + \left(\lambda^2 x^2 - \nu^2\right)J_\nu(\lambda x)\right]$$

$$= \sqrt{x} \cdot 0 \qquad \text{(since } J_n \text{ is a solution of Bessel's equation)}$$

$$= 0.$$

Therefore, $\sqrt{x}\, J_\nu(\lambda x)$ is a solution of the original equation.

13. From Problem 10 with $n = 1/2$ we find $y = x^{1/2}J_{1/2}(x)$. From Problem 11 with $n = -1/2$ we find $y = x^{1/2}J_{-1/2}(x)$.

14. From Problem 10 with $n = 1$ we find $y = xJ_1(x)$. From Problem 11 with $n = -1$ we find $y = xJ_{-1}(x) = -xJ_1(x)$.

15. From Problem 10 with $n = -1$ we find $y = x^{-1}J_{-1}(x)$. From Problem 11 with $n = 1$ we find $y = x^{-1}J_1(x) = -x^{-1}J_{-1}(x)$.

16. From Problem 12 with $\lambda = 2$ and $\nu = 0$ we find $y = \sqrt{x}\, J_0(2x)$.

17. From Problem 12 with $\lambda = 1$ and $\nu = \pm 3/2$ we find $y = \sqrt{x}\, J_{3/2}(x)$ and $y = \sqrt{x}\, J_{-3/2}(x)$.

18. From Problem 10 with $n = 3$ we find $y = x^3 J_3(x)$. From Problem 11 with $n = -3$ we find $y = x^3 J_{-3}(x) = -x^3 J_3(x)$.

19. The recurrence relation follows from

$$-\nu J_\nu(x) + x J_{\nu-1}(x) = -\sum_{n=0}^{\infty} \frac{(-1)^n \nu}{n!\Gamma(1+\nu+n)} \left(\frac{x}{2}\right)^{2n+\nu} + x\sum_{n=0}^{\infty} \frac{(-1)^n}{n!\Gamma(\nu+n)} \left(\frac{x}{2}\right)^{2n+\nu-1}$$

$$= -\sum_{n=0}^{\infty} \frac{(-1)^n \nu}{n!\Gamma(1+\nu+n)} \left(\frac{x}{2}\right)^{2n+\nu} + \sum_{n=0}^{\infty} \frac{(-1)^n(\nu+n)}{n!\Gamma(1+\nu+n)} \cdot 2\left(\frac{x}{2}\right)\left(\frac{x}{2}\right)^{2n+\nu-1}$$

$$= \sum_{n=0}^{\infty} \frac{(-1)^n(2n+\nu)}{n!\Gamma(1+\nu+n)} \left(\frac{x}{2}\right)^{2n+\nu} = xJ_\nu'(x).$$

20. Using

$$J_\nu(x) = \sum_{n=0}^{\infty} \frac{(-1)^n}{n!\Gamma(1+\nu+n)} \left(\frac{x}{2}\right)^{2n+\nu}$$

$$J_\nu'(x) = \sum_{n=0}^{\infty} \frac{(2n+\nu)(-1)^n}{2n!\Gamma(1+\nu+n)} \left(\frac{x}{2}\right)^{2n+\nu-1}$$

$$J_{\nu-1}(x) = \sum_{n=0}^{\infty} \frac{(-1)^n}{n!\Gamma(\nu+n)} \left(\frac{x}{2}\right)^{2n+\nu-1}$$

we obtain

$$\frac{d}{dx}[x^\nu J_\nu(x)] = x^\nu J_\nu'(x) + \nu x^{\nu-1} J_\nu(x)$$

$$= x^\nu \sum_{n=0}^{\infty} \frac{(2n+\nu)(-1)^n}{2n!\Gamma(1+\nu+n)} \left(\frac{x}{2}\right)^{2n+\nu-1}$$

$$+ \nu x^{\nu-1} \sum_{n=0}^{\infty} \frac{(-1)^n}{n!\Gamma(1+\nu+n)} \left(\frac{x}{2}\right)^{2n+\nu}$$

239

$$= x^\nu \sum_{n=0}^\infty \frac{(2n+\nu)(-1)^n}{2n!(\nu+n)\Gamma(\nu+n)} \left(\frac{x}{2}\right)^{2n+\nu-1}$$

$$+ x^\nu \sum_{n=0}^\infty \frac{\nu(-1)^n 2^{-1}}{n!(\nu+n)\Gamma(\nu+n)} \left(\frac{x}{2}\right)^{-1}\left(\frac{x}{2}\right)^{2n+\nu}$$

$$= x^\nu \left[\sum_{n=0}^\infty \frac{(2n+\nu)(-1)^n}{2n!(\nu+n)\Gamma(\nu+n)} \left(\frac{x}{2}\right)^{2n+\nu-1} \right.$$

$$\left. + \sum_{n=0}^\infty \frac{\nu(-1)^n}{2n!(\nu+n)\Gamma(\nu+n)} \left(\frac{x}{2}\right)^{2n+\nu-1} \right]$$

$$= x^\nu \sum_{n=0}^\infty \frac{(2n+2\nu)(-1)^n}{2n!(\nu+n)\Gamma(\nu+n)} \left(\frac{x}{2}\right)^{2n+\nu-1}$$

$$= x^\nu \sum_{n=0}^\infty \frac{(-1)^n}{n!\Gamma(\nu+n)} \left(\frac{x}{2}\right)^{2n+\nu-1}$$

$$= x^\nu J_{\nu-1}(x).$$

Alternatively, we can note that the formula in Problem 19 is a linear first-order differential equation in $J_\nu(x)$. An integrating factor for this equation is x^ν, so $\dfrac{d}{dx}[x^\nu J_\nu(x)] = x^\nu J_{\nu-1}(x)$.

21. The recurrence relation follows from

$$xJ_{\nu+1}(x) + xJ_{\nu-1}(x) = \sum_{n=0}^\infty \frac{(-1)^{n-1}2n}{n!\Gamma(1+\nu+n)} \left(\frac{x}{2}\right)^{2n+\nu} + \sum_{n=0}^\infty \frac{(-1)^n 2(\nu+n)}{n!\Gamma(1+\nu+n)} \left(\frac{x}{2}\right)^{2n+\nu}$$

$$= \sum_{n=0}^\infty \frac{(-1)^n 2\nu}{n!\Gamma(1+\nu+n)} \left(\frac{x}{2}\right)^{2n+\nu} = 2\nu J_\nu(x).$$

22. The recurrence relation follows from Example 4 in the text and Problem 21:

$$2J_\nu'(x) = \frac{1}{x}[2\nu J_\nu(x) - 2J_{\nu+1}(x)] = \frac{1}{x}[xJ_{\nu+1}(x) + xJ_{\nu-1}(x)] - 2J_{\nu+1}(x) = J_{\nu-1}(x) - J_{\nu+1}(x).$$

23. By Problem 20 $\dfrac{d}{dx}[xJ_1(x)] = xJ_0(x)$ so that $\displaystyle\int_0^x rJ_0(r)\,dr = rJ_1(r)\Big|_{r=0}^{r=x} = xJ_1(x)$.

24. By Problem 19 we obtain $J_0'(x) = J_{-1}(x)$ and by Problem 22

$$2J_0'(x) = J_{-1}(x) - J_1(x) = J_0'(x) - J_1(x)$$

so that $J_0'(x) = -J_1(x)$.

25. Using Problems 20 and 24 and integration by parts we have

$$\int x^n J_0(x)\, dx = \int x^{n-1}(xJ_0(x))\, dx = \int x^{n-1}\frac{d}{dx}(xJ_1(x))\, dx$$

$$= x^{n-1}xJ_1(x) - (n-1)\int x^{n-2}xJ_1(x)\, dx$$

$$= x^n J_1(x) - (n-1)\int x^{n-1}(-J_0'(x))\, dx$$

$$= x^n J_1(x) + (n-1)x^{n-1}J_0(x) - (n-1)^2\int x^{n-2}J_0(x)\, dx.$$

26. Using Problem 25 with $n=3$ and Problem 23 we have

$$\int x^3 J_0(x)\, dx = x^3 J_1(x) + 2x^2 J_0(x) - 4\int x J_0(x)\, dx$$

$$= x^3 J_1(x) + 2x^2 J_0(x) - 4x J_1(x) + c.$$

27. Since

$$\Gamma\left(1 - \frac{1}{2} + n\right) = \frac{(2n-1)!}{(n-1)!2^{2n-1}}$$

we obtain

$$J_{-1/2}(x) = \sum_{n=0}^{\infty} \frac{(-1)^n 2^{1/2} x^{-1/2}}{2n(2n-1)!\sqrt{\pi}} x^{2n} = \sqrt{\frac{2}{\pi x}}\cos x.$$

28. By Problem 21 we obtain $J_{1/2}(x) = xJ_{3/2}(x) + xJ_{-1/2}(x)$ so that

$$J_{3/2}(x) = \sqrt{\frac{2}{\pi x}}\left(\frac{\sin x}{x} - \cos x\right).$$

29. By Problem 21 we obtain $-J_{-1/2}(x) = xJ_{1/2}(x) + xJ_{-3/2}(x)$ so that

$$J_{-3/2}(x) = -\sqrt{\frac{2}{\pi x}}\left(\frac{\cos x}{x} + \sin x\right).$$

30. By Problem 21 we obtain $3J_{3/2}(x) = xJ_{5/2}(x) + xJ_{1/2}(x)$ so that

$$J_{5/2}(x) = \sqrt{\frac{2}{\pi x}}\left(\frac{3\sin x}{x^2} - \frac{3\cos x}{x} - \sin x\right).$$

31. By Problem 21 we obtain $-3J_{-3/2}(x) = xJ_{-1/2}(x) + xJ_{-5/2}(x)$ so that

$$J_{-5/2}(x) = \sqrt{\frac{2}{\pi x}}\left(\frac{3\cos x}{x^2} + \frac{3\sin x}{x} - \cos x\right).$$

32. By Problem 21 we obtain $5J_{5/2}(x) = xJ_{7/2}(x) + xJ_{3/2}(x)$ so that

$$J_{7/2}(x) = \sqrt{\frac{2}{\pi x}}\left(\frac{15\sin x}{x^3} - \frac{15\cos x}{x^2} - \frac{6\sin x}{x} + \cos x\right).$$

Exercises 6.4

33. By Problem 21 we obtain $-5J_{-5/2}(x) = xJ_{-3/2}(x) + xJ_{-7/2}(x)$ so that

$$J_{-7/2}(x) = \sqrt{\frac{2}{\pi x}} \left(\frac{-15\cos x}{x^3} - \frac{15\sin x}{x^2} + \frac{6\cos x}{x} + \sin x \right).$$

34. Since

$$i^{-\nu} J_\nu(ix) = i^{-\nu} i^\nu \sum_{n=0}^{\infty} \frac{(-1)^n i^{2n}}{n!\Gamma(1+\nu+n)} \left(\frac{x}{2}\right)^{2n+\nu} = \sum_{n=0}^{\infty} \frac{1}{n!\Gamma(1+\nu+n)} \left(\frac{x}{2}\right)^{2n+\nu},$$

the function is real.

35. If $y_1 = I_\nu(x) = i^{-\nu} J_\nu(ix)$ then

$$y_1' = i^{-\nu+1} J_\nu'(ix),$$

$$y_1'' = -i^{-\nu} J_\nu''(ix),$$

and

$$x^2 y_1'' + x y_1' - \left(x^2 + \nu^2\right) y_1 = i^{-\nu} \left[(ix)^2 J_\nu''(ix) + (ix)J_\nu'(ix) + \left((ix)^2 - \nu^2\right) J_\nu(ix) \right]$$

$$= i^{-\nu} \cdot 0 = 0.$$

Similarly, $y_2 = I_{-\nu}(x) = i^\nu J_{-\nu}(ix)$ satisfies the differential equation, and the general solution is $y = c_1 I_\nu(x) + c_2 I_{-\nu}(x)$.

36. If $y_1 = J_0(x)$ then using the formula for the second solution of a linear homogeneous second-order differential equation gives

$$y_2 = J_0(x) \int \frac{e^{-\int dx/x}}{(J_0(x))^2} \, dx$$

$$= J_0(x) \int \frac{dx}{x \left(1 - \dfrac{x^2}{4} + \dfrac{x^4}{64} - \dfrac{x^6}{2304} + \cdots\right)^2} \, dx$$

$$= J_0(x) \int \left(\frac{1}{x} + \frac{x}{2} + \frac{5x^3}{32} + \frac{23x^5}{576} + \cdots \right) dx$$

$$= J_0(x) \left(\ln x + \frac{x^2}{4} + \frac{5x^4}{128} + \frac{23x^6}{3456} + \cdots \right)$$

$$= J_0(x) \ln x + \left(1 - \frac{x^2}{4} + \frac{x^4}{64} - \frac{x^6}{2304} + \cdots \right) \left(\frac{x^2}{4} + \frac{5x^4}{128} + \frac{23x^6}{3456} + \cdots \right)$$

$$= J_0(x) \ln x + \frac{x^2}{4} - \frac{3x^4}{128} + \frac{11x^6}{13824} - \cdots .$$

37. Using (8) with $\nu = m$ we have

$$J_{-m}(x) = \sum_{n=0}^{\infty} \frac{(-1)^n}{n!\Gamma(1-m+n)} \left(\frac{x}{2}\right)^{2n-m} = \sum_{n=m}^{\infty} \frac{(-1)^n}{n!\Gamma(1-m+n)} \left(\frac{x}{2}\right)^{2n-m}$$

$$= \sum_{j=0}^{\infty} \frac{(-1)^j(-1)^m}{(j+m)!\Gamma(1+j)} \left(\frac{x}{2}\right)^{2j+m} = (-1)^m \sum_{j=0}^{\infty} \frac{(-1)^j}{j!\Gamma(1+m+j)} \left(\frac{x}{2}\right)^{2j+m}$$

$$= (-1)^m J_m(x).$$

38. Using (7) with $\nu = m$ we have

$$J_m(-x) = \sum_{n=0}^{\infty} \frac{(-1)^n}{n!\Gamma(1+m+n)} \left(-\frac{x}{2}\right)^{2n+m} = (-1)^m \sum_{n=0}^{\infty} \frac{(-1)^n}{n!\Gamma(1+m+n)} \left(\frac{x}{2}\right)^{2n+m} = (-1)^m J_m(x).$$

39. Letting

$$s = \frac{2}{\alpha}\sqrt{\frac{k}{m}}\, e^{-\alpha t/2},$$

we have

$$\frac{dx}{dt} = \frac{dx}{ds}\frac{ds}{dt} = \frac{dx}{dt}\left[\frac{2}{\alpha}\sqrt{\frac{k}{m}}\left(-\frac{\alpha}{2}\right)e^{-\alpha t/2}\right]$$

$$= \frac{dx}{ds}\left(-\sqrt{\frac{k}{m}}\, e^{-\alpha t/2}\right)$$

and

$$\frac{d^2x}{dt^2} = \frac{d}{dt}\left(\frac{dx}{dt}\right) = \frac{dx}{ds}\left(\frac{\alpha}{2}\sqrt{\frac{k}{m}}\, e^{-\alpha t/2}\right) + \frac{d}{dt}\left(\frac{dx}{ds}\right)\left(-\sqrt{\frac{k}{m}}\, e^{-\alpha t/2}\right)$$

$$= \frac{dx}{ds}\left(\frac{\alpha}{2}\sqrt{\frac{k}{m}}\, e^{-\alpha t/2}\right) + \frac{d^2x}{ds^2}\frac{ds}{dt}\left(-\sqrt{\frac{k}{m}}\, e^{-\alpha t/2}\right)$$

$$= \frac{dx}{ds}\left(\frac{\alpha}{2}\sqrt{\frac{k}{m}}\, e^{-\alpha t/2}\right) + \frac{d^2x}{ds^2}\left(\frac{k}{m}\, e^{-\alpha t}\right).$$

Then

$$m\frac{d^2x}{dt^2} + ke^{-\alpha t}x = ke^{-\alpha t}\frac{d^2x}{ds^2} + \frac{m\alpha}{2}\sqrt{\frac{k}{m}}\, e^{-\alpha t/2}\frac{dx}{dl} + ke^{-\alpha t}x = 0.$$

Multiplying by $2^2/\alpha^2 m$ we have

$$\frac{2^2}{\alpha^2}\frac{k}{m}\, e^{-\alpha t}\frac{d^2x}{ds^2} + \frac{2}{\alpha}\sqrt{\frac{k}{m}}\, e^{-\alpha t/2}\frac{dx}{dt} + \frac{2}{\alpha^2}\frac{k}{m}\, e^{-\alpha t}x = 0$$

or, since $s = (2/\alpha)\sqrt{k/m}\,e^{-\alpha t/2}$,

$$s^2\frac{d^2x}{ds^2} + s\frac{dx}{ds} + s^2x = 0.$$

40. (a) We identify $m = 4$, $k = 1$, and $\alpha = 0.1$. Then

$$x(t) = c_1 J_0(10e^{-0.05t}) + c_2 Y_0(10e^{-0.05t})$$

and

$$x'(t) = -0.5c_1 J_0'(10e^{-0.05t}) - 0.5c_2 Y_0'(10e^{-0.05t}).$$

Now $x(0) = 1$ and $x'(0) = -1/2$ imply

$$c_1 J_0(10) + c_2 Y_0(10) = 1$$

$$c_1 J_0'(10) + c_2 Y_0'(10) = 1.$$

Using Cramer's rule we obtain

$$c_1 = \frac{Y_0'(10) - Y_0(10)}{J_0(10)Y_0'(10) - J_0'(10)Y_0(10)}$$

and

$$c_2 = \frac{J_0(10) - J_0'(10)}{J_0(10)Y_0'(10) - J_0'(10)Y_0(10)}.$$

Using $Y_0' = -Y_1$ and $J_0' = -J_1$ and Table 6.1 we find $c_1 = -4.7860$ and $c_2 = -3.1803$. Thus

$$x(t) = -4.7860 J_0(10e^{-0.05t}) - 3.1803 Y_0(10e^{-0.05t}).$$

(b) In Problem 15 of Section 5.1 the conjecture was that the restoring force on the spring would decay to the point at which the spring would be incapable of returning the mass to the equilibrium position. This is corroborated by the graph.

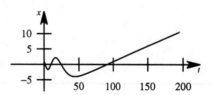

41. Differentiating $y = x^{1/2}w(\frac{2}{3}\alpha x^{3/2})$ with respect to $\frac{2}{3}\alpha x^{3/2}$ we obtain

$$y' = x^{1/2}w'\left(\frac{2}{3}\alpha x^{3/2}\right)\alpha x^{1/2} + \frac{1}{2}x^{-1/2}w\left(\frac{2}{3}\alpha x^{3/2}\right)$$

and

$$y'' = \alpha x w''\left(\frac{2}{3}\alpha x^{3/2}\right)\alpha x^{1/2} + \alpha w'\left(\frac{2}{3}\alpha x^{3/2}\right)$$

$$+ \frac{1}{2}\alpha w'\left(\frac{2}{3}\alpha x^{3/2}\right) - \frac{1}{4}x^{-3/2}w\left(\frac{2}{3}\alpha x^{3/2}\right).$$

Then, after combining terms and simplifying, we have

$$y'' + \alpha^2 xy = \alpha\left[\alpha x^{3/2}w'' + \frac{3}{2}w' + \left(\alpha x^{3/2} - \frac{1}{4\alpha x^{3/2}}\right)w\right] = 0.$$

Letting $t = \frac{2}{3}\alpha x^{3/2}$ or $\alpha x^{3/2} = \frac{3}{2}t$ this differential equation becomes

$$\frac{3}{2}\frac{\alpha}{t}\left[t^2 w''(t) + t w'(t) + \left(t^2 - \frac{1}{9}\right)w(t)\right] = 0,$$

for $t > 0$.

42. The general solution of Bessel's equation is

$$w(t) = c_1 J_{1/3}(t) + c_2 J_{-1/3}(t), \qquad t > 0.$$

Thus, the general solution of Airy's equation for $x > 0$ is

$$y = x^{1/2} w\left(\frac{2}{3}\alpha x^{3/2}\right) = c_1 x^{1/2} J_{1/3}\left(\frac{2}{3}\alpha x^{3/2}\right) + c_2 x^{1/2} J_{-1/3}\left(\frac{2}{3}\alpha x^{3/2}\right).$$

43. (a) Identifying $\alpha = \frac{1}{2}$, the general solution of $x'' + \frac{1}{4}tx = 0$ is

$$x(t) = c_1 x^{1/2} J_{1/3}\left(\frac{1}{3}x^{3/2}\right) + c_2 x^{1/2} J_{-1/3}\left(\frac{1}{3}x^{3/2}\right).$$

Solving the system $x(0.1) = 1$, $x'(0.1) = -\frac{1}{2}$ we find $c_1 = -0.809264$ and $c_2 = 0.782397$.

(b) In Problem 16 of Section 5.1 the conjecture was that the oscillations would become periodic and the spring would oscillate more rapidly. This is corroborated by the graph.

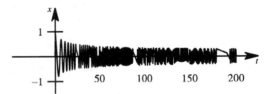

44. (a) Letting $t = L - x$, the boundary-value problem becomes

$$\frac{d^2\theta}{dt^2} + \alpha^2 t\theta = 0, \qquad \theta'(0) = 0, \quad \theta(L) = 0,$$

where $\alpha^2 = \delta g/EI$. This is Airy's differential equation, so by Problem 42 its solution is

$$y = c_1 t^{1/2} J_{1/3}\left(\frac{2}{3}\alpha t^{3/2}\right) + c_2 t^{1/2} J_{-1/3}\left(\frac{2}{3}\alpha t^{3/2}\right)$$

$$= c_1 \theta_1(t) + c_2 \theta_2(t).$$

(b) Looking at the series forms of θ_1 and θ_2 we see that $\theta_1'(0) \neq 0$, while $\theta_2'(0) = 0$. Thus, the boundary condition $\theta'(0) = 0$ implies $c_1 = 0$, and so

$$\theta(t) = c_2 \sqrt{t}\, J_{-1/3}\left(\frac{2}{3}\alpha t^{3/2}\right).$$

From $\theta(L) = 0$ we have

$$c_2 \sqrt{L}\, J_{-1/3}\left(\frac{2}{3}\alpha L^{3/2}\right) = 0,$$

so either $c_2 = 0$, in which case $\theta(t) = 0$, or $J_{-1/3}(\frac{2}{3}\alpha L^{3/2}) = 0$. The column will just start to bend when L is the length corresponding to the smallest positive zero of $J_{-1/3}$. Using *Mathematica*, the first positive root of $J_{-1/3}(x)$ is $x_1 \approx 1.86635$. Thus $\frac{2}{3}\alpha L^{3/2} = 1.86635$ implies

$$L = \left(\frac{3(1.86635)}{2\alpha}\right)^{2/3} = \left[\frac{9EI}{4\delta g}(1.86635)^2\right]^{1/3}$$

$$= \left[\frac{9(2.6 \times 10^7)\pi(0.05)^4/4}{4(0.28)\pi(0.05)^2}(1.86635)^2\right]^{1/3}$$

$$\approx 76.9 \text{ in.}$$

45. (a) Using the expressions for the two linearly independent power series solutions, $y_1(x)$ and $y_2(x)$, given in the text we obtain

$$P_6(x) = \frac{1}{16}\left(231x^6 - 315x^4 + 105x^2 - 5\right)$$

and

$$P_7(x) = \frac{1}{16}\left(429x^7 - 693x^5 + 315x^3 - 35x\right).$$

(b) $P_6(x)$ satisfies $\left(1 - x^2\right)y'' - 2xy' + 42y = 0$ and $P_7(x)$ satisfies $\left(1 - x^2\right)y'' - 2xy' + 56y = 0$.

46. We use the product rule for differentiation:

$$\frac{d}{dx}\left[(1 - x^2)\frac{dy}{dx}\right] + n(n+1)y = (1 - x^2)\frac{d^2y}{dx^2} + (-2x)\frac{dy}{dx} + n(n+1)y$$

$$= (1 - x^2)y'' - 2xy' + n(n+1)y = 0.$$

47. If $x = \cos\theta$ then

$$\frac{dy}{d\theta} = -\sin\theta\frac{dy}{dx},$$

$$\frac{d^2y}{d\theta^2} = \sin^2\theta\frac{d^2y}{dx^2} - \cos\theta\frac{dy}{dx},$$

and

$$\sin\theta\frac{d^2y}{d\theta^2} + \cos\theta\frac{dy}{d\theta} + n(n+1)(\sin\theta)y = \sin\theta\left[\left(1 - \cos^2\theta\right)\frac{d^2y}{dx^2} - 2\cos\theta\frac{dy}{dx} + n(n+1)y\right] = 0.$$

That is,

$$\left(1 - x^2\right)\frac{d^2y}{dx^2} - 2x\frac{dy}{dx} + n(n+1)y = 0.$$

48. The polynomials are shown in (19) on page 251 in the text.

49. By the binomial theorem we have

$$\left[1 + \left(t^2 - 2xt\right)\right]^{-1/2} = 1 - \frac{1}{2}\left(t^2 - 2xt\right) + \frac{3}{8}\left(t^2 - 2xt\right)^2 + \cdots = 1 + xt + \frac{1}{2}\left(3x^2 - 1\right)t^2 + \cdots .$$

50. Letting $x = 1$ in $(1 - 2xt + t^2)^{-1/2}$, we have

$$(1 - 2t + t^2)^{-1/2} = (1 - t)^{-1} = \frac{1}{1-t} = 1 + t + t^2 + t^3 + \ldots \qquad (|t| < 1)$$

$$= \sum_{n=0}^{\infty} t^n.$$

From Problem 49 we have

$$\sum_{n=0}^{\infty} P_n(1)t^n = (1 - 2t + t^2)^{-1/2} = \sum_{n=0}^{\infty} t^n.$$

Equating the coefficients of corresponding terms in the two series, we see that $P_n(1) = 1$. Similarly, letting $x = -1$ we have

$$(1 + 2t + t^2)^{-1/2} = (1 + t)^{-1} = \frac{1}{1+t} = 1 - t + t^2 - 3t^3 + \ldots \qquad (|t| < 1)$$

$$= \sum_{n=0}^{\infty} (-1)^n t^n = \sum_{n=0}^{\infty} P_n(-1)t^n,$$

so that $P_n(-1) = (-1)^n$.

51. The recurrence relation can be written

$$P_{k+1}(x) = \frac{2k+1}{k+1}xP_k(x) - \frac{k}{k+1}P_{k-1}(x), \qquad k = 2,\ 3,\ 4,\ \ldots .$$

$k = 1$: $\ P_2(x) = \frac{3}{2}x^2 - \frac{1}{2}$

$k = 2$: $\ P_3(x) = \frac{5}{3}x\left(\frac{3}{2}x^2 - \frac{1}{2}\right) - \frac{2}{3}x = \frac{5}{2}x^3 - \frac{3}{2}x$

$k = 3$: $\ P_4(x) = \frac{7}{4}x\left(\frac{5}{2}x^3 - \frac{3}{2}x\right) - \frac{3}{4}\left(\frac{3}{2}x^2 - \frac{1}{2}\right) = \frac{35}{8}x^4 - \frac{30}{8}x^2 + \frac{3}{8}$

$k = 4$: $\ P_5(x) = \frac{9}{5}x\left(\frac{35}{8}x^4 - \frac{30}{8}x^2 + \frac{3}{8}\right) - \frac{4}{5}\left(\frac{5}{2}x^3 - \frac{3}{2}x\right) = \frac{63}{8}x^5 - \frac{35}{4}x^3 + \frac{15}{8}x$

$k = 5$: $\ P_6(x) = \frac{11}{6}x\left(\frac{63}{8}x^5 - \frac{35}{4}x^3 + \frac{15}{8}x\right) - \frac{5}{6}\left(\frac{35}{8}x^4 - \frac{30}{8}x^2 + \frac{3}{8}\right) = \frac{231}{16}x^6 - \frac{315}{16}x^4 + \frac{105}{16}x - \frac{5}{16}$

52. $n = 0$: $\ P_0(x) = 1$

$n = 1$: $\ P_1(x) = \frac{1}{2}\frac{d}{dx}(x^2 - 1) = x$

$n = 2$: $\ P_2(x) = \frac{1}{8}\frac{d^2}{dx^2}(x^2 - 1)^2 = \frac{1}{8}\frac{d^2}{dx^2}(x^4 - 2x^2 + 1) = \frac{1}{8}(12x^2 - 4) = \frac{3}{2}x^2 - \frac{1}{2}$

$$n = 3: \quad P_3(x) = \frac{1}{48}\frac{d^3}{dx^3}(x^2-1)^3 = \frac{1}{48}\frac{d^3}{dx^3}(x^6 - 3x^4 + 3x^2 - 3) = \frac{1}{48}(120x^3 - 72x) = \frac{5}{2}x^3 - \frac{3}{2}x$$

53. From $y = u/\sqrt{x}$ we find

$$y' = \frac{2xu' - u}{2x^{3/2}} \qquad \text{and} \qquad y'' = \frac{4x^{5/2} - 4x^{3/2}u' + 3x^{1/2}u}{4x^3}.$$

Then

$$x^2 y'' + xy' + (x^2 - \nu^2)y = \frac{4x^{5/2}u'' - 4x^{3/2}u' + 3x^{1/2}u}{4x} + \frac{2xu' - u}{2x^{1/2}} + \frac{(x^2\nu^2)u}{x^{1/2}}$$

$$= x^{3/2}u'' + \frac{1}{4}x^{-1/2}u + x^{3/2}u - \nu^2 x^{-1/2}u = 0.$$

Multiplying by $x^{1/2}$ we obtain

$$x^2 u'' + \frac{1}{4}u + x^2 u - \nu^2 u = 0$$

or

$$\frac{d^2 u}{dx^2} + \left(1 - \frac{\nu^2 - 1/4}{x^2}\right)u = 0.$$

For large values of x this equation can be approximated by $u'' + u = 0$ whose solutions are $\sin x$ and $\cos x$. Thus, we expect the solutions to oscillate with increasing frequency for larger x.

54. Multiplying the first identity by P_m and the second identity by P_n we obtain

$$P_m \frac{d}{dx}[(1-x^2)P_n'] = -n(n+1)P_m P_n$$

and

$$P_n \frac{d}{dx}[(1-x^2)P_m'] = -m(m+1)P_m P_n.$$

Assuming $m \neq n$ and subtracting these equations gives

$$P_m \frac{d}{dx}[(1-x^2)P_n'] - P_n \frac{d}{dx}[(1-x^2)P_m'] = [m(m+1) - n(n+1)]P_m P_n.$$

Integrating the left-hand side by parts from -1 to 1 we have

$$(1-x^2)P_m P_n' \Big|_{-1}^{1} - \int_{-1}^{1}(1-x^2)P_m' P_n'\, dx - (1-x^2)P_m' P_n \Big|_{-1}^{1} + \int_{-1}^{1}(1-x^2)P_m' P_n' = 0.$$

Thus, $\int_{-1}^{1} P_m(x)P_n(x)dx = 0$.

Chapter 6 Review Exercises

1. Since
$$P(x) = \frac{-2x}{(x-2)(x^2+2x+4)} \quad \text{and} \quad Q(x) = \frac{1}{(x-2)(x^2+2x+4)}$$
the singular points are $x = 0$, $x = -1 + \sqrt{3}\,i$, and $x = -1 - \sqrt{3}\,i$. All others are ordinary points.

2. Since
$$P(x) = 0 \quad \text{and} \quad Q(x) = \frac{2}{(x^2-4)(x^2+4)}$$
the singular points are $x = 2$, $x = -2$, $x = 2i$, and $x = -2i$. All others are ordinary points.

3. Since
$$P(x) = \frac{1}{x(x-5)^2} \quad \text{and} \quad Q(x) = 0$$
the regular singular point is $x - 0$ and the irregular singular point is $x = 5$.

4. Since
$$P(x) = 0 \quad \text{and} \quad Q(x) = \frac{1}{x(x-5)^2}$$
the regular singular points are $x = 0$ and $x = 5$. There are no irregular singular points.

5. Since
$$P(x) = -\frac{1}{x^2(x^2-9)} \quad \text{and} \quad Q(x) = \frac{1}{x(x^2-9)^2}$$
the regular singular points are $x = 3$ and $x = -3$. The irregular singular point is $x = 0$.

6. Since
$$P(x) = \frac{1}{x(x^2+1)^3} \quad \text{and} \quad Q(x) = -\frac{8}{(x^2+1)^3}$$
the regular singular point is $x = 0$. The irregular singular points are $x = i$ and $x = -i$.

7. Since $P(x) = -x$ and $Q(x) = 6$ the interval of convergence is $-\infty < x < \infty$.

8. Since $P(x) = -2x/(x^2-4)$ and $Q(x) = 9/(x^2-4)$ an interval of convergence is $-2 < x < 2$.

9. Substituting $y = \sum_{n=0}^{\infty} c_n x^n$ into the differential equation we have

$$y'' - xy' - y = \underbrace{\sum_{n=2}^{\infty} n(n-1)c_n x^{n-2}}_{k=n-2} - \underbrace{\sum_{n=1}^{\infty} nc_n x^n}_{k=n} - \underbrace{\sum_{n=0}^{\infty} c_n x^n}_{k=n}$$

$$= \sum_{k=0}^{\infty} (k+2)(k+1)c_{k+2}x^k - \sum_{k=1}^{\infty} kc_k x^k - \sum_{k=0}^{\infty} c_k x^k$$

$$= 2c_2 - c_0 + \sum_{k=1}^{\infty} [(k+2)(k+1)c_{k+2} - (k+1)c_k]x^k = 0.$$

Thus
$$2c_2 - c_0 = 0$$

$$(k+2)(k+1)c_{k+2} - (k+1)c_k = 0$$

and
$$c_2 = \frac{1}{2}c_0$$

$$c_{k+2} = \frac{1}{k+2}c_k, \quad k = 1, 2, 3, \ldots .$$

Choosing $c_0 = 1$ and $c_1 = 0$ we find

$$c_2 = \frac{1}{2}$$

$$c_3 = c_5 = c_7 = \cdots = 0$$

$$c_4 = \frac{1}{8}$$

$$c_6 = \frac{1}{48}$$

and so on. For $c_0 = 0$ and $c_1 = 1$ we obtain

$$c_2 = c_4 = c_6 = \cdots = 0$$

$$c_3 = \frac{1}{3}$$

$$c_5 = \frac{1}{15}$$

$$c_7 = \frac{1}{105}$$

and so on. Thus, two solutions are

$$y_1 = 1 + \frac{1}{2}x^2 + \frac{1}{8}x^4 + \frac{1}{48}x^6 + \cdots$$

and

$$y_2 = x + \frac{1}{3}x^3 + \frac{1}{15}x^5 + \frac{1}{105}x^7 + \cdots .$$

10. Substituting $y = \sum_{n=0}^{\infty} c_n x^n$ into the differential equation we obtain

$$y'' - x^2 y' + xy = 2c_2 + (6c_3 + c_0)x + \sum_{k=1}^{\infty} [(k+3)(k+2)c_{k+3} - (k-1)c_k]x^{k+1} = 0$$

which implies $c_2 = 0$, $c_3 = -c_0/6$, and

$$c_{k+3} = \frac{k-1}{(k+3)(k+2)}c_k, \quad k = 1, 2, 3, \ldots .$$

Choosing $c_0 = 1$ and $c_1 = 0$ we find

$$c_3 = -\frac{1}{6}$$

$$c_4 = c_7 = c_{10} = \cdots = 0$$

$$c_5 = c_8 = c_{11} = \cdots = 0$$

$$c_6 = -\frac{1}{90}$$

and so on. For $c_0 = 0$ and $c_1 = 1$ we obtain

$$c_3 = c_6 = c_9 = \cdots = 0$$

$$c_4 = c_7 = c_{10} = \cdots = 0$$

$$c_5 = c_8 = c_{11} = \cdots = 0$$

and so on. Thus, two solutions are

$$y_1 = c_0 \left(1 - \frac{1}{6}x^3 - \frac{1}{90}x^6 - \cdots \right) \quad \text{and} \quad y_2 = c_1 x.$$

11. Substituting $y = \sum_{n=0}^{\infty} c_n x^n$ into the differential equation we obtain

$$(x - 1)y'' + 3y = (-2c_2 + 3c_0) + \sum_{k=3}^{\infty} (k - 1)(k - 2)c_{k-1} - k(k - 1)c_k + 3c_{k-2}]x^{k-2} = 0$$

which implies $c_2 = 3c_0/2$ and

$$c_k = \frac{(k - 1)(k - 2)c_{k-1} + 3c_{k-2}}{k(k - 1)}, \quad k = 3, 4, 5, \ldots .$$

Choosing $c_0 = 1$ and $c_1 = 0$ we find

$$c_2 = \frac{3}{2}, \qquad c_3 = \frac{1}{2}, \qquad c_4 = \frac{5}{8}$$

and so on. For $c_0 = 0$ and $c_1 = 1$ we obtain

$$c_2 = 0, \qquad c_3 = \frac{1}{2}, \qquad c_4 = \frac{1}{4}$$

and so on. Thus, two solutions are

$$y_1 = C_1 \left(1 + \frac{3}{2}x^2 + \frac{1}{2}x^3 + \frac{5}{8}x^4 + \cdots \right)$$

and

$$y_2 = C_2 \left(x + \frac{1}{2}x^3 + \frac{1}{4}x^4 + \cdots \right).$$

251

12. Substituting $y = \sum_{n=0}^{\infty} c_n x^n$ into the differential equation we have

$$(\cos x)y'' + y = \left(1 - \frac{1}{2}x^2 + \frac{1}{24}x^4 - \frac{1}{720}x^6 + \cdots\right)\left(2c_2 + 6c_3 x + 12c_4 x^2 + 20c_5 x^3 + 30c_6 x^4 + \cdots\right)$$

$$+ \sum_{n=0}^{\infty} c_n x^n$$

$$= \left[2c_2 + 6c_3 x + (12c_4 - c_2)x^2 + (20c_5 - 3c_3)x^3 + \left(30c_6 - 6c_4 + \frac{1}{12}c_2\right)x^4 + \cdots\right]$$

$$+ \left[c_0 + c_1 x + c_2 x^2 + c_3 x^3 + c_4 x^4 + \cdots\right]$$

$$= (c_0 + 2c_2) + (c_1 + 6c_3)x + 12c_4 x^2 + (20c_5 - 2c_3)x^3 + \left(30c_6 - 5c_4 + \frac{1}{12}c_2\right)x^4 + \cdots$$

$$= 0.$$

Thus

$$c_0 + 2c_2 = 0$$

$$c_1 + 6c_3 = 0$$

$$12c_4 = 0$$

$$20c_5 - 2c_3 = 0$$

$$30c_6 - 5c_4 + \frac{1}{12}c_2 = 0$$

and

$$c_2 = -\frac{1}{2}c_0$$

$$c_3 = -\frac{1}{6}c_1$$

$$c_4 = 0$$

$$c_5 = \frac{1}{10}c_3$$

$$c_6 = \frac{1}{6}c_4 - \frac{1}{360}c_2.$$

Choosing $c_0 = 1$ and $c_1 = 0$ we find

$$c_2 = -\frac{1}{2}, \quad c_3 = 0, \quad c_4 = 0, \quad c_5 = 0, \quad c_6 = \frac{1}{720}$$

and so on. For $c_0 = 0$ and $c_1 = 1$ we find

$$c_2 = 0, \quad c_3 = -\frac{1}{6}, \quad c_4 = 0, \quad c_5 = -\frac{1}{60}, \quad c_6 = 0$$

and so on. Thus, two solutions are

$$y_1 = 1 - \frac{1}{2}x^2 + \frac{1}{720}x^6 + \cdots \quad \text{and} \quad y_2 = x - \frac{1}{6}x^3 - \frac{1}{60}x^5 + \cdots.$$

13. Substituting $y = \sum_{n=0}^{\infty} c_n x^n$ into the differential equation we have

$$y'' + xy' + 2y = \underbrace{\sum_{n=2}^{\infty} n(n-1)c_n x^{n-2}}_{k=n-2} + \underbrace{\sum_{n=1}^{\infty} nc_n x^n}_{k=n} + 2\underbrace{\sum_{n=0}^{\infty} c_n x^n}_{k=n}$$

$$= \sum_{k=0}^{\infty} (k+2)(k+1)c_{k+2} x^k + \sum_{k=1}^{\infty} kc_k x^k + 2\sum_{k=0}^{\infty} c_k x^k$$

$$= 2c_2 + 2c_0 + \sum_{k=1}^{\infty} [(k+2)(k+1)c_{k+2} + (k+2)c_k]x^k = 0.$$

Thus
$$2c_2 + 2c_0 = 0$$

$$(k+2)(k+1)c_{k+2} + (k+2)c_k = 0$$

and
$$c_2 = -c_0$$

$$c_{k+2} = -\frac{1}{k+1} c_k, \quad k = 1, 2, 3, \ldots.$$

Choosing $c_0 = 1$ and $c_1 = 0$ we find

$$c_2 = -1$$

$$c_3 = c_5 = c_7 = \cdots = 0$$

$$c_4 = \frac{1}{3}$$

$$c_6 = -\frac{1}{15}$$

and so on. For $c_0 = 0$ and $c_1 = 1$ we obtain

$$c_2 = c_4 = c_6 = \cdots = 0$$

$$c_3 = -\frac{1}{2}$$

$$c_5 = \frac{1}{8}$$

$$c_7 = -\frac{1}{48}$$

and so on. Thus, the general solution is

$$y = c_0 \left(1 - x^2 + \frac{1}{3}x^4 - \frac{1}{15}x^6 + \cdots\right) + c_1 \left(x - \frac{1}{2}x^3 + \frac{1}{8}x^5 - \frac{1}{48}x^7 + \cdots\right)$$

and

$$y' = c_0 \left(-2x + \frac{4}{3}x^3 - \frac{2}{5}x^5 + \cdots\right) + c_1 \left(1 - \frac{3}{2}x^2 + \frac{5}{8}x^4 - \frac{7}{48}x^6 + \cdots\right).$$

Setting $y(0) = 3$ and $y'(0) = -2$ we find $c_0 = 3$ and $c_1 = -2$. Therefore, the solution of the initial-value problem is

$$y = 3 - 2x - 3x^2 + x^3 + x^4 - \frac{1}{4}x^5 - \frac{1}{5}x^6 + \frac{1}{24}x^7 + \cdots.$$

14. Substituting $y = \sum_{n=0}^{\infty} c_n x^n$ into the differential equation we have

$$(x+2)y'' + 3y = \sum_{n=2}^{\infty} n(n-1)c_n x^{n-1} + 2\sum_{n=2}^{\infty} n(n-1)c_n x^{n-2} + 3\sum_{n=0}^{\infty} c_n x^n$$
$$\underbrace{\phantom{\sum_{n=2}^{\infty} n(n-1)c_n x^{n-1}}}_{k=n-1} \quad \underbrace{\phantom{2\sum_{n=2}^{\infty} n(n-1)c_n x^{n-2}}}_{k=n-2} \quad \underbrace{\phantom{3\sum_{n=0}^{\infty} c_n x^n}}_{k=n}$$

$$= \sum_{k=1}^{\infty} (k+1)k c_{k+1} x^k + 2\sum_{k=0}^{\infty} (k+2)(k+1)c_{k+2} x^k + 3\sum_{k=0}^{\infty} c_k x^k$$

$$= 4c_2 + 3c_0 + \sum_{k=1}^{\infty} [(k+1)k c_{k+1} + 2(k+2)(k+1)c_{k+2} + 3c_k]x^k = 0.$$

Thus
$$4c_2 + 3c_0 = 0$$

$$(k+1)k c_{k+1} + 2(k+2)(k+1)c_{k+2} + 3c_k = 0$$

and
$$c_2 = -\frac{3}{4}c_0$$

$$c_{k+2} = -\frac{k}{2(k+2)}c_{k+1} - \frac{3}{2(k+2)(k+1)}c_k, \quad k = 1, 2, 3, \ldots.$$

Choosing $c_0 = 1$ and $c_1 = 0$ we find

$$c_2 = -\frac{3}{4}$$

$$c_3 = \frac{1}{8}$$

$$c_4 = \frac{1}{16}$$

$$c_5 = -\frac{9}{320}$$

and so on. For $c_0 = 0$ and $c_1 = 1$ we obtain

$$c_2 = 0$$

$$c_3 = -\frac{1}{4}$$

$$c_4 = \frac{1}{16}$$

$$c_5 = 0$$

254

and so on. Thus, the general solution is

$$y = c_0\left(1 - \frac{3}{4}x^2 + \frac{1}{8}x^3 + \frac{1}{16}x^4 - \frac{9}{320}x^5 + \cdots\right) + c_1\left(x - \frac{1}{4}x^3 + \frac{1}{16}x^4 + \cdots\right)$$

and

$$y' = c_0\left(-\frac{3}{2}x + \frac{3}{8}x^2 + \frac{1}{4}x^3 - \frac{9}{64}x^4 + \cdots\right) + c_1\left(1 - \frac{3}{4}x^2 + \frac{1}{4}x^3 + \cdots\right).$$

Setting $y(0) = 0$ and $y'(0) = 1$ we find $c_0 = 0$ and $c_1 = 1$. Therefore, the solution of the initial-value problem is

$$y = x - \frac{1}{4}x^3 + \frac{1}{16}x^4 + \cdots.$$

15. Substituting $y = \sum_{n=0}^{\infty} c_n x^{n+r}$ into the differential equation we obtain

$$2x^2 y'' + xy' - (x+1)y$$

$$= \left(2r^2 - r - 1\right)c_0 x^r + \sum_{k=1}^{\infty}[2(k+r)(k+r-1)c_k + (k+r)c_k - c_k - c_{k-1}]x^{k+r}$$

$$= 0$$

which implies

$$2r^2 - r - 1 = (2r+1)(r-1) = 0$$

and

$$[(k+r)(2k+2r-1) - 1]c_k - c_{k-1} = 0$$

The indicial roots are $r = 1$ and $r = -1/2$. For $r = 1$ the recurrence relation is

$$c_k = \frac{c_{k-1}}{k(2k+3)}, \quad k = 1, 2, 3, \ldots,$$

so

$$c_1 = \frac{1}{5}c_0, \qquad c_2 = \frac{1}{70}c_0, \qquad c_3 = \frac{1}{1,890}c_0.$$

For $r = -1/2$ the recurrence relation is

$$c_k = \frac{c_{k-1}}{k(2k-3)}, \quad k = 1, 2, 3, \ldots,$$

so

$$c_1 = -c_0, \qquad c_2 = -\frac{1}{2}c_0, \qquad c_3 = -\frac{1}{18}c_0.$$

Two linearly independent solutions are

$$y_1 = C_1 x\left(1 + \frac{1}{5}x + \frac{1}{70}x^2 + \frac{1}{1,890}x^3 + \cdots\right)$$

and

$$y_2 = C_2 x^{-1/2}\left(1 - x - \frac{1}{2}x^2 - \frac{1}{18}x^3 - \cdots\right).$$

16. Substituting $y = \sum_{n=0}^{\infty} c_n x^{n+r}$ into the differential equation we obtain

$$2xy'' + y' + y = \left(2r^2 - r\right) c_0 x^{r-1} + \sum_{k=1}^{\infty} [2(k+r)(k+r-1)c_k + (k+r)c_k + c_{k-1}]x^{k+r-1} = 0$$

which implies $$2r^2 - r = r(2r - 1) = 0$$

and $$(k+r)(2k + 2r - 1)c_k + c_{k-1} = 0.$$

The indicial roots are $r = 0$ and $r = 1/2$. For $r = 0$ the recurrence relation is

$$c_k = -\frac{c_{k-1}}{k(2k-1)}, \quad k = 1, 2, 3, \ldots,$$

so $$c_1 = -c_0, \quad c_2 = \frac{1}{6}c_0, \quad c_3 = -\frac{1}{90}c_0.$$

For $r = 1/2$ the recurrence relation is

$$c_k = -\frac{c_{k-1}}{k(2k+1)}, \quad k = 1, 2, 3, \ldots,$$

so $$c_1 = -\frac{1}{3}c_0, \quad c_2 = \frac{1}{30}c_0, \quad c_3 = -\frac{1}{630}c_0.$$

Two linearly independent solutions are

$$y_1 = C_1 x \left(1 - x + \frac{1}{6}x^2 - \frac{1}{90}x^3 + \cdots\right)$$

and

$$y_2 = C_2 x^{1/2} \left(1 - \frac{1}{3}x + \frac{1}{30}x^2 - \frac{1}{630}x^3 + \cdots\right).$$

17. Substituting $y = \sum_{n=0}^{\infty} c_n x^{n+r}$ into the differential equation we obtain

$$x(1-x)y'' - 2y' + y = \left(r^2 - 3r\right) c_0 x^{r-1} + \sum_{k=1}^{\infty} [(k+r)(k+r-1)c_k - 2(k+r)c_k$$

$$- (k+r-1)(k+r-2)c_{k-1} + c_{k-1}]x^{k+r-1}$$

$$= 0$$

which implies $$r^2 - 3r = r(r - 3) = 0$$

and $$(k+r)(k+r-3)c_k - [(k+r-1)(k+r-2) - 1]c_{k-1} = 0.$$

The indicial roots are $r_1 = 3$ and $r_2 = 0$. For $r_1 = 3$ the recurrence relation is

$$c_k = \frac{\left(k^2 + 3k + 1\right)c_{k-1}}{k(k+3)}, \quad k = 1, 2, 3, \ldots,$$

so
$$c_1 = \frac{5}{4}c_0, \qquad c_2 = \frac{11}{8}c_0, \qquad c_3 = \frac{209}{144}c_0.$$

One solution is
$$y_1 = c_0 x^3 \left(1 + \frac{5}{4}x + \frac{11}{8}x^2 + \frac{209}{144}x^3 + \cdots\right).$$

A second solution is
$$y_2 = y_1 \int \frac{e^{\int [2/x(1-x)]dx}}{y_1^2}\,dx = y_1 \int \frac{x^2 dx}{(1-x)^2 x^6 \left(1 + \frac{5}{2}x + \frac{69}{16}x^2 + \frac{913}{144}x^3 + \cdots\right)}$$

$$= y_1 \int \frac{dx}{x^4 \left(1 - 2x + x^2\right)\left(1 + \frac{5}{2}x + \frac{69}{16}x^2 + \frac{913}{144}x^3 + \cdots\right)}$$

$$= y_1 \int \frac{dx}{x^4 \left(1 + \frac{1}{2}x + \frac{5}{16}x^2 + \frac{31}{144}x^3 + \cdots\right)} = y_1 \int \frac{1}{x^4}\left(1 - \frac{1}{2}x - \frac{1}{16}x^2 - \frac{1}{36}x^3 - \cdots\right)dx$$

$$= y_1 \int \left(\frac{1}{x^4} - \frac{1}{2x^3} - \frac{1}{16x^2} - \frac{1}{36x} - \cdots\right)dx = y_1 \left[-\frac{1}{3x^3} + \frac{1}{4x^2} + \frac{1}{16x} - \frac{1}{36}\ln x + \cdots\right]$$

$$= -\frac{1}{36}y_1 \ln x + y_1\left(-\frac{1}{3x^3} + \frac{1}{4x^2} + \frac{1}{16x} + \cdots\right).$$

18. Substituting $y = \sum_{n=0}^{\infty} c_n x^{n+r}$ into the differential equation we obtain
$$x^2 y'' - xy' + \left(x^2 + 1\right)y = \left(r^2 - 2r + 1\right)c_0 x^r + r^2 c_1 x^{r+1}$$

$$+ \sum_{k=2}^{\infty}[(k+r)(k+r-1)c_k - (k+r)c_k + c_k + c_{k-2}]x^{k+r}$$

$$= 0$$

which implies
$$r^2 - 2r + 1 = (r-1)^2 = 0$$

$$r^2 c_1 = 0$$

$$[(k+r)(k+r-2) + 1]c_k + c_{k-2} = 0.$$

The indicial roots are $r_1 = r_2 = 1$, so $c_1 = 0$ and
$$c_k = -\frac{c_{k-2}}{k^2}, \quad k = 2, 3, 4, \ldots.$$

Thus
$$c_2 = -\frac{1}{4}c_0$$

$$c_3 = c_5 = c_7 = \cdots = 0$$

$$c_4 = \frac{1}{64}c_0$$

$$c_6 = -\frac{1}{2,304}c_0$$

and one solution is

$$y_1 = c_0 x \left(1 - \frac{1}{4}x^2 + \frac{1}{64}x^4 - \frac{1}{2,304}x^6 + \cdots\right).$$

A second solution is

$$y_2 = y_1 \int \frac{e^{dx/x}}{y_1^2}\, dx = y_1 \int \frac{x\, dx}{x^2 \left(1 - \frac{1}{4}x^2 + \frac{1}{64}x^4 - \frac{1}{2304}x^6 + \cdots\right)^2}$$

$$= y_1 \int \frac{dx}{x \left(1 - \frac{1}{2}x^2 + \frac{3}{32}x^4 - \frac{5}{576}x^6 + \cdots\right)}$$

$$= y_1 \int \frac{1}{x}\left(1 + \frac{1}{2}x^2 + \frac{5}{32}x^4 + \frac{23}{576}x^6 + \cdots\right) dx$$

$$= y_1 \int \left(\frac{1}{x} + \frac{1}{2}x + \frac{5}{32}x^3 + \frac{23}{576}x^5 + \cdots\right) dx$$

$$= y_1 \ln x + y_1 \left(\frac{1}{4}x^2 + \frac{5}{128}x^4 + \frac{23}{3,456}x^6 + \cdots\right).$$

19. Substituting $y = \sum_{n=0}^{\infty} c_n x^{n+r}$ into the differential equation we obtain

$$xy'' - (2x - 1)y' + (x - 1)y = r^2 c_0 x^{r-1} + \left[\left(r^2 + 2r + 1\right)c_1 - (2r + 1)c_0\right]x^r$$

$$+ \sum_{k=2}^{\infty} [(k + r)(k + r - 1)c_k + (k + r)c_k - 2(k + r - 1)c_{k-1} - c_{k-1} + c_{k-2}]x^{k+r-1}$$

$$= 0$$

which implies

$$r^2 = 0$$

$$(r + 1)^2 c_1 - (2r + 1)c_0 = 0$$

and

$$(k + r)^2 c_k - (2k + 2r - 1)c_{k-1} + c_{k-2} = 0.$$

The indicial roots are $r_1 = r_2 = 0$, so $c_1 = c_0$ and

$$c_k = \frac{(2k - 1)c_{k-1} - c_{k-2}}{k^2}, \quad k = 2, 3, 4, \ldots.$$

Thus

$$c_2 = \frac{1}{2}c_0, \qquad c_3 = \frac{1}{3!}c_0, \qquad c_4 = \frac{1}{4!}c_0$$

and one solution is

$$y_1 = c_0 \left(1 + x + \frac{1}{2}x^2 + \frac{1}{3!}x^3 + \frac{1}{4!}x^4 + \cdots\right) = c_0 e^x.$$

A second solution is

$$y_2 = e^x \int \frac{e^{\int (2 - 1/x)dx}}{e^{2x}} \, dx = e^x \int \frac{e^{2x} dx}{xe^{2x}} = e^x \int \frac{1}{x} \, dx = e^x \ln x.$$

20. Substituting $y = \sum_{n=0}^{\infty} c_n x^{n+r}$ into the differential equation we obtain

$$x^2 y'' - x^2 y' + \left(x^2 - 2\right) y = \left(r^2 - r - 2\right) c_0 x^r + \left[\left(r^2 + r - 2\right) c_1 - rc_0\right] x^{r+1}$$

$$+ \sum_{k=2}^{\infty} [(k+r)(k+r-1)c_k - 2c_k - (k+r-1)c_{k-1} + c_{k-2}] x^{k+r}$$

$$= 0$$

which implies
$$r^2 - r - 2 = (r-2)(r+1) = 0$$

$$\left(r^2 + r - 2\right) c_1 - rc_0 = 0$$

and
$$[(k+r)(k+r-1) - 2]c_k - (k+r-1)c_{k-1} + c_{k-2} = 0.$$

The indicial roots are $r_1 = 2$ and $r_2 = -1$. For $r_2 = -1$,

$$-2c_1 + c_0 = 0 \quad \text{and} \quad k(k-3)c_k - (k-2)c_{k-1} + c_{k-2} = 0, \quad k = 2, 3, 4, \ldots .$$

Thus
$$c_1 = \frac{1}{2} c_0$$

$$-2c_2 - 0 \cdot c_1 + c_0 = 0 \quad \Rightarrow \quad c_2 = \frac{1}{2} c_0$$

$$0 \cdot c_3 - c_2 + c_1 = 0 \quad \Rightarrow \quad c_2 = c_1 = \frac{1}{2} c_0 \text{ and } c_3 \text{ is arbitrary}$$

and
$$c_k = \frac{(k-2)c_{k-1} - c_{k-2}}{k(k-3)}, \quad k = 4, 5, 6, \ldots .$$

Taking $c_0 = 1$ and $c_3 = 0$ we have

$$c_1 = \frac{1}{2}, \qquad c_2 = \frac{1}{2}, \qquad c_4 = -\frac{1}{8}, \qquad c_5 = -\frac{3}{80},$$

and so on. Choosing $c_0 = 0$ and $c_3 = 1$ we have

$$c_1 = c_2 = 0, \qquad c_4 = \frac{1}{2}, \qquad c_5 = \frac{3}{20},$$

and so on. Two solutions are

$$y_1 = C_1 x^{-1} \left(1 + \frac{1}{2} x + \frac{1}{2} x^2 - \frac{1}{8} x^4 - \frac{3}{80} x^5 + \cdots\right)$$

and

$$y_2 = C_2 x^{-1} \left(x^3 + \frac{1}{2} x^4 + \frac{3}{20} x^5 + \cdots\right).$$

7 The Laplace Transform

1. $\mathscr{L}\{f(t)\} = \int_0^1 -e^{-st}dt + \int_1^\infty e^{-st}dt = \dfrac{1}{s}e^{-st}\Big|_0^1 - \dfrac{1}{s}e^{-st}\Big|_1^\infty$

   $= \dfrac{1}{s}e^{-s} - \dfrac{1}{s} - \left(0 - \dfrac{1}{s}e^{-s}\right) = \dfrac{2}{s}e^{-s} - \dfrac{1}{s}, \quad s > 0$

2. $\mathscr{L}\{f(t)\} = \int_0^2 4e^{-st}dt = -\dfrac{4}{s}e^{-st}\Big|_0^2 = -\dfrac{4}{s}(e^{-2s} - 1), \quad s > 0$

3. $\mathscr{L}\{f(t)\} = \int_0^1 te^{-st}dt + \int_1^\infty e^{-st}dt = \left(-\dfrac{1}{s}te^{-st} - \dfrac{1}{s^2}e^{-st}\right)\Big|_0^1 - \dfrac{1}{s}e^{-st}\Big|_1^\infty$

   $= \left(-\dfrac{1}{s}e^{-s} - \dfrac{1}{s^2}e^{-s}\right) - \left(0 - \dfrac{1}{s^2}\right) - \dfrac{1}{s}(0 - e^{-s}) = \dfrac{1}{s^2}(1 - e^{-s}), \quad s > 0$

4. $\mathscr{L}\{f(t)\} = \int_0^1 (2t + 1)e^{-st}dt = \left(-\dfrac{2}{s}te^{-st} - \dfrac{2}{s^2}e^{-st} - \dfrac{1}{s}e^{-st}\right)\Big|_0^1$

   $= \left(-\dfrac{2}{s}e^{-s} - \dfrac{2}{s^2}e^{-s} - \dfrac{1}{s}e^{-s}\right) - \left(0 - \dfrac{2}{s^2} - \dfrac{1}{s}\right) = \dfrac{1}{s}(1 - 3e^{-s}) + \dfrac{2}{s^2}(1 - e^{-s}), \quad s > 0$

5. $\mathscr{L}\{f(t)\} = \int_0^\pi (\sin t)e^{-st}dt = \left(-\dfrac{s}{s^2 + 1}e^{-st}\sin t - \dfrac{1}{s^2 + 1}e^{-st}\cos t\right)\Big|_0^\pi$

   $= \left(0 + \dfrac{1}{s^2 + 1}e^{-\pi s}\right) - \left(0 - \dfrac{1}{s^2 + 1}\right) = \dfrac{1}{s^2 + 1}(e^{-\pi s} + 1), \quad s > 0$

6. $\mathscr{L}\{f(t)\} = \int_{\pi/2}^\infty (\cos t)e^{-st}dt = \left(-\dfrac{s}{s^2 + 1}e^{-st}\cos t + \dfrac{1}{s^2 + 1}e^{-st}\sin t\right)\Big|_{\pi/2}^\infty$

   $= 0 - \left(0 + \dfrac{1}{s^2 + 1}e^{-\pi s/2}\right) = -\dfrac{1}{s^2 + 1}e^{-\pi s/2}, \quad s > 0$

7. $f(t) = \begin{cases} 0, & 0 < t < 1 \\ t, & t > 1 \end{cases}$

   $\mathscr{L}\{f(t)\} = \int_1^\infty te^{-st}\,dt = \left(-\dfrac{1}{s}te^{-st} - \dfrac{1}{s^2}e^{-st}\right)\Big|_1^\infty = \dfrac{1}{s}e^{-s} + \dfrac{1}{s^2}e^{-s}, \quad s > 0$

8. $f(t) = \begin{cases} 0, & 0 < t < 1 \\ 2t - 2, & t > 1 \end{cases}$

   $\mathscr{L}\{f(t)\} = 2\int_1^\infty (t - 1)e^{-st}\,dt = 2\left(-\dfrac{1}{s}(t - 1)e^{-st} - \dfrac{1}{s^2}e^{-st}\right)\Big|_1^\infty = \dfrac{2}{s^2}e^{-s}, \quad s > 0$

9. $f(t) = \begin{cases} 1 - t, & 0 < t < 1 \\ 0, & t > 0 \end{cases}$

$$\mathscr{L}\{f(t)\} = \int_0^1 (1 - t)e^{-st}\,dt = \left(-\frac{1}{s}(1 - t)e^{-st} + \frac{1}{s^2}e^{-st}\right)\Big|_0^1 = \frac{1}{s^2}e^{-s} + \frac{1}{s} - \frac{1}{s^2}, \quad s > 0$$

10. $f(t) = \begin{cases} 0, & 0 < t < a \\ c, & a < t < b; \\ 0, & t > b \end{cases}$ $\quad \mathscr{L}\{f(t)\} = \int_a^b ce^{-st}\,dt = -\frac{c}{s}e^{-st}\Big|_a^b = \frac{c}{s}(e^{-sa} - e^{-sb}), \quad s > 0$

11. $\mathscr{L}\{f(t)\} = \int_0^\infty e^{t+7}e^{-st}\,dt = e^7 \int_0^\infty e^{(1-s)t}\,dt = \frac{e^7}{1-s}e^{(1-s)t}\Big|_0^\infty = 0 - \frac{e^7}{1-s} = \frac{e^7}{s-1}, \quad s > 1$

12. $\mathscr{L}\{f(t)\} = \int_0^\infty e^{-2t-5}e^{-st}\,dt = e^{-5}\int_0^\infty e^{-(s+2)t}\,dt = -\frac{e^{-5}}{s+2}e^{-(s+2)t}\Big|_0^\infty = \frac{e^{-5}}{s+2}, \quad s > -2$

13. $\mathscr{L}\{f(t)\} = \int_0^\infty te^{4t}e^{-st}\,dt = \int_0^\infty te^{(4-s)t}\,dt = \left(\frac{1}{4-s}te^{(4-s)t} - \frac{1}{(4-s)^2}e^{(4-s)t}\right)\Big|_0^\infty$

$$= \frac{1}{(4-s)^2}, \quad s > 4$$

14. $\mathscr{L}\{f(t)\} = \int_0^\infty t^2 e^{3t}e^{-st}\,dt = \int_0^\infty t^2 e^{(3-s)t}\,dt$

$$= \left(\frac{1}{3-s}t^2 e^{(3-s)t} - \frac{2}{(3-s)^2}te^{(3-s)t} + \frac{2}{(3-s)^3}e^{(3-s)t}\right)\Big|_0^\infty$$

$$= -\frac{2}{(3-s)^3} = \frac{2}{(s-3)^3}, \quad s > 3$$

15. $\mathscr{L}\{f(t)\} = \int_0^\infty e^{-t}(\sin t)e^{-st}\,dt = \int_0^\infty (\sin t)e^{-(s+1)t}\,dt$

$$= \left(\frac{-(s+1)}{(s+1)^2 + 1}e^{-(s+1)t}\sin t - \frac{1}{(s+1)^2 + 1}e^{-(s+1)t}\cos t\right)\Big|_0^\infty$$

$$= \frac{1}{(s+1)^2 + 1} = \frac{1}{s^2 + 2s + 2}, \quad s > -1$$

16. $\mathscr{L}\{f(t)\} = \int_0^\infty e^t(\cos t)e^{-st}\,dt = \int_0^\infty (\cos t)e^{(1-s)t}\,dt$

$$= \left(\frac{1-s}{(1-s)^2 + 1}e^{(1-s)t}\cos t + \frac{1}{(1-s)^2 + 1}e^{(1-s)t}\sin t\right)\Big|_0^\infty$$

$$= -\frac{1-s}{(1-s)^2 + 1} = \frac{s-1}{s^2 - 2s + 2}, \quad s > 1$$

Exercises 7.1

17. $\mathcal{L}\{f(t)\} = \displaystyle\int_0^\infty t(\cos t)e^{-st}\,dt$

$$= \left[\left(-\frac{st}{s^2+1} - \frac{s^2-1}{(s^2+1)^2}\right)(\cos t)e^{-st} + \left(\frac{t}{s^2+1} + \frac{2s}{(s^2+1)^2}\right)(\sin t)e^{-st}\right]_0^\infty$$

$$= \frac{s^2-1}{(s^2+1)^2}, \quad s > 0$$

18. $\mathcal{L}\{f(t)\} = \displaystyle\int_0^\infty t(\sin t)e^{-st}\,dt$

$$= \left[\left(-\frac{t}{s^2+1} - \frac{2s}{(s^2+1)^2}\right)(\cos t)e^{-st} - \left(\frac{st}{s^2+1} + \frac{s^2-1}{(s^2+1)^2}\right)(\sin t)e^{-st}\right]_0^\infty$$

$$= \frac{2s}{(s^2+1)^2}, \quad s > 0$$

19. $\mathcal{L}\{2t^4\} = 2\dfrac{4!}{s^5}$

20. $\mathcal{L}\{t^5\} = \dfrac{5!}{s^6}$

21. $\mathcal{L}\{4t - 10\} = \dfrac{4}{s^2} - \dfrac{10}{s}$

22. $\mathcal{L}\{7t + 3\} = \dfrac{7}{s^2} + \dfrac{3}{s}$

23. $\mathcal{L}\{t^2 + 6t - 3\} = \dfrac{2}{s^3} + \dfrac{6}{s^2} - \dfrac{3}{s}$

24. $\mathcal{L}\{-4t^2 + 16t + 9\} = -4\dfrac{2}{s^3} + \dfrac{16}{s^2} + \dfrac{9}{s}$

25. $\mathcal{L}\{t^3 + 3t^2 + 3t + 1\} = \dfrac{3!}{s^4} + 3\dfrac{2}{s^3} + \dfrac{3}{s^2} + \dfrac{1}{s}$

26. $\mathcal{L}\{8t^3 - 12t^2 + 6t - 1\} = 8\dfrac{3!}{s^4} - 12\dfrac{2}{s^3} + \dfrac{6}{s^2} - \dfrac{1}{s}$

27. $\mathcal{L}\{1 + e^{4t}\} = \dfrac{1}{s} + \dfrac{1}{s-4}$

28. $\mathcal{L}\{t^2 - e^{-9t} + 5\} = \dfrac{2}{s^3} - \dfrac{1}{s+9} + \dfrac{5}{s}$

29. $\mathcal{L}\{1 + 2e^{2t} + e^{4t}\} = \dfrac{1}{s} + \dfrac{2}{s-2} + \dfrac{1}{s-4}$

30. $\mathcal{L}\{e^{2t} - 2 + e^{-2t}\} = \dfrac{1}{s-2} - \dfrac{2}{s} + \dfrac{1}{s+2}$

31. $\mathcal{L}\{4t^2 - 5\sin 3t\} = 4\dfrac{2}{s^3} - 5\dfrac{3}{s^2+9}$

32. $\mathcal{L}\{\cos 5t + \sin 2t\} = \dfrac{s}{s^2+25} + \dfrac{2}{s^2+4}$

33. $\mathcal{L}\{\sinh kt\} = \dfrac{k}{s^2-k^2}$

34. $\mathcal{L}\{\cosh kt\} = \dfrac{s}{s^2-k^2}$

35. $\mathcal{L}\{e^t \sinh t\} = \mathcal{L}\left\{e^t\dfrac{e^t - e^{-t}}{2}\right\} = \mathcal{L}\left\{\dfrac{1}{2}e^{2t} - \dfrac{1}{2}\right\} = \dfrac{1}{2(s-2)} - \dfrac{1}{2s}$

36. $\mathcal{L}\{e^{-t} \cosh t\} = \mathcal{L}\left\{e^{-t}\dfrac{e^t + e^{-t}}{2}\right\} = \mathcal{L}\left\{\dfrac{1}{2} + \dfrac{1}{2}e^{-2t}\right\} = \dfrac{1}{2s} + \dfrac{1}{2(s+2)}$

37. $\mathscr{L}\{\sin 2t \cos 2t\} = \mathscr{L}\left\{\dfrac{1}{2}\sin 4t\right\} = \dfrac{2}{s^2+16}$

38. $\mathscr{L}\{\cos^2 t\} = \mathscr{L}\left\{\dfrac{1}{2}+\dfrac{1}{2}\cos 2t\right\} = \dfrac{1}{2s}+\dfrac{1}{2}\dfrac{s}{s^2+4}$

39. Let $u = st$ so that $du = s\,dt$ and $\mathscr{L}\{t^\alpha\} = \displaystyle\int_0^\infty e^{-st}t^\alpha dt = \int_0^\infty e^{-u}\left(\dfrac{u}{s}\right)^\alpha \dfrac{1}{s}\,du = \dfrac{1}{s^{\alpha+1}}\Gamma(\alpha+1)$

for $\alpha > -1$.

40. $\mathscr{L}\{t^{-1/2}\} = \dfrac{\Gamma(1/2)}{s^{1/2}} = \sqrt{\dfrac{\pi}{s}}$

41. $\mathscr{L}\{t^{1/2}\} = \dfrac{\Gamma(3/2)}{s^{3/2}} = \dfrac{\sqrt{\pi}}{2s^{3/2}}$

42. $\mathscr{L}\{t^{3/2}\} = \dfrac{\Gamma(5/2)}{s^{5/2}} = \dfrac{3\sqrt{\pi}}{4s^{5/2}}$

43. If we attempt to compute the Laplace transform of $1/t^2$ we obtain

$$\mathscr{L}\{1/t^2\} = \int_0^1 \dfrac{1}{t^2}e^{-st}dt + \int_1^\infty \dfrac{1}{t^2}e^{-st}dt.$$

If $s = 0$ then

$$\int_0^1 \dfrac{1}{t^2}e^{-st}dt = \int_0^1 \dfrac{1}{t^2}\,dt,$$

which diverges. If $s < 0$ then

$$\int_0^1 \dfrac{1}{t^2}e^{-st}dt > \int_0^1 \dfrac{1}{t^2}\,dt,$$

which diverges. If $s > 0$ then

$$\int_0^1 \dfrac{1}{t^2}e^{-st}dt > e^{-s}\int_0^1 \dfrac{1}{t^2}e^{-st}dt,$$

which diverges. Thus, the Laplace transform of $1/t^2$ does not exist.

44. Let $F(t) = t^{1/3}$. Then $F(t)$ is of exponential order, but $f(t) = F'(t) = \frac{1}{3}r^{-2/3}$ t bounded near $t = 0$ and hence is not of exponential order.

Let $f(t) = 2te^{t^2}\cos e^{t^2} = \frac{d}{dt}\sin e^{t^2}$. This function is not of exponential order, but we show that its Laplace transform exists. Using integration by parts we have

$$\mathscr{L}\{2te^{t^2}\cos e^{t^2}\} = \int_0^\infty e^{-st}\left(\dfrac{d}{dt}\sin e^{t^2}\right)dt = \lim_{a\to\infty}\left[e^{-st}\sin e^{t^2}\Big|_0^a + s\int_0^a e^{-st}\sin e^{t^2}\,dt\right]$$

$$= s\int_0^\infty e^{-st}\sin e^{t^2}\,dt = s\,\mathscr{L}\{\sin e^{t^2}\}.$$

Since $\sin e^{t^2}$ is continuous and of exponential order, $\mathscr{L}\{\sin e^{t^2}\}$ exists, and therefore $\mathscr{L}\{2te^{t^2}\cos e^{t^2}\}$ exists.

Exercises 7.2

1. $\mathcal{L}^{-1}\left\{\dfrac{1}{s^3}\right\} = \dfrac{1}{2}\mathcal{L}^{-1}\left\{\dfrac{2}{s^3}\right\} = \dfrac{1}{2}t^2$

2. $\mathcal{L}^{-1}\left\{\dfrac{1}{s^4}\right\} = \dfrac{1}{6}\mathcal{L}^{-1}\left\{\dfrac{3!}{s^4}\right\} = \dfrac{1}{6}t^3$

3. $\mathcal{L}^{-1}\left\{\dfrac{1}{s^2} - \dfrac{48}{s^5}\right\} = \mathcal{L}^{-1}\left\{\dfrac{1}{s^2} - \dfrac{48}{24}\cdot\dfrac{4!}{s^5}\right\} = t - 2t^4$

4. $\mathcal{L}^{-1}\left\{\left(\dfrac{2}{s} - \dfrac{1}{s^3}\right)^2\right\} = \mathcal{L}^{-1}\left\{4\cdot\dfrac{1}{s^2} - \dfrac{4}{6}\cdot\dfrac{3!}{s^4} + \dfrac{1}{120}\cdot\dfrac{5!}{s^6}\right\} = 4t - \dfrac{2}{3}t^3 + \dfrac{1}{120}t^5$

5. $\mathcal{L}^{-1}\left\{\dfrac{(s+1)^3}{s^4}\right\} = \mathcal{L}^{-1}\left\{\dfrac{1}{s} + 3\cdot\dfrac{1}{s^2} + \dfrac{3}{2}\cdot\dfrac{2}{s^3} + \dfrac{1}{6}\cdot\dfrac{3!}{s^4}\right\} = 1 + 3t + \dfrac{3}{2}t^2 + \dfrac{1}{6}t^3$

6. $\mathcal{L}^{-1}\left\{\dfrac{(s+2)^2}{s^3}\right\} = \mathcal{L}^{-1}\left\{\dfrac{1}{s} + 4\cdot\dfrac{1}{s^2} + 2\cdot\dfrac{2}{s^3}\right\} = 1 + 4t + 2t^2$

7. $\mathcal{L}^{-1}\left\{\dfrac{1}{s^2} - \dfrac{1}{s} + \dfrac{1}{s-2}\right\} = t - 1 + e^{2t}$

8. $\mathcal{L}^{-1}\left\{\dfrac{4}{s} + \dfrac{6}{s^5} - \dfrac{1}{s+8}\right\} = \mathcal{L}^{-1}\left\{4\cdot\dfrac{1}{s} + \dfrac{1}{4}\cdot\dfrac{4!}{s^5} - \dfrac{1}{s+8}\right\} = 4 + \dfrac{1}{4}t^4 - e^{-8t}$

9. $\mathcal{L}^{-1}\left\{\dfrac{1}{4s+1}\right\} = \mathcal{L}^{-1}\left\{\dfrac{1}{4}\cdot\dfrac{1}{s+1/4}\right\} = \dfrac{1}{4}e^{-t/4}$

10. $\mathcal{L}^{-1}\left\{\dfrac{1}{5s-2}\right\} = \mathcal{L}^{-1}\left\{\dfrac{1}{5}\cdot\dfrac{1}{s-2/5}\right\} = \dfrac{1}{5}e^{2t/5}$

11. $\mathcal{L}^{-1}\left\{\dfrac{5}{s^2+49}\right\} = \mathcal{L}^{-1}\left\{\dfrac{5}{7}\cdot\dfrac{7}{s^2+49}\right\} = \dfrac{5}{7}\sin 7t$

12. $\mathcal{L}^{-1}\left\{\dfrac{10s}{s^2+16}\right\} = 10\cos 4t$

13. $\mathcal{L}^{-1}\left\{\dfrac{4s}{4s^2+1}\right\} = \mathcal{L}^{-1}\left\{\dfrac{s}{s^2+1/4}\right\} = \cos\dfrac{1}{2}t$

14. $\mathcal{L}^{-1}\left\{\dfrac{1}{4s^2+1}\right\} = \mathcal{L}^{-1}\left\{\dfrac{1}{2}\cdot\dfrac{1/2}{s^2+1/4}\right\} = \dfrac{1}{2}\sin\dfrac{1}{2}t$

15. $\mathcal{L}^{-1}\left\{\dfrac{1}{s^2-16}\right\} = \mathcal{L}^{-1}\left\{\dfrac{1/8}{s-4} - \dfrac{1/8}{s+4}\right\} = \dfrac{1}{8}e^{4t} - \dfrac{1}{8}e^{-4t} = \dfrac{1}{4}\sinh 4t$

16. $\mathscr{L}^{-1}\left\{\dfrac{10s}{s^2-25}\right\} = 10\cosh 5t$

17. $\mathscr{L}^{-1}\left\{\dfrac{2s-6}{s^2+9}\right\} = \mathscr{L}^{-1}\left\{2\cdot\dfrac{s}{s^2+9} - 2\cdot\dfrac{3}{s^2+9}\right\} = 2\cos 3t - 2\sin 3t$

18. $\mathscr{L}^{-1}\left\{\dfrac{s+1}{s^2+2}\right\} = \mathscr{L}^{-1}\left\{\dfrac{s}{s^2+2} + \dfrac{1}{\sqrt{2}}\cdot\dfrac{\sqrt{2}}{s^2+2}\right\} = \cos\sqrt{2}\,t + \dfrac{1}{\sqrt{2}}\sin\sqrt{2}\,t$

19. $\mathscr{L}^{-1}\left\{\dfrac{1}{s^2+3s}\right\} = \mathscr{L}^{-1}\left\{\dfrac{1}{3}\cdot\dfrac{1}{s} - \dfrac{1}{3}\cdot\dfrac{1}{s+3}\right\} = \dfrac{1}{3} - \dfrac{1}{3}e^{-3t}$

20. $\mathscr{L}^{-1}\left\{\dfrac{s+1}{s^2-4s}\right\} = \mathscr{L}^{-1}\left\{-\dfrac{1}{4}\cdot\dfrac{1}{s} + \dfrac{5}{4}\cdot\dfrac{1}{s-4}\right\} = -\dfrac{1}{4} + \dfrac{5}{4}e^{4t}$

21. $\mathscr{L}^{-1}\left\{\dfrac{s}{s^2+2s-3}\right\} = \mathscr{L}^{-1}\left\{\dfrac{1}{4}\cdot\dfrac{1}{s-1} + \dfrac{3}{4}\cdot\dfrac{1}{s+3}\right\} = \dfrac{1}{4}e^{t} + \dfrac{3}{4}e^{-3t}$

22. $\mathscr{L}^{-1}\left\{\dfrac{1}{s^2+s-20}\right\} = \mathscr{L}^{-1}\left\{\dfrac{1}{9}\cdot\dfrac{1}{s-4} - \dfrac{1}{9}\cdot\dfrac{1}{s+5}\right\} = \dfrac{1}{9}e^{4t} - \dfrac{1}{9}e^{-5t}$

23. $\mathscr{L}^{-1}\left\{\dfrac{0.9s}{(s-0.1)(s+0.2)}\right\} = \mathscr{L}^{-1}\left\{(0.3)\cdot\dfrac{1}{s-0.1} + (0.6)\cdot\dfrac{1}{s+0.2}\right\} = 0.3e^{0.1t} + 0.6e^{-0.2t}$

24. $\mathscr{L}^{-1}\left\{\dfrac{s-3}{(s-\sqrt{3})(s+\sqrt{3})}\right\} = \mathscr{L}^{-1}\left\{\dfrac{s}{s^2-3} - \sqrt{3}\cdot\dfrac{\sqrt{3}}{s^2-3}\right\} = \cosh\sqrt{3}\,t - \sqrt{3}\sinh\sqrt{3}\,t$

25. $\mathscr{L}^{-1}\left\{\dfrac{s}{(s-2)(s-3)(s-6)}\right\} = \mathscr{L}^{-1}\left\{\dfrac{1}{2}\cdot\dfrac{1}{s-2} - \dfrac{1}{s-3} + \dfrac{1}{2}\cdot\dfrac{1}{s-6}\right\} = \dfrac{1}{2}e^{2t} - e^{3t} + \dfrac{1}{2}e^{6t}$

26. $\mathscr{L}^{-1}\left\{\dfrac{s^2+1}{s(s-1)(s+1)(s-2)}\right\} = \mathscr{L}^{-1}\left\{\dfrac{1}{2}\cdot\dfrac{1}{s} - \dfrac{1}{s-1} - \dfrac{1}{3}\cdot\dfrac{1}{s+1} + \dfrac{5}{6}\cdot\dfrac{1}{s-2}\right\}$

$$= \dfrac{1}{2} - e^{t} - \dfrac{1}{3}e^{-t} + \dfrac{5}{6}e^{2t}$$

27. $\mathscr{L}^{-1}\left\{\dfrac{2s+4}{(s-2)(s^2+4s+3)}\right\} = \mathscr{L}^{-1}\left\{\dfrac{8}{15}\cdot\dfrac{1}{s-2} - \dfrac{1}{3}\cdot\dfrac{1}{s+1} - \dfrac{1}{5}\cdot\dfrac{1}{s+3}\right\} = \dfrac{8}{15}e^{2t} - \dfrac{1}{3}e^{-t} - \dfrac{1}{5}e^{-3t}$

28. $\mathscr{L}^{-1}\left\{\dfrac{s+1}{(s^2-4s)(s+5)}\right\} = \mathscr{L}^{-1}\left\{-\dfrac{1}{20}\cdot\dfrac{1}{s} + \dfrac{5}{36}\cdot\dfrac{1}{s-4} - \dfrac{4}{45}\cdot\dfrac{1}{s+5}\right\} = -\dfrac{1}{20} + \dfrac{5}{36}e^{4t} - \dfrac{4}{45}e^{-5t}$

29. $\mathscr{L}^{-1}\left\{\dfrac{1}{s^2(s^2+4)}\right\} = \mathscr{L}^{-1}\left\{\dfrac{1}{4}\cdot\dfrac{1}{s^2} - \dfrac{1}{8}\cdot\dfrac{2}{s^2+4}\right\} = \dfrac{1}{4}t - \dfrac{1}{8}\sin 2t$

30. $\mathscr{L}^{-1}\left\{\dfrac{s-1}{s^2(s^2+1)}\right\} = \mathscr{L}^{-1}\left\{\dfrac{1}{s} - \dfrac{1}{s^2} - \dfrac{s}{s^2+1} + \dfrac{1}{s^2+1}\right\} = 1 - t - \cos t + \sin t$

31. $\mathcal{L}^{-1}\left\{\dfrac{s}{(s^2+4)(s+2)}\right\} = \mathcal{L}^{-1}\left\{\dfrac{1}{4}\cdot\dfrac{s}{s^2+4} + \dfrac{1}{4}\cdot\dfrac{2}{s^2+4} - \dfrac{1}{4}\cdot\dfrac{1}{s+2}\right\} = \dfrac{1}{4}\cos 2t + \dfrac{1}{4}\sin 2t - \dfrac{1}{4}e^{-2t}$

32. $\mathcal{L}^{-1}\left\{\dfrac{1}{s^4-9}\right\} = \mathcal{L}^{-1}\left\{\dfrac{1}{6\sqrt{3}}\cdot\dfrac{\sqrt{3}}{s^2-3} - \dfrac{1}{6\sqrt{3}}\cdot\dfrac{\sqrt{3}}{s^2-3}\right\} = \dfrac{1}{6\sqrt{3}}\sinh\sqrt{3}\,t - \dfrac{1}{6\sqrt{3}}\sin\sqrt{3}\,t$

33. $\mathcal{L}^{-1}\left\{\dfrac{1}{(s^2+1)(s^2+4)}\right\} = \mathcal{L}^{-1}\left\{\dfrac{1}{3}\cdot\dfrac{1}{s^2+1} - \dfrac{1}{6}\cdot\dfrac{2}{s^2+4}\right\} = \dfrac{1}{3}\sin t - \dfrac{1}{6}\sin 2t$

34. $\mathcal{L}^{-1}\left\{\dfrac{6s+3}{(s^2+1)(s^2+4)}\right\} = \mathcal{L}^{-1}\left\{2\cdot\dfrac{s}{s^2+1} + \dfrac{1}{s^2+1} - 2\cdot\dfrac{s}{s^2+4} - \dfrac{1}{2}\cdot\dfrac{2}{s^2+4}\right\}$

$$= 2\cos t + \sin t - 2\cos 2t - \dfrac{1}{2}\sin 2t$$

35. Let $f(t) = 1$ and $g(t) = \begin{cases} 1, & t \geq 0, \ t \neq 1 \\ 0, & t = 1 \end{cases}$. Then $\mathcal{L}\{f(t)\} = \mathcal{L}\{g(t)\} = 1$, but $f(t) \neq g(t)$.

Exercises 7.3

1. $\mathcal{L}\left\{te^{10t}\right\} = \dfrac{1}{(s-10)^2}$

2. $\mathcal{L}\left\{te^{-6t}\right\} = \dfrac{1}{(s+6)^2}$

3. $\mathcal{L}\left\{t^3 e^{-2t}\right\} = \dfrac{3!}{(s+2)^4}$

4. $\mathcal{L}\left\{t^{10}e^{-7t}\right\} = \dfrac{10!}{(s+7)^{11}}$

5. $\mathcal{L}\left\{e^t \sin 3t\right\} = \dfrac{3}{(s-1)^2+9}$

6. $\mathcal{L}\left\{e^{-2t}\cos 4t\right\} = \dfrac{s+2}{(s+2)^2+16}$

7. $\mathcal{L}\left\{e^{5t}\sinh 3t\right\} = \dfrac{3}{(s-5)^2-9}$

8. $\mathcal{L}\left\{e^{-t}\cosh t\right\} = \dfrac{s+1}{(s+1)^2-1}$

9. $\mathcal{L}\left\{t\left(e^t + e^{2t}\right)^2\right\} = \mathcal{L}\left\{te^{2t} + 2te^{3t} + te^{4t}\right\} = \dfrac{1}{(s-2)^2} + \dfrac{2}{(s-3)^2} + \dfrac{1}{(s-4)^2}$

10. $\mathcal{L}\left\{e^{2t}(t-1)^2\right\} = \mathcal{L}\left\{t^2 e^{2t} - 2te^{2t} + e^{2t}\right\} = \dfrac{2}{(s-2)^3} - \dfrac{2}{(s-2)^2} + \dfrac{1}{s-2}$

11. $\mathcal{L}\left\{e^{-t}\sin^2 t\right\} = \mathcal{L}\left\{\dfrac{1}{2}e^{-t} - \dfrac{1}{2}e^{-t}\cos 2t\right\} = \dfrac{1}{2}\dfrac{1}{s+1} - \dfrac{1}{2}\dfrac{s+1}{(s+1)^2+4}$

12. $\mathscr{L}\left\{e^t\cos^2 3t\right\} = \mathscr{L}\left\{\dfrac{1}{2}e^t + \dfrac{1}{2}e^t\cos 6t\right\} = \dfrac{1}{2}\dfrac{1}{s-1} + \dfrac{1}{2}\dfrac{s-1}{(s-1)^2+36}$

13. $\mathscr{L}^{-1}\left\{\dfrac{1}{(s+2)^3}\right\} = \mathscr{L}^{-1}\left\{\dfrac{1}{2}\dfrac{2}{(s+2)^3}\right\} = \dfrac{1}{2}t^2 e^{-2t}$

14. $\mathscr{L}^{-1}\left\{\dfrac{1}{(s-1)^4}\right\} = \mathscr{L}^{-1}\left\{\dfrac{1}{6}\dfrac{3!}{(s-1)^4}\right\} = \dfrac{1}{6}t^3 e^t$

15. $\mathscr{L}^{-1}\left\{\dfrac{1}{s^2-6s+10}\right\} = \mathscr{L}^{-1}\left\{\dfrac{1}{(s-3)^2+1^2}\right\} = e^{3t}\sin t$

16. $\mathscr{L}^{-1}\left\{\dfrac{1}{s^2+2s+5}\right\} = \mathscr{L}^{-1}\left\{\dfrac{1}{2}\dfrac{2}{(s+1)^2+2^2}\right\} = \dfrac{1}{2}e^{-t}\sin 2t$

17. $\mathscr{L}^{-1}\left\{\dfrac{s}{s^2+4s+5}\right\} = \mathscr{L}^{-1}\left\{\dfrac{(s+2)}{(s+2)^2+1^2} - 2\dfrac{1}{(s+2)^2+1^2}\right\} = e^{-2t}\cos t - 2e^{-2t}\sin t$

18. $\mathscr{L}^{-1}\left\{\dfrac{2s+5}{s^2+6s+34}\right\} = \mathscr{L}^{-1}\left\{2\dfrac{(s+3)}{(s+3)^2+5^2} - \dfrac{1}{5}\dfrac{5}{(s+3)^2+5^2}\right\} = 2e^{-3t}\cos 5t - \dfrac{1}{5}e^{-3t}\sin 5t$

19. $\mathscr{L}^{-1}\left\{\dfrac{s}{(s+1)^2}\right\} = \mathscr{L}^{-1}\left\{\dfrac{s+1-1}{(s+1)^2}\right\} = \mathscr{L}^{-1}\left\{\dfrac{1}{s+1} - \dfrac{1}{(s+1)^2}\right\} = e^{-t} - te^{-t}$

20. $\mathscr{L}^{-1}\left\{\dfrac{5s}{(s-2)^2}\right\} = \mathscr{L}^{-1}\left\{\dfrac{5(s-2)+10}{(s-2)^2}\right\} = \mathscr{L}^{-1}\left\{\dfrac{5}{s-2} + \dfrac{10}{(s-2)^2}\right\} = 5e^{2t} + 10te^{2t}$

21. $\mathscr{L}^{-1}\left\{\dfrac{2s-1}{s^2(s+1)^3}\right\} = \mathscr{L}^{-1}\left\{\dfrac{5}{s} - \dfrac{1}{s^2} - \dfrac{5}{s+1} - \dfrac{4}{(s+1)^2} - \dfrac{3}{2}\dfrac{2}{(s+1)^3}\right\} = 5-t-5e^{-t}-4te^{-t}-\dfrac{3}{2}t^2e^{-t}$

22. $\mathscr{L}^{-1}\left\{\dfrac{(s+1)^2}{(s+2)^4}\right\} = \mathscr{L}^{-1}\left\{\dfrac{1}{(s+2)^2} - \dfrac{2}{(s+2)^3} + \dfrac{1}{6}\dfrac{3!}{(s+2)^4}\right\} = te^{-2t} - t^2 e^{-2t} + \dfrac{1}{6}t^3 e^{-2t}$

23. $\mathscr{L}\{(t-1)\,\mathscr{U}(t-1)\} = \dfrac{e^{-s}}{s^2}$

24. $\mathscr{L}\{e^{2-t}\,\mathscr{U}(t-2)\} = \mathscr{L}\left\{e^{-(t-2)}\,\mathscr{U}(t-2)\right\} = \dfrac{e^{-2s}}{s+1}$

25. $\mathscr{L}\{t\,\mathscr{U}(t-2)\} = \mathscr{L}\{(t-2)\,\mathscr{U}(t-2) + 2\,\mathscr{U}(t-2)\} = \dfrac{e^{-2s}}{s^2} + \dfrac{2e^{-2s}}{s}$

26. $\mathscr{L}\{(3t+1)\,\mathscr{U}(t-3)\} = 3\,\mathscr{L}\left\{\left(t+\dfrac{1}{3}\right)\mathscr{U}(t-3)\right\} = 3\,\mathscr{L}\left\{(t-3)\,\mathscr{U}(t-3) + \dfrac{10}{3}\,\mathscr{U}(t-3)\right\}$

$$= \dfrac{3e^{-3s}}{s^2} + \dfrac{10e^{-3s}}{s}$$

27. $\mathcal{L}\{\cos 2t\,\mathcal{U}(t-\pi)\} = \mathcal{L}\{\cos 2(t-\pi)\,\mathcal{U}(t-\pi)\} = \dfrac{se^{-\pi s}}{s^2+4}$

28. $\mathcal{L}\left\{\sin t\,\mathcal{U}\left(t-\dfrac{\pi}{2}\right)\right\} = \mathcal{L}\left\{\cos\left(t-\dfrac{\pi}{2}\right)\mathcal{U}\left(t-\dfrac{\pi}{2}\right)\right\} = \dfrac{se^{-\pi s}}{s^2+1}$

29. $\mathcal{L}\{(t-1)^3 e^{t-1}\,\mathcal{U}(t-1)\} = \dfrac{6e^{-s}}{(s-1)^4}$

30. $\mathcal{L}\{te^{t-5}\,\mathcal{U}(t-5)\} = \mathcal{L}\{(t-5)e^{t-5}\,\mathcal{U}(t-5) + 5e^{t-5}\,\mathcal{U}(t-5)\} = \dfrac{e^{-5s}}{(s-1)^2} + \dfrac{5e^{-5s}}{s-1}$

31. $\mathcal{L}^{-1}\left\{\dfrac{e^{-2s}}{s^3}\right\} = \mathcal{L}^{-1}\left\{\dfrac{1}{2}\cdot\dfrac{2}{s^3}e^{-2s}\right\} = \dfrac{1}{2}(t-2)^2\,\mathcal{U}(t-2)$

32. $\mathcal{L}^{-1}\left\{\dfrac{(1+e^{-2s})^2}{s+2}\right\} = \mathcal{L}^{-1}\left\{\dfrac{1}{s+2} + \dfrac{2e^{-2s}}{s+2} + \dfrac{e^{-4s}}{s+2}\right\} = e^{-2t} + 2e^{-2(t-2)}\,\mathcal{U}(t-2) + e^{-2(t-4)}\,\mathcal{U}(t-4)$

33. $\mathcal{L}^{-1}\left\{\dfrac{e^{-\pi s}}{s^2+1}\right\} = \sin(t-\pi)\,\mathcal{U}(t-\pi)$

34. $\mathcal{L}^{-1}\left\{\dfrac{se^{-\pi s/2}}{s^2+4}\right\} = \cos 2\left(t-\dfrac{\pi}{2}\right)\mathcal{U}\left(t-\dfrac{\pi}{2}\right)$

35. $\mathcal{L}^{-1}\left\{\dfrac{e^{-s}}{s(s+1)}\right\} = \mathcal{L}^{-1}\left\{\dfrac{e^{-s}}{s} - \dfrac{e^{-s}}{s+1}\right\} = \mathcal{U}(t-1) - e^{-(t-1)}\,\mathcal{U}(t-1)$

36. $\mathcal{L}^{-1}\left\{\dfrac{e^{-2s}}{s^2(s-1)}\right\} = \mathcal{L}^{-1}\left\{-\dfrac{e^{-2s}}{s} - \dfrac{e^{-2s}}{s^2} + \dfrac{e^{-2s}}{s-1}\right\} = -\mathcal{U}(t-2) - (t-2)\,\mathcal{U}(t-2) + e^{t-2}\,\mathcal{U}(t-2)$

37. $\mathcal{L}\{t\cos 2t\} = -\dfrac{d}{ds}\left(\dfrac{s}{s^2+4}\right) = \dfrac{s^2-4}{(s^2+4)^2}$

38. $\mathcal{L}\{t\sinh 3t\} = -\dfrac{d}{ds}\left(\dfrac{3}{s^2-9}\right) = \dfrac{6s}{(s^2-9)^2}$

39. $\mathcal{L}\{t^2\sinh t\} = \dfrac{d^2}{ds^2}\left(\dfrac{1}{s^2-1}\right) = \dfrac{6s^2+2}{(s^2-1)^3}$

40. $\mathcal{L}\{t^2\cos t\} = \dfrac{d^2}{ds^2}\left(\dfrac{s}{s^2+1}\right) = \dfrac{d}{ds}\left(\dfrac{1-s^2}{(s^2+1)^2}\right) = \dfrac{2s\left(s^2-3\right)}{(s^2+1)^3}$

41. $\mathcal{L}\{te^{2t}\sin 6t\} = -\dfrac{d}{ds}\left(\dfrac{6}{(s-2)^2+36}\right) = \dfrac{12(s-2)}{[(s-2)^2+36]^2}$

42. $\mathcal{L}\{te^{-3t}\cos 3t\} = -\dfrac{d}{ds}\left(\dfrac{s+3}{(s+3)^2+9}\right) = \dfrac{(s+3)^2-9}{[(s+3)^2+9]^2}$

43. $\mathcal{L}^{-1}\left\{\dfrac{s}{(s^2+1)^2}\right\} = \mathcal{L}^{-1}\left\{-\dfrac{1}{2}\cdot\dfrac{-2s}{(s^2+1)^2}\right\} = \mathcal{L}^{-1}\left\{\dfrac{1}{2}(-1)\dfrac{d}{ds}\left(\dfrac{1}{s^2+1}\right)\right\} = \dfrac{1}{2}t\sin t$

44. $\mathcal{L}^{-1}\left\{\dfrac{s+1}{(s^2+2s+2)^2}\right\} = \mathcal{L}^{-1}\left\{-\dfrac{1}{2}\cdot\dfrac{-2(s+1)}{[(s+1)^2+1]^2}\right\} = \mathcal{L}^{-1}\left\{\dfrac{1}{2}\cdot\dfrac{-d}{ds}\left(\dfrac{1}{(s+1)^2+1}\right)\right\}$

$\qquad = \dfrac{1}{2}te^{-t}\sin t$

45. (c)   **46.** (e)   **47.** (f)   **48.** (b)   **49.** (a)   **50.** (d)

51. $\mathcal{L}\{2-4\mathcal{U}(t-3)\} = \dfrac{2}{s} - \dfrac{4}{s}e^{-3s}$

52. $\mathcal{L}\{1-\mathcal{U}(t-4)+\mathcal{U}(t-5)\} = \dfrac{1}{s} - \dfrac{e^{-4s}}{s} + \dfrac{e^{-5s}}{s}$

53. $\mathcal{L}\{t^2\mathcal{U}(t-1)\} = \mathcal{L}\{[(t-1)^2+2t-1]\mathcal{U}(t-1)\} = \mathcal{L}\{[(t-1)^2+2(t-1)-1]\mathcal{U}(t-1)\}$

$\qquad = \left(\dfrac{2}{s^3} + \dfrac{2}{s^2} + \dfrac{1}{s}\right)e^{-s}$

54. $\mathcal{L}\left\{\sin t\,\mathcal{U}\left(t-\dfrac{3\pi}{2}\right)\right\} = \mathcal{L}\left\{-\cos\left(t-\dfrac{3\pi}{2}\right)\mathcal{U}\left(t-\dfrac{3\pi}{2}\right)\right\} = -\dfrac{se^{-3\pi s/2}}{s^2+1}$

55. $\mathcal{L}\{t-t\,\mathcal{U}(t-2)\} = \mathcal{L}\{t-(t-2)\mathcal{U}(t-2)-2\mathcal{U}(t-2)\} = \dfrac{1}{s^2} - \dfrac{e^{-2s}}{s^2} - \dfrac{2e^{-2s}}{s}$

56. $\mathcal{L}\{\sin t-\sin t\,\mathcal{U}(t-2\pi)\} = \mathcal{L}\{\sin t-\sin(t-2\pi)\mathcal{U}(t-2\pi)\} = \dfrac{1}{s^2+1} - \dfrac{e^{-2\pi s}}{s^2+1}$

57. $\mathcal{L}\{f(t)\} = \mathcal{L}\{\mathcal{U}(t-a)-\mathcal{U}(t-b)\} = \dfrac{e^{-as}}{s} - \dfrac{e^{-bs}}{s}$

58. $\mathcal{L}\{f(t)\} = \mathcal{L}\{\mathcal{U}(t-1)+\mathcal{U}(t-2)+\mathcal{U}(t-3)+\cdots\} = \dfrac{e^{-s}}{s} + \dfrac{e^{-2s}}{s} + \dfrac{e^{-3s}}{s} + \cdots = \dfrac{1}{s}\dfrac{e^{-s}}{1-e^{-s}}$

59. $\mathcal{L}^{-1}\left\{\dfrac{1}{s^2} - \dfrac{e^{-s}}{s^2}\right\} = t-(t-1)\mathcal{U}(t-1) = \begin{cases} t, & 0\le t<1 \\ 1, & t\ge 1 \end{cases}$

Exercises 7.3

60. $\mathcal{L}^{-1}\left\{\dfrac{2}{s} - \dfrac{3e^{-s}}{s^2} + \dfrac{5e^{-2s}}{s^2}\right\} = 2 - 3(t-1)\,\mathcal{U}(t-1) + 5(t-2)\,\mathcal{U}(t-2)$

$$= \begin{cases} 2, & 0 \le t < 1 \\ -3t + 5, & 1 \le t < 2 \\ 2t - 5, & t \ge 2 \end{cases}$$

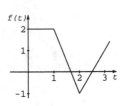

61. $f(t) = -\dfrac{1}{t}\mathcal{L}^{-1}\left\{\dfrac{d}{ds}[\ln(s-3) - \ln(s+1)]\right\} = -\dfrac{1}{t}\mathcal{L}^{-1}\left\{\dfrac{1}{s-3} - \dfrac{1}{s+1}\right\} = -\dfrac{1}{t}\left(e^{3t} - e^{-t}\right)$

62. $f(t) = -\dfrac{1}{t}\mathcal{L}^{-1}\left\{\dfrac{d}{ds}\left[\ln\left(s^2+1\right) - \ln\left(s^2+4\right)\right]\right\} = -\dfrac{1}{t}\mathcal{L}^{-1}\left\{\dfrac{2s}{s^2+1} - \dfrac{2s}{s^2+2^2}\right\}$

$$= -\dfrac{1}{t}(2\cos t - 2\cos 2t)$$

63. Since

$$t^2 - 3t = t(t-3) = (t - 2 + 2)(t - 2 - 1) = (t-2)^2 + (t-2) - 2$$

we have

$$\mathcal{L}\{(t^2 - 3t)\,\mathcal{U}(t-2)\} = \mathcal{L}\{(t-2)^2\,\mathcal{U}(t-2) + (t-2)\,\mathcal{U}(t-2) - 2\,\mathcal{U}(t-2)\}$$

$$= \dfrac{2}{s^3}e^{-2s} + \dfrac{1}{s^2}e^{-2s} - \dfrac{2}{s}e^{-2s}.$$

Using the alternative form of the second translation theorem we obtain

$$\mathcal{L}\{(t^2 - 3t)\,\mathcal{U}(t-2)\} = e^{-2s}\,\mathcal{L}\{(t+2)^2 - 3(t+2)\}$$

$$= e^{-2s}\,\mathcal{L}\{t^2 + t - 2\} = e^{-2s}\left(\dfrac{2}{s^3} + \dfrac{1}{s^2} - \dfrac{2}{s}\right).$$

64. (a) $2t + 1 = 2(t - 1 + 1) + 1 = 2(t - 1) + 3$

(b) $e^t = e^{t-5+5} = e^5 e^{t-5}$

(c) $\cos t = -\cos(t - \pi)$

(d) $t\sin t = (t - 2\pi + 2\pi)\sin(t - 2\pi) = (t - 2\pi)\sin(t - 2\pi) + 2\pi\sin(t - 2\pi)$

Exercises 7.4

1. $\mathcal{L}\{e^t\} = \mathcal{L}\left\{\dfrac{d}{dt}e^t\right\} = s\mathcal{L}\{e^t\} - e^0 = \dfrac{s}{s-1} - 1 = \dfrac{1}{s-1}$

2. $\mathcal{L}\{-\sin 2t\} = \mathcal{L}\left\{\dfrac{d}{dt}\cos^2 t\right\} = s\mathcal{L}\{\cos^2 t\} - \cos^2 0 = s\mathcal{L}\{\cos^2 t\} - 1$

Solving for $\mathcal{L}\{\cos^2 t\}$ we obtain

$$\mathcal{L}\{\cos^2 t\} = \frac{1}{s}\mathcal{L}\{-\sin 2t\} + \frac{1}{s} = \frac{1}{s}\frac{-2}{s^2+4} + \frac{1}{s} = \frac{s^2+2}{s(s^2+4)}.$$

3. $\mathcal{L}\{y'' + 3y'\} = \mathcal{L}\{y''\} + 3\mathcal{L}\{y'\} = s^2 Y(s) - sy(0) - y'(0) + 3[sY(s) - y(0)] = (s^2 + 3s)Y(s) - s - 2$

4. $\mathcal{L}\{y'' - 4y' + 5y\} = \mathcal{L}\{y''\} - 4\mathcal{L}\{y'\} + 5\mathcal{L}\{y\}$

$$= s^2 Y(s) - sy(0) - y'(0) - 4[sY(s) - y(0)] + 5Y(s) = (s^2 - 4s + 5)Y(s) - s + 5$$

5. We solve $\mathcal{L}\{y'' - 2y' + y\} = \mathcal{L}\{0\} = 0$.

$$s^2 Y(s) - sy(0) - y'(0) - 2[sY(s) - y(0)] + Y(s) = 0$$

$$(s^2 - 2s + 1)Y(s) - 2s + 1 = 0$$

$$Y(s) = \frac{2s-1}{(s-1)^2}$$

6. We solve $\mathcal{L}\{y'' + y\} = \mathcal{L}\{1\} = 1/s$.

$$s^2 Y(s) - sy(0) - y'(0) + Y(s) = \frac{1}{s}$$

$$(s^2 + 1)Y(s) - 2s - 3 = \frac{1}{s}$$

$$Y(s) = \frac{1}{s(s^2+1)} + \frac{2s+3}{s^2+1}$$

7. $\mathcal{L}\left\{\displaystyle\int_0^t e^\tau \, d\tau\right\} = \dfrac{1}{s}\mathcal{L}\{e^t\} = \dfrac{1}{s(s-1)}$

8. $\mathcal{L}\left\{\displaystyle\int_0^t \cos\tau \, d\tau\right\} = \dfrac{1}{s}\mathcal{L}\{\cos t\} = \dfrac{s}{s(s^2+1)} = \dfrac{1}{s^2+1}$

9. $\mathcal{L}\left\{\displaystyle\int_0^t e^{-\tau}\cos\tau \, d\tau\right\} = \dfrac{1}{s}\mathcal{L}\{e^{-t}\cos t\} = \dfrac{1}{s}\dfrac{s+1}{(s+1)^2+1} = \dfrac{s+1}{s(s^2+2s+2)}$

271

10. $\mathscr{L}\left\{\int_0^t \tau \sin \tau \, d\tau\right\} = \frac{1}{s} \mathscr{L}\{t \sin t\} = \frac{1}{s}\left(-\frac{d}{ds}\frac{1}{s^2+1}\right) = -\frac{1}{s}\frac{-2s}{(s^2+1)^2} = \frac{2}{(s^2+1)^2}$

11. $\mathscr{L}\left\{\int_0^t \tau e^{t-\tau} \, d\tau\right\} = \mathscr{L}\{t\}\,\mathscr{L}\{e^t\} = \frac{1}{s^2(s-1)}$

12. $\mathscr{L}\left\{\int_0^t \sin \tau \cos(t-\tau) \, d\tau\right\} = \mathscr{L}\{\sin t\}\,\mathscr{L}\{\cos t\} = \frac{s}{(s^2+1)^2}$

13. $\mathscr{L}\left\{t\int_0^t \sin \tau \, d\tau\right\} = -\frac{d}{ds}\mathscr{L}\left\{\int_0^t \sin \tau \, d\tau\right\} = -\frac{d}{ds}\left(\frac{1}{s}\frac{1}{s^2+1}\right) = \frac{3s^2+1}{s^2(s^2+1)^2}$

14. $\mathscr{L}\left\{t\int_0^t \tau e^{-\tau} d\tau\right\} = -\frac{d}{ds}\mathscr{L}\left\{t\int_0^t \tau e^{-\tau} d\tau\right\} = -\frac{d}{ds}\left(\frac{1}{s}\frac{1}{(s+1)^2}\right) = \frac{3s+1}{s^2(s+1)^3}$

15. $\mathscr{L}\left\{1*t^3\right\} = \frac{1}{s}\frac{3!}{s^4} = \frac{6}{s^5}$

16. $\mathscr{L}\left\{1*e^{-2t}\right\} = \frac{1}{s(s+2)}$

17. $\mathscr{L}\left\{t^2*t^4\right\} = \frac{2}{s^3}\frac{4!}{s^5} = \frac{48}{s^8}$

18. $\mathscr{L}\left\{t^2*te^t\right\} = \frac{2}{s^3(s-1)^2}$

19. $\mathscr{L}\left\{e^{-t}*e^t\cos t\right\} = \frac{s-1}{(s+1)\left[(s-1)^2+1\right]}$

20. $\mathscr{L}\left\{e^{2t}*\sin t\right\} = \frac{1}{(s-2)(s^2+1)}$

21. $\mathscr{L}^{-1}\left\{\frac{1}{s+5}F(s)\right\} = e^{-5t}*f(t) = \int_0^t f(\tau)e^{-5(t-\tau)}d\tau$

22. $\mathscr{L}^{-1}\left\{\frac{s}{s^2+4}F(s)\right\} = \cos 2t * f(t) = \int_0^t f(\tau)\cos 2(t-\tau)\,d\tau$

23. $\mathscr{L}^{-1}\left\{\frac{1}{s(s+1)}\right\} = 1*e^{-t} = \int_0^t e^{-(t-\tau)}d\tau = e^{-(t-\tau)}\Big|_0^t = 1 - e^{-t}$

$\mathscr{L}^{-1}\left(\frac{1}{s\,(s+1)}\right) =$

24. We use repeated applications of $\int_0^t f(\tau)\,d\tau = \mathscr{L}^{-1}\{F(s)/s\}$.

#19

$$\mathscr{L}^{-1}\left\{\frac{1}{s(s-1)}\right\} = \int_0^t e^\tau \, d\tau = e^t - 1$$

$$\mathscr{L}^{-1}\left\{\frac{1}{s^2(s-1)}\right\} = \int_0^t (e^\tau - 1)\, d\tau = e^t - t - 1$$

$$\mathscr{L}^{-1}\left\{\frac{1}{s^3(s-1)}\right\} = \int_0^t (e^\tau - \tau - 1)\, d\tau = e^t - \frac{1}{2}t^2 - t - 1$$

25. $\mathscr{L}^{-1}\left\{\frac{1}{(s+1)(s-2)}\right\} = e^{-t}*e^{2t} = \int_0^t e^{-\tau}e^{2(t-\tau)}d\tau = \int_0^t e^{2t-3\tau}d\tau = -\frac{1}{3}e^{2t-3\tau}\Big|_0^t = \frac{1}{3}\left(e^{2t}-e^{-t}\right)$

26. $\mathcal{L}^{-1}\left\{\dfrac{1}{(s+1)^2}\right\} = e^{-t} * e^{-t} = \displaystyle\int_0^t e^{-\tau}e^{-(t-\tau)}d\tau = e^{-t}\int_0^t d\tau = te^{-t}$

27. $\mathcal{L}^{-1}\left\{\dfrac{s}{(s^2+4)^2}\right\} = \cos 2t * \dfrac{1}{2}\sin 2t = \dfrac{1}{2}\displaystyle\int_0^t \cos 2\tau \sin 2(t-\tau)\,d\tau$

$= \dfrac{1}{2}\displaystyle\int_0^t \cos 2\tau(\sin 2t\cos 2\tau - \cos 2t\sin 2\tau)\,d\tau = \dfrac{1}{2}\left[\sin 2t\int_0^t \cos^2 2\tau\,d\tau - \cos 2t\int_0^t \dfrac{1}{2}\sin 4\tau\,d\tau\right]$

$= \dfrac{1}{2}\sin 2t\left[\dfrac{1}{2}\tau + \dfrac{1}{8}\sin 4\tau\right]_0^t - \dfrac{1}{4}\cos 2t\left[-\dfrac{1}{4}\cos 4\tau\right]_0^t$

$= \dfrac{1}{2}\sin 2t\left(\dfrac{1}{2}t + \dfrac{1}{8}\sin 4t\right) + \dfrac{1}{16}\cos 2t(\cos 4t - 1)$

$= \dfrac{1}{4}t\sin 2t + \dfrac{1}{16}\sin 2t\sin 4t + \dfrac{1}{16}\cos 2t\cos 4t - \dfrac{1}{16}\cos 2t$

$= \dfrac{1}{4}t\sin 2t + \dfrac{1}{16}\left[\sin 2t(2\sin 2t\cos 2t) + \cos 2t\left(\cos^2 2t - \sin^2 2t\right) - \cos 2t\right]$

$= \dfrac{1}{4}t\sin 2t + \dfrac{1}{16}\cos 2t\left[2\sin^2 2t + \cos^2 2t - \sin^2 2t - 1\right] = \dfrac{1}{4}t\sin 2t$

28. $\mathcal{L}^{-1}\left\{\dfrac{1}{(s^2+4s+5)^2}\right\} = \mathcal{L}^{-1}\left\{\dfrac{1}{\left[(s+2)^2+1\right]^2}\right\} = e^{-2t}\sin t * e^{-2t}\sin t$

$= \displaystyle\int_0^t e^{-2\tau}\sin\tau e^{-2(t-\tau)}\sin(t-\tau)\,d\tau = e^{-2t}\int_0^t \sin\tau(\sin t\cos\tau - \cos t\sin\tau)\,d\tau$

$= e^{-2t}\left[\sin t\displaystyle\int_0^t \dfrac{1}{2}\sin 2\tau\,d\tau - \cos t\int_0^t \dfrac{1}{2}(1-\cos 2\tau)\,d\tau\right]$

$= e^{-2t}\left[-\dfrac{1}{4}\sin t(\cos 2\tau)\Big|_0^t - \dfrac{1}{2}\cos t\left(\tau - \dfrac{1}{2}\sin 2\tau\right)\Big|_0^t\right]$

$= e^{-2t}\left[-\dfrac{1}{4}\sin t(\cos 2t - 1) - \dfrac{1}{2}\cos t\left(t - \dfrac{1}{2}\sin 2t\right)\right]$

$= \dfrac{1}{2}e^{-2t}\left[-\dfrac{1}{2}\sin t\left(\cos^2 t - \sin^2 t - 1\right) - t\cos t + \dfrac{1}{2}\cos t(2\sin t\cos t)\right]$

$= \dfrac{1}{2}e^{-2t}\left[\dfrac{1}{2}\sin t\left(-\cos^2 t + \sin^2 t + 1 + 2\cos^2 t\right) - t\cos t\right] = \dfrac{1}{2}e^{-2t}(\sin t - t\cos t)$

29. Let $u = t - \tau$ so that $du = d\tau$ and

$$f * g = \int_0^t f(\tau)g(t-\tau)\,d\tau = -\int_t^0 f(t-u)g(u)\,du = g * f.$$

30. $f * (g + h) = \int_0^t f(\tau)[g(t - \tau) + h(t - \tau)] \, d\tau = \int_0^t f(\tau) g(t - \tau) \, d\tau + \int_0^t f(\tau) h(t - \tau) \, d\tau$

$$= \int_0^t f(\tau)[g(t - \tau) + h(t - \tau)] \, d\tau = f * g + f * h$$

31. $\mathscr{L}\{f(t)\} = \dfrac{1}{1 - e^{-2as}} \left[\int_0^a e^{-st} dt - \int_a^{2a} e^{-st} dt \right] = \dfrac{(1 - e^{-as})^2}{s(1 - e^{-2as})} = \dfrac{1 - e^{-as}}{s(1 + e^{-as})}$

32. $\mathscr{L}\{f(t)\} = \dfrac{1}{1 - e^{-2as}} \int_0^a e^{-st} dt = \dfrac{1}{s(1 + e^{-as})}$

33. $\mathscr{L}\{f(t)\} = \dfrac{1}{1 - e^{-bs}} \int_0^b \dfrac{a}{b} t e^{-st} dt = \dfrac{a}{s} \left(\dfrac{1}{bs} - \dfrac{1}{e^{bs} - 1} \right)$

34. $\mathscr{L}\{f(t)\} = \dfrac{1}{1 - e^{-2s}} \left[\int_0^1 t e^{-st} dt + \int_1^2 (2 - t) e^{-st} dt \right] = \dfrac{1 - e^{-s}}{s^2 (1 - e^{-2s})}$

35. $\mathscr{L}\{f(t)\} = \dfrac{1}{1 - e^{-\pi s}} \int_0^\pi e^{-st} \sin t \, dt = \dfrac{1}{s^2 + 1} \cdot \dfrac{e^{\pi s/2} + e^{-\pi s/2}}{e^{\pi s/2} - e^{-\pi s/2}} = \dfrac{1}{s^2 + 1} \coth \dfrac{\pi s}{2}$

36. $\mathscr{L}\{f(t)\} = \dfrac{1}{1 - e^{-2\pi s}} \int_0^\pi e^{-st} \sin t \, dt = \dfrac{1}{s^2 + 1} \cdot \dfrac{1}{1 - e^{-\pi s}}$

37. $\mathscr{L}\{f(t)\} = \dfrac{1}{1 - e^{-2\pi s}} \int_0^{2\pi} e^{-st} \sin t \, dt = \dfrac{1}{s^2 + 1}$

38. $\mathscr{L}\{f(t)\} = \dfrac{1}{1 - e^{-2\pi s}} \int_0^{2\pi} e^{-st} \cos t \, dt = \dfrac{s}{s^2 + 1}$

39. By definition, $t * \mathscr{U}(t - a) = \int_0^t (t - \tau) \mathscr{U}(\tau - a) \, d\tau$. We consider separately the cases when $a < t$ and when $a > t$. When $a < t$, then

$$\int_0^t (t - \tau) \mathscr{U}(\tau - a) \, d\tau = \int_a^t (t - \tau) \, d\tau = -\dfrac{(t - \tau)^2}{2} \Big|_a^t = \dfrac{1}{2}(t - a)^2.$$

When $a > t$, then $\mathscr{U}(\tau - a) = 0$ since $\tau < t < a$ and

$$\int_0^t (t - \tau) \mathscr{U}(\tau - a) \, d\tau = 0.$$

Therefore

$$t * \mathscr{U}(t - a) = \begin{cases} \frac{1}{2}(t - a)^2, & t > a \\ 0, & t < a \end{cases} = \dfrac{1}{2}(t - a)^2 \, \mathscr{U}(t - a).$$

40. First method: By the definition of the Laplace transform and integration by parts we have

$$\mathscr{L}\left\{\int_0^t f(\tau)\,d\tau\right\} = \int_0^\infty e^{-st}\left(\int_0^t f(\tau)\,d\tau\right) dt$$

$$= -\frac{e^{-st}}{s}\int_0^t f(\tau)\,d\tau\,\Big|_0^\infty + \frac{1}{s}\int_0^\infty e^{-st}f(t)\,dt$$

$$= \frac{1}{s}\int_0^\infty e^{-st}f(t)\,dt = \frac{F(s)}{s}.$$

Second Method: Let $g(t) = \int_0^t f(\tau)\,d\tau$; then $g'(t) = f(t)$ and $g(0) = 0$. By Theorem 7.8

$$\mathscr{L}\{g'(t)\} = s\,\mathscr{L}\{g(t)\} - g(0) = s\,\mathscr{L}\{g(t)\},$$

so

$$\mathscr{L}\{f(t)\} = s\,\mathscr{L}\left\{\int_0^t f(\tau)\,d\tau\right\},$$

and

$$\mathscr{L}\left\{\int_0^t f(\tau)\,d\tau\right\} = \frac{1}{s}\mathscr{L}\{f(t)\} = \frac{F(s)}{s}.$$

———— Exercises 7.5 ————

1. The Laplace transform of the differential equation is

$$s\,\mathscr{L}\{y\} - y(0) - \mathscr{L}\{y\} = \frac{1}{s}.$$

Solving for $\mathscr{L}\{y\}$ we obtain

$$\mathscr{L}\{y\} = -\frac{1}{s} + \frac{1}{s-1}.$$

Thus

$$y = -1 + e^t.$$

2. The Laplace transform of the differential equation is

$$s\,\mathscr{L}\{y\} - y(0) + 2\,\mathscr{L}\{y\} = \frac{1}{s^2}.$$

Solving for $\mathscr{L}\{y\}$ we obtain

$$\mathscr{L}\{y\} = \frac{1-s^2}{s^2(s+2)} = -\frac{1}{4}\frac{1}{s} + \frac{1}{2}\frac{1}{s^2} - \frac{3}{4}\frac{1}{s+2}.$$

Thus

$$y = -\frac{1}{4} + \frac{1}{2}t - \frac{3}{4}e^{-2t}.$$

3. The Laplace transform of the differential equation is

$$s\mathscr{L}\{y\} - y(0) + 4\mathscr{L}\{y\} = \frac{1}{s+4}.$$

Solving for $\mathscr{L}\{y\}$ we obtain $\mathscr{L}\{y\} = \dfrac{1}{(s+4)^2} + \dfrac{2}{s+4}$.

Thus
$$y = te^{-4t} + 2e^{-4t}.$$

4. The Laplace transform of the differential equation is

$$s\mathscr{L}\{y\} - y(0) - \mathscr{L}\{y\} = \frac{1}{s^2+1}.$$

Solving for $\mathscr{L}\{y\}$ we obtain

$$\mathscr{L}\{y\} = \frac{1}{(s^2+1)(s-1)} = -\frac{1}{2}\frac{s+1}{s^2+1} + \frac{1}{2}\frac{1}{s-1}.$$

Thus
$$y = \frac{1}{2}e^t - \frac{1}{2}(\cos t + \sin t).$$

5. The Laplace transform of the differential equation is

$$s^2\mathscr{L}\{y\} - sy(0) - y'(0) + 5\left[s\mathscr{L}\{y\} - y(0)\right] + 4\mathscr{L}\{y\} = 0.$$

Solving for $\mathscr{L}\{y\}$ we obtain

$$\mathscr{L}\{y\} = \frac{s+5}{s^2+5s+4} = \frac{4}{3}\frac{1}{s+1} - \frac{1}{3}\frac{1}{s+4}.$$

Thus
$$y = \frac{4}{3}e^{-t} - \frac{1}{3}e^{-4t}.$$

6. The Laplace transform of the differential equation is

$$s^2\mathscr{L}\{y\} - sy(0) - y'(0) - 6\left[s\mathscr{L}\{y\} - y(0)\right] + 13\mathscr{L}\{y\} = 0.$$

Solving for $\mathscr{L}\{y\}$ we obtain

$$\mathscr{L}\{y\} = -\frac{3}{s^2-6s+13} = -\frac{3}{2}\frac{2}{(s-3)^2+2^2}.$$

Thus
$$y = -\frac{3}{2}e^{3t}\sin 2t.$$

7. The Laplace transform of the differential equation is

$$s^2\mathscr{L}\{y\} - sy(0) - y'(0) - 6\left[s\mathscr{L}\{y\} - y(0)\right] + 9\mathscr{L}\{y\} = \frac{1}{s^2}.$$

Solving for $\mathcal{L}\{y\}$ we obtain

$$\mathcal{L}\{y\} = \frac{1+s^2}{s^2(s-3)^2} = \frac{2}{27}\frac{1}{s} + \frac{1}{9}\frac{1}{s^2} - \frac{2}{27}\frac{1}{s-3} + \frac{10}{9}\frac{1}{(s-3)^2}.$$

Thus

$$y = \frac{2}{27} + \frac{1}{9}t - \frac{2}{27}e^{3t} + \frac{10}{9}te^{3t}.$$

8. The Laplace transform of the differential equation is

$$s^2\,\mathcal{L}\{y\} - sy(0) - y'(0) - 4\,[s\,\mathcal{L}\{y\} - y(0)] + 4\mathcal{L}\{y\} = \frac{6}{s^4}.$$

Solving for $\mathcal{L}\{y\}$ we obtain

$$\mathcal{L}\{y\} = \frac{s^5 - 4s^4 + 6}{s^4(s-2)^2} = \frac{3}{4}\frac{1}{s} + \frac{9}{8}\frac{1}{s^2} + \frac{3}{4}\frac{2}{s^3} + \frac{1}{4}\frac{3!}{s^4} + \frac{1}{4}\frac{1}{s-2} - \frac{13}{8}\frac{1}{(s-2)^2}.$$

Thus

$$y = \frac{3}{4} + \frac{9}{8}t + \frac{3}{4}t^2 + \frac{1}{4}t^3 + \frac{1}{4}e^{2t} - \frac{13}{8}te^{2t}.$$

9. The Laplace transform of the differential equation is

$$s^2\,\mathcal{L}\{y\} - sy(0) - y'(0) - 4\,[s\,\mathcal{L}\{y\} - y(0)] + 4\mathcal{L}\{y\} = \frac{6}{(s-2)^4}.$$

Solving for $\mathcal{L}\{y\}$ we obtain $\mathcal{L}\{y\} = \frac{1}{20}\frac{5!}{(s-2)^6}$. Thus, $y = \frac{1}{20}t^5e^{2t}$.

10. The Laplace transform of the differential equation is

$$s^2\,\mathcal{L}\{y\} - sy(0) - y'(0) - 2\,[s\,\mathcal{L}\{y\} - y(0)] + 5\mathcal{L}\{y\} = \frac{1}{s} + \frac{1}{s^2}.$$

Solving for $\mathcal{L}\{y\}$ we obtain

$$\mathcal{L}\{y\} = \frac{4s^2 + s + 1}{s^2(s^2 - 2s + 5)} = \frac{7}{25}\frac{1}{s} + \frac{1}{5}\frac{1}{s^2} + \frac{-7s/25 + 109/25}{s^2 - 2s + 5}$$

$$= \frac{7}{25}\frac{1}{s} + \frac{1}{5}\frac{1}{s^2} - \frac{7}{25}\frac{s-1}{(s-1)^2 + 2^2} + \frac{51}{25}\frac{2}{(s-1)^2 + 2^2}.$$

Thus

$$y = \frac{7}{25} + \frac{1}{5}t - \frac{7}{25}e^t\cos 2t + \frac{51}{25}e^t\sin 2t.$$

11. The Laplace transform of the differential equation is

$$s^2\,\mathcal{L}\{y\} - sy(0) - y'(0) + \mathcal{L}\{y\} = \frac{1}{s^2+1}.$$

Solving for $\mathcal{L}\{y\}$ we obtain

$$\mathcal{L}\{y\} = \frac{s^3 - s^2 + s}{(s^2+1)^2} = \frac{s}{s^2+1} - \frac{1}{s^2+1} + \frac{1}{(s^2+1)^2}.$$

Thus
$$y = \cos t - \frac{1}{2}\sin t - \frac{1}{2}t\cos t.$$

12. The Laplace transform of the differential equation is

$$s^2 \mathscr{L}\{y\} - sy(0) - y'(0) + 16\,\mathscr{L}\{y\} = \frac{1}{s}.$$

Solving for $\mathscr{L}\{y\}$ we obtain

$$\mathscr{L}\{y\} = \frac{s^2 + 2s + 1}{s(s^2 + 16)} = \frac{1}{16}\frac{1}{s} + \frac{15}{16}\frac{s}{s^2 + 4^2} + \frac{1}{2}\frac{4}{s^2 + 4^2}.$$

Thus
$$y = \frac{1}{16} + \frac{15}{16}\cos 4t + \frac{1}{2}\sin 4t.$$

13. The Laplace transform of the differential equation is

$$s^2 \mathscr{L}\{y\} - sy(0) - y'(0) - [s\,\mathscr{L}\{y\} - y(0)] = \frac{s - 1}{(s - 1)^2 + 1}.$$

Solving for $\mathscr{L}\{y\}$ we obtain

$$\mathscr{L}\{y\} = \frac{1}{s(s^2 - 2s + 2)} = \frac{1}{2}\frac{1}{s} - \frac{1}{2}\frac{s - 1}{(s - 1)^2 + 1} + \frac{1}{2}\frac{1}{(s - 1)^2 + 1}.$$

Thus
$$y = \frac{1}{2} - \frac{1}{2}e^t\cos t + \frac{1}{2}e^t\sin t.$$

14. The Laplace transform of the differential equation is

$$s^2 \mathscr{L}\{y\} - sy(0) - y'(0) - 2\,[s\,\mathscr{L}\{y\} - y(0)] = \frac{1}{(s - 1)^2 - 1}.$$

Solving for $\mathscr{L}\{y\}$ we obtain

$$\mathscr{L}\{y\} = \frac{1}{s^2(s - 2)^2} = \frac{1}{4}\frac{1}{s} + \frac{1}{4}\frac{1}{s^2} - \frac{1}{4}\frac{1}{s - 2} + \frac{1}{4}\frac{1}{(s - 2)^2}.$$

Thus
$$y = \frac{1}{4} + \frac{1}{4}t - \frac{1}{4}e^{2t} + \frac{1}{4}te^{2t}.$$

15. The Laplace transform of the differential equation is

$$2\left[s^3 \mathscr{L}\{y\} - s^2(0) - sy'(0) - y''(0)\right] + 3[s^2\,\mathscr{L}\{y\} - sy(0) - y'(0)] - 3[s\,\mathscr{L}\{y\} - y(0)] - 2\,\mathscr{L}\{y\} = \frac{1}{s + 1}.$$

Solving for $\mathscr{L}\{y\}$ we obtain

$$\mathscr{L}\{y\} = \frac{2s + 3}{(s + 1)(s - 1)(2s + 1)(s + 2)} = \frac{1}{2}\frac{1}{s + 1} + \frac{5}{18}\frac{1}{s - 1} - \frac{8}{9}\frac{1}{s + 1/2} + \frac{1}{9}\frac{1}{s + 2}.$$

Thus
$$y = \frac{1}{2}e^{-t} + \frac{5}{18}e^t - \frac{8}{9}e^{-t/2} + \frac{1}{9}e^{-2t}.$$

16. The Laplace transform of the differential equation is

$$s^3 \mathscr{L}\{y\} - s^2(0) - sy'(0) - y''(0) + 2[s^2 \mathscr{L}\{y\} - sy(0) - y'(0)] - [s\mathscr{L}\{y\} - y(0)] - 2\mathscr{L}\{y\} = \frac{3}{s^2 + 9}.$$

Solving for $\mathscr{L}\{y\}$ we obtain

$$\mathscr{L}\{y\} = \frac{s^2 + 12}{(s-1)(s+1)(s+2)(s^2+9)}$$

$$= \frac{13}{60}\frac{1}{s-1} - \frac{13}{20}\frac{1}{s+1} + \frac{16}{39}\frac{1}{s+2} + \frac{3}{130}\frac{s}{s^2+9} - \frac{1}{65}\frac{3}{s^2+9}.$$

Thus

$$y = \frac{13}{60}e^t - \frac{13}{20}e^{-t} + \frac{16}{39}e^{-2t} + \frac{3}{130}\cos 3t - \frac{1}{65}\sin 3t.$$

17. The Laplace transform of the differential equation is

$$s^4 \mathscr{L}\{y\} - s^3 y(0) - s^2 y'(0) - sy''(0) - y'''(0) - \mathscr{L}\{y\} = 0.$$

Solving for $\mathscr{L}\{y\}$ we obtain $\mathscr{L}\{y\} = \dfrac{s}{s^2+1}$. Thus, $y = \cos t$.

18. The Laplace transform of the differential equation is

$$s^4 \mathscr{L}\{y\} - s^3 y(0) - s^2 y'(0) - sy''(0) - y'''(0) - \mathscr{L}\{y\} = \frac{1}{s^2}.$$

Solving for $\mathscr{L}\{y\}$ we obtain

$$\mathscr{L}\{y\} = \frac{1}{s^2(s^4-1)} = -\frac{1}{s^2} + \frac{1}{4}\frac{1}{s-1} - \frac{1}{4}\frac{1}{s+1} + \frac{1}{2}\frac{1}{s^2+1}.$$

Thus

$$y = -t + \frac{1}{4}e^t - \frac{1}{4}e^{-t} + \frac{1}{2}\sin t.$$

19. The Laplace transform of the differential equation is

$$s\mathscr{L}\{y\} - y(0) + \mathscr{L}\{y\} = \frac{5}{s}e^{-s}.$$

Solving for $\mathscr{L}\{y\}$ we obtain

$$\mathscr{L}\{y\} = \frac{5e^{-s}}{s(s+1)} = 5e^{-s}\left[\frac{1}{s} - \frac{1}{s+1}\right].$$

Thus

$$y = 5\,\mathscr{U}(t-1) - 5e^{-(t-1)}\,\mathscr{U}(t-1).$$

20. The Laplace transform of the differential equation is

$$s\mathscr{L}\{y\} - y(0) + \mathscr{L}\{y\} = \frac{1}{s} - \frac{2}{s}e^{-s}.$$

279

Solving for $\mathscr{L}\{y\}$ we obtain

$$\mathscr{L}\{y\} = \frac{1}{s(s+1)} - \frac{2e^{-s}}{s(s+1)} = \frac{1}{s} - \frac{1}{s+1} - 2e^{-s}\left[\frac{1}{s} - \frac{1}{s+1}\right].$$

Thus

$$y = 1 - e^{-t} - 2\left[1 - e^{-(t-1)}\right]\mathcal{U}(t-1).$$

21. The Laplace transform of the differential equation is

$$s\,\mathscr{L}\{y\} - y(0) + 2\,\mathscr{L}\{y\} = \frac{1}{s^2} - e^{-s}\frac{s+1}{s^2}.$$

Solving for $\mathscr{L}\{y\}$ we obtain

$$\mathscr{L}\{y\} = \frac{1}{s^2(s+2)} - e^{-s}\frac{s+1}{s^2(s+1)} = -\frac{1}{4}\frac{1}{s} + \frac{1}{2}\frac{1}{s^2} + \frac{1}{4}\frac{1}{s+2} - e^{-s}\left[\frac{1}{4}\frac{1}{s} + \frac{1}{2}\frac{1}{s^2} - \frac{1}{4}\frac{1}{s+2}\right].$$

Thus

$$y = -\frac{1}{4} + \frac{1}{2}t + \frac{1}{4}e^{-2t} - \left[\frac{1}{4} + \frac{1}{2}(t-1) - \frac{1}{4}e^{-2(t-1)}\right]\mathcal{U}(t-1).$$

22. The Laplace transform of the differential equation is

$$s^2\,\mathscr{L}\{y\} - sy(0) - y'(0) + 4\,\mathscr{L}\{y\} = \frac{1}{s} - \frac{e^{-s}}{s}.$$

Solving for $\mathscr{L}\{y\}$ we obtain

$$\mathscr{L}\{y\} = \frac{1-s}{s(s^2+4)} - e^{-s}\frac{1}{s(s^2+4)} = \frac{1}{4}\frac{1}{s} - \frac{1}{4}\frac{s}{s^2+4} - \frac{1}{2}\frac{2}{s^2+4} - e^{-s}\left[\frac{1}{4}\frac{1}{s} - \frac{1}{4}\frac{s}{s^2+4}\right].$$

Thus

$$y = \frac{1}{4} - \frac{1}{4}\cos 2t - \frac{1}{2}\sin 2t - \left[\frac{1}{4} - \frac{1}{4}\cos 2(t-1)\right]\mathcal{U}(t-1).$$

23. The Laplace transform of the differential equation is

$$s^2\,\mathscr{L}\{y\} - sy(0) - y'(0) + 4\,\mathscr{L}\{y\} = e^{-2\pi s}\frac{1}{s^2+1}.$$

Solving for $\mathscr{L}\{y\}$ we obtain

$$\mathscr{L}\{y\} = \frac{s}{s^2+4} + e^{-2\pi s}\left[\frac{1}{3}\frac{1}{s^2+1} - \frac{1}{6}\frac{2}{s^2+4}\right].$$

Thus

$$y = \cos 2t + \left[\frac{1}{3}\sin(t-2\pi) - \frac{1}{6}\sin 2(t-2\pi)\right]\mathcal{U}(t-2\pi).$$

24. The Laplace transform of the differential equation is

$$s^2\,\mathscr{L}\{y\} - sy(0) - y'(0) - 5\left[s\,\mathscr{L}\{y\} - y(0)\right] + 6\,\mathscr{L}\{y\} = \frac{e^{-s}}{s}.$$

Solving for $\mathcal{L}\{y\}$ we obtain

$$\mathcal{L}\{y\} = e^{-s}\frac{1}{s(s-2)(s-3)} + \frac{1}{(s-2)(s-3)}$$

$$= e^{-s}\left[\frac{1}{6}\frac{1}{s} - \frac{1}{2}\frac{1}{s-2} + \frac{1}{3}\frac{1}{s-3}\right] - \frac{1}{s-2} + \frac{1}{s-3}.$$

Thus

$$y = \left[\frac{1}{6} - \frac{1}{2}e^{2(t-1)} + \frac{1}{3}e^{3(t-1)}\right]\mathcal{U}(t-1) + e^{3t} - e^{2t}.$$

25. The Laplace transform of the differential equation is

$$s^2\,\mathcal{L}\{y\} - sy(0) - y'(0) + \mathcal{L}\{y\} = \frac{e^{-\pi s}}{s} - \frac{e^{-2\pi s}}{s}.$$

Solving for $\mathcal{L}\{y\}$ we obtain

$$\mathcal{L}\{y\} = e^{-\pi s}\left[\frac{1}{s} - \frac{s}{s^2+1}\right] - e^{-2\pi s}\left[\frac{1}{s} - \frac{s}{s^2+1}\right] + \frac{1}{s^2+1}.$$

Thus

$$y = [1 - \cos(t-\pi)]\mathcal{U}(t-\pi) - [1 - \cos(t-2\pi)]\mathcal{U}(t-2\pi) + \sin t.$$

26. The Laplace transform of the differential equation is

$$s^2\,\mathcal{L}\{y\} - sy(0) - y'(0) + 4[s\,\mathcal{L}\{y\} - y(0)] + 3\mathcal{L}\{y\} = \frac{1}{s} - \frac{e^{-2s}}{s} - \frac{e^{-4s}}{s} + \frac{e^{-6s}}{s}.$$

Solving for $\mathcal{L}\{y\}$ we obtain

$$\mathcal{L}\{y\} = \frac{1}{3}\frac{1}{s} - \frac{1}{2}\frac{1}{s+1} + \frac{1}{6}\frac{1}{s+3} - e^{-2s}\left[\frac{1}{3}\frac{1}{s} - \frac{1}{2}\frac{1}{s+1} + \frac{1}{6}\frac{1}{s+3}\right]$$

$$- e^{-4s}\left[\frac{1}{3}\frac{1}{s} - \frac{1}{2}\frac{1}{s+1} + \frac{1}{6}\frac{1}{s+3}\right] + e^{-6s}\left[\frac{1}{3}\frac{1}{s} - \frac{1}{2}\frac{1}{s+1} + \frac{1}{6}\frac{1}{s+3}\right].$$

Thus

$$y = \frac{1}{3} - \frac{1}{2}e^{-t} + \frac{1}{6}e^{-3t} - \left[\frac{1}{3} - \frac{1}{2}e^{-(t-2)} + \frac{1}{6}e^{-3(t-2)}\right]\mathcal{U}(t-2)$$

$$- \left[\frac{1}{3} - \frac{1}{2}e^{-(t-4)} + \frac{1}{6}e^{-3(t-4)}\right]\mathcal{U}(t-4) + \left[\frac{1}{3} - \frac{1}{2}e^{-(t-6)} + \frac{1}{6}e^{-3(t-6)}\right]\mathcal{U}(t-6).$$

27. Taking the Laplace transform of both sides of the differential equation and letting $c = y(0)$ we

281

obtain

$$\mathcal{L}\{y''\} + \mathcal{L}\{2y'\} + \mathcal{L}\{y\} = 0$$

$$s^2 \mathcal{L}\{y\} - sy(0) - y'(0) + 2s\,\mathcal{L}\{y\} - 2y(0) + \mathcal{L}\{y\} = 0$$

$$s^2 \mathcal{L}\{y\} - cs - 2 + 2s\,\mathcal{L}\{y\} - 2c + \mathcal{L}\{y\} = 0$$

$$\left(s^2 + 2s + 1\right)\mathcal{L}\{y\} = cs + 2c + 2$$

$$\mathcal{L}\{y\} = \frac{cs}{(s+1)^2} + \frac{2c+2}{(s+1)^2}$$

$$= c\frac{s+1-1}{(s+1)^2} + \frac{2c+2}{(s+1)^2}$$

$$= \frac{c}{s+1} + \frac{c+2}{(s+1)^2}.$$

Therefore,

$$y(t) = c\mathcal{L}^{-1}\left\{\frac{1}{s+1}\right\} + (c+2)\,\mathcal{L}^{-1}\left\{\frac{1}{(s+1)^2}\right\} = ce^{-t} + (c+2)te^{-t}.$$

To find c we let $y(1) = 2$. Then $2 = ce^{-1} + (c+2)e^{-1} = 2(c+1)e^{-1}$ and $c = e - 1$.

Thus
$$y(t) = (e-1)e^{-t} + (e+1)te^{-t}.$$

28. Taking the Laplace transform of both sides of the differential equation and letting $c = y'(0)$ we
obtain

$$\mathcal{L}\{y''\} - \mathcal{L}\{9y'\} + \mathcal{L}\{20y\} = \mathcal{L}\{1\}$$

$$s^2\,\mathcal{L}\{y\} - sy(0) - y'(0) - 9s\,\mathcal{L}\{y\} + 9y(0) + 20\,\mathcal{L}\{y\} = \frac{1}{s}$$

$$s^2\,\mathcal{L}\{y\} - c - 9s\,\mathcal{L}\{y\} + 20\,\mathcal{L}\{y\} = \frac{1}{s}$$

$$(s^2 - 9s + 20)\,\mathcal{L}\{y\} = \frac{1}{s} + c.$$

Solving for $\mathcal{L}\{y\}$ we obtain

$$\mathcal{L}\{y\} = \frac{1}{s(s^2 - 9s + 20)} + \frac{c}{s^2 - 9s + 20}$$

$$= \frac{1}{s(s-4)(s-5)} + \frac{c}{(s-4)(s-5)}$$

$$= \frac{1/20}{s} - \frac{1/4}{s-4} + \frac{1/5}{s-5} - \frac{c}{s-4} + \frac{c}{s-5}.$$

Therefore

$$y(t) = \frac{1}{20}\mathscr{L}^{-1}\left\{\frac{1}{s}\right\} - \frac{1}{4}\mathscr{L}^{-1}\left\{\frac{1}{s-4}\right\} + \frac{1}{5}\mathscr{L}^{-1}\left\{\frac{1}{s-5}\right\} - c\mathscr{L}^{-1}\left\{\frac{1}{s-4}\right\} + c\mathscr{L}^{-1}\left\{\frac{1}{s-5}\right\}$$

$$= \frac{1}{20} - \frac{1}{4}e^{4t} + \frac{1}{5}e^{5t} - c\left(e^{4t} - e^{5t}\right).$$

To find c we compute

$$y'(t) = -e^{4t} + e^{5t} - c\left(4e^{4t} - 5e^{5t}\right)$$

and let $y'(1) = 0$. Then

$$0 = -e^4 + e^5 - c\left(4e^4 - 5e^5\right)$$

and

$$c = \frac{e^5 - e^4}{4e^4 - 5e^5} = \frac{e-1}{4-5e}.$$

Thus

$$y(t) = \frac{1}{20} - \frac{1}{4}e^{4t} + \frac{1}{5}e^{5t} - \frac{e-1}{4-5e}\left(e^{4t} - e^{5t}\right)$$

$$= \frac{1}{20} + \frac{e}{4(4-5e)}e^{4t} - \frac{1}{5(4-5e)}e^{5t}.$$

29. The Laplace transform of the given equation is

$$\mathscr{L}\{f\} + \mathscr{L}\{t\}\mathscr{L}\{f\} = \mathscr{L}\{t\}.$$

Solving for $\mathscr{L}\{f\}$ we obtain $\mathscr{L}\{f\} = \dfrac{1}{s^2+1}$. Thus, $f(t) = \sin t$.

30. The Laplace transform of the given equation is

$$\mathscr{L}\{f\} = \mathscr{L}\{2t\} - 4\mathscr{L}\{\sin t\}\mathscr{L}\{f\}.$$

Solving for $\mathscr{L}\{f\}$ we obtain

$$\mathscr{L}\{f\} = \frac{2s^2+2}{s^2(s^2+5)} = \frac{2}{5}\frac{1}{s^2} + \frac{8}{5\sqrt{5}}\frac{\sqrt{5}}{s^2+5}.$$

Thus

$$f(t) = \frac{2}{5}t + \frac{8}{5\sqrt{5}}\sin\sqrt{5}\,t.$$

31. The Laplace transform of the given equation is

$$\mathscr{L}\{f\} = \mathscr{L}\{te^t\} + \mathscr{L}\{t\}\mathscr{L}\{f\}.$$

Solving for $\mathscr{L}\{f\}$ we obtain

$$\mathscr{L}\{f\} = \frac{s^2}{(s-1)^3(s+1)} = \frac{1}{8}\frac{1}{s-1} + \frac{3}{4}\frac{1}{(s-1)^2} + \frac{1}{4}\frac{2}{(s-1)^3} - \frac{1}{8}\frac{1}{s+1}.$$

Thus
$$f(t) = \frac{1}{8}e^t + \frac{3}{4}te^t + \frac{1}{4}t^2e^t - \frac{1}{8}e^{-t}$$

32. The Laplace transform of the given equation is

$$\mathscr{L}\{f\} + 2\mathscr{L}\{\cos t\}\mathscr{L}\{f\} = 4\mathscr{L}\{e^{-t}\} + \mathscr{L}\{\sin t\}.$$

Solving for $\mathscr{L}\{f\}$ we obtain

$$\mathscr{L}\{f\} = \frac{4s^2 + s + 5}{(s+1)^3} = \frac{4}{s+1} - \frac{7}{(s+1)^2} + 4\frac{2}{(s+1)^3}.$$

Thus
$$f(t) = 4e^{-t} - 7te^{-t} + 4t^2e^{-t}.$$

33. The Laplace transform of the given equation is

$$\mathscr{L}\{f\} + \mathscr{L}\{1\}\mathscr{L}\{f\} = \mathscr{L}\{1\}.$$

Solving for $\mathscr{L}\{f\}$ we obtain $\mathscr{L}\{f\} = \dfrac{1}{s+1}$. Thus, $f(t) = e^{-t}$.

34. The Laplace transform of the given equation is

$$\mathscr{L}\{f\} = \mathscr{L}\{\cos t\} + \mathscr{L}\{e^{-t}\}\mathscr{L}\{f\}.$$

Solving for $\mathscr{L}\{f\}$ we obtain

$$\mathscr{L}\{f\} = \frac{s}{s^2+1} + \frac{1}{s^2+1}.$$

Thus
$$f(t) = \cos t + \sin t.$$

35. The Laplace transform of the given equation is

$$\mathscr{L}\{f\} = \mathscr{L}\{1\} + \mathscr{L}\{t\} + \frac{8}{3}\mathscr{L}\{t^3\}\mathscr{L}\{f\}.$$

Solving for $\mathscr{L}\{f\}$ we obtain

$$\mathscr{L}\{f\} = \frac{s^2(s+1)}{s^4 - 16} = \frac{3}{8}\frac{1}{s-2} + \frac{1}{8}\frac{1}{s+2} + \frac{1}{2}\frac{s}{s^2+4} + \frac{1}{4}\frac{2}{s^2+4}.$$

Thus
$$f(t) = \frac{3}{8}e^{2t} + \frac{1}{8}e^{-2t} + \frac{1}{2}\cos 2t + \frac{1}{4}\sin 2t.$$

36. The Laplace transform of the given equation is

$$\mathscr{L}\{t\} - 2\mathscr{L}\{f\} = \mathscr{L}\{e^t - e^{-t}\}\mathscr{L}\{f\}.$$

Solving for $\mathscr{L}\{f\}$ we obtain

$$\mathscr{L}\{f\} = \frac{s^2 - 1}{2s^4} = \frac{1}{2}\frac{1}{s^2} - \frac{1}{12}\frac{3!}{s^4}.$$

Thus
$$f(t) = \frac{1}{2}t - \frac{1}{12}t^3.$$

37. The Laplace transform of the given equation is

$$s\mathcal{L}\{y\} - y(0) = \mathcal{L}\{1\} - \mathcal{L}\{\sin t\} - \mathcal{L}\{1\}\mathcal{L}\{y\}.$$

Solving for $\mathcal{L}\{f\}$ we obtain

$$\mathcal{L}\{y\} = \frac{s^3 - s^2 + s}{s(s^2 + 1)^2} = \frac{1}{s^2 + 1} + \frac{1}{2}\frac{2s}{(s^2 + 1)^2}.$$

Thus
$$y = \sin t - \frac{1}{2}t\sin t.$$

38. The Laplace transform of the given equation is

$$s\mathcal{L}\{y\} - y(0) + 6\mathcal{L}\{y\} + 9\mathcal{L}\{1\}\mathcal{L}\{y\} = \mathcal{L}\{1\}.$$

Solving for $\mathcal{L}\{f\}$ we obtain $\mathcal{L}\{y\} = \dfrac{1}{(s+3)^2}$. Thus, $y = te^{-3t}$.

39. From equation (3) in the text the differential equation is

$$0.005\frac{di}{dt} + i + 50\int_0^t i(\tau)\,d\tau = 100[1 - \mathcal{U}(t-1)], \quad i(0) = 0.$$

The Laplace transform of this equation is

$$0.005[s\mathcal{L}\{i\} - i(0)] + \mathcal{L}\{i\} + 50\frac{1}{s}\mathcal{L}\{i\} = 100\left[\frac{1}{s} - \frac{1}{s}e^{-s}\right].$$

Solving for $\mathcal{L}\{i\}$ we obtain

$$\mathcal{L}\{i\} = \frac{20{,}000}{(s+100)^2}(1 - e^{-s}).$$

Thus
$$i(t) = 20{,}000te^{-100t} - 20{,}000(t-1)e^{-100(t-1)}\mathcal{U}(t-1).$$

40. From equation (3) in the text the differential equation is

$$0.005\frac{di}{dt} + i + \frac{1}{0.02}\int_0^t i(\tau)\,d\tau = 100[t - (t-1) - \mathcal{U}(t-1)], \quad i(0) = 0$$

or
$$\frac{di}{dt} + 200i + 10{,}000\int_0^t i(\tau)\,d\tau = 20{,}000[t - (t-1)\mathcal{U}(t-1)], \quad i(0) = 0.$$

The Laplace transform of this equation is

$$s\mathcal{L}\{i\} + 200\,\mathcal{L}\{i\} + 10{,}000\frac{1}{s}\mathcal{L}\{i\} = 20{,}000\left[\frac{1}{s^2} - \frac{1}{s^2}e^{-s}\right].$$

Solving for $\mathcal{L}\{i\}$ we obtain

$$\mathcal{L}\{i\} = \frac{20{,}000}{s(s+100)^2}(1 - e^{-s}) = \left[\frac{2}{s} - \frac{2}{s+100} - \frac{200}{(s+100)^2}\right](1 - e^{-s}).$$

Thus

$$i(t) = 2 - 2e^{-100t} - 200te^{-100t} - 2\mathcal{U}(t-1) + 2e^{-100(t-1)}\mathcal{U}(t-1) + 200(t-1)e^{-100(t-1)}\mathcal{U}(t-1).$$

41. The differential equation is

$$R\frac{dq}{dt} + \frac{1}{C}q = E_0e^{-kt}, \quad q(0) = 0.$$

The Laplace transform of this equation is

$$R\mathcal{L}\{q\} + \frac{1}{C}\mathcal{L}\{q\} = E_0\frac{1}{s+k}.$$

Solving for $\mathcal{L}\{q\}$ we obtain

$$\mathcal{L}\{q\} = \frac{E_0C}{(s+k)(RC_s+1)} = \frac{E_0/R}{(s+k)(s+1/RC)}.$$

When $1/RC \neq k$ we have by partial fractions

$$\mathcal{L}\{q\} = \frac{E_0}{R}\left(\frac{1/(1/RC-k)}{s+k} - \frac{1/(1/RC-k)}{s+1/RC}\right) = \frac{E_0}{R}\frac{1}{1/RC-k}\left(\frac{1}{s+k} - \frac{1}{s+1/RC}\right).$$

Thus

$$q(t) = \frac{E_0C}{1-kRC}\left(e^{-kt} - e^{-t/RC}\right).$$

When $1/RC = k$ we have

$$\mathcal{L}\{q\} = \frac{E_0}{R}\frac{1}{(s+k)^2}.$$

Thus

$$q(t) = \frac{E_0}{R}te^{-kt} = \frac{E_0}{R}te^{-t/RC}.$$

42. The differential equation is

$$10\frac{dq}{dt} + 10q = 30e^t - 30e^t\mathcal{U}(t-1.5).$$

The Laplace transform of this equation is

$$s\mathcal{L}\{q\} - q_0 + \mathcal{L}\{q\} = \frac{3}{s-1} - \frac{3e^{1.5}}{s-1.5}e^{-1.5s}.$$

Solving for $\mathcal{L}\{q\}$ we obtain

$$\mathcal{L}\{q\} = \left(q_0 - \frac{3}{2}\right)\cdot\frac{1}{s+1} + \frac{3}{2}\cdot\frac{1}{s-1} - 3e^{1.5}\left(\frac{-2/5}{s+1} + \frac{2/5}{s-1.5}\right)e^{-1.55}.$$

Thus

$$q(t) = \left(q_0 - \frac{3}{2}\right)e^{-t} + \frac{3}{2}e^t + \frac{6}{5}e^{1.5}\left(e^{-(t-1.5)} - e^{1.5(t-1.5)}\right)\mathcal{U}(t-1.5).$$

43. The differential equation is

$$2.5\frac{dq}{dt} + 12.5q = 5\mathcal{U}(t-3).$$

The Laplace transform of this equation is

$$s\mathscr{L}\{q\} + 5\mathscr{L}\{q\} = \frac{2}{s}e^{-3s}.$$

Solving for $\mathscr{L}\{q\}$ we obtain

$$\mathscr{L}\{q\} = \frac{2}{s(s+5)}e^{-3s} = \left(\frac{2}{5}\cdot\frac{1}{s} - \frac{2}{5}\cdot\frac{1}{s+5}\right)e^{-3s}.$$

Thus

$$q(t) = \frac{2}{5}\mathscr{U}(t-3) - \frac{2}{5}e^{-5(t-3)}\mathscr{U}(t-3).$$

44. (a) The differential equation is

$$50\frac{dq}{dt} + \frac{1}{0.01}q = E_0[\mathscr{U}(t-1) - \mathscr{U}(t-3)], \quad q(0) = 0$$

or

$$50\frac{dq}{dt} + 100q = E_0[\mathscr{U}(t-1) - \mathscr{U}(t-3)], \quad q(0) = 0.$$

The Laplace transform of this equation is

$$50s\,\mathscr{L}\{q\} + 100\,\mathscr{L}\{q\} = E_0\left(\frac{1}{s}e^{-s} - \frac{1}{s}e^{-3s}\right).$$

Solving for $\mathscr{L}\{q\}$ we obtain

$$\mathscr{L}\{q\} = \frac{E_0}{50}\left[\frac{e^{-s}}{s(s+2)} - \frac{e^{-3s}}{s(s+2)}\right] = \frac{E_0}{50}\left[\frac{1}{2}\left(\frac{1}{s} - \frac{1}{s+2}\right)e^{-s} - \frac{1}{2}\left(\frac{1}{s} - \frac{1}{s+2}\right)e^{-3s}\right].$$

Thus

$$q(t) = \frac{E_0}{100}\left[\left(1 - e^{-2(t-1)}\right)\mathscr{U}(t-1) - \left(1 - e^{-2(t-3)}\right)\mathscr{U}(t-3)\right].$$

(b)

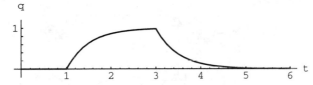

The maximum value of $q(t)$ is approximately 1 at $t - 3$.

45. (a) The differential equation is

$$\frac{di}{dt} + 10i = \sin t + \cos\left(t - \frac{3\pi}{2}\right)\mathscr{U}\left(t - \frac{3\pi}{2}\right), \quad i(0) = 0.$$

The Laplace transform of this equation is

$$s\mathscr{L}\{i\} + 10\mathscr{L}\{i\} = \frac{1}{s^2+1} + \frac{se^{-3\pi s/2}}{s^2+1}.$$

287

Solving for $\mathscr{L}\{i\}$ we obtain

$$\mathscr{L}\{i\} = \frac{1}{(s^2+1)(s+10)} + \frac{s}{(s^2+1)(s+10)} e^{-3\pi s/2}$$

$$= \frac{1}{101}\left(\frac{1}{s+10} - \frac{s}{s^2+1} + \frac{10}{s^2+1}\right) + \frac{1}{101}\left(\frac{-10}{s+10} + \frac{10s}{s^2+1} + \frac{1}{s^2+1}\right) e^{-3\pi s/2}.$$

Thus

$$i(t) = \frac{1}{101}\left(e^{-10t} - \cos t + 10\sin t\right)$$

$$+ \frac{1}{101}\left[-10e^{-10(t-3\pi/2)} + 10\cos\left(t - \frac{3\pi}{2}\right) + \sin\left(t - \frac{3\pi}{2}\right)\right]\mathscr{U}\left(t - \frac{3\pi}{2}\right).$$

(b)

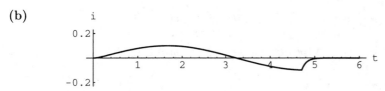

The maximum value of $i(t)$ is approximately 0.1 at $t = 1.7$, the minimum is approximately -0.1 at 4.7.

46. The differential equation is

$$\frac{di}{dt} + \frac{R}{L}i = \frac{1}{L}E(t), \quad i(0) = 0.$$

The Laplace transform of this equation is

$$s\mathscr{L}\{i\} + \frac{R}{L}\mathscr{L}\{i\} = \frac{1}{L}\mathscr{L}\{E(t)\}.$$

From Problem 31, Exercises 7.4, we have

$$\mathscr{L}\{E(t)\} = \frac{1 - e^{-s}}{s(1 + e^{-s})}.$$

Thus

$$\left(s + \frac{R}{L}\right)\mathscr{L}\{i\} = \frac{1}{L}\frac{1 - e^{-s}}{s(1 + e^{-s})}$$

and

$$\mathscr{L}\{i\} = \frac{1}{L}\frac{1 - e^{-s}}{s(s + R/L)(1 + e^{-s})} = \frac{1}{L}\frac{1 - e^{-s}}{s(s + R/L)}\frac{1}{1 + e^{-s}}$$

$$= \frac{1}{L}\left[\frac{L/R}{s} - \frac{L/R}{s + R/L}\right](1 - e^{-s})(1 - e^{-s} + e^{-2s} - e^{-3s} + \cdots)$$

$$= \frac{1}{R}\left[\frac{1}{s} - \frac{1}{s + R/L}\right](1 - 2e^{-s} + 2e^{-2s} - 2e^{-3s} + \cdots).$$

Therefore

$$i(t) = \frac{1}{R}[1 - 2\,\mathcal{U}(t-1) + 2\,\mathcal{U}(t-2) - 2\,\mathcal{U}(t-3) + \cdots]$$

$$- \frac{1}{R}\Big[e^{-Rt/L} + 2e^{-R(t-1)/L}\,\mathcal{U}(t-1) - 2e^{-R(t-2)/L}\,\mathcal{U}(t-2)$$

$$+ 2e^{-R(t-3)/L}\,\mathcal{U}(t-3) + \cdots\Big]$$

$$= \frac{1}{R}\left(1 - e^{-Rt/L}\right) + \frac{2}{R}\sum_{n=1}^{\infty}(-1)^n\left(1 - e^{-R(t-n)/L}\right)\mathcal{U}(t-n).$$

47. The differential equation is

$$\frac{di}{dt} + \frac{R}{L}i = \frac{1}{L}E(t), \quad i(0) = 0.$$

The Laplace transform of this equation is

$$s\,\mathscr{L}\{i\} + \frac{R}{L}\,\mathscr{L}\{i\} = \frac{1}{L}\,\mathscr{L}\{E(t)\}.$$

From Problem 33, Exercises 7.4, we have

$$\mathscr{L}\{E(t)\} = \frac{1}{s}\left(\frac{1}{s} - \frac{1}{e^s - 1}\right) = \frac{1}{s^2} + \frac{1}{s}\frac{1}{1 - e^s}.$$

Thus

$$\left(s + \frac{R}{L}\right)\mathscr{L}\{i\} = \frac{1}{L}\frac{1}{s^2} + \frac{1}{L}\frac{1}{s}\frac{1}{1 - e^s}$$

and

$$\mathscr{L}\{i\} = \frac{1}{L}\frac{1}{s^2(s + R/L)} + \frac{1}{L}\frac{1}{s(s + R/L)}\frac{1}{1 - e^s}$$

$$= \frac{1}{L}\left(\frac{L/R}{s^2} - \frac{L^2/R^2}{s} + \frac{L^2/R^2}{s + R/L}\right) + \frac{1}{L}\left(\frac{L/R}{s} - \frac{L/R}{s + R/L}\right)\frac{1}{1 - e^s}$$

$$= \frac{1}{R}\left[\frac{1}{s} - \frac{L/R}{s} + \frac{L/R}{s + R/L}\right] + \frac{1}{R}\left(\frac{1}{s} - \frac{1}{s + R/L}\right)\left(1 + e^s + e^{2s} + e^{3s} + \cdots\right).$$

Thus

$$i(t) = \frac{1}{R}\left(t - \frac{L}{R} + \frac{L}{R}e^{-Rt/L}\right) + \frac{1}{R}\sum_{n=1}^{\infty}\left(1 - e^{-R(t-n)/L}\right)\mathcal{U}(t-n).$$

For $0 \le t < 2$ we have

$$i(t) = \frac{1}{R}\left(t - \frac{L}{R} + \frac{L}{R}e^{-Rt/L}\right) + \frac{1}{R}\left(1 - e^{-R(t-1)/L}\right)\mathcal{U}(t-1)$$

$$= \begin{cases} \frac{1}{R}\left(t - \frac{L}{R} + \frac{L}{R}e^{-Rt/L}\right), & 0 \le t < 1 \\ \frac{1}{R}\left(t - \frac{L}{R} + \frac{L}{R}e^{-Rt/L}\right) + \frac{1}{R}\left(1 - e^{-R(t-1)/L}\right), & 1 \le t < 2. \end{cases}$$

48. The differential equation is

$$\frac{d^2q}{dt^2} + 20\frac{dq}{dt} + 200q = 150, \quad q(0) = q'(0) = 0.$$

The Laplace transform of this equation is

$$s^2\mathscr{L}\{q\} + 20s\mathscr{L}\{q\} + 200\mathscr{L}\{q\} = \frac{150}{s}.$$

Solving for $\mathscr{L}\{q\}$ we obtain

$$\mathscr{L}\{q\} = \frac{150}{s(s^2 + 20s + 200)} = \frac{3}{4}\frac{1}{s} - \frac{3}{4}\frac{s+10}{(s+10)^2 + 10^2} - \frac{3}{4}\frac{10}{(s+10)^2 + 10^2}.$$

Thus

$$q(t) = \frac{3}{4} - \frac{3}{4}e^{-10t}\cos 10t - \frac{3}{4}e^{-10t}\sin 10t$$

and

$$i(t) = q'(t) = 15e^{-10t}\sin 10t.$$

If $E(t) = 150 - 150\,\mathcal{U}(t-2)$, then

$$\mathscr{L}\{q\} = \frac{150}{s(s^2 + 20s + 200)}\left(1 - e^{-2s}\right)$$

$$q(t) = \frac{3}{4} - \frac{3}{4}e^{-10t}\cos 10t - \frac{3}{4}e^{-10t}\sin 10t - \left[\frac{3}{4} - \frac{3}{4}e^{-10(t-2)}\cos 10(t-2)\right.$$

$$\left. - \frac{3}{4}e^{-10(t-2)}\sin 10(t-2)\right]\mathcal{U}(t-2).$$

49. The differential equation is

$$\frac{d^2q}{dt^2} + 20\frac{dq}{dt} + 100q = 120\sin 10t.$$

The Laplace transform of this equation is

$$s^2\mathscr{L}\{q\} + 20s\mathscr{L}\{q\} + 100\mathscr{L}\{q\} = \frac{120}{s^2 + 100}.$$

Solving for $\mathscr{L}\{q\}$ we obtain

$$\mathscr{L}\{q\} = \frac{1200}{(s+10)^2(s^2+100)} = \frac{3}{5}\frac{1}{s+10} + 6\frac{1}{(s+10)^2} - \frac{3}{5}\frac{s}{s^2+10^2}.$$

Thus

$$q(t) = \frac{3}{5}e^{-10t} + 6te^{-10t} - \frac{3}{5}\cos 10t$$

and

$$i(t) = q'(t) = -60te^{-10t} + 6\sin 10t.$$

The steady-state current is $6\sin 10t$.

50. The differential equation is

$$\frac{d^2q}{dt^2} + 2\lambda\frac{dq}{dt} + \omega^2 q = \frac{E_0}{L}, \quad q(0) = q'(0) = 0.$$

The Laplace transform of this equation is

$$s^2\mathscr{L}\{q\} + 2\lambda s\mathscr{L}\{q\} + \omega^2\mathscr{L}\{q\} = \frac{E_0}{L}\frac{1}{s}$$

or

$$\left(s^2 + 2\lambda s + \omega^2\right)\mathscr{L}\{q\} = \frac{E_0}{L}\frac{1}{s}.$$

Solving for $\mathscr{L}\{q\}$ and using partial fractions we obtain

$$\mathscr{L}\{q\} = \frac{E_0}{L}\left(\frac{1/\omega^2}{s} - \frac{(1/\omega^2)s + 2\lambda/\omega^2}{s^2 + 2\lambda s + \omega^2}\right) = \frac{E_0}{L\omega^2}\left(\frac{1}{s} - \frac{s + 2\lambda}{s^2 + 2\lambda s + \omega^2}\right).$$

For $\lambda > \omega$ we write $s^2 + 2\lambda s + \omega^2 = (s + \lambda)^2 - \left(\lambda^2 - \omega^2\right)$, so (recalling that $\omega^2 = 1/LC$,)

$$\mathscr{L}\{q\} = E_0 C\left(\frac{1}{s} - \frac{s + \lambda}{(s + \lambda)^2 - (\lambda^2 - \omega^2)} - \frac{\lambda}{(s + \lambda)^2 - (\lambda^2 - \omega^2)}\right).$$

Thus for $\lambda > \omega$,

$$q(t) = E_0 C\left(1 - e^{-\lambda t}\cosh\sqrt{\lambda^2 - \omega^2}\,t - \frac{\lambda}{\sqrt{\lambda^2 - \omega^2}}\sinh\sqrt{\lambda^2 - \omega^2}\,t\right).$$

For $\lambda < \omega$ we write $s^2 + 2\lambda s + \omega^2 = (s + \lambda)^2 + \left(\omega^2 - \lambda^2\right)$, so

$$\mathscr{L}\{q\} = E_0\left(\frac{1}{s} - \frac{s + \lambda}{(s + \lambda)^2 + (\omega^2 - \lambda^2)} - \frac{\lambda}{(s + \lambda)^2 + (\omega^2 - \lambda^2)}\right).$$

Thus for $\lambda < \omega$,

$$q(t) = E_0 C\left(1 - e^{-\lambda t}\cos\sqrt{\omega^2 - \lambda^2}\,t - \frac{\lambda}{\sqrt{\lambda^2 - \omega^2}}\sin\sqrt{\omega^2 - \lambda^2}\,t\right).$$

For $\lambda = \omega$, $s^2 + 2\lambda + \omega^2 = (s + \lambda)^2$ and

$$\mathscr{L}\{q\} = \frac{E_0}{L}\frac{1}{s(s + \lambda)^2} = \frac{E_0}{L}\left(\frac{1/\lambda^2}{s} - \frac{1/\lambda^2}{s + \lambda} - \frac{1/\lambda}{(s + \lambda)^2}\right) = \frac{E_0}{L\lambda^2}\left(\frac{1}{s} - \frac{1}{s + \lambda} - \frac{\lambda}{(s + \lambda)^2}\right).$$

Thus for $\lambda = \omega$,

$$q(t) = E_0 C\left(1 - e^{-\lambda t} - \lambda t e^{-\lambda t}\right).$$

51. The differential equation is

$$\frac{d^2q}{dt^2} + \frac{1}{LC}q = \frac{E_0}{L}e^{-kt}, \quad q(0) = q'(0) = 0.$$

The Laplace transform of this equation is

$$s^2 \mathscr{L}\{q\} + \frac{1}{LC}\mathscr{L}\{q\} = \frac{E_0}{L}\frac{1}{s+k}.$$

Solving for $\mathscr{L}\{q\}$ we obtain

$$\mathscr{L}\{q\} = \frac{E_0}{L}\frac{1}{(s+k)(s^2+1/LC)} = \frac{E_0}{L}\left(\frac{1/(k^2+1/LC)}{s+k} - \frac{s/(k^2+1/LC)}{s^2+1/LC} + \frac{k/(k^2+1/LC)}{s^2+1/LC}\right).$$

Thus
$$q(t) = \frac{E_0}{L(k^2+1/LC)}\left[e^{-kt} - \cos\left(t/\sqrt{LC}\right) + k\sqrt{LC}\,\sin\left(t/\sqrt{LC}\right)\right].$$

52. Recall from Chapter 5 that $mx'' = -kx + f(t)$. Now $m = W/g = 32/32 = 1$ slug, and $32 = 2k$ so that $k = 16$ lb/ft. Thus, the differential equation is $x'' + 16x = f(t)$. The initial conditions are $x(0) = 0$, $x'(0) = 0$. Also, since

$$f(t) = \begin{cases} \sin t, & 0 \le t < 2\pi \\ 0, & t \ge 2\pi \end{cases}$$

and $\sin t = \sin(t - 2\pi)$ we can write

$$f(t) = \sin t - \sin(t - 2\pi)\mathscr{U}(t - 2\pi).$$

The Laplace transform of the differential equation is

$$s^2 \mathscr{L}\{x\} + 16\mathscr{L}\{x\} = \frac{1}{s^2+1} - \frac{1}{s^2+1}e^{-2\pi s}.$$

Solving for $\mathscr{L}\{x\}$ we obtain

$$\mathscr{L}\{x\} = \frac{1}{(s^2+16)(s^2+1)} - \frac{1}{(s^2+16)(s^2+1)}e^{-2\pi s}$$

$$= \frac{-1/15}{s^2+16} + \frac{1/15}{s^2+1} - \left[\frac{-1/15}{s^2+16} + \frac{1/15}{s^2+1}\right]e^{-2\pi s}.$$

Thus

$$x(t) = -\frac{1}{60}\sin 4t + \frac{1}{15}\sin t + \frac{1}{60}\sin 4(t - 2\pi)\mathscr{U}(t - 2\pi) - \frac{1}{15}\sin(t - 2\pi)\mathscr{U}(t - 2\pi)$$

$$= \begin{cases} -\frac{1}{60}\sin 4t + \frac{1}{15}\sin t, & 0 \le t < 2\pi \\ 0, & t \ge 2\pi. \end{cases}$$

53. Recall from Chapter 5 that $mx'' = -kx - \beta x'$. Now $m = W/g = 4/32 = \frac{1}{8}$ slug, and $4 = 2k$ so that $k = 2$ lb./ft. Thus, the differential equation is $x'' + 7x' + 16x = 0$. The initial conditions are $x(0) = -3/2$ and $x'(0) = 0$. The Laplace transform of the differential equation is

$$s^2 \mathscr{L}\{x\} + \frac{3}{2}s + 7s\mathscr{L}\{x\} + \frac{21}{2} + 16\mathscr{L}\{x\} = 0.$$

Solving for $\mathscr{L}\{x\}$ we obtain

$$\mathscr{L}\{x\} = \frac{-3s/2 - 21/2}{s^2 + 7s + 16} = -\frac{3}{2} \frac{s + 7/2}{(s + 7/2)^2 + (\sqrt{15}/2)^2} - \frac{7\sqrt{15}}{10} \frac{\sqrt{15}/2}{(s + 7/2)^2 + (\sqrt{15}/2)^2}.$$

Thus
$$x = -\frac{3}{2} e^{-7t/2} \cos \frac{\sqrt{15}}{2} t - \frac{7\sqrt{15}}{10} e^{-7t/2} \sin \frac{\sqrt{15}}{2} t.$$

54. Recall from Chapter 5 that $mx'' = -kx + f(t)$. Now $m = W/g = 16/32 = 1/2$ slug, and $k = 4.5$, so the differential equation is

$$\frac{1}{2} x'' + 4.5x = 4 \sin 3t + 2 \cos 3t \quad \text{or} \quad x'' + 9x = 8 \sin 3t + 4 \cos 3t.$$

The initial conditions are $x(0) = x'(0) = 0$. The Laplace transform of the differential equation is

$$s^2 \mathscr{L}\{x\} + 9\mathscr{L}\{x\} = \frac{24}{s^2 + 9} + \frac{4s}{s^2 + 9}.$$

Solving for $\mathscr{L}\{x\}$ we obtain

$$\mathscr{L}\{x\} = \frac{4s + 24}{(s^2 + 9)^2} = \frac{2}{3} \frac{2(3)s}{(s^2 + 9)^2} + \frac{12}{27} \frac{2(3)^3}{(s^2 + 9)^2}.$$

Thus
$$x(t) = \frac{2}{3} t \sin 3t + \frac{4}{9} (\sin 3t - 3t \cos 3t) = \frac{2}{3} t \sin 3t + \frac{4}{9} \sin 3t - \frac{4}{3} t \cos 3t.$$

55. The differential equation is

$$EI \frac{d^4 y}{dx^4} = w_0 [1 - \mathscr{U}(x - L/2)].$$

Taking the Laplace transform of both sides and using $y(0) = y'(0) = 0$ we obtain

$$s^4 \mathscr{L}\{y\} - sy''(0) - y'''(0) = \frac{w_0}{EI} \frac{1}{s} \left(1 - e^{-Ls/2}\right).$$

Letting $y''(0) = c_1$ and $y'''(0) = c_2$ we have

$$\mathscr{L}\{y\} = \frac{c_1}{s^3} + \frac{c_2}{s^4} + \frac{w_0}{EI} \frac{1}{s^5} \left(1 - e^{-Ls/2}\right)$$

so that
$$y(x) = \frac{1}{2} c_1 x^2 + \frac{1}{6} c_2 x^3 + \frac{1}{24} \frac{w_0}{EI} \left[x^4 - \left(x - \frac{L}{2}\right)^4 \mathscr{U}\left(x - \frac{L}{2}\right)\right].$$

To find c_1 and c_2 we compute

$$y''(x) = c_1 + c_2 x + \frac{1}{2} \frac{w_0}{EI} \left[x^2 - \left(x - \frac{L}{2}\right)^2 \mathscr{U}\left(x - \frac{L}{2}\right)\right]$$

and
$$y'''(x) = c_2 + \frac{w_0}{EI} \left[x - \left(x - \frac{L}{2}\right)\mathscr{U}\left(x - \frac{L}{2}\right)\right].$$

293

Then $y''(L) = y'''(L) = 0$ yields the system

$$c_1 + c_2 L + \frac{1}{2}\frac{w_0}{EI}\left[L^2 - \left(\frac{L}{2}\right)^2\right] = c_1 + c_2 L + \frac{3}{8}\frac{w_0 L^2}{EI} = 0$$

$$c_2 + \frac{w_0}{EI}\left(\frac{L}{2}\right) = c_2 + \frac{1}{2}\frac{w_0 L}{EI} = 0.$$

Solving for c_1 and c_2 we obtain $c_1 = \frac{1}{8}w_0 L^2/EI$ and $c_2 = -\frac{1}{2}w_0 L/EI$. Thus

$$y(x) = \frac{w_0}{EI}\left(\frac{1}{16}L^2 x^2 - \frac{1}{12}Lx^3 + \frac{1}{24}x^4 - \frac{1}{24}\left(x - \frac{L}{2}\right)^4 \mathscr{U}\left(x - \frac{L}{2}\right)\right).$$

56. The differential equation is

$$EI\frac{d^4 y}{dx^4} = w_0[\mathscr{U}(x - L/3) - \mathscr{U}(x - 2L/3)].$$

Taking the Laplace transform of both sides and using $y(0) = y'(0) = 0$ we obtain

$$s^4 \mathscr{L}\{y\} - sy''(0) - y'''(0) = \frac{w_0}{EI}\frac{1}{s}\left(e^{-Ls/3} - e^{-2Ls/3}\right).$$

Letting $y''(0) = c_1$ and $y'''(0) = c_2$ we have

$$\mathscr{L}\{y\} = \frac{c_1}{s^3} + \frac{c_2}{s^4} + \frac{w_0}{EI}\frac{1}{s^5}\left(e^{-Ls/3} - e^{-2Ls/3}\right)$$

so that

$$y(x) = \frac{1}{2}c_1 x^2 + \frac{1}{6}c_2 x^3 + \frac{1}{24}\frac{w_0}{EI}\left[\left(x - \frac{L}{3}\right)^4 \mathscr{U}\left(x - \frac{L}{3}\right) - \left(x - \frac{2L}{3}\right)^4 \mathscr{U}\left(x - \frac{2L}{3}\right)\right].$$

To find c_1 and c_2 we compute

$$y''(x) = c_1 + c_2 x + \frac{1}{2}\frac{w_0}{EI}\left[\left(x - \frac{L}{3}\right)^2 \mathscr{U}\left(x - \frac{L}{3}\right) - \left(x - \frac{2L}{3}\right)^2 \mathscr{U}\left(x - \frac{2L}{3}\right)\right]$$

and $\qquad y'''(x) = c_2 + \frac{w_0}{EI}\left[\left(x - \frac{L}{3}\right)\mathscr{U}\left(x - \frac{L}{3}\right) - \left(x - \frac{2L}{3}\right)\mathscr{U}\left(x - \frac{2L}{3}\right)\right].$

Then $y''(L) = y'''(L) = 0$ yields the system

$$c_1 + c_2 L + \frac{1}{2}\frac{w_0}{EI}\left[\left(\frac{2L}{3}\right)^2 - \left(\frac{L}{3}\right)^2\right] = c_1 + c_2 L + \frac{1}{6}\frac{w_0 L^2}{EI} = 0$$

$$c_2 + \frac{w_0}{EI}\left[\frac{2L}{3} - \frac{L}{3}\right] = c_2 + \frac{1}{3}\frac{w_0 L}{EI} = 0.$$

Solving for c_1 and c_2 we obtain $c_1 = \frac{1}{6}w_0 L^2/EI$ and $c_2 = -\frac{1}{3}w_0 L/EI$. Thus

$$y(x) = \frac{w_0}{EI}\left(\frac{1}{12}L^2 x^2 - \frac{1}{18}Lx^3 + \frac{1}{24}\left[\left(x - \frac{L}{3}\right)^4 \mathscr{U}\left(x - \frac{L}{3}\right) - \left(x - \frac{2L}{3}\right)^4 \mathscr{U}\left(x - \frac{2L}{3}\right)\right]\right).$$

57. The differential equation is

$$EI \frac{d^4y}{dx^4} = \frac{2w_0}{L} \left[\frac{L}{2} - x + \left(x - \frac{L}{2} \right) \mathcal{U} \left(x - \frac{L}{2} \right) \right].$$

Taking the Laplace transform of both sides and using $y(0) = y'(0) = 0$ we obtain

$$s^4 \mathcal{L}\{y\} - sy''(0) - y'''(0) = \frac{2w_0}{EIL} \left[\frac{L}{2s} - \frac{1}{s^2} + \frac{1}{s^2} e^{-Ls/2} \right].$$

Letting $y''(0) = c_1$ and $y'''(0) = c_2$ we have

$$\mathcal{L}\{y\} = \frac{c_1}{s^3} + \frac{c_2}{s^4} + \frac{2w_0}{EIL} \left[\frac{L}{2s^5} - \frac{1}{s^6} + \frac{1}{s^6} e^{-Ls/2} \right]$$

so that

$$y(x) = \frac{1}{2}c_1 x^2 + \frac{1}{6}c_2 x^3 + \frac{2w_0}{EIL} \left[\frac{L}{48}x^4 - \frac{1}{120}x^5 + \frac{1}{120} \left(x - \frac{L}{2} \right)^5 \mathcal{U} \left(x - \frac{L}{2} \right) \right]$$

$$= \frac{1}{2}c_1 x^2 + \frac{1}{6}c_2 x^3 + \frac{w_0}{60EIL} \left[\frac{5L}{2}x^4 - x^5 + \left(x - \frac{L}{2} \right)^5 \mathcal{U} \left(x - \frac{L}{2} \right) \right].$$

To find c_1 and c_2 we compute

$$y''(x) = c_1 + c_2 x + \frac{w_0}{60EIL} \left[30Lx^2 - 20x^3 + 20 \left(x - \frac{L}{2} \right)^3 \mathcal{U} \left(x - \frac{L}{2} \right) \right]$$

and

$$y'''(x) = c_2 + \frac{w_0}{60EIL} \left[60Lx - 60x^2 + 60 \left(x - \frac{L}{2} \right)^2 \mathcal{U} \left(x - \frac{L}{2} \right) \right].$$

Then $y''(L) = y'''(L) = 0$ yields the system

$$c_1 + c_2 L + \frac{w_0}{60EIL} \left[30L^3 - 20L^3 + \frac{5}{2}L^3 \right] = c_1 + c_2 L + \frac{5w_0 L^2}{24EI} = 0$$

$$c_2 + \frac{w_0}{60EIL}[60L^2 - 60L^2 + 15L^2] = c_2 + \frac{w_0 L}{4EI} = 0.$$

Solving for c_1 and c_2 we obtain $c_1 = w_0 L^2/24EI$ and $c_2 = -w_0 L/4EI$. Thus

$$y(x) = \frac{w_0 L^2}{48EI}x^2 - \frac{w_0 L}{24EI} + \frac{w_0}{60EIL} \left[\frac{5L}{2}x^4 - x^5 + \left(x - \frac{L}{2} \right)^5 \mathcal{U} \left(x - \frac{L}{2} \right) \right].$$

58. The differential equation is

$$EI\frac{d^4y}{dx^4} = w_0[1 - \mathcal{U}(x - L/2)].$$

Taking the Laplace transform of both sides and using $y(0) = y'(0) = 0$ we obtain

$$s^4 \mathcal{L}\{y\} - sy''(0) - y'''(0) = \frac{w_0}{EI} \frac{1}{s} \left(1 - e^{-Ls/2} \right).$$

Letting $y''(0) = c_1$ and $y'''(0) = c_2$ we have

$$\mathcal{L}\{y\} = \frac{c_1}{s^3} + \frac{c_2}{s^4} + \frac{w_0}{EI} \frac{1}{s^5} \left(1 - e^{-Ls/2} \right)$$

295

so that
$$y(x) = \frac{1}{2}c_1 x^2 + \frac{1}{6}c_2 x^3 + \frac{1}{24}\frac{w_0}{EI}\left[x^4 - \left(x - \frac{L}{2}\right)^4 \mathcal{U}\left(x - \frac{L}{2}\right)\right].$$

To find c_1 and c_2 we compute
$$y''(x) = c_1 + c_2 x + \frac{1}{2}\frac{w_0}{EI}\left[x^2 - \left(x - \frac{L}{2}\right)^2 \mathcal{U}\left(x - \frac{L}{2}\right)\right].$$

Then $y(L) = y''(L) = 0$ yields the system
$$\frac{1}{2}c_1 L^2 + \frac{1}{6}c_2 L^3 + \frac{1}{24}\frac{w_0}{EI}\left[L^4 - \left(\frac{L}{2}\right)^4\right] = \frac{1}{2}c_1 L^2 + \frac{1}{6}c_2 L^3 + \frac{5w_0}{128EI}L^4 = 0$$

$$c_1 + c_2 L + \frac{1}{2}\frac{w_0}{EI}\left[L^2 - \left(\frac{L}{2}\right)^2\right] = c_1 + c_2 L + \frac{3w_0}{8EI}L^2 = 0.$$

Solving for c_1 and c_2 we obtain $c_1 = \frac{9}{128}w_0 L^2/EI$ and $c_2 = -\frac{57}{128}w_0 L/EI$. Thus
$$y(x) = \frac{w_0}{EI}\left(\frac{9}{256}L^2 x^2 - \frac{19}{256}Lx^3 + \frac{1}{24}x^4 - \frac{1}{24}\left(x - \frac{L}{2}\right)^4 \mathcal{U}\left(x - \frac{L}{2}\right)\right).$$

59. The Laplace transform of the differential equation is
$$-\frac{d}{ds}[s^2 Y - y'(0)] - 2\frac{d}{ds}[sY] + 2Y = \frac{1}{s^2}.$$

Then
$$-s^2\left(\frac{d}{ds}Y\right) - 2sY - 2s\left(\frac{d}{ds}Y\right) - 2Y + 2Y = \frac{1}{s^2}$$

and
$$\frac{d}{ds}Y + \frac{2}{s+2}Y = -\frac{1}{s^3(s+2)}.$$

This is a linear differential equation with integrating factor $(s+2)^2$. Thus we have
$$\frac{d}{ds}[(s+2)^s Y] = -\frac{s+2}{s^3} = -\frac{1}{s^2} - \frac{2}{s^3}$$

$$(s+2)^2 Y = \frac{1}{s} + \frac{1}{s^2} + c = \frac{s^2+1}{s^2} + c_1$$

$$Y = \frac{s^2+1}{s^2(s+2)^2} + \frac{c_1}{(s+2)^2}.$$

Using partial fractions we obtain
$$Y = \frac{1}{4s^2} - \frac{1}{4s} + \frac{5}{4(s+2)^2} + \frac{1}{4(s+2)} + \frac{c_1}{(s+2)^2}$$

so
$$y(t) = \frac{1}{4}t - \frac{1}{4} + \frac{5}{4}te^{-2t} + \frac{1}{4}e^{-2t} + c_1 te^{-2t}$$

$$= \frac{1}{4}t - \frac{1}{4} + \frac{1}{4}e^{-2t} + c_2 te^{-2t}.$$

60. (a) The output for the first three lines of the program are

$$9y[t] + 6y'[t] + y''[t] == t \sin[t]$$

$$1 - 2s + 9Y + s^2Y + 6(-2 + sY) == \frac{2s}{(1+s^2)^2}$$

$$Y \rightarrow -\left(\frac{-11 - 4s - 22s^2 - 4s^3 - 11s^4 - 2s^5}{(1+s^2)^2(9 + 6s + s^2)}\right)$$

The fourth line is the same as the third line with $Y \rightarrow$ removed. The final line of output shows a solution involving complex coefficients of e^{it} and e^{-it}. To get the solution in more standard form write the last line as two lines:

euler = {E^(I t)-> Cos[t] + I Sin[t], E^(-I t)-> Cos[t] - I Sin[t]}
InverseLaplaceTransform[Y,s,t]/.euler//Expand

We see that the solution is

$$y(t) = \left(\frac{487}{250} + \frac{247}{50}t\right)e^{-3t} + \frac{1}{250}(13\cos t - 15t\cos t - 9\sin t + 20t\sin t).$$

(b) The solution is

$$y(t) = \frac{1}{6}e^t - \frac{1}{6}e^{-t/2}\cos\sqrt{15}\,t - \frac{\sqrt{3/5}}{6}e^{-t/2}\sin\sqrt{15}\,t.$$

(c) The solution is

$$q(t) = 1 - \cos t + (6 - 6\cos t)\,\mathscr{U}(t - 3\pi) - (4 + 4\cos t)\,\mathscr{U}(t - \pi).$$

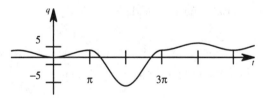

Exercises 7.6
---

1. The Laplace transform of the differential equation yields

$$\mathscr{L}\{y\} = \frac{1}{s-3}e^{-2s}$$

so that

$$y = e^{3(t-2)}\mathscr{U}(t - 2).$$

2. The Laplace transform of the differential equation yields

$$\mathscr{L}\{y\} = \frac{2}{s+1} + \frac{e^{-s}}{s+1}$$

so that

$$y = 2e^{-t} + e^{-(t-1)}\mathscr{U}(t-1).$$

3. The Laplace transform of the differential equation yields

$$\mathscr{L}\{y\} = \frac{1}{s^2+1}\left(1 + e^{-2\pi s}\right)$$

so that

$$y = \sin t + \sin t\,\mathscr{U}(t - 2\pi).$$

4. The Laplace transform of the differential equation yields

$$\mathscr{L}\{y\} = \frac{1}{4}\frac{4}{s^2+16}e^{-2\pi s}$$

so that

$$y = \frac{1}{4}\sin 4(t - 2\pi)\,\mathscr{U}(t - 2\pi).$$

5. The Laplace transform of the differential equation yields

$$\mathscr{L}\{y\} = \frac{1}{s^2+1}\left(e^{-\pi s/2} + e^{-3\pi s/2}\right)$$

so that

$$y = \sin\left(t - \frac{\pi}{2}\right)\mathscr{U}\left(t - \frac{\pi}{2}\right) + \sin\left(t - \frac{3\pi}{2}\right)\mathscr{U}\left(t - \frac{3\pi}{2}\right)$$

$$= -\cos t\,\mathscr{U}\left(t - \frac{\pi}{2}\right) + \cos t\,\mathscr{U}\left(t - \frac{\pi}{2}\right).$$

6. The Laplace transform of the differential equation yields

$$\mathscr{L}\{y\} = \frac{s}{s^2+1} + \frac{1}{s^2+1}(e^{-2\pi s} + e^{-4\pi s})$$

so that

$$y = \cos t + \sin t[\,\mathscr{U}(t - 2\pi) + \mathscr{U}(t - 4\pi)].$$

7. The Laplace transform of the differential equation yields

$$\mathscr{L}\{y\} = \frac{1}{s^2+2s}(1 + e^{-s}) = \left[\frac{1}{2}\frac{1}{s} - \frac{1}{2}\frac{1}{s+2}\right](1 + e^{-s})$$

so that

$$y = \frac{1}{2} - \frac{1}{2}e^{-2t} + \left[\frac{1}{2} - \frac{1}{2}e^{-2(t-1)}\right]\mathscr{U}(t - 1).$$

8. The Laplace transform of the differential equation yields

$$\mathscr{L}\{y\} = \frac{s+1}{s^2(s-2)} + \frac{1}{s(s-2)}e^{-2s} = \frac{3}{4}\frac{1}{s-2} - \frac{3}{4}\frac{1}{s} - \frac{1}{2}\frac{1}{s^2} + \left[\frac{1}{2}\frac{1}{s-2} - \frac{1}{2}\frac{1}{s}\right]e^{-2s}$$

so that
$$y = \frac{3}{4}e^{2t} - \frac{3}{4} - \frac{1}{2}t + \left[\frac{1}{2}e^{2(t-2)} - \frac{1}{2}\right]\mathcal{U}(t-2).$$

9. The Laplace transform of the differential equation yields
$$\mathcal{L}\{y\} = \frac{1}{(s+2)^2+1}e^{-2\pi s}$$

so that
$$y = e^{-2(t-2\pi)}\sin t\, \mathcal{U}(t-2\pi).$$

10. The Laplace transform of the differential equation yields
$$\mathcal{L}\{y\} = \frac{1}{(s+1)^2}e^{-s}$$

so that
$$y = (t-1)e^{-(t-1)}\mathcal{U}(t-1).$$

11. The Laplace transform of the differential equation yields
$$\mathcal{L}\{y\} = \frac{4+s}{s^2+4s+13} + \frac{e^{-\pi s}+e^{-3\pi s}}{s^2+4s+13}$$

$$= \frac{2}{3}\frac{3}{(s+2)^2+3^2} + \frac{s+2}{(s+2)^2+3^2} + \frac{1}{3}\frac{3}{(s+2)^2+3^2}\left(e^{-\pi s}+e^{-3\pi s}\right)$$

so that
$$y = \frac{2}{3}e^{-2t}\sin 3t + e^{-2t}\cos 3t + \frac{1}{3}e^{-2(t-\pi)}\sin 3(t-\pi)\mathcal{U}(t-\pi)$$

$$+ \frac{1}{3}e^{-2(t-3\pi)}\sin 3(t-3\pi)\mathcal{U}(t-3\pi).$$

12. The Laplace transform of the differential equation yields
$$\mathcal{L}\{y\} = \frac{1}{(s-1)^2(s-6)} + \frac{e^{-2s}+e^{-4s}}{(s-1)(s-6)}$$

$$= -\frac{1}{25}\frac{1}{s-1} - \frac{1}{5}\frac{1}{(s-1)^2} + \frac{1}{25}\frac{1}{s-6} + \left[-\frac{1}{5}\frac{1}{s-1} + \frac{1}{5}\frac{1}{s-6}\right]\left(e^{-2s}+e^{-4s}\right)$$

so that
$$y = -\frac{1}{25}e^t - \frac{1}{5}te^t + \frac{1}{25}e^{6t} + \left[-\frac{1}{5}e^{t-2} + \frac{1}{5}e^{6(t-2)}\right]\mathcal{U}(t-2)$$

$$+ \left[-\frac{1}{5}e^{t-4} + \frac{1}{5}e^{6(t-4)}\right]\mathcal{U}(t-4).$$

13. The Laplace transform of the differential equation yields
$$\mathcal{L}\{y\} = \frac{1}{2}\frac{2}{s^3}y''(0) + \frac{1}{6}\frac{3!}{s^4}y'''(0) + \frac{1}{6}\frac{P_0}{EI}\frac{3!}{s^4}e^{-Ls/2}$$

299

so that
$$y = \frac{1}{2}y''(0)x^2 + \frac{1}{6}y'''(0)x^3 + \frac{1}{6}\frac{P_0}{EI}\left(X - \frac{L}{2}\right)^3 \mathcal{U}\left(x - \frac{L}{2}\right).$$

Using $y''(L) = 0$ and $y'''(L) = 0$ we obtain

$$y = \frac{1}{4}\frac{P_0 L}{EI}x^2 - \frac{1}{6}\frac{P_0}{EI}x^3 + \frac{1}{6}\frac{P_0}{EI}\left(x - \frac{L}{2}\right)^3 \mathcal{U}\left(x - \frac{L}{2}\right)$$

$$= \begin{cases} \frac{P_0}{EI}\left(\frac{L}{4}x^2 - \frac{1}{6}x^3\right), & 0 \le x < \frac{L}{2} \\ \frac{P_0 L^2}{4EI}\left(\frac{1}{2}x - \frac{L}{12}\right), & \frac{L}{2} \le x \le L. \end{cases}$$

14. From Problem 13 we know that

$$y = \frac{1}{2}y''(0)x^2 + \frac{1}{6}y'''(0)x^3 + \frac{1}{6}\frac{P_0}{EI}\left(X - \frac{L}{2}\right)^3 \mathcal{U}\left(x - \frac{L}{2}\right).$$

Using $y(L) = 0$ and $y'(L) = 0$ we obtain

$$y = \frac{1}{16}\frac{P_0 L}{EI}x^2 - \frac{1}{12}\frac{P_0}{EI}x^3 + \frac{1}{6}\frac{P_0}{EI}\left(x - \frac{L}{2}\right)^3 \mathcal{U}\left(x - \frac{L}{2}\right)$$

$$= \begin{cases} \frac{P_0}{EI}\left(\frac{L}{16}x^2 - \frac{1}{12}x^3\right), & 0 \le x < \frac{L}{2} \\ \frac{P_0}{EI}\left(\frac{L}{16}x^2 - \frac{1}{12}x^3\right) + \frac{1}{6}\frac{P_0}{EI}\left(x - \frac{L}{2}\right)^3, & \frac{L}{2} \le x \le L. \end{cases}$$

15. The Laplace transform of the differential equation yields

$$\mathcal{L}\{y\} = \frac{1}{s^2 + \omega^2}$$

so that
$$y(t) = \sin \omega t.$$

Note that $y'(0) = 1$, even though the initial condition was $y'(0) = 0$.

Exercises 7.7

1. Taking the Laplace transform of the system gives

$$s\mathcal{L}\{x\} = -\mathcal{L}\{x\} + \mathcal{L}\{y\}$$

$$s\mathcal{L}\{y\} - 1 = 2\mathcal{L}\{x\}$$

so that
$$\mathcal{L}\{x\} = \frac{1}{(s-1)(s+2)} = \frac{1}{3}\frac{1}{s-1} - \frac{1}{3}\frac{1}{s+2}$$

and
$$\mathcal{L}\{y\} = \frac{1}{s} + \frac{2}{s(s-1)(s+2)} = \frac{2}{3}\frac{1}{s-1} + \frac{1}{3}\frac{1}{s+2}.$$

Then
$$x = \frac{1}{3}e^t - \frac{1}{3}e^{-2t} \quad \text{and} \quad y = \frac{2}{3}e^t + \frac{1}{3}e^{-2t}.$$

2. Taking the Laplace transform of the system gives

$$s\mathcal{L}\{x\} - 1 = 2\mathcal{L}\{y\} + \frac{1}{s-1}$$

$$s\mathcal{L}\{y\} - 1 = 8\mathcal{L}\{x\} - \frac{1}{s^2}$$

so that
$$\mathcal{L}\{y\} = \frac{s^3 + 7s^2 - s + 1}{s(s-1)(s^2-16)} = \frac{1}{16}\frac{1}{s} - \frac{8}{15}\frac{1}{s-1} + \frac{173}{96}\frac{1}{s-4} - \frac{53}{160}\frac{1}{s+4}$$

and
$$y = \frac{1}{16} - \frac{8}{15}e^t + \frac{173}{96}e^{4t} - \frac{53}{160}e^{-4t}.$$

Then
$$x = \frac{1}{8}y' + \frac{1}{8}t = \frac{1}{8}t - \frac{1}{15}e^t + \frac{173}{192}e^{4t} + \frac{53}{320}e^{-4t}.$$

3. Taking the Laplace transform of the system gives

$$s\mathcal{L}\{x\} + 1 = \mathcal{L}\{x\} - 2\mathcal{L}\{y\}$$

$$s\mathcal{L}\{y\} - 2 = 5\mathcal{L}\{x\} - \mathcal{L}\{y\}$$

so that
$$\mathcal{L}\{x\} = \frac{-s-5}{s^2+9} = -\frac{s}{s^2+9} - \frac{5}{3}\frac{3}{s^2+9}$$

and
$$x = -\cos 3t - \frac{5}{3}\sin 3t.$$

Then
$$y = \frac{1}{2}x - \frac{1}{2}x' = 2\cos 3t - \frac{7}{3}\sin 3t.$$

4. Taking the Laplace transform of the system gives

$$(s+3)\mathcal{L}\{x\} + s\mathcal{L}\{y\} = \frac{1}{s}$$

$$(s-1)\mathcal{L}\{x\} + (s-1)\mathcal{L}\{y\} = \frac{1}{s-1}$$

so that
$$\mathcal{L}\{y\} = \frac{5s-1}{3s(s-1)^2} = -\frac{1}{3}\frac{1}{s} + \frac{1}{3}\frac{1}{s-1} + \frac{4}{3}\frac{1}{(s-1)^2}$$

and
$$\mathcal{L}\{x\} = \frac{1-2s}{3s(s-1)^2} = \frac{1}{3}\frac{1}{s} - \frac{1}{3}\frac{1}{s-1} - \frac{1}{3}\frac{1}{(s-1)^2}.$$

Then
$$x = \frac{1}{3} - \frac{1}{3}e^t - \frac{1}{3}te^t \quad \text{and} \quad y = -\frac{1}{3} + \frac{1}{3}e^t + \frac{4}{3}te^t.$$

Exercises 7.7

5. Taking the Laplace transform of the system gives

$$(2s - 2)\mathcal{L}\{x\} + s\mathcal{L}\{y\} = \frac{1}{s}$$

$$(s - 3)\mathcal{L}\{x\} + (s - 3)\mathcal{L}\{y\} = \frac{2}{s}$$

so that

$$\mathcal{L}\{x\} = \frac{-s - 3}{s(s - 2)(s - 3)} = -\frac{1}{2}\frac{1}{s} + \frac{5}{2}\frac{1}{s - 2} - \frac{2}{s - 3}$$

and

$$\mathcal{L}\{y\} = \frac{3s - 1}{s(s - 2)(s - 3)} = -\frac{1}{6}\frac{1}{s} - \frac{5}{2}\frac{1}{s - 2} + \frac{8}{3}\frac{1}{s - 3}.$$

Then

$$x = -\frac{1}{2} + \frac{5}{2}e^{2t} - 2e^{3t} \quad \text{and} \quad y = -\frac{1}{6} - \frac{5}{2}e^{2t} + \frac{8}{3}e^{3t}.$$

6. Taking the Laplace transform of the system gives

$$(s + 1)\mathcal{L}\{x\} - (s - 1)\mathcal{L}\{y\} = -1$$

$$s\mathcal{L}\{x\} + (s + 2)\mathcal{L}\{y\} = 1$$

so that

$$\mathcal{L}\{y\} = \frac{s + 1/2}{s^2 + s + 1} = \frac{s + 1/2}{(s + 1/2)^2 + (\sqrt{3}/2)^2}$$

and

$$\mathcal{L}\{x\} = \frac{-3/2}{s^2 + s + 1} = \frac{-3/2}{(s + 1/2)^2 + (\sqrt{3}/2)^2}.$$

Then

$$y = e^{-t/2}\cos\frac{\sqrt{3}}{2}t \quad \text{and} \quad x = e^{-t/2}\sin\frac{\sqrt{3}}{2}t.$$

7. Taking the Laplace transform of the system gives

$$(s^2 + 1)\mathcal{L}\{x\} - \mathcal{L}\{y\} = -2$$

$$-\mathcal{L}\{x\} + (s^2 + 1)\mathcal{L}\{y\} = 1$$

so that

$$\mathcal{L}\{x\} = \frac{-2s^2 - 1}{s^4 + 2s^2} = -\frac{1}{2}\frac{1}{s^2} - \frac{3}{2}\frac{1}{s^2 + 2}$$

and

$$x = -\frac{1}{2}t - \frac{3}{2\sqrt{2}}\sin\sqrt{2}t.$$

Then

$$y = x'' + x = -\frac{1}{2}t + \frac{3}{2\sqrt{2}}\sin\sqrt{2}t.$$

8. Taking the Laplace transform of the system gives

$$(s+1)\mathcal{L}\{x\} + \mathcal{L}\{y\} = 1$$

$$4\mathcal{L}\{x\} - (s+1)\mathcal{L}\{y\} = 1$$

so that

$$\mathcal{L}\{x\} = \frac{s+2}{s^2 + 2s + 5} = \frac{s+1}{(s+1)^2 + 2^2} + \frac{1}{2}\frac{2}{(s+1)^2 + 2^2}$$

and

$$\mathcal{L}\{y\} = \frac{-s+3}{s^2 + 2s + 5} = -\frac{s+1}{(s+1)^2 + 2^2} + 2\frac{2}{(s+1)^2 + 2^2}.$$

Then

$$x = e^{-t}\cos 2t + \frac{1}{2}e^{-t}\sin 2t \quad \text{and} \quad y = -e^{-t}\cos 2t + 2e^{-t}\sin 2t.$$

9. Adding the equations and then subtracting them gives

$$\frac{d^2x}{dt^2} = \frac{1}{2}t^2 + 2t$$

$$\frac{d^2y}{dt^2} = \frac{1}{2}t^2 - 2t.$$

Taking the Laplace transform of the system gives

and

$$\mathcal{L}\{x\} = 8\frac{1}{s} + \frac{1}{24}\frac{4!}{s^5} + \frac{1}{3}\frac{3!}{s^4}$$

$$\mathcal{L}\{y\} = \frac{1}{24}\frac{4!}{s^5} - \frac{1}{3}\frac{3!}{s^4}$$

so that

$$x = 8 + \frac{1}{24}t^4 + \frac{1}{3}t^3 \quad \text{and} \quad y = \frac{1}{24}t^4 - \frac{1}{3}t^3.$$

10. Taking the Laplace transform of the system gives

$$(s-4)\mathcal{L}\{x\} + s^3\mathcal{L}\{y\} = \frac{6}{s^2 + 1}$$

$$(s+2)\mathcal{L}\{x\} - 2s^3\mathcal{L}\{y\} = 0$$

so that $\quad \mathcal{L}\{x\} = \dfrac{4}{(s-2)(s^2 + 1)} = \dfrac{4}{5}\dfrac{1}{s-2} - \dfrac{4}{5}\dfrac{s}{s^2 + 1} - \dfrac{8}{5}\dfrac{1}{s^2 + 1}$

and $\quad \mathcal{L}\{y\} = \dfrac{2s + 4}{s^3(s-2)(s^2 + 1)} = \dfrac{1}{s} - \dfrac{2}{s^2} - 2\dfrac{2}{s^3} + \dfrac{1}{5}\dfrac{1}{s-2} - \dfrac{6}{5}\dfrac{s}{s^2 + 1} + \dfrac{8}{5}\dfrac{1}{s^2 + 1}.$

Then

$$x = \frac{4}{5}e^{2t} - \frac{4}{5}\cos t - \frac{8}{5}\sin t$$

and

$$y = 1 - 2t - 2t^2 + \frac{1}{5}e^{2t} - \frac{6}{5}\cos t + \frac{8}{5}\sin t.$$

11. Taking the Laplace transform of the system gives

$$s^2 \mathcal{L}\{x\} + 3(s+1)\mathcal{L}\{y\} = 2$$

$$s^2 \mathcal{L}\{x\} + 3\mathcal{L}\{y\} = \frac{1}{(s+1)^2}$$

so that
$$\mathcal{L}\{x\} = -\frac{2s+1}{s^3(s+1)} = \frac{1}{s} + \frac{1}{s^2} + \frac{1}{2}\frac{2}{s^3} - \frac{1}{s+1}.$$

Then
$$x = 1 + t + \frac{1}{2}t^2 - e^{-t}$$

and
$$y = \frac{1}{3}te^{-t} - \frac{1}{3}x'' = \frac{1}{3}te^{-t} + \frac{1}{3}e^{-t} - \frac{1}{3}.$$

12. Taking the Laplace transform of the system gives

$$(s-4)\,\mathcal{L}\{x\} + 2\mathcal{L}\{y\} = \frac{2e^{-s}}{s}$$

$$-3\,\mathcal{L}\{x\} + (s+1)\,\mathcal{L}\{y\} = \frac{1}{2} + \frac{e^{-s}}{s}$$

so that
$$\mathcal{L}\{x\} = \frac{-1/2}{(s-1)(s-2)} + e^{-s}\frac{1}{(s-1)(s-2)}$$

$$= \left[\frac{1}{2}\frac{1}{s-1} - \frac{1}{2}\frac{1}{s-2}\right] + e^{-s}\left[-\frac{1}{s-1} + \frac{1}{s-2}\right]$$

and
$$\mathcal{L}\{y\} = \frac{e^{-s}}{s} + \frac{s/4-1}{(s-1)(s-2)} + e^{-s}\frac{-s/2+2}{(s-1)(s-2)}$$

$$= \frac{3}{4}\frac{1}{s-1} - \frac{1}{2}\frac{1}{s-2} + e^{-s}\left[\frac{1}{s} - \frac{3}{2}\frac{1}{s-1} + \frac{1}{s-2}\right].$$

Then
$$x = \frac{1}{2}e^t - \frac{1}{2}e^{2t} + \left[-e^{t-1} + e^{2(t-1)}\right]\mathcal{U}(t-1)$$

and
$$y = \frac{3}{4}e^t - \frac{1}{2}e^{2t} + \left[1 - \frac{3}{2}e^{t-1} + e^{2(t-1)}\right]\mathcal{U}(t-1).$$

13. The system is

$$x_1'' = -3x_1 + 2(x_2 - x_1)$$

$$x_2'' = -2(x_2 - x_1)$$

$$x_1(0) = 0$$

$$x_1'(0) = 1$$

$$x_2(0) = 1$$

$$x_2'(0) = 0.$$

Taking the Laplace transform of the system gives

$$(s^2 + 5)\mathscr{L}\{x_1\} - 2\mathscr{L}\{x_2\} = 1$$

$$-2\mathscr{L}\{x_1\} + (s^2 + 2)\mathscr{L}\{x_2\} = s$$

so that $\quad \mathscr{L}\{x_1\} = \dfrac{s^2 + 2s + 2}{s^4 + 7s^2 + 6} = \dfrac{2}{5}\dfrac{s}{s^2+1} + \dfrac{1}{5}\dfrac{1}{s^2+1} - \dfrac{2}{5}\dfrac{s}{s^2+6} + \dfrac{4}{5\sqrt{6}}\dfrac{\sqrt{6}}{s^2+6}$

and $\quad \mathscr{L}\{x_2\} = \dfrac{s^3 + 5s + 2}{(s^2+1)(s^2+6)} = \dfrac{4}{5}\dfrac{s}{s^2+1} + \dfrac{2}{5}\dfrac{1}{s^2+1} + \dfrac{1}{5}\dfrac{s}{s^2+6} - \dfrac{2}{5\sqrt{6}}\dfrac{\sqrt{6}}{s^2+6}.$

Then $\qquad\qquad x_1 = \dfrac{2}{5}\cos t + \dfrac{1}{5}\sin t - \dfrac{2}{5}\cos\sqrt{6}\,t + \dfrac{4}{5\sqrt{6}}\sin\sqrt{6}\,t$

and $\qquad\qquad x_2 = \dfrac{4}{5}\cos t + \dfrac{2}{5}\sin t + \dfrac{1}{5}\cos\sqrt{6}\,t - \dfrac{2}{5\sqrt{6}}\sin\sqrt{6}\,t.$

14. In this system x_1 and x_2 represent displacements of masses m_1 and m_2 from their equilibrium positions. Since the net forces acting on m_1 and m_2 are

$$-k_1x_1 + k_2(x_2 - x_1) \quad\text{and}\quad -k_2(x_2 - x_1) - k_3x_2,$$

respectively, Newton's second law of motion gives

$$m_1 x_1'' = -k_1 x_1 + k_2(x_2 - x_1)$$

$$m_2 x_2'' = -k_2(x_2 - x_1) - k_3 x_2.$$

Using $k_1 = k_2 = k_3 = 1$, $m_1 = m_2 = 1$, $x_1(0) = 0$, $x_1(0) = -1$, $x_2(0) = 0$, and $x_2'(0) = 1$, and taking the Laplace transform of the system, we obtain

$$(2 + s^2)\mathscr{L}\{x_1\} - \mathscr{L}\{x_2\} = -1$$

$$\mathscr{L}\{x_1\} - (2 + s^2)\mathscr{L}\{x_2\} = -1$$

so that
$$\mathscr{L}\{x_1\} = -\frac{1}{s^2 + 3} \quad \text{and} \quad \mathscr{L}\{x_2\} = \frac{1}{s^2 + 3}.$$

Then
$$x_1 = -\frac{1}{\sqrt{3}} \sin \sqrt{3}\,t \quad \text{and} \quad x_2 = \frac{1}{\sqrt{3}} \sin \sqrt{3}\,t.$$

15. **(a)** By Kirchoff's first law we have $i_1 = i_2 + i_3$. By Kirchoff's second law, on each loop we have $E(t) = Ri_1 + L_1 i_2'$ and $E(t) = Ri_1 + L_2 i_3'$ or $L_1 i_2' + Ri_2 + Ri_3 = E(t)$ and $L_2 i_3' + Ri_2 + Ri_3 = E(t)$.

(b) Taking the Laplace transform of the system
$$0.01 i_2' + 5 i_2 + 5 i_3 = 100$$

$$0.0125 i_3' + 5 i_2 + 5 i_3 = 100$$

gives
$$(s + 500)\,\mathscr{L}\{i_2\} + 500 \mathscr{L}\{i_3\} = \frac{10,000}{s}$$

$$400 \mathscr{L}\{i_2\} + (s + 400)\,\mathscr{L}\{i_3\} = \frac{8,000}{s}$$

so that
$$\mathscr{L}\{i_3\} = \frac{8,000}{s^2 + 900s} = \frac{80}{9}\frac{1}{s} - \frac{80}{9}\frac{1}{s + 900}.$$

Then
$$i_3 = \frac{80}{9} - \frac{80}{9} e^{-900t} \quad \text{and} \quad i_2 = 20 - 0.0025 i_3' - i_3 = \frac{100}{9} - \frac{100}{9} e^{-900t}.$$

(c) $i_1 = i_2 + i_3 = 20 - 20 e^{-900t}$

16. **(a)** Taking the Laplace transform of the system
$$i_2' + i_3' + 10 i_2 = 120 - 120\,\mathscr{U}(t - 2)$$

$$-10 i_2' + 5 i_3' + 5 i_3 = 0$$

gives
$$(s + 10)\,\mathscr{L}\{i_2\} + s \mathscr{L}\{i_3\} = \frac{120}{s}\left(1 - e^{-2s}\right)$$

$$-10s \mathscr{L}\{i_2\} + 5(s + 1)\,\mathscr{L}\{i_3\} = 0$$

so that
$$\mathscr{L}\{i_2\} = \frac{120(s + 1)}{(3s^2 + 11s + 10)s}\left(1 - e^{-2s}\right) = \left[\frac{48}{s + 5/3} - \frac{60}{s + 2} + \frac{12}{s}\right]\left(1 - e^{-2s}\right)$$

and
$$\mathscr{L}\{i_3\} = \frac{240}{3s^2 + 11s + 10}\left(1 - e^{-2s}\right) = \left[\frac{240}{s + 5/3} - \frac{240}{s + 2}\right]\left(1 - e^{-2s}\right).$$

306

Then $\quad i_2 = 12 + 48e^{-5t/3} - 60e^{-2t} - \left[12 + 48e^{-5(t-2)/3} - 60e^{-2(t-2)}\right]\mathcal{U}(t-2)$

and $\quad i_3 = 240e^{-5t/3} - 240e^{-2t} - \left[240e^{-5(t-2)/3} - 240e^{-2(t-2)}\right]\mathcal{U}(t-2).$

(b) $i_1 = i_2 + i_3 = 12 + 288e^{-5t/3} - 300e^{-2t} - \left[12 + 288e^{-5(t-2)/3} - 300e^{-2(t-2)}\right]\mathcal{U}(t-2)$

17. Taking the Laplace transform of the system

$$i_2' + 11i_2 + 6i_3 = 50\sin t$$

$$i_3' + 6i_2 + 6i_3 = 50\sin t$$

gives $$(s+11)\mathcal{L}\{i_2\} + 6\mathcal{L}\{i_3\} = \frac{50}{s^2+1}$$

$$6\mathcal{L}\{i_2\} + (s+6)\mathcal{L}\{i_3\} = \frac{50}{s^2+1}$$

so that

$$\mathcal{L}\{i_2\} = \frac{50s}{(s+2)(s+15)(s^2+1)} = -\frac{20}{13}\frac{1}{s+2} + \frac{375}{1469}\frac{1}{s+15} + \frac{145}{113}\frac{s}{s^2+1} + \frac{85}{113}\frac{1}{s^2+1}.$$

Then $$i_2 = -\frac{20}{13}e^{-2t} + \frac{375}{1469}e^{-15t} + \frac{145}{113}\cos t + \frac{85}{113}\sin t$$

and $$i_3 = \frac{25}{3}\sin t - \frac{1}{6}i_2' - \frac{11}{6}i_2 = \frac{30}{13}e^{-2t} + \frac{250}{1469}e^{-15t} - \frac{280}{113}\cos t + \frac{810}{113}\sin t.$$

18. Taking the Laplace transform of the system

$$0.5i_1' + 50i_2 = 60$$

$$0.005i_2' + i_2 - i_1 = 0$$

gives $$s\,\mathcal{L}\{i_1\} + 100\,\mathcal{L}\{i_2\} = \frac{120}{s}$$

$$-200\,\mathcal{L}\{i_1\} + (s+200)\,\mathcal{L}\{i_2\} = 0$$

so that

$$\mathcal{L}\{i_2\} = \frac{24{,}000}{s(s^2+200s+20{,}000)} = \frac{6}{5}\frac{1}{s} - \frac{6}{5}\frac{s+100}{(s+100)^2+100^2} - \frac{6}{5}\frac{100}{(s+100)^2+100^2}.$$

Then $$i_2 = \frac{6}{5} - \frac{6}{5}e^{-100t}\cos 100t - \frac{6}{5}e^{-100t}\sin 100t$$

and $$i_1 = 0.005i_2' + i_2 = \frac{6}{5} - \frac{6}{5}e^{-100t}\cos 100t.$$

307

19. Taking the Laplace transform of the system

$$2i_1' + 50i_2 = 60$$

$$0.005i_2' + i_2 - i_1 = 0$$

gives

$$2s\mathcal{L}\{i_1\} + 50\mathcal{L}\{i_2\} = \frac{60}{s}$$

$$-200\mathcal{L}\{i_1\} + (s + 200)\mathcal{L}\{i_2\} = 0$$

so that

$$\mathcal{L}\{i_2\} = \frac{6{,}000}{s(s^2 + 200s + 5{,}000)}$$

$$= \frac{6}{5}\frac{1}{s} - \frac{6}{5}\frac{s + 100}{(s + 100)^2 - (50\sqrt{2})^2} - \frac{6\sqrt{2}}{5}\frac{50\sqrt{2}}{(s + 100)^2 - (50\sqrt{2})^2}.$$

Then

$$i_2 = \frac{6}{5} - \frac{6}{5}e^{-100t}\cosh 50\sqrt{2}\,t - \frac{6\sqrt{2}}{5}e^{-100t}\sinh 50\sqrt{2}\,t$$

and

$$i_1 = 0.005i_2' + i_2 = \frac{6}{5} - \frac{6}{5}e^{-100t}\cosh 50\sqrt{2}\,t - \frac{9\sqrt{2}}{10}e^{-100t}\sinh 50\sqrt{2}\,t.$$

20. (a) Using Kirchoff's first law we write $i_1 = i_2 + i_3$. Since $i_2 = dq/dt$ we have $i_1 - i_3 = dq/dt$. Using Kirchoff's second law and summing the voltage drops across the shorter loop gives

$$E(t) = iR_1 + \frac{1}{C}q, \tag{1}$$

so that

$$i_1 = \frac{1}{R_1}E(t) - \frac{1}{R_1 C}q.$$

Then

$$\frac{dq}{dt} = i_1 - i_3 = \frac{1}{R_1}E(t) - \frac{1}{R_1 C}q - i_3$$

and

$$R_1\frac{dq}{dt} + \frac{1}{C}q + R_1 i_3 = E(t).$$

Summing the voltage drops across the longer loop gives

$$E(t) = i_1 R_1 + L\frac{di_3}{dt} + R_2 i_3.$$

Combining this with (1) we obtain

$$i_1 R_1 + L\frac{di_3}{dt} + R_2 i_3 = i_1 R_1 + \frac{1}{C}q$$

or

$$L\frac{di_3}{dt} + R_2 i_3 - \frac{1}{C}q = 0.$$

308

(b) Using $L = R_1 = R_2 = C - 1$, $E(t) - 50e^{-t}\mathcal{U}(t \quad 1) - 50e^{-1}e^{-(t-1)}\mathcal{U}(t - 1)$, $q(0) = i_3(0) = 0$, and taking the Laplace transform of the system we obtain

$$(s + 1)\mathcal{L}\{q\} + \mathcal{L}\{i_3\} = \frac{50e^{-1}}{s + 1}e^{-s}$$

$$(s + 1)\mathcal{L}\{i_3\} - \mathcal{L}\{q\} = 0,$$

so that
$$\mathcal{L}\{q\} = \frac{50e^{-1}e^{-s}}{(s + 1)^2 + 1}$$

and $\quad q(t) = 50e^{-1}e^{-(t-1)}\sin(t - 1)\mathcal{U}(t - 1) = 50e^{-t}\sin(t - 1)\mathcal{U}(t - 1)$.

21. Taking the Laplace transform of the system

$$4\theta_1'' + \theta_2'' + 8\theta_1 = 0$$

$$\theta_1'' + \theta_2'' + 2\theta_2 = 0$$

gives
$$4\left(s^2 + 2\right)\mathcal{L}\{\theta_1\} + s^2\mathcal{L}\{\theta_2\} = 3s$$

$$s^2\mathcal{L}\{\theta_1\} + \left(s^2 + 2\right)\mathcal{L}\{\theta_2\} = 0$$

so that
$$\left(3s^2 + 4\right)\left(s^2 + 4\right)\mathcal{L}\{\theta_2\} = -3s^3$$

or
$$\mathcal{L}\{\theta_2\} = \frac{1}{2}\frac{s}{s^2 + 4/3} - \frac{3}{2}\frac{s}{s^2 + 4}.$$

Then
$$\theta_2 = \frac{1}{2}\cos\frac{2}{\sqrt{3}}t - \frac{3}{2}\cos 2t \quad \text{and} \quad \theta_1'' = -\theta_2'' - 2\theta_2$$

so that
$$\theta_1 = \frac{1}{4}\cos\frac{2}{\sqrt{3}}t + \frac{3}{4}\cos 2t.$$

Chapter 7 Review Exercises

1. $\mathcal{L}\{f(t)\} = \int_0^1 te^{-st}dt + \int_1^\infty (2-t)e^{-st}dt = \dfrac{1}{s^2} - \dfrac{2}{s^2}e^{-s}$

2. $\mathcal{L}\{f(t)\} = \int_2^4 e^{-st}dt = \dfrac{1}{s}\left(e^{-2s} - e^{-4s}\right)$

3. False; consider $f(t) = t^{-1/2}$.

4. False, since $f(t) = (e^t)^{10} = e^{10t}$.

5. True, since $\lim_{s\to\infty} F(s) = 1 \neq 0$. (See Theorem 7.4 in the text.)

6. False; consider $f(t) = 1$ and $g(t) = 1$.

7. $\mathcal{L}\{e^{-7t}\} = \dfrac{1}{s+7}$

8. $\mathcal{L}\{te^{-7t}\} = \dfrac{1}{(s+7)^2}$

9. $\mathcal{L}\{\sin 2t\} = \dfrac{2}{s^2+4}$

10. $\mathcal{L}\{e^{-3t}\sin 2t\} = \dfrac{2}{(s+3)^2+4}$

11. $\mathcal{L}\{t\sin 2t\} = -\dfrac{d}{ds}\left[\dfrac{2}{s^2+4}\right] = \dfrac{4s}{(s^2+4)^2}$

12. $\mathcal{L}\{\sin 2t\,\mathcal{U}(t-\pi)\} = \mathcal{L}\{\sin 2(t-\pi)\mathcal{U}(t-\pi)\} = \dfrac{2}{s^2+4}e^{-\pi s}$

13. $\mathcal{L}^{-1}\left\{\dfrac{20}{s^6}\right\} = \mathcal{L}^{-1}\left\{\dfrac{1}{6}\dfrac{5!}{s^6}\right\} = \dfrac{1}{6}t^5$

14. $\mathcal{L}^{-1}\left\{\dfrac{1}{3s-1}\right\} = \mathcal{L}^{-1}\left\{\dfrac{1}{3}\dfrac{1}{s-1/3}\right\} = \dfrac{1}{3}e^{t/3}$

15. $\mathcal{L}^{-1}\left\{\dfrac{1}{(s-5)^3}\right\} = \mathcal{L}^{-1}\left\{\dfrac{1}{2}\dfrac{2}{(s-5)^3}\right\} = \dfrac{1}{2}t^2e^{5t}$

16. $\mathcal{L}^{-1}\left\{\dfrac{1}{s^2-5}\right\} = \mathcal{L}^{-1}\left\{-\dfrac{1}{\sqrt{5}}\dfrac{1}{s+\sqrt{5}} + \dfrac{1}{\sqrt{5}}\dfrac{1}{s-\sqrt{5}}\right\} = -\dfrac{1}{\sqrt{5}}e^{-\sqrt{5}t} + \dfrac{1}{\sqrt{5}}e^{\sqrt{5}t}$

17. $\mathcal{L}^{-1}\left\{\dfrac{s}{s^2-10s+29}\right\} = \mathcal{L}^{-1}\left\{\dfrac{s-5}{(s-5)^2+2^2} + \dfrac{5}{2}\dfrac{2}{(s-5)^2+2^2}\right\} = e^{5t}\cos 2t + \dfrac{5}{2}e^{5t}\sin 2t$

18. $\mathcal{L}^{-1}\left\{\dfrac{1}{s^2}e^{-5s}\right\} = (t-5)\mathcal{U}(t-5)$

19. $\mathcal{L}^{-1}\left\{\dfrac{s+\pi}{s^2+\pi^2}\,e^{-s}\right\} = \mathcal{L}^{-1}\left\{\dfrac{s}{s^2+\pi^2}\,e^{-s} + \dfrac{\pi}{s^2+\pi^2}\,e^{-s}\right\}$

$\qquad\qquad = \cos\pi(t-1)\mathcal{U}(t-1) + \sin\pi(t-1)\mathcal{U}(t-1)$

20. $\mathcal{L}^{-1}\left\{\dfrac{1}{L^2 s^2 + n^2\pi^2}\right\} = \dfrac{1}{L^2}\dfrac{L}{n\pi}\mathcal{L}^{-1}\left\{\dfrac{n\pi/L}{s^2+(n^2\pi^2)/L^2}\right\} = \dfrac{1}{Ln\pi}\sin\dfrac{n\pi}{L}t$

21. $\mathcal{L}\left\{e^{-5t}\right\}$ exists for $s > -5$.

22. $\mathcal{L}\left\{te^{8t}f(t)\right\} = -\dfrac{d}{ds}F(s-8)$.

23. $\mathcal{L}\{e^{at}f(t-k)\,\mathcal{U}(t-k)\} = e^{-ks}\,\mathcal{L}\{e^{a(t+k)}f(t)\} = e^{-ks}e^{ak}\,\mathcal{L}\{e^{at}f(t)\}$

$\qquad\qquad = e^{-k(s-a)}F(s-a)$

24. $1 * 1 = \displaystyle\int_0^t d\tau = t$

25. (a) $f(t) = t - [(t-1)+1]\mathcal{U}(t-1) + \mathcal{U}(t-1) - \mathcal{U}(t-4) = t - (t-1)\mathcal{U}(t-1) - \mathcal{U}(t-4)$

(b) $\mathcal{L}\{f(t)\} = \dfrac{1}{s^2} - \dfrac{1}{s^2}e^{-s} - \dfrac{1}{s}e^{-4s}$

(c) $\mathcal{L}\{e^t f(t)\} = \dfrac{1}{(s-1)^2} - \dfrac{1}{(s-1)^2}e^{-(s-1)} - \dfrac{1}{s-1}e^{-4(s-1)}$

26. (a) $f(t) = \sin t\,\mathcal{U}(t-\pi) - \sin t\,\mathcal{U}(t-3\pi) = -\sin(t-\pi)\mathcal{U}(t-\pi) + \sin(t-3\pi)\mathcal{U}(t-3\pi)$

(b) $\mathcal{L}\{f(t)\} = -\dfrac{1}{s^2+1}e^{-\pi s} + \dfrac{1}{s^2+1}e^{-3\pi s}$

(c) $\mathcal{L}\{e^t f(t)\} = -\dfrac{1}{(s-1)^2+1}e^{-\pi(s-1)} + \dfrac{1}{(s-1)^2+1}e^{-3\pi(s-1)}$

27. (a) $f(t) = 2 - 2\mathcal{U}(t-2) + [(t-2)+2]\mathcal{U}(t-2) = 2 + (t-2)\mathcal{U}(t-2)$

(b) $\mathcal{L}\{f(t)\} = \dfrac{2}{s} + \dfrac{1}{s^2}e^{-2s}$

(c) $\mathcal{L}\{e^t f(t)\} = \dfrac{2}{s-1} + \dfrac{1}{(s-1)^2}e^{-2(s-1)}$

28. (a) $f(t) = t - t\mathcal{U}(t-1) + (2-t)\mathcal{U}(t-1) - (2-t)\mathcal{U}(t-2)$

$\qquad\qquad = t - 2(t-1)\mathcal{U}(t-1) + (t-2)\mathcal{U}(t-2)$

(b) $\mathcal{L}\{f(t)\} = \dfrac{1}{s^2} - \dfrac{2}{s^2}e^{-s} + \dfrac{1}{s^2}e^{-2s}$

(c) $\mathcal{L}\{e^t f(t)\} = \dfrac{1}{(s-1)^2} - \dfrac{2}{(s-1)^2}e^{-(s-1)} + \dfrac{1}{(s-1)^2}e^{-2(s-1)}$

29. Taking the Laplace transform of the differential equation we obtain

$$\mathcal{L}\{y\} = \frac{5}{(s-1)^2} + \frac{1}{2}\frac{2}{(s-1)^3}$$

so that

$$y = 5te^t + \frac{1}{2}t^2 e^t.$$

30. Taking the Laplace transform of the differential equation we obtain

$$\mathcal{L}\{y\} = \frac{1}{(s-1)^2(s^2 - 8s + 20)}$$

$$= \frac{6}{169}\frac{1}{s-1} + \frac{1}{13}\frac{1}{(s-1)^2} - \frac{6}{169}\frac{s-4}{(s-4)^2 + 2^2} + \frac{5}{338}\frac{2}{(s-4)^2 + 2^2}$$

so that

$$y = \frac{6}{169}e^t + \frac{1}{13}te^t - \frac{6}{169}e^{4t}\cos 2t + \frac{5}{338}e^{4t}\sin 2t.$$

31. Taking the Laplace transform of the given differential equation we obtain

$$\mathcal{L}\{y\} = \frac{30}{s(s^2 - 4s + 6)}e^{-\pi s} = \left(5\cdot\frac{1}{s} - 5\cdot\frac{s-2}{(s-2)^2 + 2} + 5\sqrt{2}\cdot\frac{\sqrt{2}}{(s-2)^2 + 2}\right)e^{-\pi s}$$

so that

$$y = 5\,\mathcal{U}(t-\pi) - 5e^{2(t-\pi)}\cos\sqrt{2}\,(t-\pi)\,\mathcal{U}(t-\pi) + 5\sqrt{2}\,e^{2(t-\pi)}\sin\sqrt{2}\,(t-\pi)\,\mathcal{U}(t-\pi).$$

32. Taking the Laplace transform of the given differential equation we obtain

$$\mathcal{L}\{y\} = \frac{s^3 + 6s^2 + 1}{s^2(s+1)(s+5)} - \frac{1}{s^2(s+1)(s+5)}e^{-2s} - \frac{2}{s(s+1)(s+5)}e^{-2s}$$

$$= -\frac{6}{25}\cdot\frac{1}{s} + \frac{1}{5}\cdot\frac{1}{s^2} + \frac{3}{2}\cdot\frac{1}{s+1} - \frac{13}{50}\cdot\frac{1}{s+5}$$

$$- \left(-\frac{6}{25}\cdot\frac{1}{s} + \frac{1}{5}\cdot\frac{1}{s^2} + \frac{1}{4}\cdot\frac{1}{s+1} - \frac{1}{100}\cdot\frac{1}{s+5}\right)e^{-2s}$$

$$- \left(\frac{2}{5}\cdot\frac{1}{5} - \frac{1}{2}\cdot\frac{1}{s+1} + \frac{1}{10}\cdot\frac{1}{s+5}\right)e^{-2s}$$

so that

$$y = -\frac{6}{25} + \frac{1}{5}t^2 + \frac{3}{2}e^{-t} - \frac{13}{50}e^{-5t} - \frac{4}{25}\mathcal{U}(t-2) - \frac{1}{5}(t-2)^2\mathcal{U}(t-2)$$

$$+ \frac{1}{4}e^{-(t-2)}\mathcal{U}(t-2) - \frac{9}{100}e^{-5(t-2)}\mathcal{U}(t-2).$$

33. Taking the Laplace transform of the differential equation we obtain

$$\mathscr{L}\{y\} = \frac{s^3 + 2}{s^3(s-5)} - \frac{2 + 2s + s^2}{s^3(s-5)} e^{-s}$$

$$= -\frac{2}{125}\frac{1}{s} - \frac{2}{25}\frac{1}{s^2} - \frac{1}{5}\frac{2}{s^3} + \frac{127}{125}\frac{1}{s-5} - \left[-\frac{37}{125}\frac{1}{s} - \frac{12}{25}\frac{1}{s^2} - \frac{1}{5}\frac{2}{s^3} + \frac{37}{125}\frac{1}{s-5}\right] e^{-s}$$

so that

$$y = -\frac{2}{125} - \frac{2}{25}t - \frac{1}{5}t^2 + \frac{127}{125}e^{5t} - \left[-\frac{37}{125} - \frac{12}{25}(t-1) - \frac{1}{5}(t-1)^2 + \frac{37}{125}e^{5(t-1)}\right]\mathscr{U}(t-1).$$

34. Taking the Laplace transform of the integral equation we obtain

$$\mathscr{L}\{f\} = \frac{s+3}{s(s+5)} = \frac{3}{5}\frac{1}{s} + \frac{2}{5}\frac{1}{s+5}$$

so that

$$f(t) = \frac{3}{5} + \frac{2}{5}e^{-5t}.$$

35. Taking the Laplace transform of the integral equation we obtain

$$\mathscr{L}\{y\} = \frac{1}{s} + \frac{1}{s^2} + \frac{1}{2}\frac{2}{s^3}$$

so that

$$y(t) = 1 + t + \frac{1}{2}t^2.$$

36. Taking the Laplace transform of the integral equation we obtain

$$(\mathscr{L}\{f\})^2 = 6 \cdot \frac{6}{s^4} \quad \text{or} \quad \mathscr{L}\{f\} = \pm 6 \cdot \frac{1}{s^2}$$

so that $f(t) = \pm 6t$.

37. Taking the Laplace transform of the system gives

$$s\mathscr{L}\{x\} + \mathscr{L}\{y\} = \frac{1}{s^2} + 1$$

$$4\mathscr{L}\{x\} + s\mathscr{L}\{y\} = 2$$

so that

$$\mathscr{L}\{x\} = \frac{s^2 - 2s + 1}{s(s-2)(s+2)} = -\frac{1}{4}\frac{1}{s} + \frac{1}{8}\frac{1}{s-2} + \frac{9}{8}\frac{1}{s+2}.$$

Then

$$x = -\frac{1}{4} + \frac{1}{8}e^{2t} + \frac{9}{8}e^{-2t} \quad \text{and} \quad y = -x' + t = \frac{9}{4}e^{-2t} - \frac{1}{4}e^{2t} + t.$$

38. Taking the Laplace transform of the system gives

$$s^2\mathscr{L}\{x\} + s^2\mathscr{L}\{y\} = \frac{1}{s-2}$$

$$2s\mathscr{L}\{x\} + s^2\mathscr{L}\{y\} = -\frac{1}{s-2}$$

so that
$$\mathcal{L}\{x\} = \frac{2}{s(s-2)^2} = \frac{1}{2}\frac{1}{s} - \frac{1}{2}\frac{1}{s-2} + \frac{1}{(s-2)^2}$$

and
$$\mathcal{L}\{y\} = \frac{-s-2}{s^2(s-2)^2} = -\frac{3}{4}\frac{1}{s} - \frac{1}{2}\frac{1}{s^2} + \frac{3}{4}\frac{1}{s-2} - \frac{1}{(s-2)^2}.$$

Then
$$x = \frac{1}{2} - \frac{1}{2}e^{2t} + te^{2t} \quad \text{and} \quad y = -\frac{3}{4} - \frac{1}{2}t + \frac{3}{4}e^{2t} - te^{2t}.$$

39. The integral equation is

$$10i + 2\int_0^t i(\tau)\,d\tau = 2t^2 + 2t.$$

Taking the Laplace transform we obtain

$$\mathcal{L}\{i\} = \left(\frac{4}{s^3} + \frac{2}{s^2}\right)\frac{s}{10s+2} = \frac{s+2}{s^2(5s+2)} = -\frac{9}{s} + \frac{2}{s^2} + \frac{45}{5s+1} = -\frac{9}{s} + \frac{2}{s^2} + \frac{9}{s+1/5}.$$

Thus
$$i(t) = -9 + 2t + 9e^{-t/5}.$$

40. The differential equation is

$$\frac{1}{2}\frac{d^2q}{dt^2} + 10\frac{dq}{dt} + 100q = 10 - 10\,\mathcal{U}(t-5).$$

Taking the Laplace transform we obtain

$$\mathcal{L}\{q\} = \frac{20}{2(s^2+20s+200)}\left(1 - e^{-5s}\right)$$

$$= \left[\frac{1}{10}\frac{1}{s} - \frac{1}{10}\frac{s+10}{(s+10)^2 + 10^2} - \frac{1}{10}\frac{10}{(s+10)^2 + 10^2}\right]\left(1 - e^{-5s}\right)$$

so that

$$q(t) = \frac{1}{10} - \frac{1}{10}e^{-10t}\cos 10t - \frac{1}{10}e^{-10t}\sin 10t$$

$$- \left[\frac{1}{10} - \frac{1}{10}e^{-10(t-5)}\cos 10(t-5) - \frac{1}{10}e^{-10(t-5)}\sin 10(t-5)\right]\mathcal{U}(t-5).$$

41. Taking the Laplace transform of the given differential equation we obtain

$$\mathcal{L}\{y\} = \frac{2w_0}{EIL}\left(\frac{L}{48}\cdot\frac{4!}{s^5} - \frac{1}{120}\cdot\frac{5!}{s^6} + \frac{1}{120}\cdot\frac{5!}{s^6}e^{-sL/2}\right) + \frac{c_1}{2}\cdot\frac{2!}{s^3} + \frac{c_2}{6}\cdot\frac{3!}{s^4}$$

so that

$$y = \frac{2w_0}{EIL}\left[\frac{L}{48}x^4 - \frac{1}{120}x^5 + \frac{1}{120}\left(x - \frac{L}{2}\right)^5\mathcal{U}\left(x - \frac{L}{2}\right) + \frac{c_1}{2}x^2 + \frac{c_2}{6}x^3\right]$$

where $y''(0) = c_1$ and $y'''(0) = c_2$. Using $y''(L) = 0$ and $y'''(L) = 0$ we find

$$c_1 = w_0L^2/24EI, \qquad c_2 = -w_0L/4EI.$$

Hence

$$y = \frac{w_0}{12EIL} \left[-\frac{1}{5}x^5 + \frac{L}{2}x^4 - \frac{L^2}{2}x^3 + \frac{L^3}{4}x^2 + \frac{1}{5}\left(x - \frac{L}{2}\right)^5 \mathcal{U}\left(x - \frac{L}{2}\right) \right].$$

42. Taking the Laplace transform of the given differential equation we obtain

$$\mathcal{L}\{y\} = \frac{c_1}{2} \cdot \frac{2s}{s^4 + 4} + \frac{c_2}{4} \cdot \frac{4}{s^4 + 4} + \frac{w_0}{4EI} \cdot \frac{4}{s^4 + 4} e^{-s\pi/2}$$

so that

$$y = \frac{c_1}{2} \sin x \sinh x + \frac{c_2}{4}(\sin x \cosh x - \cos x \sinh x)$$

$$+ \frac{w_0}{4EI} \left[\sin\left(x - \frac{\pi}{2}\right) \cosh\left(x - \frac{\pi}{2}\right) - \cos\left(x - \frac{\pi}{2}\right) \sinh\left(x - \frac{\pi}{2}\right) \right] \mathcal{U}\left(x - \frac{\pi}{2}\right)$$

where $y''(0) = c_1$ and $y'''(0) = c_2$. Using $y(\pi) = 0$ and $y'(\pi) = 0$ we find

$$c_1 = \frac{w_0}{EI} \frac{\sinh \frac{\pi}{2}}{\sinh \pi}, \qquad c_2 = -\frac{w_0}{EI} \frac{\cosh \frac{\pi}{2}}{\sinh \pi}.$$

Hence

$$y = \frac{w_0}{2EI} \frac{\sinh \frac{\pi}{2}}{\sinh \pi} \sin x \sinh x - \frac{w_0}{4EI} \frac{\cosh \frac{\pi}{2}}{\sinh \pi}(\sin x \cosh x - \cos x \sinh x)$$

$$+ \frac{w_0}{4EI} \left[\sin\left(x - \frac{\pi}{2}\right) \cosh\left(x - \frac{\pi}{2}\right) - \cos\left(x - \frac{\pi}{2}\right) \sinh\left(x - \frac{\pi}{2}\right) \right] \mathcal{U}\left(x - \frac{\pi}{2}\right).$$

8 Systems of Linear First-Order Differential Equations

───────── **Exercises 8.1** ─────────

1. Let $\mathbf{X} = \begin{pmatrix} x \\ y \end{pmatrix}$. Then

$$\mathbf{X}' = \begin{pmatrix} 3 & -5 \\ 4 & 8 \end{pmatrix} \mathbf{X}.$$

2. Let $\mathbf{X} = \begin{pmatrix} x \\ y \end{pmatrix}$. Then

$$\mathbf{X}' = \begin{pmatrix} 4 & -7 \\ 5 & 0 \end{pmatrix} \mathbf{X}.$$

3. Let $\mathbf{X} = \begin{pmatrix} x \\ y \\ z \end{pmatrix}$. Then

$$\mathbf{X}' = \begin{pmatrix} -3 & 4 & -9 \\ 6 & -1 & 0 \\ 10 & 4 & 3 \end{pmatrix} \mathbf{X}.$$

4. Let $\mathbf{X} = \begin{pmatrix} x \\ y \\ z \end{pmatrix}$. Then

$$\mathbf{X}' = \begin{pmatrix} 1 & -1 & 0 \\ 1 & 0 & 2 \\ -1 & 0 & 1 \end{pmatrix} \mathbf{X}.$$

5. Let $\mathbf{X} = \begin{pmatrix} x \\ y \\ z \end{pmatrix}$. Then

$$\mathbf{X}' = \begin{pmatrix} 1 & -1 & 1 \\ 2 & 1 & -1 \\ 1 & 1 & 1 \end{pmatrix} \mathbf{X} + \begin{pmatrix} 0 \\ -3t^2 \\ t^2 \end{pmatrix} + \begin{pmatrix} t \\ 0 \\ -t \end{pmatrix} + \begin{pmatrix} -1 \\ 0 \\ 2 \end{pmatrix}.$$

6. Let $\mathbf{X} = \begin{pmatrix} x \\ y \end{pmatrix}$. Then

$$\mathbf{X}' = \begin{pmatrix} -3 & 4 \\ 5 & 9 \end{pmatrix} \mathbf{X} + \begin{pmatrix} e^{-t} \sin 2t \\ 4e^{-t} \cos 2t \end{pmatrix}.$$

7. $\dfrac{dx}{dt} = 4x + 2y + e^t; \quad \dfrac{dy}{dt} = -x + 3y - e^t$

8. $\dfrac{dx}{dt} = 7x + 5y - 9z - 8e^{-2t}; \quad \dfrac{dy}{dt} = 4x + y + z + 2e^{5t}; \quad \dfrac{dz}{dt} = -2y + 3z + e^{5t} - 3e^{-2t}$

9. $\dfrac{dx}{dt} = x - y + 2z + e^{-t} - 3t; \quad \dfrac{dy}{dt} = 3x - 4y + z + 2e^{-t} + t; \quad \dfrac{dz}{dt} = -2x + 5y + 6z + 2e^{-t} - t$

10. $\dfrac{dx}{dt} = 3x - 7y + 4\sin t + (t-4)e^{4t}; \quad \dfrac{dy}{dt} = x + y + 8\sin t + (2t+1)e^{4t}$

11. Since

$$\mathbf{X}' = \begin{pmatrix} -5 \\ -10 \end{pmatrix} e^{-5t} \quad \text{and} \quad \begin{pmatrix} 3 & -4 \\ 4 & -7 \end{pmatrix} \mathbf{X} = \begin{pmatrix} -5 \\ -10 \end{pmatrix} e^{-5t}$$

we see that
$$\mathbf{X}' = \begin{pmatrix} 3 & -4 \\ 4 & -7 \end{pmatrix} \mathbf{X}.$$

12. Since

$$\mathbf{X}' = \begin{pmatrix} 5\cos t - 5\sin t \\ 2\cos t - 4\sin t \end{pmatrix} e^t \quad \text{and} \quad \begin{pmatrix} -2 & 5 \\ -2 & 4 \end{pmatrix} \mathbf{X} = \begin{pmatrix} 5\cos t - 5\sin t \\ 2\cos t - 4\sin t \end{pmatrix} e^t$$

we see that
$$\mathbf{X}' = \begin{pmatrix} -2 & 5 \\ -2 & 4 \end{pmatrix} \mathbf{X}.$$

13. Since

$$\mathbf{X}' = \begin{pmatrix} 3/2 \\ -3 \end{pmatrix} e^{-3t/2} \quad \text{and} \quad \begin{pmatrix} -1 & 1/4 \\ 1 & -1 \end{pmatrix} \mathbf{X} = \begin{pmatrix} 3/2 \\ -3 \end{pmatrix} e^{-3t/2}$$

we see that
$$\mathbf{X}' = \begin{pmatrix} -1 & 1/4 \\ 1 & -1 \end{pmatrix} \mathbf{X}.$$

14. Since

$$\mathbf{X}' = \begin{pmatrix} 5 \\ -1 \end{pmatrix} e^t + \begin{pmatrix} 4 \\ -4 \end{pmatrix} te^t \quad \text{and} \quad \begin{pmatrix} 2 & 1 \\ -1 & 0 \end{pmatrix} \mathbf{X} = \begin{pmatrix} 5 \\ -1 \end{pmatrix} e^t + \begin{pmatrix} 4 \\ -4 \end{pmatrix} te^t$$

we see that
$$\mathbf{X}' = \begin{pmatrix} 2 & 1 \\ -1 & 0 \end{pmatrix} \mathbf{X}.$$

15. Since

$$\mathbf{X}' = \begin{pmatrix} 0 \\ 0 \\ 0 \end{pmatrix} \quad \text{and} \quad \begin{pmatrix} 1 & 2 & 1 \\ 6 & -1 & 0 \\ -1 & -2 & -1 \end{pmatrix} \mathbf{X} = \begin{pmatrix} 0 \\ 0 \\ 0 \end{pmatrix}$$

we see that
$$\mathbf{X}' = \begin{pmatrix} 1 & 2 & 1 \\ 6 & -1 & 0 \\ -1 & -2 & -1 \end{pmatrix} \mathbf{X}.$$

16. Since

$$\mathbf{X}' = \begin{pmatrix} \cos t \\ \frac{1}{2}\sin t - \frac{1}{2}\cos t \\ -\cos t - \sin t \end{pmatrix} \quad \text{and} \quad \begin{pmatrix} 1 & 0 & 1 \\ 1 & 1 & 0 \\ -2 & 0 & -1 \end{pmatrix} \mathbf{X} = \begin{pmatrix} \cos t \\ \frac{1}{2}\sin t - \frac{1}{2}\cos t \\ -\cos t - \sin t \end{pmatrix}$$

we see that
$$\mathbf{X}' = \begin{pmatrix} 1 & 0 & 1 \\ 1 & 1 & 0 \\ -2 & 0 & -1 \end{pmatrix} \mathbf{X}.$$

17. Yes, since $W(\mathbf{X}_1, \mathbf{X}_2) = -2e^{-8t} \neq 0$ and $\mathbf{X}_1$ and $\mathbf{X}_2$ are linearly independent on $-\infty < t < \infty$.

18. Yes, since $W(\mathbf{X}_1, \mathbf{X}_2) = 8e^{2t} \neq 0$ and $\mathbf{X}_1$ and $\mathbf{X}_2$ are linearly independent on $-\infty < t < \infty$.

19. No, since $W(\mathbf{X}_1, \mathbf{X}_2, \mathbf{X}_3) = 0$ and $\mathbf{X}_1$, $\mathbf{X}_2$, and $\mathbf{X}_3$ are linearly dependent on $-\infty < t < \infty$.

20. Yes, since $W(\mathbf{X}_1, \mathbf{X}_2, \mathbf{X}_3) = -84e^{-t} \neq 0$ and $\mathbf{X}_1$, $\mathbf{X}_2$, and $\mathbf{X}_3$ are linearly independent on $-\infty < t < \infty$.

21. Since

$$\mathbf{X}'_p = \begin{pmatrix} 2 \\ -1 \end{pmatrix} \quad \text{and} \quad \begin{pmatrix} 1 & 4 \\ 3 & 2 \end{pmatrix} \mathbf{X}_p + \begin{pmatrix} 2 \\ -4 \end{pmatrix} t + \begin{pmatrix} -7 \\ -18 \end{pmatrix} = \begin{pmatrix} 2 \\ -1 \end{pmatrix}$$

we see that
$$\mathbf{X}'_p = \begin{pmatrix} 1 & 4 \\ 3 & 2 \end{pmatrix} \mathbf{X}_p + \begin{pmatrix} 2 \\ -4 \end{pmatrix} t + \begin{pmatrix} -7 \\ -18 \end{pmatrix}.$$

22. Since

$$\mathbf{X}'_p = \begin{pmatrix} 0 \\ 0 \end{pmatrix} \quad \text{and} \quad \begin{pmatrix} 2 & 1 \\ 1 & -1 \end{pmatrix} \mathbf{X}_p + \begin{pmatrix} -5 \\ 2 \end{pmatrix} = \begin{pmatrix} 0 \\ 0 \end{pmatrix}$$

we see that
$$\mathbf{X}'_p = \begin{pmatrix} 2 & 1 \\ 1 & -1 \end{pmatrix} \mathbf{X}_p + \begin{pmatrix} -5 \\ 2 \end{pmatrix}.$$

23. Since

$$\mathbf{X}'_p = \begin{pmatrix} 2 \\ 0 \end{pmatrix} e^t + \begin{pmatrix} 1 \\ -1 \end{pmatrix} te^t \quad \text{and} \quad \begin{pmatrix} 2 & 1 \\ 3 & 4 \end{pmatrix} \mathbf{X}_p - \begin{pmatrix} 1 \\ 7 \end{pmatrix} e^t = \begin{pmatrix} 2 \\ 0 \end{pmatrix} e^t + \begin{pmatrix} 1 \\ -1 \end{pmatrix} te^t$$

we see that
$$\mathbf{X}'_p = \begin{pmatrix} 2 & 1 \\ 3 & 4 \end{pmatrix} \mathbf{X}_p - \begin{pmatrix} 1 \\ 7 \end{pmatrix} e^t.$$

24. Since

$$\mathbf{X}'_p = \begin{pmatrix} 3\cos 3t \\ 0 \\ -3\sin 3t \end{pmatrix} \quad \text{and} \quad \begin{pmatrix} 1 & 2 & 3 \\ -4 & 2 & 0 \\ -6 & 1 & 0 \end{pmatrix} \mathbf{X}_p + \begin{pmatrix} -1 \\ 4 \\ 3 \end{pmatrix} \sin 3t = \begin{pmatrix} 3\cos 3t \\ 0 \\ -3\sin 3t \end{pmatrix}$$

we see that
$$\mathbf{X}'_p = \begin{pmatrix} 1 & 2 & 3 \\ -4 & 2 & 0 \\ -6 & 1 & 0 \end{pmatrix} \mathbf{X}_p + \begin{pmatrix} -1 \\ 4 \\ 3 \end{pmatrix} \sin 3t.$$

25. Let

$$\mathbf{X}_1 = \begin{pmatrix} 6 \\ -1 \\ -5 \end{pmatrix} e^{-t}, \quad \mathbf{X}_2 = \begin{pmatrix} -3 \\ 1 \\ 1 \end{pmatrix} e^{-2t}, \quad \mathbf{X}_3 = \begin{pmatrix} 2 \\ 1 \\ 1 \end{pmatrix} e^{3t}, \quad \text{and} \quad \mathbf{A} = \begin{pmatrix} 0 & 6 & 0 \\ 1 & 0 & 1 \\ 1 & 1 & 0 \end{pmatrix}.$$

Then
$$\mathbf{X}'_1 = \begin{pmatrix} -6 \\ 1 \\ 5 \end{pmatrix} e^{-t} = \mathbf{A}\mathbf{X}_1,$$

$$\mathbf{X}'_2 = \begin{pmatrix} 6 \\ -2 \\ -2 \end{pmatrix} e^{-2t} = \mathbf{A}\mathbf{X}_2,$$

$$\mathbf{X}'_3 = \begin{pmatrix} 6 \\ 3 \\ 3 \end{pmatrix} e^{3t} = \mathbf{A}\mathbf{X}_3,$$

and $W(\mathbf{X}_1, \mathbf{X}_2, \mathbf{X}_3) = 20 \neq 0$ so that $\mathbf{X}_1$, $\mathbf{X}_2$, and $\mathbf{X}_3$ form a fundamental set for $\mathbf{X}' = \mathbf{A}\mathbf{X}$ on $-\infty < t < \infty$.

26. Let

$$\mathbf{X}_1 = \begin{pmatrix} 1 \\ -1 - \sqrt{2} \end{pmatrix} e^{\sqrt{2}t},$$

$$\mathbf{X}_2 = \begin{pmatrix} 1 \\ -1 + \sqrt{2} \end{pmatrix} e^{-\sqrt{2}t},$$

$$\mathbf{X}_p = \begin{pmatrix} 1 \\ 0 \end{pmatrix} t^2 + \begin{pmatrix} -2 \\ 4 \end{pmatrix} t + \begin{pmatrix} 1 \\ 0 \end{pmatrix},$$

and
$$\mathbf{A} = \begin{pmatrix} -1 & -1 \\ -1 & 1 \end{pmatrix}.$$

Then
$$\mathbf{X}'_1 = \begin{pmatrix} \sqrt{2} \\ -2 - \sqrt{2} \end{pmatrix} e^{\sqrt{2}t} = \mathbf{A}\mathbf{X}_1,$$

$$\mathbf{X}'_2 = \begin{pmatrix} -\sqrt{2} \\ -2 - \sqrt{2} \end{pmatrix} e^{-\sqrt{2}t} = \mathbf{A}\mathbf{X}_2,$$

$$\mathbf{X}_p' = \begin{pmatrix} 2 \\ 0 \end{pmatrix} t + \begin{pmatrix} -2 \\ 4 \end{pmatrix} = \mathbf{AX}_p + \begin{pmatrix} 1 \\ 1 \end{pmatrix} t^2 + \begin{pmatrix} 4 \\ -6 \end{pmatrix} t + \begin{pmatrix} -1 \\ 5 \end{pmatrix},$$

and $W(\mathbf{X}_1, \mathbf{X}_2) = 2\sqrt{2} \neq 0$ so that $\mathbf{X}_p$ is a particular solution and $\mathbf{X}_1$ and $\mathbf{X}_2$ form a fundamental set on $-\infty < t < \infty$.

Exercises 8.2

1. The system is

$$\mathbf{X}' = \begin{pmatrix} 1 & 2 \\ 4 & 3 \end{pmatrix} \mathbf{X}$$

and $\det(\mathbf{A} - \lambda\mathbf{I}) = (\lambda - 5)(\lambda + 1) = 0$. For $\lambda_1 = 5$ we obtain

$$\begin{pmatrix} -4 & 2 & | & 0 \\ 4 & -2 & | & 0 \end{pmatrix} \implies \begin{pmatrix} 1 & -1/2 & | & 0 \\ 0 & 0 & | & 0 \end{pmatrix} \quad \text{so that} \quad \mathbf{K}_1 = \begin{pmatrix} 1 \\ 2 \end{pmatrix}.$$

For $\lambda_2 = -1$ we obtain

$$\begin{pmatrix} 2 & 2 & | & 0 \\ 4 & 4 & | & 0 \end{pmatrix} \implies \begin{pmatrix} 1 & 1 & | & 0 \\ 0 & 0 & | & 0 \end{pmatrix} \quad \text{so that} \quad \mathbf{K}_2 = \begin{pmatrix} -1 \\ 1 \end{pmatrix}.$$

Then $\qquad\qquad\qquad\qquad \mathbf{X} = c_1 \begin{pmatrix} 1 \\ 2 \end{pmatrix} e^{5t} + c_2 \begin{pmatrix} -1 \\ 1 \end{pmatrix} e^{-t}.$

2. The system is

$$\mathbf{X}' = \begin{pmatrix} 0 & 2 \\ 8 & 0 \end{pmatrix} \mathbf{X}$$

and $\det(\mathbf{A} - \lambda\mathbf{I}) = (\lambda - 4)(\lambda + 4) = 0$. For $\lambda_1 = 4$ we obtain

$$\begin{pmatrix} -4 & 2 & | & 0 \\ 8 & -4 & | & 0 \end{pmatrix} \implies \begin{pmatrix} 1 & -1/2 & | & 0 \\ 0 & 0 & | & 0 \end{pmatrix} \quad \text{so that} \quad \mathbf{K}_1 = \begin{pmatrix} 1 \\ 2 \end{pmatrix}.$$

For $\lambda_2 = -4$ we obtain

$$\begin{pmatrix} 4 & 2 & | & 0 \\ 8 & 4 & | & 0 \end{pmatrix} \implies \begin{pmatrix} 1 & 1/2 & | & 0 \\ 0 & 0 & | & 0 \end{pmatrix} \quad \text{so that} \quad \mathbf{K}_2 = \begin{pmatrix} -1 \\ 2 \end{pmatrix}.$$

Then $\qquad\qquad\qquad\qquad \mathbf{X} = c_1 \begin{pmatrix} 1 \\ 2 \end{pmatrix} e^{4t} + c_2 \begin{pmatrix} -1 \\ 2 \end{pmatrix} e^{-4t}.$

3. The system is

$$\mathbf{X}' = \begin{pmatrix} -4 & 2 \\ -5/2 & 2 \end{pmatrix} \mathbf{X}$$

and $\det(\mathbf{A} - \lambda\mathbf{I}) = (\lambda - 1)(\lambda + 3) = 0$. For $\lambda_1 = 1$ we obtain

$$\begin{pmatrix} -5 & 2 & | & 0 \\ -5/2 & 1 & | & 0 \end{pmatrix} \Longrightarrow \begin{pmatrix} -5 & 2 & | & 0 \\ 0 & 0 & | & 0 \end{pmatrix} \quad \text{so that} \quad \mathbf{K}_1 = \begin{pmatrix} 2 \\ 5 \end{pmatrix}.$$

For $\lambda_2 = -3$ we obtain

$$\begin{pmatrix} -1 & 2 & | & 0 \\ -5/2 & 5 & | & 0 \end{pmatrix} \Longrightarrow \begin{pmatrix} -1 & 2 & | & 0 \\ 0 & 0 & | & 0 \end{pmatrix} \quad \text{so that} \quad \mathbf{K}_2 = \begin{pmatrix} 2 \\ 1 \end{pmatrix}.$$

Then

$$\mathbf{X} = c_1 \begin{pmatrix} 2 \\ 5 \end{pmatrix} e^t + c_2 \begin{pmatrix} 2 \\ 1 \end{pmatrix} e^{-3t}.$$

4. The system is

$$\mathbf{X}' = \begin{pmatrix} 1/2 & 9 \\ 1/2 & 2 \end{pmatrix} \mathbf{X}$$

and $\det(\mathbf{A} - \lambda\mathbf{I}) = (\lambda - 7/2)(\lambda + 1) = 0$. For $\lambda_1 = 7/2$ we obtain

$$\begin{pmatrix} -3 & 9 & | & 0 \\ 1/2 & -1/2 & | & 0 \end{pmatrix} \Longrightarrow \begin{pmatrix} 1 & -3 & | & 0 \\ 0 & 0 & | & 0 \end{pmatrix} \quad \text{so that} \quad \mathbf{K}_1 = \begin{pmatrix} 3 \\ 1 \end{pmatrix}.$$

For $\lambda_2 = -1$ we obtain

$$\begin{pmatrix} 3/2 & 9 & | & 0 \\ 1/2 & 3 & | & 0 \end{pmatrix} \Longrightarrow \begin{pmatrix} 1 & 6 & | & 0 \\ 0 & 0 & | & 0 \end{pmatrix} \quad \text{so that} \quad \mathbf{K}_2 = \begin{pmatrix} -6 \\ 1 \end{pmatrix}.$$

Then

$$\mathbf{X} = c_1 \begin{pmatrix} 3 \\ 1 \end{pmatrix} e^{7t/2} + c_2 \begin{pmatrix} -6 \\ 1 \end{pmatrix} e^{-t}.$$

5. The system is

$$\mathbf{X}' = \begin{pmatrix} 10 & -5 \\ 8 & -12 \end{pmatrix} \mathbf{X}$$

and $\det(\mathbf{A} - \lambda\mathbf{I}) = (\lambda - 8)(\lambda + 10) = 0$. For $\lambda_1 = 8$ we obtain

$$\begin{pmatrix} 2 & -5 & | & 0 \\ 8 & -20 & | & 0 \end{pmatrix} \Longrightarrow \begin{pmatrix} 1 & -5/2 & | & 0 \\ 0 & 0 & | & 0 \end{pmatrix} \quad \text{so that} \quad \mathbf{K}_1 = \begin{pmatrix} 5 \\ 2 \end{pmatrix}.$$

For $\lambda_2 = -10$ we obtain

$$\begin{pmatrix} 20 & -5 & | & 0 \\ 8 & -2 & | & 0 \end{pmatrix} \Longrightarrow \begin{pmatrix} 1 & -1/4 & | & 0 \\ 0 & 0 & | & 0 \end{pmatrix} \quad \text{so that} \quad \mathbf{K}_2 = \begin{pmatrix} 1 \\ 4 \end{pmatrix}.$$

Then

$$\mathbf{X} = c_1 \begin{pmatrix} 5 \\ 2 \end{pmatrix} e^{8t} + c_2 \begin{pmatrix} 1 \\ 4 \end{pmatrix} e^{-10t}.$$

6. The system is

$$\mathbf{X}' = \begin{pmatrix} -6 & 2 \\ -3 & 1 \end{pmatrix} \mathbf{X}$$

and $\det(\mathbf{A} - \lambda\mathbf{I}) = \lambda(\lambda + 5) = 0$. For $\lambda_1 = 0$ we obtain

$$\begin{pmatrix} -6 & 2 & | & 0 \\ -3 & 1 & | & 0 \end{pmatrix} \implies \begin{pmatrix} 1 & -1/3 & | & 0 \\ 0 & 0 & | & 0 \end{pmatrix} \quad \text{so that} \quad \mathbf{K}_1 = \begin{pmatrix} 1 \\ 3 \end{pmatrix}.$$

For $\lambda_2 = -5$ we obtain

$$\begin{pmatrix} -1 & 2 & | & 0 \\ -3 & 6 & | & 0 \end{pmatrix} \implies \begin{pmatrix} 1 & -2 & | & 0 \\ 0 & 0 & | & 0 \end{pmatrix} \quad \text{so that} \quad \mathbf{K}_2 = \begin{pmatrix} 2 \\ 1 \end{pmatrix}.$$

Then

$$\mathbf{X} = c_1 \begin{pmatrix} 1 \\ 3 \end{pmatrix} + c_2 \begin{pmatrix} 2 \\ 1 \end{pmatrix} e^{-5t}.$$

7. The system is

$$\mathbf{X}' = \begin{pmatrix} 1 & 1 & -1 \\ 0 & 2 & 0 \\ 0 & 1 & -1 \end{pmatrix} \mathbf{X}$$

and $\det(\mathbf{A} - \lambda\mathbf{I}) = (\lambda - 1)(2 - \lambda)(\lambda + 1) = 0$. For $\lambda_1 = 1$, $\lambda_2 = 2$, and $\lambda_3 = -1$ we obtain

$$\mathbf{K}_1 = \begin{pmatrix} 1 \\ 0 \\ 0 \end{pmatrix}, \quad \mathbf{K}_2 = \begin{pmatrix} 2 \\ 3 \\ 1 \end{pmatrix}, \quad \text{and} \quad \mathbf{K}_3 = \begin{pmatrix} 1 \\ 0 \\ 2 \end{pmatrix},$$

so that

$$\mathbf{X} = c_1 \begin{pmatrix} 1 \\ 0 \\ 0 \end{pmatrix} e^t + c_2 \begin{pmatrix} 2 \\ 3 \\ 1 \end{pmatrix} e^{2t} + c_3 \begin{pmatrix} 1 \\ 0 \\ 2 \end{pmatrix} e^{-t}.$$

8. The system is

$$\mathbf{X}' = \begin{pmatrix} 2 & -7 & 0 \\ 5 & 10 & 4 \\ 0 & 5 & 2 \end{pmatrix} \mathbf{X}$$

and $\det(\mathbf{A} - \lambda\mathbf{I}) = (2 - \lambda)(\lambda - 5)(\lambda - 7) = 0$. For $\lambda_1 = 2$, $\lambda_2 = 5$, and $\lambda_3 = 7$ we obtain

$$\mathbf{K}_1 = \begin{pmatrix} 4 \\ 0 \\ -5 \end{pmatrix}, \quad \mathbf{K}_2 = \begin{pmatrix} -7 \\ 3 \\ 5 \end{pmatrix}, \quad \text{and} \quad \mathbf{K}_3 = \begin{pmatrix} -7 \\ 5 \\ 5 \end{pmatrix},$$

so that

$$\mathbf{X} = c_1 \begin{pmatrix} 4 \\ 0 \\ -5 \end{pmatrix} e^{2t} + c_2 \begin{pmatrix} -7 \\ 3 \\ 5 \end{pmatrix} e^{5t} + c_3 \begin{pmatrix} -7 \\ 5 \\ 5 \end{pmatrix} e^{7t}.$$

9. We have $\det(\mathbf{A} - \lambda\mathbf{I}) = -(\lambda + 1)(\lambda - 3)(\lambda + 2) = 0$. For $\lambda_1 = -1$, $\lambda_2 = 3$, and $\lambda_3 = -2$ we obtain

$$\mathbf{K}_1 = \begin{pmatrix} -1 \\ 0 \\ 1 \end{pmatrix}, \quad \mathbf{K}_2 = \begin{pmatrix} 1 \\ 4 \\ 3 \end{pmatrix}, \quad \text{and} \quad \mathbf{K}_3 = \begin{pmatrix} 1 \\ -1 \\ 3 \end{pmatrix},$$

so that

$$\mathbf{X} = c_1 \begin{pmatrix} -1 \\ 0 \\ 1 \end{pmatrix} e^{-t} + c_2 \begin{pmatrix} 1 \\ 4 \\ 3 \end{pmatrix} e^{3t} + c_3 \begin{pmatrix} 1 \\ -1 \\ 3 \end{pmatrix} e^{-2t}.$$

10. We have $\det(\mathbf{A} - \lambda\mathbf{I}) = -\lambda(\lambda - 1)(\lambda - 2) = 0$. For $\lambda_1 = 0$, $\lambda_2 = 1$, and $\lambda_3 = 2$ we obtain

$$\mathbf{K}_1 = \begin{pmatrix} 1 \\ 0 \\ -1 \end{pmatrix}, \quad \mathbf{K}_2 = \begin{pmatrix} 0 \\ 1 \\ 0 \end{pmatrix}, \quad \text{and} \quad \mathbf{K}_3 = \begin{pmatrix} 1 \\ 0 \\ 1 \end{pmatrix},$$

so that

$$\mathbf{X} = c_1 \begin{pmatrix} 1 \\ 0 \\ -1 \end{pmatrix} + c_2 \begin{pmatrix} 0 \\ 1 \\ 0 \end{pmatrix} e^{t} + c_3 \begin{pmatrix} 1 \\ 0 \\ 1 \end{pmatrix} e^{2t}.$$

11. We have $\det(\mathbf{A} - \lambda\mathbf{I}) = -(\lambda + 1)(\lambda + 1/2)(\lambda + 3/2) = 0$. For $\lambda_1 = -1$, $\lambda_2 = -1/2$, and $\lambda_3 = -3/2$ we obtain

$$\mathbf{K}_1 = \begin{pmatrix} 4 \\ 0 \\ -1 \end{pmatrix}, \quad \mathbf{K}_2 = \begin{pmatrix} -12 \\ 6 \\ 5 \end{pmatrix}, \quad \text{and} \quad \mathbf{K}_3 = \begin{pmatrix} 4 \\ 2 \\ -1 \end{pmatrix},$$

so that

$$\mathbf{X} = c_1 \begin{pmatrix} 4 \\ 0 \\ -1 \end{pmatrix} e^{-t} + c_2 \begin{pmatrix} -12 \\ 6 \\ 5 \end{pmatrix} e^{-t/2} + c_3 \begin{pmatrix} 4 \\ 2 \\ -1 \end{pmatrix} e^{-3t/2}.$$

12. We have $\det(\mathbf{A} - \lambda\mathbf{I}) = (\lambda - 3)(\lambda + 5)(6 - \lambda) = 0$. For $\lambda_1 = 3$, $\lambda_2 = -5$, and $\lambda_3 = 6$ we obtain

$$\mathbf{K}_1 = \begin{pmatrix} 1 \\ 1 \\ 0 \end{pmatrix}, \quad \mathbf{K}_2 = \begin{pmatrix} 1 \\ -1 \\ 0 \end{pmatrix}, \quad \text{and} \quad \mathbf{K}_3 = \begin{pmatrix} 2 \\ -2 \\ 11 \end{pmatrix},$$

so that

$$\mathbf{X} = c_1 \begin{pmatrix} 1 \\ 1 \\ 0 \end{pmatrix} e^{3t} + c_2 \begin{pmatrix} 1 \\ -1 \\ 0 \end{pmatrix} e^{-5t} + c_3 \begin{pmatrix} 2 \\ -2 \\ 11 \end{pmatrix} e^{6t}.$$

13. We have $\det(\mathbf{A} - \lambda\mathbf{I}) = (\lambda + 1/2)(\lambda - 1/2) = 0$. For $\lambda_1 = -1/2$ and $\lambda_2 = 1/2$ we obtain

$$\mathbf{K}_1 = \begin{pmatrix} 0 \\ 1 \end{pmatrix} \quad \text{and} \quad \mathbf{K}_2 = \begin{pmatrix} 1 \\ 1 \end{pmatrix},$$

so that
$$\mathbf{X} = c_1 \begin{pmatrix} 0 \\ 1 \end{pmatrix} e^{-t/2} + c_2 \begin{pmatrix} 1 \\ 1 \end{pmatrix} e^{t/2}.$$

If
$$\mathbf{X}(0) = \begin{pmatrix} 3 \\ 5 \end{pmatrix}$$

then $c_1 = 2$ and $c_2 = 3$.

14. We have $\det(\mathbf{A} - \lambda \mathbf{I}) = (2 - \lambda)(\lambda - 3)(\lambda + 1) = 0$. For $\lambda_1 = 2$, $\lambda_2 = 3$, and $\lambda_3 = -1$ we obtain

$$\mathbf{K}_1 = \begin{pmatrix} 5 \\ -3 \\ 2 \end{pmatrix}, \quad \mathbf{K}_2 = \begin{pmatrix} 2 \\ 0 \\ 1 \end{pmatrix}, \quad \text{and} \quad \mathbf{K}_3 = \begin{pmatrix} -2 \\ 0 \\ 1 \end{pmatrix},$$

so that
$$\mathbf{X} = c_1 \begin{pmatrix} 5 \\ -3 \\ 2 \end{pmatrix} e^{2t} + c_2 \begin{pmatrix} 2 \\ 0 \\ 1 \end{pmatrix} e^{3t} + c_3 \begin{pmatrix} -2 \\ 0 \\ 1 \end{pmatrix} e^{-t}.$$

If
$$\mathbf{X}(0) = \begin{pmatrix} 1 \\ 3 \\ 0 \end{pmatrix}$$

then $c_1 = -1$, $c_2 = 5/2$, and $c_3 = -1/2$.

15. $\mathbf{X} = c_1 \begin{pmatrix} 0.382175 \\ 0.851161 \\ 0.359815 \end{pmatrix} e^{8.58979t} + c_2 \begin{pmatrix} 0.405188 \\ -0.676043 \\ 0.615458 \end{pmatrix} e^{2.25684t} + c_3 \begin{pmatrix} -0.923562 \\ -0.132174 \\ 0.35995 \end{pmatrix} e^{-0.0466321t}$

16. $\mathbf{X} = c_1 \begin{pmatrix} 0.0312209 \\ 0.949058 \\ 0.239535 \\ 0.195825 \\ 0.0508861 \end{pmatrix} e^{5.05452t} + c_2 \begin{pmatrix} -0.280232 \\ -0.836611 \\ -0.275304 \\ 0.176045 \\ 0.338775 \end{pmatrix} e^{4.09561t} + c_3 \begin{pmatrix} 0.262219 \\ -0.162664 \\ -0.826218 \\ -0.346439 \\ 0.31957 \end{pmatrix} e^{-2.92362t}$

$+ c_4 \begin{pmatrix} 0.313235 \\ 0.64181 \\ 0.31754 \\ 0.173787 \\ -0.599108 \end{pmatrix} e^{2.02882t} + c_5 \begin{pmatrix} -0.301294 \\ 0.466599 \\ 0.222136 \\ 0.0534311 \\ -0.799567 \end{pmatrix} e^{-0.155338t}$

17. We have $\det(\mathbf{A} - \lambda\mathbf{I}) = \lambda^2 = 0$. For $\lambda_1 = 0$ we obtain

$$\mathbf{K} = \begin{pmatrix} 1 \\ 3 \end{pmatrix}.$$

A solution of $(\mathbf{A} - \lambda_1\mathbf{I})\mathbf{P} = \mathbf{K}$ is

$$\mathbf{P} = \begin{pmatrix} 1 \\ 2 \end{pmatrix}$$

so that

$$\mathbf{X} = c_1 \begin{pmatrix} 1 \\ 3 \end{pmatrix} + c_2 \left[\begin{pmatrix} 1 \\ 3 \end{pmatrix} t + \begin{pmatrix} 1 \\ 2 \end{pmatrix} \right].$$

18. We have $\det(\mathbf{A} - \lambda\mathbf{I}) = (\lambda + 1)^2 = 0$. For $\lambda_1 = -1$ we obtain

$$\mathbf{K} = \begin{pmatrix} 1 \\ 1 \end{pmatrix}.$$

A solution of $(\mathbf{A} - \lambda_1\mathbf{I})\mathbf{P} = \mathbf{K}$ is

$$\mathbf{P} = \begin{pmatrix} 0 \\ 1/5 \end{pmatrix}$$

so that

$$\mathbf{X} = c_1 \begin{pmatrix} 1 \\ 1 \end{pmatrix} e^{-t} + c_2 \left[\begin{pmatrix} 1 \\ 1 \end{pmatrix} te^{-t} + \begin{pmatrix} 0 \\ 1/5 \end{pmatrix} e^{-t} \right].$$

19. We have $\det(\mathbf{A} - \lambda\mathbf{I}) = (\lambda - 2)^2 = 0$. For $\lambda_1 = 2$ we obtain

$$\mathbf{K} = \begin{pmatrix} 1 \\ 1 \end{pmatrix}.$$

A solution of $(\mathbf{A} - \lambda_1\mathbf{I})\mathbf{P} = \mathbf{K}$ is

$$\mathbf{P} = \begin{pmatrix} -1/3 \\ 0 \end{pmatrix}$$

so that

$$\mathbf{X} = c_1 \begin{pmatrix} 1 \\ 1 \end{pmatrix} e^{2t} + c_2 \left[\begin{pmatrix} 1 \\ 1 \end{pmatrix} te^{2t} + \begin{pmatrix} -1/3 \\ 0 \end{pmatrix} e^{2t} \right].$$

20. We have $\det(\mathbf{A} - \lambda\mathbf{I}) = (\lambda - 6)^2 = 0$. For $\lambda_1 = 6$ we obtain

$$\mathbf{K} = \begin{pmatrix} 3 \\ 2 \end{pmatrix}.$$

A solution of $(\mathbf{A} - \lambda_1\mathbf{I})\mathbf{P} = \mathbf{K}$ is

$$\mathbf{P} = \begin{pmatrix} 1/2 \\ 0 \end{pmatrix}$$

so that
$$\mathbf{X} = c_1 \begin{pmatrix} 3 \\ 2 \end{pmatrix} e^{6t} + c_2 \left[\begin{pmatrix} 3 \\ 2 \end{pmatrix} te^{6t} + \begin{pmatrix} 1/2 \\ 0 \end{pmatrix} e^{6t} \right].$$

21. We have $\det(\mathbf{A} - \lambda\mathbf{I}) = (1 - \lambda)(\lambda - 2)^2 = 0$. For $\lambda_1 = 1$ we obtain
$$\mathbf{K}_1 = \begin{pmatrix} 1 \\ 1 \\ 1 \end{pmatrix}.$$

For $\lambda_2 = 2$ we obtain
$$\mathbf{K}_2 = \begin{pmatrix} 1 \\ 0 \\ 1 \end{pmatrix} \quad \text{and} \quad \mathbf{K}_3 = \begin{pmatrix} 1 \\ 1 \\ 0 \end{pmatrix}.$$

Then
$$\mathbf{X} = c_1 \begin{pmatrix} 1 \\ 1 \\ 1 \end{pmatrix} e^{t} + c_2 \begin{pmatrix} 1 \\ 0 \\ 1 \end{pmatrix} e^{2t} + c_3 \begin{pmatrix} 1 \\ 1 \\ 0 \end{pmatrix} e^{2t}.$$

22. We have $\det(\mathbf{A} - \lambda\mathbf{I}) = (\lambda - 8)(\lambda + 1)^2 = 0$. For $\lambda_1 = 8$ we obtain
$$\mathbf{K}_1 = \begin{pmatrix} 2 \\ 1 \\ 2 \end{pmatrix}.$$

For $\lambda_2 = -1$ we obtain
$$\mathbf{K}_2 = \begin{pmatrix} 0 \\ -2 \\ 1 \end{pmatrix} \quad \text{and} \quad \mathbf{K}_3 = \begin{pmatrix} 1 \\ -2 \\ 0 \end{pmatrix}.$$

Then
$$\mathbf{X} = c_1 \begin{pmatrix} 2 \\ 1 \\ 2 \end{pmatrix} e^{8t} + c_2 \begin{pmatrix} 0 \\ -2 \\ 1 \end{pmatrix} e^{-t} + c_3 \begin{pmatrix} 1 \\ -2 \\ 0 \end{pmatrix} e^{-t}.$$

23. We have $\det(\mathbf{A} - \lambda\mathbf{I}) = -\lambda(5 - \lambda)^2 = 0$. For $\lambda_1 = 0$ we obtain
$$\mathbf{K}_1 = \begin{pmatrix} -4 \\ -5 \\ 2 \end{pmatrix}.$$

For $\lambda_2 = 5$ we obtain
$$\mathbf{K} = \begin{pmatrix} -2 \\ 0 \\ 1 \end{pmatrix}.$$

A solution of $(\mathbf{A} - \lambda_1\mathbf{I})\mathbf{P} = \mathbf{K}$ is

$$\mathbf{P} = \begin{pmatrix} 5/2 \\ 1/2 \\ 0 \end{pmatrix}$$

so that

$$\mathbf{X} = c_1 \begin{pmatrix} -4 \\ -5 \\ 2 \end{pmatrix} + c_2 \begin{pmatrix} -2 \\ 0 \\ 1 \end{pmatrix} e^{5t} + c_3 \left[\begin{pmatrix} -2 \\ 0 \\ 1 \end{pmatrix} te^{5t} + \begin{pmatrix} 5/2 \\ 1/2 \\ 0 \end{pmatrix} e^{5t} \right].$$

24. We have $\det(\mathbf{A} - \lambda\mathbf{I}) = (1 - \lambda)(\lambda - 2)^2 = 0$. For $\lambda_1 = 1$ we obtain

$$\mathbf{K}_1 = \begin{pmatrix} 1 \\ 0 \\ 0 \end{pmatrix}.$$

For $\lambda_2 = 2$ we obtain

$$\mathbf{K} = \begin{pmatrix} 0 \\ -1 \\ 1 \end{pmatrix}.$$

A solution of $(\mathbf{A} - \lambda_2\mathbf{I})\mathbf{P} = \mathbf{K}$ is

$$\mathbf{P} = \begin{pmatrix} 0 \\ -1 \\ 0 \end{pmatrix}$$

so that

$$\mathbf{X} = c_1 \begin{pmatrix} 1 \\ 0 \\ 0 \end{pmatrix} e^t + c_2 \begin{pmatrix} 0 \\ -1 \\ 1 \end{pmatrix} e^{2t} + c_3 \left[\begin{pmatrix} 0 \\ -1 \\ 1 \end{pmatrix} te^{2t} + \begin{pmatrix} 0 \\ -1 \\ 0 \end{pmatrix} e^{2t} \right].$$

25. We have $\det(\mathbf{A} - \lambda\mathbf{I}) = -(\lambda - 1)^3 = 0$. For $\lambda_1 = 1$ we obtain

$$\mathbf{K} = \begin{pmatrix} 0 \\ 1 \\ 1 \end{pmatrix}.$$

Solutions of $(\mathbf{A} - \lambda_1\mathbf{I})\mathbf{P} = \mathbf{K}$ and $(\mathbf{A} - \lambda_1\mathbf{I})\mathbf{Q} = \mathbf{P}$ are

$$\mathbf{P} = \begin{pmatrix} 0 \\ 1 \\ 0 \end{pmatrix} \quad \text{and} \quad \mathbf{Q} = \begin{pmatrix} 1/2 \\ 0 \\ 0 \end{pmatrix}$$

so that

$$\mathbf{X} = c_1 \begin{pmatrix} 0 \\ 1 \\ 1 \end{pmatrix} + c_2 \left[\begin{pmatrix} 0 \\ 1 \\ 1 \end{pmatrix} te^t + \begin{pmatrix} 0 \\ 1 \\ 0 \end{pmatrix} e^t \right] + c_3 \left[\begin{pmatrix} 0 \\ 1 \\ 1 \end{pmatrix} \frac{t^2}{2} e^t + \begin{pmatrix} 0 \\ 1 \\ 0 \end{pmatrix} te^t + \begin{pmatrix} 1/2 \\ 0 \\ 0 \end{pmatrix} e^t \right].$$

26. We have $\det(\mathbf{A} - \lambda\mathbf{I}) = (\lambda - 4)^3 = 0$. For $\lambda_1 = 4$ we obtain

$$\mathbf{K} = \begin{pmatrix} 1 \\ 0 \\ 0 \end{pmatrix}.$$

Solutions of $(\mathbf{A} - \lambda_1\mathbf{I})\mathbf{P} = \mathbf{K}$ and $(\mathbf{A} - \lambda_1\mathbf{I})\mathbf{Q} = \mathbf{P}$ are

$$\mathbf{P} = \begin{pmatrix} 0 \\ 1 \\ 0 \end{pmatrix} \quad \text{and} \quad \mathbf{Q} = \begin{pmatrix} 0 \\ 0 \\ 1 \end{pmatrix}$$

so that $\quad \mathbf{X} = c_1 \begin{pmatrix} 1 \\ 0 \\ 0 \end{pmatrix} e^{4t} + c_2 \left[\begin{pmatrix} 1 \\ 0 \\ 0 \end{pmatrix} te^{4t} + \begin{pmatrix} 0 \\ 1 \\ 0 \end{pmatrix} e^{4t} \right] + c_3 \left[\begin{pmatrix} 1 \\ 0 \\ 0 \end{pmatrix} \frac{t^2}{2} e^{4t} + \begin{pmatrix} 0 \\ 1 \\ 0 \end{pmatrix} te^{4t} + \begin{pmatrix} 0 \\ 0 \\ 1 \end{pmatrix} e^{4t} \right].$

27. We have $\det(\mathbf{A} - \lambda\mathbf{I}) = (\lambda - 4)^2 = 0$. For $\lambda_1 = 4$ we obtain

$$\mathbf{K} = \begin{pmatrix} 2 \\ 1 \end{pmatrix}.$$

A solution of $(\mathbf{A} - \lambda_1\mathbf{I})\mathbf{P} = \mathbf{K}$ is

$$\mathbf{P} = \begin{pmatrix} 1 \\ 1 \end{pmatrix}$$

so that $$\mathbf{X} = c_1 \begin{pmatrix} 2 \\ 1 \end{pmatrix} e^{4t} + c_2 \left[\begin{pmatrix} 2 \\ 1 \end{pmatrix} te^{4t} + \begin{pmatrix} 1 \\ 1 \end{pmatrix} e^{4t} \right].$$

If $$\mathbf{X}(0) = \begin{pmatrix} -1 \\ 6 \end{pmatrix}$$

then $c_1 = -7$ and $c_2 = 13$.

28. We have $\det(\mathbf{A} - \lambda\mathbf{I}) = -(\lambda + 1)(\lambda - 1)^2 = 0$. For $\lambda_1 = -1$ we obtain

$$\mathbf{K}_1 = \begin{pmatrix} -1 \\ 0 \\ 1 \end{pmatrix}.$$

For $\lambda_2 = 1$ we obtain

$$\mathbf{K}_2 = \begin{pmatrix} 1 \\ 0 \\ 1 \end{pmatrix} \quad \text{and} \quad \mathbf{K}_3 = \begin{pmatrix} 0 \\ 1 \\ 0 \end{pmatrix}$$

so that
$$\mathbf{X} = c_1 \begin{pmatrix} -1 \\ 0 \\ 1 \end{pmatrix} e^{-t} + c_2 \begin{pmatrix} 1 \\ 0 \\ 1 \end{pmatrix} e^{t} + c_3 \begin{pmatrix} 0 \\ 1 \\ 0 \end{pmatrix} e^{t}.$$

If
$$\mathbf{X}(0) = \begin{pmatrix} 1 \\ 2 \\ 5 \end{pmatrix}$$

then $c_1 = 2$, $c_2 = 3$, and $c_3 = 2$.

29. In this case $\det(\mathbf{A} - \lambda\mathbf{I}) = (2 - \lambda)^5$, and $\lambda_1 = 2$ is an eigenvalue of multiplicity 5. Linearly independent eigenvectors are

$$\mathbf{K}_1 = \begin{pmatrix} 1 \\ 0 \\ 0 \\ 0 \\ 0 \end{pmatrix}, \qquad \mathbf{K}_2 = \begin{pmatrix} 0 \\ 0 \\ 1 \\ 0 \\ 0 \end{pmatrix}, \qquad \text{and} \qquad \mathbf{K}_3 = \begin{pmatrix} 0 \\ 0 \\ 0 \\ 1 \\ 0 \end{pmatrix}.$$

30. The system of differential equations is

$$x_1' = 2x_1 + x_2$$
$$x_2' = 2x_2$$
$$x_3' = 2x_3$$
$$x_4' = 2x_4 + x_5$$
$$x_5' = 2x_5.$$

We see immediately that $x_2 = c_2 e^{2t}$, $x_3 = c_3 e^{2t}$, and $x_5 = c_5 e^{2t}$. Then

$$x_1' = 2x_1 + c_2 e^{2t} \implies x_1 = c_2 t e^{2t} + c_1 e^{2t}$$

and
$$x_4' = 2x_4 + c_5 e^{2t} \implies x_4 = c_5 t e^{2t} + c_4 e^{2t}.$$

The general solution of the system is

$$\mathbf{X} = \begin{pmatrix} c_2 t e^{2t} + c_1 e^{2t} \\ c_2 e^{2t} \\ c_3 e^{2t} \\ c_5 t e^{2t} + c_4 e^{2t} \\ c_5 e^{2t} \end{pmatrix}$$

329

$$= c_1 \begin{pmatrix} 1 \\ 0 \\ 0 \\ 0 \\ 0 \end{pmatrix} e^{2t} + c_2 \left[\begin{pmatrix} 1 \\ 0 \\ 0 \\ 0 \\ 0 \end{pmatrix} te^{2t} + \begin{pmatrix} 0 \\ 1 \\ 0 \\ 0 \\ 0 \end{pmatrix} e^{2t} \right]$$

$$+ c_3 \begin{pmatrix} 0 \\ 0 \\ 1 \\ 0 \\ 0 \end{pmatrix} e^{2t} + c_4 \begin{pmatrix} 0 \\ 0 \\ 0 \\ 1 \\ 0 \end{pmatrix} e^{2t} + c_5 \left[\begin{pmatrix} 0 \\ 0 \\ 0 \\ 1 \\ 0 \end{pmatrix} te^{2t} + \begin{pmatrix} 0 \\ 0 \\ 0 \\ 0 \\ 1 \end{pmatrix} e^{2t} \right]$$

$$= c_1 \mathbf{K}_1 e^{2t} + c_2 \left[\mathbf{K}_1 te^{2t} + \begin{pmatrix} 0 \\ 1 \\ 0 \\ 0 \\ 0 \end{pmatrix} e^{2t} \right]$$

$$+ c_3 \mathbf{K}_2 e^{2t} + c_4 \mathbf{K}_3 e^{2t} + c_5 \left[\mathbf{K}_3 te^{2t} + \begin{pmatrix} 0 \\ 0 \\ 0 \\ 0 \\ 1 \end{pmatrix} e^{2t} \right].$$

There are three solutions of the form $\mathbf{X} = \mathbf{K}e^{2t}$, where $\mathbf{K}$ is an eigenvector, and two solutions of the form $\mathbf{X} = \mathbf{K}te^{2t} + \mathbf{P}e^{2t}$. See (12) in Section 8.2. From (13) and (14) in Section 8.2

$$(\mathbf{A} - 2\mathbf{I})\mathbf{K}_1 = \mathbf{0}$$

and

$$(\mathbf{A} - 2\mathbf{I})\mathbf{K}_2 = \mathbf{K}_1.$$

This implies

$$\begin{pmatrix} 0 & 1 & 0 & 0 & 0 \\ 0 & 0 & 0 & 0 & 0 \\ 0 & 0 & 0 & 0 & 0 \\ 0 & 0 & 0 & 0 & 1 \\ 0 & 0 & 0 & 0 & 0 \end{pmatrix} \begin{pmatrix} p_1 \\ p_2 \\ p_3 \\ p_4 \\ p_5 \end{pmatrix} = \begin{pmatrix} 1 \\ 0 \\ 0 \\ 0 \\ 0 \end{pmatrix},$$

so $p_2 = 1$ and $p_5 = 0$, while p_1, p_3, and p_4 are arbitrary. Choosing $p_1 = p_3 = p_4 = 0$ we have

$$P = \begin{pmatrix} 1 \\ 0 \\ 0 \\ 0 \\ 0 \end{pmatrix}.$$

Therefore a solution is

$$X = \begin{pmatrix} 1 \\ 0 \\ 0 \\ 0 \\ 0 \end{pmatrix} te^{2t} + \begin{pmatrix} 1 \\ 0 \\ 0 \\ 0 \\ 0 \end{pmatrix} e^{2t}.$$

Repeating for $\mathbf{K}_3$ we find

$$P = \begin{pmatrix} 0 \\ 0 \\ 0 \\ 0 \\ 1 \end{pmatrix},$$

so another solution is

$$X = \begin{pmatrix} 0 \\ 0 \\ 0 \\ 1 \\ 0 \end{pmatrix} te^{2t} + \begin{pmatrix} 0 \\ 0 \\ 0 \\ 0 \\ 1 \end{pmatrix} e^{2t}.$$

In Problems 31-44 the form of the answer will vary according to the choice of eigenvector. For example, in Problem 31, if $\mathbf{K}_1$ is chosen to be $\begin{pmatrix} 1 \\ -2i \end{pmatrix}$ the solution has the form

$$X = c_1 \begin{pmatrix} \cos t \\ 2\cos t + \sin t \end{pmatrix} e^{4t} + c_2 \begin{pmatrix} \sin t \\ 2\sin t - \cos t \end{pmatrix} e^{4t}.$$

31. We have $\det(\mathbf{A} - \lambda\mathbf{I}) = \lambda^2 - 8\lambda + 17 = 0$. For $\lambda_1 = 4 + i$ we obtain

$$\mathbf{K}_1 = \begin{pmatrix} 2 + i \\ 5 \end{pmatrix}$$

so that $\quad X_1 = \begin{pmatrix} 2 + i \\ 5 \end{pmatrix} e^{(4+i)t} = \begin{pmatrix} 2\cos t - \sin t \\ 5\cos t \end{pmatrix} e^{4t} + i \begin{pmatrix} \cos t + 2\sin t \\ 5\sin t \end{pmatrix} e^{4t}.$

Then
$$\mathbf{X} = c_1 \begin{pmatrix} 2\cos t - \sin t \\ 5\cos t \end{pmatrix} e^{4t} + c_2 \begin{pmatrix} 2\sin t + \cos t \\ 5\sin t \end{pmatrix} e^{4t}.$$

32. We have $\det(\mathbf{A} - \lambda\mathbf{I}) = \lambda^2 + 1 = 0$. For $\lambda_1 = i$ we obtain

$$\mathbf{K}_1 = \begin{pmatrix} -1 - i \\ 2 \end{pmatrix}$$

so that
$$\mathbf{X}_1 = \begin{pmatrix} -1 - i \\ 2 \end{pmatrix} e^{it} = \begin{pmatrix} \sin t - \cos t \\ 2\cos t \end{pmatrix} + i \begin{pmatrix} -\cos t - \sin t \\ 2\sin t \end{pmatrix}.$$

Then
$$\mathbf{X} = c_1 \begin{pmatrix} \sin t - \cos t \\ 2\cos t \end{pmatrix} + c_2 \begin{pmatrix} -\cos t - \sin t \\ 2\sin t \end{pmatrix}.$$

33. We have $\det(\mathbf{A} - \lambda\mathbf{I}) = \lambda^2 - 8\lambda + 17 = 0$. For $\lambda_1 = 4 + i$ we obtain

$$\mathbf{K}_1 = \begin{pmatrix} -1 - i \\ 2 \end{pmatrix}$$

so that
$$\mathbf{X}_1 = \begin{pmatrix} -1 - i \\ 2 \end{pmatrix} e^{(4+i)t} = \begin{pmatrix} \sin t - \cos t \\ 2\cos t \end{pmatrix} e^{4t} + i \begin{pmatrix} -\sin t - \cos t \\ 2\sin t \end{pmatrix} e^{4t}.$$

Then
$$\mathbf{X} = c_1 \begin{pmatrix} \sin t - \cos t \\ 2\cos t \end{pmatrix} e^{4t} + c_2 \begin{pmatrix} -\sin t - \cos t \\ 2\sin t \end{pmatrix} e^{4t}.$$

34. We have $\det(\mathbf{A} - \lambda\mathbf{I}) = \lambda^2 - 10\lambda + 34 = 0$. For $\lambda_1 = 5 + 3i$ we obtain

$$\mathbf{K}_1 = \begin{pmatrix} 1 - 3i \\ 2 \end{pmatrix}$$

so that
$$\mathbf{X}_1 = \begin{pmatrix} 1 - 3i \\ 2 \end{pmatrix} e^{(5+3i)t} = \begin{pmatrix} \cos 3t + 3\sin 3t \\ 2\cos 3t \end{pmatrix} e^{5t} + i \begin{pmatrix} \sin 3t - 3\cos 3t \\ 2\cos 3t \end{pmatrix} e^{5t}.$$

Then
$$\mathbf{X} = c_1 \begin{pmatrix} \cos 3t + 3\sin 3t \\ 2\cos 3t \end{pmatrix} e^{5t} + c_2 \begin{pmatrix} \sin 3t - 3\cos 3t \\ 2\cos 3t \end{pmatrix} e^{5t}.$$

35. We have $\det(\mathbf{A} - \lambda\mathbf{I}) = \lambda^2 + 9 = 0$. For $\lambda_1 = 3i$ we obtain

$$\mathbf{K}_1 = \begin{pmatrix} 4 + 3i \\ 5 \end{pmatrix}$$

so that
$$\mathbf{X}_1 = \begin{pmatrix} 4 + 3i \\ 5 \end{pmatrix} e^{3it} = \begin{pmatrix} 4\cos 3t - 3\sin 3t \\ 5\cos 3t \end{pmatrix} + i \begin{pmatrix} 4\sin 3t + 3\cos 3t \\ 5\sin 3t \end{pmatrix}.$$

Then
$$\mathbf{X} = c_1 \begin{pmatrix} 4\cos 3t - 3\sin 3t \\ 5\cos 3t \end{pmatrix} + c_2 \begin{pmatrix} 4\sin 3t + 3\cos 3t \\ 5\sin 3t \end{pmatrix}.$$

36. We have $\det(\mathbf{A} - \lambda\mathbf{I}) = \lambda^2 + 2\lambda + 5 = 0$. For $\lambda_1 = -1 + 2i$ we obtain

$$\mathbf{K}_1 = \begin{pmatrix} 2 + 2i \\ 1 \end{pmatrix}$$

so that
$$\mathbf{X}_1 = \begin{pmatrix} 2 + 2i \\ 1 \end{pmatrix} e^{(-1+2i)t}$$

$$= (2\cos 2t - 2\sin 2t \cos 2t) e^{-t} + i \begin{pmatrix} 2\cos 2t + 2\sin 2t \\ \sin 2t \end{pmatrix} e^{-t}.$$

Then
$$\mathbf{X} = c_1 \begin{pmatrix} 2\cos 2t - 2\sin 2t \\ \cos 2t \end{pmatrix} e^{-t} + c_2 \begin{pmatrix} 2\cos 2t + 2\sin 2t \\ \sin 2t \end{pmatrix} e^{-t}.$$

37. We have $\det(\mathbf{A} - \lambda\mathbf{I}) = -\lambda\left(\lambda^2 + 1\right) = 0$. For $\lambda_1 = 0$ we obtain

$$\mathbf{K}_1 = \begin{pmatrix} 1 \\ 0 \\ 0 \end{pmatrix}.$$

For $\lambda_2 = i$ we obtain

$$\mathbf{K}_2 = \begin{pmatrix} -i \\ i \\ 1 \end{pmatrix}$$

so that
$$\mathbf{X}_2 = \begin{pmatrix} -i \\ i \\ 1 \end{pmatrix} e^{it} = \begin{pmatrix} \sin t \\ -\sin t \\ \cos t \end{pmatrix} + i \begin{pmatrix} -\cos t \\ \cos t \\ \sin t \end{pmatrix}.$$

Then
$$\mathbf{X} = c_1 \begin{pmatrix} 1 \\ 0 \\ 0 \end{pmatrix} + c_2 \begin{pmatrix} \sin t \\ -\sin t \\ \cos t \end{pmatrix} + c_3 \begin{pmatrix} -\cos t \\ \cos t \\ \sin t \end{pmatrix}.$$

38. We have $\det(\mathbf{A} - \lambda\mathbf{I}) = (\lambda + 3)(\lambda^2 - 2\lambda + 5) = 0$. For $\lambda_1 = -3$ we obtain

$$\mathbf{K}_1 = \begin{pmatrix} 0 \\ -2 \\ 1 \end{pmatrix}.$$

For $\lambda_2 = 1 + 2i$ we obtain

$$\mathbf{K}_2 = \begin{pmatrix} -2 - i \\ -2 - 3i \\ 2 \end{pmatrix}$$

so that
$$\mathbf{X}_2 = \begin{pmatrix} -2\cos 2t + \sin 2t \\ -2\cos 2t + 3\sin 2t \\ 2\cos 2t \end{pmatrix} e^t + i \begin{pmatrix} -\cos 2t - 2\sin 2t \\ -3\cos 2t - 2\sin 2t \\ 2\sin 2t \end{pmatrix} e^t.$$

Then
$$\mathbf{X} = c_1 \begin{pmatrix} 0 \\ -2 \\ 1 \end{pmatrix} e^{-3t} + c_2 \begin{pmatrix} -2\cos 2t + \sin 2t \\ -2\cos 2t + 3\sin 2t \\ 2\cos 2t \end{pmatrix} e^t + c_3 \begin{pmatrix} -\cos 2t - 2\sin 2t \\ -3\cos 2t - 2\sin 2t \\ 2\sin 2t \end{pmatrix} e^t.$$

39. We have $\det(\mathbf{A} - \lambda\mathbf{I}) = (1 - \lambda)(\lambda^2 - 2\lambda + 2) = 0$. For $\lambda_1 = 1$ we obtain

$$\mathbf{K}_1 = \begin{pmatrix} 0 \\ 2 \\ 1 \end{pmatrix}.$$

For $\lambda_2 = 1 + i$ we obtain

$$\mathbf{K}_2 = \begin{pmatrix} 1 \\ i \\ i \end{pmatrix}$$

so that
$$\mathbf{X}_2 = \begin{pmatrix} 1 \\ i \\ i \end{pmatrix} e^{(1+i)t} = \begin{pmatrix} \cos t \\ -\sin t \\ -\sin t \end{pmatrix} e^t + i \begin{pmatrix} \sin t \\ \cos t \\ \cos t \end{pmatrix} e^t.$$

Then
$$\mathbf{X} = c_1 \begin{pmatrix} 0 \\ 2 \\ 1 \end{pmatrix} e^t + c_2 \begin{pmatrix} \cos t \\ -\sin t \\ -\sin t \end{pmatrix} e^t + c_3 \begin{pmatrix} \sin t \\ \cos t \\ \cos t \end{pmatrix} e^t.$$

40. We have $\det(\mathbf{A} - \lambda\mathbf{I}) = -(\lambda - 6)(\lambda^2 - 8\lambda + 20) = 0$. For $\lambda_1 = 6$ we obtain

$$\mathbf{K}_1 = \begin{pmatrix} 0 \\ 1 \\ 0 \end{pmatrix}.$$

For $\lambda_2 = 4 + 2i$ we obtain

$$\mathbf{K}_2 = \begin{pmatrix} -i \\ 0 \\ 2 \end{pmatrix}$$

so that
$$\mathbf{X}_2 = \begin{pmatrix} -i \\ 0 \\ 2 \end{pmatrix} e^{(4+2i)t} = \begin{pmatrix} \sin 2t \\ 0 \\ 2\cos 2t \end{pmatrix} e^{4t} + i \begin{pmatrix} -\cos 2t \\ 0 \\ 2\sin 2t \end{pmatrix} e^{4t}.$$

Then
$$\mathbf{X} = c_1 \begin{pmatrix} 0 \\ 1 \\ 0 \end{pmatrix} e^{6t} + c_2 \begin{pmatrix} \sin 2t \\ 0 \\ 2\cos 2t \end{pmatrix} e^{4t} + c_3 \begin{pmatrix} -\cos 2t \\ 0 \\ 2\sin 2t \end{pmatrix} e^{4t}.$$

41. We have $\det(\mathbf{A} - \lambda\mathbf{I}) = (2 - \lambda)(\lambda^2 + 4\lambda + 13) = 0$. For $\lambda_1 = 2$ we obtain

$$\mathbf{K}_1 = \begin{pmatrix} 28 \\ -5 \\ 25 \end{pmatrix}.$$

For $\lambda_2 = -2 + 3i$ we obtain

$$\mathbf{K}_2 = \begin{pmatrix} 4 + 3i \\ -5 \\ 0 \end{pmatrix}$$

so that
$$\mathbf{X}_2 = \begin{pmatrix} 4 + 3i \\ -5 \\ 0 \end{pmatrix} e^{(-2+3i)t}$$

$$= \begin{pmatrix} 4\cos 3t - 3\sin 3t \\ -5\cos 3t \\ 0 \end{pmatrix} e^{-2t} + i \begin{pmatrix} 4\sin 3t + 3\cos 3t \\ -5\sin 3t \\ 0 \end{pmatrix} e^{-2t}.$$

Then
$$\mathbf{X} = c_1 \begin{pmatrix} 28 \\ -5 \\ 25 \end{pmatrix} e^{2t} + c_2 \begin{pmatrix} 4\cos 3t - 3\sin 3t \\ -5\cos 3t \\ 0 \end{pmatrix} e^{-2t} + c_3 \begin{pmatrix} 4\sin 3t + 3\cos 3t \\ -5\sin 3t \\ 0 \end{pmatrix} e^{-2t}.$$

42. We have $\det(\mathbf{A} - \lambda\mathbf{I}) = -(\lambda + 2)(\lambda^2 + 4) = 0$. For $\lambda_1 = -2$ we obtain

$$\mathbf{K}_1 = \begin{pmatrix} 0 \\ -1 \\ 1 \end{pmatrix}.$$

For $\lambda_2 = 2i$ we obtain

$$\mathbf{K}_2 = \begin{pmatrix} -2 - 2i \\ 1 \\ 1 \end{pmatrix}$$

so that
$$\mathbf{X}_2 = \begin{pmatrix} -2 - 2i \\ 1 \\ 1 \end{pmatrix} e^{2it} = \begin{pmatrix} -2\cos 2t + 2\sin 2t \\ \cos 2t \\ \cos 2t \end{pmatrix} + i \begin{pmatrix} -2\cos 2t - 2\sin 2t \\ \sin 2t \\ \sin 2t \end{pmatrix}.$$

Then
$$\mathbf{X} = c_1 \begin{pmatrix} 0 \\ -1 \\ 1 \end{pmatrix} e^{-2t} + c_2 \begin{pmatrix} -2\cos 2t + 2\sin 2t \\ \cos 2t \\ \cos 2t \end{pmatrix} + c_3 \begin{pmatrix} -2\cos 2t - 2\sin 2t \\ \sin 2t \\ \sin 2t \end{pmatrix}.$$

43. We have $\det(\mathbf{A} - \lambda\mathbf{I}) = (1 - \lambda)(\lambda^2 + 25) = 0$. For $\lambda_1 = 1$ we obtain
$$\mathbf{K}_1 = \begin{pmatrix} 25 \\ -7 \\ 6 \end{pmatrix}.$$

For $\lambda_2 = 5i$ we obtain
$$\mathbf{K}_2 = \begin{pmatrix} 1 + 5i \\ 1 \\ 1 \end{pmatrix}$$

so that
$$\mathbf{X}_2 = \begin{pmatrix} 1 + 5i \\ 1 \\ 1 \end{pmatrix} e^{5it} = \begin{pmatrix} \cos 5t - 5\sin 5t \\ \cos 5t \\ \cos 5t \end{pmatrix} + i \begin{pmatrix} \sin 5t + 5\cos 5t \\ \sin 5t \\ \sin 5t \end{pmatrix}.$$

Then
$$\mathbf{X} = c_1 \begin{pmatrix} 25 \\ -7 \\ 6 \end{pmatrix} e^{t} + c_2 \begin{pmatrix} \cos 5t - 5\sin 5t \\ \cos 5t \\ \cos 5t \end{pmatrix} + c_3 \begin{pmatrix} \sin 5t + 5\cos 5t \\ \sin 5t \\ \sin 5t \end{pmatrix}.$$

If
$$\mathbf{X}(0) = \begin{pmatrix} 4 \\ 6 \\ -7 \end{pmatrix}$$

then $c_1 = c_2 = -1$ and $c_3 = 6$.

44. We have $\det(\mathbf{A} - \lambda\mathbf{I}) = \lambda^2 - 10\lambda + 29 = 0$. For $\lambda_1 = 5 + 2i$ we obtain
$$\mathbf{K}_1 = \begin{pmatrix} 1 \\ 1 - 2i \end{pmatrix}$$

so that
$$\mathbf{X}_1 = \begin{pmatrix} 1 \\ 1 - 2i \end{pmatrix} e^{(5+2i)t} = \begin{pmatrix} \cos 2t \\ \cos 2t + 2\sin 2t \end{pmatrix} e^{5t} + i \begin{pmatrix} \sin 2t \\ \sin 2t - 2\cos 2t \end{pmatrix} e^{5t}.$$

and
$$\mathbf{X} = c_1 \begin{pmatrix} \cos 2t \\ \cos 2t + 2\sin 2t \end{pmatrix} e^{5t} + c_3 \begin{pmatrix} \sin 2t \\ \sin 2t - 2\cos 2t \end{pmatrix} e^{5t}.$$

If

$$\mathbf{X}(0) = \begin{pmatrix} -2 \\ 8 \end{pmatrix}$$

then $c_1 = -2$ and $c_2 = 5$.

Exercises 8.3

1. From

$$\mathbf{X}' = \begin{pmatrix} 3 & -3 \\ 2 & -2 \end{pmatrix} \mathbf{X} + \begin{pmatrix} 4 \\ -1 \end{pmatrix}$$

we obtain

$$\mathbf{X}_c = c_1 \begin{pmatrix} 1 \\ 1 \end{pmatrix} + c_2 \begin{pmatrix} 3 \\ 2 \end{pmatrix} e^t.$$

Then

$$\boldsymbol{\Phi} = \begin{pmatrix} 1 & 3e^t \\ 1 & 2e^t \end{pmatrix} \quad \text{and} \quad \boldsymbol{\Phi}^{-1} = \begin{pmatrix} -2 & 3 \\ e^{-t} & -e^{-t} \end{pmatrix}$$

so that

$$\mathbf{U} = \int \boldsymbol{\Phi}^{-1}\mathbf{F}\,dt = \int \begin{pmatrix} -11 \\ 5e^{-t} \end{pmatrix} dt = \begin{pmatrix} -11t \\ -5e^{-t} \end{pmatrix}$$

and

$$\mathbf{X}_p = \boldsymbol{\Phi}\mathbf{U} = \begin{pmatrix} -11 \\ -11 \end{pmatrix} t + \begin{pmatrix} -15 \\ -10 \end{pmatrix}.$$

2. From

$$\mathbf{X}' = \begin{pmatrix} 2 & -1 \\ 3 & -2 \end{pmatrix} \mathbf{X} + \begin{pmatrix} 0 \\ 4 \end{pmatrix} t$$

we obtain

$$\mathbf{X}_c = c_1 \begin{pmatrix} 1 \\ 1 \end{pmatrix} e^t + c_2 \begin{pmatrix} 1 \\ 3 \end{pmatrix} e^{-t}.$$

Then

$$\boldsymbol{\Phi} = \begin{pmatrix} e^t & e^{-t} \\ e^t & 3e^{-t} \end{pmatrix} \quad \text{and} \quad \boldsymbol{\Phi}^{-1} = \begin{pmatrix} \frac{3}{2}e^{-t} & -\frac{1}{2}e^{-t} \\ -\frac{1}{2}e^t & \frac{1}{2}e^t \end{pmatrix}$$

so that

$$\mathbf{U} = \int \boldsymbol{\Phi}^{-1}\mathbf{F}\,dt = \int \begin{pmatrix} -2te^{-t} \\ 2te^t \end{pmatrix} dt = \begin{pmatrix} 2te^{-t} + 2e^{-t} \\ 2te^t - 2e^t \end{pmatrix}$$

and

$$\mathbf{X}_p = \boldsymbol{\Phi}\mathbf{U} = \begin{pmatrix} 4 \\ 8 \end{pmatrix} t + \begin{pmatrix} 0 \\ -4 \end{pmatrix}.$$

3. From

$$\mathbf{X}' = \begin{pmatrix} 3 & -5 \\ 3/4 & -1 \end{pmatrix} \mathbf{X} + \begin{pmatrix} 1 \\ -1 \end{pmatrix} e^{t/2}$$

337

we obtain
$$\mathbf{X}_c = c_1 \begin{pmatrix} 10 \\ 3 \end{pmatrix} e^{3t/2} + c_2 \begin{pmatrix} 2 \\ 1 \end{pmatrix} e^{t/2}.$$

Then
$$\mathbf{\Phi} = \begin{pmatrix} 10e^{3t/2} & 2e^{t/2} \\ 3e^{3t/2} & e^{t/2} \end{pmatrix} \quad \text{and} \quad \mathbf{\Phi}^{-1} = \begin{pmatrix} \frac{1}{4}e^{-3t/2} & -\frac{1}{2}e^{-3t/2} \\ -\frac{3}{4}e^{-t/2} & \frac{5}{2}e^{-t/2} \end{pmatrix}$$

so that
$$\mathbf{U} = \int \mathbf{\Phi}^{-1}\mathbf{F}\, dt = \int \begin{pmatrix} \frac{3}{4}e^{-t} \\ -\frac{13}{4} \end{pmatrix} dt = \begin{pmatrix} -\frac{3}{4}e^{-t} \\ -\frac{13}{4}t \end{pmatrix}$$

and
$$\mathbf{X}_p = \mathbf{\Phi}\mathbf{U} = \begin{pmatrix} -13/2 \\ -13/4 \end{pmatrix} te^{t/2} + \begin{pmatrix} -15/2 \\ -9/4 \end{pmatrix} e^{t/2}.$$

4. From
$$\mathbf{X}' = \begin{pmatrix} 2 & -1 \\ 4 & 2 \end{pmatrix} \mathbf{X} + \begin{pmatrix} \sin 2t \\ 2\cos 2t \end{pmatrix}$$

we obtain
$$\mathbf{X}_c = c_1 \begin{pmatrix} -\sin 2t \\ 2\cos 2t \end{pmatrix} e^{2t} + c_2 \begin{pmatrix} \cos 2t \\ 2\sin 2t \end{pmatrix} e^{2t}.$$

Then
$$\mathbf{\Phi} = \begin{pmatrix} -e^{2t}\sin 2t & e^{2t}\cos 2t \\ 2e^{2t}\cos 2t & 2e^{2t}\sin 2t \end{pmatrix} \quad \text{and} \quad \mathbf{\Phi}^{-1} = \begin{pmatrix} -\frac{1}{2}e^{-2t}\sin 2t & \frac{1}{4}e^{-2t}\cos 2t \\ \frac{1}{2}e^{-2t}\cos 2t & \frac{1}{4}e^{-2t}\sin 2t \end{pmatrix}$$

so that
$$\mathbf{U} = \int \mathbf{\Phi}^{-1}\mathbf{F}\, dt = \int \begin{pmatrix} \frac{1}{2}\cos 4t \\ \frac{1}{2}\sin 4t \end{pmatrix} dt = \begin{pmatrix} \frac{1}{8}\sin 4t \\ -\frac{1}{8}\cos 4t \end{pmatrix}$$

and
$$\mathbf{X}_p = \mathbf{\Phi}\mathbf{U} = \begin{pmatrix} -\frac{1}{8}\sin 2t\cos 4t - \frac{1}{8}\cos 2t\cos 4t \\ \frac{1}{4}\cos 2t\sin 4t - \frac{1}{4}\sin 2t\cos 4t \end{pmatrix} e^{2t}.$$

5. From
$$\mathbf{X}' = \begin{pmatrix} 0 & 2 \\ -1 & 3 \end{pmatrix} \mathbf{X} + \begin{pmatrix} 1 \\ -1 \end{pmatrix} e^{t}$$

we obtain
$$\mathbf{X}_c = c_1 \begin{pmatrix} 2 \\ 1 \end{pmatrix} e^{t} + c_2 \begin{pmatrix} 1 \\ 1 \end{pmatrix} e^{2t}.$$

Then
$$\mathbf{\Phi} = \begin{pmatrix} 2e^{t} & e^{2t} \\ e^{t} & e^{2t} \end{pmatrix} \quad \text{and} \quad \mathbf{\Phi}^{-1} = \begin{pmatrix} e^{-t} & -e^{-t} \\ -e^{-2t} & 2e^{-2t} \end{pmatrix}$$

so that
$$\mathbf{U} = \int \mathbf{\Phi}^{-1}\mathbf{F}\, dt = \int \begin{pmatrix} 2 \\ -3e^{-t} \end{pmatrix} dt = \begin{pmatrix} 2t \\ 3e^{-t} \end{pmatrix}$$

and
$$\mathbf{X}_p = \mathbf{\Phi}\mathbf{U} = \begin{pmatrix} 4 \\ 2 \end{pmatrix} te^{t} + \begin{pmatrix} 3 \\ 3 \end{pmatrix} e^{t}.$$

6. From

$$\mathbf{X}' = \begin{pmatrix} 0 & 2 \\ -1 & 3 \end{pmatrix} \mathbf{X} + \begin{pmatrix} 2 \\ e^{-3t} \end{pmatrix}$$

we obtain

$$\mathbf{X}_c = c_1 \begin{pmatrix} 2 \\ 1 \end{pmatrix} e^t + c_2 \begin{pmatrix} 1 \\ 1 \end{pmatrix} e^{2t}.$$

Then

$$\mathbf{\Phi} = \begin{pmatrix} 2e^t & e^{2t} \\ e^t & e^{2t} \end{pmatrix} \quad \text{and} \quad \mathbf{\Phi}^{-1} = \begin{pmatrix} e^{-t} & -e^{-t} \\ -e^{-2t} & 2e^{-2t} \end{pmatrix}$$

so that

$$\mathbf{U} = \int \mathbf{\Phi}^{-1} \mathbf{F}\, dt = \int \begin{pmatrix} 2e^{-t} - e^{-4t} \\ -2e^{-2t} + 2e^{-5t} \end{pmatrix} dt = \begin{pmatrix} -2e^{-t} + \frac{1}{4}e^{-4t} \\ e^{-2t} - \frac{2}{5}e^{-5t} \end{pmatrix}$$

and

$$\mathbf{X}_p = \mathbf{\Phi}\mathbf{U} = \begin{pmatrix} \frac{1}{10}e^{-3t} - 3 \\ -\frac{3}{20}e^{-3t} - 1 \end{pmatrix}.$$

7. From

$$\mathbf{X}' = \begin{pmatrix} 1 & 8 \\ 1 & -1 \end{pmatrix} \mathbf{X} + \begin{pmatrix} 12 \\ 12 \end{pmatrix} t$$

we obtain

$$\mathbf{X}_c = c_1 \begin{pmatrix} 4 \\ 1 \end{pmatrix} e^{3t} + c_2 \begin{pmatrix} -2 \\ 1 \end{pmatrix} e^{-3t}.$$

Then

$$\mathbf{\Phi} = \begin{pmatrix} 4e^{3t} & -2e^{-3t} \\ e^{3t} & e^{-3t} \end{pmatrix} \quad \text{and} \quad \mathbf{\Phi}^{-1} = \begin{pmatrix} \frac{1}{6}e^{-3t} & \frac{1}{3}e^{-3t} \\ -\frac{1}{6}e^{3t} & \frac{2}{3}e^{3t} \end{pmatrix}$$

so that

$$\mathbf{U} = \int \mathbf{\Phi}^{-1} \mathbf{F}\, dt = \int \begin{pmatrix} 6te^{-3t} \\ 6te^{3t} \end{pmatrix} dt = \begin{pmatrix} -2te^{-3t} - \frac{2}{3}e^{-3t} \\ 2te^{3t} - \frac{2}{3}e^{3t} \end{pmatrix}$$

and

$$\mathbf{X}_p = \mathbf{\Phi}\mathbf{U} = \begin{pmatrix} -12 \\ 0 \end{pmatrix} t + \begin{pmatrix} -4/3 \\ -4/3 \end{pmatrix}.$$

8. From

$$\mathbf{X}' = \begin{pmatrix} 1 & 8 \\ 1 & -1 \end{pmatrix} \mathbf{X} + \begin{pmatrix} e^{-t} \\ te^t \end{pmatrix}$$

we obtain

$$\mathbf{X}_c = c_1 \begin{pmatrix} 4 \\ 1 \end{pmatrix} e^{3t} + c_2 \begin{pmatrix} -2 \\ 1 \end{pmatrix} e^{-3t}.$$

Then

$$\mathbf{\Phi} = \begin{pmatrix} 4e^{3t} & -2e^{3t} \\ e^{3t} & e^{-3t} \end{pmatrix} \quad \text{and} \quad \mathbf{\Phi}^{-1} = \begin{pmatrix} \frac{1}{6}e^{-3t} & \frac{1}{3}e^{-3t} \\ -\frac{1}{6}e^{3t} & \frac{2}{3}e^{3t} \end{pmatrix}$$

so that
$$\mathbf{U} = \int \mathbf{\Phi}^{-1}\mathbf{F}\,dt = \int \begin{pmatrix} \frac{1}{6}e^{-4t} + \frac{1}{3}te^{-2t} \\ -\frac{1}{6}e^{2t} + \frac{2}{3}te^{4t} \end{pmatrix} dt = \begin{pmatrix} -\frac{1}{24}e^{-4t} - \frac{1}{6}te^{-2t} - \frac{1}{12}e^{-2t} \\ -\frac{1}{12}e^{2t} + \frac{1}{6}te^{4t} - \frac{1}{24}e^{4t} \end{pmatrix}$$

and
$$\mathbf{X}_p = \mathbf{\Phi}\mathbf{U} = \begin{pmatrix} -te^t - \frac{1}{4}e^t \\ -\frac{1}{8}e^{-t} - \frac{1}{8}e^t \end{pmatrix}.$$

9. From
$$\mathbf{X}' = \begin{pmatrix} 3 & 2 \\ -2 & -1 \end{pmatrix}\mathbf{X} + \begin{pmatrix} 2 \\ 1 \end{pmatrix}e^{-t}$$

we obtain
$$\mathbf{X}_c = c_1 \begin{pmatrix} 1 \\ -1 \end{pmatrix}e^t + c_2 \left[\begin{pmatrix} 1 \\ -1 \end{pmatrix}te^t + \begin{pmatrix} 0 \\ 1/2 \end{pmatrix}e^t\right].$$

Then
$$\mathbf{\Phi} = \begin{pmatrix} e^t & te^t \\ -e^t & \frac{1}{2}e^t - te^t \end{pmatrix} \quad \text{and} \quad \mathbf{\Phi}^{-1} = \begin{pmatrix} e^{-t} - 2te^{-t} & -2te^{-t} \\ 2e^{-t} & 2e^{-t} \end{pmatrix}$$

so that
$$\mathbf{U} = \int \mathbf{\Phi}^{-1}\mathbf{F}\,dt = \int \begin{pmatrix} 2e^{-2t} - 6te^{-2t} \\ 6e^{-2t} \end{pmatrix} dt = \begin{pmatrix} \frac{1}{2}e^{-2t} + 3te^{-2t} \\ -3e^{-2t} \end{pmatrix}$$

and
$$\mathbf{X}_p = \mathbf{\Phi}\mathbf{U} = \begin{pmatrix} 1/2 \\ -2 \end{pmatrix}e^{-t}.$$

10. From
$$\mathbf{X}' = \begin{pmatrix} 3 & 2 \\ -2 & -1 \end{pmatrix}\mathbf{X} + \begin{pmatrix} 1 \\ 1 \end{pmatrix}$$

we obtain
$$\mathbf{X}_c = c_1 \begin{pmatrix} 1 \\ -1 \end{pmatrix}e^t + c_2 \left[\begin{pmatrix} 1 \\ -1 \end{pmatrix}te^t + \begin{pmatrix} 0 \\ 1/2 \end{pmatrix}e^t\right].$$

Then
$$\mathbf{\Phi} = \begin{pmatrix} e^t & te^t \\ -e^t & \frac{1}{2}e^t - te^t \end{pmatrix} \quad \text{and} \quad \mathbf{\Phi}^{-1} = \begin{pmatrix} e^{-t} - 2te^{-t} & -2te^{-t} \\ 2e^{-t} & 2e^{-t} \end{pmatrix}$$

so that
$$\mathbf{U} = \int \mathbf{\Phi}^{-1}\mathbf{F}\,dt = \int \begin{pmatrix} e^{-t} - 4te^{-t} \\ 2e^{-t} \end{pmatrix} dt = \begin{pmatrix} 3e^{-t} + 4te^{-t} \\ -2e^{-t} \end{pmatrix}$$

and
$$\mathbf{X}_p = \mathbf{\Phi}\mathbf{U} = \begin{pmatrix} 3 \\ -5 \end{pmatrix}.$$

11. From
$$\mathbf{X}' = \begin{pmatrix} 0 & -1 \\ 1 & 0 \end{pmatrix}\mathbf{X} + \begin{pmatrix} \sec t \\ 0 \end{pmatrix}$$

we obtain
$$\mathbf{X}_c = c_1 \begin{pmatrix} \cos t \\ \sin t \end{pmatrix} + c_2 \begin{pmatrix} \sin t \\ -\cos t \end{pmatrix}.$$

Then
$$\mathbf{\Phi} = \begin{pmatrix} \cos t & \sin t \\ \sin t & -\cos t \end{pmatrix} \quad \text{and} \quad \mathbf{\Phi}^{-1} = \begin{pmatrix} \cos t & \sin t \\ \sin t & -\cos t \end{pmatrix}$$

so that
$$\mathbf{U} = \int \mathbf{\Phi}^{-1}\mathbf{F}\, dt = \int \begin{pmatrix} 1 \\ \tan t \end{pmatrix} dt = \begin{pmatrix} t \\ \ln|\sec t| \end{pmatrix}$$

and
$$\mathbf{X}_p = \mathbf{\Phi U} = \begin{pmatrix} t\cos t + \sin t \ln|\sec t| \\ t\sin t - \cos t \ln|\sec t| \end{pmatrix}.$$

12. From
$$\mathbf{X}' = \begin{pmatrix} 1 & -1 \\ 1 & 1 \end{pmatrix}\mathbf{X} + \begin{pmatrix} 3 \\ 3 \end{pmatrix} e^t$$

we obtain
$$\mathbf{X}_c = c_1 \begin{pmatrix} -\sin t \\ \cos t \end{pmatrix} e^t + c_2 \begin{pmatrix} \cos t \\ \sin t \end{pmatrix} e^t.$$

Then
$$\mathbf{\Phi} = \begin{pmatrix} -\sin t & \cos t \\ \cos t & \sin t \end{pmatrix} e^t \quad \text{and} \quad \mathbf{\Phi}^{-1} = \begin{pmatrix} -\sin t & \cos t \\ \cos t & \sin t \end{pmatrix} e^{-t}$$

so that
$$\mathbf{U} = \int \mathbf{\Phi}^{-1}\mathbf{F}\, dt = \int \begin{pmatrix} -3\sin t + 3\cos t \\ 3\cos t + 3\sin t \end{pmatrix} dt = \begin{pmatrix} 3\cos t + 3\sin t \\ 3\sin t - 3\cos t \end{pmatrix}$$

and
$$\mathbf{X}_p = \mathbf{\Phi U} = \begin{pmatrix} -3 \\ 3 \end{pmatrix} e^t.$$

13. From
$$\mathbf{X}' = \begin{pmatrix} 1 & -1 \\ 1 & 1 \end{pmatrix}\mathbf{X} + \begin{pmatrix} \cos t \\ \sin t \end{pmatrix} e^t$$

we obtain
$$\mathbf{X}_c = c_1 \begin{pmatrix} -\sin t \\ \cos t \end{pmatrix} c^t + c_2 \begin{pmatrix} \cos t \\ \sin t \end{pmatrix} e^t.$$

Then
$$\mathbf{\Phi} = \begin{pmatrix} -\sin t & \cos t \\ \cos t & \sin t \end{pmatrix} e^t \quad \text{and} \quad \mathbf{\Phi}^{-1} = \begin{pmatrix} -\sin t & \cos t \\ \cos t & \sin t \end{pmatrix} e^{-t}$$

so that
$$\mathbf{U} = \int \mathbf{\Phi}^{-1}\mathbf{F}\, dt = \int \begin{pmatrix} 0 \\ 1 \end{pmatrix} dt = \begin{pmatrix} 0 \\ t \end{pmatrix}$$

and
$$\mathbf{X}_p = \mathbf{\Phi U} = \begin{pmatrix} \cos t \\ \sin t \end{pmatrix} t e^t.$$

341

14. From

$$X' = \begin{pmatrix} 2 & -2 \\ 8 & -6 \end{pmatrix} X + \begin{pmatrix} 1 \\ 3 \end{pmatrix} \frac{1}{t} e^{-2t}$$

we obtain

$$X_c = c_1 \begin{pmatrix} 1 \\ 2 \end{pmatrix} e^{-2t} + c_2 \left[\begin{pmatrix} 1 \\ 2 \end{pmatrix} t e^{-2t} + \begin{pmatrix} 1/2 \\ 1/2 \end{pmatrix} e^{-2t} \right].$$

Then

$$\Phi = \begin{pmatrix} 1 & 2t+1 \\ 2 & 4t+1 \end{pmatrix} e^{-2t} \quad \text{and} \quad \Phi^{-1} = \begin{pmatrix} -(4t+1) & 2t+1 \\ 2 & -1 \end{pmatrix} e^{2t}$$

so that

$$U = \int \Phi^{-1} F \, dt = \int \begin{pmatrix} 2t + 2\ln t \\ -\ln t \end{pmatrix} dt$$

and

$$X_p = \Phi U = \begin{pmatrix} 2t + \ln t - 2t\ln t \\ 4t + 3\ln t - 4t\ln t \end{pmatrix} e^{-2t}.$$

15. From

$$X' = \begin{pmatrix} 0 & 1 \\ -1 & 0 \end{pmatrix} X + \begin{pmatrix} 0 \\ \sec t \tan t \end{pmatrix}$$

we obtain

$$X_c = c_1 \begin{pmatrix} \cos t \\ -\sin t \end{pmatrix} + c_2 \begin{pmatrix} \sin t \\ \cos t \end{pmatrix}.$$

Then

$$\Phi = \begin{pmatrix} \cos t & \sin t \\ -\sin t & \cos t \end{pmatrix} t \quad \text{and} \quad \Phi^{-1} = \begin{pmatrix} \cos t & -\sin t \\ \sin t & \cos t \end{pmatrix}$$

so that

$$U = \int \Phi^{-1} F \, dt = \int \begin{pmatrix} -\tan^2 t \\ \tan t \end{pmatrix} dt = \begin{pmatrix} t - \tan t \\ \ln|\sec t| \end{pmatrix}$$

and

$$X_p = \Phi U = \begin{pmatrix} \cos t \\ -\sin t \end{pmatrix} t + \begin{pmatrix} -\sin t \\ \sin t \tan t \end{pmatrix} + \begin{pmatrix} \sin t \\ \cos t \end{pmatrix} \ln|\sec t|.$$

16. From

$$X' = \begin{pmatrix} 0 & 1 \\ -1 & 0 \end{pmatrix} X + \begin{pmatrix} 1 \\ \cot t \end{pmatrix}$$

we obtain

$$X_c = c_1 \begin{pmatrix} \cos t \\ -\sin t \end{pmatrix} + c_2 \begin{pmatrix} \sin t \\ \cos t \end{pmatrix}.$$

Then

$$\Phi = \begin{pmatrix} \cos t & \sin t \\ -\sin t & \cos t \end{pmatrix} \quad \text{and} \quad \Phi^{-1} = \begin{pmatrix} \cos t & -\sin t \\ \sin t & \cos t \end{pmatrix}$$

so that

$$U = \int \Phi^{-1} F \, dt = \int \begin{pmatrix} 0 \\ \csc t \end{pmatrix} dt = \begin{pmatrix} 0 \\ \ln|\csc t - \cot t| \end{pmatrix}$$

and
$$\mathbf{X}_p = \boldsymbol{\Phi}\mathbf{U} = \begin{pmatrix} \sin t \ln|\csc t - \cot t| \\ \cos t \ln|\csc t - \cot t| \end{pmatrix}.$$

17. From
$$\mathbf{X}' = \begin{pmatrix} 1 & 2 \\ -1/2 & 1 \end{pmatrix}\mathbf{X} + \begin{pmatrix} \csc t \\ \sec t \end{pmatrix}e^t$$

we obtain
$$\mathbf{X}_c = c_1 \begin{pmatrix} 2\sin t \\ \cos t \end{pmatrix}e^t + c_2 \begin{pmatrix} 2\cos t \\ -\sin t \end{pmatrix}e^t.$$

Then
$$\boldsymbol{\Phi} = \begin{pmatrix} 2\sin t & 2\cos t \\ \cos t & -\sin t \end{pmatrix}e^t \quad \text{and} \quad \boldsymbol{\Phi}^{-1} = \begin{pmatrix} \frac{1}{2}\sin t & \cos t \\ \frac{1}{2}\cos t & -\sin t \end{pmatrix}e^{-t}$$

so that
$$\mathbf{U} = \int \boldsymbol{\Phi}^{-1}\mathbf{F}\,dt = \int \begin{pmatrix} \frac{3}{2} \\ \frac{1}{2}\cos t - \tan t \end{pmatrix}dt = \begin{pmatrix} \frac{3}{2}t \\ \frac{1}{2}\ln|\sin t| - \ln|\sec t| \end{pmatrix}$$

and
$$\mathbf{X}_p = \boldsymbol{\Phi}\mathbf{U} = \begin{pmatrix} 3\sin t \\ \frac{3}{2}\cos t \end{pmatrix}te^t + \begin{pmatrix} \cos t \\ -\frac{1}{2}\sin t \end{pmatrix}\ln|\sin t| + \begin{pmatrix} -2\cos t \\ \sin t \end{pmatrix}\ln|\sec t|.$$

18. From
$$\mathbf{X}' = \begin{pmatrix} 1 & -2 \\ 1 & -1 \end{pmatrix}\mathbf{X} + \begin{pmatrix} \tan t \\ 1 \end{pmatrix}$$

we obtain
$$\mathbf{X}_c = c_1 \begin{pmatrix} \cos t - \sin t \\ \cos t \end{pmatrix} + c_2 \begin{pmatrix} \cos t + \sin t \\ \sin t \end{pmatrix}.$$

Then
$$\boldsymbol{\Phi} = \begin{pmatrix} \cos t - \sin t & \cos t + \sin t \\ \cos t & \sin t \end{pmatrix} \quad \text{and} \quad \boldsymbol{\Phi}^{-1} = \begin{pmatrix} -\sin t & \cos t + \sin t \\ \cos t & \sin t - \cos t \end{pmatrix}$$

so that
$$\mathbf{U} = \int \boldsymbol{\Phi}^{-1}\mathbf{F}\,dt = \int \begin{pmatrix} 2\cos t + \sin t - \sec t \\ 2\sin t - \cos t \end{pmatrix}dt = \begin{pmatrix} 2\sin t - \cos t - \ln|\sec t + \tan t| \\ -2\cos t - \sin t \end{pmatrix}$$

and
$$\mathbf{X}_p = \boldsymbol{\Phi}\mathbf{U} = \begin{pmatrix} 3\sin t \cos t - \cos^2 t - 2\sin^2 t + (\sin t - \cos t)\ln|\sec t + \tan t| \\ \sin^2 t - \cos^2 t - \cos t(\ln|\sec t + \tan t|) \end{pmatrix}.$$

19. From
$$\mathbf{X}' = \begin{pmatrix} 1 & 1 & 0 \\ 1 & 1 & 0 \\ 0 & 0 & 3 \end{pmatrix}\mathbf{X} + \begin{pmatrix} e^t \\ e^{2t} \\ te^{3t} \end{pmatrix}$$

we obtain
$$\mathbf{X}_c = c_1 \begin{pmatrix} 1 \\ -1 \\ 0 \end{pmatrix} + c_2 \begin{pmatrix} 1 \\ 1 \\ 0 \end{pmatrix}e^{2t} + c_3 \begin{pmatrix} 0 \\ 0 \\ 1 \end{pmatrix}e^{3t}.$$

343

Then
$$\Phi = \begin{pmatrix} 1 & e^{2t} & 0 \\ -1 & e^{2t} & 0 \\ 0 & 0 & e^{3t} \end{pmatrix} \quad \text{and} \quad \Phi^{-1} = \begin{pmatrix} \frac{1}{2} & -\frac{1}{2} & 0 \\ \frac{1}{2}e^{-2t} & \frac{1}{2}e^{-2t} & 0 \\ 0 & 0 & e^{-3t} \end{pmatrix}$$

so that
$$U = \int \Phi^{-1}F \, dt = \int \begin{pmatrix} \frac{1}{2}e^t - \frac{1}{2}e^{2t} \\ \frac{1}{2}e^{-t} + \frac{1}{2} \\ t \end{pmatrix} dt = \begin{pmatrix} \frac{1}{2}e^t - \frac{1}{4}e^{2t} \\ -\frac{1}{2}e^{-t} + \frac{1}{2}t \\ \frac{1}{2}t^2 \end{pmatrix}$$

and
$$X_p = \Phi U = \begin{pmatrix} -\frac{1}{4}e^{2t} + \frac{1}{2}te^{2t} \\ -e^t + \frac{1}{4}e^{2t} + \frac{1}{2}te^{2t} \\ \frac{1}{2}t^2 e^{3t} \end{pmatrix}.$$

20. From
$$X' = \begin{pmatrix} 3 & -1 & -1 \\ 1 & 1 & -1 \\ 1 & -1 & 1 \end{pmatrix} X + \begin{pmatrix} 0 \\ t \\ 2e^t \end{pmatrix}$$

we obtain
$$X_c = c_1 \begin{pmatrix} 1 \\ 1 \\ 1 \end{pmatrix} e^t + c_2 \begin{pmatrix} 1 \\ 1 \\ 0 \end{pmatrix} e^{2t} + c_3 \begin{pmatrix} 1 \\ 0 \\ 1 \end{pmatrix} e^{2t}.$$

Then
$$\Phi = \begin{pmatrix} e^t & e^{2t} & e^{2t} \\ e^t & e^{2t} & 0 \\ e^t & 0 & e^{2t} \end{pmatrix} \quad \text{and} \quad \Phi^{-1} = \begin{pmatrix} -e^{-t} & e^{-t} & e^{-t} \\ e^{-2t} & 0 & -e^{-2t} \\ e^{-2t} & -e^{-2t} & 0 \end{pmatrix}$$

so that
$$U = \int \Phi^{-1}F \, dt = \int \begin{pmatrix} te^{-t} + 2 \\ -2e^{-t} \\ -te^{-2t} \end{pmatrix} dt = \begin{pmatrix} -te^{-t} - e^{-t} + 2t \\ 2e^{-t} \\ \frac{1}{2}te^{-2t} + \frac{1}{4}e^{-2t} \end{pmatrix}$$

and
$$X_p = \Phi U = \begin{pmatrix} -1/2 \\ -1 \\ -1/2 \end{pmatrix} t + \begin{pmatrix} -3/4 \\ -1 \\ -3/4 \end{pmatrix} + \begin{pmatrix} 2 \\ 2 \\ 0 \end{pmatrix} e^t + \begin{pmatrix} 2 \\ 2 \\ 2 \end{pmatrix} te^t.$$

21. From
$$X' = \begin{pmatrix} 3 & -1 \\ -1 & 3 \end{pmatrix} X + \begin{pmatrix} 4e^{2t} \\ 4e^{4t} \end{pmatrix}$$

we obtain
$$\Phi = \begin{pmatrix} -e^{4t} & e^{2t} \\ e^{4t} & e^{2t} \end{pmatrix}, \quad \Phi^{-1} = \begin{pmatrix} -\frac{1}{2}e^{-4t} & \frac{1}{2}e^{-4t} \\ \frac{1}{2}e^{-2t} & \frac{1}{2}e^{-2t} \end{pmatrix},$$

and
$$X = \Phi\Phi^{-1}(0)X(0) + \Phi\int_0^t \Phi^{-1}F\,ds = \Phi \cdot \begin{pmatrix} 0 \\ 1 \end{pmatrix} + \Phi \cdot \begin{pmatrix} e^{-2t} + 2t - 1 \\ e^{2t} + 2t - 1 \end{pmatrix}$$

$$= \begin{pmatrix} 2 \\ 2 \end{pmatrix} te^{2t} + \begin{pmatrix} -1 \\ 1 \end{pmatrix} e^{2t} + \begin{pmatrix} -2 \\ 2 \end{pmatrix} te^{4t} + \begin{pmatrix} 2 \\ 0 \end{pmatrix} e^{4t}.$$

22. From
$$X' = \begin{pmatrix} 1 & -1 \\ 1 & -1 \end{pmatrix} X + \begin{pmatrix} 1/t \\ 1/t \end{pmatrix}$$

we obtain
$$\Phi = \begin{pmatrix} 1 & 1+t \\ 1 & t \end{pmatrix}, \qquad \Phi^{-1} = \begin{pmatrix} -t & 1+t \\ 1 & -1 \end{pmatrix},$$

and
$$X = \Phi\Phi^{-1}(1)X(1) + \Phi\int_1^t \Phi^{-1}F\,ds = \Phi \cdot \begin{pmatrix} -4 \\ 3 \end{pmatrix} + \Phi \cdot \begin{pmatrix} \ln t \\ 0 \end{pmatrix} = \begin{pmatrix} 3 \\ 3 \end{pmatrix} t - \begin{pmatrix} 1 \\ 4 \end{pmatrix} + \begin{pmatrix} 1 \\ 1 \end{pmatrix} \ln t.$$

23. Let $I = \begin{pmatrix} i_1 \\ i_2 \end{pmatrix}$ so that

$$I' = \begin{pmatrix} -11 & 3 \\ 3 & -3 \end{pmatrix} I + \begin{pmatrix} 100\sin t \\ 0 \end{pmatrix}$$

and
$$X_c = c_1 \begin{pmatrix} 1 \\ 3 \end{pmatrix} e^{-2t} + c_2 \begin{pmatrix} 3 \\ -1 \end{pmatrix} e^{-12t}.$$

Then
$$\Phi = \begin{pmatrix} e^{-2t} & 3e^{-12t} \\ 3e^{-2t} & -e^{-12t} \end{pmatrix}, \qquad \Phi^{-1} = \begin{pmatrix} \frac{1}{10}e^{2t} & \frac{3}{10}e^{2t} \\ \frac{3}{10}e^{12t} & -\frac{1}{10}e^{12t} \end{pmatrix},$$

$$U = \int \Phi^{-1}F\,dt = \int \begin{pmatrix} 10e^{2t}\sin t \\ 30e^{12t}\sin t \end{pmatrix} dt = \begin{pmatrix} 2e^{2t}(2\sin t - \cos t) \\ \frac{6}{29}e^{12t}(12\sin t - \cos t) \end{pmatrix},$$

and
$$I_p = \Phi U = \begin{pmatrix} \frac{332}{29}\sin t - \frac{76}{29}\cos t \\ \frac{276}{29}\sin t - \frac{168}{29}\cos t \end{pmatrix}$$

so that
$$I = c_1 \begin{pmatrix} 1 \\ 3 \end{pmatrix} e^{-2t} + c_2 \begin{pmatrix} 3 \\ -1 \end{pmatrix} e^{-12t} + I_p.$$

If $I(0) = \begin{pmatrix} 0 \\ 0 \end{pmatrix}$ then $c_1 = 2$ and $c_2 = \frac{6}{29}$.

345

24. (a) The eigenvalues are 0, 1, 3, and 4, with corresponding eigenvectors

$$\begin{pmatrix} -6 \\ -4 \\ 1 \\ 2 \end{pmatrix}, \quad \begin{pmatrix} 2 \\ 1 \\ 0 \\ 0 \end{pmatrix}, \quad \begin{pmatrix} 3 \\ 1 \\ 2 \\ 1 \end{pmatrix}, \quad \text{and} \quad \begin{pmatrix} -1 \\ 1 \\ 0 \\ 0 \end{pmatrix}.$$

(b) $\Phi = \begin{pmatrix} -6 & 2e^t & 3e^{3t} & -e^{4t} \\ -4 & e^t & e^{3t} & e^{4t} \\ 1 & 0 & 2e^{3t} & 0 \\ 2 & 0 & e^{3t} & 0 \end{pmatrix}$, $\quad \Phi^{-1} = \begin{pmatrix} 0 & 0 & -\frac{1}{3} & \frac{2}{3} \\ \frac{1}{3}e^{-t} & \frac{1}{3}e^{-t} & -2e^{-t} & \frac{8}{3}e^{-t} \\ 0 & 0 & \frac{2}{3}e^{-3t} & -\frac{1}{3}e^{-3t} \\ -\frac{1}{3}e^{-4t} & \frac{2}{3}e^{-4t} & 0 & \frac{1}{3}e^{-4t} \end{pmatrix}$

(c) $\Phi^{-1}(t)\mathbf{F}(t) = \begin{pmatrix} \frac{2}{3} - \frac{1}{3}e^{2t} \\ \frac{1}{3}e^{-2t} + \frac{8}{3}e^{-t} - 2e^t + \frac{1}{3}t \\ -\frac{1}{3}e^{-3t} + \frac{2}{3}e^{-t} \\ \frac{2}{3}e^{-5t} + \frac{1}{3}e^{-4t} - \frac{1}{3}te^{-3t} \end{pmatrix},$

$$\int \Phi^{-1}(t)\mathbf{F}(t)\,dt = \begin{pmatrix} -\frac{1}{6}e^{2t} + \frac{2}{3}t \\ -\frac{1}{6}e^{-2t} - \frac{8}{3}e^{-t} - 2e^t + \frac{1}{6}t^2 \\ \frac{1}{9}e^{-3t} - \frac{2}{3}e^{-t} \\ -\frac{2}{15}e^{-5t} - \frac{1}{12}e^{-4t} + \frac{1}{27}e^{-3t} + \frac{1}{9}te^{-3t} \end{pmatrix},$$

$$\Phi(t)\int \Phi^{-1}(t)\mathbf{F}(t)\,dt = \begin{pmatrix} -5e^{2t} - \frac{1}{5}e^{-t} - \frac{1}{27}e^t - \frac{1}{9}te^t + \frac{1}{3}t^2e^t - 4t - \frac{59}{12} \\ -2e^{2t} - \frac{3}{10}e^{-t} + \frac{1}{27}e^t + \frac{1}{9}te^t + \frac{1}{6}t^2e^t - \frac{8}{3}t - \frac{95}{36} \\ -\frac{3}{2}e^{2t} + \frac{2}{3}t + \frac{2}{9} \\ -e^{2t} + \frac{4}{3}t - \frac{1}{9} \end{pmatrix},$$

$$\Phi(t)\mathbf{C} = \begin{pmatrix} -6c_1 + 2c_2e^t + 3c_3e^{3t} - c_4e^{4t} \\ -4c_1 + c_2e^t + c_3e^{3t} + c_4e^{4t} \\ c_1 + 2c_3e^{3t} \\ 2c_1 + c_3e^{3t} \end{pmatrix},$$

$$\Phi(t)\mathbf{C} + \Phi(t)\int \Phi^{-1}(t)\mathbf{F}(t)\,dt$$

$$= \begin{pmatrix} -6c_1 + 2c_2e^t + 3c_3e^{3t} - c_4e^{4t} \\ -4c_1 + c_2e^t + c_3e^{3t} + c_4e^{4t} \\ c_1 + 2c_3e^{3t} \\ 2c_1 + c_3e^{3t} \end{pmatrix} + \begin{pmatrix} -5e^{2t} - \frac{1}{5}e^{-t} - \frac{1}{27}e^t - \frac{1}{9}te^t + \frac{1}{3}t^2e^t - 4t - \frac{59}{12} \\ -2e^{2t} - \frac{3}{10}e^{-t} + \frac{1}{27}e^t + \frac{1}{9}te^t + \frac{1}{6}t^2e^t - \frac{8}{3}t - \frac{95}{36} \\ -\frac{3}{2}e^{2t} + \frac{2}{3}t + \frac{2}{9} \\ -e^{2t} + \frac{4}{3}t - \frac{1}{9} \end{pmatrix}$$

(d) $\mathbf{X}(t) = c_1 \begin{pmatrix} -6 \\ -4 \\ 1 \\ 2 \end{pmatrix} + c_2 \begin{pmatrix} 2 \\ 1 \\ 0 \\ 0 \end{pmatrix} + c_3 \begin{pmatrix} 3 \\ 1 \\ 2 \\ 1 \end{pmatrix} + c_4 \begin{pmatrix} -1 \\ 1 \\ 0 \\ 0 \end{pmatrix}$

$$+ \begin{pmatrix} -5e^{2t} - \frac{1}{5}e^{-t} - \frac{1}{27}e^t - \frac{1}{9}te^t + \frac{1}{3}t^2e^t - 4t - \frac{59}{12} \\ -2e^{2t} - \frac{3}{10}e^{-t} + \frac{1}{27}e^t + \frac{1}{9}te^t + \frac{1}{6}t^2e^t - \frac{8}{3}t - \frac{95}{36} \\ -\frac{3}{2}e^{2t} + \frac{2}{3}t + \frac{2}{9} \\ -e^{2t} + \frac{4}{3}t - \frac{1}{9} \end{pmatrix}$$

———— Exercises 8.4 ————

1. For $\mathbf{A} = \begin{pmatrix} 1 & 0 \\ 0 & 2 \end{pmatrix}$ we have

$$\mathbf{A}^2 = \begin{pmatrix} 1 & 0 \\ 0 & 2 \end{pmatrix}\begin{pmatrix} 1 & 0 \\ 0 & 2 \end{pmatrix} = \begin{pmatrix} 1 & 0 \\ 0 & 4 \end{pmatrix},$$

$$\mathbf{A}^3 = \mathbf{A}\mathbf{A}^2 = \begin{pmatrix} 1 & 0 \\ 0 & 2 \end{pmatrix}\begin{pmatrix} 1 & 0 \\ 0 & 4 \end{pmatrix} = \begin{pmatrix} 1 & 0 \\ 0 & 8 \end{pmatrix},$$

$$\mathbf{A}^4 = \mathbf{A}\mathbf{A}^3 = \begin{pmatrix} 1 & 0 \\ 0 & 2 \end{pmatrix}\begin{pmatrix} 1 & 0 \\ 0 & 8 \end{pmatrix} = \begin{pmatrix} 1 & 0 \\ 0 & 16 \end{pmatrix},$$

and so on. In general

$$\mathbf{A}^k = \begin{pmatrix} 1 & 0 \\ 0 & 2^k \end{pmatrix} \quad \text{for} \quad k = 1, 2, 3, \ldots.$$

Thus $\qquad e^{t\mathbf{A}} = \mathbf{I} + \frac{\mathbf{A}}{1!}t + \frac{\mathbf{A}^2}{2!}t^2 + \frac{\mathbf{A}^3}{3!}t^3 + \cdots$

$$= \begin{pmatrix} 1 & 0 \\ 0 & 1 \end{pmatrix} + \frac{1}{1!}\begin{pmatrix} 1 & 0 \\ 0 & 2 \end{pmatrix}t + \frac{1}{2!}\begin{pmatrix} 1 & 0 \\ 0 & 4 \end{pmatrix}t^2 + \frac{1}{3!}\begin{pmatrix} 1 & 0 \\ 0 & 8 \end{pmatrix}t^3 + \cdots$$

$$= \begin{pmatrix} 1 + t + \frac{t^2}{2!} + \frac{t^3}{3!} + \cdots & 0 \\ 0 & 1 + t + \frac{(2t)^2}{2!} + \frac{(2t)^3}{3!} + \cdots \end{pmatrix} = \begin{pmatrix} e^t & 0 \\ 0 & e^{2t} \end{pmatrix}$$

and $\qquad\qquad\qquad\qquad\qquad e^{-t\mathbf{A}} = \begin{pmatrix} e^{-t} & 0 \\ 0 & e^{-2t} \end{pmatrix}.$

2. For $\mathbf{A} = \begin{pmatrix} 0 & 1 \\ 1 & 0 \end{pmatrix}$ we have

$$\mathbf{A}^2 = \begin{pmatrix} 0 & 1 \\ 1 & 0 \end{pmatrix}\begin{pmatrix} 0 & 1 \\ 1 & 0 \end{pmatrix} = \begin{pmatrix} 1 & 0 \\ 0 & 1 \end{pmatrix} = \mathbf{I}$$

$$\mathbf{A}^3 = \mathbf{A}\mathbf{A}^2 = \begin{pmatrix} 0 & 1 \\ 1 & 0 \end{pmatrix}\mathbf{I} = \begin{pmatrix} 0 & 1 \\ 1 & 0 \end{pmatrix} = \mathbf{A}$$

$$\mathbf{A}^4 = (\mathbf{A}^2)^2 = \mathbf{I}$$

$$\mathbf{A}^5 = \mathbf{A}\mathbf{A}^4 = \mathbf{A}\mathbf{I} = \mathbf{A}$$

and so on. In general

$$\mathbf{A}^k = \begin{cases} \mathbf{A}, & k = 1, 3, 5, \ldots \\ \mathbf{I}, & k = 2, 4, 6, \ldots \end{cases}.$$

Thus

$$e^{t\mathbf{A}} = \mathbf{I} + \frac{\mathbf{A}}{1!}t + \frac{\mathbf{A}^2}{2!}t^2 + \frac{\mathbf{A}^3}{3!}t^3 + \cdots$$

$$= \mathbf{I} + \mathbf{A}t + \frac{1}{2!}\mathbf{I}t^2 + \frac{1}{3!}\mathbf{A}t^3 + \cdots$$

$$= \mathbf{I}\left(1 + \frac{1}{2!}t^2 + \frac{1}{4!}t^4 + \cdots\right) + \mathbf{A}\left(t + \frac{1}{3!}t^3 + \frac{1}{5!}t^5 + \cdots\right)$$

$$= \mathbf{I}\cosh t + \mathbf{A}\sinh t = \begin{pmatrix} \cosh t & \sinh t \\ \sinh t & \cosh t \end{pmatrix}$$

and

$$e^{-t\mathbf{A}} = \begin{pmatrix} \cosh(-t) & \sinh(-t) \\ \sinh(-t) & \cosh(-t) \end{pmatrix} = \begin{pmatrix} \cosh t & -\sinh t \\ -\sinh t & \cosh t \end{pmatrix}.$$

3. For

$$\mathbf{A} = \begin{pmatrix} 1 & 1 & 1 \\ 1 & 1 & 1 \\ -2 & -2 & -2 \end{pmatrix}$$

we have

$$\mathbf{A}^2 = \begin{pmatrix} 1 & 1 & 1 \\ 1 & 1 & 1 \\ -2 & -2 & -2 \end{pmatrix}\begin{pmatrix} 1 & 1 & 1 \\ 1 & 1 & 1 \\ -2 & -2 & -2 \end{pmatrix} = \begin{pmatrix} 0 & 0 & 0 \\ 0 & 0 & 0 \\ 0 & 0 & 0 \end{pmatrix}.$$

Thus, $\mathbf{A}^3 = \mathbf{A}^4 = \mathbf{A}^5 = \cdots = \mathbf{0}$ and

$$e^{\mathbf{A}t} = \mathbf{I} + \mathbf{A}t = \begin{pmatrix} 1 & 0 & 0 \\ 0 & 1 & 0 \\ 0 & 0 & 1 \end{pmatrix} + \begin{pmatrix} t & t & t \\ t & t & t \\ -2t & -2t & -2t \end{pmatrix} = \begin{pmatrix} t+1 & t & t \\ t & t+1 & t \\ -2t & -2t & -2t+1 \end{pmatrix}.$$

4. For

$$\mathbf{A} = \begin{pmatrix} 0 & 0 & 0 \\ 3 & 0 & 0 \\ 5 & 1 & 0 \end{pmatrix}$$

we have

$$\mathbf{A}^2 = \begin{pmatrix} 0 & 0 & 0 \\ 3 & 0 & 0 \\ 5 & 1 & 0 \end{pmatrix}\begin{pmatrix} 0 & 0 & 0 \\ 3 & 0 & 0 \\ 5 & 1 & 0 \end{pmatrix} = \begin{pmatrix} 0 & 0 & 0 \\ 0 & 0 & 0 \\ 3 & 0 & 0 \end{pmatrix}$$

$$\mathbf{A}^3 = \mathbf{A}\mathbf{A}^2 = \begin{pmatrix} 0 & 0 & 0 \\ 3 & 0 & 0 \\ 5 & 1 & 0 \end{pmatrix}\begin{pmatrix} 0 & 0 & 0 \\ 0 & 0 & 0 \\ 3 & 0 & 0 \end{pmatrix} = \begin{pmatrix} 0 & 0 & 0 \\ 0 & 0 & 0 \\ 0 & 0 & 0 \end{pmatrix}.$$

Thus, $\mathbf{A}^4 = \mathbf{A}^5 = \mathbf{A}^6 = \cdots = \mathbf{0}$ and

$$e^{\mathbf{A}t} = \mathbf{I} + \mathbf{A}t + \frac{1}{2}\mathbf{A}^2 t^2$$

$$= \begin{pmatrix} 1 & 0 & 0 \\ 0 & 1 & 0 \\ 0 & 0 & 1 \end{pmatrix} + \begin{pmatrix} 0 & 0 & 0 \\ 3t & 0 & 0 \\ 5t & t & 0 \end{pmatrix} + \begin{pmatrix} 0 & 0 & 0 \\ 0 & 0 & 0 \\ \frac{3}{2}t^2 & 0 & 0 \end{pmatrix} = \begin{pmatrix} 1 & 0 & 0 \\ 3t & 1 & 0 \\ \frac{3}{2}t^2 + 5t & t & 1 \end{pmatrix}.$$

5. Using the result of Problem 1

$$\mathbf{X} = \begin{pmatrix} e^t & 0 \\ 0 & c^{2t} \end{pmatrix}\begin{pmatrix} c_1 \\ c_2 \end{pmatrix} = c_1 \begin{pmatrix} e^t \\ 0 \end{pmatrix} + c_2 \begin{pmatrix} 0 \\ e^t \end{pmatrix}.$$

6. Using the result of Problem 2

$$\mathbf{X} = \begin{pmatrix} \cosh t & \sinh t \\ \sinh t & \cosh t \end{pmatrix}\begin{pmatrix} c_1 \\ c_2 \end{pmatrix} = c_1 \begin{pmatrix} \cosh t \\ \sinh t \end{pmatrix} + c_2 \begin{pmatrix} \sinh t \\ \cosh t \end{pmatrix}.$$

7. Using the result of Problem 3

$$\mathbf{X} = \begin{pmatrix} t+1 & t & t \\ t & t+1 & t \\ -2t & -2t & -2t+1 \end{pmatrix}\begin{pmatrix} c_1 \\ c_2 \\ c_3 \end{pmatrix} = c_1 \begin{pmatrix} t+1 \\ t \\ -2t \end{pmatrix} + c_2 \begin{pmatrix} t \\ t+1 \\ -2t \end{pmatrix} + c_3 \begin{pmatrix} t \\ t \\ -2t+1 \end{pmatrix}.$$

8. Using the result of Problem 4

$$\mathbf{X} = \begin{pmatrix} 1 & 0 & 0 \\ 3t & 1 & 0 \\ \frac{3}{2}t^2 + 5t & t & 1 \end{pmatrix}\begin{pmatrix} c_1 \\ c_2 \\ c_3 \end{pmatrix} = c_1 \begin{pmatrix} 1 \\ 3t \\ \frac{3}{2}t^2 + 5t \end{pmatrix} + c_2 \begin{pmatrix} 0 \\ 1 \\ t \end{pmatrix} + c_3 \begin{pmatrix} 0 \\ 0 \\ 1 \end{pmatrix}.$$

9. To solve

$$\mathbf{X}' = \begin{pmatrix} 1 & 0 \\ 0 & 2 \end{pmatrix}\mathbf{X} + \begin{pmatrix} 3 \\ -1 \end{pmatrix}$$

349

we identify $t_0 = 0$, $\mathbf{F}(s) = \begin{pmatrix} 3 \\ -1 \end{pmatrix}$, and use the results of Problem 1 and equation (5) in the text.

$$\mathbf{X}(t) = e^{t\mathbf{A}}\mathbf{C} + e^{t\mathbf{A}} \int_{t_0}^{t} e^{-s\mathbf{A}}\mathbf{F}(s)\, ds$$

$$= \begin{pmatrix} e^t & 0 \\ 0 & e^{2t} \end{pmatrix} \begin{pmatrix} c_1 \\ c_2 \end{pmatrix} + \begin{pmatrix} e^t & 0 \\ 0 & e^{2t} \end{pmatrix} \int_0^t \begin{pmatrix} e^{-s} & 0 \\ 0 & e^{-2s} \end{pmatrix} \begin{pmatrix} 3 \\ -1 \end{pmatrix} ds$$

$$= \begin{pmatrix} c_1 e^t \\ c_2 e^{2t} \end{pmatrix} + \begin{pmatrix} e^t & 0 \\ 0 & e^{2t} \end{pmatrix} \int_0^t \begin{pmatrix} 3e^{-s} \\ -e^{-2s} \end{pmatrix} ds$$

$$= \begin{pmatrix} c_1 e^t \\ c_2 e^{2t} \end{pmatrix} + \begin{pmatrix} e^t & 0 \\ 0 & e^{2t} \end{pmatrix} \begin{pmatrix} -3e^{-s} \\ \frac{1}{2}e^{-2s} \end{pmatrix} \Big|_0^t$$

$$= \begin{pmatrix} c_1 e^t \\ c_2 e^{2t} \end{pmatrix} + \begin{pmatrix} e^t & 0 \\ 0 & e^{2t} \end{pmatrix} \begin{pmatrix} -3e^{-t} - 3 \\ \frac{1}{2}e^{-2t} - \frac{1}{2} \end{pmatrix}$$

$$= \begin{pmatrix} c_1 e^t \\ c_2 e^{2t} \end{pmatrix} + \begin{pmatrix} -3 - 3e^t \\ \frac{1}{2} - \frac{1}{2}e^{2t} \end{pmatrix} = c_3 \begin{pmatrix} 1 \\ 0 \end{pmatrix} e^t + c_4 \begin{pmatrix} 0 \\ 1 \end{pmatrix} e^{2t} + \begin{pmatrix} -3 \\ \frac{1}{2} \end{pmatrix}.$$

10. To solve

$$\mathbf{X}' = \begin{pmatrix} 1 & 0 \\ 0 & 2 \end{pmatrix} \mathbf{X} + \begin{pmatrix} t \\ e^{4t} \end{pmatrix}$$

we identify $t_0 = 0$, $\mathbf{F}(s) = \begin{pmatrix} t \\ e^{4t} \end{pmatrix}$, and use the results of Problem 1 and equation (5) in the text.

$$\mathbf{X}(t) = e^{t\mathbf{A}}\mathbf{C} + e^{t\mathbf{A}} \int_{t_0}^{t} e^{-s\mathbf{A}}\mathbf{F}(s)\, ds$$

$$= \begin{pmatrix} e^t & 0 \\ 0 & e^{2t} \end{pmatrix} \begin{pmatrix} c_1 \\ c_2 \end{pmatrix} + \begin{pmatrix} e^t & 0 \\ 0 & e^{2t} \end{pmatrix} \int_0^t \begin{pmatrix} e^{-s} & 0 \\ 0 & e^{-2s} \end{pmatrix} \begin{pmatrix} s \\ e^{4s} \end{pmatrix} ds$$

$$= \begin{pmatrix} c_1 e^t \\ c_2 e^{2t} \end{pmatrix} + \begin{pmatrix} e^t & 0 \\ 0 & e^{2t} \end{pmatrix} \int_0^t \begin{pmatrix} se^{-s} \\ e^{2s} \end{pmatrix} ds$$

$$= \begin{pmatrix} c_1 e^t \\ c_2 e^{2t} \end{pmatrix} + \begin{pmatrix} e^t & 0 \\ 0 & e^{2t} \end{pmatrix} \begin{pmatrix} -se^{-s} - e^{-s} \\ \frac{1}{2}e^{2s} \end{pmatrix} \Big|_0^t$$

$$= \begin{pmatrix} c_1 e^t \\ c_2 e^{2t} \end{pmatrix} + \begin{pmatrix} e^t & 0 \\ 0 & e^{2t} \end{pmatrix} \begin{pmatrix} -te^{-t} - e^{-t} + 1 \\ \frac{1}{2}e^{2t} - \frac{1}{2} \end{pmatrix}$$

$$= \begin{pmatrix} c_1 e^t \\ c_2 e^{2t} \end{pmatrix} + \begin{pmatrix} -t - 1 + e^t \\ \frac{1}{2} e^{4t} - \frac{1}{2} e^{2t} \end{pmatrix} = c_3 \begin{pmatrix} 1 \\ 0 \end{pmatrix} e^t + c_4 \begin{pmatrix} 0 \\ 1 \end{pmatrix} e^{2t} + \begin{pmatrix} -t - 1 \\ \frac{1}{2} e^{4t} \end{pmatrix}.$$

11. To solve

$$\mathbf{X}' = \begin{pmatrix} 0 & 1 \\ 1 & 0 \end{pmatrix} \mathbf{X} + \begin{pmatrix} 1 \\ 1 \end{pmatrix}$$

we identify $t_0 = 0$, $\mathbf{F}(s) = \begin{pmatrix} 1 \\ 1 \end{pmatrix}$, and use the results of Problem 2 and equation (5) in the text.

$$\mathbf{X}(t) = e^{t\mathbf{A}} \mathbf{C} + e^{t\mathbf{A}} \int_{t_0}^{t} e^{-s\mathbf{A}} \mathbf{F}(s)\, ds$$

$$= \begin{pmatrix} \cosh t & \sinh t \\ \sinh t & \cosh t \end{pmatrix} \begin{pmatrix} c_1 \\ c_2 \end{pmatrix} + \begin{pmatrix} \cosh t & \sinh t \\ \sinh t & \cosh t \end{pmatrix} \int_0^t \begin{pmatrix} \cosh s & -\sinh s \\ -\sinh s & \cosh s \end{pmatrix} \begin{pmatrix} 1 \\ 1 \end{pmatrix} ds$$

$$= \begin{pmatrix} c_1 \cosh t + c_2 \sinh t \\ c_1 \sinh t + c_2 \cosh t \end{pmatrix} + \begin{pmatrix} \cosh t & \sinh t \\ \sinh t & \cosh t \end{pmatrix} \int_0^t \begin{pmatrix} \cosh s & -\sinh s \\ -\sinh s & \cosh s \end{pmatrix} ds$$

$$= \begin{pmatrix} c_1 \cosh t + c_2 \sinh t \\ c_1 \sinh t + c_2 \cosh t \end{pmatrix} + \begin{pmatrix} \cosh t & \sinh t \\ \sinh t & \cosh t \end{pmatrix} \begin{pmatrix} \sinh s - \cosh s \\ -\cosh s + \sinh s \end{pmatrix} \Big|_0^t$$

$$= \begin{pmatrix} c_1 \cosh t + c_2 \sinh t \\ c_1 \sinh t + c_2 \cosh t \end{pmatrix} + \begin{pmatrix} \cosh t & \sinh t \\ \sinh t & \cosh t \end{pmatrix} \begin{pmatrix} \sinh t - \cosh t \\ -\cosh t + \sinh t \end{pmatrix}$$

$$= \begin{pmatrix} c_1 \cosh t + c_2 \sinh t \\ c_1 \sinh t + c_2 \cosh t \end{pmatrix} + \begin{pmatrix} \sinh^2 t - \cosh^2 t \\ \sinh^2 t - \cosh^2 t \end{pmatrix} = c_1 \begin{pmatrix} \cosh t \\ \sinh t \end{pmatrix} + c_2 \begin{pmatrix} \sinh t \\ \cosh t \end{pmatrix} - \begin{pmatrix} 1 \\ 1 \end{pmatrix}.$$

12. To solve

$$\mathbf{X}' = \begin{pmatrix} 0 & 1 \\ 1 & 0 \end{pmatrix} \mathbf{X} + \begin{pmatrix} \cosh t \\ \sinh t \end{pmatrix}$$

we identify $t_0 = 0$, $\mathbf{F}(s) = \begin{pmatrix} \cosh t \\ \sinh t \end{pmatrix}$, and use the results of Problem 2 and equation (5) in the text.

$$\mathbf{X}(t) = e^{t\mathbf{A}} \mathbf{C} + e^{t\mathbf{A}} \int_{t_0}^{t} e^{-s\mathbf{A}} \mathbf{F}(s)\, ds$$

$$= \begin{pmatrix} \cosh t & \sinh t \\ \sinh t & \cosh t \end{pmatrix} \begin{pmatrix} c_1 \\ c_2 \end{pmatrix} + \begin{pmatrix} \cosh t & \sinh t \\ \sinh t & \cosh t \end{pmatrix} \int_0^t \begin{pmatrix} \cosh s & -\sinh s \\ -\sinh s & \cosh s \end{pmatrix} \begin{pmatrix} \cosh s \\ \sinh s \end{pmatrix} ds$$

$$= \begin{pmatrix} c_1 \cosh t + c_2 \sinh t \\ c_1 \sinh t + c_2 \cosh t \end{pmatrix} + \begin{pmatrix} \cosh t & \sinh t \\ \sinh t & \cosh t \end{pmatrix} \int_0^t \begin{pmatrix} 1 \\ 0 \end{pmatrix} ds$$

$$= \begin{pmatrix} c_1 \cosh t + c_2 \sinh t \\ c_1 \sinh t + c_2 \cosh t \end{pmatrix} + \begin{pmatrix} \cosh t & \sinh t \\ \sinh t & \cosh t \end{pmatrix} \begin{pmatrix} s \\ 0 \end{pmatrix} \Big|_0^t$$

$$= \begin{pmatrix} c_1 \cosh t + c_2 \sinh t \\ c_1 \sinh t + c_2 \cosh t \end{pmatrix} + \begin{pmatrix} \cosh t & \sinh t \\ \sinh t & \cosh t \end{pmatrix} \begin{pmatrix} t \\ 0 \end{pmatrix}$$

$$= \begin{pmatrix} c_1 \cosh t + c_2 \sinh t \\ c_1 \sinh t + c_2 \cosh t \end{pmatrix} + \begin{pmatrix} t \cosh t \\ t \sinh t \end{pmatrix} = c_1 \begin{pmatrix} \cosh t \\ \sinh t \end{pmatrix} + c_2 \begin{pmatrix} \sinh t \\ \cosh t \end{pmatrix} + t \begin{pmatrix} \cosh t \\ \sinh t \end{pmatrix}.$$

13. We have

$$\mathbf{X}(0) = c_1 \begin{pmatrix} 1 \\ 0 \\ 0 \end{pmatrix} + c_2 \begin{pmatrix} 0 \\ 1 \\ 0 \end{pmatrix} + c_3 \begin{pmatrix} 0 \\ 0 \\ 1 \end{pmatrix} = \begin{pmatrix} c_1 \\ c_2 \\ c_3 \end{pmatrix} = \begin{pmatrix} 1 \\ -4 \\ 6 \end{pmatrix}.$$

Thus, the solution of the initial-value problem is

$$\mathbf{X} = \begin{pmatrix} t+1 \\ t \\ -2t \end{pmatrix} - 4 \begin{pmatrix} t \\ t+1 \\ -2t \end{pmatrix} + 6 \begin{pmatrix} t \\ t \\ -2t+1 \end{pmatrix}.$$

14. We have

$$\mathbf{X}(0) = c_3 \begin{pmatrix} 1 \\ 0 \end{pmatrix} + c_4 \begin{pmatrix} 0 \\ 1 \end{pmatrix} + \begin{pmatrix} -3 \\ \frac{1}{2} \end{pmatrix} = \begin{pmatrix} c_3 - 3 \\ c_4 + \frac{1}{2} \end{pmatrix} = \begin{pmatrix} 4 \\ 3 \end{pmatrix}.$$

Thus, $c_3 = 7$ and $c_4 = \frac{5}{2}$, so

$$\mathbf{X} = 7 \begin{pmatrix} 1 \\ 0 \end{pmatrix} e^t + \frac{5}{2} \begin{pmatrix} 0 \\ 1 \end{pmatrix} e^{2t} + \begin{pmatrix} -3 \\ \frac{1}{2} \end{pmatrix}.$$

15. Solving

$$\begin{vmatrix} 2-\lambda & 1 \\ -3 & 6-\lambda \end{vmatrix} = \lambda^2 - 8\lambda + 15 = (\lambda - 3)(\lambda - 5) = 0$$

we find eigenvalues $\lambda_1 = 3$ and $\lambda_2 = 5$. Corresponding eigenvectors are

$$\mathbf{K}_1 = \begin{pmatrix} 1 \\ 1 \end{pmatrix} \quad \text{and} \quad \mathbf{K}_2 = \begin{pmatrix} 1 \\ 3 \end{pmatrix}.$$

Then

$$\mathbf{P} = \begin{pmatrix} 1 & 1 \\ 1 & 3 \end{pmatrix}, \quad \mathbf{P}^{-1} = \begin{pmatrix} 3/2 & -1/2 \\ -1/2 & 1/2 \end{pmatrix}, \quad \text{and} \quad \mathbf{D} = \begin{pmatrix} 3 & 0 \\ 0 & 5 \end{pmatrix},$$

so

$$\mathbf{PDP}^{-1} = \begin{pmatrix} 2 & 1 \\ -3 & 6 \end{pmatrix}.$$

16. Solving

$$\begin{vmatrix} 2-\lambda & 1 \\ 1 & 2-\lambda \end{vmatrix} = \lambda^2 - 4\lambda + 3 = (\lambda - 1)(\lambda - 3) = 0$$

we find eigenvalues $\lambda_1 = 1$ and $\lambda_2 = 3$. Corresponding eigenvectors are

$$\mathbf{K}_1 = \begin{pmatrix} -1 \\ 1 \end{pmatrix} \quad \text{and} \quad \mathbf{K}_2 = \begin{pmatrix} 1 \\ 1 \end{pmatrix}.$$

Then $\qquad \mathbf{P} = \begin{pmatrix} -1 & 1 \\ 1 & 1 \end{pmatrix}, \quad \mathbf{P}^{-1} = \begin{pmatrix} -1/2 & 1/2 \\ 1/2 & 1/2 \end{pmatrix}, \quad \text{and} \quad \mathbf{D} = \begin{pmatrix} 1 & 0 \\ 0 & 3 \end{pmatrix}$

so $\qquad\qquad\qquad\qquad\qquad \mathbf{PDP}^{-1} = \begin{pmatrix} 2 & 1 \\ 1 & 2 \end{pmatrix}.$

17. From equation (3) in the text

$$e^{t\mathbf{A}} = e^{t\mathbf{PDP}^{-1}} = \mathbf{I} + t(\mathbf{PDP}^{-1}) + \frac{1}{2!}t^2(\mathbf{PDP}^{-1})^2 + \frac{1}{3!}t^3(\mathbf{PDP}^{-1})^3 + \cdots$$

$$= \mathbf{P}\left[\mathbf{I} + t\mathbf{D} + \frac{1}{2!}(t\mathbf{D})^2 + \frac{1}{3!}(t\mathbf{D})^3 + \cdots\right]\mathbf{P}^{-1} = \mathbf{P}e^{t\mathbf{D}}\mathbf{P}^{-1}.$$

18. From equation (3) in the text

$$e^{t\mathbf{D}} = \begin{pmatrix} 1 & 0 & \cdots & 0 \\ 0 & 1 & \cdots & 0 \\ \vdots & \vdots & \ddots & \vdots \\ 0 & 0 & \cdots & 1 \end{pmatrix} + \begin{pmatrix} \lambda_1 & 0 & \cdots & 0 \\ 0 & \lambda_2 & \cdots & 0 \\ \vdots & \vdots & \ddots & \vdots \\ 0 & 0 & \cdots & \lambda_n \end{pmatrix} + \frac{1}{2!}t^2\begin{pmatrix} \lambda_1^2 & 0 & \cdots & 0 \\ 0 & \lambda_2^2 & \cdots & 0 \\ \vdots & \vdots & \ddots & \vdots \\ 0 & 0 & \cdots & \lambda_n^2 \end{pmatrix}$$

$$+ \frac{1}{3!}t^3\begin{pmatrix} \lambda_1^3 & 0 & \cdots & 0 \\ 0 & \lambda_2^3 & \cdots & 0 \\ \vdots & \vdots & \ddots & \vdots \\ 0 & 0 & \cdots & \lambda_n^3 \end{pmatrix} + \cdots$$

$$= \begin{pmatrix} 1 + \lambda_1 t + \frac{1}{2!}(\lambda_1 t)^2 + \cdots & 0 & \cdots & 0 \\ 0 & 1 + \lambda_2 t + \frac{1}{2!}(\lambda_2 t)^2 + \cdots & \cdots & 0 \\ \vdots & \vdots & \ddots & \vdots \\ 0 & 0 & \cdots & 1 + \lambda_n t + \frac{1}{2!}(\lambda_n t)^2 + \cdots \end{pmatrix}$$

$$= \begin{pmatrix} e^{\lambda_1 t} & 0 & \cdots & 0 \\ 0 & e^{\lambda_2 t} & \cdots & 0 \\ \vdots & \vdots & \ddots & \vdots \\ 0 & 0 & \cdots & e^{\lambda_n t} \end{pmatrix}.$$

19. From Problems 15, 17, and 18, and equation (1) in the text

$$\mathbf{X} = e^{t\mathbf{A}}\mathbf{C} = \mathbf{P}e^{t\mathbf{D}}\mathbf{P}^{-1}\mathbf{C}$$

$$= \begin{pmatrix} e^{3t} & e^{5t} \\ e^{3t} & 3e^{5t} \end{pmatrix} \begin{pmatrix} e^{3t} & 0 \\ 0 & e^{5t} \end{pmatrix} \begin{pmatrix} \frac{3}{2}e^{-3t} & -\frac{1}{2}e^{-3t} \\ -\frac{1}{2}e^{-5t} & \frac{1}{2}e^{-5t} \end{pmatrix} \begin{pmatrix} c_1 \\ c_2 \end{pmatrix}$$

$$= \begin{pmatrix} \frac{3}{2}e^{3t} - \frac{1}{2}e^{5t} & -\frac{1}{2}e^{3t} + \frac{1}{2}e^{5t} \\ \frac{3}{2}e^{3t} - \frac{3}{2}e^{5t} & -\frac{1}{2}e^{3t} + \frac{3}{2}e^{5t} \end{pmatrix} \begin{pmatrix} c_1 \\ c_2 \end{pmatrix}.$$

20. From Problems 16-18 and equation (1) in the text

$$\mathbf{X} = e^{t\mathbf{A}}\mathbf{C} = \mathbf{P}e^{t\mathbf{D}}\mathbf{P}^{-1}\mathbf{C}$$

$$= \begin{pmatrix} -e^{t} & e^{3t} \\ e^{t} & e^{3t} \end{pmatrix} \begin{pmatrix} e^{t} & 0 \\ 0 & e^{3t} \end{pmatrix} \begin{pmatrix} -\frac{1}{2}e^{-t} & \frac{1}{2}e^{-t} \\ \frac{1}{2}e^{3t} & \frac{1}{2}e^{-3t} \end{pmatrix} \begin{pmatrix} c_1 \\ c_2 \end{pmatrix}$$

$$= \begin{pmatrix} \frac{1}{2}e^{t} + \frac{1}{2}e^{9t} & -\frac{1}{2}e^{t} + \frac{1}{2}e^{3t} \\ -\frac{1}{2}e^{t} + \frac{1}{2}e^{9t} & \frac{1}{2}e^{t} + \frac{1}{2}e^{3t} \end{pmatrix} \begin{pmatrix} c_1 \\ c_2 \end{pmatrix}.$$

———— Chapter 8 Review Exercises ————

1. If

$$\mathbf{X} = c_1 \begin{pmatrix} 2\cos 3t \\ \cos 3t + 3\sin 3t \end{pmatrix} e^{3t} + c_2 \begin{pmatrix} 2\sin 3t \\ \sin 3t - 3\cos 3t \end{pmatrix} e^{3t},$$

then

$$\mathbf{X}' = 3c_1 \begin{pmatrix} 2\cos 3t \\ \cos 3t + 3\sin 3t \end{pmatrix} e^{3t} + c_1 \begin{pmatrix} -6\sin 3t \\ -3\sin 3t + 9\cos 3t \end{pmatrix} e^{3t}$$

$$+ 3c_2 \begin{pmatrix} 2\sin 3t \\ \sin 3t - 3\cos 3t \end{pmatrix} e^{3t} + c_2 \begin{pmatrix} 6\cos 3t \\ 3\cos 3t + 9\sin 3t \end{pmatrix} e^{3t}$$

$$= c_1 \begin{pmatrix} 6\cos 3t - 6\sin 3t \\ 12\cos 3t + 6\sin 3t \end{pmatrix} e^{3t} + c_2 \begin{pmatrix} 6\sin 3t + 6\cos 3t \\ 12\sin 3t - 6\cos 3t \end{pmatrix} e^{3t}$$

and

$$\begin{pmatrix} 4 & -2 \\ 5 & 2 \end{pmatrix} \mathbf{X} = c_1 \begin{pmatrix} 4 & -2 \\ 5 & 2 \end{pmatrix} \begin{pmatrix} 2\cos 3t \\ \cos 3t + 3\sin 3t \end{pmatrix} e^{3t} + c_2 \begin{pmatrix} 4 & -2 \\ 5 & 2 \end{pmatrix} \begin{pmatrix} 2\sin 3t \\ \sin 3t - 3\cos 3t \end{pmatrix} e^{3t}$$

$$= c_1 \begin{pmatrix} 6\cos 3t - 6\sin 3t \\ 12\cos 3t + 6\sin 3t \end{pmatrix} e^{3t} + c_2 \begin{pmatrix} 6\sin 3t + 6\cos 3t \\ 12\sin 3t - 6\cos 3t \end{pmatrix} e^{3t}.$$

Therefore
$$\mathbf{X}' = \begin{pmatrix} 4 & -2 \\ 5 & 2 \end{pmatrix} \mathbf{X}.$$

2. If
$$x = c_1 e^t + c_2 t e^t + \sin t \quad \text{and} \quad y = c_1 e^t + c_2 (t e^t + e^t) + \cos t$$

then
$$x' = c_1 e^t + c_2 (t e^t + e^t) + \cos t = y$$

and
$$y' = c_1 e^t + c_2 (t e^t + 2 e^t) - \sin t$$
$$= -(c_1 e^t + c_2 t e^t + \sin t) + 2(c_1 e^t + c_2 (t e^t + e^t) + \cos t) - 2 \cos t$$
$$= -x + 2y - 2 \cos t.$$

3. We have $\det(\mathbf{A} - \lambda\mathbf{I}) = (\lambda - 1)^2 = 0$ and $\mathbf{K} = \begin{pmatrix} 1 \\ -1 \end{pmatrix}$. A solution to $(\mathbf{A} - \lambda\mathbf{I})\mathbf{P} = \mathbf{K}$ is $\mathbf{P} = \begin{pmatrix} 0 \\ 1 \end{pmatrix}$

so that
$$\mathbf{X} = c_1 \begin{pmatrix} 1 \\ -1 \end{pmatrix} e^t + c_2 \left[\begin{pmatrix} 1 \\ -1 \end{pmatrix} t e^t + \begin{pmatrix} 0 \\ 1 \end{pmatrix} e^t \right].$$

4. We have $\det(\mathbf{A} - \lambda\mathbf{I}) = (\lambda + 6)(\lambda + 2) = 0$ so that
$$\mathbf{X} = c_1 \begin{pmatrix} 1 \\ -1 \end{pmatrix} e^{-6t} + c_2 \begin{pmatrix} 1 \\ 1 \end{pmatrix} e^{-2t}.$$

5. We have $\det(\mathbf{A} - \lambda\mathbf{I}) = \lambda^2 - 2\lambda + 5 = 0$. For $\lambda = 1 + 2i$ we obtain $\mathbf{K}_1 = \begin{pmatrix} 1 \\ i \end{pmatrix}$ and
$$\mathbf{X}_1 = \begin{pmatrix} 1 \\ i \end{pmatrix} e^{(1+2i)t} = \begin{pmatrix} \cos 2t \\ -\sin 2t \end{pmatrix} e^t + i \begin{pmatrix} \sin 2t \\ \cos 2t \end{pmatrix} e^t.$$

Then
$$\mathbf{X} = c_1 \begin{pmatrix} \cos 2t \\ -\sin 2t \end{pmatrix} e^t + c_2 \begin{pmatrix} \sin 2t \\ \cos 2t \end{pmatrix} e^t.$$

6. We have $\det(\mathbf{A} - \lambda\mathbf{I}) = \lambda^2 - 2\lambda + 2 = 0$. For $\lambda = 1 + i$ we obtain $\mathbf{K}_1 = \begin{pmatrix} 3 - i \\ 2 \end{pmatrix}$ and
$$\mathbf{X}_1 = \begin{pmatrix} 3 - i \\ 2 \end{pmatrix} e^{(1+i)t} = \begin{pmatrix} 3\cos t + \sin t \\ 2\cos t \end{pmatrix} e^t + i \begin{pmatrix} -\cos t + 3\sin t \\ 2\sin t \end{pmatrix} e^t.$$

Then
$$\mathbf{X} = c_1 \begin{pmatrix} 3\cos t + \sin t \\ 2\cos t \end{pmatrix} e^t + c_2 \begin{pmatrix} -\cos t + 3\sin t \\ 2\sin t \end{pmatrix} e^t.$$

7. We have $\det(\mathbf{A} - \lambda\mathbf{I}) = \lambda^2(3 - \lambda) = 0$ so that
$$\mathbf{X} = c_1 \begin{pmatrix} -1 \\ 1 \\ 0 \end{pmatrix} + c_2 \begin{pmatrix} -1 \\ 0 \\ 1 \end{pmatrix} + c_3 \begin{pmatrix} 1 \\ 1 \\ 1 \end{pmatrix} e^{3t}.$$

8. We have $\det(\mathbf{A} - \lambda\mathbf{I}) = -(\lambda - 2)(\lambda - 4)(\lambda + 3) = 0$ so that

$$\mathbf{X} = c_1 \begin{pmatrix} -2 \\ 3 \\ 1 \end{pmatrix} e^{2t} + c_2 \begin{pmatrix} 0 \\ 1 \\ 1 \end{pmatrix} e^{4t} + c_3 \begin{pmatrix} 7 \\ 12 \\ -16 \end{pmatrix} e^{-3t}.$$

9. We have

$$\mathbf{X}_c = c_1 \begin{pmatrix} 1 \\ 0 \end{pmatrix} e^{2t} + c_2 \begin{pmatrix} 4 \\ 1 \end{pmatrix} e^{4t}.$$

Then
$$\mathbf{\Phi} = \begin{pmatrix} e^{2t} & 4e^{4t} \\ 0 & e^{4t} \end{pmatrix}, \quad \mathbf{\Phi}^{-1} = \begin{pmatrix} e^{-2t} & -4e^{-2t} \\ 0 & e^{-4t} \end{pmatrix},$$

and
$$\mathbf{U} = \int \mathbf{\Phi}^{-1}\mathbf{F}\, dt = \int \begin{pmatrix} 2e^{-2t} - 64te^{-2t} \\ 16te^{-4t} \end{pmatrix} dt = \begin{pmatrix} 15e^{-2t} + 32te^{-2t} \\ -e^{-4t} - 4te^{-4t} \end{pmatrix},$$

so that
$$\mathbf{X}_p = \mathbf{\Phi}\mathbf{U} = \begin{pmatrix} 11 + 16t \\ -1 - 4t \end{pmatrix}.$$

10. We have

$$\mathbf{X}_c = c_1 \begin{pmatrix} 2\cos t \\ -\sin t \end{pmatrix} e^t + c_2 \begin{pmatrix} 2\sin t \\ \cos t \end{pmatrix} e^t.$$

Then
$$\mathbf{\Phi} = \begin{pmatrix} 2\cos t & 2\sin t \\ -\sin t & \cos t \end{pmatrix} e^t, \quad \mathbf{\Phi}^{-1} = \begin{pmatrix} \frac{1}{2}\cos t & -\sin t \\ \frac{1}{2}\sin t & \cos t \end{pmatrix} e^{-t},$$

and
$$\mathbf{U} = \int \mathbf{\Phi}^{-1}\mathbf{F}\, dt = \int \begin{pmatrix} \cos t - \sec t \\ \sin t \end{pmatrix} dt = \begin{pmatrix} \sin t - \ln|\sec t + \tan t| \\ -\cos t \end{pmatrix},$$

so that
$$\mathbf{X}_p = \mathbf{\Phi}\mathbf{U} = \begin{pmatrix} -2\cos t \ln|\sec t + \tan t| \\ -1 + \sin t \ln|\sec t + \tan t| \end{pmatrix}.$$

11. We have

$$\mathbf{X}_c = c_1 \begin{pmatrix} \cos t + \sin t \\ 2\cos t \end{pmatrix} + c_2 \begin{pmatrix} \sin t - \cos t \\ 2\sin t \end{pmatrix}.$$

Then
$$\mathbf{\Phi} = \begin{pmatrix} \cos t + \sin t & \sin t - \cos t \\ 2\cos t & 2\sin t \end{pmatrix}, \quad \mathbf{\Phi}^{-1} = \begin{pmatrix} \sin t & \frac{1}{2}\cos t - \frac{1}{2}\sin t \\ -\cos t & \frac{1}{2}\cos t + \frac{1}{2}\sin t \end{pmatrix},$$

and
$$\mathbf{U} = \int \mathbf{\Phi}^{-1}\mathbf{F}\,dt = \int \begin{pmatrix} \frac{1}{2}\sin t - \frac{1}{2}\cos t + \frac{1}{2}\csc t \\ -\frac{1}{2}\sin t - \frac{1}{2}\cos t + \frac{1}{2}\csc t \end{pmatrix} dt$$

$$= \begin{pmatrix} -\frac{1}{2}\cos t - \frac{1}{2}\sin t + \frac{1}{2}\ln|\csc t - \cot t| \\ \frac{1}{2}\cos t - \frac{1}{2}\sin t + \frac{1}{2}\ln|\csc t - \cot t| \end{pmatrix},$$

so that
$$\mathbf{X}_p = \mathbf{\Phi}\mathbf{U} = \begin{pmatrix} -1 \\ -1 \end{pmatrix} + \begin{pmatrix} \sin t \\ \sin t + \cos t \end{pmatrix} \ln|\csc t - \cot t|.$$

12. We have
$$\mathbf{X}_c = c_1 \begin{pmatrix} 1 \\ -1 \end{pmatrix} e^{2t} + c_2 \left[\begin{pmatrix} 1 \\ -1 \end{pmatrix} t e^{2t} + \begin{pmatrix} 1 \\ 0 \end{pmatrix} e^{2t} \right].$$

Then
$$\mathbf{\Phi} = \begin{pmatrix} e^{2t} & t e^{2t} + e^{2t} \\ -e^{2t} & -t e^{2t} \end{pmatrix}, \quad \mathbf{\Phi}^{-1} = \begin{pmatrix} -t e^{-2t} & -t e^{-2t} - e^{-2t} \\ e^{-2t} & e^{-2t} \end{pmatrix},$$

and
$$\mathbf{U} = \int \mathbf{\Phi}^{-1}\mathbf{F}\,dt = \int \begin{pmatrix} t - 1 \\ -1 \end{pmatrix} dt = \begin{pmatrix} \frac{1}{2}t^2 - t \\ -t \end{pmatrix},$$

so that
$$\mathbf{X}_p = \mathbf{\Phi}\mathbf{U} = \begin{pmatrix} -1/2 \\ 1/2 \end{pmatrix} t^2 e^{2t} + \begin{pmatrix} -2 \\ 1 \end{pmatrix} t e^{2t}.$$

9 Numerical Methods for Ordinary Differential Equations

1.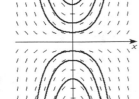

2.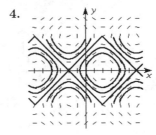

3.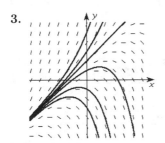

4.

5. Setting $x = c$ we see that the isoclines form a family of vertical lines.

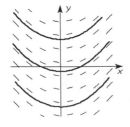

6. Setting $x + y = c$ we obtain the isoclines $y = -x + c$.

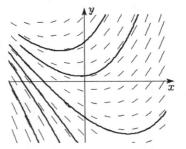

7. Setting $-x/y = c$ we obtain the isoclines $y = -x/c$.

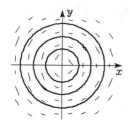

8. Setting $1/y = c$ we obtain the isoclines $y = 1/c$.

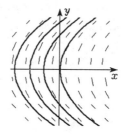

9. Setting $0.2x^2 + y = c$ we obtain the isoclines $y = c - 0.2x^2$.

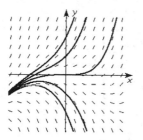

10. Setting $xe^y = c$ we obtain the isoclines $y = \ln c - \ln x$.

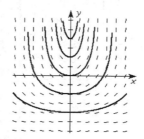

11. Setting $y - \cos\frac{\pi}{2}x = c$ we obtain the isoclines $y = \cos\frac{\pi}{2}x + c$.

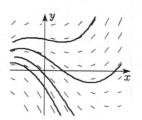

12. Setting $1 - y/x = c$ we obtain the isoclines $y = (1-c)x$.

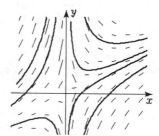

Exercises 9.2

All tables in this chapter were constructed in a spreadsheet program which does not support subscripts. Consequently, x_n and y_n will be indicated as $x(n)$ and $y(n)$, respectively.

1. (a) Let $u = x + y - 1$ so that $y' = (x + y - 1)^2$ becomes

$$\frac{1}{u^2 + 1} du = dx$$

and $\tan^{-1} u = x + c$. Then $y = 1 - x + \tan(x + c)$ and $y(0) = 2$ gives $c = \pi/4$, so

$$y = 1 - x + \tan\left(x + \frac{\pi}{4}\right).$$

(b)

x(n)	h = 0.1 y(n}	exact
0.00	2.0000	2.0000
0.10	2.1000	2.1230
0.20	2.2440	2.3085
0.30	2.4525	2.5958
0.40	2.7596	3.0650
0.50	3.2261	3.9082

x(n)	h = 0.05 y(n}	exact
0.00	2.0000	2.0000
0.05	2.0500	2.0554
0.10	2.1105	2.1230
0.15	2.1838	2.2061
0.20	2.2727	2.3085
0.25	2.3812	2.4358
0.30	2.5142	2.5958
0.35	2.6788	2.7997
0.40	2.8845	3.0650
0.45	3.1455	3.4189
0.50	3.4823	3.9082

2.

x(n)	h = 0.1 y(n}	exact
0.00	2.0000	2.0000
0.10	2.1220	2.1230
0.20	2.3049	2.3085
0.30	2.5858	2.5958
0.40	3.0378	3.0650
0.50	3.8254	3.9082

x(n)	h = 0.05 y(n}	exact
0.00	2.0000	2.0000
0.05	2.0553	2.0554
0.10	2.1228	2.1230
0.15	2.2056	2.2061
0.20	2.3075	2.3085
0.25	2.4342	2.4358
0.30	2.5931	2.5958
0.35	2.7953	2.7997
0.40	3.0574	3.0650
0.45	3.4057	3.4189
0.50	3.8840	3.9082

3.

h = 0.1	
x(n)	y(n)
1.00	5.0000
1.10	3.8000
1.20	2.9800
1.30	2.4260
1.40	2.0582
1.50	1.8207

h = 0.05	
x(n)	y(n)
1.00	5.0000
1.05	4.4000
1.10	3.8950
1.15	3.4708
1.20	3.1151
1.25	2.8179
1.30	2.5702
1.35	2.3647
1.40	2.1950
1.45	2.0557
1.50	1.9424

4.

h = 0.1	
x(n)	y(n)
0.00	2.0000
0.10	1.6000
0.20	1.3200
0.30	1.1360
0.40	1.0288
0.50	0.9830

h = 0.05	
x(n)	y(n)
0.00	2.0000
0.05	1.8000
0.10	1.6300
0.15	1.4870
0.20	1.3683
0.25	1.2715
0.30	1.1943
0.35	1.1349
0.40	1.0914
0.45	1.0623
0.50	1.0460

5.

h = 0.1	
x(n)	y(n)
0.00	0.0000
0.10	0.1000
0.20	0.2010
0.30	0.3050
0.40	0.4143
0.50	0.5315

h = 0.05	
x(n)	y(n)
0.00	0.0000
0.05	0.0500
0.10	0.1001
0.15	0.1506
0.20	0.2018
0.25	0.2538
0.30	0.3070
0.35	0.3617
0.40	0.4183
0.45	0.4770
0.50	0.5384

Exercises 9.2

6.

h = 0.1		h = 0.05	
x(n)	y(n)	x(n)	y(n)
0.00	1.0000	0.00	1.0000
0.10	1.1000	0.05	1.0500
0.20	1.2220	0.10	1.1053
0.30	1.3753	0.15	1.1668
0.40	1.5735	0.20	1.2360
0.50	1.8371	0.25	1.3144
		0.30	1.4039
		0.35	1.5070
		0.40	1.6267
		0.45	1.7670
		0.50	1.9332

7.

h = 0.1		h = 0.05	
x(n)	y(n)	x(n)	y(n)
0.00	0.0000	0.00	0.0000
0.10	0.1000	0.05	0.0500
0.20	0.1905	0.10	0.0976
0.30	0.2731	0.15	0.1429
0.40	0.3492	0.20	0.1863
0.50	0.4198	0.25	0.2278
		0.30	0.2676
		0.35	0.3058
		0.40	0.3427
		0.45	0.3782
		0.50	0.4124

8.

h = 0.1		h = 0.05	
x(n)	y(n)	x(n)	y(n)
0.00	0.0000	0.00	0.0000
0.10	0.0000	0.05	0.0000
0.20	0.0100	0.10	0.0025
0.30	0.0300	0.15	0.0075
0.40	0.0601	0.20	0.0150
0.50	0.1005	0.25	0.0250
		0.30	0.0375
		0.35	0.0526
		0.40	0.0703
		0.45	0.0905
		0.50	0.1134

9.

h = 0.1	
x(n)	y(n)
0.00	0.5000
0.10	0.5250
0.20	0.5431
0.30	0.5548
0.40	0.5613
0.50	0.5639

h = 0.05	
x(n)	y(n)
0.00	0.5000
0.05	0.5125
0.10	0.5232
0.15	0.5322
0.20	0.5395
0.25	0.5452
0.30	0.5496
0.35	0.5527
0.40	0.5547
0.45	0.5559
0.50	0.5565

10.

h = 0.1	
x(n)	y(n)
0.00	1.0000
0.10	1.1000
0.20	1.2159
0.30	1.3505
0.40	1.5072
0.50	1.6902

h = 0.05	
x(n)	y(n)
0.00	1.0000
0.05	1.0500
0.10	1.1039
0.15	1.1619
0.20	1.2245
0.25	1.2921
0.30	1.3651
0.35	1.4440
0.40	1.5293
0.45	1.6217
0.50	1.7219

11.

h = 0.1	
x(n)	y(n)
1.00	1.0000
1.10	1.0000
1.20	1.0191
1.30	1.0588
1.40	1.1231
1.50	1.2194

h = 0.05	
x(n)	y(n)
1.00	1.0000
1.05	1.0000
1.10	1.0049
1.15	1.0147
1.20	1.0298
1.25	1.0506
1.30	1.0775
1.35	1.1115
1.40	1.1538
1.45	1.2057
1.50	1.2696

12.

h = 0.1	
x(n)	y(n}
0.00	0.5000
0.10	0.5250
0.20	0.5499
0.30	0.5747
0.40	0.5991
0.50	0.6231

h = 0.05	
x(n)	y(n}
0.00	0.5000
0.05	0.5125
0.10	0.5250
0.15	0.5375
0.20	0.5499
0.25	0.5623
0.30	0.5746
0.35	0.5868
0.40	0.5989
0.45	0.6109
0.50	0.6228

13. (a)

h = 0.1	
x	y
1.00	5.0000
1.10	3.9900
1.20	3.2546
1.30	2.7236
1.40	2.3451
1.50	2.0801

h = 0.05	
x(n)	y(n)
1.00	5.0000
1.05	4.4475
1.10	3.9763
1.15	3.5751
1.20	3.2342
1.25	2.9452
1.30	2.7009
1.35	2.4952
1.40	2.3226
1.45	2.1786
1.50	2.0592

(b)

h = 0.1	
x(n)	y(n)
0.00	0.0000
0.10	0.1005
0.20	0.2030
0.30	0.3098
0.40	0.4234
0.50	0.5470

h = 0.05	
x(n)	y(n)
0.00	0.0000
0.05	0.0501
0.10	0.1004
0.15	0.1512
0.20	0.2028
0.25	0.2554
0.30	0.3095
0.35	0.3652
0.40	0.4230
0.45	0.4832
0.50	0.5465

(c)

h = 0.1	
x(n)	y(n)
0.00	0.0000
0.10	0.0952
0.20	0.1822
0.30	0.2622
0.40	0.3363
0.50	0.4053

h = 0.05	
x(n)	y(n)
0.00	0.0000
0.05	0.0488
0.10	0.0953
0.15	0.1397
0.20	0.1823
0.25	0.2231
0.30	0.2623
0.35	0.3001
0.40	0.3364
0.45	0.3715
0.50	0.4054

(d)

h = 0.1	
x(n)	y(n)
0.00	0.5000
0.10	0.5215
0.20	0.5362
0.30	0.5449
0.40	0.5490
0.50	0.5503

h = 0.05	
x(n)	y(n)
0.00	0.5000
0.05	0.5116
0.10	0.5214
0.15	0.5294
0.20	0.5359
0.25	0.5408
0.30	0.5444
0.35	0.5469
0.40	0.5484
0.45	0.5492
0.50	0.5495

(e)

h = 0.1	
x(n)	y(n)
1.00	1.0000
1.10	1.0095
1.20	1.0404
1.30	1.0967
1.40	1.1866
1.50	1.3260

h = 0.05	
x(n)	y(n)
1.00	1.0000
1.05	1.0024
1.10	1.0100
1.15	1.0228
1.20	1.0414
1.25	1.0663
1.30	1.0984
1.35	1.1389
1.40	1.1895
1.45	1.2526
1.50	1.3315

Exercises 9.2

14. (a)

h = 0.1	
x(n)	y(n)
0.00	2.0000
0.10	1.6600
0.20	1.4172
0.30	1.2541
0.40	1.1564
0.50	1.1122

h = 0.05	
x(n)	y(n)
0.00	2.0000
0.05	1.8150
0.10	1.6571
0.15	1.5237
0.20	1.4124
0.25	1.3212
0.30	1.2482
0.35	1.1916
0.40	1.1499
0.45	1.1217
0.50	1.1056

(b)

h = 0.1	
x(n)	y(n)
0.00	1.0000
0.10	1.1110
0.20	1.2515
0.30	1.4361
0.40	1.6880
0.50	2.0488

h = 0.05	
x(n)	y(n)
0.00	1.0000
0.05	1.0526
0.10	1.1113
0.15	1.1775
0.20	1.2526
0.25	1.3388
0.30	1.4387
0.35	1.5556
0.40	1.6939
0.45	1.8598
0.50	2.0619

(c)

h = 0.1	
x(n)	y(n)
0.00	0.0000
0.10	0.0050
0.20	0.0200
0.30	0.0451
0.40	0.0805
0.50	0.1266

h = 0.05	
x(n)	y(n)
0.00	0.0000
0.05	0.0013
0.10	0.0050
0.15	0.0113
0.20	0.0200
0.25	0.0313
0.30	0.0451
0.35	0.0615
0.40	0.0805
0.45	0.1022
0.50	0.1266

(d)

h = 0.1		h = 0.05	
x(n)	y(n)	x(n)	y(n)
0.00	1.0000	0.00	1.0000
0.10	1.1079	0.05	1.0519
0.20	1.2337	0.10	1.1079
0.30	1.3806	0.15	1.1684
0.40	1.5529	0.20	1.2337
0.50	1.7557	0.25	1.3043
		0.30	1.3807
		0.35	1.4634
		0.40	1.5530
		0.45	1.6503
		0.50	1.7560

(e)

h = 0.1		h = 0.05	
x(n)	y(n)	x(n)	y(n)
0.00	0.5000	0.00	0.5000
0.10	0.5250	0.05	0.5125
0.20	0.5498	0.10	0.5250
0.30	0.5744	0.15	0.5374
0.40	0.5986	0.20	0.5498
0.50	0.6224	0.25	0.5622
		0.30	0.5744
		0.35	0.5866
		0.40	0.5987
		0.45	0.6106
		0.50	0.6224

15. (a)

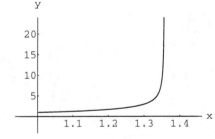

(b)

h=0.1	EULER	IMPROVED EULER
x(n)	y(n)	y(n)
1.00	1.0000	1.0000
1.10	1.2000	1.2469
1.20	1.4938	1.6668
1.30	1.9711	2.6427
1.40	2.9060	8.7988

16. (a) With $k = 2$ the Taylor polynomial is

$$y(x) = y(a) + y'(a)(x - a) + \frac{1}{2}y''(a)(x - a)^2 + \frac{1}{6}y'''(c)(x - a)^3.$$

Differentiating we obtain

$$y'(x) = y'(a) + y''(a)(x - a) + \frac{1}{2}y'''(c)(x - a)^2.$$

Identifying $a = x_n$ and $x = x_{n+1} = x_n + h$ we have

$$y'(x_{n+1}) = y'(x_n) + y''(x_n)h + \frac{1}{2}y'''(c)h^2$$

or

$$y''(x_n) = \frac{y'(x_{n+1}) - y'(x_n)}{h} - \frac{1}{2}hy'''(c).$$

(b) Using the Taylor polynomial in part **(a)** with $a = x_n$ and $x = x_{n+1} = x_n + h$ we have

$$y(x_{n+1}) = y(x_n) + hy'(x_n) + \frac{1}{2}h^2y''(x_n) + \frac{1}{6}h^3y'''(c)$$

$$= y(x_n) + hy'(x_n) + \frac{1}{2}h^2\left(\frac{y'(x_{n+1}) - y'(x_n)}{h} - \frac{1}{2}hy'''(c)\right) + \frac{1}{6}h^3y'''(c)$$

$$= y(x_n) + hy'(x_n) + \frac{1}{2}hy'(x_{n+1}) - \frac{1}{2}hy'(x_n) + 0(h^3)$$

$$= y(x_n) + h\frac{y'(x_n) + y'(x_{n+1})}{2} + 0(h^3).$$

17. (a) Using the Euler method we obtain $y(0.1) \approx y_1 = 1.2$.

(b) Using $y'' = 4e^{2x}$ we see that the local truncation error is

$$y''(c)\frac{h^2}{2} = 4e^{2c}\frac{(0.1)^2}{2} = 0.02e^{2c}.$$

Since e^{2x} is an increasing function, $e^{2c} \le e^{2(0.1)} = e^{0.2}$ for $0 \le c \le 0.1$. Thus an upper bound for the local truncation error is $0.02e^{0.2} = 0.0244$.

(c) Since $y(0.1) = e^{0.2} = 1.2214$, the actual error is $y(0.1) - y_1 = 0.0214$, which is less than 0.0244.

(d) Using the Euler method with $h = 0.05$ we obtain $y(0.1) \approx y_2 = 1.21$.

(e) The error in (d) is $1.2214 - 1.21 = 0.0114$. With global truncation error $O(h)$, when the step size is halved we expect the error for $h = 0.05$ to be one-half the error when $h = 0.1$. Comparing 0.0114 with 0.214 we see that this is the case.

18. (a) Using the improved Euler method we obtain $y(0.1) \approx y_1 = 1.22$.

(b) Using $y''' = 8e^{2x}$ we see that the local truncation error is

$$y'''(c)\frac{h^3}{6} = 8e^{2c}\frac{(0.1)^3}{6} = 0.001333e^{2c}.$$

Since e^{2x} is an increasing function, $e^{2c} < e^{2(0.1)} = e^{0.2}$ for $0 \le c \le 0.1$. Thus an upper bound for the local truncation error is $0.001333e^{0.2} = 0.001628$.

(c) Since $y(0.1) = e^{0.2} = 1.221403$, the actual error is $y(0.1) - y_1 = 0.001403$ which is less than 0.001628.

(d) Using the improved Euler method with $h = 0.05$ we obtain $y(0.1) \approx y_2 = 1.221025$.

(e) The error in (d) is $1.221403 - 1.221025 = 0.000378$. With global truncation error $O(h^2)$, when the step size is halved we expect the error for $h = 0.05$ to be one-fourth the error for $h = 0.1$. Comparing 0.000378 with 0.001403 we see that this is the case.

19. (a) Using the Euler method we obtain $y(0.1) \approx y_1 = 0.8$.

(b) Using $y'' = 5e^{-2x}$ we see that the local truncation error is

$$5e^{-2c} \frac{(0.1)^2}{2} = 0.025e^{-2c}$$

Since e^{-2x} is a decreasing function, $e^{-2c} \le e^0 = 1$ for $0 \le c \le 0.1$. Thus an upper bound for the local truncation error is $0.025(1) = 0.025$.

(c) Since $y(0.1) = 0.8234$, the actual error is $y(0.1) - y_1 = 0.0234$, which is less than 0.025.

(d) Using the Euler method with $h = 0.05$ we obtain $y(0.1) \approx y_2 = 0.8125$.

(e) The error in (d) is $0.8234 - 0.8125 = 0.0109$. With global truncation error $O(h)$, when the step size is halved we expect the error for $h = 0.05$ to be one-half the error when $h = 0.1$. Comparing 0.0109 with 0.0234 we see that this is the case.

20. (a) Using the improved Euler method we obtain $y(0.1) \approx y_1 = 0.825$.

(b) Using $y''' = -10e^{-2x}$ we see that the local truncation error is

$$10e^{-2c} \frac{(0.1)^3}{6} = 0.001667e^{-2c}.$$

Since e^{-2x} is a decreasing function, $e^{-2c} \le e^0 = 1$ for $0 \le c \le 0.1$. Thus an upper bound for the local truncation error is $0.001667(1) = 0.001667$.

(c) Since $y(0.1) = 0.823413$, the actual error is $y(0.1) - y_1 = 0.001587$, which is less than 0.001667.

(d) Using the improved Euler method with $h = 0.05$ we obtain $y(0.1) \approx y_2 = 0.823781$.

(e) The error in (d) is $|0.823413 - 0.823781| = 0.000305$. With global truncation error $O(h^2)$, when the step size is halved we expect the error for $h = 0.05$ to be one-fourth the error when $h = 0.1$. Comparing 0.000305 with 0.001587 we see that this is the case.

21. (a) Using $y'' = 38e^{-3(x-1)}$ we see that the local truncation error is

$$y''(c) \frac{h^2}{2} = 38e^{-3(c-1)} \frac{h^2}{2} = 19h^2 e^{-3(c-1)}.$$

Exercises 9.2

(b) Since $e^{-3(x-1)}$ is a decreasing function for $1 \le x \le 1.5$, $e^{-3(c-1)} \le e^{-3(1-1)} = 1$ for $1 \le c \le 1.5$ and

$$y''(c)\frac{h^2}{2} \le 19(0.1)^2(1) = 0.19.$$

(c) Using the Euler method with $h = 0.1$ we obtain $y(1.5) \approx 1.8207$. With $h = 0.05$ we obtain $y(1.5) \approx 1.9424$.

(d) Since $y(1.5) = 2.0532$, the error for $h = 0.1$ is $E_{0.1} = 0.2325$, while the error for $h = 0.05$ is $E_{0.05} = 0.1109$. With global truncation error $O(h)$ we expect $E_{0.1}/E_{0.05} \approx 2$. We actually have $E_{0.1}/E_{0.05} = 2.10$.

22. (a) Using $y''' = -114e^{-3(x-1)}$ we see that the local truncation error is

$$\left| y'''(c)\frac{h^3}{6} \right| = 114e^{-3(x-1)}\frac{h^3}{6} = 19h^3 e^{-3(c-1)}.$$

(b) Since $e^{-3(x-1)}$ is a decreasing function for $1 \le x \le 1.5$, $e^{-3(c-1)} \le e^{-3(1-1)} = 1$ for $1 \le c \le 1.5$ and

$$\left| y'''(c)\frac{h^3}{6} \right| \le 19(0.1)^3(1) = 0.019.$$

(c) Using the improved Euler method with $h = 0.1$ we obtain $y(1.5) \approx 2.080108$. With $h = 0.05$ we obtain $y(1.5) \approx 2.059166$.

(d) Since $y(1.5) = 2.053216$, the error for $h = 0.1$ is $E_{0.1} = 0.026892$, while the error for $h = 0.05$ is $E_{0.05} = 0.005950$. With global truncation error $O(h^2)$ we expect $E_{0.1}/E_{0.05} \approx 4$. We actually have $E_{0.1}/E_{0.05} = 4.52$.

23. (a) Using $y'' = -\dfrac{1}{(x+1)^2}$ we see that the local truncation error is

$$\left| y''(c)\frac{h^2}{2} \right| = \frac{1}{(c+1)^2}\frac{h^2}{2}.$$

(b) Since $\dfrac{1}{(x+1)^2}$ is a decreasing function for $0 \le x \le 0.5$, $\dfrac{1}{(c+1)^2} \le \dfrac{1}{(0+1)^2} = 1$ for $0 \le c \le 0.5$ and

$$\left| y''(c)\frac{h^2}{2} \right| \le (1)\frac{(0.1)^2}{2} = 0.005.$$

(c) Using the Euler method with $h = 0.1$ we obtain $y(0.5) \approx 0.4198$. With $h = 0.05$ we obtain $y(0.5) \approx 0.4124$.

(d) Since $y(0.5) = 0.4055$, the error for $h = 0.1$ is $E_{0.1} = 0.0143$, while the error for $h = 0.05$ is $E_{0.05} = 0.0069$. With global truncation error $O(h)$ we expect $E_{0.1}/E_{0.05} \approx 2$. We actually have $E_{0.1}/E_{0.05} = 2.06$.

24. (a) Using $y''' = \dfrac{2}{(x+1)^3}$ we see that the local truncation error is

$$y'''(c)\frac{h^3}{6} = \frac{1}{(c+1)^3}\frac{h^3}{3}.$$

(b) Since $\dfrac{1}{(x+1)^3}$ is a decreasing function for $0 \le x \le 0.5$, $\dfrac{1}{(c+1)^3} \le \dfrac{1}{(0+1)^3} = 1$ for $0 \le c \le 0.5$ and

$$y'''(c)\frac{h^3}{6} \le (1)\frac{(0.1)^3}{3} = 0.000333.$$

(c) Using the improved Euler method with $h = 0.1$ we obtain $y(0.5) \approx 0.405281$. With $h = 0.05$ we obtain $y(0.5) \approx 0.405419$.

(d) Since $y(0.5) = 0.405465$, the error for $h = 0.1$ is $E_{0.1} = 0.000184$, while the error for $h = 0.05$ is $E_{0.05} = 0.000046$. With global truncation error $O(h^2)$ we expect $E_{0.1}/E_{0.05} \approx 4$. We actually have $E_{0.1}/E_{0.05} = 3.98$.

Exercises 9.3

1.

h = 0.1		
x(n)	y(n)	exact
0.00	2.0000	2.0000
0.10	2.1230	2.1230
0.20	2.3085	2.3085
0.30	2.5958	2.5958
0.40	3.0649	3.0650
0.50	3.9078	3.9082

2. Setting $\alpha = 1/4$ we find $b = 2$, $a = -1$ and $\beta = 1/4$. The resulting second-order Runge-Kutta method is

$$y_{n+1} = y_n - k_1 + 2k_2$$

$$k_1 = hf(x_n, y_n)$$

$$k_2 = hf(x_n + h/4, y_n + k_1/4).$$

h = 0.1	R-K	IMP EULER
x(n)	y(n)	y(n)
0.00	2.0000	2.0000
0.10	2.1205	2.1220
0.20	2.3006	2.3049
0.30	2.5759	2.5858
0.40	3.0152	3.0378
0.50	3.7693	3.8254

3.

x(n)	y(n)
1.00	5.0000
1.10	3.9724
1.20	3.2284
1.30	2.6945
1.40	2.3163
1.50	2.0533

4.

x(n)	y(n)
0.00	2.0000
0.10	1.6562
0.20	1.4110
0.30	1.2465
0.40	1.1480
0.50	1.1037

5.

x(n)	y(n)
0.00	0.0000
0.10	0.1003
0.20	0.2027
0.30	0.3093
0.40	0.4228
0.50	0.5463

6.

x(n)	y(n)
0.00	1.0000
0.10	1.1115
0.20	1.2530
0.30	1.4397
0.40	1.6961
0.50	2.0670

7.

x(n)	y(n)
0.00	0.0000
0.10	0.0953
0.20	0.1823
0.30	0.2624
0.40	0.3365
0.50	0.4055

8.

x(n)	y(n)
0.00	0.0000
0.10	0.0050
0.20	0.0200
0.30	0.0451
0.40	0.0805
0.50	0.1266

9.

x(n)	y(n)
0.00	0.5000
0.10	0.5213
0.20	0.5358
0.30	0.5443
0.40	0.5482
0.50	0.5493

10.

x(n)	y(n)
0.00	1.0000
0.10	1.1079
0.20	1.2337
0.30	1.3807
0.40	1.5531
0.50	1.7561

11.

x(n)	y(n)
1.00	1.0000
1.10	1.0101
1.20	1.0417
1.30	1.0989
1.40	1.1905
1.50	1.3333

12.

x(n)	y(n)
0.00	0.5000
0.10	0.5250
0.20	0.5498
0.30	0.5744
0.40	0.5987
0.50	0.6225

13. **(a)** Write the equation in the form

$$\frac{dv}{dt} = 32 - 0.025v^2 = f(t, v).$$

t (n)	v (n)
0.0	0.0000
1.0	25.2570
2.0	32.9390
3.0	34.9772
4.0	35.5503
5.0	35.7128

(b)

v(t)

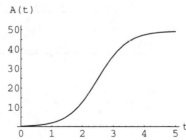

(c) Separating variables and using partial fractions we have

$$\frac{1}{2\sqrt{32}} \left(\frac{1}{\sqrt{32} - \sqrt{0.025}\, v} + \frac{1}{\sqrt{32} + \sqrt{0.025}\, v} \right) dv = dt$$

and

$$\frac{1}{2\sqrt{32}\sqrt{0.025}} \left(\ln |\sqrt{32} + \sqrt{0.025}\, v| - \ln |\sqrt{32} - \sqrt{0.025}\, v| \right) = t + c.$$

Since $v(0) = 0$ we find $c = 0$. Solving for v we obtain

$$v(t) = \frac{16\sqrt{5}\, (e^{\sqrt{3.2}\, t} - 1)}{e^{\sqrt{3.2}\, t} + 1}$$

and $v(5) \approx 35.7678$.

14. **(a)** See the table in part (c) of this problem.

(b) From the graph we estimate $A(1) \approx 1.68$, $A(2) \approx 13.2$, $A(3) \approx 36.8$, $A(4) \approx 46.9$, and $A(5) \approx 48.9$.

A(t)

(c) Let $\alpha = 2.128$ and $\beta = 0.0432$. Separating variables we obtain

$$\frac{dA}{A(\alpha - \beta A)} = dt$$

$$\frac{1}{\alpha} \left(\frac{1}{A} + \frac{\beta}{\alpha - \beta A} \right) dA = dt$$

$$\frac{1}{\alpha}[\ln A - \ln(\alpha - \beta A)] = t + c$$

$$\ln \frac{A}{\alpha - \beta A} = \alpha(t + c)$$

$$\frac{A}{\alpha - \beta A} = e^{\alpha(t+c)}$$

$$A = \alpha e^{\alpha(t+c)} - \beta A e^{\alpha(t+c)}$$

$$\left[1 + \beta e^{\alpha(t+c)} \right] A = \alpha e^{\alpha(t+c)}.$$

Thus

$$A(t) = \frac{\alpha e^{\alpha(t+c)}}{1 + \beta e^{\alpha(t+c)}} = \frac{\alpha}{\beta + e^{-\alpha(t+c)}} = \frac{\alpha}{\beta + e^{-\alpha c}e^{-\alpha t}}.$$

From $A(0) = 0.24$ we obtain

$$0.24 = \frac{\alpha}{\beta + e^{-\alpha c}}$$

so that $e^{-\alpha c} = \alpha/0.24 - \beta \approx 8.8235$ and

$$A(t) \approx \frac{2.128}{0.0432 + 8.8235 e^{-2.128t}}.$$

t (days)	1	2	3	4	5
A (observed)	2.78	13.53	36.30	47.50	49.40
A (approximated)	1.93	12.50	36.46	47.23	49.00
A (exact)	1.95	12.64	36.63	47.32	49.02

15. (a)

	h = 0.05	h = 0.1
x(n)	y(n)	y(n)
1.00	1.0000	1.0000
1.05	1.1112	
1.10	1.2511	1.2511
1.15	1.4348	
1.20	1.6934	1.6934
1.25	2.1047	
1.30	2.9560	2.9425
1.35	7.8981	
1.40	1.06E+15	903.0282

(b)

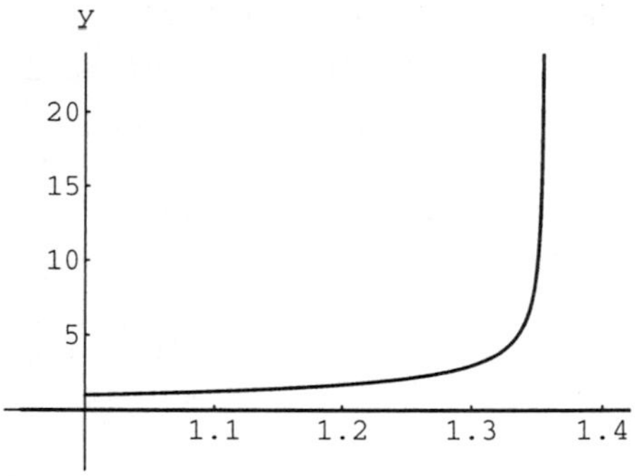

16. (a) Using the fourth-order Runge-Kutta method we obtain $y(0.1) \approx y_1 = 1.2214$.

(b) Using $y^{(5)}(x) = 32e^{2x}$ we see that the local truncation error is

$$y^{(5)}(c)\,\frac{h^5}{120} = 32e^{2c}\,\frac{(0.1)^5}{120} = 0.000002667e^{2c}.$$

Since e^{2x} is an increasing function, $e^{2c} \le e^{2(0.1)} = e^{0.2}$ for $0 \le c \le 0.1$. Thus an upper bound for the local truncation error is $0.000002667e^{0.2} = 0.000003257$.

(c) Since $y(0.1) = e^{0.2} = 1.221402758$, the actual error is $y(0.1) - y_1 = 0.000002758$ which is less than 0.000003257.

(d) Using the fourth-order Runge-Kutta formula with $h = 0.05$ we obtain $y(0.1) \approx y_2 = 1.221402571$.

(e) The error in (d) is $1.221402758 - 1.221402571 = 0.000000187$. With global truncation error $O(h^4)$, when the step size is halved we expect the error for $h = 0.05$ to be one-sixteenth the error for $h = 0.1$. Comparing 0.000000187 with 0.000002758 we see that this is the case.

17. (a) Using the fourth-order Runge-Kutta method we obtain $y(0.1) \approx y_1 = 0.823416667$.

(b) Using $y^{(5)}(x) = -40e^{-2x}$ we see that the local truncation error is

$$40e^{-2c}\,\frac{(0.1)^5}{120} = 0.000003333.$$

Since e^{-2x} is a decreasing function, $e^{-2c} \le e^0 = 1$ for $0 \le c \le 0.1$. Thus an upper bound for the local truncation error is $0.000003333(1) = 0.000003333$.

(c) Since $y(0.1) = 0.823413441$, the actual error is $|y(0.1) - y_1| = 0.000003225$, which is less than 0.000003333.

(d) Using the fourth-order Runge-Kutta method with $h = 0.05$ we obtain $y(0.1) \approx y_2 = 0.823413627$.

(e) The error in (d) is $|0.823413441 - 0.823413627| = 0.000000185$. With global truncation error $O(h^4)$, when the step size is halved we expect the error for $h = 0.05$ to be one-sixteenth the error when $h = 0.1$. Comparing 0.000000185 with 0.000003225 we see that this is the case.

18. (a) Using $y^{(5)} = -1026e^{-3(x-1)}$ we see that the local truncation error is

$$\left| y^{(5)}(c) \frac{h^5}{120} \right| = 8.55h^5e^{-3(c-1)}.$$

(b) Since $e^{-3(x-1)}$ is a decreasing function for $1 \le x \le 1.5$, $e^{-3(c-1)} \le e^{-3(1-1)} = 1$ for $1 \le c \le 1.5$ and

$$y^{(5)}(c) \frac{h^5}{120} \le 8.55(0.1)^5(1) = 0.0000855.$$

(c) Using the fourth-order Runge-Kutta method with $h = 0.1$ we obtain $y(1.5) \approx 2.053338827$. With $h = 0.05$ we obtain $y(1.5) \approx 2.053222989$.

19. (a) Using $y^{(5)} = \dfrac{24}{(x+1)^5}$ we see that the local truncation error is

$$y^{(5)}(c) \frac{h^5}{120} = \frac{1}{(c+1)^5} \frac{h^5}{5}.$$

(b) Since $\dfrac{1}{(x+1)^5}$ is a decreasing function for $0 \le x \le 0.5$, $\dfrac{1}{(c+1)^5} \le \dfrac{1}{(0+1)^5} = 1$ for $0 \le c \le 0.5$ and

$$y^{(5)}(c) \frac{h^5}{5} \le (1) \frac{(0.1)^5}{5} = 0.000002.$$

(c) Using the fourth-order Runge-Kutta method with $h = 0.1$ we obtain $y(0.5) \approx 0.405465168$. With $h = 0.05$ we obtain $y(0.5) \approx 0.405465111$.

20. (a) For $y' + y = 10\sin 3x$ an integrating factor is e^x so that

$$\frac{d}{dx}[e^x y] = 10e^x \sin 3x \implies e^x y = e^x \sin 3x - 3e^x \cos 3x + c$$

$$\implies y = \sin 3x - 3\cos 3x + ce^{-x}.$$

When $x = 0$, $y = 0$, so $0 = -3 + c$ and $c = 3$. The solution is $y = \sin 3x - 3\cos 3x + 3e^{-x}$. Using Newton's method we find that $x = 1.53235$ is the only positive root in $[0, 2]$.

(b) Using the fourth-order Runge-Kutta method with $h = 0.1$ we obtain the table of values shown. These values are used to obtain an interpolating function in *Mathematica*. The graph of the interpolating function is shown. Using *Mathematica*'s root finding capability we see that the only positive root in $[0, 2]$ is $x = 1.53236$.

x(n)	y(n)	x(n)	y(n)
0.0	0.0000	1.0	4.2147
0.1	0.1440	1.1	3.8033
0.2	0.5448	1.2	3.1513
0.3	1.1409	1.3	2.3076
0.4	1.8559	1.4	1.3390
0.5	2.6049	1.5	0.3243
0.6	3.3019	1.6	-0.6530
0.7	3.8675	1.7	-1.5117
0.8	4.2356	1.8	-2.1809
0.9	4.3593	1.9	-2.6061
1.0	4.2147	2.0	-2.7539

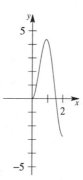

21. Each step of the Euler method requires only 1 function evaluation, while each step of the improved Euler method requires 2 function evaluations – once at (x_n, y_n) and again at (x_{n+1}, y_{n+1}^*). The second-order Runge-Kutta methods require 2 function evaluations per step, while the fourth-order Runge-Kutta method requires 4 function evaluations per step. To compare the methods we approximate the solution of $y' = (x + y - 1)^2$, $y(0) = 2$, at $x = 0.2$ using $h = 0.1$ for the Runge-Kutta method, $h = 0.05$ for the improved Euler method, and $h = 0.025$ for the Euler method. For each method a total of 8 function evaluations is required. By comparing with the exact solution we see that the fourth-order Runge-Kutta method appears to still give the most accurate result.

	EULER	IMP EUL	R-K	
	h = 0.025	h = 0.05	h = 0.1	
x(n)	y(n)	y(n)	y(n)	exact
0.000	2.0000	2.0000		2.0000
0.025	2.0250			2.0263
0.050	2.0526	2.0553		2.0554
0.075	2.0830			2.0875
0.100	2.1165	2.1228	2.1230	2.1230
0.125	2.1535			2.1624
0.150	2.1943	2.2056		2.2061
0.175	2.2395			2.2546
0.200	2.2895	2.3075	2.3085	2.3085

—————— **Exercises 9.4** ——————

1. For $y' - y = x - 1$ an integrating factor is $e^{-\int dx} = e^{-x}$, so that

$$\frac{d}{dx}[e^{-x}y] = (x - 1)e^{-x}$$

and

$$y = e^x(-xe^{-x} + c) = -x + ce^x.$$

From $y(0) = 1$ we find $c = 1$ and $y = -x + e^x$. Comparing exact values with approximations obtained in Example 1, we find $y(0.2) \approx 1.02140276$ compared to $y_1 = 1.02140000$, $y(0.4) \approx 1.09182470$ compared to $y_2 = 1.09181796$, $y(0.6) \approx 1.22211880$ compared to $y_3 = 1.22210646$, and $y(0.8) \approx 1.42554093$ compared to $y_4 = 1.42552788$.

2. 100 **REM** ADAMS-BASHFORTH/ADAMS-MOULTON
   110 **REM** METHOD TO SOLVE Y'=FNF(X,Y)
   120 **REM** DEFINE FNF(X,Y) HERE
   130 **REM** GET INPUTS
   140 **PRINT**
   150 **INPUT** "STEP SIZE=", H
   160 **INPUT** "NUMBER OF STEPS (AT LEAST 4)=",N
   170 **IF** N<4 **GOTO** 160
   180 **INPUT** "X0 =",X
   190 **INPUT** "Y0 =",Y
   200 **PRINT**
   210 **REM** SET UP TABLE
   220 **PRINT** "X","Y"
   230 **PRINT**
   240 **REM** COMPUTE 3 ITERATES USING RUNGE-KUTTA
   250 **DIM** Z(4)
   260 Z(1)=Y
   270 **FOR** I=1 **TO** 3
   280 K1=H*FNF(X,Y)
   290 K2=H*FNF(X+H/2,Y+K1/2)
   300 K3=H*FNF(X+H/2,Y+K2/2)
   310 K4=H*FNF(X+H,Y+K3)
   320 Y=Y+(K1+2*K2+2*K3+K4)/6
   330 Z(I+1)+Y
   340 X=X+H
   350 **PRINT** X,Y
   360 **NEXT** I
   370 **REM** COMPUTE REMAINING X AND Y VALUES
   380 **FOR** I=4 **TO** N
   390 YP=Y+H*(55*FNF(X,Z(4))-59*FNF(X-H,Z(3))+37*FNF(X-2*H,Z(2))
       -9*FNF(X-3*H,Z(1)))/24

400 Y=Y+H*(9*FNF(X+H,YP)+19*FNF(X,Z(4))-5*FNF(X-H,Z(3))+FNF(X-2*H,Z(2)))/24

410 X=X+H

420 **PRINT** X,Y

430 Z(1)=Z(2)

440 Z(2)=Z(3)

450 Z(3)=Z(4)

460 Z(4)=Y

470 **NEXT** I

480 **END**

3.

x(n)	y(n)	
0.00	1.0000	initial condition
0.20	0.7328	Runge-Kutta
0.40	0.6461	Runge-Kutta
0.60	0.6585	Runge-Kutta
	0.7332	*predictor*
0.80	0.7232	corrector

4.

x(n)	y(n)	
0.00	2.0000	initial condition
0.20	1.4112	Runge-Kutta
0.40	1.1483	Runge-Kutta
0.60	1.1039	Runge-Kutta
	1.2109	*predictor*
0.80	1.2049	corrector

5.

x(n)	y(n)	
0.00	0.0000	initial condition
0.20	0.2027	Runge-Kutta
0.40	0.4228	Runge-Kutta
0.60	0.6841	Runge-Kutta
	1.0234	*predictor*
0.80	1.0297	corrector
	1.5376	*predictor*
1.00	1.5569	corrector

x(n)	y(n)	
0.00	0.0000	initial condition
0.10	0.1003	Runge-Kutta
0.20	0.2027	Runge-Kutta
0.30	0.3093	Runge-Kutta
	0.4227	*predictor*
0.40	0.4228	corrector
	0.5462	*predictor*
0.50	0.5463	corrector
	0.6840	*predictor*
0.60	0.6842	corrector
	0.8420	*predictor*
0.70	0.8423	corrector
	1.0292	*predictor*
0.80	1.0297	corrector
	1.2592	*predictor*
0.90	1.2603	corrector
	1.5555	*predictor*
1.00	1.5576	corrector

379

6.

x(n)	y(n)	
0.00	1.0000	initial condition
0.20	1.4414	Runge-Kutta
0.40	1.9719	Runge-Kutta
0.60	2.6028	Runge-Kutta
	3.3483	predictor
0.80	3.3486	corrector
	4.2276	predictor
1.00	4.2280	corrector

x(n)	y(n)	
0.00	1.0000	initial condition
0.10	1.2102	Runge-Kutta
0.20	1.4414	Runge-Kutta
0.30	1.6949	Runge-Kutta
	1.9719	predictor
0.40	1.9719	corrector
	2.2740	predictor
0.50	2.2740	corrector
	2.6028	predictor
0.60	2.6028	corrector
	2.9603	predictor
0.70	2.9603	corrector
	3.3486	predictor
0.80	3.3486	corrector
	3.7703	predictor
0.90	3.7703	corrector
	4.2280	predictor
1.00	4.2280	corrector

7.

x(n)	y(n)	
0.00	0.0000	initial condition
0.20	0.0026	Runge-Kutta
0.40	0.0201	Runge-Kutta
0.60	0.0630	Runge-Kutta
	0.1362	predictor
0.80	0.1360	corrector
	0.2379	predictor
1.00	0.2385	corrector

x(n)	y(n)	
0.00	0.0000	initial condition
0.10	0.0003	Runge-Kutta
0.20	0.0026	Runge-Kutta
0.30	0.0087	Runge-Kutta
	0.0201	predictor
0.40	0.0200	corrector
	0.0379	predictor
0.50	0.0379	corrector
	0.0630	predictor
0.60	0.0629	corrector
	0.0956	predictor
0.70	0.0956	corrector
	0.1359	predictor
0.80	0.1360	corrector
	0.1837	predictor
0.90	0.1837	corrector
	0.2384	predictor
1.00	0.2384	corrector

8.

x(n)	y(n)	
0.00	1.0000	initial condition
0.20	1.2337	Runge-Kutta
0.40	1.5531	Runge-Kutta
0.60	1.9961	Runge-Kutta
	2.6180	predictor
0.80	2.6214	corrector
	3.5151	predictor
1.00	3.5208	corrector

x(n)	y(n)	
0.00	1.0000	initial condition
0.10	1.1079	Runge-Kutta
0.20	1.2337	Runge-Kutta
0.30	1.3807	Runge-Kutta
	1.5530	predictor
0.40	1.5531	corrector
	1.7560	predictor
0.50	1.7561	corrector
	1.9960	predictor
0.60	1.9961	corrector
	2.2811	predictor
0.70	2.2812	corrector
	2.6211	predictor
0.80	2.6213	corrector
	3.0289	predictor
0.90	3.0291	corrector
	3.5203	predictor
1.00	3.5207	corrector

Exercises 9.5

1. The substitution $y' = u$ leads to the iteration formulas

$$y_{n+1} = y_n + hu_n, \qquad u_{n+1} = u_n + h(4u_n - 4y_n).$$

The initial conditions are $y_0 = -2$ and $u_0 = 1$. Then

$$y_1 = y_0 + 0.1u_0 = -2 + 0.1(1) = -1.9$$

$$u_1 = u_0 + 0.1(4u_0 - 4y_0) = 1 + 0.1(4 + 8) = 2.2$$

$$y_2 = y_1 + 0.1u_1 = -1.9 + 0.1(2.2) = -1.68.$$

The general solution of the differential equation is $y = c_1 e^{2x} + c_2 x e^{2x}$. From the initial conditions we find $c_1 = -2$ and $c_2 = 5$. Thus $y = -2e^{2x} + 5xe^{2x}$ and $y(0.2) \approx 1.4918$.

2. The substitution $y' = u$ leads to the iteration formulas

$$y_{n+1} = y_n + hu_n, \qquad u_{n+1} = u_n + h\left(\frac{2}{x}u_n - \frac{2}{x^2}y_n\right).$$

The initial conditions are $y_0 = 4$ and $u_0 = 9$. Then

$$y_1 = y_0 + 0.1u_0 = 4 + 0.1(9) = 4.9$$

$$u_1 = u_0 + 0.1\left(\frac{2}{1}u_0 - \frac{2}{1}y_0\right) = 9 + 0.1[2(9) - 2(4)] = 10$$

$$y_2 = y_1 + 0.1u_1 = 4.9 + 0.1(10) = 5.9.$$

381

Exercises 9.5

The general solution of the Cauchy-Euler differential equation is $y = c_1 x + c_2 x^2$. From the initial conditions we find $c_1 = -1$ and $c_2 = 5$. Thus $y = -x + 5x^2$ and $y(1.2) = 6$.

3. The substitution $y' = u$ leads to the system

$$y' = u, \qquad u' = 4u - 4y.$$

Using formulas (5) and (6) in the text with x corresponding to t, y corresponding to x, and u corresponding to y, we obtain

Runge-Kutta method with h=0.2

m1	m2	m3	m4	k1	k2	k3	k4	x	y	u
								0.00	-2.0000	1.0000
0.2000	0.4400	0.5280	0.9072	2.4000	3.2800	3.5360	4.8064	0.20	-1.4928	4.4731

Runge-Kutta method with h=0.1

m1	m2	m3	m4	k1	k2	k3	k4	x	y	u
								0.00	-2.0000	1.0000
0.1000	0.1600	0.1710	0.2452	1.2000	1.4200	1.4520	1.7124	0.10	-1.8321	2.4427
0.2443	0.3298	0.3444	0.4487	1.7099	2.0031	2.0446	2.3900	0.20	-1.4919	4.4753

4. The substitution $y' = u$ leads to the system

$$y' = u, \qquad u' = \frac{2}{x} u - \frac{2}{x^2} y.$$

Using formulas (5) and (6) in the text with x corresponding to t, y corresponding to x, and u corresponding to y, we obtain

Runge-Kutta method with h=0.2

m1	m2	m3	m4	k1	k2	k3	k4	x	y	u
								1.00	4.0000	9.0000
1.8000	2.0000	2.0017	2.1973	2.0000	2.0165	1.9865	1.9950	1.20	6.0001	11.0002

Runge-Kutta method with h=0.1

m1	m2	m3	m4	k1	k2	k3	k4	x	y	u
								1.00	4.0000	9.0000
0.9000	0.9500	0.9501	0.9998	1.0000	1.0023	0.9979	0.9996	1.10	4.9500	10.0000
1.0000	1.0500	1.0501	1.0998	1.0000	1.0019	0.9983	0.9997	1.20	6.0000	11.0000

5. The substitution $y' = u$ leads to the system

$$y' = u, \qquad u' = 2u - 2y + e^t \cos t.$$

Using formulas (5) and (6) in the text with y corresponding to x and u corresponding to y, we obtain

Runge-Kutta method with h=0.2

m1	m2	m3	m4	k1	k2	k3	k4	t	y	u
								0.00	1.0000	2.0000
0.4000	0.4600	0.4660	0.5320	0.6000	0.6599	0.6599	0.7170	0.20	1.4640	2.6594

Runge-Kutta method with h=0.1

m1	m2	m3	m4	k1	k2	k3	k4	t	y	u
								0.00	1.0000	2.0000
0.2000	0.2150	0.2157	0.2315	0.3000	0.3150	0.3150	0.3298	0.10	1.2155	2.3150
0.2315	0.2480	0.2487	0.2659	0.3299	0.3446	0.3444	0.3587	0.20	1.4640	2.6594

6.

Runge-Kutta method with h=0.1

m1	m2	m3	m4	k1	k2	k3	k4	t	i1	i2
								0.00	0.0000	0.0000
10.0000	0.0000	12.5000	-20.0000	0.0000	5.0000	-5.0000	22.5000	0.10	2.5000	3.7500
8.7500	-2.5000	13.4375	-28.7500	-5.0000	4.3750	-10.6250	29.6875	0.20	2.8125	5.7813
10.1563	-4.3750	17.0703	-40.0000	-8.7500	5.0781	-16.0156	40.3516	0.30	2.0703	7.4023
13.2617	-6.3672	22.9443	-55.1758	-12.7344	6.6309	-22.5488	55.3076	0.40	0.6104	9.1919
17.9712	-8.8867	31.3507	-75.9326	-17.7734	8.9856	-31.2024	75.9821	0.50	-1.5619	11.4877

7.

Runge-Kutta method with h=0.2

m1	m2	m3	m4	k1	k2	k3	k4	t	x	y
								0.00	6.0000	2.0000
2.0000	2.2800	2.3160	2.6408	1.2000	1.4000	1.4280	1.6632	0.20	8.3055	3.4199

Runge-Kutta method with h=0.1

m1	m2	m3	m4	k1	k2	k3	k4	t	x	y
								0.00	6.0000	2.0000
1.0000	1.0700	1.0745	1.1496	0.6000	0.6500	0.6535	0.7075	0.10	7.0731	2.6524
1.1494	1.2289	1.2340	1.3193	0.7073	0.7648	0.7688	0.8307	0.20	8.3055	3.4199

8.

Runge-Kutta method with h=0.2

m1	m2	m3	m4	k1	k2	k3	k4	t	x	y
								0.00	1.0000	1.0000
0.6000	0.9400	1.1060	1.7788	1.4000	2.0600	2.3940	3.7212	0.20	2.0785	3.3382

Runge-Kutta method with h=0.1

m1	m2	m3	m4	k1	k2	k3	k4	t	x	y
								0.00	1.0000	1.0000
0.3000	0.3850	0.4058	0.5219	0.7000	0.8650	0.9068	1.1343	0.10	1.4006	1.8963
0.5193	0.6582	0.6925	0.8828	1.1291	1.4024	1.4711	1.8474	0.20	2.0845	3.3502

383

9.

										Runge-Kutta method with h=0.2
m1	m2	m3	m4	k1	k2	k3	k4	t	x	y
								0.00	-3.0000	5.0000
-1.0000	-0.9200	-0.9080	-0.8176	-0.6000	-0.7200	-0.7120	-0.8216	0.20	-3.9123	4.2857

										Runge-Kutta method with h=0.1
m1	m2	m3	m4	k1	k2	k3	k4	t	x	y
								0.00	-3.0000	5.0000
-0.5000	-0.4800	-0.4785	-0.4571	-0.3000	-0.3300	-0.3290	-0.3579	0.10	-3.4790	4.6707
-0.4571	-0.4342	-0.4328	-0.4086	-0.3579	-0.3858	-0.3846	-0.4112	0.20	-3.9123	4.2857

10.

										Runge-Kutta method with h=0.2
m1	m2	m3	m4	k1	k2	k3	k4	t	x	y
								0.00	0.5000	0.2000
0.6400	1.2760	1.7028	3.3558	1.3200	1.7720	2.1620	3.5794	0.20	2.1589	2.3279

										Runge-Kutta method with h=0.1
m1	m2	m3	m4	k1	k2	k3	k4	t	x	y
								0.00	0.5000	0.2000
0.3200	0.4790	0.5324	0.7816	0.6600	0.7730	0.8218	1.0195	0.10	1.0207	1.0115
0.7736	1.0862	1.1929	1.6862	1.0117	1.2682	1.3692	1.7996	0.20	2.1904	2.3592

11. Solving for x' and y' we obtain the system

$$x' = -2x + y + 5t$$

$$y' = 2x + y - 2t.$$

										Runge-Kutta method with h=0.2
m1	m2	m3	m4	k1	k2	k3	k4	t	x	y
								0.00	1.0000	-2.0000
-0.8000	-0.5400	-0.6120	-0.3888	0.0000	-0.2000	-0.1680	-0.3584	0.20	0.4179	-2.1824

										Runge-Kutta method with h=0.1
m1	m2	m3	m4	k1	k2	k3	k4	t	x	y
								0.00	1.0000	-2.0000
-0.4000	-0.3350	-0.3440	-0.2858	0.0000	-0.0500	-0.0460	-0.0934	0.10	0.6594	-2.0476
-0.2866	-0.2376	-0.2447	-0.2010	-0.0929	-0.1362	-0.1335	-0.1752	0.20	0.4173	-2.1821

12. Solving for x' and y' we obtain the system

$$x' = \frac{1}{2}y - 3t^2 + 2t - 5$$

$$y' = -\frac{1}{2}y + 3t^2 + 2t + 5.$$

| | | | | | | | | Runge-Kutta method with h=0.2 | | |
m1	m2	m3	m4	k1	k2	k3	k4	t	x	y
								0.00	3.0000	-1.0000
-1.1000	-1.0110	-1.0115	-0.9349	1.1000	1.0910	1.0915	1.0949	0.20	1.9867	0.0933

| | | | | | | | | Runge-Kutta method with h=0.1 | | |
m1	m2	m3	m4	k1	k2	k3	k4	t	x	y
								0.00	3.0000	-1.0000
-0.5500	-0.5270	-0.5271	-0.5056	0.5500	0.5470	0.5471	0.5456	0.10	2.4727	-0.4527
-0.5056	-0.4857	-0.4857	-0.4673	0.5456	0.5457	0.5457	0.5473	0.20	1.9867	0.0933

13. (a) From the graph we see that θ is a good approximation to $\sin\theta$ for $0 \le \theta \le 0.5$.

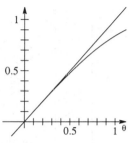

(b) The auxiliary equation is $m^2 + g/l = 0$ so the general solution is $\theta = c_1 \cos\sqrt{g/l}\,t + c_2 \sin\sqrt{g/l}\,t$. From the initial conditions we find $c_1 = \theta_0\sqrt{l/g}$ and $c_2 = -\sqrt{l/g}$. Thus, the exact solution is

$$\theta = \theta_0 \cos\sqrt{\frac{g}{l}}\,t - \sqrt{\frac{l}{g}}\,\sin\sqrt{\frac{g}{l}}\,t.$$

(c) In order to use the fourth-order Runge-Kutta method (or any numerical procedure) we must assign numerical values to the parameters g and l. In this case we take $g/l = 9$, so that the differential equation is $d^2\theta/dt^2 + 9\sin\theta = 0$. We then apply the fourth-order Runge-Kutta method with $h = 0.1$. Tables are shown on the next page for $\theta_0 = 0.2$, $\theta_0 = 0.5$, and $\theta_0 = 1$. Values obtained using $\theta = \theta_0 \cos 3t - \frac{1}{3}\sin 3t$ are also shown. The conjecture in part (a) seems valid.

x	nonlinear R-K	linear exact	x	nonlinear R-K	linear exact	x	nonlinear R-K	linear exact
0.0	0.2000	0.2000	0.0	0.5000	0.5000	0.0	1.0000	1.0000
0.1	0.0926	0.0926	0.1	0.3799	0.3792	0.1	0.8631	0.8568
0.2	-0.0230	-0.0231	0.2	0.2267	0.2245	0.2	0.6584	0.6371
0.3	-0.1366	-0.1368	0.3	0.0535	0.0497	0.3	0.3992	0.3605
0.4	-0.2381	-0.2382	0.4	-0.1245	-0.1295	0.4	0.1055	0.0517
0.5	-0.3184	-0.3184	0.5	-0.2914	-0.2971	0.5	-0.1976	-0.2618
0.6	-0.3709	-0.3701	0.6	-0.4327	-0.4382	0.6	-0.4833	-0.5518
0.7	-0.3909	-0.3887	0.7	-0.5365	-0.5402	0.7	-0.7276	-0.7926
0.8	-0.3769	-0.3726	0.8	-0.5947	-0.5939	0.8	-0.9127	-0.9625
0.9	-0.3300	-0.3233	0.9	-0.6028	-0.5945	0.9	-1.0270	-1.0465
1.0	-0.2541	-0.2450	1.0	-0.5602	-0.5420	1.0	-1.0647	-1.0370
1.1	-0.1558	-0.1449	1.1	-0.4701	-0.4412	1.1	-1.0239	-0.9349
1.2	-0.0437	-0.0318	1.2	-0.3395	-0.3009	1.2	-0.9066	-0.7493
1.3	0.0723	0.0841	1.3	-0.1793	-0.1337	1.3	-0.7189	-0.4967
1.4	0.1819	0.1925	1.4	-0.0031	0.0454	1.4	-0.4725	-0.1997
1.5	0.2754	0.2837	1.5	0.1733	0.2204	1.5	-0.1857	0.1150
1.6	0.3445	0.3496	1.6	0.3344	0.3758	1.6	0.1175	0.4196
1.7	0.3835	0.3842	1.7	0.4662	0.4976	1.7	0.4103	0.6866
1.8	0.3891	0.3845	1.8	0.5578	0.5749	1.8	0.6676	0.8923
1.9	0.3607	0.3505	1.9	0.6022	0.6009	1.9	0.8699	1.0183
2.0	0.3008	0.2852	2.0	0.5958	0.5732	2.0	1.0038	1.0533
2.1	0.2145	0.1944	2.1	0.5393	0.4943	2.1	1.0622	0.9943
2.2	0.1092	0.0862	2.2	0.4369	0.3713	2.2	1.0423	0.8464
2.3	-0.0059	-0.0297	2.3	0.2968	0.2150	2.3	0.9450	0.6229
2.4	-0.1204	-0.1429	2.4	0.1305	0.0396	2.4	0.7752	0.3438
2.5	-0.2242	-0.2433	2.5	-0.0473	-0.1393	2.5	0.5430	0.0340
2.6	-0.3082	-0.3221	2.6	-0.2209	-0.3059	2.6	0.2648	-0.2789
2.7	-0.3651	-0.3720	2.7	-0.3750	-0.4451	2.7	-0.0366	-0.5668
2.8	-0.3900	-0.3887	2.8	-0.4964	-0.5445	2.8	-0.3348	-0.8042
2.9	-0.3810	-0.3707	2.9	-0.5753	-0.5953	2.9	-0.6037	-0.9696
3.0	-0.3388	-0.3196	3.0	-0.6055	-0.5929	3.0	-0.8222	-1.0485

Exercises 9.6

1. We identify $P(x) = 0$, $Q(x) = 9$, $f(x) = 0$, and $h = (2-0)/4 = 0.5$. Then the finite difference equation is

$$y_{i+1} + 0.25y_i + y_{i-1} = 0.$$

The solution of the corresponding linear system gives

x	0.0	0.5	1.0	1.5	2.0
y	4.0000	-5.6774	-2.5807	6.3226	1.0000

2. We identify $P(x) = 0$, $Q(x) = -1$, $f(x) = x^2$, and $h = (1-0)/4 = 0.25$. Then the finite difference equation is

$$y_{i+1} - 2.0625y_i + y_{i-1} = 0.0625x_i^2.$$

386

The solution of the corresponding linear system gives

x	0.00	0.25	0.50	0.75	1.00
y	0.0000	-0.0172	-0.0316	-0.0324	0.0000

3. We identify $P(x) = 2$, $Q(x) = 1$, $f(x) = 5x$, and $h = (1 - 0)/5 = 0.2$. Then the finite difference equation is

$$1.2y_{i+1} - 1.96y_i + 0.8y_{i-1} = 0.04(5x_i).$$

The solution of the corresponding linear system gives

x	0.0	0.2	0.4	0.6	0.8	1.0
y	0.0000	-0.2259	-0.3356	-0.3308	-0.2167	0.0000

4. We identify $P(x) = -10$, $Q(x) = 25$, $f(x) = 1$, and $h = (1 - 0)/5 = 0.2$. Then the finite difference equation is

$$-y_i + 2y_{i-1} = 0.04.$$

The solution of the corresponding linear system gives

x	0.0	0.2	0.4	0.6	0.8	1.0
y	1.0000	1.9600	3.8800	7.7200	15.4000	0.0000

5. We identify $P(x) = -4$, $Q(x) = 4$, $f(x) = (1+x)e^{2x}$, and $h = (1 - 0)/6 = 0.1667$. Then the finite difference equation is

$$0.6667y_{i+1} - 1.8889y_i + 1.3333y_{i-1} = 0.2778(1 + x_i)e^{2x_i}.$$

The solution of the corresponding linear system gives

x	0.0000	0.1667	0.3333	0.5000	0.6667	0.8333	1.0000
y	3.0000	3.3751	3.6306	3.6448	3.2355	2.1411	0.0000

6. We identify $P(x) = 5$, $Q(x) = 0$, $f(x) = 4\sqrt{x}$, and $h = (2 - 1)/6 = 0.1667$. Then the finite difference equation is

$$1.4167y_{i+1} - 2y_i + 0.5833y_{i-1} = 0.2778(4\sqrt{x_i}).$$

The solution of the corresponding linear system gives

x	1.0000	1.1667	1.3333	1.5000	1.6667	1.8333	2.0000
y	1.0000	-0.5918	-1.1626	-1.3070	-1.2704	-1.1541	-1.0000

7. We identify $P(x) = 3/x$, $Q(x) = 3/x^2$, $f(x) = 0$, and $h = (2 - 1)/8 = 0.125$. Then the finite difference equation is

$$\left(1 + \frac{0.1875}{x_i}\right) y_{i+1} + \left(-2 + \frac{0.0469}{x_i^2}\right) y_i + \left(1 - \frac{0.1875}{x_i}\right) y_{i-1} = 0.$$

The solution of the corresponding linear system gives

x	1.000	1.125	1.250	1.375	1.500	1.625	1.750	1.875	2.000
y	5.0000	3.8842	2.9640	2.2064	1.5826	1.0681	0.6430	0.2913	0.0000

8. We identify $P(x) = -1/x$, $Q(x) = x^{-2}$, $f(x) = \ln x/x^2$, and $h = (2-1)/8 = 0.125$. Then the finite difference equation is

$$\left(1 - \frac{0.0625}{x_i}\right) y_{i+1} + \left(-2 + \frac{0.0156}{x_i^2}\right) y_i + \left(1 + \frac{0.0625}{x_i}\right) y_{i-1} = 0.0156 \ln x_i.$$

The solution of the corresponding linear system gives

x	1.000	1.125	1.250	1.375	1.500	1.625	1.750	1.875	2.000
y	0.0000	-0.1988	-0.4168	-0.6510	-0.8992	-1.1594	-1.4304	-1.7109	-2.0000

9. We identify $P(x) = 1 - x$, $Q(x) = x$, $f(x) = x$, and $h = (1-0)/10 = 0.1$. Then the finite difference equation is

$$[1 + 0.05(1 - x_i)]y_{i+1} + [-2 + 0.01x_i]y_i + [1 - 0.05(1 - x_i)]y_{i-1} = 0.01x_i.$$

The solution of the corresponding linear system gives

x	0.0	0.1	0.2	0.3	0.4	0.5	0.6
y	0.0000	0.2660	0.5097	0.7357	0.9471	1.1465	1.3353

0.7	0.8	0.9	1.0
1.5149	1.6855	1.8474	2.0000

10. We identify $P(x) = x$, $Q(x) = 1$, $f(x) = x$, and $h = (1-0)/10 = 0.1$. Then the finite difference equation is

$$(1 + 0.05x_i)y_{i+1} - 1.99y_i + (1 - 0.05x_i)y_{i-1} = 0.01x_i.$$

The solution of the corresponding linear system gives

x	0.0	0.1	0.2	0.3	0.4	0.5	0.6
y	1.0000	0.8929	0.7789	0.6615	0.5440	0.4296	0.3216

0.7	0.8	0.9	1.0
0.2225	0.1347	0.0601	0.0000

11. We identify $P(x) = 0$, $Q(x) = -4$, $f(x) = 0$, and $h = (1-0)/8 = 0.125$. Then the finite difference equation is

$$y_{i+1} - 2.0625y_i + y_{i-1} = 0.$$

The solution of the corresponding linear system gives

x	0.000	0.125	0.250	0.375	0.500	0.625	0.750	0.875	1.000
y	0.0000	0.3492	0.7202	1.1363	1.6233	2.2118	2.9386	3.8490	5.0000

12. We identify $P(r) = 2/r$, $Q(r) = 0$, $f(r) = 0$, and $h = (4-1)/6 = 0.5$. Then the finite difference equation is

$$\left(1 + \frac{0.5}{r_i}\right) u_{i+1} - 2u_i + \left(1 - \frac{0.5}{r_i}\right) u_{i-1} = 0.$$

The solution of the corresponding linear system gives

r	1.0	1.5	2.0	2.5	3.0	3.5	4.0
u	50.0000	72.2222	83.3333	90.0000	94.4444	97.6190	100.0000

13. (a) The difference equation

$$\left(1 + \frac{h}{2}P_i\right)y_{i+1} + (-2 + h^2 Q_i)y_i + \left(1 - \frac{h}{2}P_i\right)y_{i-1} = h^2 f_i$$

is the same as the one derived on page 383 in the text. The equations are the same because the derivation was based only on the differential equation, not the boundary conditions. If we allow i to range from 0 to $n-1$ we obtain n equations in the $n+1$ unknowns $y_{-1}, y_0, y_1, \ldots, y_{n-1}$. Since y_n is one of the given boundary conditions, it is not an unknown.

(b) Identifying $y_0 = y(0)$, $y_{-1} = y(0-h)$, and $y_1 = y(0+h)$ we have from (5) in the text

$$\frac{1}{2h}[y_1 - y_{-1}] = y'(0) = 1 \quad \text{or} \quad y_1 - y_{-1} = 2h.$$

The difference equation corresponding to $i = 0$,

$$\left(1 + \frac{h}{2}P_0\right)y_1 + (-2 + h^2 Q_0)y_0 + \left(1 - \frac{h}{2}P_0\right)y_{-1} = h^2 f_0$$

becomes, with $y_{-1} = y_1 - 2h$,

$$\left(1 + \frac{h}{2}P_0\right)y_1 + (-2 + h^2 Q_0)y_0 + \left(1 - \frac{h}{2}P_0\right)(y_1 - 2h) = h^2 f_0$$

or

$$2y_1 + (-2 + h^2 Q_0)y_0 = h^2 f_0 + 2h - P_0.$$

Alternatively, we may simply add the equation $y_1 - y_{-1} = 2h$ to the list of n difference equations obtaining $n+1$ equations in the $n+1$ unknowns $y_{-1}, y_0, y_1, \ldots, y_{n-1}$.

(c) Using $n = 5$ we obtain

x	0.0	0.2	0.4	0.6	0.8	1.0
y	-2.2755	-2.0755	-1.8589	-1.6126	-1.3275	-1.0000

14. Using $h = 0.1$ and, after shooting a few times, $y'(0) = 0.43535$ we obtain the following table with the fourth-order Runge-Kutta method.

x	0.0	0.1	0.2	0.3	0.4	0.5	0.6
y	1.00000	1.04561	1.09492	1.14714	1.20131	1.25633	1.31096

	0.7	0.8	0.9	1.0
	1.36392	1.41388	1.45962	1.50003

Chapter 9 Review Exercises

1.

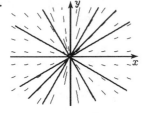

2.

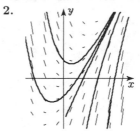

3.

h=0.1		IMPROVED	RUNGE
x(n)	EULER	EULER	KUTTA
1.00	2.0000	2.0000	2.0000
1.10	2.1386	2.1549	2.1556
1.20	2.3097	2.3439	2.3454
1.30	2.5136	2.5672	2.5695
1.40	2.7504	2.8246	2.8278
1.50	3.0201	3.1157	3.1197

h=0.05		IMPROVED	RUNGE
x(n)	EULER	EULER	KUTTA
1.00	2.0000	2.0000	2.0000
1.05	2.0693	2.0735	2.0736
1.10	2.1469	2.1554	2.1556
1.15	2.2328	2.2459	2.2462
1.20	2.3272	2.3450	2.3454
1.25	2.4299	2.4527	2.4532
1.30	2.5409	2.5689	2.5695
1.35	2.6604	2.6937	2.6944
1.40	2.7883	2.8269	2.8278
1.45	2.9245	2.9686	2.9696
1.50	3.0690	3.1187	3.1197

4.

h=0.1		IMPROVED	RUNGE
x(n)	EULER	EULER	KUTTA
0.00	0.0000	0.0000	0.0000
0.10	0.1000	0.1005	0.1003
0.20	0.2010	0.2030	0.2026
0.30	0.3049	0.3092	0.3087
0.40	0.4135	0.4207	0.4201
0.50	0.5279	0.5382	0.5376

h=0.05		IMPROVED	RUNGE
x(n)	EULER	EULER	KUTTA
0.00	0.0000	0.0000	0.0000
0.05	0.0500	0.0501	0.0500
0.10	0.1001	0.1004	0.1003
0.15	0.1506	0.1512	0.1511
0.20	0.2017	0.2027	0.2026
0.25	0.2537	0.2552	0.2551
0.30	0.3067	0.3088	0.3087
0.35	0.3610	0.3638	0.3637
0.40	0.4167	0.4202	0.4201
0.45	0.4739	0.4782	0.4781
0.50	0.5327	0.5378	0.5376

5.

h=0.1		IMPROVED	RUNGE
x(n)	EULER	EULER	KUTTA
0.50	0.5000	0.5000	0.5000
0.60	0.6000	0.6048	0.6049
0.70	0.7095	0.7191	0.7194
0.80	0.8283	0.8427	0.8431
0.90	0.9559	0.9752	0.9757
1.00	1.0921	1.1163	1.1169

h=0.05		IMPROVED	RUNGE
x(n)	EULER	EULER	KUTTA
0.50	0.5000	0.5000	0.5000
0.55	0.5500	0.5512	0.5512
0.60	0.6024	0.6049	0.6049
0.65	0.6573	0.6609	0.6610
0.70	0.7144	0.7193	0.7194
0.75	0.7739	0.7800	0.7801
0.80	0.8356	0.8430	0.8431
0.85	0.8996	0.9082	0.9083
0.90	0.9657	0.9755	0.9757
0.95	1.0340	1.0451	1.0452
1.00	1.1044	1.1168	1.1169

6.

h=0.1		IMPROVED	RUNGE
x(n)	EULER	EULER	KUTTA
1.00	1.0000	1.0000	1.0000
1.10	1.2000	1.2380	1.2415
1.20	1.4760	1.5910	1.6036
1.30	1.8710	2.1524	2.1909
1.40	2.4643	3.1458	3.2745
1.50	3.4165	5.2510	5.8338

h=0.05		IMPROVED	RUNGE
x(n)	EULER	EULER	KUTTA
1.00	1.0000	1.0000	1.0000
1.05	1.1000	1.1091	1.1095
1.10	1.2183	1.2405	1.2415
1.15	1.3595	1.4010	1.4029
1.20	1.5300	1.6001	1.6036
1.25	1.7389	1.8523	1.8586
1.30	1.9988	2.1799	2.1911
1.35	2.3284	2.6197	2.6401
1.40	2.7567	3.2360	3.2755
1.45	3.3296	4.1528	4.2363
1.50	4.1253	5.6404	5.8446

7. Using

$$y_{n+1} = y_n + hu_n, \qquad y_0 = 3$$

$$u_{n+1} = u_n + h(2x_n + 1)y_n, \qquad u_0 = 1$$

we obtain (when $h = 0.2$) $y_1 = y(0.2) = y_0 + hu_0 = 3 + (0.2)1 = 3.2$. When $h = 0.1$ we have

$$y_1 = y_0 + 0.1u_0 = 3 + (0.1)1 = 3.1$$

$$u_1 = u_0 + 0.1(2x_0 + 1)y_0 = 1 + 0.1(1)3 = 1.3$$

$$y_2 = y_1 + 0.1u_1 = 3.1 + 0.1(1.3) = 3.23.$$

8.

x(n)	y(n)	
0.00	2.0000	initial condition
0.10	2.4734	Runge-Kutta
0.20	3.1781	Runge-Kutta
0.30	4.3925	Runge-Kutta
	6.7689	*predictor*
0.40	7.0783	corrector

391

9. Using $x_0 = 1$, $y_0 = 2$, and $h = 0.1$ we have

$$x_1 = x_0 + h(x_0 + y_0) = 1 + 0.1(1 + 2) = 1.3$$

$$y_1 = y_0 + h(x_0 - y_0) = 2 + 0.1(1 - 2) = 1.9$$

and

$$x_2 = x_1 + h(x_1 + y_1) = 1.3 + 0.1(1.3 + 1.9) = 1.62$$

$$y_2 = y_1 + h(x_1 - y_1) = 1.9 + 0.1(1.3 - 1.9) = 1.84.$$

Thus, $x(0.2) \approx 1.62$ and $y(0.2) \approx 1.84$.

10. We identify $P(x) = 0$, $Q(x) = 6.55(1 + x)$, $f(x) = 1$, and $h = (1 - 0)/10 = 0.1$. Then the finite difference equation is

$$y_{i+1} + [-2 + 0.0655(1 + x_i)]y_i + y_{i-1} = 0.001$$

or

$$y_{i+1} + (0.0655x_i - 1.9345)y_i + y_{i-1} = 0.001.$$

The solution of the corresponding linear system gives

x	0.0	0.1	0.2	0.3	0.4	0.5	0.6
y	0.0000	4.1987	8.1049	11.3840	13.7038	14.7770	14.4083

0.7	0.8	0.9	1.0
12.5396	9.2847	4.9450	0.0000

10 Plane Autonomous Systems and Stability

———————— **Exercises 10.1** ————————————————

1. The corresponding plane autonomous system is

$$x' = y, \quad y' = -9\sin x.$$

If (x, y) is a critical point, $y = 0$ and $-9\sin x = 0$. Therefore $x = \pm n\pi$ and so the critical points are $(\pm n\pi, 0)$ for $n = 0, 1, 2, \ldots$.

2. The corresponding plane autonomous system is

$$x' = y, \quad y' = -2x - y^2.$$

If (x, y) is a critical point, then $y = 0$ and so $-2x - y^2 = -2x = 0$. Therefore $(0, 0)$ is the sole critical point.

3. The corresponding plane autonomous system is

$$x' = y, \quad y' = x^2 - y(1 - x^3).$$

If (x, y) is a critical point, $y = 0$ and so $x^2 - y(1 - x^3) = x^2 = 0$. Therefore $(0, 0)$ is the sole critical point.

4. The corresponding plane autonomous system is

$$x' = y, \quad y' = -4\frac{x}{1 + x^2} - 2y.$$

If (x, y) is a critical point, $y = 0$ and so $-4\dfrac{x}{1 + x^2} - 2(0) = 0$. Therefore $x = 0$ and so $(0, 0)$ is the sole critical point.

5. The corresponding plane autonomous system is

$$x' = y, \quad y' = -x + \epsilon x^3.$$

If (x, y) is a critical point, $y = 0$ and $-x + \epsilon x^3 = 0$. Hence $x(-1 + \epsilon x^2) = 0$ and so $x = 0, \sqrt{1/\epsilon}$, $-\sqrt{1/\epsilon}$. The critical points are $(0, 0)$, $(\sqrt{1/\epsilon}, 0)$ and $(-\sqrt{1/\epsilon}, 0)$.

6. The corresponding plane autonomous system is

$$x' = y, \quad y' = -x + \epsilon x|x|.$$

393

If (x, y) is a critical point, $y = 0$ and $-x + \epsilon x|x| = x(-1 + \epsilon|x|) = 0$. Hence $x = 0, 1/\epsilon, -1/\epsilon$. The critical points are $(0, 0)$, $(1/\epsilon, 0)$ and $(-1/\epsilon, 0)$.

7. From $x + xy = 0$ we have $x(1 + y) = 0$. Therefore $x = 0$ or $y = -1$. If $x = 0$, then, substituting into $-y - xy = 0$, we obtain $y = 0$, Likewise, if $y = -1$, $1 + x = 0$ or $x = -1$. We may conclude that $(0, 0)$ and $(-1, -1)$ are critical points of the system.

8. From $y^2 - x = 0$ we have $x = y^2$. Substituting into $x^2 - y = 0$, we obtain $y^4 - y = 0$ or $y(y^3 - 1) = 0$. It follows that $y = 0$, 1 and so $(0, 0)$ and $(1, 1)$ are the critical points of the system.

9. From $x - y = 0$ we have $y = x$. Substituting into $3x^2 - 4y = 0$ we obtain $3x^2 - 4x = x(3x - 4) = 0$. It follows that $(0, 0)$ and $(4/3, 4/3)$ are the critical points of the system.

10. From $x^3 - y = 0$ we have $y = x^3$. Substituting into $x - y^3 = 0$ we obtain $x - x^9 = 0$ or $x(1 - x^8)$. Therefore $x = 0$, 1, -1 and so the critical points of the system are $(0, 0)$, $(1, 1)$, and $(-1, -1)$.

11. From $x(10 - x - \frac{1}{2}y) = 0$ we obtain $x = 0$ or $x + \frac{1}{2}y = 10$. Likewise $y(16 - y - x) = 0$ implies that $y = 0$ or $x + y = 16$. We therefore have four cases. If $x = 0$, $y = 0$ or $y = 16$. If $x + \frac{1}{2}y = 10$, we may conclude that $y(-\frac{1}{2}y + 6) = 0$ and so $y = 0$, 12. Therefore the critical points of the system are $(0, 0)$, $(0, 16)$, $(10, 0)$, and $(4, 12)$.

12. Adding the two equations we obtain $10 - 15\dfrac{y}{y + 5} = 0$. It follows that $y = 10$, and from $-2x + y + 10 = 0$ we may conclude that $x = 10$. Therefore $(10, 10)$ is the sole critical point of the system.

13. From $x^2 e^y = 0$ we have $x = 0$. Since $e^x - 1 = e^0 - 1 = 0$, the second equation is satisfied for an arbitrary value of y. Therefore any point of the form $(0, y)$ is a critical point.

14. From $\sin y = 0$ we have $y = \pm n\pi$. From $e^{x-y} = 1$, we may conclude that $x - y = 0$ or $x = y$. The critical points of the system are therefore $(\pm n\pi, \pm n\pi)$ for $n = 0, 1, 2, \ldots$.

15. From $x(1 - x^2 - 3y^2) = 0$ we have $x = 0$ or $x^2 + 3y^2 = 1$. If $x = 0$, then substituting into $y(3 - x^2 - 3y^2)$ gives $y(3 - 3y^2) = 0$. Therefore $y = 0$, 1, -1. Likewise $x^2 = 1 - 3y^2$ yields $2y = 0$ so that $y = 0$ and $x^2 = 1 - 3(0)^2 = 1$. The critical points of the system are therefore $(0, 0)$, $(0, 1)$, $(0, -1)$, $(1, 0)$, and $(-1, 0)$.

16. From $-x(4 - y^2) = 0$ we obtain $x = 0$, $y = 2$, or $y = -2$. If $x = 0$, then substituting into $4y(1 - x^2)$ yields $y = 0$. Likewise $y = 2$ gives $8(1 - x^2) = 0$ or $x = 1$, -1. Finally $y = -2$ yields $-8(1 - x^2) = 0$ or $x = 1$, -1. The critical points of the system are therefore $(0, 0)$, $(1, 2)$, $(-1, 2)$, $(1, -2)$, and $(-1, -2)$.

17. (a) From Exercises 8.2, Problem 1, $x = c_1 e^{5t} - c_2 e^{-t}$ and $y = 2c_1 e^{5t} + c_2 e^{-t}$.

(b) From $\mathbf{X}(0) = (2, -1)$ it follows that $c_1 = 0$ and $c_2 = 2$. Therefore $x = -2e^{-t}$ and $y = 2e^{-t}$.

(c)

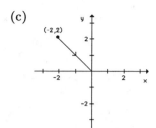

18. (a) From Exercises 8.2, Problem 6, $x = c_1 + 2c_2 e^{-5t}$ and $y = 3c_1 + c_2 e^{-5t}$.

(b) From $\mathbf{X}(0) = (3, 4)$ it follows that $c_1 = c_2 = 1$. Therefore $x = 1 + 2e^{-5t}$ and $y = 3 + e^{-5t}$.

(c)

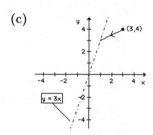

19. (a) From Exercises 8.2, Problem 35, $x = c_1(4\cos 3t - 3\sin 3t) + c_2(4\sin 3t + 3\cos 3t)$ and $y = c_1(5\cos 3t) + c_2(5\sin 3t)$. All solutions are one periodic with $p = 2\pi/3$.

(b) From $\mathbf{X}(0) = (4, 5)$ it follows that $c_1 = 1$ and $c_2 = 0$. Therefore $x = 4\cos 3t - 3\sin 3t$ and $y = 5\cos 3t$.

(c)

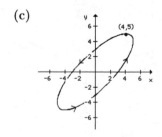

20. (a) From Exercises 8.2, Problem 32, $x = c_1(\sin t - \cos t) + c_2(-\cos t - \sin t)$ and $y = 2c_1 \cos t + 2c_2 \sin t$. All solutions are periodic with $p = 2\pi$.

(b) From $\mathbf{X}(0) = (-2, 2)$ it follows that $c_1 = c_2 = 1$. Therefore $x = -2\cos t$ and $y = 2\cos t + 2\sin t$.

(c)

21. **(a)** From Exercises 8.2, Problem 33, $x = c_1(\sin t - \cos t)e^{4t} + c_2(-\sin t - \cos t)e^{4t}$ and $y = 2c_1(\cos t)\,e^{4t} + 2c_2(\sin t)\,e^{4t}$. Because of the presence of e^{4t}, there are no periodic solutions.

(b) From $\mathbf{X}(0) = (-1, 2)$ it follows that $c_1 = 1$ and $c_2 = 0$. Therefore $x = (\sin t - \cos t)e^{4t}$ and $y = 2(\cos t)\,e^{4t}$.

(c)

22. **(a)** From Exercises 8.2, Problem 36, $x = c_1 e^{-t}(2\cos 2t - 2\sin 2t) + c_2 e^{-t}(2\cos 2t + 2\sin 2t)$ and $y = c_1 e^{-t}\cos 2t + c_2 e^{-t}\sin 2t$. Because of the presence of e^{-t}, there are no periodic solutions.

(b) From $\mathbf{X}(0) = (2, 1)$ it follows that $c_1 = 1$ and $c_2 = 0$. Therefore $x = e^{-t}(2\cos 2t - 2\sin 2t)$ and $y = e^{-t}\cos 2t$.

(c)

23. Switching to polar coordinates,

$$\frac{dr}{dt} = \frac{1}{r}\left(x\frac{dx}{dt} + y\frac{dy}{dt}\right) = \frac{1}{r}(-xy - x^2 r^4 + xy - y^2 r^4) = -r^5$$

$$\frac{d\theta}{dt} = \frac{1}{r^2}\left(-y\frac{dx}{dt} + x\frac{dy}{dt}\right) = \frac{1}{r^2}(y^2 + xyr^4 + x^2 - xyr^4) = 1\,.$$

If we use separation of variables on $\dfrac{dr}{dt} = -r^5$ we obtain

$$r = \left(\frac{1}{4t + c_1}\right)^{1/4} \quad \text{and} \quad \theta = t + c_2.$$

Since $\mathbf{X}(0) = (4, 0)$, $r = 4$ and $\theta = 0$ when $t = 0$. It follows that $c_2 = 0$ and $c_1 = \dfrac{1}{256}$. The final solution may be written as

$$r = \frac{4}{\sqrt[4]{1024t + 1}}, \quad \theta = t$$

and so the solution spirals toward the origin as t increases.

24. Switching to polar coordinates,

$$\frac{dr}{dt} = \frac{1}{r}\left(x\frac{dx}{dt} + y\frac{dy}{dt}\right) = \frac{1}{r}(xy - x^2r^2 - xy + y^2r^2) = r^3$$

$$\frac{d\theta}{dt} = \frac{1}{r^2}\left(-y\frac{dx}{dt} + x\frac{dy}{dt}\right) = \frac{1}{r^2}(-y^2 - xyr^2 - x^2 + xyr^2) = -1.$$

If we use separation of variables, it follows that

$$r = \frac{1}{\sqrt{-2t + c_1}} \quad \text{and} \quad \theta = -t + c_2.$$

Since $\mathbf{X}(0) = (4, 0)$, $r = 4$ and $\theta = 0$ when $t = 0$. It follows that $c_2 = 0$ and $c_1 = \frac{1}{16}$. The final solution may be written as

$$r = \frac{4}{\sqrt{1 - 32t}}, \quad \theta = -t.$$

Note that $r \to \infty$ as $t \to \left(\frac{1}{32}\right)^-$. Because $0 \le t \le \frac{1}{32}$, the curve is not a spiral.

25. Switching to polar coordinates,

$$\frac{dr}{dt} = \frac{1}{r}\left(x\frac{dx}{dt} + y\frac{dy}{dt}\right) = \frac{1}{r}[-xy + x^2(1 - r^2) + xy + y^2(1 - r^2)] = r(1 - r^2)$$

$$\frac{d\theta}{dt} = \frac{1}{r^2}\left(-y\frac{dx}{dt} + x\frac{dy}{dt}\right) = \frac{1}{r^2}[y^2 - xy(1 - r^2) + x^2 + xy(1 - r^2)] = 1.$$

Now $\dfrac{dr}{dt} = r - r^3$ or $\dfrac{dr}{dt} - r = -r^3$ is a Bernoulli differential equation. Following the procedure in Section 2.4 of the text, we let $w = r^{-2}$ so that $w' = -2r^{-3}\dfrac{dr}{dt}$. Therefore $w' + 2w = 2$, a linear first order differential equation. It follows that $w = 1 + c_1e^{-2t}$ and so $r^2 = \dfrac{1}{1 + c_1e^{-2t}}$. The general solution may be written as

$$r = \frac{1}{\sqrt{1 + c_1e^{-2t}}}, \quad \theta = t + c_2.$$

If $\mathbf{X}(0) = (1, 0)$, $r = 1$ and $\theta = 0$ when $t = 0$. Therefore $c_1 = 0 = c_2$ and so $x = r \cos t = \cos t$ and $y = r \sin t = \sin t$. This solution generates the circle $r = 1$. If $\mathbf{X}(0) = (2, 0)$, $r = 2$ and $\theta = 0$ when $t = 0$. Therefore $c_1 = -3/4$, $c_2 = 0$ and so

$$r = \frac{1}{\sqrt{1 - \frac{3}{4} e^{-2t}}}, \qquad \theta = t.$$

This solution spirals toward the circle $r = 1$ as t increases.

26. Switching to polar coordinates,

$$\frac{dr}{dt} = \frac{1}{r}\left(x \frac{dx}{dt} + y \frac{dy}{dt}\right) = \frac{1}{r}\left[xy - \frac{x^2}{r}(4 - r^2) - xy - \frac{y^2}{r}(4 - r^2)\right] = r^2 - 4$$

$$\frac{d\theta}{dt} = \frac{1}{r^2}\left(-y \frac{dx}{dt} + x \frac{dy}{dt}\right) = \frac{1}{r^2}\left[-y^2 + \frac{xy}{r}(4 - r^2) - x^2 - \frac{xy}{r}(4 - r^2)\right] = -1.$$

From Example 4, Section 2.1,

$$r = 2 \frac{1 + c_1 e^{4t}}{1 - c_1 e^{4t}} \quad \text{and} \quad \theta = -t + c_2.$$

If $\mathbf{X}(0) = (1, 0)$, $r = 1$ and $\theta = 0$ when $t = 0$. It follows that $c_2 = 0$ and $c_1 = -\frac{1}{3}$. Therefore

$$r = 2 \frac{1 - \frac{1}{3} e^{4t}}{1 + \frac{1}{3} e^{4t}} \quad \text{and} \quad \theta = -t.$$

Note that $r = 0$ when $e^{4t} = 3$ or $t = \dfrac{\ln 3}{4}$ and $r \to -2$ as $t \to \infty$. The solution therefore approaches the circle $r = 2$. If $\mathbf{X}(0) = (2, 0)$, it follows that $c_1 = c_2 = 0$. Therefore $r = 2$ and $\theta = -t$ so that the solution generates the circle $r = 2$ traversed in the clockwise direction. Note also that the original system is not defined at $(0, 0)$ but the corresponding polar system is defined for $r = 0$. If the Runge-Kutta method is applied to the original system, the solution corresponding to $\mathbf{X}(0) = (1, 0)$ will stall at the origin.

27. The system has no critical points, so there are no periodic solutions.

28. From $x(6y - 1) = 0$ and $y(2 - 8x) = 0$ we see that $(0, 0)$ and $(1/4, 1/6)$ are critical points. From the graph we see that there are periodic solutions around $(1/4, 1/6)$.

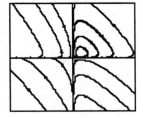

29. The only critical point is $(0,0)$. There appears to be a single periodic solution around $(0,0)$.

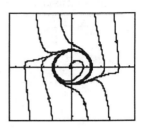

30. The system has no critical points, so there are no periodic solutions.

31. If $\mathbf{X}(t) = (x(t), y(t))$ is a solution,

$$\frac{d}{dt} f(x(t), y(t)) = \frac{\partial f}{\partial x} \frac{dx}{dt} + \frac{\partial f}{\partial y} \frac{dy}{dt} = QP - PQ = 0,$$

using the chain rule. Therefore $f(x(t), y(t)) = c$ for some constant c, and the solution lies on a level curve of f.

Exercises 10.2

1. (a) If $\mathbf{X}(0) = \mathbf{X}_0$ lies on the line $y = 2x$, then $\mathbf{X}(t)$ approaches $(0,0)$ along this line. For all other initial conditions, $\mathbf{X}(t)$ approaches $(0,0)$ from the direction determined by the line $y = -x/2$.

(b)

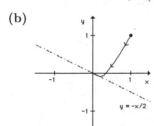

2. (a) If $\mathbf{X}(0) = \mathbf{X}_0$ lies on the line $y = -x$, then $\mathbf{X}(t)$ becomes unbounded along this line. For all other initial conditions, $\mathbf{X}(t)$ becomes unbounded and $y = -3x/2$ serves as an asymptote.

(b)

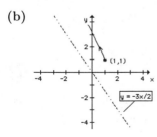

Exercises 10.2

3. (a) All solutions are unstable spirals which become unbounded as t increases.

(b)

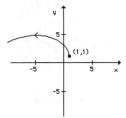

4. (a) All solutions are spirals which approach the origin.

(b)

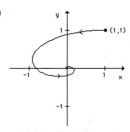

5. (a) All solutions approach $(0,0)$ from the direction specified by the line $y = x$.

(b)

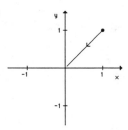

6. (a) All solutions become unbounded and $y = x/2$ serves as the asymptote.

(b)

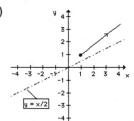

7. (a) If $\mathbf{X}(0) = \mathbf{X}_0$ lies on the line $y = 3x$, then $\mathbf{X}(t)$ approaches $(0,0)$ along this line. For all other intial conditions, $\mathbf{X}(t)$ becomes unbounded and $y = x$ serves as the asymptote.

(b)

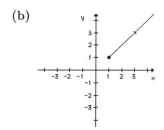

8. (a) The solutions are ellipses which encircle the origin.

(b)

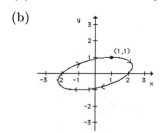

9. Since $\Delta = -41 < 0$, we may conclude from Figure 10.18 that $(0,0)$ is a saddle point.

10. Since $\Delta = 29$ and $\tau = -12$, $\tau^2 - 4\Delta > 0$ and so from Figure 10.18, $(0,0)$ is a stable node.

11. Since $\Delta = -19 < 0$, we may conclude from Figure 10.18 that $(0,0)$ is a saddle point.

12. Since $\Delta = 1$ and $\tau = -1$, $\tau^2 - 4\Delta = -3$ and so from Figure 10.18, $(0,0)$ is a stable spiral point.

13. Since $\Delta = 1$ and $\tau = -2$, $\tau^2 - 4\Delta = 0$ and so from Figure 10.18, $(0,0)$ is a degenerate stable node.

14. Since $\Delta = 1$ and $\tau = 2$, $\tau^2 - 4\Delta = 0$ and so from Figure 10.18, $(0,0)$ is a degenerate unstable node.

15. Since $\Delta = 0.01$ and $\tau = -0.03$, $\tau^2 - 4\Delta < 0$ and so from Figure 10.18, $(0,0)$ is a stable spiral point.

16. Since $\Delta = 0.0016$ and $\tau = 0.08$, $\tau^2 - 4\Delta = 0$ and so from Figure 10.18, $(0,0)$ is a degenerate unstable node.

17. $\Delta = 1 - \mu^2$, $\tau = 0$, and so we need $\Delta = 1 - \mu^2 > 0$ for $(0,0)$ to be a center. Therefore $|\mu| < 1$.

18. Note that $\Delta = 1$ and $\tau - \mu$. Therefore we need both $\tau = \mu < 0$ and $\tau^2 - 4\Delta = \mu^2 - 4 < 0$ for $(0,0)$ to be a stable spiral point. These two conditions may be written as $-2 < \mu < 0$.

19. Note that $\Delta = \mu + 1$ and $\tau = \mu + 1$ and so $\tau^2 - 4\Delta = (\mu+1)^2 - 4(\mu+1) = (\mu+1)(\mu-3)$. It follows that $\tau^2 - 4\Delta < 0$ if and only if $-1 < \mu < 3$. We may conclude that $(0,0)$ will be a saddle point when $\mu < -1$. Likewise $(0,0)$ will be an unstable spiral point when $\tau = \mu + 1 > 0$ and $\tau^2 - 4\Delta < 0$. This condition reduces to $-1 < \mu < 3$.

20. $\tau = 2\alpha$, $\Delta = \alpha^2 + \beta^2 > 0$, and $\tau^2 - 4\Delta = -4\beta < 0$. If $\alpha < 0$, $(0,0)$ is a stable spiral point. If $\alpha > 0$, $(0,0)$ is an unstable spiral point. Therefore $(0,0)$ cannot be a node or saddle point.

21. $AX_1 + F = 0$ implies that $AX_1 = -F$ or $X_1 = -A^{-1}F$. Since $X_p(t) = -A^{-1}F$ is a particular solution, it follows from Theorem 8.6 that $X(t) = X_c(t) + X_1$ is the general solution to $X' = AX + F$. If $\tau < 0$ and $\Delta > 0$ then $X_c(t)$ approaches $(0,0)$ by Theorem 10.1(a). It follows that $X(t)$ approaches X_1 as $t \to \infty$.

22. If $bc < 1$, $\Delta = ad\hat{x}\hat{y}(1 - bc) > 0$ and $\tau^2 - 4\Delta = (a\hat{x} - d\hat{y})^2 + 4abcd\hat{x}\hat{y} > 0$. Therefore $(0,0)$ is a stable node.

23. (a) The critical point is $X_1 = (-3, 4)$.

(b) From the graph, X_1 appears to be an unstable node or a saddle point.

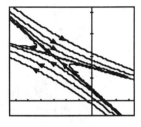

(c) Since $\Delta = -1$, $(0,0)$ is a saddle point.

24. (a) The critical point is $X_1 = (-1, -2)$.

(b) From the graph, X_1 appears to be a stable node or a degenerate stable node.

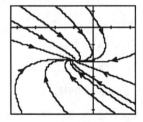

(c) Since $\tau = -16$, $\Delta = 64$, and $\tau^2 - 4\Delta = 0$, $(0,0)$ is a degenerate stable node.

25. (a) The critical point is $X_1 = (0.5, 2)$.

(b) From the graph, X_1 appears to be an unstable spiral point.

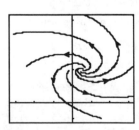

(c) Since $\tau = 0.2$, $\Delta = 0.03$, and $\tau 2 - 4\Delta = -0.08$, $(0,0)$ is an unstable spiral point.

26. (a) The critical point is $X_1 = (1, 1)$.

(b) From the graph, X_1 appears to be a center.

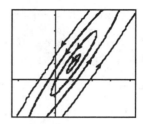

(c) Since $\tau = 0$ and $\Delta = 1$, $(0, 0)$ is a center.

Exercises 10.3

1. Switching to polar coordinates,

$$\frac{dr}{dt} = \frac{1}{r}\left(x\frac{dx}{dt} + y\frac{dy}{dt}\right) = \frac{1}{r}(\alpha x^2 - \beta xy + xy^2 + \beta xy + \alpha y^2 - xy^2) = \frac{1}{r}\alpha r^2 = \alpha r.$$

Therefore $r = ce^{\alpha t}$ and so $r \to 0$ if and only if $\alpha < 0$.

2. The differential equation $\dfrac{dr}{dt} = \alpha r(5 - r)$ is a logistic differential equation. [See Section 3.2, (4) and (5).] It follows that

$$r = \frac{5}{1 + c_1 e^{-5\alpha t}} \quad \text{and} \quad \theta = -t + c_2.$$

If $\alpha > 0$, $r \to 5$ as $t \to +\infty$ and so the critical point $(0, 0)$ is unstable. If $\alpha < 0$, $r \to 0$ as $t \to +\infty$ and so $(0, 0)$ is asymptotically stable.

3. The critical points are $x = 0$ and $x = n + 1$. Since $g'(x) = k(n + 1) - 2kx$, $g'(0) = k(n + 1) > 0$ and $g'(n + 1) = -k(n + 1) < 0$. Therefore $x = 0$ is unstable while $x = n + 1$ is asymptotically stable. See Theorem 10.2.

4. Note that $x = k$ is the only critical point since $\ln(x/k)$ is not defined at $x = 0$. Since $g'(x) = -k - k\ln(x/k)$, $g'(k) = -k < 0$. Therefore $x = k$ is an asymptotically stable critical point by Theorem 10.2.

5. The only critical point is $T = T_0$. Since $g'(T) = k$, $g'(T_0) = k > 0$. Therefore $T = T_0$ is unstable by Theorem 10.2.

6. The only critical point is $v = mg/k$. Now $g(v) = g - (k/m)v$ and so $g'(v) = -k/m < 0$. Therefore $v = mg/k$ is an asymptotically stable critical point by Theorem 10.2.

7. Critical points occur at $x = \alpha$, β. Since $g'(x) = k(-\alpha - \beta + 2x)$, $g'(\alpha) = k(\alpha - \beta)$ and $g'(\beta) = k(\beta - \alpha)$. Since $\alpha > \beta$, $g'(\alpha) > 0$ and so $x = \alpha$ is unstable. Likewise $x = \beta$ is asymptotically stable.

Exercises 10.3

8. Critical points occur at $x = \alpha, \beta, \gamma$. Since

$$g'(x) = k(\alpha - x)(-\beta - \gamma - 2x) + k(\beta - x)(\gamma - x)(-1),$$

$g'(\alpha) = -k(\beta - \alpha)(\gamma - \alpha) < 0$ since $\alpha > \beta > \gamma$. Therefore $x = \alpha$ is asymptotically stable. Similarly $g'(\beta) > 0$ and $g'(\gamma) < 0$. Therefore $x = \beta$ is unstable while $x = \gamma$ is asymptotically stable.

9. Critical points occur at $P = a/b$, c but not at $P = 0$. Since $g'(P) = (a - bP) + (P - c)(-b)$,

$$g'(a/b) = (a/b - c)(-b) = -a + bc \quad \text{and} \quad g'(c) = a - bc.$$

Since $a < bc$, $-a + bc > 0$ and $a - bc < 0$. Therefore $P = a/b$ is unstable while $P = c$ is asymptotically stable.

10. Since $A > 0$, the only critical point is $A = K^2$. Since $g'(A) = \frac{1}{2}kKA^{-1/2} - k$, $g'(K^2) = -k/2 < 0$. Therefore $A = K^2$ is asymptotically stable.

11. The sole critical point is $(1/2, 1)$ and

$$\mathbf{g}'(\mathbf{X}) = \begin{pmatrix} -2y & -2x \\ 2y & 2x - 1 \end{pmatrix}.$$

Computing $\mathbf{g}'((1/2, 1))$ we find that $\tau = -2$ and $\Delta = 2$ so that $\tau^2 - 4\Delta = -4 < 0$. Therefore $(1/2, 1)$ is a stable spiral point.

12. Critical points are $(1, 0)$ and $(-1, 0)$, and

$$\mathbf{g}'(\mathbf{X}) = \begin{pmatrix} 2x & -2y \\ 0 & 2 \end{pmatrix}.$$

At $\mathbf{X} = (1, 0)$, $\tau = 4$, $\Delta = 4$, and so $\tau^2 - 4\Delta = 0$. We may conclude that $(1, 0)$ is unstable but we are unable to classify this critical point any further. At $\mathbf{X} = (-1, 0)$, $\Delta = -4 < 0$ and so $(-1, 0)$ is a saddle point.

13. $y' = 2xy - y = y(2x - 1)$. Therefore if (x, y) is a caritical point, either $x = 1/2$ or $y = 0$. The case $x = 1/2$ and $y - x^2 + 2 = 0$ implies that $(x, y) = (1/2, -7/4)$. The case $y = 0$ leads to the critical points $(\sqrt{2}, 0)$ and $(-\sqrt{2}, 0)$. We next use the Jacobian matrix

$$\mathbf{g}'(\mathbf{X}) = \begin{pmatrix} -2x & 1 \\ 2y & 2x - 1 \end{pmatrix}$$

to classify these three critical points. For $\mathbf{X} = (\sqrt{2}, 0)$ or $(-\sqrt{2}, 0)$, $\tau = -1$ and $\Delta < 0$. Therefore both critical points are saddle points. For $\mathbf{X} = (1/2, -7/4)$, $\tau = -1$, $\Delta = 7/2$ and so $\tau^2 - 4\Delta = -13 < 0$. Therefore $(1/2, -7/4)$ is a stable spiral point.

14. $y' = -y + xy = y(-1 + x)$. Therefore if (x, y) is a critical point, either $y = 0$ or $x = 1$. The case $y = 0$ and $2x - y^2 = 0$ implies that $(x, y) = (0, 0)$. The case $x = 1$ leads to the critical points

$(1, \sqrt{2})$ and $(1, -\sqrt{2})$. We next use the Jacobian matrix

$$\mathbf{g}'(\mathbf{X}) = \begin{pmatrix} 2 & -2y \\ y & x-1 \end{pmatrix}$$

to classify these critical points. For $\mathbf{X} = (0,0)$, $\Delta = -2 < 0$ and so $(0,0)$ is a saddle point. For either $(1, \sqrt{2})$ or $(1, -\sqrt{2})$, $\tau = 2$, $\Delta = 4$, and so $\tau^2 - 4\Delta = -12$. Therefore $(1, \sqrt{2})$ and $(1, -\sqrt{2})$ are unstable spiral points.

15. Since $x^2 - y^2 = 0$, $y^2 = x^2$ and so $x^2 - 3x + 2 = (x-1)(x-2) = 0$. It follows that the critical points are $(1,1)$, $(1,-1)$, $(2,2)$, and $(2,-2)$. We next use the Jacobian

$$\mathbf{g}'(\mathbf{X}) = \begin{pmatrix} -3 & 2y \\ 2x & -2y \end{pmatrix}$$

to classify these four critical points. For $\mathbf{X} = (1,1)$, $\tau = -5$, $\Delta = 2$, and so $\tau^2 - 4\Delta = 17 > 0$. Therefore $(1,1)$ is a stable node. For $\mathbf{X} = (1,-1)$, $\Delta = -2 < 0$ and so $(1,-1)$ is a saddle point. For $\mathbf{X} = (2,2)$, $\Delta = -4 < 0$ and so we have another saddle point. Finally, if $\mathbf{X} = (2,-2)$, $\tau = 1$, $\Delta = 4$, and so $\tau^2 - 4\Delta = -15 < 0$. Therefore $(2,-2)$ is an unstable spiral point.

16. From $y^2 - x^2 = 0$, $y = x$ or $y = -x$. The case $y = x$ leads to $(4,4)$ and $(-1,1)$ but the case $y = -x$ leads to $x^2 - 3x + 4 = 0$ which has no real solutions. Therefore $(4,4)$ and $(-1,1)$ are the only critical points. We next use the Jacobian matrix

$$\mathbf{g}'(\mathbf{X}) = \begin{pmatrix} y & x-3 \\ -2x & 2y \end{pmatrix}$$

to classify these two critical points. For $\mathbf{X} = (4,4)$, $\tau = 12$, $\Delta = 40$, and so $\tau^2 - 4\Delta < 0$. Therefore $(4,4)$ is an unstable spiral point. For $\mathbf{X} = (-1,1)$, $\tau = -3$, $\Delta = 10$, and so $x^2 - 4\Delta < 0$. It follows that $(-1,-1)$ is a stable spiral piont.

17. Since $x' = -2xy = 0$, either $x = 0$ or $y = 0$. If $x = 0$, $y(1-y^2) = 0$ and so $(0,0)$, $(0,1)$, and $(0,-1)$ are critical points. The case $y = 0$ leads to $x = 0$. We next use the Jacobian matrix

$$\mathbf{g}'(\mathbf{X}) = \begin{pmatrix} -2y & -2x \\ -1+y & 1+x-3y^2 \end{pmatrix}$$

to classify these three critical points. For $\mathbf{X} = (0,0)$, $\tau = 1$ and $\Delta = 0$ and so the test is inconclusive. For $\mathbf{X} = (0,1)$, $\tau = -4$, $\Delta = 4$ and so $\tau^2 - 4\Delta = 0$. We can conclude that $(0,1)$ is a stable critical point but we are unable to classify this critical point further in this borderline case. For $\mathbf{X} = (0,-1)$, $\Delta = -4 < 0$ and so $(0,-1)$ is a saddle point.

18. We found that $(0,0)$, $(0,1)$, $(0,-1)$, $(1,0)$ and $(-1,0)$ were the critical points in Exercise 15, Section 10.1. The Jacobian is

$$\mathbf{g}'(\mathbf{X}) = \begin{pmatrix} 1 - 3x^2 - 3y^2 & -6xy \\ -2xy & 3 - x^2 - 9y^2 \end{pmatrix}.$$

405

For $\mathbf{X} = (0,0)$, $\tau = 4$, $\Delta = 3$ and so $\tau^2 - 4\Delta = 4 > 0$. Therefore $(0,0)$ is an unstable node. Both $(0,1)$ and $(0,-1)$ give $\tau = -8$, $\Delta = 12$, and $\tau^2 - 4\Delta = 16 > 0$. These two critical points are therefore stable nodes. For $\mathbf{X} = (1,0)$ or $(-1,0)$, $\Delta = -4 < 0$ and so saddle points occur.

19. We found the critical points $(0,0)$, $(10,0)$, $(0,16)$ and $(4,12)$ in Exercise 11, Section 10.1. Since the Jacobian is

$$\mathbf{g}'(\mathbf{X}) = \begin{pmatrix} 10 - 2x - \frac{1}{2}y & -\frac{1}{2}x \\ -y & 16 - 2y - x \end{pmatrix}$$

we may classify the critical points as follows:

$\mathbf{X}$	τ	Δ	$\tau^2 - 4\Delta$	Conclusion
$(0,0)$	26	160	36	unstable node
$(10,0)$	-4	-60	$-$	saddle point
$(0,16)$	-14	-32	$-$	saddle point
$(4,12)$	-16	24	160	stable node

20. We found the sole critical point $(10,10)$ in Exercise 12, Section 10.1. The Jacobian is

$$\mathbf{g}'(\mathbf{X}) = \begin{pmatrix} -2 & 1 \\ 2 & -1 - \frac{15}{(y+5)^2} \end{pmatrix},$$

$\mathbf{g}'((10,10))$ has trace $\tau = -46/15$, $\Delta = 2/15$, and $\tau^2 - 4\Delta > 0$. Therefore $(0,0)$ is a stable node.

21. The corresponding plane autonomous system is

$$\theta' = y, \quad y' = (\cos\theta - \frac{1}{2})\sin\theta.$$

Since $|\theta| < \pi$, it follows that critical points are $(0,0)$, $(\pi/3, 0)$ and $(-\pi/3, 0)$. The Jacobian matrix is

$$\mathbf{g}'(\mathbf{X}) = \begin{pmatrix} 0 & 1 \\ \cos 2\theta - \frac{1}{2}\cos\theta & 0 \end{pmatrix}$$

and so at $(0,0)$, $\tau = 0$ and $\Delta = -1/2$. Therefore $(0,0)$ is a saddle point. For $\mathbf{X} = (\pm\pi/3, 0)$, $\tau = 0$ and $\Delta = 3/4$. It is not possible to classify either critical point in this borderline case.

22. The corresponding plane autonomous system is

$$x' = y, \quad y' = -x + (\frac{1}{2} - 3y^2)y - x^2.$$

If (x,y) is a critical point, $y = 0$ and so $-x - x^2 = -x(1+x) = 0$. Therefore $(0,0)$ and $(-1,0)$ are the only two critical points. We next use the Jacobian matrix

$$\mathbf{g}'(\mathbf{X}) = \begin{pmatrix} 0 & 1 \\ -1 - 2x & \frac{1}{2} - 9y^2 \end{pmatrix}$$

to classify these critical points. For $\mathbf{X} = (0,0)$, $\tau = 1/2$, $\Delta = 1$, and $\tau^2 - 4\Delta < 0$. Therefore $(0,0)$ is an unstable spiral point. For $\mathbf{X} = (-1,0)$, $\tau = 1/2$, $\Delta = -1$ and so $(-1,0)$ is a saddle point.

23. The corresponding plane autonomous system is

$$x' = y, \quad y' = x^2 - y(1 - x^3)$$

and the only critical point is $(0,0)$. Since the Jacobian matrix is

$$g'(\mathbf{X}) = \begin{pmatrix} 0 & 1 \\ 2x + 3x^2y & x^3 - 1 \end{pmatrix},$$

$\tau = -1$ and $\Delta = 0$, and we are unable to classify the critical point in this borderline case.

24. The corresponding plane autonomous system is

$$x' = y, \quad y' = -\frac{4x}{1 + x^2} - 2y$$

and the only critical point is $(0,0)$. Since the Jacobian matrix is

$$g'(\mathbf{X}) = \begin{pmatrix} 0 & 1 \\ -4\frac{1-x^2}{(1+x^2)^2} & -2 \end{pmatrix},$$

$\tau = -2$, $\Delta = 4$, $\tau^2 - 4\Delta = -12$, and so $(0,0)$ is a stable spiral point.

25. In Exercise 5, Section 10.1, we showed that $(0,0)$, $(\sqrt{1/\epsilon}, 0)$ and $(-\sqrt{1/\epsilon}, 0)$ are the critical points. We will use the Jacobian matrix

$$g'(\mathbf{X}) = \begin{pmatrix} 0 & 1 \\ -1 + 3\epsilon x^2 & 0 \end{pmatrix}$$

to classify these three critical points. For $\mathbf{X} = (0,0)$, $\tau = 0$ and $\Delta = 1$ and we are unable to classify this critical point. For $(\pm\sqrt{1/\epsilon}, 0)$, $\tau = 0$ and $\Delta = -2$ and so both of these critical points are saddle points.

26. In Exercise 6, Section 10.1, we showed that $(0,0)$, $(1/\epsilon, 0)$, and $(-1/\epsilon, 0)$ are the critical points. Since $D_x x|x| = 2|x|$, the Jacobian matrix is

$$g'(\mathbf{X}) = \begin{pmatrix} 0 & 1 \\ 2\epsilon|x| - 1 & 0 \end{pmatrix}.$$

For $\mathbf{X} = (0,0)$, $\tau = 0$, $\Delta = 1$ and we are unable to classify this critical point. For $(\pm 1/\epsilon, 0)$, $\tau = 0$, $\Delta = -1$, and so both of these critical points are saddle points.

27. The corresponding plane autonomous system is

$$x' = y, \quad y' = -\frac{(\beta + \alpha^2 y^2)x}{1 + \alpha^2 x^2}$$

and the Jacobian matrix is

$$g'(\mathbf{X}) = \begin{pmatrix} 0 & 1 \\ \frac{(\beta + \alpha y^2)(\alpha^2 x^2 - 1)}{(1 + \alpha^2 x^2)^2} & \frac{-2\alpha^2 yx}{1 + \alpha^2 x^2} \end{pmatrix}.$$

For $\mathbf{X} = (0,0)$, $\tau = 0$ and $\Delta = \beta$. Since $\beta < 0$, we may conclude that $(0,0)$ is a saddle point.

407

28. From $x' = -\alpha x + xy = x(-\alpha + y) = 0$, either $x = 0$ or $y = \alpha$. If $x = 0$, then $1 - \beta y = 0$ and so $y = 1/\beta$. The case $y = \alpha$ implies that $1 - \beta\alpha - x^2 = 0$ or $x^2 = 1 - \alpha\beta$. Since $\alpha\beta > 1$, this equation has no real solutions. It follows that $(0, 1/\beta)$ is the unique critical point. Since the Jacobian matrix is

$$g'(X) = \begin{pmatrix} -\alpha + y & x \\ -2x & -\beta \end{pmatrix},$$

$\tau = -\alpha - \beta + \dfrac{1}{\beta} = -\beta + \dfrac{1 - \alpha\beta}{\beta} < 0$ and $\Delta = \alpha\beta - 1 > 0$. Therefore $(0, 1/\beta)$ is a stable critical point.

29. (a) The graphs of $-x + y - x^3 = 0$ and $-x - y + y^2 = 0$ are shown in the figure. The Jacobian matrix is

$$g'(X) = \begin{pmatrix} -1 - 3x^2 & 1 \\ -1 & -1 + 2y \end{pmatrix}.$$

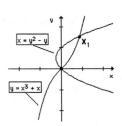

For $X = (0, 0)$, $\tau = -2$, $\Delta = 2$, $\tau^2 - 4\Delta = -4$, and so $(0, 0)$ is a stable spiral point.

(b) For X_1, $\Delta = -6.07 < 0$ and so a saddle point occurs at X_1.

30. (a) The corresponding plane autonomous system is

$$x' = y, \quad y' = \epsilon(y - \tfrac{1}{3}y^3) - x$$

and so the only critical point is $(0, 0)$. Since the Jacobian matrix is

$$g'(X) = \begin{pmatrix} 0 & 1 \\ -1 & \epsilon(1 - y^2) \end{pmatrix},$$

$\tau = \epsilon$, $\Delta = 1$, and so $\tau^2 - 4\Delta = \epsilon^2 - 4$ at the critical point $(0, 0)$.

(b) When $\tau = \epsilon > 0$, $(0, 0)$ is an unstable critical point.

(c) When $\epsilon < 0$ and $\tau^2 - 4\Delta = \epsilon^2 - 4 < 0$, $(0, 0)$ is a stable spiral point. These two requirements can be written as $-2 < \epsilon < 0$.

(d) When $\epsilon = 0$, $x'' + x = 0$ and so $x = c_1 \cos t + c_2 \sin t$. Therefore all solutions are periodic (with period 2π) and so $(0, 0)$ is a center.

31. $\dfrac{dy}{dx} = \dfrac{y'}{x'} = \dfrac{-2x^3}{y}$ may be solved by separating variales. It follows that $y^2 + x^4 = c$. If $X(0) = (x_0, 0)$ where $x_0 > 0$, then $c = x_0^4$ so that $y^2 = x_0^4 - x^4$. Therefore if $-x_0 < x < x_0$, $y^2 > 0$ and so there are two values of y corresponding to each value of x. Therefore the solution $X(t)$ with $X(0) = (x_0, 0)$ is periodic and so $(0, 0)$ is a center.

32. $\dfrac{dy}{dx} = \dfrac{y'}{x'} = \dfrac{x^2 - 2x}{y}$ may be solved by separating variables. It follows that $\dfrac{y^2}{2} = \dfrac{x^3}{3} - x^2 + c$ and since $\mathbf{X}(0) = (x(0), x'(0)) = (1, 0)$, $c = \dfrac{2}{3}$. Therefore

$$\frac{y^2}{2} = \frac{x^3 - 3x^2 + 2}{3} = \frac{(x-1)(x^2 - 2x - 2)}{3}.$$

But $(x-1)(x^2 - 2x - 2) > 0$ for $1 - \sqrt{3} < x < 1$ and so each x in this interval has 2 corresponding values of y. therefore $\mathbf{X}(t)$ is a periodic solution.

33. (a) $x' = 2xy = 0$ implies that either $x = 0$ or $y = 0$. If $x = 0$, then from $1 - x^2 + y^2 = 0$, $y^2 = -1$ and there are no real solutions. If $y = 0$, $1 - x^2 = 0$ and so $(1, 0)$ and $(-1, 0)$ are critical points. The Jacobian matrix is

$$\mathbf{g}'(\mathbf{X}) = \begin{pmatrix} 2y & 2x \\ -2x & 2y \end{pmatrix}$$

and so $\tau = 0$ and $\Delta = 4$ at either $\mathbf{X} = (1, 0)$ or $(-1, 0)$. We obtain no information about these critical points in this borderline case.

(b) $\dfrac{dy}{dx} = \dfrac{y'}{x'} = \dfrac{1 - x^2 + y^2}{2xy}$ or $2xy\dfrac{dy}{dx} = 1 - x^2 + y^2$. Letting $\mu = \dfrac{y^2}{x}$, it follows that $\dfrac{d\mu}{dx} = \dfrac{1}{x^2} - 1$ and so $\mu = -\dfrac{1}{x} - x + 2c$. Therefore $\dfrac{y^2}{x} = -\dfrac{1}{x} - x + 2c$ which can be put in the form

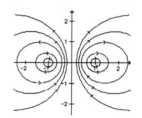

$$(x - c)^2 + y^2 = c^2 - 1.$$

The solution curves are shown and so both $(1, 0)$ and $(-1, 0)$ are centers.

34. (a) $\dfrac{dy}{dx} = \dfrac{y'}{x'} = \dfrac{-x - y^2}{y} = -\dfrac{x}{y} - y$ and so $\dfrac{dy}{dx} + y = -xy^{-1}$.

(b) Let $w = y^{1-n} = y^2$. It follows that $\dfrac{dw}{dx} + 2w = -2x$, a linear first order differential equation whose solution is

$$y^2 = w = ce^{-2x} + \left(\frac{1}{2} - x\right).$$

Since $x(0) = \frac{1}{2}$ and $y(0) = x'(0) = 0$, $0 = c$ and so

$$y^2 = \frac{1}{2} - x,$$

a parabola with vertex at $(1/2, 0)$. Therefore the solution $\mathbf{X}(t)$ with $\mathbf{X}(0) = (1/2, 0)$ is not periodic.

Exercises 10.3

35. $\dfrac{dy}{dx} = \dfrac{y'}{x'} = \dfrac{x^3 - x}{y}$ and so $\dfrac{y^2}{2} = \dfrac{x^4}{4} - \dfrac{x^2}{2} + c$ or $y^2 = \dfrac{x^4}{2} - x^2 + c_1$. Since $x(0) = 0$ and

$y(0) = x'(0) = v_0$, it follows that $c_1 = v_0^2$ and so

$$y^2 = \frac{1}{2}x^4 - x^2 + v_0^2 = \frac{(x^2 - 1)^2 + 2v_0^2 - 1}{2}.$$

The x-intercepts on this graph satisfy

$$x^2 = 1 \pm \sqrt{1 - 2v_0^2}$$

and so we must require that $1 - 2v_0^2 \geq 0$ (or $|v_0| \leq \frac{1}{2}\sqrt{2}$) for real solutions to exist. If $x_0^2 = 1 - \sqrt{1 - 2v_0^2}$ and $-x_0 < x < x_0$, then $(x^2 - 1)^2 + 2v_0^2 - 1 > 0$ and so there are two corresponding values of y. Therefore $\mathbf{X}(t)$ with $\mathbf{X}(0) = (0, v_0)$ is periodic provided that $|v_0| \leq \frac{1}{2}\sqrt{2}$.

36. The corresponding plane autonomous system is

$$x' = y, \quad y' = \epsilon x^2 - x + 1$$

and so the critical points must satisfy $y = 0$ and

$$x = \frac{1 \pm \sqrt{1 - 4\epsilon}}{2\epsilon}.$$

Therefore we must require that $\epsilon \leq \frac{1}{4}$ for real solutions to exist. We will use the Jacobian matrix

$$\mathbf{g}'(\mathbf{X}) = \begin{pmatrix} 0 & 1 \\ 2\epsilon x - 1 & 0 \end{pmatrix}$$

to attempt to classify $((1 \pm \sqrt{1 - 4\epsilon})/2\epsilon, 0)$ when $\epsilon \leq 1/4$. Note that $\tau = 0$ and $\Delta = \mp\sqrt{1 - 4\epsilon}$. For $\mathbf{X} = ((1 + \sqrt{1 - 4\epsilon})/2\epsilon, 0)$ and $\epsilon < 1/4$, $\Delta < 0$ and so a saddle point occurs. For $\mathbf{X} = ((1 - \sqrt{1 - 4\epsilon})/2\epsilon, 0)$ $\Delta \geq 0$ and we are not able to classify this critical point using linearization.

37. The corresponding plane autonomous system is

$$x' = y, \quad y' = -\frac{\alpha}{L}x - \frac{\beta}{L}x^3 - \frac{R}{L}y$$

where $x = q$ and $y = q'$. If $\mathbf{X} = (x, y)$ is a critical point, $y = 0$ and $-\alpha x - \beta x^3 = -x(\alpha + \beta x^2) = 0$. If $\beta > 0$, $\alpha + \beta x^2 = 0$ has no real solutions and so $(0, 0)$ is the only critical point. Since

$$\mathbf{g}'(\mathbf{X}) = \begin{pmatrix} 0 & 1 \\ \frac{-\alpha - 3\beta x^2}{L} & -\frac{R}{L} \end{pmatrix},$$

$\tau = -R/L < 0$ and $\Delta = \alpha/L > 0$. Therefore $(0, 0)$ is a stable critical point. If $\beta < 0$, $(0, 0)$ and $(\pm\hat{x}, 0)$, where $\hat{x}^2 = -\alpha/\beta$ are critical points. At $\mathbf{X}(\pm\hat{x}, 0)$, $\tau = -R/L < 0$ and $\Delta = -2\alpha/L < 0$. Therefore both critical points are saddles.

410

38. If we let $dx/dt = y$, then $dy/dt = -x^3 - x$. From this we obtain the first-order differential equation

$$\frac{dy}{dx} = \frac{dy/dt}{dx/dt} = -\frac{x^3 + x}{y}.$$

Separating variables and integrating we obtain

$$\int y\,dy = -\int (x^3 + x)\,dx$$

and

$$\frac{1}{2}y^2 = -\frac{1}{4}x^4 - \frac{1}{2}x^2 + c_1.$$

Completing the square we can write the solution as $y^2 = -\frac{1}{2}(x^2 + 1)^2 + c_2$. If $\mathbf{X}(0) = (x_0, 0)$, then $c_2 = \frac{1}{2}(x_0^2 + 1)^2$ and so

$$y^2 = -\frac{1}{2}(x^2 + 1)^2 + \frac{1}{2}(x_0^2 + 1)^2 = \frac{x_0^4 + 2x_0^2 + 1 - x^4 - 2x^2 - 1}{2}$$

$$= \frac{(x_0^2 + x^2)(x_0^2 - x^2) + 2(x_0^2 - x^2)}{2} = \frac{(x_0^2 + x^2 + 2)(x_0^2 - x^2)}{2}.$$

Note that $y = 0$ when $x = -x_0$. In addition, the right-hand side is positive for $-x_0 < x < x_0$, and so there are two corresponding values of y for each x between $-x_0$ and x_0. The solution $\mathbf{X} = \mathbf{X}(t)$ that satisfies $\mathbf{X}(0) = (x_0, 0)$ is therefore periodic, and so $(0, 0)$ is a center.

39. (a) Letting $x = \theta$ and $y = x'$ we obtain the system $x' = y$ and $y' = \frac{1}{2} - \sin x$. Since $\sin \pi/6 = \sin 5\pi/6 = \frac{1}{2}$ we see that $(\pi/6, 0)$ and $(5\pi/6, 0)$ are critical points of the system.

(b) The Jacobian matrix is

$$\mathbf{g}'(\mathbf{X}) = \begin{pmatrix} 0 & 1 \\ -\cos x & 0 \end{pmatrix}$$

and so

$$\mathbf{A}_1 = \mathbf{g}' = ((\pi/6, 0)) = \begin{pmatrix} 0 & 1 \\ -\sqrt{3}/2 & 0 \end{pmatrix} \quad \text{and} \quad \mathbf{A}_2 = \mathbf{g}' = ((5\pi/6, 0)) = \begin{pmatrix} 0 & 1 \\ \sqrt{3}/2 & 0 \end{pmatrix}.$$

Since $\det \mathbf{A}_1 > 0$ and the trace of $\mathbf{A}_1$ is 0, no conclusion can be drawn regarding the critical point $(\pi/6, 0)$. Since $\det \mathbf{A}_2 < 0$, we see that $(5\pi/6, 0)$ is a saddle point.

(c) From the system in part (a) we obtain the first-order differential equation

$$\frac{dy}{dx} = \frac{1/2 - \sin x}{y}.$$

Separating variables and integrating we obtain

$$\int y\,dy = \int \left(\frac{1}{2} - \sin x\right) dx$$

and

411

$$\frac{1}{2}y^2 = \frac{1}{2}x + \cos x + c_1$$

or

$$y^2 = x + 2\cos x + c_2.$$

For x_0 near $\pi/6$, if $\mathbf{X}(0) = (x_0, 0)$ then $c_2 = -x_0 - 2\cos x_0$ and $y^2 = x + 2\cos x - x_0 - 2\cos x_0$. Thus, there are two values of y for each x in a sufficiently small interval around $\pi/6$. Therefore $(\pi/6, 0)$ is a center.

40. (a) Writing the system as $x' = x(x^3 - 2y^3)$ and $y' = y(2x^3 - y^3)$ we see that $(0, 0)$ is a critical point. Setting $x^3 - 2y^3 = 0$ we have $x^3 = 2y^3$ and $2x^3 - y^3 = 4y^3 - y^3 = 3y^3$. Thus, $(0, 0)$ is the only critical point of the system.

(b) From the system we obtain the first-order differential equation

$$\frac{dy}{dx} = \frac{2x^3 y - y^4}{x^4 - 2xy^3}$$

or

$$(2x^3 y - y^4)\,dx + (2xy^3 - x^4)\,dy = 0$$

which is homogeneous. If we let $y = ux$ it follows that

$$(2x^4 u - x^4 u^4)\,dx + (2x^4 u^3 - x^4)(u\,dx + x\,du) = 0$$

$$x^4 u(1 + u^3)\,dx + x^5(2u^3 - 1)\,du = 0$$

$$\frac{1}{x}\,dx + \frac{2u^3 - 1}{u(u^3 + 1)}\,du = 0$$

$$\frac{1}{x}\,dx + \left(\frac{1}{u+1} - \frac{1}{u} + \frac{2u - 1}{u^2 - u + 1}\right)du = 0.$$

Integrating gives

$$\ln|x| + \ln|u + 1| - \ln|u| + \ln|u^2 - u + 1| = c_1$$

or

$$x\left(\frac{u+1}{u}\right)(u^2 - u + 1) = c_2$$

$$x\left(\frac{y+x}{y}\right)\left(\frac{y^2}{x^2} - \frac{y}{x} + 1\right) = c_2$$

$$(xy + x^2)(y^2 - xy + x^2) = c_2 x^2 y$$

$$xy^3 + x^4 = c_2 x^2 y$$

$$x^3 + y^2 = 3c_3 xy.$$

412

(c) We see from the graph that $(0,0)$ is unstable. It is not possible to classify the critical point as a node, saddle, center, or spiral point.

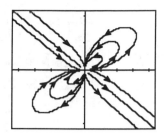

Exercises 10.4

1. We are given that $x(0) = \theta(0) = \dfrac{\pi}{3}$ and $y(0) = \theta'(0) = w_0$. Since $y^2 = \dfrac{2g}{l}\cos x + c$,

$w_0^2 = \dfrac{2g}{l}\cos\dfrac{\pi}{3} + c = \dfrac{g}{l} + c$ and so $c = w_0^2 - \dfrac{g}{l}$. Therefore

$$y^2 = \frac{2g}{l}\left(\cos x - \frac{1}{2} + \frac{l}{2g}w_0^2\right)$$

and the x-intercepts occur where $\cos x = \dfrac{1}{2} - \dfrac{l}{2g}w_0^2$ and so $\dfrac{1}{2} - \dfrac{l}{2g}w_0^2$ must be greater than -1 for

solutions to exist. This condition is equivalent to $|w_0| < \sqrt{\dfrac{3g}{l}}$.

2. (a) Since $y^2 = \dfrac{2g}{l}\cos x + c$, $x(0) = \theta(0) = \theta_0$ and $y(0) = \theta'(0) = 0$, $c = -\dfrac{2g}{l}\cos\theta_0$ and so

$y^2 = \dfrac{2g}{l}(\cos\theta - \cos\theta_0)$. When $\theta = -\theta_0$, $y^2 = \dfrac{2g}{l}(\cos(-\theta_0) - \cos(\theta_0)) = 0$. Therefore $y = \dfrac{d\theta}{dt} = 0$
when $\theta = \theta_0$.

(b) Since $y = \dfrac{d\theta}{dt}$ and θ is decreasing between the time when $\theta = \theta_0$, $t = 0$, and $\theta = -\theta_0$, that is,
$t = T$,

$$\frac{d\theta}{dt} = -\sqrt{\frac{2g}{l}}\sqrt{\cos\theta - \cos\theta_0}.$$

Therefore $\dfrac{dt}{d\theta} = -\sqrt{\dfrac{l}{2g}}\dfrac{1}{\sqrt{\cos\theta - \cos\theta_0}}$ and so

$$T = -\sqrt{\frac{l}{2g}}\int_{\theta=\theta_0}^{\theta=-\theta_0}\frac{1}{\sqrt{\cos\theta - \cos\theta_0}}\,d\theta = \sqrt{\frac{l}{2g}}\int_{-\theta_0}^{\theta_0}\frac{1}{\sqrt{\cos\theta - \cos\theta_0}}\,d\theta.$$

3. The corresponding plane autonomous system is

$$x' = y, \quad y' = -g\,\frac{f'(x)}{1 + [f'(x)]^2} - \frac{\beta}{m}y$$

and

$$\frac{\partial}{\partial x}\left(-g\frac{f'(x)}{1+[f'(x)]^2}-\frac{\beta}{m}y\right)=-g\frac{(1+[f'(x)]^2)f''(x)-f'(x)2f'(x)f''(x)}{(1+[f'(x)]^2)^2}.$$

If $X_1=(x_1,y_1)$ is a critical point, $y_1=0$ and $f'(x_1)=0$. The Jacobian at this critical point is therefore

$$g'(X_1)=\begin{pmatrix}0 & 1 \\ -gf''(x_1) & -\frac{\beta}{m}\end{pmatrix}.$$

4. When $\beta=0$ the Jacobian matrix is

$$\begin{pmatrix}0 & 1 \\ -gf''(x_1) & 0\end{pmatrix}$$

which has complex eigenvalues $\lambda=\pm\sqrt{gf''(x_1)}\,i$. The approximating linear system with $x'(0)=0$ has solution

$$x(t)=x(0)\cos\sqrt{gf''(x_1)}\,t$$

and period $2\pi/\sqrt{gf''(x_1)}$. Therefore $p\approx 2\pi/\sqrt{gf''(x_1)}$ for the actual solution.

5. **(a)** If $f(x)=\dfrac{x^2}{2}$, $f'(x)=x$ and so

$$\frac{dy}{dx}=\frac{y'}{x'}=-g\frac{x}{1+x^2}\frac{1}{y}.$$

We may separate variables to show that $y^2=-g\ln(1+x^2)+c$. But $x(0)=x_0$ and $y(0)=x'(0)=v_0$. Therefore $c=v_0^2+g\ln(1+x_0^2)$ and so

$$y^2=v_0^2-g\ln\left(\frac{1+x^2}{1+x_0^2}\right).$$

Now

$$v_0^2-g\ln\left(\frac{1+x^2}{1+x_0^2}\right)\geq 0 \quad\text{if and only if}\quad x^2\leq e^{v_0^2/g}(1+x_0^2)-1.$$

Therefore, if $|x|\leq [e^{v_0^2/g}(1+x_0^2)-1]^{1/2}$, there are two values of y for a given value of x and so the solution is periodic.

(b) Since $z=\dfrac{x^2}{2}$, the maximum height occurs at the largest value of x on the cycle. From (a), $x_{\max}=[e^{v_0^2/g}(1+x_0^2)-1]^{1/2}$ and so

$$z_{\max}=\frac{x_{\max}^2}{2}=\frac{1}{2}[e^{v_0^2/g}(1+x_0^2)-1].$$

414

6. (a) If $f(x) = \cosh x$, $f'(x) = \sinh x$ and $[f'(x)]^2 + 1 = \sinh^2 x + 1 = \cosh^2 x$. Therefore

$$\frac{dy}{dx} = \frac{y'}{x'} = -g \frac{\sinh x}{\cosh^2 x} \frac{1}{y}.$$

We may separate variables to show that $y^2 = \dfrac{2g}{\cosh x} + c$. But $x(0) = x_0$ and $y(0) = x'(0) = v_0$.

Therefore $c = v_0^2 - \dfrac{2g}{\cosh x_0}$ and so

$$y^2 = \frac{2g}{\cosh x} - \frac{2g}{\cosh x_0} + v_0^2.$$

Now

$$\frac{2g}{\cosh x} - \frac{2g}{\cosh x_0} + v_0^2 \geq 0 \quad \text{if and only if} \quad \cosh x \leq \frac{2g \cosh x_0}{2g - v_0^2 \cosh x_0}$$

and the solution to this inequality is an interval $[-a, a]$. Therefore each x in $(-a, a)$ has two corresponding values of y and so the solution is periodic.

(b) Since $z = \cosh x$, the maximum height occurs at the largest value of x on the cycle. From (a), $x_{\max} = a$ where $\cosh a = \dfrac{2g \cosh x_0}{2g - v_0^2 \cosh x_0}$. Therefore

$$z_{\max} = \frac{2g \cosh x_0}{2g - v_0^2 \cosh x_0}.$$

7. If $x_m < x_1 < x_n$, then $F(x_1) > F(x_m) = F(x_n)$. Letting $x = x_1$,

$$G(y) = \frac{c_0}{F(x_1)} = \frac{F(x_m)G(a/b)}{F(x_1)} < G(a/b).$$

Therefore from Figure 10.35(b), $G(y) = \dfrac{c_0}{F(x_1)}$ has two solutions y_1 and y_2 that satisfy $y_1 < a/b < y_2$.

8. From (i) when $y = a/b$, x_n is taken on at some time t. From (iii) if $x > x_n$ there is no corresponding value of y. Therefore the maximum number of predators is x_n and x_n occurs when $y = a/b$.

9. (a) In the Lotka-Volterra Model the average number of predators is d/c and the average number of prey is a/b. But

$$x' = -ax + bxy - \epsilon_1 x = -(a + \epsilon_1)x + bxy$$

$$y' = -cxy + dy - \epsilon_2 y = -cxy + (d - \epsilon_2)y$$

and so the new critical point in the first quadrant is $(d/c - \epsilon_2/c, a/b + \epsilon_1/b)$.

(b) The average number of predators $d/c - \epsilon_2/c$ has decreased while the average number of prey $a/b + \epsilon_1/b$ has increased. The fishery science model is consistent with Volterra's principle.

10. **(a)** Solving

$$x(-0.1 + 0.02y) = 0$$

$$y(0.2 - 0.025x) = 0$$

in the first quadrant we obtain the critical
point $(8, 5)$. The graphs are plotted using
$x(0) = 7$ and $y(0) = 4$.

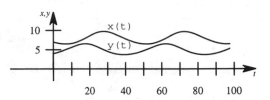

(b) The graph in part (a) was obtained using **NDSolve** in *Mathematica*. We see that the period
is around 40. Since $x(0) = 7$, we use the **FindRoot** equation solver in *Mathematica* to
approximate the solution of $x(t) = 7$ for t near 40. From this we see that the period is more
closely approximated by $t = 44.65$.

11. Solving

$$x(20 - 0.4x - 0.3y) = 0$$

$$y(10 - 0.1y - 0.3x) = 0$$

we see that critical points are $(0, 0)$, $(0, 100)$, $(50, 0)$, and $(20, 40)$. The Jacobian matrix is

$$\mathbf{g}'(\mathbf{X}) = \begin{pmatrix} 0.08(20 - 0.8x - 0.3y) & -0.024x \\ -0.018y & 0.06(10 - 0.2y - 0.3x) \end{pmatrix}$$

and so

$$\mathbf{A}_1 = \mathbf{g}'((0,0)) = \begin{pmatrix} 1.6 & 0 \\ 0 & 0.6 \end{pmatrix} \qquad \mathbf{A}_2 = \mathbf{g}'((0,100)) = \begin{pmatrix} -0.8 & 0 \\ -1.8 & -0.6 \end{pmatrix}$$

$$\mathbf{A}_3 = \mathbf{g}'((50,0)) = \begin{pmatrix} -1.6 & -1.2 \\ 0 & -0.3 \end{pmatrix} \qquad \mathbf{A}_4 = \mathbf{g}'((20,40)) = \begin{pmatrix} -0.64 & -0.48 \\ -0.72 & -0.24 \end{pmatrix}.$$

Since $\det(\mathbf{A}_1) = \Delta_1 = 0.96 > 0$, $\tau = 2.2 > 0$, and $\tau_1^2 - 4\Delta_1 = 1 > 0$, we see that $(0, 0)$ is an unstable
node. Since $\det(\mathbf{A}_2) = \Delta_2 = 0.48 > 0$, $\tau = -1.4 < 0$, and $\tau_2^2 - 4\Delta_2 = 0.04 > 0$, we see that $(0, 100)$
is a stable node. Since $\det(\mathbf{A}_3) = \Delta_3 = 0.48 > 0$, $\tau = -1.9 < 0$, and $\tau_3^2 - 4\Delta_3 = 1.69 > 0$, we see
that $(50, 0)$ is a stable node. Since $\det(\mathbf{A}_4) = -0.192 < 0$ we see that $(20, 40)$ is a saddle point.

12. $\Delta = r_1 r_2$, $\tau = r_1 + r_2$ and $\tau^2 - 4\Delta = (r_1 + r_2)^2 - 4r_1 r_2 = (r_1 - r_2)^2$. Therefore when $r_1 \neq r_2$, $(0, 0)$
is an unstable node.

13. For $\mathbf{X} = (K_1, 0)$, $\tau = -r_1 + r_2 \left(1 - \dfrac{K_1}{K_2}\alpha_{21}\right)$ and $\Delta = -r_1 r_2 \left(1 - \dfrac{K_1}{K_2}\alpha_{21}\right)$. If we let $c = 1 - \dfrac{K_1}{K_2}\alpha_{21}$,

$\tau^2 - 4\Delta = (cr_2 + r_1)^2 > 0$. Now if $k_1 > \dfrac{K_2}{\alpha_{21}}$, $c < 0$ and so $\tau < 0$, $\Delta > 0$. Therefore $(K_1, 0)$ is a

stable node. If $K_1 < \dfrac{K_2}{\alpha_{21}}$, $c > 0$ and so $\Delta < 0$. In this case $(K_1, 0)$ is a saddle point.

14. $(\hat{x}, \hat{y})$ is a stable node if and only if $\dfrac{K_1}{\alpha_{12}} > K_2$ and $\dfrac{K_2}{\alpha_{21}} > K_1$. [See Figure 10.38(a) in the text.]

From Problem 12, (0.0) is an unstable node and from Problem 13, since $K_1 < \dfrac{K_2}{\alpha_{21}}$, $(K_1, 0)$ is a

saddle point. Finally, when $K_2 < \dfrac{K_1}{\alpha_{12}}$, $(0, K_2)$ is a saddle point. This is Problem 12 with the roles

of 1 and 2 interchanged. Therefore $(0,0)$, $(K_1, 0)$, and $(0, K_2)$ are unstable.

15. $\dfrac{K_1}{\alpha_{12}} < K_2 < K_1\alpha_{21}$ and so $\alpha_{12}\alpha_{21} > 1$. Therefore $\Delta = (1 - \alpha_{12}\alpha_{21})\hat{x}\hat{y}\,\dfrac{r_1 r_2}{K_1 K_2} < 0$ and so $(\hat{x}, \hat{y})$ is

a saddle point.

16. (a) The corresponding plane autonomous system is

$$x' = y, \quad y' = \frac{-g}{l}\sin x - \frac{\beta}{ml}y$$

and so critical points must satisfy both $y = 0$ and $\sin x = 0$. Therefore $(\pm n\pi, 0)$ are critical
points.

(b) The Jacobian matrix

$$\begin{pmatrix} 0 & 1 \\ -\frac{g}{l}\cos x & -\frac{\beta}{ml} \end{pmatrix}$$

has trace $\tau = -\dfrac{\beta}{ml}$ and determinant $\Delta = \dfrac{g}{l} > 0$ at $(0, 0)$. Therefore

$$\tau^2 - 4\Delta = \frac{\beta^2}{m^2 l^2} - 4\frac{g}{l} = \frac{\beta^2 - 4glm^2}{m^2 l^2}.$$

We may conclude that $(0, 0)$ is a stable spiral point provided $\beta^2 - 4glm^2 < 0$ or $\beta < 2m\sqrt{gl}$.

17. (a) The corresponding plane autonomous system is

$$x = y, \quad y' = -\frac{\beta}{m}y|y| - \frac{k}{m}x$$

and so a critical point must satisfy both $y = 0$ and $x = 0$. Therefore $(0, 0)$ is the unique critical
point.

(b) The Jacobian matrix is

$$\begin{pmatrix} 0 & 1 \\ -\frac{k}{m} & -\frac{\beta}{m}2|y| \end{pmatrix}$$

and so $\tau = 0$ and $\Delta = \dfrac{k}{m} > 0$. Therefore $(0, 0)$ is a center, stable spiral point, or an unstable

spiral point. Physical considerations suggest that $(0, 0)$ must be asymptotically stable and so
$(0, 0)$ must be a stable spiral point.

417

18. (a) The magnitude of the frictional force between the bead and the wire is $\mu(mg\cos\theta)$ for some $\mu > 0$. The component of this frictional force in the x-direction is

$$(\mu mg\cos\theta)\cos\theta = \mu mg\cos^2\theta.$$

But

$$\cos\theta = \frac{1}{\sqrt{1 + [f'(x)]^2}} \quad \text{and so} \quad \mu mg\cos^2\theta = \frac{\mu mg}{1 + [f'(x)]^2}.$$

It follows from Newton's Second Law that

$$mx'' = -mg\frac{f'(x)}{1 + [f'(x)]^2} - \beta x' + mg\frac{\mu}{1 + [f'(x)]^2}$$

and so

$$x'' = g\frac{\mu - f'(x)}{1 + [f'(x)]^2} - \frac{\beta}{m}x'.$$

(b) A critical point (x, y) must satisfy $y = 0$ and $f'(x) = \mu$. Therefore critical points occur at $(x_1, 0)$ where $f'(x_1) = \mu$. The Jacobian matrix of the plane autonomous system is

$$\mathbf{g}'(\mathbf{X}) = \begin{pmatrix} 0 & 1 \\ g\frac{(1+[f'(x)]^2)(-f''(x))-(\mu-f'(x))2f'(x)f''(x)}{(1+[f'(x)]^2)^2} & -\frac{\beta}{m} \end{pmatrix}$$

and so at a critical point $\mathbf{X}_1$,

$$\mathbf{g}'(\mathbf{X}) = \begin{pmatrix} 0 & 1 \\ \frac{-gf''(x_1)}{1+\mu^2} & -\frac{\beta}{m} \end{pmatrix}.$$

Therefore $\tau = -\frac{\beta}{m} < 0$ and $\Delta = \frac{gf''(x_1)}{1 + \mu^2}$. When $f''(x_1) < 0$, $\Delta < 0$ and so a saddle point occurs. When $f''(x_1) > 0$ and

$$\tau^2 - 4\Delta = \frac{\beta^2}{m^2} - 4g\frac{f''(x_1)}{1 + \mu^2} < 0,$$

$(x_1, 0)$ is a stable spiral point. This condition may also be written as

$$\beta^2 < 4gm^2\frac{f''(x_1)}{1 + \mu^2}.$$

19. $\dfrac{dy}{dx} = \dfrac{y'}{x'} = -\dfrac{f(x)}{y}$ and so using separation of variables, $\dfrac{y^2}{2} = -\displaystyle\int_0^x f(\mu)\,d\mu + c$ or $y^2 + 2F(x) = c$. We may conclude that for a given value of x there are at most two corresponding values of y. If $(0, 0)$ were a stable spiral point there would exist an x with more than two corresponding values of y. Note that the condition $f(0) = 0$ is required for $(0, 0)$ to be a critical point of the corresponding plane autonomous system $x' = y$, $y' = -f(x)$.

20. (a) $x' = x(-a + by) = 0$ implies that $x = 0$ or $y = a/b$. If $x = 0$, then, from

$$-cxy + \frac{r}{K} y(K - y) = 0,$$

$y = 0$ or K. Therefore $(0,0)$ and $(0, K)$ are critical points. If $\hat{y} = a/b$, then

$$\hat{y}\left[-cx + \frac{r}{K}(K - \hat{y})\right] = 0.$$

The corresponding value of x, $x = \hat{x}$, therefore satisfies the equation $c\hat{x} = \frac{r}{K}(K - \hat{y})$.

(b) The Jacobian matrix is

$$\mathbf{g}'(\mathbf{X}) = \begin{pmatrix} -a + by & bx \\ -cy & -cx + \frac{r}{K}(K - 2y) \end{pmatrix}$$

and so at $\mathbf{X}_1 = (0,0)$, $\Delta = -ar < 0$. For $\mathbf{X}_1 = (0, K)$, $\Delta = n(Kb - a) = -rb\left(K - \frac{a}{b}\right)$. Since we are given that $K > \frac{a}{b}$, $\Delta < 0$ in this case. Therefore $(0,0)$ and $(0, K)$ are each saddle points. For $\mathbf{X}_1 = (\hat{x}, \hat{y})$ where $\hat{y} = \frac{a}{b}$ and $c\hat{x} = \frac{r}{K}(K - \hat{y})$, we may write the Jacobian matrix as

$$\mathbf{g}'((\hat{x}, \hat{y})) = \begin{pmatrix} 0 & b\hat{x} \\ -c\hat{y} & -\frac{r}{K}\hat{y} \end{pmatrix}$$

and so $\tau = -\frac{r}{K}\hat{y} < 0$ and $\Delta = bc\hat{x}\hat{y} > 0$. Therefore $(\hat{x}, \hat{y})$ is a stable critical point and so it is either a stable node (perhaps degenerate) or a stable spiral point.

(c) Write

$$\tau^2 - 4\Delta = \frac{r^2}{K^2}\hat{y}^2 - 4bc\hat{x}\hat{y} = \hat{y}\left[\frac{r^2}{K^2}\hat{y} - 4bc\hat{x}\right] = \hat{y}\left[\frac{r^2}{K^2}\hat{y} - 4b\frac{r}{K}(K - \hat{y})\right]$$

using

$$c\hat{x} = \frac{r}{K}(K - \hat{y}) = \frac{r}{K}\hat{y}\left[\left(\frac{r}{K} + 4b\right)\hat{y} - 4bK\right].$$

Therefore $\tau^2 - 4\Delta < 0$ if and only if

$$\hat{y} < \frac{4bK}{\frac{r}{K} + 4b} = \frac{4bK^2}{r + 4bK}.$$

Note that

$$\frac{4bK^2}{r + 4bK} = \frac{4bK}{r + 4bK} \cdot K \approx K$$

where K is large, and $\hat{y} = \frac{a}{b} < K$. Therefore $\tau^2 - 4\Delta < 0$ when K is large and a stable spiral point will result.

21. The equation

$$x' = \alpha \frac{y}{1+y} x - x = x\left(\frac{\alpha y}{1+y} - 1\right) = 0$$

implies that $x = 0$ or $y = \dfrac{1}{\alpha - 1}$. When $\alpha > 0$, $\hat{y} = \dfrac{1}{\alpha - 1} > 0$. If $x = 0$, then from the differential

equation for y', $y = \beta$. On the other hand, if $\hat{y} = \dfrac{1}{\alpha - 1}$, $\dfrac{\hat{y}}{1+\hat{y}} = \dfrac{1}{\alpha}$ and so $\dfrac{1}{\alpha}\hat{x} - \dfrac{1}{\alpha - 1} + \beta = 0$.

It follows that

$$\hat{x} = \alpha\left(\beta - \frac{1}{\alpha - 1}\right) = \frac{\alpha}{\alpha - 1}[(\alpha - 1)\beta - 1]$$

and if $\beta(\alpha - 1) > 1$, $\hat{x} > 0$. Therefore $(\hat{x}, \hat{y})$ is the unique critical point in the first quadrant. The
Jacobian matrix is

$$\mathbf{g}'(\mathbf{X}) = \begin{pmatrix} \alpha\frac{y}{y+1} - 1 & \frac{\alpha x}{(1+y)^2} \\ -\frac{y}{1+y} & \frac{-x}{(1+y)^2} - 1 \end{pmatrix}$$

and for $\mathbf{X} = (\hat{x}, \hat{y})$, the Jacobian can be written in the form

$$\mathbf{g}'((\hat{x}, \hat{y})) = \begin{pmatrix} 0 & \frac{(\alpha-1)^2}{\alpha}\hat{x} \\ -\frac{1}{\alpha} & -\frac{(\alpha-1)^2}{\alpha^2} - 1 \end{pmatrix}.$$

It follows that

$$\tau = -\left[\frac{(\alpha - 1)^2}{\alpha^2}\hat{x} + 1\right] < 0, \quad \Delta = \frac{(\alpha - 1)^2}{\alpha^2}\hat{x}$$

and so $\tau = -(\Delta + 1)$. Therefore $\tau^2 - 4\Delta = (\Delta + 1)^2 - 4\Delta = (\Delta - 1)^2 > 0$. Therefore $(\hat{x}, \hat{y})$ is a
stable node.

22. Letting $y = x'$ we obtain the plane autonomous system

$$x' = y$$

$$y' = -8x + 6x^3 - x^5.$$

Solving $x^5 - 6x^3 + 8x = x(x^2 - 4)(x^2 - 2) = 0$ we see that
critical points are $(0,0)$, $(0,-2)$, $(0,2)$, $(0,-\sqrt{2})$, and $(0,\sqrt{2})$.
The Jacobian matrix is

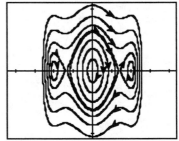

$$\mathbf{g}'(\mathbf{X}) = \begin{pmatrix} 0 & 1 \\ -8 + 18x^2 - 5x^4 & 0 \end{pmatrix}$$

and we see that $\det(\mathbf{g}'(\mathbf{X})) = 5x^4 - 18x^2 + 8$ and the trace of $\mathbf{g}'(\mathbf{X})$ is 0. Since $\det(\mathbf{g}'((\pm\sqrt{2},0))) = -8 < 0$, $(\pm\sqrt{2},0)$ are saddle points. For the other critical points the determinant is positive and
linearization discloses no information. The graph of the phase plane suggests that $(0,0)$ and $(\pm 2,0)$
are centers.

——————— **Chapter 10 Review Exercises** ———————

1. True

2. True

3. a center or a saddle point

4. complex with negative real parts

5. False; there are initial conditions for which $\lim\limits_{t\to\infty} \mathbf{X}(t) = (0,0)$.

6. True

7. False; this is a borderline case. See Figure 10.25 in the text.

8. False; see Figure 10.29 in the text.

9. The system is linear and we identify $\Delta = -\alpha$ and $\tau = \alpha + 1$. Since a critical point will be a center when $\Delta > 0$ and $\tau = 0$ we see that for $\alpha = -1$ critical points will be centers and solutions will be periodic. Note also that when $\alpha = -1$ the system is

$$x' = -x - 2y$$

$$y' = x + y,$$

which does have an isolated critical point at $(0,0)$.

10. We identify $g(x) = \sin x$ in Theorem 10.2. Then $x_1 = n\pi$ is a critical point for n an integer and $g'(n\pi) = \cos n\pi < 0$ when n is an odd integer. Thus, $n\pi$ is an asymptotically stable critical point when n is an odd integer.

11. Switching to polar coordinates,

$$\frac{dr}{dt} = \frac{1}{r}\left(x\frac{dx}{dt} + y\frac{dy}{dt}\right) = \frac{1}{r}(-xy - x^2r^3 + xy - y^2r^3) = -r^4$$

$$\frac{d\theta}{dt} = \frac{1}{r^2}\left(-y\frac{dx}{dt} + x\frac{dy}{dt}\right) = \frac{1}{r^2}(y^2 + xyr^3 + x^2 - xyr^3) = 1.$$

Using separation of variables it follows that $r = \dfrac{1}{\sqrt[3]{3t + c_1}}$ and $\theta = t + c_2$. Since $\mathbf{X}(0) = (1,0)$, $r = 1$ and $\theta = 0$. It follows that $c_1 = 1$, $c_2 = 0$, and so

$$r = \frac{1}{\sqrt[3]{3t + 1}}, \quad \theta = t.$$

As $t \to \infty$, $r \to 0$ and the solution spirals toward the origin.

12. (a) If $\mathbf{X}(0) = \mathbf{X}_0$ lies on the line $y = -2x$, then $\mathbf{X}(t)$ approaches $(0,0)$ along this line. For all other initial conditions, $\mathbf{X}(t)$ approaches $(0,0)$ from the direction determined by the line $y = x$.

(b) If $\mathbf{X}(0) = \mathbf{X}_0$ lies on the line $y = -x$, then $\mathbf{X}(t)$ approaches $(0,0)$ along this line. For all other initial conditions, $\mathbf{X}(t)$ becomes unbounded and $y = 2x$ serves as an asymptote.

13. (a) $\tau = 0$, $\Delta = 11 > 0$ and so $(0,0)$ is a center.

(b) $\tau = -2$, $\Delta = 1$, $\tau^2 - 4\Delta = 0$ and so $(0,0)$ is a degenerate stable node.

14. From $x' = x(1 + y - 3x) = 0$, either $x = 0$ or $1 + y - 3x = 0$. If $x = 0$, then, from $y(4 - 2x - y) = 0$ we obtain $y(4 - y) = 0$. It follows that $(0,0)$ and $(0,4)$ are critical points. If $1 + y - 3x = 0$, then $y(5 - 5x) = 0$. Therefore $(1/3, 0)$ and $(1,2)$ are the remaining critical points. We will use the Jacobian matrix

$$\mathbf{g}'(\mathbf{X}) = \begin{pmatrix} 1 + y - 6x & x \\ -2y & 4 - 2x - 2y \end{pmatrix}$$

to classify these four critical points. The results are as follows:

$\mathbf{X}$	τ	Δ	$\tau^2 - 4\Delta$	Conclusion
$(0,0)$	5	4	9	unstable node
$(0,4)$	$-$	-20	$-$	saddle point
$(\frac{1}{3}, 0)$	$-$	$-\frac{10}{3}$	$-$	saddle point
$(1,2)$	-5	10	-15	stable spiral point

15. From $x = r\cos\theta$, $y = r\sin\theta$ we have

$$\frac{dx}{dt} = -r\sin\theta\,\frac{d\theta}{dt} + \frac{dr}{dt}\cos\theta$$

$$\frac{dy}{dt} = r\cos\theta\,\frac{d\theta}{dt} + \frac{dr}{dt}\sin\theta.$$

Then $r' = \alpha r$, $\theta' = 1$ gives

$$\frac{dx}{dt} = -r\sin\theta + \alpha r\cos\theta$$

$$\frac{dy}{dt} = r\cos\theta + \alpha r\sin\theta.$$

We see that $r = 0$, which corresponds to $\mathbf{X} = (0,0)$, is a critical point. Solving $r' = \alpha r$ we have $r = c_1 e^{\alpha t}$. Thus, when $\alpha < 0$, $\lim_{t\to\infty} r(t) = 0$ and $(0,0)$ is a stable critical point. When $\alpha = 0$, $r' = 0$ and $r = c_1$. In this case $(0,0)$ is a center, which is stable. Therefore, $(0,0)$ is a stable critical point for the system when $\alpha \leq 0$.

16. The corresponding plane autonomous system is $x' = y$, $y' = \mu(1 - x^2) - x$ and so the Jacobian at the critical point $(0,0)$ is

$$\mathbf{g}'((0,0)) = \begin{pmatrix} 0 & 1 \\ -1 & \mu \end{pmatrix}.$$

Therefore $\tau = \mu$, $\Delta = 1$ and $\tau^2 - 4\Delta = \mu^2 - 4$. Now $\mu^2 - 4 < 0$ if and only if $-2 < \mu < 2$. We may therefore conclude that $(0,0)$ is a stable node for $\mu < -2$, a stable spiral point for $-2 < \mu < 0$, an unstable spiral point for $0 < \mu < 2$, and an unstable node for $\mu > 2$.

17. Critical points occur at $x = \pm 1$. Since

$$g'(x) = -\frac{1}{2}e^{-x/2}(x^2 - 4x - 1),$$

$g'(1) > 0$ and $g'(-1) < 0$. Therefore $x = 1$ is unstable and $x = -1$ is asymptotically stable.

18. $\dfrac{dy}{dx} = \dfrac{y'}{x'} = \dfrac{-2x\sqrt{y^2 + 1}}{y}$. We may separate variables to show that $\sqrt{y^2 + 1} = -x^2 + c$. But $x(0) = x_0$ and $y(0) = x'(0) = 0$. It follows that $c = 1 + x_0^2$ so that

$$y^2 = (1 + x_0^2 - x^2)^2 - 1.$$

Note that $1 + x_0^2 - x^2 > 1$ for $-x_0 < x < x_0$ and $y = 0$ for $x = \pm x_0$. Each x with $-x_0 < x < x_0$ has two corresponding values of y and so the solution $\mathbf{X}(t)$ with $\mathbf{X}(0) = (x_0, 0)$ is periodic.

19. The corresponding plane autonomous system

$$x' = y, \quad y' = -\frac{\beta}{m}y - \frac{k}{m}(s+x)^3 + g$$

and so the Jacobian is

$$\mathbf{g}'(\mathbf{X}) = \begin{pmatrix} 0 & 1 \\ -\frac{3k}{m}(s+x)^2 & -\frac{\beta}{m} \end{pmatrix}.$$

For $\mathbf{X} = (0,0)$, $\tau = -\dfrac{\beta}{m} < 0$, $\Delta = \dfrac{3k}{m}s^2 > 0$. Therefore

$$\tau^2 - 4\Delta = \frac{\beta^2}{m^2} - \frac{12k}{m}s^2 = \frac{1}{m^2}(\beta^2 - 12kms^2).$$

Therefore $(0,0)$ is a stable node if $\beta^2 > 12kms^2$ and a stable spiral point provided $\beta^2 < 12kms^2$, where $ks^3 = mg$.

20. (a) If (x, y) is a critical point, $y = 0$ and so $\sin x(\omega^2 \cos x - g/l) = 0$. Either $\sin x = 0$ (in which case $x = 0$) of $\cos x = g/\omega^2 l$. But if $\omega^2 < g/l$, $g/\omega^2 l > 1$ and so the latter equation has no real solutions. Therefore $(0,0)$ is the only critical point if $\omega^2 < g/l$. The Jacobian matrix is

$$\mathbf{g}'(\mathbf{X}) = \begin{pmatrix} 0 & 1 \\ \omega^2 \cos 2x - \frac{g}{l}\cos x & -\frac{\beta}{ml} \end{pmatrix}$$

and so $\tau = -\beta/ml < 0$ and $\Delta = g/l - \omega^2 > 0$ for $\mathbf{X} = (0,0)$. It follows that $(0,0)$ is asymptotically stable and so after a small displacement, the pendulum will return to $\theta = 0$, $\theta' = 0$.

(b) If $\omega^2 > g/l$, $\cos x = g/\omega^2 l$ will have two solutions $x = \pm\hat{x}$ that satisfy $-\pi < x < \pi$. Therefore $(\pm\hat{x}, 0)$ are two additional critical points. If $\mathbf{X}_1 = (0, 0)$, $\Delta = g/l - \omega^2 < 0$ and so $(0, 0)$ is a saddle point. If $\mathbf{X}_1 = (\pm\hat{x}, 0)$, $\tau = -\beta/ml < 0$ and

$$\Delta = \frac{g}{l}\cos\hat{x} - \omega^2\cos 2\hat{x} = \frac{g^2}{\omega^2 l^2} - \omega^2\left(2\frac{g^2}{\omega^4 l^2} - 1\right) = \omega^2 - \frac{g^2}{\omega^2 l^2} > 0.$$

Therefore $(\hat{x}, 0)$ and $(-\hat{x}, 0)$ are each stable. When $\theta(0) = \theta_0$, $\theta'(0) = 0$ and θ_0 is small we expect the pendulum to reach one of these two stable equilibrium positions.

11 Orthogonal Functions and Fourier Series

---------- **Exercises 11.1** ----------

1. $\int_{-2}^{2} xx^2dx = \frac{1}{4}x^4 \Big|_{-2}^{2} = 0$

2. $\int_{-1}^{1} x^3(x^2+1)dx = \frac{1}{6}x^6 \Big|_{-1}^{1} + \frac{1}{4}x^4 \Big|_{-1}^{1} = 0$

3. $\int_{0}^{2} e^x(xe^{-x} - e^{-x})dx = \int_{0}^{2}(x-1)dx = \left(\frac{1}{2}x^2 - x\right)\Big|_{0}^{2} = 0$

4. $\int_{0}^{\pi} \cos x \sin^2 x \, dx = \frac{1}{3}\sin^3 x \Big|_{0}^{\pi} = 0$

5. $\int_{-\pi/2}^{\pi/2} x \cos 2x \, dx = \frac{1}{2}\left(\frac{1}{2}\cos 2x + x \sin 2x\right)\Big|_{-\pi/2}^{\pi/2} = 0$

6. $\int_{\pi/4}^{5\pi/4} e^x \sin x \, dx = \left(\frac{1}{2}e^x \sin x - \frac{1}{2}e^x \cos x\right)\Big|_{\pi/4}^{5\pi/4} = 0$

7. For $m \neq n$

$$\int_{0}^{\pi/2} \sin(2n+1)x \, \sin(2m+1)x \, dx$$

$$= \frac{1}{2}\int_{0}^{\pi/2}[\cos 2(n-m)x - \cos 2(n+m+1)x]\,dx$$

$$= \frac{1}{4(n-m)}\sin 2(n-m)x \Big|_{0}^{\pi/2} - \frac{1}{4(n+m+1)}\sin 2(n+m+1)x \Big|_{0}^{\pi/2}$$

$$= 0.$$

For $m = n$

$$\int_{0}^{\pi/2} \sin^2(2n+1)x \, dx = \int_{0}^{\pi/2}\left(\frac{1}{2} - \frac{1}{2}\cos 2(2n+1)x\right)dx$$

$$= \frac{1}{2}x \Big|_{0}^{\pi/2} - \frac{1}{4(2n+1)}\sin 2(2n+1)x \Big|_{0}^{\pi/2}$$

$$= \frac{\pi}{4}$$

so that

$$\| \sin(2n + 1)x \| = \frac{1}{2}\sqrt{\pi} \, .$$

8. For $m \neq n$

$$\int_0^{\pi/2} \cos(2n + 1)x \, \cos(2m + 1)x \, dx$$

$$= \frac{1}{2} \int_0^{\pi/2} [\cos 2(n - m)x + \cos 2(n + m + 1)x] \, dx$$

$$= \frac{1}{4(n - m)} \sin 2(n - m)x \, \bigg|_0^{\pi/2} + \frac{1}{4(n + m + 1)} \sin 2(n + m + 1)x \, \bigg|_0^{\pi/2}$$

$$= 0.$$

For $m = n$

$$\int_0^{\pi/2} \cos^2(2n + 1)x \, dx = \int_0^{\pi/2} \left(\frac{1}{2} + \frac{1}{2} \cos 2(2n + 1)x \right) dx$$

$$= \frac{1}{2}x \, \bigg|_0^{\pi/2} + \frac{1}{4(2n + 1)} \sin 2(2n + 1)x \, \bigg|_0^{\pi/2}$$

$$= \frac{\pi}{4}$$

so that

$$\| \cos(2n + 1)x \| = \frac{1}{2}\sqrt{\pi} \, .$$

9. For $m \neq n$

$$\int_0^{\pi} \sin nx \sin mx \, dx = \frac{1}{2} \int_0^{\pi} [\cos(n - m)x - \cos(n + m)x] \, dx$$

$$= \frac{1}{2(n - m)} \sin(n - m)x \, \bigg|_0^{\pi} - \frac{1}{2(n + m)} \sin 2(n + m)x \, \bigg|_0^{\pi}$$

$$= 0.$$

For $m = n$

$$\int_0^{\pi} \sin^2 nx \, dx = \int_0^{\pi} \left[\frac{1}{2} - \frac{1}{2} \cos 2nx \right] dx = \frac{1}{2}x \, \bigg|_0^{\pi} - \frac{1}{4n} \sin 2nx \, \bigg|_0^{\pi} = \frac{\pi}{2}$$

so that

$$\| \sin nx \| = \sqrt{\frac{\pi}{2}} \, .$$

426

10. For $m \neq n$

$$\int_0^p \sin \frac{n\pi}{p} x \sin \frac{m\pi}{p} x \, dx = \frac{1}{2} \int_0^p \left(\cos \frac{(n-m)\pi}{p} x - \cos \frac{(n+m)\pi}{p} x \right) dx$$

$$= \frac{p}{2(n-m)\pi} \sin \frac{(n-m)\pi}{p} x \Big|_0^p - \frac{p}{2(n+m)\pi} \sin \frac{(n+m)\pi}{p} x \Big|_0^p$$

$$= 0.$$

For $m = n$

$$\int_0^p \sin^2 \frac{n\pi}{p} x \, dx = \int_0^p \left[\frac{1}{2} - \frac{1}{2} \cos \frac{2n\pi}{p} x \right] dx = \frac{1}{2} x \Big|_0^p - \frac{p}{4n\pi} \sin \frac{2n\pi}{p} x \Big|_0^p = \frac{p}{2}$$

so that

$$\left\| \sin \frac{n\pi}{p} x \right\| = \sqrt{\frac{p}{2}} \, .$$

11. For $m \neq n$

$$\int_0^p \cos \frac{n\pi}{p} x \cos \frac{m\pi}{p} x \, dx = \frac{1}{2} \int_0^p \left(\cos \frac{(n-m)\pi}{p} x + \cos \frac{(n+m)\pi}{p} x \right) dx$$

$$= \frac{p}{2(n-m)\pi} \sin \frac{(n-m)\pi}{p} x \Big|_0^p + \frac{p}{2(n+m)\pi} \sin \frac{(n+m)\pi}{p} x \Big|_0^p$$

$$= 0.$$

For $m = n$

$$\int_0^p \cos^2 \frac{n\pi}{p} x \, dx = \int_0^p \left(\frac{1}{2} + \frac{1}{2} \cos \frac{2n\pi}{p} x \right) dx = \frac{1}{2} x \Big|_0^p + \frac{p}{4n\pi} \sin \frac{2n\pi}{p} x \Big|_0^p = \frac{p}{2} \, .$$

Also

$$\int_0^p 1 \cdot \cos \frac{n\pi}{p} x \, dx = \frac{p}{n\pi} \sin \frac{n\pi}{p} x \Big|_0^p = 0 \quad \text{and} \quad \int_0^p 1^2 dx = p$$

so that

$$\|1\| = \sqrt{p} \quad \text{and} \quad \left\| \cos \frac{n\pi}{p} x \right\| = \sqrt{\frac{p}{2}} \, .$$

12. For $m \neq n$, we use Problems 11 and 10:

$$\int_{-p}^p \cos \frac{n\pi}{p} x \cos \frac{m\pi}{p} x \, dx = 2 \int_0^p \cos \frac{n\pi}{p} x \cos \frac{m\pi}{p} x \, dx = 0$$

$$\int_{-p}^p \sin \frac{n\pi}{p} x \sin \frac{m\pi}{p} x \, dx = 2 \int_0^p \sin \frac{n\pi}{p} x \sin \frac{m\pi}{p} x \, dx = 0.$$

427

Also

$$\int_{-p}^{p} \sin\frac{n\pi}{p}x \cos\frac{m\pi}{p}x \, dx = \frac{1}{2}\int_{-p}^{p}\left(\sin\frac{(n-m)\pi}{p}x + \sin\frac{(n+m)\pi}{p}x\right)dx = 0,$$

$$\int_{-p}^{p} 1 \cdot \cos\frac{n\pi}{p}x \, dx = \frac{p}{n\pi}\sin\frac{n\pi}{p}x \, \Big|_{-p}^{p} = 0,$$

$$\int_{-p}^{p} 1 \cdot \sin\frac{n\pi}{p}x \, dx = -\frac{p}{n\pi}\cos\frac{n\pi}{p}x \, \Big|_{-p}^{p} = 0,$$

and

$$\int_{-p}^{p} \sin\frac{n\pi}{p}x \cos\frac{n\pi}{p}x \, dx = \int_{-p}^{p}\frac{1}{2}\sin\frac{2n\pi}{p}x \, dx = -\frac{p}{4n\pi}\cos\frac{2n\pi}{p}x \, \Big|_{-p}^{p} = 0.$$

For $m = n$

$$\int_{-p}^{p} \cos^2\frac{n\pi}{p}x \, dx = \int_{-p}^{p}\left(\frac{1}{2} + \frac{1}{2}\cos\frac{2n\pi}{p}x\right)dx = p,$$

$$\int_{-p}^{p} \sin^2\frac{n\pi}{p}x \, dx = \int_{-p}^{p}\left(\frac{1}{2} - \frac{1}{2}\cos\frac{2n\pi}{p}x\right)dx = p,$$

and

$$\int_{-p}^{p} 1^2 dx = 2p$$

so that

$$\|1\| = \sqrt{2p}, \quad \left\|\cos\frac{n\pi}{p}x\right\| = \sqrt{p}, \quad \text{and} \quad \left\|\sin\frac{n\pi}{p}x\right\| = \sqrt{p}.$$

13. Since

$$\int_{-\infty}^{\infty} e^{-x^2} \cdot 1 \cdot 2x \, dx = -e^{-x^2}\Big|_{-\infty}^{0} - e^{-x^2}\Big|_{0}^{\infty} = 0,$$

$$\int_{-\infty}^{\infty} e^{-x^2} \cdot 1 \cdot (4x^2 - 2) \, dx = 2\int_{-\infty}^{\infty} x\left(2xe^{-x^2}\right)dx - 2\int_{-\infty}^{\infty} e^{-x^2}dx$$

$$= 2\left(-xe^{-x^2}\Big|_{-\infty}^{\infty} + \int_{-\infty}^{\infty} e^{-x^2}dx\right) - 2\int_{-\infty}^{\infty} e^{-x^2}dx$$

$$= 2\left(-xe^{-x^2}\Big|_{-\infty}^{0} - xe^{-x^2}\Big|_{0}^{\infty}\right) = 0,$$

and

$$\int_{-\infty}^{\infty} e^{-x^2} \cdot 2x \cdot (4x^2 - 2)\, dx = 4 \int_{-\infty}^{\infty} x^2 \left(2xe^{-x^2}\right) dx - 4 \int_{-\infty}^{\infty} xe^{-x^2} dx$$

$$= 4 \left(-x^2 e^{-x^2} \Big|_{-\infty}^{\infty} + 2 \int_{-\infty}^{\infty} xe^{-x^2} dx \right) - 4 \int_{-\infty}^{\infty} xe^{-x^2} dx$$

$$= 4 \left(-x^2 e^{-x^2} \Big|_{-\infty}^{0} - x^2 e^{-x^2} \Big|_{0}^{\infty} \right) + 2 \int_{-\infty}^{\infty} 2xe^{-x^2} dx = 0,$$

the functions are orthogonal.

14. Since

$$\int_{0}^{\infty} e^{-x^2} \cdot 1(1 - x)\, dx = (x - 1)e^{-x} \Big|_{0}^{\infty} - \int_{0}^{\infty} e^{-x} dx = 0,$$

$$\int_{0}^{\infty} e^{-x} \cdot 1 \cdot \left(\frac{1}{2}x^2 - 2x + 1 \right) dx = \left(2x - 1 - \frac{1}{2}x^2 \right) e^{-x} \Big|_{0}^{\infty} + \int_{0}^{\infty} e^{-x}(x - 2)\, dx$$

$$= 1 + (2 - x)e^{-x} \Big|_{0}^{\infty} + \int_{0}^{\infty} e^{-x} dx = 0,$$

and

$$\int_{0}^{\infty} e^{-x} \cdot (1 - x) \left(\frac{1}{2}x^2 - 2x + 1 \right) dx$$

$$= \int_{0}^{\infty} e^{-x} \left(-\frac{1}{2}x^3 + \frac{5}{2}x^2 - 3x + 1 \right) dx$$

$$= e^{-x} \left(\frac{1}{2}x^3 - \frac{5}{2}x^2 + 3x - 1 \right) \Big|_{0}^{\infty} + \int_{0}^{\infty} e^{-x} \left(-\frac{3}{2}x^2 + 5x - 3 \right) dx$$

$$= 1 + e^{-x} \left(\frac{3}{2}x^2 - 5x + 3 \right) \Big|_{0}^{\infty} + \int_{0}^{\infty} e^{-x}(5 - 3x)\, dx$$

$$= 1 - 3 + e^{-x}(3x - 5) \Big|_{0}^{\infty} - 3 \int_{0}^{\infty} e^{-x} dx = 0,$$

the functions are orthogonal

15. By orthogonality $\int_{a}^{b} \phi_0(x)\phi_n(x)dx = 0$ for $n = 1, 2, 3, \ldots$; that is, $\int_{a}^{b} \phi_n(x)dx = 0$ for $n = 1, 2, 3, \ldots$.

Exercises 11.1

16. Using the facts that ϕ_0 and ϕ_1 are orthogonal to ϕ_n for $n > 1$, we have

$$\int_a^b (\alpha x + \beta)\phi_n(x)\,dx = \alpha \int_a^b x\phi_n(x)\,dx + \beta \int_a^b 1 \cdot \phi_n(x)\,dx$$

$$= \alpha \int_a^b \phi_1(x)\phi_n(x)\,dx + \beta \int_a^b \phi_0(x)\phi_n(x)\,dx$$

$$= \alpha \cdot 0 + \beta \cdot 0 = 0$$

for $n = 2, 3, 4, \ldots$.

17. Using the fact that ϕ_n and ϕ_m are orthogonal for $n \neq m$ we have

$$\|\phi_m(x) + \phi_n(x)\|^2 = \int_a^b [\phi_m(x) + \phi_n(x)]^2 dx \int_a^b \left[\phi_m^2(x) + 2\phi_m(x)\phi_n(x) + \phi_n^2(x)\right] dx$$

$$= \int_a^b \phi_m^2(x)\,dx + 2\int_a^b \phi_m(x)\phi_n(x)\,dx + \int_a^b \phi_n^2(x)\,dx$$

$$= \|\phi_m(x)\|^2 + \|\phi_n(x)\|^2 .$$

18. Setting

$$0 = \int_{-2}^2 f_3(x)f_1(x)\,dx = \int_{-2}^2 \left(x^2 + c_1 x^3 + c_2 x^4\right) dx = \frac{16}{3} + \frac{64}{5}c_2$$

and

$$0 = \int_{-2}^2 f_3(x)f_2(x)\,dx = \int_{-2}^2 \left(x^3 + c_1 x^4 + c_2 x^5\right) dx = \frac{64}{5}c_1$$

we obtain $c_1 = 0$ and $c_2 = -5/12$.

19. Since $\sin nx$ is an odd function on $[-\pi, \pi]$,

$$(1, \sin nx) = \int_{-\pi}^{\pi} \sin nx\,dx = 0$$

and $f(x) = 1$ is orthogonal to every member of $\{\sin nx\}$. Thus $\{\sin nx\}$ is not complete.

20. $(f_1 + f_2, f_3) = \int_a^b [f_1(x) + f_2(x)]f_3(x)\,dx = \int_a^b f_1(x)f_3(x)\,dx + \int_a^b f_2(x)f_3(x)\,dx = (f_1, f_3) + (f_2, f_3)$

Exercises 11.2

1. $a_0 = \dfrac{1}{\pi}\displaystyle\int_{-\pi}^{\pi} f(x)\,dx = \dfrac{1}{\pi}\displaystyle\int_0^{\pi} 1\,dx = 1$

$a_n = \dfrac{1}{\pi}\displaystyle\int_{-\pi}^{\pi} f(x)\cos\dfrac{n\pi}{\pi}x\,dx = \dfrac{1}{\pi}\displaystyle\int_0^{\pi}\cos nx\,dx = 0$

$b_n = \dfrac{1}{\pi}\displaystyle\int_{-\pi}^{\pi} f(x)\sin\dfrac{n\pi}{\pi}x\,dx = \dfrac{1}{\pi}\displaystyle\int_0^{\pi}\sin nx\,dx = \dfrac{1}{n\pi}(1-\cos n\pi) = \dfrac{1}{n\pi}[1-(-1)^n]$

$f(x) = \dfrac{1}{2} + \dfrac{1}{\pi}\displaystyle\sum_{n=1}^{\infty}\dfrac{1-(-1)^n}{n}\sin nx$

2. $a_0 = \dfrac{1}{\pi}\displaystyle\int_{-\pi}^{\pi} f(x)\,dx = \dfrac{1}{\pi}\displaystyle\int_{-\pi}^{0} -1\,dx + \dfrac{1}{\pi}\displaystyle\int_0^{\pi} 2\,dx = 1$

$a_n = \dfrac{1}{\pi}\displaystyle\int_{-\pi}^{\pi} f(x)\cos nx\,dx = \dfrac{1}{\pi}\displaystyle\int_{-\pi}^{0} -\cos nx\,dx + \dfrac{1}{\pi}\displaystyle\int_0^{\pi} 2\cos nx\,dx = 0$

$b_n = \dfrac{1}{\pi}\displaystyle\int_{-\pi}^{\pi} f(x)\sin nx\,dx = \dfrac{1}{\pi}\displaystyle\int_{-\pi}^{0} -\sin nx\,dx + \dfrac{1}{\pi}\displaystyle\int_0^{\pi} 2\sin nx\,dx = \dfrac{3}{n\pi}[1-(-1)^n]$

$f(x) = \dfrac{1}{2} + \dfrac{3}{\pi}\displaystyle\sum_{n=1}^{\infty}\dfrac{1-(-1)^n}{n}\sin nx$

3. $a_0 = \displaystyle\int_{-1}^{1} f(x)\,dx = \displaystyle\int_{-1}^{0} 1\,dx + \displaystyle\int_0^{1} x\,dx = \dfrac{3}{2}$

$a_n = \displaystyle\int_{-1}^{1} f(x)\cos n\pi x\,dx = \displaystyle\int_{-1}^{0}\cos n\pi x\,dx + \displaystyle\int_0^{1} x\cos n\pi x\,dx = \dfrac{1}{n^2\pi^2}[(-1)^n-1]$

$b_n = \displaystyle\int_{-1}^{1} f(x)\sin n\pi x\,dx = \displaystyle\int_{-1}^{0}\sin n\pi x\,dx + \displaystyle\int_0^{1} x\sin n\pi x\,dx = -\dfrac{1}{n\pi}$

$f(x) = \dfrac{3}{4} + \displaystyle\sum_{n=1}^{\infty}\left[\dfrac{(-1)^n-1}{n^2\pi^2}\cos n\pi x - \dfrac{1}{n\pi}\sin n\pi x\right]$

4. $a_0 = \displaystyle\int_{-1}^{1} f(x)\,dx = \displaystyle\int_0^{1} x\,dx = \dfrac{1}{2}$

$a_n = \displaystyle\int_{-1}^{1} f(x)\cos n\pi x\,dx - \displaystyle\int_0^{1} x\cos n\pi x\,dx = \dfrac{1}{n^2\pi^2}[(-1)^n-1]$

$b_n = \displaystyle\int_{-1}^{1} f(x)\sin n\pi x\,dx = \displaystyle\int_0^{1} x\sin n\pi x\,dx = \dfrac{(-1)^{n+1}}{n\pi}$

$f(x) = \dfrac{3}{4} + \displaystyle\sum_{n=1}^{\infty}\left[\dfrac{(-1)^n-1}{n^2\pi^2}\cos n\pi x + \dfrac{(-1)^{n+1}}{n\pi}\sin n\pi x\right]$

5. $a_0 = \dfrac{1}{\pi} \displaystyle\int_{-\pi}^{\pi} f(x)\, dx = \dfrac{1}{\pi} \int_0^{\pi} x^2\, dx = \dfrac{1}{3}\pi^2$

$a_n = \dfrac{1}{\pi} \displaystyle\int_{-\pi}^{\pi} f(x) \cos nx\, dx = \dfrac{1}{\pi} \int_0^{\pi} x^2 \cos nx\, dx = \dfrac{1}{\pi}\left(\dfrac{x^2}{\pi} \sin nx \,\Big|_0^{\pi} - \dfrac{2}{n} \int_0^{\pi} x \sin nx\, dx \right) = \dfrac{2(-1)^n}{n^2}$

$b_n = \dfrac{1}{\pi} \displaystyle\int_0^{\pi} x^2 \sin nx\, dx = \dfrac{1}{\pi}\left(-\dfrac{x^2}{n} \cos nx \,\Big|_0^{\pi} + \dfrac{2}{n} \int_0^{\pi} x \cos nx\, dx \right) = \dfrac{\pi}{n}(-1)^{n+1} + \dfrac{2}{n^3 \pi}[(-1)^n - 1]$

$f(x) = \dfrac{\pi^2}{6} + \displaystyle\sum_{n=1}^{\infty} \left[\dfrac{2(-1)^n}{n^2} \cos nx + \left(\dfrac{\pi}{n}(-1)^{n+1} + \dfrac{2[(-1)^n - 1]}{n^3 \pi} \right) \sin nx \right]$

6. $a_0 = \dfrac{1}{\pi} \displaystyle\int_{-\pi}^{\pi} f(x)\, dx = \dfrac{1}{\pi} \int_{-\pi}^{0} \pi^2\, dx + \dfrac{1}{\pi} \int_0^{\pi} \left(\pi^2 - x^2 \right) dx = \dfrac{5}{3}\pi^2$

$a_n = \dfrac{1}{\pi} \displaystyle\int_{-\pi}^{\pi} f(x) \cos nx\, dx = \dfrac{1}{\pi} \int_{-\pi}^{0} \pi^2 \cos nx\, dx + \dfrac{1}{\pi} \int_0^{\pi} \left(\pi^2 - x^2 \right) \cos nx\, dx$

$= \dfrac{1}{\pi}\left(\dfrac{\pi^2 - x^2}{n} \sin nx \,\Big|_0^{\pi} + \dfrac{2}{n} \int_0^{\pi} x \sin nx\, dx \right) = \dfrac{2}{n^2}(-1)^{n+1}$

$b_n = \dfrac{1}{\pi} \displaystyle\int_{-\pi}^{\pi} f(x) \sin nx\, dx = \dfrac{1}{\pi} \int_{-\pi}^{0} \pi^2 \sin nx\, dx + \dfrac{1}{\pi} \int_0^{\pi} \left(\pi^2 - x^2 \right) \sin nx\, dx$

$= \dfrac{\pi}{n}[(-1)^n - 1] + \dfrac{1}{\pi}\left(\dfrac{x^2 - \pi^2}{n} \cos nx \,\Big|_0^{\pi} - \dfrac{2}{n} \int_0^{\pi} x \cos nx\, dx \right) = \dfrac{\pi}{n}(-1)^n + \dfrac{2}{n^3 \pi}[1 - (-1)^n]$

$f(x) = \dfrac{5\pi^2}{6} + \displaystyle\sum_{n=1}^{\infty} \left[\dfrac{2}{n^2}(-1)^{n+1} \cos nx + \left(\dfrac{\pi}{n}(-1)^n + \dfrac{2[1 - (-1^n]}{n^3 \pi} \right) \sin nx \right]$

7. $a_0 = \dfrac{1}{\pi} \displaystyle\int_{-\pi}^{\pi} f(x)\, dx = \dfrac{1}{\pi} \int_{-\pi}^{\pi} (x + \pi)\, dx = 2\pi$

$a_n = \dfrac{1}{\pi} \displaystyle\int_{-\pi}^{\pi} f(x) \cos nx\, dx = \dfrac{1}{\pi} \int_{-\pi}^{\pi} (x + \pi) \cos nx\, dx = 0$

$b_n = \dfrac{1}{\pi} \displaystyle\int_{-\pi}^{\pi} f(x) \sin nx\, dx = \dfrac{2}{n}(-1)^{n+1}$

$f(x) = \pi + \displaystyle\sum_{n=1}^{\infty} \dfrac{2}{n}(-1)^{n+1} \sin nx$

8. $a_0 = \dfrac{1}{\pi} \displaystyle\int_{-\pi}^{\pi} f(x)\, dx = \dfrac{1}{\pi} \int_{-\pi}^{\pi} (3 - 2x)\, dx = 6$

$a_n = \dfrac{1}{\pi} \displaystyle\int_{-\pi}^{\pi} f(x) \cos nx\, dx = \dfrac{1}{\pi} \int_{-\pi}^{\pi} (3 - 2x) \cos nx\, dx = 0$

$b_n = \dfrac{1}{\pi} \displaystyle\int_{-\pi}^{\pi} (3 - 2x) \sin nx\, dx = \dfrac{4}{n}(-1)^n$

$$f(x) = 3 + 4 \sum_{n=1}^{\infty} \frac{(-1)^n}{n} \sin nx$$

9. $a_0 = \dfrac{1}{\pi} \displaystyle\int_{-\pi}^{\pi} f(x)\, dx = \dfrac{1}{\pi} \int_0^{\pi} \sin x\, dx = \dfrac{2}{\pi}$

$a_n = \dfrac{1}{\pi} \displaystyle\int_{-\pi}^{\pi} f(x) \cos nx\, dx = \dfrac{1}{\pi} \int_0^{\pi} \sin x \cos nx\, dx = \dfrac{1}{2\pi} \int_0^{\pi} \left[\sin(n+1)x + \sin(1-n)x \right] dx$

$\quad = \dfrac{1 + (-1)^n}{\pi(1 - n^2)} \quad \text{for } n = 2, 3, 4, \ldots$

$a_1 = \dfrac{1}{2\pi} \displaystyle\int_0^{\pi} \sin 2x\, dx = 0$

$b_n = \dfrac{1}{\pi} \displaystyle\int_{-\pi}^{\pi} f(x) \sin nx\, dx = \dfrac{1}{\pi} \int_0^{\pi} \sin x \sin nx\, dx$

$\quad = \dfrac{1}{2\pi} \displaystyle\int_0^{\pi} \left[\cos(1-n)x - \cos(1+n)x \right] dx = 0 \quad \text{for } n = 2, 3, 4, \ldots$

$b_1 = \dfrac{1}{2\pi} \displaystyle\int_0^{\pi} (1 - \cos 2x)\, dx = \dfrac{1}{2}$

$f(x) = \dfrac{1}{\pi} + \dfrac{1}{2} \sin x + \displaystyle\sum_{n=2}^{\infty} \dfrac{1 + (-1)^n}{\pi(1 - n^2)} \cos nx$

10. $a_0 = \dfrac{2}{\pi} \displaystyle\int_{-\pi/2}^{\pi/2} f(x)\, dx = \dfrac{2}{\pi} \int_0^{\pi/2} \cos x\, dx = \dfrac{2}{\pi}$

$a_n = \dfrac{2}{\pi} \displaystyle\int_{-\pi/2}^{\pi/2} f(x) \cos 2nx\, dx = \dfrac{2}{\pi} \int_0^{\pi/2} \cos x \cos 2nx\, dx = \dfrac{1}{\pi} \int_0^{\pi/2} \left[\cos(2n-1)x + \cos(2n+1)x \right] dx$

$\quad = \dfrac{2(-1)^{n+1}}{\pi(4n^2 - 1)}$

$b_n = \dfrac{2}{\pi} \displaystyle\int_{-\pi/2}^{\pi/2} f(x) \sin 2nx\, dx = \dfrac{2}{\pi} \int_0^{\pi/2} \cos x \sin 2nx\, dx = \dfrac{1}{\pi} \int_0^{\pi/2} \left[\sin(2n-1)x + \sin(2n+1)x \right] dx$

$\quad = \dfrac{4n}{\pi(4n^2 - 1)}$

$f(x) = \dfrac{1}{\pi} + \displaystyle\sum_{n=1}^{\infty} \left[\dfrac{2(-1)^{n+1}}{\pi(4n^2 - 1)} \cos 2nx + \dfrac{4n}{\pi(4n^2 - 1)} \sin 2nx \right]$

11. $a_0 = \dfrac{1}{2} \displaystyle\int_{-2}^{2} f(x)\, dx = \dfrac{1}{2} \left(\int_{-1}^0 -2\, dx + \int_0^1 1\, dx \right) = -\dfrac{1}{2}$

$a_n = \dfrac{1}{2} \displaystyle\int_{-2}^{2} f(x) \cos \dfrac{n\pi}{2} x\, dx = \dfrac{1}{2} \left(\int_{-1}^0 \left[-2 \cos \dfrac{n\pi}{2} x \right] dx + \int_0^1 \cos \dfrac{n\pi}{2} x\, dx \right) = -\dfrac{1}{n\pi} \sin \dfrac{n\pi}{2}$

433

$$b_n = \frac{1}{2}\int_{-2}^{2} f(x)\sin\frac{n\pi}{2}x\,dx = \frac{1}{2}\left(\int_{-1}^{0}\left[-2\sin\frac{n\pi}{2}x\right]dx + \int_{0}^{1}\sin\frac{n\pi}{2}x\,dx\right) = \frac{3}{n\pi}\left(1 - \cos\frac{n\pi}{2}\right)$$

$$f(x) = -\frac{1}{4} + \sum_{n=1}^{\infty}\left[-\frac{1}{n\pi}\sin\frac{n\pi}{2}\cos\frac{n\pi}{2}x + \frac{3}{n\pi}\left(1 - \cos\frac{n\pi}{2}\right)\sin\frac{n\pi}{2}x\right]$$

12. $a_0 = \dfrac{1}{2}\displaystyle\int_{-2}^{2} f(x)\,dx = \dfrac{1}{2}\left(\displaystyle\int_{0}^{1} x\,dx + \displaystyle\int_{1}^{2} 1\,dx\right) = \dfrac{3}{4}$

$$a_n = \frac{1}{2}\int_{-2}^{2} f(x)\cos\frac{n\pi}{2}x\,dx = \frac{1}{2}\left(\int_{0}^{1} x\cos\frac{n\pi}{2}x\,dx + \int_{1}^{2}\cos\frac{n\pi}{2}x\,dx\right) = \frac{2}{n^2\pi^2}\left(\cos\frac{n\pi}{2} - 1\right)$$

$$b_n = \frac{1}{2}\int_{-2}^{2} f(x)\sin\frac{n\pi}{2}x\,dx = \frac{1}{2}\left(\int_{0}^{1} x\sin\frac{n\pi}{2}x\,dx + \int_{1}^{2}\sin\frac{n\pi}{2}x\,dx\right)$$

$$= \frac{2}{n^2\pi^2}\left(\sin\frac{n\pi}{2} + \frac{n\pi}{2}(-1)^{n+1}\right)$$

$$f(x) = \frac{3}{8} + \sum_{n=1}^{\infty}\left[\frac{2}{n^2\pi^2}\left(\cos\frac{n\pi}{2} - 1\right)\cos\frac{n\pi}{2}x + \frac{2}{n^2\pi^2}\left(\sin\frac{n\pi}{2} + \frac{n\pi}{2}(-1)^{n+1}\right)\sin\frac{n\pi}{2}x\right]$$

13. $a_0 = \dfrac{1}{5}\displaystyle\int_{-5}^{5} f(x)\,dx = \dfrac{1}{5}\left(\displaystyle\int_{-5}^{0} 1\,dx + \displaystyle\int_{0}^{5}(1+x)\,dx\right) = \dfrac{9}{2}$

$$a_n = \frac{1}{5}\int_{-5}^{5} f(x)\cos\frac{n\pi}{5}x\,dx = \frac{1}{5}\left(\int_{-5}^{0}\cos\frac{n\pi}{5}x\,dx + \int_{0}^{5}(1+x)\cos\frac{n\pi}{5}x\,dx\right) = \frac{5}{n^2\pi^2}[(-1)^n - 1]$$

$$b_n = \frac{1}{5}\int_{-5}^{5} f(x)\sin\frac{n\pi}{5}x\,dx = \frac{1}{5}\left(\int_{-5}^{0}\sin\frac{n\pi}{5}x\,dx + \int_{0}^{5}(1+x)\cos\frac{n\pi}{5}x\,dx\right) = \frac{5}{n\pi}(-1)^{n+1}$$

$$f(x) = \frac{9}{4} + \sum_{n=1}^{\infty}\left[\frac{5}{n^2\pi^2}[(-1)^n - 1]\cos\frac{n\pi}{5}x + \frac{5}{n\pi}(-1)^{n+1}\sin\frac{n\pi}{5}x\right]$$

14. $a_0 = \dfrac{1}{2}\displaystyle\int_{-2}^{2} f(x)\,dx = \dfrac{1}{2}\left(\displaystyle\int_{-2}^{0}(2+x)\,dx + \displaystyle\int_{0}^{2} 2\,dx\right) = 3$

$$a_n = \frac{1}{2}\int_{-2}^{2} f(x)\cos\frac{n\pi}{2}x\,dx = \frac{1}{2}\left(\int_{-2}^{0}(2+x)\cos\frac{n\pi}{2}x\,dx + \int_{0}^{2} 2\cos\frac{n\pi}{2}x\,dx\right) = \frac{2}{n^2\pi^2}[1 - (-1)^n]$$

$$b_n = \frac{1}{2}\int_{-2}^{2} f(x)\sin\frac{n\pi}{2}x\,dx = \frac{1}{2}\left(\int_{-2}^{0}(2+x)\sin\frac{n\pi}{2}x\,dx + \int_{0}^{2} 2\sin\frac{n\pi}{2}x\,dx\right) = \frac{2}{n\pi}(-1)^{n+1}$$

$$f(x) = \frac{3}{2} + \sum_{n=1}^{\infty}\left[\frac{2}{n^2\pi^2}[1 - (-1)^n]\cos\frac{n\pi}{2}x + \frac{2}{n\pi}(-1)^{n+1}\sin\frac{n\pi}{2}x\right]$$

15. $a_0 = \dfrac{1}{\pi}\displaystyle\int_{-\pi}^{\pi} f(x)\,dx = \dfrac{1}{\pi}\displaystyle\int_{-\pi}^{\pi} e^x\,dx = \dfrac{1}{\pi}(e^\pi - e^{-\pi})$

$$a_n = \frac{1}{\pi}\int_{-\pi}^{\pi} f(x)\cos nx\,dx = \frac{(-1)^n(e^\pi - e^{-\pi})}{\pi(1+n^2)}$$

$$b_n = \frac{1}{\pi}\int_{-\pi}^{\pi} f(x)\sin nx\,dx = \frac{1}{\pi}\int_{-\pi}^{\pi} e^x \sin nx\,dx = \frac{(-1)^n n(e^{-\pi} - e^\pi)}{\pi(1+n^2)}$$

$$f(x) = \frac{e^\pi - e^{-\pi}}{2\pi} + \sum_{n=1}^{\infty}\left[\frac{(-1)^n(e^\pi - e^{-\pi})}{\pi(1+n^2)}\cos nx + \frac{(-1)^n n(e^{-\pi} - e^\pi)}{\pi(1+n^2)}\sin nx\right]$$

16. $a_0 = \dfrac{1}{\pi}\displaystyle\int_{-\pi}^{\pi} f(x)\,dx = \dfrac{1}{\pi}\displaystyle\int_{0}^{\pi}(e^x - 1)\,dx = \dfrac{1}{\pi}(e^\pi - \pi - 1)$

$$a_n = \frac{1}{\pi}\int_{-\pi}^{\pi} f(x)\cos nx\,dx = \frac{1}{\pi}\int_{0}^{\pi}(e^x - 1)\cos nx\,dx = \frac{[e^\pi(-1)^n - 1]}{\pi(1+n^2)}$$

$$b_n = \frac{1}{\pi}\int_{-\pi}^{\pi} f(x)\sin nx\,dx = \frac{1}{\pi}\int_{0}^{\pi}(e^x - 1)\sin nx\,dx = \frac{1}{\pi}\left(\frac{ne^\pi(-1)^{n+1}}{1+n^2} + \frac{n}{1+n^2} + \frac{(-1)^n}{n} - \frac{1}{n}\right)$$

$$f(x) = \frac{e^\pi - \pi - 1}{2\pi} + \sum_{n=1}^{\infty}\left[\frac{e^\pi(-1)^n - 1}{\pi(1+n^2)}\cos nx + \left(\frac{n}{1+n^2}[e^\pi(-1)^{n+1} + 1] + \frac{(-1)^n - 1}{n}\right)\sin nx\right]$$

17. The function in Problem 5 is discontinuous at $x = \pi$, so the corresponding Fourier series converges to $\pi^2/2$ at $x = \pi$. That is,

$$\frac{\pi^2}{2} = \frac{\pi^2}{6} + \sum_{n=1}^{\infty}\left[\frac{2(-1)^n}{n^2}\cos n\pi + \left(\frac{\pi}{n}(-1)^{n+1} + \frac{2[(-1)^n - 1]}{n^3\pi}\right)\sin n\pi\right]$$

$$= \frac{\pi^2}{6} + \sum_{n=1}^{\infty}\frac{2(-1)^n}{n^2}(-1)^n = \frac{\pi^2}{6} + \sum_{n=1}^{\infty}\frac{2}{n^2} = \frac{\pi^2}{6} + 2\left(1 + \frac{1}{2^2} + \frac{1}{3^2} + \cdots\right)$$

and

$$\frac{\pi^2}{6} = \frac{1}{2}\left(\frac{\pi^2}{2} - \frac{\pi^2}{6}\right) = 1 + \frac{1}{2^2} + \frac{1}{3^2} + \cdots.$$

At $x = 0$ the series converges to 0 and

$$0 = \frac{\pi^2}{6} + \sum_{n=1}^{\infty}\frac{2(-1)^n}{n^2} = \frac{\pi^2}{6} + 2\left(-1 + \frac{1}{2^2} - \frac{1}{3^2} + \frac{1}{4^2} - \cdots\right)$$

so

$$\frac{\pi^2}{12} = 1 - \frac{1}{2^2} + \frac{1}{3^2} - \frac{1}{4^2} + \cdots.$$

18. From Problem 17

$$\frac{\pi^2}{8} = \frac{1}{2}\left(\frac{\pi^2}{6} + \frac{\pi^2}{12}\right) = \frac{1}{2}\left(2 + \frac{2}{3^2} + \frac{2}{5^2} + \cdots\right) = 1 + \frac{1}{3^2} + \frac{1}{5^2} + \cdots.$$

19. The function in Problem 7 is continuous at $x = \pi/2$ so

$$\frac{3\pi}{2} = f\left(\frac{\pi}{2}\right) = \pi + \sum_{n=1}^{\infty} \frac{2}{n}(-1)^{n+1} \sin \frac{n\pi}{2}$$

$$= \pi + 2\left(1 - \frac{1}{3} + \frac{1}{5} - \frac{1}{7} + \cdots\right)$$

and

$$\frac{\pi}{4} = 1 - \frac{1}{3} + \frac{1}{5} - \frac{1}{7} + \cdots.$$

20. The function in Problem 9 is continuous at $x = \pi/2$ so

$$1 = f\left(\frac{\pi}{2}\right) = \frac{1}{\pi} + \frac{1}{2} + \sum_{n=2}^{\infty} \frac{1 + (-1)^n}{\pi(1 - n^2)} \cos \frac{n\pi}{2}$$

$$= \frac{1}{\pi} + \frac{1}{2} + \frac{2}{3\pi} - \frac{2}{3 \cdot 5\pi} + \frac{2}{5 \cdot 7\pi} - \cdots$$

and

$$\pi = 1 + \frac{\pi}{2} + \frac{2}{3} - \frac{2}{3 \cdot 5} + \frac{2}{5 \cdot 7} - \cdots$$

or

$$\frac{\pi}{4} = \frac{1}{2} + \frac{1}{1 \cdot 3} - \frac{1}{3 \cdot 5} + \frac{1}{5 \cdot 7} - \cdots.$$

21. (a) Letting $c_0 = a_0/2$, $c_n = (a_n - ib_n)$, and $c_{-n} = (a_n + ib_n)/2$ we have

$$f(x) = \frac{a_0}{2} + \sum_{n=1}^{\infty} \left(a_n \cos \frac{n\pi}{p}x + b_n \sin \frac{n\pi}{p}x\right)$$

$$= c_0 + \sum_{n=1}^{\infty} \left(a_n \frac{e^{in\pi x/p} + e^{-in\pi x/p}}{2} + b_n \frac{e^{in\pi x/p} - e^{-in\pi x/p}}{2i}\right)$$

$$= c_0 + \sum_{n=1}^{\infty} \left(a_n \frac{e^{in\pi x/p} + e^{-in\pi x/p}}{2} - b_n \frac{ie^{in\pi x/p} - ie^{-in\pi x/p}}{2}\right)$$

$$= c_0 + \sum_{n=1}^{\infty} \left(\frac{a_n - ib_n}{2}e^{in\pi x/p} + \frac{a_n + ib_n}{2}e^{-in\pi x/p}\right)$$

$$= c_0 + \sum_{n=1}^{\infty} \left(c_n e^{in\pi x/p} + c_{-n}e^{i(-n)\pi x/p}\right) = \sum_{n=-\infty}^{\infty} c_n e^{in\pi x/p}.$$

436

(b) Multiplying both sides of the expression in (a) by $e^{-im\pi x/p}$ and integrating we obtain

$$\int_{-p}^{p} f(x)e^{-im\pi x/p}dx = \int_{-p}^{p}\left(\sum_{n=-\infty}^{\infty} c_n e^{in\pi x/p}e^{-im\pi x/p}\right)dx$$

$$= \sum_{n=-\infty}^{\infty} c_n \int_{-p}^{p} e^{i(n-m)\pi x/p}dx$$

$$= \sum_{n\neq m} c_n \int_{-p}^{p} e^{i(n-m)\pi x/p}dx + c_m \int_{-p}^{p} e^{i(m-n)\pi x/p}dx$$

$$= \sum_{n\neq m} c_n \int_{-p}^{p} e^{i(n-m)\pi x/p}dx + c_m \int_{-p}^{p} dx$$

$$= \sum_{n\neq m} c_n \int_{-p}^{p} e^{i(n-m)\pi x/p}dx + 2pc_m.$$

Recalling that

$$e^{iy} = \cos y + i\sin y \quad \text{and} \quad e^{-iy} = \cos y - i\sin y$$

we have for $n - m$ an integer and $n \neq m$

$$\int_{-p}^{p} e^{i(n-m)\pi x/p}dx = \frac{p}{i(n-m)\pi}e^{i(n-m)\pi x/p}\bigg|_{-p}^{p}$$

$$= \frac{p}{i(n-m)\pi}\left(e^{i(n-m)\pi} - e^{-i(n-m)\pi}\right)$$

$$= \frac{p}{i(n-m)\pi}[\cos(n-m)\pi + i\sin(n-m)\pi - \cos(n-m)\pi + i\sin(n-m)\pi]$$

$$= 0.$$

Thus

$$\int_{-p}^{p} f(x)e^{-im\pi x/p}dx = 2pc_m$$

and

$$c_m = \frac{1}{2p}\int_{-p}^{p} f(x)e^{-im\pi x/p}dx.$$

22. Identifying $f(x) = e^{-x}$ and $p = \pi$, we have

$$c_n = \frac{1}{2\pi}\int_{-\pi}^{\pi} e^{-x}e^{-inx}dx = \frac{1}{2\pi}\int_{-\pi}^{\pi} e^{-(in+1)x}dx$$

$$= -\frac{1}{2(in+1)\pi}e^{-(in+1)x}\bigg|_{-\pi}^{\pi}$$

$$= -\frac{1}{2(in+1)\pi}\left[e^{-(in+1)\pi} - e^{(in+1)\pi}\right]$$

$$= \frac{e^{(in+1)\pi} - e^{-(in+1)\pi}}{2(in+1)\pi}$$

$$= \frac{e^{\pi}(\cos n\pi + i\sin n\pi) - e^{-\pi}(\cos n\pi - i\sin n\pi)}{2(in+1)\pi}$$

$$= \frac{(e^{\pi} - e^{-\pi}\cos n\pi}{2(in+1)\pi} = \frac{(e\pi - e^{-\pi})(-1)^n}{2(in+1)\pi}.$$

Thus

$$f(x) = \sum_{n=-\infty}^{\infty}(-1)^n \frac{e^p i - e^{-\pi}}{2(in+1)\pi}e^{inx}.$$

Exercises 11.3

1. Since

$$f(-x) = \sin(-3x) = -\sin 3x = -f(x),$$

$f(x)$ is an odd function.

2. Since

$$f(-x) = -x\,\cos(-x) = -x\,\cos x = -f(x),$$

$f(x)$ is an odd function.

3. Since

$$f(-x) = (-x)^2 - x = x^2 - x,$$

$f(x)$ is neither even nor odd.

4. Since

$$f(-x) = (-x)^3 + 4x = -(x^3 - 4x) = -f(x),$$

$f(x)$ is an odd function.

5. Since

$$f(-x) = e^{|-x|} = e^{|x|} = f(x),$$

$f(x)$ is an even function.

6. Since

$$f(-x) = \left|(-x)^5\right| = \left|x^5\right| = f(x),$$

$f(x)$ is an even function.

7. For $0 < x < 1$

$$f(-x) = (-x)^2 = x^2 = -f(x),$$

$f(x)$ is an odd function.

8. For $0 \leq x < 2$

$$f(-x) = -x + 5 = f(x),$$

$f(x)$ is an even function.

9. Since $f(x)$ is not defined for $x < 0$, it is neither even nor odd.

10. Since

$$f(-x) = 2|-x| - 1 = 2|x| - 1 = f(x),$$

$f(x)$ is an even function.

11. Since $f(x)$ is an odd function, we expand in a sine series:

$$b_n = \frac{2}{\pi} \int_0^\pi 1 \cdot \sin nx \, dx = \frac{2}{n\pi}[1 - (-1)^n].$$

Thus

$$f(x) = \sum_{n=1}^\infty \frac{2}{n\pi}[1 - (-1)^n] \sin nx.$$

12. Since $f(x)$ is an even function, we expand in a cosine series:

$$a_0 = \cdot \int_1^2 1 \, dx = 1$$

$$a_n = \int_1^2 \cos \frac{n\pi}{2} x \, dx = -\frac{2}{n\pi} \sin \frac{n\pi}{2}.$$

Thus

$$f(x) = \frac{1}{2} + \sum_{n=1}^\infty \frac{-2}{n\pi} \sin \frac{n\pi}{2} \cos \frac{n\pi}{2} x.$$

13. Since $f(x)$ is an even function, we expand in a cosine series:

$$a_0 = \frac{2}{\pi} \int_0^\pi x \, dx = \pi$$

$$a_n = \frac{2}{\pi} \int_0^\pi x \cos nx \, dx = \frac{2}{n^2\pi}[(-1)^n - 1].$$

Thus

$$f(x) = \frac{\pi}{2} + \sum_{n=1}^\infty \frac{2}{n^2\pi}[(-1)^n - 1] \cos nx.$$

14. Since $f(x)$ is an odd function, we expand in a sine series:

$$b_n = \frac{2}{\pi} \int_0^\pi x \sin nx \, dx = \frac{2}{n}(-1)^{n+1}.$$

439

Thus

$$f(x) = \sum_{n=1}^{\infty} \frac{2}{n}(-1)^{n+1} \sin nx.$$

15. Since $f(x)$ is an even function, we expand in a cosine series:

$$a_0 = 2\int_0^1 x^2\, dx = \frac{2}{3}$$

$$a_n = 2\int_0^1 x^2 \cos n\pi x\, dx = 2\left(\frac{x^2}{n\pi}\sin n\pi x \,\Big|_0^1 - \frac{2}{n\pi}\int_0^1 x \sin n\pi x\, dx \right) = \frac{4}{n^2\pi^2}(-1)^n.$$

Thus

$$f(x) = \frac{1}{3} + \sum_{n=1}^{\infty} \frac{4}{n^2\pi^2}(-1)^n \cos n\pi x.$$

16. Since $f(x)$ is an odd function, we expand in a sine series:

$$b_n = 2\int_0^1 x^2 \sin n\pi x\, dx = 2\left(-\frac{x^2}{n\pi}\cos n\pi x \,\Big|_0^1 + \frac{2}{n\pi}\int_0^1 x \cos n\pi x\, dx \right)$$

$$= \frac{2(-1)^{n+1}}{n\pi} + \frac{4}{n^3\pi^3}[(-1)^n - 1].$$

Thus

$$f(x) = \sum_{n=1}^{\infty} \left(\frac{2(-1)^{n+1}}{n\pi} + \frac{4}{n^3\pi^3}[(-1)^n - 1] \right) \sin n\pi x.$$

17. Since $f(x)$ is an even function, we expand in a cosine series:

$$a_0 = \frac{2}{\pi}\int_0^\pi (\pi^2 - x^2)\, dx = \frac{4}{3}\pi^2$$

$$a_n = \frac{2}{\pi}\int_0^\pi (\pi^2 - x^2)\cos nx\, dx = \frac{2}{\pi}\left(\frac{\pi^2 - x^2}{n}\sin nx \,\Big|_0^\pi + \frac{2}{n}\int_0^\pi x \sin nx\, dx \right) = \frac{4}{n^2}(-1)^{n+1}.$$

Thus

$$f(x) = \frac{2}{3}\pi^2 + \sum_{n=1}^{\infty} \frac{4}{n^2}(-1)^{n+1}\cos nx\, dx.$$

18. Since $f(x)$ is an odd function, we expand in a sine series:

$$b_n = \frac{2}{\pi}\int_0^\pi x^3 \sin nx\, dx = \frac{2}{\pi}\left(-\frac{x^3}{n}\cos nx \,\Big|_0^\pi + \frac{3}{n}\int_0^\pi x^2 \cos nx\, dx \right)$$

$$= \frac{2\pi^2}{n}(-1)^{n+1} - \frac{12}{n^2\pi}\int_0^\pi x \sin nx\, dx$$

$$= \frac{2\pi^2}{n}(-1)^{n+1} - \frac{12}{n^2\pi}\left(-\frac{x}{n}\cos nx \,\Big|_0^\pi + \frac{1}{n}\int_0^\pi \cos nx\, dx \right) = \frac{2\pi^2}{n}(-1)^{n+1} + \frac{12}{n^3}(-1)^n.$$

Thus

$$f(x) = \sum_{n=1}^{\infty} \left(\frac{2\pi^2}{n}(-1)^{n+1} + \frac{12}{n^3}(-1)^n \right) \sin nx.$$

19. Since $f(x)$ is an odd function, we expand in a sine series:

$$b_n = \frac{2}{\pi} \int_0^{\pi} (x+1) \sin nx \, dx = \frac{2(\pi+1)}{n\pi}(-1)^{n+1} + \frac{2}{n\pi}.$$

Thus

$$f(x) = \sum_{n=1}^{\infty} \left(\frac{2(\pi+1)}{n\pi}(-1)^{n+1} + \frac{2}{n\pi} \right) \sin nx.$$

20. Since $f(x)$ is an odd function, we expand in a sine series:

$$b_n = 2 \int_0^1 (x-1) \sin n\pi x \, dx = 2 \left[\int_0^1 x \sin n\pi x \, dx - \int_0^1 \sin n\pi x \, dx \right]$$

$$= 2 \left[\frac{1}{n^2\pi^2} \sin n\pi x - \frac{x}{n\pi} \cos n\pi x + \frac{1}{n\pi} \cos n\pi x \right]_0^1 = -\frac{2}{n\pi}.$$

Thus

$$f(x) = -\sum_{n=1}^{\infty} \frac{2}{n\pi} \sin n\pi x.$$

21. Since $f(x)$ is an even function, we expand in a cosine series:

$$a_0 = \int_0^1 x \, dx + \int_1^2 1 \, dx = \frac{3}{2}$$

$$a_n = \int_0^1 x \cos \frac{n\pi}{2} x \, dx + \int_1^2 \cos \frac{n\pi}{2} x \, dx = \frac{4}{n^2\pi^2} \left(\cos \frac{n\pi}{2} - 1 \right).$$

Thus

$$f(x) = \frac{3}{4} + \sum_{n=1}^{\infty} \frac{4}{n^2\pi^2} \left(\cos \frac{n\pi}{2} - 1 \right) \cos \frac{n\pi}{2} x.$$

22. Since $f(x)$ is an odd function, we expand in a sine series:

$$b_n = \frac{1}{\pi} \int_0^{\pi} x \sin \frac{n}{2} x \, dx + \int_{\pi}^{2\pi} \pi \sin \frac{n}{2} x \, dx = \frac{4}{n^2\pi} \sin \frac{n\pi}{2} + \frac{2}{n}(-1)^{n+1}.$$

Thus

$$f(x) = \sum_{n=1}^{\infty} \left(\frac{4}{n^2\pi} \sin \frac{n\pi}{2} + \frac{2}{n}(-1)^{n+1} \right) \sin \frac{n}{2} x.$$

441

23. Since $f(x)$ is an even function, we expand in a cosine series:

$$a_0 = \frac{2}{\pi} \int_0^\pi \sin x \, dx = \frac{4}{\pi}$$

$$a_n = \frac{2}{\pi} \int_0^\pi \sin x \, \cos nx \, dx = \frac{1}{\pi} \int_0^\pi [\sin(n+1)x + \sin(1-n)x] \, dx$$

$$= \frac{2}{\pi(1-n^2)}[1 + (-1)^n] \quad \text{for } n = 2, 3, 4, \dots$$

$$a_1 = \frac{1}{\pi} \int_0^\pi \sin 2x \, dx = 0.$$

Thus

$$f(x) = \frac{2}{\pi} + \sum_{n=2}^\infty \frac{2[1 + (-1)^n]}{\pi(1-n^2)} \cos nx.$$

24. Since $f(x)$ is an even function, we expand in a cosine series. [See the solution of Problem 10 in Exercise 11.2 for the computation of the integrals.]

$$a_0 = \frac{2}{\pi/2} \int_0^{\pi/2} \cos x \, dx = \frac{4}{\pi}$$

$$a_n = \frac{2}{\pi/2} \int_0^{\pi/2} \cos x \cos \frac{n\pi}{\pi/2} x \, dx = \frac{4(-1)^{n+1}}{\pi(4n^2-1)}$$

Thus

$$f(x) = \frac{2}{\pi} + \sum_{n=1}^\infty \frac{4(-1)^{n+1}}{\pi(4n^2-1)} \cos 2nx.$$

25. $a_0 = 2 \int_0^{1/2} 1 \, dx = 1$

$$a_n = 2 \int_0^{1/2} 1 \cdot \cos n\pi x \, dx = \frac{2}{n\pi} \sin \frac{n\pi}{2}$$

$$b_n = 2 \int_0^{1/2} 1 \cdot \sin n\pi x \, dx = \frac{2}{n\pi} \left(1 - \cos \frac{n\pi}{2}\right)$$

$$f(x) = \frac{1}{2} + \sum_{n=1}^\infty \frac{2}{n\pi} \sin \frac{n\pi}{2} \cos n\pi x$$

$$f(x) = \sum_{n=1}^\infty \frac{2}{n\pi} \left(1 - \cos \frac{n\pi}{2}\right) \sin n\pi x$$

26. $a_0 = 2 \int_{1/2}^1 1 \, dx = 1$

$$a_n = 2 \int_{1/2}^{1} 1 \cdot \cos n\pi x \, dx = -\frac{2}{n\pi} \sin \frac{n\pi}{2}$$

$$b_n = 2 \int_{1/2}^{1} 1 \cdot \sin n\pi x \, dx = \frac{2}{n\pi} \left(\cos \frac{n\pi}{2} + (-1)^{n+1} \right)$$

$$f(x) = \frac{1}{2} + \sum_{n=1}^{\infty} \left(-\frac{2}{n\pi} \sin \frac{n\pi}{2} \cos n\pi x \right)$$

$$f(x) = \sum_{n=1}^{\infty} \frac{2}{n\pi} \left(\cos \frac{n\pi}{2} + (-1)^{n+1} \right) \sin n\pi x$$

27. $a_0 = \dfrac{4}{\pi} \displaystyle\int_{0}^{\pi/2} \cos x \, dx = \dfrac{4}{\pi}$

$$a_n = \frac{4}{\pi} \int_{0}^{\pi/2} \cos x \cos 2nx \, dx = \frac{2}{\pi} \int_{0}^{\pi/2} [\cos(2n+1)x + \cos(2n-1)x] \, dx = \frac{4(-1)^n}{\pi(1-4n^2)}$$

$$b_n = \frac{4}{\pi} \int_{0}^{\pi/2} \cos x \sin 2nx \, dx = \frac{2}{\pi} \int_{0}^{\pi/2} [\sin(2n+1)x + \sin(2n-1)x] \, dx = \frac{8n}{\pi(4n^2-1)}$$

$$f(x) = \frac{2}{\pi} + \sum_{n=1}^{\infty} \frac{4(-1)^n}{\pi(1-4n^2)} \cos 2nx$$

$$f(x) = \sum_{n=1}^{\infty} \frac{8n}{\pi(4n^2-1)} \sin 2nx$$

28. $a_0 = \dfrac{2}{\pi} \displaystyle\int_{0}^{\pi} \sin x \, dx = \dfrac{4}{\pi}$

$$a_n = \frac{2}{\pi} \int_{0}^{\pi} \sin x \cos nx \, dx = \frac{1}{\pi} \int_{0}^{\pi} [\sin(n+1)x - \sin(n-1)x] \, dx$$

$$= \frac{2[(-1)^n + 1]}{\pi(1-n^2)} \quad \text{for } n = 2, 3, 4, \ldots$$

$$b_n = \frac{2}{\pi} \int_{0}^{\pi} \sin x \sin nx \, dx = \frac{1}{\pi} \int_{0}^{\pi} [\cos(n-1)x - \cos(n+1)x] \, dx = 0 \quad \text{for } n = 2, 3, 4, \ldots$$

$$a_1 = \frac{1}{\pi} \int_{0}^{\pi} \sin 2x \, dx = 0$$

$$b_1 = \frac{2}{\pi} \int_{0}^{\pi} \sin^2 x \, dx = 1$$

$$f(x) = \sin x$$

$$f(x) = \frac{2}{\pi} + \frac{2}{\pi} \sum_{n=2}^{\infty} \frac{(-1)^n + 1}{1 - n^2} \cos nx$$

29. $a_0 = \dfrac{2}{\pi}\left(\displaystyle\int_0^{\pi/2} x\,dx + \int_{\pi/2}^{\pi}(\pi - x)\,dx\right) = \dfrac{\pi}{2}$

$a_n = \dfrac{2}{\pi}\left(\displaystyle\int_0^{\pi/2} x\cos nx\,dx + \int_{\pi/2}^{\pi}(\pi - x)\cos nx\,dx\right) = \dfrac{2}{n^2\pi}\left(2\cos\dfrac{n\pi}{2} + (-1)^{n+1} - 1\right)$

$b_n = \dfrac{2}{\pi}\left(\displaystyle\int_0^{\pi/2} x\sin nx\,dx + \int_{\pi/2}^{\pi}(\pi - x)\sin nx\,dx\right) = \dfrac{4}{n^2\pi}\sin\dfrac{n\pi}{2}$

$f(x) = \dfrac{\pi}{4} + \displaystyle\sum_{n=1}^{\infty}\dfrac{2}{n^2\pi}\left(2\cos\dfrac{n\pi}{2} + (-1)^{n+1} - 1\right)\cos nx$

$f(x) = \displaystyle\sum_{n=1}^{\infty}\dfrac{4}{n^2\pi}\sin\dfrac{n\pi}{2}\sin nx$

30. $a_0 = \dfrac{1}{\pi}\displaystyle\int_{\pi}^{2\pi}(x - \pi)\,dx = \dfrac{\pi}{2}$

$a_n = \dfrac{1}{\pi}\displaystyle\int_{\pi}^{2\pi}(x - \pi)\cos\dfrac{n}{2}x\,dx = \dfrac{4}{n^2\pi}\left[(-1)^n - \cos\dfrac{n\pi}{2}\right]$

$b_n = \dfrac{1}{\pi}\displaystyle\int_{\pi}^{2\pi}(x - \pi)\sin\dfrac{n}{2}x\,dx = \dfrac{2}{n}(-1)^{n+1} - \dfrac{4}{n^2\pi}\sin\dfrac{n\pi}{2}$

$f(x) = \dfrac{\pi}{4} + \displaystyle\sum_{n=1}^{\infty}\dfrac{4}{n^2\pi}\left[(-1)^n - \cos\dfrac{n\pi}{2}\right]\cos\dfrac{n}{2}x$

$f(x) = \displaystyle\sum_{n=1}^{\infty}\left(\dfrac{2}{n}(-1)^{n+1} - \dfrac{4}{n^2\pi}\sin\dfrac{n\pi}{2}\right)\sin\dfrac{n}{2}x$

31. $a_0 = \displaystyle\int_0^1 x\,dx + \int_1^2 1\,dx = \dfrac{3}{2}$

$a_n = \displaystyle\int_0^1 x\cos\dfrac{n\pi}{2}x\,dx = \dfrac{4}{n^2\pi^2}\left(\cos\dfrac{n\pi}{2} - 1\right)$

$b_n = \displaystyle\int_0^1 x\sin\dfrac{n\pi}{2}x\,dx + \int_1^2 1\cdot\sin\dfrac{n\pi}{2}x\,dx = \dfrac{4}{n^2\pi^2}\sin\dfrac{n\pi}{2} + \dfrac{2}{n\pi}(-1)^{n+1}$

$f(x) = \dfrac{3}{4} + \displaystyle\sum_{n=1}^{\infty}\dfrac{4}{n^2\pi^2}\left(\cos\dfrac{n\pi}{2} - 1\right)\cos\dfrac{n\pi}{2}x$

$f(x) = \displaystyle\sum_{n=1}^{\infty}\left(\dfrac{4}{n^2\pi^2}\sin\dfrac{n\pi}{2} + \dfrac{2}{n\pi}(-1)^{n+1}\right)\sin\dfrac{n\pi}{2}x$

32. $a_0 = \displaystyle\int_0^1 1\,dx + \int_1^2(2 - x)\,dx = \dfrac{3}{2}$

$a_n = \displaystyle\int_0^1 1\cdot\cos\dfrac{n\pi}{2}x\,dx + \int_1^2(2 - x)\cos\dfrac{n\pi}{2}x\,dx = \dfrac{4}{n^2\pi^2}\left(\cos\dfrac{n\pi}{2} + (-1)^{n+1}\right)$

$$b_n = \int_0^1 1 \cdot \sin \frac{n\pi}{2} x \, dx + \int_1^2 (2-x) \sin \frac{n\pi}{2} x \, dx = \frac{2}{n\pi} + \frac{4}{n^2\pi^2} \sin \frac{n\pi}{2}$$

$$f(x) = \frac{3}{4} + \sum_{n=1}^{\infty} \frac{4}{n^2\pi^2} \left(\cos \frac{n\pi}{2} + (-1)^{n+1} \right) \cos \frac{n\pi}{2} x$$

$$f(x) = \sum_{n=1}^{\infty} \left(\frac{2}{n\pi} + \frac{4}{n^2\pi^2} \sin \frac{n\pi}{2} \right) \sin \frac{n\pi}{2} x$$

33. $a_0 = 2 \int_0^1 (x^2 + x) \, dx = \frac{5}{3}$

$$a_n = 2 \int_0^1 (x^2+x) \cos n\pi x \, dx = \frac{2(x^2+x)}{n\pi} \sin n\pi x \bigg|_0^1 - \frac{2}{n\pi} \int_0^1 (2x+1) \sin n\pi x \, dx = \frac{2}{n^2\pi^2}[3(-1)^n - 1]$$

$$b_n = 2 \int_0^1 (x^2 + x) \sin n\pi x \, dx = -\frac{2(x^2+x)}{n\pi} \cos n\pi x \bigg|_0^1 + \frac{2}{n\pi} \int_0^1 (2x+1) \cos n\pi x \, dx$$

$$= \frac{4}{n\pi}(-1)^{n+1} + \frac{4}{n^3\pi^3}[(-1)^n - 1]$$

$$f(x) = \frac{5}{6} + \sum_{n=1}^{\infty} \frac{2}{n^2\pi^2}[3(-1)^n - 1] \cos n\pi x$$

$$f(x) = \sum_{n=1}^{\infty} \left(\frac{4}{n\pi}(-1)^{n+1} + \frac{4}{n^3\pi^3}[(-1)^n - 1] \right) \sin n\pi x$$

34. $a_0 = \int_0^2 (2x - x^2) \, dx = \frac{4}{3}$

$$a_n = \int_0^2 (2x - x^2) \cos \frac{n\pi}{2} x \, dx = \frac{8}{n^2\pi^2}[(-1)^{n+1} - 1]$$

$$b_n = \int_0^2 (2x - x^2) \sin \frac{n\pi}{2} x \, dx = \frac{16}{n^3\pi^3}[1 - (-1)^n]$$

$$f(x) = \frac{2}{3} + \sum_{n=1}^{\infty} \frac{8}{n^2\pi^2}[(-1)^{n+1} - 1] \cos \frac{n\pi}{2} x$$

$$f(x) = \sum_{n=1}^{\infty} \frac{16}{n^3\pi^3}[1 - (-1)^n] \sin \frac{n\pi}{2} x$$

35. $a_0 = \frac{1}{\pi} \int_0^{2\pi} x^2 \, dx = \frac{8}{3}\pi^2$

$$a_n = \frac{1}{\pi} \int_0^{2\pi} x^2 \cos nx \, dx = \frac{4}{n^2}$$

$$b_n = \frac{1}{\pi} \int_0^{2\pi} x^2 \sin nx \, dx = -\frac{4\pi}{n}$$

$$f(x) = \frac{4}{3}\pi^2 + \sum_{n=1}^{\infty} \left(\frac{4}{n^2} \cos nx - \frac{4\pi}{n} \sin nx \right)$$

36. $a_0 = \frac{2}{\pi} \int_0^{\pi} x \, dx = \pi$

$a_n = \frac{2}{\pi} \int_0^{\pi} x \cos 2nx \, dx = 0$

$b_n = \frac{2}{\pi} \int_0^{\pi} x \sin 2nx \, dx = -\frac{1}{n}$

$f(x) = \frac{\pi}{2} + \sum_{n=1}^{\infty} \left(-\frac{1}{n} \sin 2nx \right)$

37. $a_0 = 2 \int_0^1 (x+1) \, dx = 3$

$a_n = 2 \int_0^1 (x+1) \cos 2n\pi x \, dx = 0$

$b_n = 2 \int_0^1 (x+1) \sin 2n\pi x \, dx = -\frac{1}{n\pi}$

$f(x) = \frac{3}{2} - \sum_{n=1}^{\infty} \frac{1}{n\pi} \sin 2n\pi x$

38. $a_0 = 2 \int_0^2 (2-x) \, dx = 2$

$a_n = 2 \int_0^2 (2-x) \cos n\pi x \, dx = 0$

$b_n = 2 \int_0^2 (2-x) \sin n\pi x \, dx = \frac{2}{n\pi}$

$f(x) = 1 + \sum_{n=1}^{\infty} \frac{2}{n\pi} \sin n\pi x$

39. We have

$$b_n = \frac{2}{\pi} \int_0^{\pi} 5 \sin nt \, dt = \frac{10}{n\pi} [1 - (-1)^n]$$

so that

$$f(t) = \sum_{n=1}^{\infty} \frac{10[1 - (-1)^n]}{n\pi} \sin nt.$$

Substituting the assumption $x_p(t) = \sum_{n=1}^{\infty} B_n \sin nt$ into the differential equation then gives

$$x_p'' + 10x_p = \sum_{n=1}^{\infty} B_n (10 - n^2) \sin nt = \sum_{n=1}^{\infty} \frac{10[1 - (-1)^n]}{n\pi} \sin nt$$

and so $B_n = \dfrac{10[1-(-1)^n]}{n\pi(10-n^2)}$. Thus

$$x_p(t) = \frac{10}{\pi}\sum_{n=1}^{\infty}\frac{1-(-1)^n}{n(10-n^2)}\sin nt.$$

40. We have

$$b_n = \frac{2}{\pi}\int_0^1 (1-t)\sin n\pi t\, dt = \frac{2}{n\pi}$$

so that

$$f(t) = \sum_{n=1}^{\infty}\frac{2}{n\pi}\sin n\pi t.$$

Substituting the assumption $x_p(t) = \displaystyle\sum_{n=1}^{\infty} B_n \sin n\pi t$ into the differential equation then gives

$$x_p'' + 10x_p = \sum_{n=1}^{\infty} B_n(10 - n^2\pi^2)\sin n\pi t = \sum_{n=1}^{\infty}\frac{2}{n\pi}\sin n\pi t$$

and so $B_n = \dfrac{2}{n\pi(10-n^2\pi^2)}$. Thus

$$x_p(t) = \frac{2}{\pi}\sum_{n=1}^{\infty}\frac{1}{n(10-n^2\pi^2)}\sin n\pi t.$$

41. We have

$$a_0 = \frac{2}{\pi}\int_0^{\pi}(2\pi t - t^2)\, dt = \frac{4}{3}\pi^2$$

$$a_n = \frac{2}{\pi}\int_0^{\pi}(2\pi t - t^2)\cos nt\, dt = -4n^2$$

so that

$$f(t) = \frac{2\pi^2}{3} + \sum_{n=1}^{\infty}\frac{(-4)}{n^2}\cos nt.$$

Substituting the assumption

$$x_p(t) = \frac{A_0}{2} + \sum_{n=1}^{\infty} A_n \cos nt$$

into the differential equation then gives

$$\frac{1}{4}x_p'' + 12x_p = 6A_0 + \sum_{n=1}^{\infty} A_n\left(-\frac{1}{4}n^2 + 12\right)\cos nt = \frac{2\pi^2}{3} + \sum_{n=1}^{\infty}\frac{(-4)}{n^2}\cos nt$$

and $A_0 = \dfrac{\pi^2}{9}$, $A_n = \dfrac{16}{n^2(n^2-48)}$. Thus

$$x_p(t) = \frac{\pi^2}{18} + 16\sum_{n=1}^{\infty}\frac{1}{n^2(n^2-48)}\cos nt.$$

447

42. We have

$$a_0 = \frac{2}{(1/2)} \int_0^{1/2} t \, dt = \frac{1}{2}$$

$$a_n = \frac{2}{(1/2)} \int_0^{1/2} t \cos 2n\pi t \, dt = \frac{1}{n^2\pi^2}[(-1)^n - 1]$$

so that

$$f(t) = \frac{1}{4} + \sum_{n=1}^{\infty} \frac{(-1)^n - 1}{n^2\pi^2} \cos 2n\pi t.$$

Substituting the assumption

$$x_p(t) = \frac{A_0}{2} + \sum_{n=1}^{\infty} A_n \cos 2n\pi t$$

into the differential equation then gives

$$\frac{1}{4} x_p'' + 12 x_p = 6 A_0 + \sum_{n=1}^{\infty} A_n (12 - n^2\pi^2) \cos 2n\pi t = \frac{1}{4} + \sum_{n=1}^{\infty} \frac{(-1)^n - 1}{n^2\pi^2} \cos 2n\pi t$$

and $A_0 = \dfrac{1}{24}$, $A_n = \dfrac{(-1)^n - 1}{n^2\pi^2(12 - n^2\pi^2)}$. Thus

$$x_p(t) = \frac{1}{48} + \frac{1}{\pi^2} \sum_{n=1}^{\infty} \frac{(-1)^n - 1}{n^2(12 - n^2\pi^2)} \cos 2n\pi t.$$

43. We have

$$b_n = \frac{2}{L} \int_0^L \frac{w_0 x}{L} \sin \frac{n\pi}{L} x \, dx = \frac{2w_0}{n\pi} (-1)^{n+1}$$

so that

$$w(x) = \sum_{n=1}^{\infty} \frac{2w_0}{n\pi} (-1)^{n+1} \sin \frac{n\pi}{L} x.$$

If we assume $y(x) = \displaystyle\sum_{n=1}^{\infty} B_n \sin \frac{n\pi}{L} x$ then

$$y^{(4)} = \sum_{n=1}^{\infty} \frac{n^4\pi^4}{L^4} B_n \sin \frac{n\pi}{L} x$$

and so the differential equation $EIy^{(4)} = w(x)$ gives

$$B_n = \frac{2w_0(-1)^{n+1}L^4}{EIn^5\pi^5}.$$

Thus

$$y(x) = \frac{2w_0 L^4}{EI\pi^5} \sum_{n=1}^{\infty} \frac{(-1)^{n+1}}{n^5} \sin \frac{n\pi}{L} x.$$

44. We have

$$b_n = \frac{2}{L} \int_{L/3}^{2L/3} w_0 \sin \frac{n\pi}{L} x \, dx = \frac{2w_0}{n\pi} \left[\cos \frac{n\pi}{3} - \cos \frac{2n\pi}{3} \right]$$

so that

$$w(x) = \sum_{n=1}^{\infty} \frac{2w_0}{n\pi} \left[\cos \frac{n\pi}{3} - \cos \frac{2n\pi}{3} \right] \sin \frac{n\pi}{L} x.$$

If we assume $y(x) = \sum_{n=1}^{\infty} B_n \sin \frac{n\pi}{L} x$ then

$$y^{(4)} = \sum_{n=1}^{\infty} \frac{n^4 \pi^4}{L^4} B_n \sin \frac{n\pi}{L} x$$

and so the differential equation $EIy^{(4)} = w(x)$ gives

$$B_n = 2w_0 L^4 \frac{\cos \frac{n\pi}{3} - \cos \frac{2n\pi}{3}}{EIn^5 \pi^5}.$$

Thus

$$y(x) = \frac{2w_0 L^4}{EI\pi^5} \sum_{n=1}^{\infty} \frac{\cos \frac{n\pi}{3} - \cos \frac{2n\pi}{3}}{n^5} \sin \frac{n\pi}{L} x.$$

45. If f and g are even and $h(x) = f(x)g(x)$ then

$$h(-x) = f(-x)g(-x) = f(x)g(x) = h(x)$$

and h is even.

46. If f is even and g is odd and $h(x) = f(x)g(x)$ then

$$h(-x) = f(-x)g(-x) = f(x)[-g(x)] = -h(x)$$

and h is odd.

47. Let $h(x) = f(x) \pm g(x)$ where f and g are even. Then

$$h(-x) = f(-x) \pm g(-x) = f(x) \pm g(x),$$

and h is an even function.

48. If f is even then

$$\int_{-a}^{a} f(x) \, dx = -\int_{a}^{0} f(-u) \, du + \int_{0}^{a} f(x) \, dx = \int_{0}^{a} f(u) \, du + \int_{0}^{a} f(x) \, dx = 2\int_{0}^{a} f(x) \, dx.$$

49. If f is odd then

$$\int_{-a}^{a} f(x) \, dx = -\int_{-a}^{0} f(-x) \, dx + \int_{0}^{a} f(x) \, dx = \int_{a}^{0} f(u) \, du + \int_{0}^{a} f(x) \, dx$$

$$= -\int_{0}^{a} f(u) \, du + \int_{0}^{a} f(x) \, dx = 0.$$

449

$$F(x) = \frac{1}{2}[f(x) + f(-x)] \quad \text{and} \quad G(x) = \frac{1}{2}[f(x) - f(-x)].$$

Then $F(-x) = F(x)$ so that F is even, $G(-x) = -G(x)$ so that G is odd, and $f(x) = F(x) + G(x)$.

51. Using Problems 13 and 14 we obtain

$$f(x) = \frac{1}{2}|x| + \frac{1}{2}x = \frac{\pi}{4} + \sum_{n=1}^{\infty}\left(\frac{(-1)^n - 1}{\pi n^2}\cos nx + \frac{(-1)^{n+1}}{n}\sin nx\right).$$

52. Identifying $b = c = \pi$ and $f(x, y) = 1$ we have

$$A_{mn} = \frac{4}{\pi^2}\int_0^\pi\int_0^\pi \sin mx \sin ny \, dx \, dy = \frac{4}{\pi^2}\int_0^\pi \sin mx \, dx \int_0^\pi \sin ny \, dy$$

$$= \frac{4}{\pi^2}\left(-\frac{1}{m}\cos mx\right)\Big|_0^\pi\left(-\frac{1}{n}\cos ny\right)\Big|_0^\pi = \frac{4}{mn\pi^2}(\cos m\pi - 1)(\cos n\pi - 1)$$

$$= \frac{4}{mn\pi^2}[(-1)^m - 1][(-1)^n - 1]$$

and

$$f(x, y) = 1 = \frac{4}{\pi^2}\sum_{m=1}^{\infty}\sum_{n=1}^{\infty}\frac{[(-1)^m - 1][(-1)^n - 1]}{mn}\sin mx \sin nx.$$

53. Identifying $b = c = 1$ and $f(x, y) = xy$ we have

$$A_{00} = \int_0^1\int_0^1 xy \, dx \, dy = \int_0^1 \frac{1}{2}x^2 y \Big|_0^1 dy = \frac{1}{2}\int_0^1 y \, dy = \frac{1}{4},$$

$$A_{m0} = 2\int_0^1\int_0^1 xy \cos m\pi x \, dx \, dy = 2\int_0^1 \frac{1}{m^2\pi^2}(\cos m\pi x + m\pi x \sin m\pi x)\Big|_0^1 y \, dy$$

$$= 2\int_0^1 \frac{\cos m\pi - 1}{m^2\pi^2}y \, dy = \frac{\cos m\pi - 1}{m^2\pi^2} = \frac{(-1)^m - 1}{m^2\pi^2},$$

$$A_{0n} = 2\int_0^1\int_0^1 xy \cos n\pi y \, dx \, dy = \frac{(-1)^n - 1}{n^2\pi^2},$$

and

$$A_{mn} = 4\int_0^1\int_0^1 xy \cos m\pi x \cos n\pi y \, dx \, dy = 4\int_0^1 x \cos m\pi x \, dx \int_0^1 y \cos n\pi y \, dy$$

$$= 4\left(\frac{(-1)^m - 1}{m^2\pi^2}\right)\left(\frac{(-1)^n - 1}{n^2\pi^2}\right).$$

Thus

$$f(x, y) = xy = \frac{1}{4} + \frac{1}{\pi^2}\sum_{m=1}^{\infty}\frac{(-1)^m - 1}{m^2}\cos m\pi x + \frac{1}{\pi^2}\sum_{n=1}^{\infty}\frac{(-1)^n - 1}{n^2}\cos n\pi y$$

$$+ \frac{4}{\pi^4}\sum_{m=1}^{\infty}\sum_{n=1}^{\infty}\frac{[(-1)^m - 1][(-1)^n - 1]}{m^2 n^2}\cos m\pi x \cos n\pi y.$$

Exercises 11.4

1. For $\lambda \leq 0$ the only solution of the boundary-value problem is $y = 0$. For $\lambda > 0$ we have

$$y = c_1 \cos \sqrt{\lambda}\, x + c_2 \sin \sqrt{\lambda}\, x.$$

Now

$$y'(x) = -c_1 \sqrt{\lambda} \sin \sqrt{\lambda}\, x + c_2 \sqrt{\lambda} \cos \sqrt{\lambda}\, x$$

and $y'(0) = 0$ implies $c_2 = 0$, so

$$y(1) + y'(1) = c_1(\cos \sqrt{\lambda} - \sqrt{\lambda} \sin \sqrt{\lambda}) = 0 \quad \text{or} \quad \cot \sqrt{\lambda} = \sqrt{\lambda}.$$

The eigenvalues are $\lambda_n = x_n^2$ where x_1, x_2, x_3, ... are the consecutive positive solutions of $\cot \sqrt{\lambda} = \sqrt{\lambda}$. The corresponding eigenfunctions are $\cos \sqrt{\lambda_n}\, x = \cos x_n x$ for $n = 1, 2, 3, \ldots$. Using a CAS we find that the first four eigenvalues are 0.7402, 11.7349, 41.4388, and 90.8082 with corresponding eigenfunctions $\cos 0.8603x$, $\cos 3.4256x$, $\cos 6.4373x$, and $\cos 9.5293x$.

2. For $\lambda < 0$ the only solution of the boundary-value problem is $y = 0$. For $\lambda = 0$ we have $y = c_1 x + c_2$. Now $y' = c_1$ and the boundary conditions both imply $c_1 + c_2 = 0$. Thus, $\lambda = 0$ is an eigenvalue with corresponding eigenfunction $y_0 = x - 1$.

For $\lambda > 0$ we have

$$y = c_1 \cos \sqrt{\lambda}\, x + c_2 \sin \sqrt{\lambda}\, x$$

and

$$y' = -c_1 \sqrt{\lambda} \sin \sqrt{\lambda}\, x + c_2 \sqrt{\lambda} \cos \sqrt{\lambda}\, x.$$

The boundary conditions imply

$$c_1 + c_2 \sqrt{\lambda} = 0$$

$$c_1 \cos \sqrt{\lambda} + c_2 \sin \sqrt{\lambda} = 0$$

which gives

$$-c_2 \sqrt{\lambda} \cos \sqrt{\lambda} + c_2 \sin \sqrt{\lambda} = 0 \quad \text{or} \quad \tan \sqrt{\lambda} = \sqrt{\lambda}.$$

The eigenvalues are $\lambda_n = x_n^2$ where x_1, x_2, x_3, ... are the consecutive positive solutions of $\tan \sqrt{\lambda} = \sqrt{\lambda}$. The corresponding eigenfunctions are $\sqrt{\lambda_n} \cos \sqrt{\lambda_n}\, x - \sin \sqrt{\lambda_n}\, x$ (obtained by taking $c_2 = -1$ in the first equation of the system.) Using a CAS we find that the first four positive eigenvalues are 20.1907, 59.6795, 118.9000, and 197.858 with corresponding eigenfunctions $4.4934 \cos 4.4934x - \sin 4.4934x$, $7.7253 \cos 7.7253x - \sin 7.7253x$, $10.9041 \cos 10.9041x - \sin 10.9041x$, and $14.0662 \cos 14.0662x - \sin 14.0662x$.

3. For $\lambda = 0$ the solution of $y'' = 0$ is $y = c_1 x + c_2$. The condition $y'(0) = 0$ implies $c_1 = 0$, so $\lambda = 0$ is an eigenvalue with corresponding eigenfunction 1. For $\lambda < 0$ we have $y = c_1 \cosh \sqrt{-\lambda}\, x + c_2 \sin \sqrt{-\lambda}\, x$

and $y' = c_1\sqrt{-\lambda}\sinh\sqrt{-\lambda}\,x + c_2\sqrt{-\lambda}\cosh\sqrt{-\lambda}\,x$. The condition $y'(0) = 0$ implies $c_2 = 0$ and so $y = c_1\cosh\sqrt{-\lambda}\,x$. Now the condition $y'(L) = 0$ implies $c_1 = 0$. Thus $y = 0$ and there are no negative eigenvalues. For $\lambda > 0$ we have $y = c_1\cos\sqrt{\lambda}\,x + c_2\sin\sqrt{\lambda}\,x$ and $y' = -c_1\sqrt{\lambda}\sin\sqrt{\lambda}\,x + c_2\sqrt{\lambda}\cos\sqrt{\lambda}\,x$. The condition $y'(0) = 0$ implies $c_2 = 0$ and so $y = c_1\cos\sqrt{\lambda}\,x$. Now the condition $y'(L) = 0$ implies $-c_1\sqrt{\lambda}\sin\sqrt{\lambda}\,L = 0$. For $c_1 \neq 0$ this condition will hold when $\sqrt{\lambda}\,L = n\pi$ or $\lambda = n^2\pi^2/L^2$, where $n = 1, 2, 3, \ldots$. These are the positive eigenvalues with corresponding eigenfunctions $\cos(n\pi/L)x$, $n = 1, 2, 3, \ldots$.

4. For $\lambda < 0$ we have
$$y = c_1\cosh\sqrt{-\lambda}\,x + c_2\sinh\sqrt{-\lambda}\,x$$
$$y' = c_1\sqrt{-\lambda}\sinh\sqrt{-\lambda}\,x + c_2\sqrt{-\lambda}\cosh\sqrt{-\lambda}\,x.$$

Using the fact that $\cosh x$ is an even function and $\sinh x$ is odd we have
$$y(-L) = c_1\cosh(-\sqrt{-\lambda}\,L) + c_2\sinh(-\sqrt{-\lambda}\,L)$$
$$= c_1\cosh\sqrt{-\lambda}\,L - c_2\sinh\sqrt{-\lambda}\,L$$

and
$$y'(-L) = c_1\sqrt{-\lambda}\sinh(-\sqrt{-\lambda}\,L) + c_2\sqrt{-\lambda}\cosh(-\sqrt{-\lambda}\,L)$$
$$= -c_1\sqrt{-\lambda}\sinh\sqrt{-\lambda}\,L + c_2\sqrt{-\lambda}\cosh\sqrt{-\lambda}\,L.$$

The boundary conditions imply
$$c_1\cosh\sqrt{-\lambda}\,L - c_2\sinh\sqrt{-\lambda}\,L = c_1\cosh\sqrt{-\lambda}\,L + c_2\sinh\sqrt{-\lambda}\,L$$

or
$$2c_2\sinh\sqrt{-\lambda}\,L = 0$$

and
$$-c_1\sqrt{-\lambda}\sinh\sqrt{-\lambda}\,L + c_2\sqrt{-\lambda}\cosh\sqrt{-\lambda}\,L = c_1\sqrt{-\lambda}\sinh\sqrt{-\lambda}\,L + c_2\sqrt{-\lambda}\cosh\sqrt{-\lambda}\,L$$

or
$$2c_1\sqrt{-\lambda}\sinh\sqrt{-\lambda}\,L = 0.$$

Since $\sqrt{-\lambda}\,L \neq 0$, $c_1 = c_2 = 0$ and the only solution of the boundary-value problem in this case is $y = 0$.

For $\lambda = 0$ we have
$$y = c_1 x + c_2$$
$$y' = c_1.$$

From $y(-L) = y(L)$ we obtain
$$-c_1 L + c_2 = c_1 L + c_2.$$

Then $c_1 = 0$ and $y = 1$ is an eigenfunction corresponding to the eigenvalue $\lambda = 0$.
For $\lambda > 0$ we have

$$y = c_1 \cos \sqrt{\lambda}\, x + c_2 \sin \sqrt{\lambda}\, x$$

$$y' = -c_1 \sqrt{\lambda} \sin \sqrt{\lambda}\, x + c_2 \sqrt{\lambda} \cos \sqrt{\lambda}\, x.$$

The first boundary condition implies

$$c_1 \cos \sqrt{\lambda}\, L - c_2 \sin \sqrt{\lambda}\, L = c_1 \cos \sqrt{\lambda}\, L + c_2 \sin \sqrt{\lambda}\, L$$

or

$$2c_2 \sin \sqrt{\lambda}\, L = 0.$$

Thus, if $c_1 = 0$ and $c_2 \neq 0$,

$$\sqrt{\lambda}\, L = n\pi \quad \text{or} \quad \lambda = \frac{n^2 \pi^2}{L^2}, \quad n = 1, 2, 3, \dots .$$

The corresponding eigenfunctions are $\sin \dfrac{n\pi}{L} x$, for $n = 1, 2, 3, \dots$. Similarly, the second
boundary condition implies

$$2c_1 \sqrt{\lambda} \sin \sqrt{\lambda}\, L = 0.$$

If $c_2 = 0$ and $c_1 \neq 0$,

$$\sqrt{\lambda}\, L = n\pi \quad \text{or} \quad \lambda = \frac{n^2 \pi^2}{L^2}, \quad n = 1, 2, 3, \dots ,$$

and the corresponding eigenfunctions are $\cos \dfrac{n\pi}{L} x$, for $n = 1, 2, 3, \dots$.

5. The eigenfunctions are $\cos \sqrt{\lambda_n}\, x$ where $\cot \sqrt{\lambda_n} = \sqrt{\lambda_n}$. Thus

$$\| \cos \sqrt{\lambda_n}\, x \|^2 = \int_0^1 \cos^2 \sqrt{\lambda_n}\, x \, dx = \frac{1}{2} \int_0^1 \left(1 + \cos 2\sqrt{\lambda_n}\, x \right) dx$$

$$= \frac{1}{2} \left(x + \frac{1}{2\sqrt{\lambda_n}} \sin 2\sqrt{\lambda_n}\, x \right) \Big|_0^1 = \frac{1}{2} \left(1 + \frac{1}{2\sqrt{\lambda_n}} \sin 2\sqrt{\lambda_n} \right)$$

$$= \frac{1}{2} \left[1 + \frac{1}{2\sqrt{\lambda_n}} \left(2 \sin \sqrt{\lambda_n} \cos \sqrt{\lambda_n} \right) \right]$$

$$= \frac{1}{2} \left[1 + \frac{1}{\sqrt{\lambda_n}} \sin \sqrt{\lambda_n} \cot \sqrt{\lambda_n} \sin \sqrt{\lambda_n} \right]$$

$$= \frac{1}{2} \left[1 + \frac{1}{\sqrt{\lambda_n}} \left(\sin \sqrt{\lambda_n} \right) \sqrt{\lambda_n} \left(\sin \sqrt{\lambda_n} \right) \right] = \frac{1}{2} \left(1 + \sin^2 \sqrt{\lambda_n} \right).$$

6. The eigenfunctions are $\sin \sqrt{\lambda_n}\, x$ where $\tan \sqrt{\lambda_n} = -\lambda_n$. Thus

$$\| \sin \sqrt{\lambda_n}\, x \|^2 = \int_0^1 \sin^2 \sqrt{\lambda_n}\, x\, dx = \frac{1}{2} \int_0^1 \left(1 - \cos 2\sqrt{\lambda_n}\, x \right) dx$$

$$= \frac{1}{2} \left(x - \frac{1}{2\sqrt{\lambda_n}} \sin 2\sqrt{\lambda_n}\, x \right) \Big|_0^1 = \frac{1}{2} \left(1 - \frac{1}{2\sqrt{\lambda_n}} \sin 2\sqrt{\lambda_n} \right)$$

$$= \frac{1}{2} \left[1 - \frac{1}{2\sqrt{\lambda_n}} \left(2\sin \sqrt{\lambda_n} \cos \sqrt{\lambda_n} \right) \right]$$

$$= \frac{1}{2} \left[1 - \frac{1}{\sqrt{\lambda_n}} \tan \sqrt{\lambda_n} \cos \sqrt{\lambda_n} \cos \sqrt{\lambda_n} \right]$$

$$= \frac{1}{2} \left[1 - \frac{1}{\sqrt{\lambda_n}} \left(-\sqrt{\lambda_n} \cos^2 \sqrt{\lambda_n} \right) \right] = \frac{1}{2} \left(1 + \cos^2 \sqrt{\lambda_n} \right).$$

7. (a) If $\lambda \leq 0$ the initial conditions imply $y = 0$. For $\lambda > 0$ the general solution of the Cauchy-Euler differential equation is $y = c_1 \cos(\sqrt{\lambda}\, \ln x) + c_2 \sin(\sqrt{\lambda}\, \ln x)$. The condition $y(1) = 0$ implies $c_1 = 0$, so that $y = c_2 \sin(\sqrt{\lambda}\, \ln x)$. The condition $y(5) = 0$ implies $\sqrt{\lambda}\, \ln 5 = n\pi$, $n = 1$, 2, 3, ... Thus, the eigenvalues are $n^2 \pi^2 / (\ln 5)^2$ for $n = 1$, 2, 3, ..., with corresponding eigenfunctions $\sin[(n\pi / \ln 5) \ln x]$.

(b) The self-adjoint form is

$$\frac{d}{dx}[xy'] + \frac{\lambda}{x} y = 0.$$

(c) An orthogonality relation is

$$\int_1^5 \sin \left(\frac{m\pi}{\ln 5} \ln x \right) \sin \left(\frac{n\pi}{\ln 5} \ln x \right) dx = 0.$$

8. (a) The roots of the auxiliary equation $m^2 + m + \lambda = 0$ are $\frac{1}{2}(-1 \pm \sqrt{1 - 4\lambda})$. When $\lambda = 0$ the general solution of the differential equation is $c_1 + c_2 e^{-x}$. The initial conditions imply $c_1 + c_2 = 0$ and $c_1 + c_2 e^{-2} = 0$. Since the determinant of the coefficients is not 0, the only solution of this homogeneous system is $c_1 = c_2 = 0$, in which case $y = 0$. Similarly, if $0 < \lambda < \frac{1}{4}$, the general solution is

$$y = c_1 e^{\frac{1}{2}(-1+\sqrt{1-4\lambda})x} + c_2 e^{\frac{1}{2}(-1-\sqrt{1-4\lambda})x}.$$

In this case the initial conditions again imply $c_1 = c_2 = 0$, and so $y = 0$. Now, for $\lambda > \frac{1}{4}$, the general solution of the differential equation is

$$y = c_1 e^{-x/2} \cos \sqrt{4\lambda - 1}\, x + c_2 e^{-x/2} \sin \sqrt{4\lambda - 1}\, x.$$

The condition $y(0) = 0$ implies $c_1 = 0$ so $y = c_2 e^{-x/2} \sin \sqrt{4\lambda - 1}\, x$. From

$$y(2) = c_2 e^{-1} \sin 2\sqrt{4\lambda - 1} = 0$$

we see that the eigenvalues are determined by $2\sqrt{4\lambda - 1} = n\pi$ for $n = 1, 2, 3, \ldots$. Thus, the eigenvalues are $n^2\pi^2/4^2 + 1/4$ for $n = 1, 2, 3, \ldots$, with corresponding eigenfunctions $e^{-x/2} \sin \dfrac{n\pi}{2} x$.

(b) The self-adjoint form is

$$\frac{d}{dx}[e^x y'] + \lambda e^x y = 0.$$

(c) An orthogonality relation is

$$\int_0^2 e^x \left(e^{-x/2} \sin \frac{m\pi}{2}x\right)\left(e^{-x/2} \cos \frac{n\pi}{2}x\right) dx = \int_0^2 \left(\sin \frac{m\pi}{2}x\right)\left(\cos \frac{n\pi}{2}x\right) dx = 0.$$

9. (a) An orthogonality relation is

$$\int_0^1 \cos x_m x \cos x_n x = 0$$

where $x_m \neq x_n$ are positive solutions of $\cot x = x$.

(b) Referring to Problem 1 we use a CAS to compute

$$\int_0^1 (\cos 0.8603x)(\cos 3.4256x)\, dx = -1.8771 \times 10^{-6}.$$

10. (a) An orthogonality relation is

$$\int_0^1 (x_m \cos x_m x - \sin x_m x)(x_n \cos x_n x - \sin x_n x)\, dx = 0$$

where $x_m \neq x_n$ are positive solutions of $\tan x = x$.

(b) Referring to Problem 2 we use a CAS to compute

$$\int_0^1 (4.4934 \cos 4.4934x - \sin 4.4934x)(7.7253 \cos 7.7253x - \sin 7.7253x)\, dx = -2.5650 \times 10^{-4}.$$

11. To obtain the self-adjoint form we note that an integrating factor is $(1/x)e^{\int(1-x)dx/x} = e^{-x}$. Thus, the differential equation is

$$xe^{-x}y'' + (1 - x)e^{-x}y' + ne^{-x}y = 0$$

and the self-adjoint form is

$$\frac{d}{dx}\left[xe^{-x}y'\right] + ne^{-x}y = 0.$$

Identifying the weight function $p(x) = e^{-x}$ and noting that since $r(x) = xe^{-x}$, $r(0) = 0$ and $\lim_{x\to\infty} r(x) = 0$, we have the orthogonality relation

$$\int_0^\infty e^{-x} L_n(x) L_m(x)\, dx = 0, \quad m \neq n.$$

455

12. To obtain the self-adjoint form we note that an integrating factor is $e^{\int -2x\,dx} = e^{-x^2}$. Thus, the differential equation is

$$e^{-x^2}y'' - 2xe^{-x^2}y' + 2ne^{-x^2}y = 0$$

and the self-adjoint form is

$$\frac{d}{dx}\left[e^{-x^2}y'\right] + 2ne^{-x^2}y = 0.$$

Identifying the weight function $p(x) = 2e^{-x^2}$ and noting that since $r(x) = e^{-x^2}$, $\lim_{x\to-\infty} r(x) = \lim_{x\to\infty} r(x) = 0$, we have the orthogonality relation

$$\int_{\infty}^{\infty} 2e^{-x^2}H_n(x)H_m(x)\,dx = 0, \ m \neq n.$$

13. (a) The differential equation is

$$(1+x^2)y'' + 2xy' + \frac{\lambda}{1+x^2}y = 0.$$

Letting $x = \tan\theta$ we have $\theta = \tan^{-1}x$ and

$$\frac{dy}{dx} = \frac{dy}{d\theta}\frac{d\theta}{dx} = \frac{1}{1+x^2}\frac{dy}{d\theta}$$

$$\frac{d^2y}{dx^2} = \frac{d}{dx}\left[\frac{1}{1+x^2}\frac{dy}{d\theta}\right] = \frac{1}{1+x^2}\left(\frac{d^2y}{d\theta^2}\frac{d\theta}{dx}\right) - \frac{2x}{(1+x^2)^2}\frac{dy}{d\theta}$$

$$= \frac{1}{(1+x^2)^2}\frac{d^2y}{d\theta^2} - \frac{2x}{(1+x^2)^2}\frac{dy}{d\theta}.$$

The differential equation can then be written in terms of $y(\theta)$ as

$$(1+x^2)\left[\frac{1}{(1+x^2)^2}\frac{d^2y}{d\theta^2} - \frac{2x}{(1+x^2)^2}\frac{dy}{d\theta}\right] + 2x\left[\frac{1}{1+x^2}\frac{dy}{d\theta}\right] + \frac{\lambda}{1+x^2}y$$

$$= \frac{1}{1+x^2}\frac{d^2y}{d\theta^2} + \frac{\lambda}{1+x^2}y = 0$$

or

$$\frac{d^2y}{d\theta^2} + \lambda y = 0.$$

The boundary conditions become $y(0) = y(\pi/4) = 0$. For $\lambda \leq 0$ the only solution of the boundary-value problem is $y = 0$. For $\lambda > 0$ the general solution of the differential equation is $y = c_1 \cos\sqrt{\lambda}\,\theta + c_2 \sin\sqrt{\lambda}\,\theta$. The condition $y(0) = 0$ implies $c_1 = 0$ so $y = c_2 \sin\sqrt{\lambda}\,\theta$. Now the condition $y(\pi/4) = 0$ implies $c_2 \sin\sqrt{\lambda}\,\pi/4 = 0$. For $c_2 \neq 0$ this condition will hold when $\sqrt{\lambda}\,\pi/4 = n\pi$ or $\lambda = 16n^2$, where $n = 1, 2, 3, \ldots$. These are the eigenvalues with corresponding eigenfunctions $\sin 4n\theta = \sin(4n\tan^{-1}x)$, for $n = 1, 2, 3, \ldots$.

(b) An orthogonality relation is

$$\int_0^1 \frac{1}{x^2+1} \sin(4m \tan^{-1} x) \sin(4n \tan^{-1} x)\, dx = 0.$$

14. (a) This is the parametric Bessel equation with $\nu = 1$. The general solution is

$$y = c_1 J_1(\lambda x) c_2 Y_1(\lambda x).$$

Since Y is bounded at 0 we must have $c_2 = 0$, so that $y = c_1 J_1(\lambda x)$. The condition $J_1(3\lambda) = 0$ defines the eigenvalues $\lambda_1, \lambda_2, \lambda_3, \dots$. (When $\lambda = 0$ the differential equation is Cauchy-Euler and the only solution satisfying the boundary condition is $y = 0$, so $\lambda = 0$ is not an eigenvalue.)

(b) From Table 6.1 in the text we see that eigenvalues are determined by $3\lambda_1 = 3.832$, $3\lambda_2 = 7.016$, $3\lambda_3 = 10.173$, and $3\lambda_4 = 13.323$. The first four eigenvalues are thus $\lambda_1 = 1.2773$, $\lambda_2 = 2.3387$, $\lambda_3 = 3.391$, and $\lambda_4 = 4.441$.

15. When $\lambda = 0$ the differential equation is $r(x)y'' + r'(x)y' = 0$. By inspection we see that $y = 1$ is a solution of the boundary-value problem. Thus, $\lambda = 0$ is an eigenvalue.

Exercises 11.5

---

1. Identifying $b = 3$, the first four eigenvalues are

$$\lambda_1 = \frac{3.832}{3} \approx 1.277$$

$$\lambda_2 = \frac{7.016}{3} \approx 2.339$$

$$\lambda_3 = \frac{10.173}{3} = 3.391$$

$$\lambda_4 = \frac{13.323}{3} \approx 4.441.$$

2. We first note from Case III in the text that 0 is an eigenvalue. Now, since $J_0'(2\lambda) = 0$ is equivalent to $J_1(2\lambda) = 0$, the next three eigenvalues are

$$\lambda_2 = \frac{3.832}{2} = 1.916$$

$$\lambda_3 = \frac{7.016}{2} = 3.508$$

$$\lambda_4 = \frac{10.173}{2} = 5.087.$$

3. The boundary condition indicates that we use (15) and (16) of Section 11.5. With $b = 2$ we obtain

$$c_i = \frac{2}{4J_1^2(2\lambda_i)} \int_0^2 x J_0(\lambda_i x)\, dx$$

$$\boxed{t = \lambda_i x \qquad dt = \lambda_i\, dx}$$

$$= \frac{1}{2J_1^2(2\lambda_i)} \cdot \frac{1}{\lambda_i^2} \int_0^{2\lambda_i} t J_0(t)\, dt$$

$$= \frac{1}{2\lambda_i^2 J_1^2(2\lambda_i)} \int_0^{2\lambda_i} \frac{d}{dt}[t J_1(t)]\, dt \qquad \text{[From (4) in the text]}$$

$$= \frac{1}{2\lambda_i^2 J_1^2(2\lambda_i)} t J_1(t) \Big|_0^{2\lambda_i}$$

$$= \frac{1}{\lambda_i J_1(2\lambda_i)} \cdot$$

Thus

$$f(x) = \sum_{i=1}^{\infty} \frac{1}{\lambda_i J_1(2\lambda_i)} J_0(\lambda_i x).$$

4. The boundary condition indicates that we use (19) and (20) of Section 11.5. With $b = 2$ we obtain

$$c_1 = \frac{2}{4} \int_0^2 x\, dx = \frac{2}{4} \frac{x^2}{2} \Big|_0^2 = 1,$$

$$c_i = \frac{2}{4J_0^2(2\lambda_i)} \int_0^2 x J_0(\lambda_i x)\, dx$$

$$\boxed{t = \lambda_i x \qquad dt = \lambda_i\, dx}$$

$$= \frac{1}{2J_0^2(2\lambda_i)} \cdot \frac{1}{\lambda_i^2} \int_0^{2\lambda_i} t J_0(t)\, dt$$

$$= \frac{1}{2\lambda_i^2 J_0^2(2\lambda_i)} \int_0^{2\lambda_i} \frac{d}{dt}[t J_1(t)]\, dt \qquad \text{[From (4) in the text]}$$

$$= \frac{1}{2\lambda_i^2 J_0^2(2\lambda_i)} t J_1(t) \Big|_0^{2\lambda_i}$$

$$= \frac{J_1(2\lambda_i)}{\lambda_i J_0^2(2\lambda_i)} \cdot$$

Now since $J_0'(2\lambda_i) = 0$ is equivalent to $J_1(2\lambda_i) = 0$ we conclude $c_i = 0$ for $i = 2, 3, 4, \ldots$. Thus the expansion of f on $0 < x < 2$ consists of a series with one nontrivial term:

$$f(x) = c_1 = 1.$$

5. The boundary condition indicates that we use (17) and (18) of Section 11.5. With $b = 2$ and $h = 1$ we obtain

$$c_i = \frac{2\lambda_i^2}{(4\lambda_i^2 + 1)J_0^2(2\lambda_i)} \int_0^2 x J_0(\lambda_i x)\, dx$$

$$\boxed{t = \lambda_i x \qquad dt = \lambda_i\, dx}$$

$$= \frac{2\lambda_i^2}{(4\lambda_i^2 + 1)J_0^2(2\lambda_i)} \cdot \frac{1}{\lambda_i^2} \int_0^{2\lambda_i} t J_0(t)\, dt$$

$$= \frac{2}{(4\lambda_i^2 + 1)J_0^2(2\lambda_i)} \int_0^{2\lambda_i} \frac{d}{dt}[t J_1(t)]\, dt \qquad \text{[From (4) in the text]}$$

$$= \frac{2}{(4\lambda_i^2 + 1)J_0^2(2\lambda_i)} t J_1(t) \Big|_0^{2\lambda_i}$$

$$= \frac{4\lambda_i J_1(2\lambda_i)}{(4\lambda_i^2 + 1)J_0^2(2\lambda_i)}.$$

Thus

$$f(x) = 4\sum_{i=1}^{\infty} \frac{\lambda_i J_1(2\lambda_i)}{(4\lambda_i^2 + 1)J_0^2(2\lambda_i)} J_0(\lambda_i x).$$

6. Writing the boundary condition in the form

$$2J_0(2\lambda) + 2\lambda J_0'(2\lambda) = 0$$

we identify $b = 2$ and $h = 2$. Using (17) and (18) of Section 11.5 we obtain

$$c_i = \frac{2\lambda_i^2}{(4\lambda_i^2 + 4)J_0^2(2\lambda_i)} \int_0^2 x J_0(\lambda_i x)\, dx$$

$$\boxed{t = \lambda_i x \qquad dt = \lambda_i\, dx}$$

$$= \frac{\lambda_i^2}{2(\lambda_i^2 + 1)J_0^2(2\lambda_i)} \cdot \frac{1}{\lambda_i^2} \int_0^{2\lambda_i} t J_0(t)\, dt$$

$$= \frac{1}{2(\lambda_i^2 + 1)J_0^2(2\lambda_i)} \int_0^{2\lambda_i} \frac{d}{dt}[t J_1(t)]\, dt \qquad \text{[From (4) in the text]}$$

$$= \frac{1}{2(\lambda_i^2 + 1)J_0^2(2\lambda_i)} t J_1(t) \Big|_0^{2\lambda_i}$$

$$= \frac{\lambda_i J_1(2\lambda_i)}{(\lambda_i^2 + 1)J_0^2(2\lambda_i)}.$$

Thus

$$f(x) = \sum_{i=1}^{\infty} \frac{\lambda_i J_1(2\lambda_i)}{(\lambda_i^2 + 1)J_0^2(2\lambda_i)} J_0(\lambda_i x).$$

7. The boundary condition indicates that we use (17) and (18) of Section 11.5. With $n = 1$, $b = 4$, and $h = 3$ we obtain

$$c_i = \frac{2\lambda_i^2}{(16\lambda_i^2 - 1 + 9)J_1^2(4\lambda_i)} \int_0^4 x J_1(\lambda_i x) 5x \, dx$$

$$\boxed{t = \lambda_i x \qquad dt = \lambda_i \, dx}$$

$$= \frac{5\lambda_i^2}{4(2\lambda_i^2 + 1)J_1^2(4\lambda_i)} \cdot \frac{1}{\lambda_i^3} \int_0^{4\lambda_i} t^2 J_1(t) \, dt$$

$$= \frac{5}{4\lambda_i(2\lambda_i^2 + 1)J_1^2(4\lambda_i)} \int_0^{4\lambda_i} \frac{d}{dt}[t^2 J_2(t)] \, dt \qquad \text{[From (4) in the text]}$$

$$= \frac{5}{4\lambda_i(2\lambda_i^2 + 1)J_1^2(4\lambda_i)} t^2 J_2(t) \Big|_0^{4\lambda_i}$$

$$= \frac{20\lambda_i J_2(4\lambda_i)}{(2\lambda_i^2 + 1)J_1^2(4\lambda_i)}.$$

Thus

$$f(x) = 20 \sum_{i=1}^{\infty} \frac{\lambda_i J_2(4\lambda_i)}{(2\lambda_i^2 + 1)J_1^2(4\lambda_i)} J_1(\lambda_i x).$$

8. The boundary condition indicates that we use (15) and (16) of Section 11.5. With $n = 2$ and $b = 1$ we obtain

$$c_1 = \frac{2}{J_3^2(\lambda_i)} \int_0^1 x J_2(\lambda_i x) x^2 \, dx$$

$$\boxed{t = \lambda_i x \qquad dt = \lambda_i \, dx}$$

$$= \frac{2}{J_3^2(\lambda_i)} \cdot \frac{1}{\lambda_i^4} \int_0^{\lambda_i} t^3 J_2(t) \, dt$$

$$= \frac{2}{\lambda_i^4 J_3^2(\lambda_i)} \int_0^{\lambda_i} \frac{d}{dt}[t^3 J_3(t)] \, dt \qquad \text{[From (4) in the text]}$$

$$= \frac{2}{\lambda_i^4 J_3^2(\lambda_i)} t^3 J_3(t) \Big|_0^{\lambda_i}$$

$$= \frac{2}{\lambda_i J_3(\lambda_i)}.$$

Thus

$$f(x) = 2 \sum_{i=1}^{\infty} \frac{1}{\lambda_i J_3(\lambda_i)} J_2(\lambda_i x).$$

9. The boundary condition indicates that we use (19) and (20) of Section 11.5. With $b = 3$ we obtain

$$c_1 = \frac{2}{9} \int_0^3 x x^2 \, dx = \frac{2}{9} \frac{x^4}{4} \Big|_0^3 = \frac{9}{2},$$

$$c_i = \frac{2}{9 J_0^2(3\lambda_i)} \int_0^3 x J_0(\lambda_i x) x^2 \, dx$$

$$\boxed{t = \lambda_i x \qquad dt = \lambda_i \, dx}$$

$$= \frac{2}{9 J_0^2(3\lambda_i)} \cdot \frac{1}{\lambda_i^4} \int_0^{3\lambda_i} t^3 J_0(t) \, dt$$

$$= \frac{2}{9 \lambda_i^4 J_0^2(3\lambda_i)} \int_0^{3\lambda_i} t^2 \frac{d}{dt} [t J_1(t)] \, dt$$

$$\boxed{\begin{array}{ll} u = t^2 & dv = \frac{d}{dt}[t J_1(t)] \, dt \\ du = 2t \, dt & v = t J_1(t) \end{array}}$$

$$= \frac{2}{9 \lambda_i^4 J_0^2(3\lambda_i)} \left(t^3 J_1(t) \Big|_0^{3\lambda_i} - 2 \int_0^{3\lambda_i} t^2 J_1(t) \, dt \right)$$

With $n = 0$ in equation (5) in Section 11.5 in the text we have $J_0'(x) = -J_1(x)$, so the boundary condition $J_0'(3\lambda_i) = 0$ implies $J_1(3\lambda_i) = 0$. Then

$$c_i = \frac{2}{9 \lambda_i^4 J_0^2(3\lambda_i)} \left(-2 \int_0^{3\lambda_i} \frac{d}{dt} [t^2 J_2(t)] \, dt \right) = \frac{2}{9 \lambda_i^4 J_0^2(3\lambda_i)} \left(-2 t^2 J_2(t) \Big|_0^{3\lambda_i} \right)$$

$$= \frac{2}{9 \lambda_i^4 J_0^2(3\lambda_i)} \left[-18 \lambda_i^2 J_2(3\lambda_i) \right] = \frac{-4 J_2(3\lambda_i)}{\lambda_i^2 J_0^2(3\lambda_i)}.$$

Thus

$$f(x) = \frac{9}{2} - 4 \sum_{i=1}^{\infty} \frac{J_2(3\lambda_i)}{\lambda_i^2 J_0^2(3\lambda_i)} J_0(\lambda_i x).$$

10. The boundary condition indicates that we use (15) and (16) of Section 11.5. With $b = 1$ it follows that

$$c_i = \frac{2}{J_1^2(\lambda_i)} \int_0^1 x \left(1 - x^2 \right) J_0(\lambda_i x) \, dx$$

$$= \frac{2}{J_1^2(\lambda_i)} \left[\int_0^1 x J_0(\lambda_i x) \, dx - \int_0^1 x^3 J_0(\lambda_i x) \, dx \right]$$

461

$$\boxed{t = \lambda_i x \qquad dt = \lambda_i \, dx}$$

$$= \frac{2}{J_1^2(\lambda_i)} \left[\frac{1}{\lambda_i^2} \int_0^{\lambda_i} t J_0(t) \, dt - \frac{1}{\lambda_i^4} \int_0^{\lambda_i} t^3 J_0(t) \, dt \right]$$

$$= \frac{2}{J_1^2(\lambda_i)} \left[\frac{1}{\lambda_i^2} \int_0^{\lambda_i} \frac{d}{dt} \left(t J_1(t) \right) dt - \frac{1}{\lambda_i^4} \int_0^{\lambda_i} t^2 \frac{d}{dt} \left[t J_1(t) \right] dt \right]$$

$$\boxed{\begin{array}{ll} u = t^2 & dv = \frac{d}{dt} \left[t J_1(t) \right] dt \\ du = 2t \, dt & v = t J_1(t) \end{array}}$$

$$= \frac{2}{J_1^2(\lambda_i)} \left[\frac{1}{\lambda_i^2} t J_1(t) \Big|_0^{\lambda_i} - \frac{1}{\lambda_i^4} \left(t^3 J_1(t) \Big|_0^{\lambda_i} - 2 \int_0^{\lambda_i} t^2 J_1(t) \, dt \right) \right]$$

$$= \frac{2}{J_1^2(\lambda_i)} \left[\frac{J_1(\lambda_i)}{\lambda_i} - \frac{J_1(\lambda_i)}{\lambda_i} + \frac{2}{\lambda_i^4} \int_0^{\lambda_i} \frac{d}{dt} \left[t^2 J_2(t) \right] dt \right]$$

$$= \frac{2}{J_1^2(\lambda_i)} \left[\frac{2}{\lambda_i^4} t^2 J_2(t) \Big|_0^{\lambda_i} \right] = \frac{4 J_2(\lambda_i)}{\lambda_i^2 J_1^2(\lambda_i)} .$$

Thus

$$f(x) = 4 \sum_{i=1}^{\infty} \frac{J_2(\lambda_i)}{\lambda_i^2 J_1^2(\lambda_i)} j_0(\lambda_i x).$$

11. Since $f(x) = x^2$ is a polynomial in x, an expansion of f in polynomials in x must terminate with the term having the same degree as f. We have

$$c_0 = \frac{1}{2} \int_{-1}^1 x^2 P_0(x) \, dx = \frac{1}{2} \int_{-1}^1 x^2 \, dx = \frac{1}{3},$$

$$c_1 = \frac{3}{2} \int_{-1}^1 x^2 P_1(x) \, dx = \frac{3}{2} \int_{-1}^1 x^3 \, dx = 0,$$

$$c_2 = \frac{5}{2} \int_{-1}^1 x^2 P_2(x) \, dx = \frac{5}{2} \int_{-1}^1 x^2 \frac{1}{2} \left(3x^2 - 1 \right) dx = \frac{2}{3}.$$

Thus

$$f(x) = c_0 P_0(x) + c_1 P_1(x) + c_2 P_2(x) = \frac{1}{3} P_0(x) + \frac{2}{3} P_2(x).$$

12. Since $f(x) = x^3$ is a polynomial in x, an expansion of f in polynomials in x must terminate with the term having the same degree as f. We have

$$c_0 = \frac{1}{2}\int_{-1}^1 x^3 P_0(x)\,dx = \frac{1}{2}\int_{-1}^1 x^3\,dx = 0,$$

$$c_1 = \frac{3}{2}\int_{-1}^1 x^3 P_1(x)\,dx = \frac{3}{2}\int_{-1}^1 x^4\,dx = \frac{3}{5},$$

$$c_2 = \frac{5}{2}\int_{-1}^1 x^3 P_2(x)\,dx = \frac{5}{2}\int_{-1}^1 x^3 \frac{1}{2}\left(3x^2 - 1\right)\,dx = 0,$$

$$c_3 = \frac{7}{2}\int_{-1}^1 x^3 P_3(x)\,dx = \frac{7}{2}\int_{-1}^1 x^3 \frac{1}{2}\left(5x^3 - 3x\right)\,dx = \frac{2}{5}.$$

Thus

$$f(x) = c_0 P_0(x) + c_1 P_1(x) + c_2 P_2(x) + c_3 P_3(x)$$

$$= \frac{3}{5}P_1(x) + \frac{2}{5}P_3(x).$$

13. We compute

$$c_0 = \frac{1}{2}\int_0^1 x P_0(x)\,dx = \frac{1}{2}\int_0^1 x\,dx = \frac{1}{4}$$

$$c_1 = \frac{3}{2}\int_0^1 x P_1(x)\,dx = \frac{3}{2}\int_0^1 x^2\,dx - \frac{1}{2}$$

$$c_2 = \frac{5}{2}\int_0^1 x P_2(x)\,dx = \frac{5}{2}\int_0^1 \frac{1}{2}(3x^3 - x)\,dx = \frac{5}{16}$$

$$c_3 = \frac{7}{2}\int_0^1 x P_3(x)\,dx = \frac{7}{2}\int_0^1 \frac{1}{2}(5x^4 - 3x^2)\,dx = 0$$

$$c_4 = \frac{9}{2}\int_0^1 x P_4(x)\,dx = \frac{9}{2}\int_0^1 \frac{1}{8}(35x^5 - 30x^3 + 3x)\,dx = -\frac{3}{32}.$$

Thus

$$f(x) = \frac{1}{4}P_0(x) + \frac{1}{2}P_1(x) + \frac{5}{16}P_2(x) - \frac{3}{32}P_4(x) + \cdots.$$

14. We compute

$$c_0 = \frac{1}{2}\int_{-1}^1 e^x P_0(x)\,dx = \frac{1}{2}\int_{-1}^1 e^x\,dx = \frac{1}{2}(e - e^{-1})$$

$$c_1 = \frac{3}{2}\int_{-1}^1 e^x P_1(x)\,dx = \frac{3}{2}\int_{-1}^1 x e^x\,dx = 3e^{-1}$$

$$c_2 = \frac{5}{2} \int_{-1}^{1} e^x P_2(x)\, dx = \frac{5}{2} \int_{-1}^{1} \frac{1}{2}(3x^2 e^x - e^x)\, dx = \frac{5}{2}(e - 7e^{-1})$$

$$c_3 = \frac{7}{2} \int_{-1}^{1} e^x P_3(x)\, dx = \frac{7}{2} \int_{-1}^{1} \frac{1}{2}(5x^3 e^x - 3xe^x)\, dx = \frac{7}{2}(-5e + 37e^{-1}).$$

Thus

$$f(x) = \frac{1}{2}(e - e^{-1})P_0(x) + 3e^{-1}P_1(x) + \frac{5}{2}(e - 7e^{-1})P_2(x) + \frac{7}{2}(-5e + 37e^{-1})P_3(x) + \cdots.$$

15. Using $\cos^2 \theta = \frac{1}{2}(\cos 2\theta + 1)$ we have

$$P_2(\cos\theta) = \frac{1}{2}(3\cos^2\theta - 1) = \frac{3}{2}\cos^2\theta - \frac{1}{2}$$

$$= \frac{3}{4}(\cos 2\theta + 1) - \frac{1}{2} = \frac{3}{4}\cos 2\theta + \frac{1}{4} = \frac{1}{4}(3\cos 2\theta + 1).$$

16. From Problem 15 we have

$$P_2(\cos 2\theta) = \frac{1}{4}(3\cos 2\theta + 1)$$

or

$$\cos 2\theta = \frac{4}{3}P_2(\cos\theta) - \frac{1}{3}.$$

Then, using $P_0(\cos\theta) = 1$,

$$F(\theta) = 1 - \cos 2\theta = 1 - \left[\frac{4}{3}P_2(\cos\theta) - \frac{1}{3}\right]$$

$$= \frac{4}{3} - \frac{4}{3}P_2(\cos\theta) = \frac{4}{3}P_0(\cos\theta) - \frac{4}{3}P_2(\cos\theta).$$

17. If f is an even function on $(-1, 1)$ then

$$\int_{-1}^{1} f(x)P_{2n}(x)\, dx = 2\int_{0}^{1} f(x)P_{2n}(x)\, dx$$

and

$$\int_{-1}^{1} f(x)P_{2n+1}(x)\, dx = 0.$$

Thus

$$c_{2n} = \frac{2(2n) + 1}{2}\int_{-1}^{1} f(x)P_{2n}(x)\, dx = \frac{4n + 1}{2}\left(2\int_{0}^{1} f(x)P_{2n}(x)\, dx\right)$$

$$= (4n + 1)\int_{0}^{1} f(x)P_{2n}(x)\, dx,$$

$c_{2n+1} = 0$, and

$$f(x) = \sum_{n=0}^{\infty} c_{2n}P_{2n}(x).$$

18. If f is an odd function on $(-1, 1)$ then

$$\int_{-1}^{1} f(x) P_{2n}(x)\, dx = 0$$

and

$$\int_{-1}^{1} f(x) P_{2n+1}(x)\, dx = 2 \int_{0}^{1} f(x) P_{2n+1}(x)\, dx.$$

Thus

$$c_{2n+1} = \frac{2(2n+1)+1}{2} \int_{-1}^{1} f(x) P_{2n+1}(x)\, dx = \frac{4n+3}{2} \left(2 \int_{0}^{1} f(x) P_{2n+1}(x)\, dx \right)$$

$$= (4n+1) \int_{0}^{1} f(x) P_{2n+1}(x)\, dx,$$

$c_{2n} = 0$, and

$$f(x) = \sum_{n=0}^{\infty} c_{2n+1} P_{2n+1}(x).$$

19. From (26) in Problem 17 in the text we find

$$c_0 = \int_{0}^{1} x P_0(x)\, dx = \int_{0}^{1} x\, dx = \frac{1}{2},$$

$$c_2 = 5 \int_{0}^{1} x P_2(x)\, dx = 5 \int_{0}^{1} \frac{1}{2}(3x^3 - x)\, dx = \frac{5}{8},$$

and

$$c_4 = 9 \int_{0}^{1} x P_4(x)\, dx = 9 \int_{0}^{1} \frac{1}{8}(35x^5 - 30x^3 + 3x)\, dx = -\frac{3}{16}.$$

Hence, from (25) in the text,

$$f(x) = \frac{1}{2} P_0(x) + \frac{5}{8} P_2(x) - \frac{3}{16} P_4(x) + \cdots.$$

On the interval $-1 < x < 1$ this series represents the even function $f(x) = |x|$.

20. From (28) in Problem 18 in the text we find

$$c_1 = 3 \int_{0}^{1} P_1(x)\, dx = 3 \int_{0}^{1} x\, dx = \frac{3}{2},$$

$$c_3 = 7 \int_{0}^{1} P_3(x)\, dx = 7 \int_{0}^{1} \frac{1}{2} \left(5x^3 - 3x \right) dx = -\frac{7}{8},$$

and

$$c_5 = 11 \int_{0}^{1} P_5(x)\, dx = 11 \int_{0}^{1} \frac{1}{8} \left(63x^5 - 70x^3 + 15x \right) dx = \frac{11}{16}.$$

Hence, from (27) in Problem 18 in the text,

$$f(x) = \frac{3}{2} P_1(x) - \frac{7}{8} P_3(x) + \frac{11}{16} P_5(x) + \cdots.$$

On the interval $-1 < x < 1$ this series represents the odd function

$$f(x) = \begin{cases} -1, & -1 < x < 0 \\ 1, & 0 < x < 1 \end{cases}.$$

Chapter 11 Review Exercises

1. True, since $\int_{-\pi}^{\pi}(x^2 - 1)x^5\,dx = 0$.

2. Even, since if f and g are odd then $h(-x) = f(-x)g(-x) = -f(x)[-g(x)] = f(x)g(x) = h(x)$.

3. cosine, since f is even.

4. False, since it has an even extension on $-2 < x < 2$.

5. $3/2$, the average of 3 and 0.

6. True

7. False; see the solution of Problem 15 in Section 11.4.

8. $\cos 5x$, since the general solution is $y = c_1 \cos\sqrt{\lambda}\,x + c_2 \sin\sqrt{\lambda}\,x$ and $y'(0) = 0$ implies $c_2 = 0$.

9. Since the coefficient of y in the differential equation is n^2, the weight function is the integrating factor

$$\frac{1}{a(x)}e^{\int (b/a)dx} = \frac{1}{1-x^2}e^{\int -\frac{x}{1-x^2}\,dx} = \frac{1}{1-x^2}e^{\frac{1}{2}\ln(1-x^2)} = \frac{\sqrt{1-x^2}}{1-x^2} = \frac{1}{\sqrt{1-x^2}}$$

on the interval $[-1, 1]$.

10. Since $P_n(x)$ is orthogonal to $P_0(x) = 1$ for $n > 0$,

$$\int_{-1}^{1} P_n(x)\,dx = \int_{-1}^{1} P_0(x)P_n(x)\,dx = 0.$$

11. For $m \neq n$

$$\int_0^L \sin\frac{(2n+1)\pi}{2L}x \sin\frac{(2m+1)\pi}{2L}x\,dx = \frac{1}{2}\int_0^L \left(\cos\frac{n-m}{L}\pi x - \cos\frac{n+m+\pi}{L}\pi x\right)dx = 0.$$

12. From

$$\int_0^L \sin^2\frac{(2n+1)\pi}{2L}x\,dx = \int_0^L \left(\frac{1}{2} - \frac{1}{2}\cos\frac{(2n+1)\pi}{2L}x\right)dx = \frac{L}{2}$$

we see that

$$\left\| \sin\frac{(2n+1)\pi}{2L}x \right\| = \sqrt{\frac{L}{2}}.$$

13. Since

$$A_0 = \int_{-1}^{0} (-2x)\, dx = 1,$$

$$A_n = \int_{-1}^{0} (-2x) \cos n\pi x\, dx = \frac{2}{n^2\pi^2}[(-1)^n - 1],$$

and

$$B_n = \int_{-1}^{0} (-2x) \sin n\pi x\, dx = \frac{4}{n\pi}(-1)^n$$

for $n = 1, 2, 3, \ldots$ we have

$$f(x) = \frac{1}{2} + \sum_{n=1}^{\infty} \left(\frac{2}{n^2\pi^2}[(-1)^n - 1] \cos n\pi x + \frac{4}{n\pi}(-1)^n \sin n\pi x \right).$$

14. Since

$$A_0 = \int_{-1}^{1} (2x^2 - 1)\, dx = -\frac{2}{3},$$

$$A_n = \int_{-1}^{1} (2x^2 - 1) \cos n\pi x\, dx = \frac{8}{n^2\pi^2}(-1)^n,$$

and

$$B_n = \int_{-1}^{1} (2x^2 - 1) \sin n\pi x\, dx = 0$$

for $n = 1, 2, 3, \ldots$ we have

$$f(x) = -\frac{1}{3} + \sum_{n=1}^{\infty} \frac{8}{n^2\pi^2}(-1)^n \cos n\pi x.$$

15. Since

$$A_0 = 2 \int_{0}^{1} e^{-x} dx$$

and

$$A_n = 2 \int_{-1}^{1} e^{-x} \cos n\pi x\, dx = \frac{2}{1 + n^2\pi^2}[(1 - (-1)^n e^{-1}]$$

for $n = 1, 2, 3, \ldots$ we have

$$f(x) = 1 - e^{-1} + 2 \sum_{n=1}^{\infty} \frac{1 - (-1)^n e^{-1}}{1 + n^2\pi^2} \cos n\pi x.$$

16. Since

$$B_n = 2 \int_{0}^{1} e^{-x} \sin n\pi x\, dx = \frac{2n\pi}{1 + n^2\pi^2}[(1 - (-1)^n e^{-1}]$$

for $n = 1, 2, 3, \ldots$ we have

$$f(x) = \sum_{n=1}^{\infty} \frac{2n\pi}{1 + n^2\pi^2}[(1 - (-1)^n e^{-1}] \sin n\pi x.$$

17. For $\lambda > 0$ a general solution of the given differential equation is

$$y = c_1 \cos(3\sqrt{\lambda}\,\ln x) + c_2 \sin(3\sqrt{\lambda}\,\ln x)$$

and

$$y' = -\frac{3c_1\sqrt{\lambda}}{x}\sin(3\sqrt{\lambda}\,\ln x) + \frac{3c_2\sqrt{\lambda}}{x}\cos(3\sqrt{\lambda}\,\ln x).$$

Since $\ln 1 = 0$, the boundary condition $y'(1) = 0$ implies $c_2 = 0$. Therefore

$$y = c_1 \cos(3\sqrt{\lambda}\,\ln x).$$

Using $\ln e = 1$ we find that $y(e) = 0$ implies $c_1 \cos 3\sqrt{\lambda} = 0$ or $3\sqrt{\lambda} = \dfrac{2n-1}{2}\pi$, for $n = 1, 2, 3, \ldots$. The eigenvalues are $\lambda = (2n-1)^2\pi^2/36$ with corresponding eigenfunctions $\cos\left(\dfrac{2n-1}{2}\pi \ln x\right)$ for $n = 1, 2, 3, \ldots$.

18. To obtain the self-adjoint form of the differential equation in Problem 17 we note that an integrating factor is $(1/x^2)e^{\int dx/x} = 1/x$. Thus the weight function is $9/x$ and an orthogonality relation is

$$\int_1^e \frac{9}{x} \cos\left(\frac{2n-1}{2}\pi \ln x\right) \cos\left(\frac{2m-1}{2}\pi \ln x\right) dx = 0, \quad m \neq n.$$

19. The boundary condition indicates that we use (15) and (16) of Section 11.5. With $b = 4$ we obtain

$$c_i = \frac{2}{16 J_1^2(4\lambda_i)} \int_0^4 x J_0(\lambda_i x) f(x)\, dx$$

$$= \frac{1}{8 J_1^2(4\lambda_i)} \int_0^2 x J_0(\lambda_i x)\, dx \qquad \boxed{t = \lambda_i x \qquad dt = \lambda_i\, dx}$$

$$= \frac{1}{8 J_1^2(4\lambda_i)} \cdot \frac{1}{\lambda_i^2} \int_0^{2\lambda_i} t J_0(t)\, dt$$

$$= \frac{1}{8 J_1^2(4\lambda_i)} \int_0^{2\lambda_i} \frac{d}{dt}[t J_1(t)]\, dt \qquad \text{[From (4) in 11.5 in the text]}$$

$$= \frac{1}{8 J_1^2(4\lambda_i)} t J_1(t)\Big|_0^{2\lambda_i} = \frac{J_1(2\lambda_i)}{4\lambda_i J_1^2(4\lambda_i)}.$$

Thus

$$f(x) = \frac{1}{4}\sum_{i=1}^{\infty} \frac{J_1(2\lambda_i)}{\lambda_i J_1^2(4\lambda_i)} J_0(\lambda_i x).$$

20. Since $f(x) = x^4$ is a polynomial in x, an expansion of f in polynomials in x must terminate with the term having the same degree as f. Using the fact that $x^4 P_1(x)$ and $x^4 P_3(x)$ are odd functions, we see immediately that $c_1 = c_3 = 0$. Now

$$c_0 = \frac{1}{2} \int_{-1}^{1} x^4 P_0(x)\, dx = \frac{1}{2} \int_{-1}^{1} x^4 dx = \frac{1}{5}$$

$$c_2 = \frac{5}{2} \int_{-1}^{1} x^4 P_2(x)\, dx = \frac{5}{2} \int_{-1}^{1} \frac{1}{2}(3x^6 - x^4)dx = \frac{4}{7}$$

$$c_4 = \frac{9}{2} \int_{-1}^{1} x^4 P_4(x)\, dx = \frac{9}{2} \int_{-1}^{1} \frac{1}{8}(35x^8 - 30x^6 + 3x^4)dx = \frac{8}{35}.$$

Thus

$$f(x) = \frac{1}{5} P_0(x) + \frac{4}{7} P_2(x) + \frac{8}{35} P_4(x).$$

12 Partial Differential Equations and Boundary-Value Problems in Rectangular Coordinates

---
Exercises 12.1
---

1. If $u = XY$ then

$$u_x = X'Y,$$
$$u_y = XY',$$
$$X'Y = XY',$$

and

$$\frac{X'}{X} = \frac{Y'}{Y} = \pm\lambda^2.$$

Then

$$X' \mp \lambda^2 X = 0 \quad \text{and} \quad Y' \mp \lambda^2 Y = 0$$

so that

$$X = A_1 e^{\pm\lambda^2 x},$$
$$Y = A_2 e^{\pm\lambda^2 y},$$

and

$$u = XY = c_1 e^{c_2(x+y)}.$$

2. If $u = XY$ then

$$u_x = X'Y,$$
$$u_y = XY',$$
$$X'Y = -3XY',$$

and

$$\frac{X'}{-3X} = \frac{Y'}{Y} = \pm\lambda^2.$$

Then

$$X' \pm 3\lambda^2 X = 0 \quad \text{and} \quad Y' \mp \lambda^2 Y = 0$$

so that

$$X = A_1 e^{\mp 3\lambda^2 x},$$

$$Y = A_2 e^{\pm \lambda^2 y},$$

and

$$u = XY = c_1 e^{c_2(y-3x)}.$$

3. If $u = XY$ then

$$u_x = X'Y,$$

$$u_y = XY',$$

$$X'Y = X(Y - Y'),$$

and

$$\frac{X'}{X} = \frac{Y - Y'}{Y} = \pm\lambda^2.$$

Then

$$X' \mp \lambda^2 X = 0 \quad \text{and} \quad Y' - (1 \mp \lambda^2)Y = 0$$

so that

$$X = A_1 e^{\pm\lambda^2 x},$$

$$Y = A_2 e^{(1\mp\lambda^2)y},$$

and

$$u = XY = c_1 e^{y + c_2(x-y)}.$$

4. If $u = XY$ then

$$u_x = X'Y,$$

$$u_y = XY',$$

$$X'Y = X(Y + Y'),$$

and

$$\frac{X'}{X} = \frac{Y + Y'}{Y} = \pm\lambda^2.$$

Then

$$X' \mp \lambda^2 X = 0 \quad \text{and} \quad Y' - (-1 \pm \lambda^2)Y = 0$$

so that

$$X = A_1 e^{\pm\lambda^2 x},$$

$$Y = A_2 e^{(-1\pm\lambda^2)y},$$

and

$$u = XY = c_1 e^{-y + c_2(x+y)}.$$

5. If $u = XY$ then

$$u_x = X'Y,$$

$$u_y = XY',$$

$$xX'Y = yXY',$$

and

$$\frac{xX'}{X} = \frac{yY'}{Y} = \pm\lambda^2.$$

Then

$$X \mp \frac{1}{x}\lambda^2 X = 0 \quad \text{and} \quad Y' \mp \frac{1}{y}\lambda^2 Y = 0$$

so that

$$X = A_1 x^{\pm\lambda^2},$$

$$Y = A_2 y^{\pm\lambda^2},$$

and

$$u = XY = c_1(xy)^{c_2}.$$

6. If $u = XY$ then

$$u_x = X'Y,$$

$$u_y = XY',$$

$$yX'Y = xXY',$$

and

$$\frac{X'}{xX} = \frac{Y'}{-yY} = \pm\lambda^2.$$

Then

$$X \mp \lambda^2 x X = 0 \quad \text{and} \quad Y' \pm \lambda^2 y Y = 0$$

so that

$$X = A_1 e^{\pm\lambda^2 x^2/2},$$

$$Y = A_2 e^{\mp\lambda^2 y^2/2},$$

and

$$u = XY = c_1 e^{c_2(x^2 - y^2)}.$$

7. If $u = XY$ then

$$u_{xx} = X''Y, \quad u_{yy} = XY'', \quad u_{yx} = X'Y',$$

and

$$X''Y + X'Y' + XY'' = 0,$$

which is not separable.

8. If $u = XY$ then

$$u_{yx} = X'Y',$$

$$yX'Y' + XY = 0,$$

and

$$\frac{X'}{-X} = \frac{Y}{yY'} = \pm\lambda^2.$$

Then

$$X' \mp \lambda^2 X = 0 \quad \text{and} \quad \pm\lambda^2 yY' - Y = 0$$

so that

$$X = A_1 e^{\mp\lambda^2 x},$$

$$Y = A_2 y^{\pm 1/\lambda^2},$$

and

$$u = XY - c_1 e^{-c_2 x} y^{1/c_2}.$$

9. If $u = XT$ then

$$u_t = XT',$$

$$u_{xx} = X''T,$$

$$kX''T - XT = XT',$$

and we choose

$$\frac{T'}{T} = \frac{kX'' - X}{X} = -1 \pm k\lambda^2$$

so that

$$T' - (-1 \pm k\lambda^2)T = 0 \quad \text{and} \quad X'' - (\pm\lambda^2)X = 0.$$

For $\lambda^2 > 0$ we obtain

$$X = A_1 \cosh \lambda x + A_2 \sinh \lambda x \quad \text{and} \quad T = A_3 e^{(-1+k\lambda^2)t}$$

so that

$$u = XT = e^{(-1+k\lambda^2)t} \left(c_1 \cosh \lambda x + c_2 \sinh \lambda x \right).$$

For $-\lambda^2 < 0$ we obtain

$$X = A_1 \cos \lambda x + A_2 \sin \lambda x \quad \text{and} \quad T = A_3 e^{(-1-k\lambda^2)t}$$

so that

$$u = XT = e^{(-1-k\lambda^2)t}(c_3 \cos \lambda x + c_4 \sin \lambda x).$$

If $\lambda^2 = 0$ then

$$X'' = 0 \quad \text{and} \quad T' + T = 0,$$

and we obtain

$$X = A_1 x + A_2 \quad \text{and} \quad T = A_3 e^{-t}.$$

In this case

$$u = XT = e^{-t}(c_5 x + c_6)$$

10. If $u = XT$ then

$$u_t = XT',$$

$$u_{xx} = X''T,$$

$$kX''T = XT',$$

and

$$\frac{X''}{X} = \frac{T'}{kT} = \pm \lambda^2$$

so that

$$X'' \mp \lambda^2 X = 0 \quad \text{and} \quad T' \mp \lambda^2 kT = 0.$$

For $\lambda^2 > 0$ we obtain

$$X = A_1 \cos \lambda x + A_2 \sin \lambda x \quad \text{and} \quad T = A_3 e^{\lambda^2 kt},$$

so that

$$u = XT = e^{-c_3^2 kt}(c_1 \cos c_3 x + c_2 \sin c_3 x).$$

For $-\lambda^2 < 0$ we obtain

$$X = A_1 e^{-\lambda x} + A_2 e^{\lambda x},$$

$$T = A_3 e^{\lambda^2 kt},$$

and

$$u = XT = e^{c_3^2 kt}\left(c_1 e^{-c_3 x} + c_2 e^{c_3 x}\right).$$

For $\lambda^2 = 0$ we obtain

$$T = A_3, \quad X = A_1 x + A_2, \quad \text{and} \quad u = XT = c_1 x + c_2.$$

11. If $u = XT$ then

$$u_{xx} = X''T,$$
$$u_{tt} = XT'',$$
$$a^2 X''T = XT'',$$

and

$$\frac{X''}{X} = \frac{T''}{a^2 T} = \pm \lambda^2$$

so that

$$X'' \mp \lambda^2 X = 0 \quad \text{and} \quad T'' \mp a^2 \lambda^2 T = 0.$$

For $\lambda^2 > 0$ we obtain

$$X = A_1 \cos \lambda x + A_2 \sin \lambda x,$$
$$T = A_3 \cos a\lambda t + A_4 \sin a\lambda t,$$

and

$$u = XT = (A_1 \cos \lambda x + A_2 \sin \lambda x)(A_3 \cos a\lambda t + A_4 \sin a\lambda t).$$

For $-\lambda^2 < 0$ we obtain

$$X = A_1 e^{\lambda x} + A_2 e^{-\lambda x},$$
$$T = A_3 e^{a\lambda t} + A_4 e^{-a\lambda t},$$

and

$$u = XT = \left(A_1 e^{\lambda x} + A_2 e^{-\lambda x} \right) \left(A_3 e^{a\lambda t} + A_4 e^{-a\lambda t} \right).$$

For $\lambda^2 = 0$ we obtain

$$X = A_1 x + A_2,$$
$$T = A_3 t + A_4,$$

and

$$u = XT = (A_1 x + A_2)(A_3 t + A_4).$$

12. If $u = XT$ then

$$u_t = XT',$$
$$u_{tt} = XT'',$$
$$u_{xx} = X''T,$$
$$a^2 X''T = XT'' + 2kXT',$$

and

475

$$\frac{X''}{X} = \frac{T'' + 2kT'}{a^2 T} = \pm\lambda^2$$

so that

$$X'' \mp \lambda^2 X = 0 \quad \text{and} \quad T'' + 2kT' \mp a^2\lambda^2 T = 0.$$

For $\lambda^2 > 0$ we obtain

$$X = A_1 e^{\lambda x} + A_2 e^{-\lambda x},$$

$$T = A_3 e^{(-k+\sqrt{k^2+a^2\lambda^2})t} + A_4 e^{(-k-\sqrt{k^2+a^2\lambda^2})t},$$

and

$$u = XT = \left(A_1 e^{\lambda x} + A_2 e^{-\lambda x}\right)\left(A_3 e^{(-k+\sqrt{k^2+a^2\lambda^2})t} + A_4 e^{(-k-\sqrt{k^2+a^2\lambda^2})t}\right).$$

For $-\lambda^2 < 0$ we obtain

$$X = A_1 \cos \lambda x + A_2 \sin \lambda x.$$

If $k^2 - a^2\lambda^2 > 0$ then

$$T = A_3 e^{(-k+\sqrt{k^2-a^2\lambda^2})t} + A_4 e^{(-k-\sqrt{k^2-a^2\lambda^2})t}.$$

If $k^2 - a^2\lambda^2 < 0$ then

$$T = e^{-kt}\left(A_3 \cos \sqrt{a^2\lambda^2 - k^2}\,t + A_4 \sin \sqrt{a^2\lambda^2 - k^2}\,t\right).$$

If $k^2 - a^2\lambda^2 = 0$ then

$$T = A_3 e^{-kt} + A_4 t e^{-kt}$$

so that

$$u = XT = (A_1 \cos \lambda x + A_2 \sin \lambda x)\left(A_3 e^{(-k+\sqrt{k^2-a^2\lambda^2})t} + A_4 e^{(-k-\sqrt{k^2-a^2\lambda^2})t}\right)$$

$$= (A_1 \cos \lambda x + A_2 \sin \lambda x)e^{-kt}\left(A_3 \cos \sqrt{a^2\lambda^2 - k^2}\,t + A_4 \sin \sqrt{a^2\lambda^2 - k^2}\,t\right)$$

$$= \left(A_1 \cos \frac{k}{a}x + A_2 \sin \frac{k}{a}x\right)\left(A_3 e^{-kt} + A_4 t e^{-kt}\right).$$

For $\lambda^2 = 0$ we obtain

$$X A_1 x + A_2,$$

$$T = A_3 + A_4 e^{-2kt},$$

and

$$u = XT = (A_1 x + A_2)(A_3 + A_4 e^{-2kt}).$$

13. If $u = XY$ then

$$u_{xx} = X''Y,$$

$$u_{yy} = XY'',$$

$$X''Y + xY'' = 0,$$

and

$$\frac{X''}{-X} = \frac{Y''}{Y} = \pm \lambda^2$$

so that

$$X'' \pm \lambda^2 X = 0 \quad \text{and} \quad Y'' \mp \lambda^2 Y = 0.$$

For $\lambda^2 > 0$ we obtain

$$X = A_1 \cos \lambda x + A_2 \sin \lambda x,$$

$$Y = A_3 e^{\lambda y} + A_4 e^{-\lambda y},$$

and

$$u = XY = (A_1 \cos \lambda x + A_2 \sin \lambda x)(A_3 e^{\lambda y} + A_4 e^{-\lambda y}).$$

For $-\lambda^2 < 0$ we obtain

$$X = A_1 e^{\lambda x} + A_2 e^{-\lambda x},$$

$$Y = A_3 \cos \lambda y + A_4 \sin \lambda y,$$

and

$$u = XY = (A_1 e^{\lambda x} + A_2 e^{-\lambda x})(A_3 \cos \lambda y + A_4 \sin \lambda y).$$

For $\lambda^2 = 0$ we obtain

$$u = XY = (A_1 x + A_2)(A_3 y + A_4).$$

14. If $u = XY$ then

$$u_{xx} = X''Y,$$

$$u_{yy} = XY'',$$

$$x^2 X''Y + xY'' = 0,$$

and

$$\frac{x^2 X''}{-X} = \frac{Y''}{Y} = \pm \lambda^2$$

so that

$$x^2 X'' \pm \lambda^2 X = 0 \quad \text{and} \quad Y'' \mp \lambda^2 Y = 0.$$

For $\lambda^2 > 0$ we obtain

$$Y = A_3 e^{\lambda y} + A_4 e^{-\lambda y}.$$

477

If $1 - 4\lambda^2 > 0$ then

$$X = A_1 x^{(1/2+\sqrt{1-4\lambda^2}/2)} + A_2 x^{(1/2-\sqrt{1-4\lambda^2}/2)}.$$

If $1 - 4\lambda^2 < 0$ then

$$X = x^{1/2} \left(A_1 \cos \frac{1}{2}\sqrt{4\lambda^2 - 1} \, \ln x + A_2 \sin \frac{1}{2}\sqrt{4\lambda^2 - 1} \, \ln x \right).$$

If $1 - 4\lambda^2 = 0$ then

$$X = A_1 x^{1/2} + A_2 x^{1/2} \ln x$$

so that

$$u = XY = \left(A_1 x^{(1/2+\sqrt{1-4\lambda^2}/2)} + A_2 x^{(1/2-\sqrt{1-4\lambda^2}/2)} \right) \left(A_3 e^{\lambda y} + A_4 e^{-\lambda y} \right)$$

$$= x^{1/2} \left(A_1 \cos \frac{1}{2}\sqrt{4\lambda^2 - 1} \, \ln x + A_2 \sin \frac{1}{2}\sqrt{4\lambda^2 - 1} \, \ln x \right) \left(A_3 e^{y/2} + A_4 e^{-y/2} \right)$$

$$= \left(A_1 x^{1/2} + A_2 x^{1/2} \ln x \right) \left(A_3 e^{y/2} + A_4 e^{-y/2} \right).$$

For $-\lambda^2 < 0$ we obtain

$$X = A_1 e^{(1/2+\sqrt{1+4\lambda^2}/2)x} + A_2 e^{(1/2-\sqrt{1+4\lambda^2}/2)x},$$

$$Y = A_3 \cos \lambda y + A_4 \sin \lambda y,$$

and

$$u = XY = \left(A_1 e^{(1/2+\sqrt{1+4\lambda^2}/2)x} + A_2 e^{(1/2-\sqrt{1+4\lambda^2}/2)x} \right) \left(A_3 \cos \lambda y + A_4 \sin \lambda y \right).$$

For $\lambda^2 = 0$ we obtain

$$X = A_1 x + A_2,$$

$$Y = A_3 y + A_4,$$

and

$$u = XY = (A_1 x + A_2)(A_3 y + A_4).$$

15. If $u = XY$ then

$$u_{xx} = X''Y,$$

$$u_{yy} = XY'',$$

$$X''Y + XY'' = XY,$$

and

$$\frac{X''}{X} = \frac{Y - Y''}{Y} = \pm\lambda^2$$

so that

$$X'' \mp \lambda^2 X = 0 \quad \text{and} \quad Y'' + (\pm\lambda^2 - 1)Y = 0.$$

For $\lambda^2 > 0$ we obtain

$$X = A_1 e^{\lambda x} + A_2 e^{-\lambda x}.$$

If $\lambda^2 - 1 > 0$ then

$$Y = A_3 \cos \sqrt{\lambda^2 - 1}\, y + A_4 \sin \sqrt{\lambda^2 - 1}\, y.$$

If $\lambda^2 - 1 < 0$ then

$$Y = A_3 e^{\sqrt{1-\lambda^2}\, y} + A_4 e^{-\sqrt{1-\lambda^2}\, y}.$$

If $\lambda^2 - 1 = 0$ then $Y = A_3 y + A_4$ so that

$$u = XY = \left(A_1 e^{\lambda x} + A_2 e^{-\lambda x} \right) \left(A_3 \cos \sqrt{\lambda^2 - 1}\, y + A_4 \sin \sqrt{\lambda^2 - 1}\, y \right),$$

$$= \left(A_1 e^{\lambda x} + A_2 e^{-\lambda x} \right) \left(A_3 e^{\sqrt{1-\lambda^2}\, y} + A_4 e^{-\sqrt{1-\lambda^2}\, y} \right)$$

$$= (A_1 e^x + A_2 e^{-x})(A_3 y + A_4).$$

For $-\lambda^2 < 0$ we obtain

$$X = A_1 \cos \lambda x + A_2 \sin \lambda x,$$

$$Y = A_3 e^{\sqrt{1+\lambda^2}\, y} + A_4 e^{-\sqrt{1+\lambda^2}\, y},$$

and

$$u = XY = (A_1 \cos \lambda x + A_2 \sin \lambda x) \left(A_3 e^{\sqrt{1+\lambda^2}\, y} + A_4 e^{-\sqrt{1+\lambda^2}\, y} \right).$$

For $\lambda^2 = 0$ we obtain

$$X = A_1 x + A_2,$$

$$Y = A_3 e^y + A_4 e^{-y},$$

and

$$u = XY = (A_1 x + A_2)(A_3 e^y + A_4 e^{-y}).$$

16. If $u = XT$ then

$$u_{tt} = XT'', \qquad u_{xx} = X''T, \quad \text{and} \quad a^2 X''T - g = XT'',$$

which is not separable.

17. Identifying $A = B = C = 1$, we compute $B^2 - 4AC = -3 < 0$. The equation is elliptic.

18. Identifying $A = 3$, $B = 5$, and $C = 1$, we compute $B^2 - 4AC = 13 > 0$. The equation is hyperbolic.

19. Identifying $A = 1$, $B = 6$, and $C = 9$, we compute $B^2 - 4AC = 0$. The equation is parabolic.

20. Identifying $A = 1$, $B = -1$, and $C = -3$, we compute $B^2 - 4AC = 13 > 0$. The equation is hyperbolic.

21. Identifying $A = 1$, $B = -9$, and $C = 0$, we compute $B^2 - 4AC = 81 > 0$. The equation is hyperbolic.

22. Identifying $A = 0$, $B = 1$, and $C = 0$, we compute $B^2 - 4AC = 1 > 0$. The equation is hyperbolic.

23. Identifying $A = 1$, $B = 2$, and $C = 1$, we compute $B^2 - 4AC = 0$. The equation is parabolic.

24. Identifying $A = 1$, $B = 0$, and $C = 1$, we compute $B^2 - 4AC = -4 < 0$. The equation is elliptic.

25. Identifying $A = a^2$, $B = 0$, and $C = -1$, we compute $B^2 - 4AC = 4a^2 > 0$. The equation is hyperbolic.

26. Identifying $A = k > 0$, $B = 0$, and $C = 0$, we compute $B^2 - 4AC = -4k < 0$. The equation is elliptic.

27. If $u = RT$ then

$$u_r = R'T,$$

$$u_{rr} = R''T,$$

$$u_t = RT',$$

$$RT' = k\left(R''T + \frac{1}{r}R'T\right),$$

and

$$\frac{r^2 R'' + rR'}{r^2 R} = \frac{T'}{kT} = \pm\lambda^2.$$

If we use $-\lambda^2 < 0$ then

$$r^2 R'' + rR' + \lambda^2 r^2 R = 0 \quad \text{and} \quad T'' \mp \lambda^2 kT = 0$$

so that

$$T = A_1 e^{-k\lambda^2 t},$$

$$R = A_2 J_0(\lambda r) + A_3 Y_0(\lambda r),$$

and

$$u = RT = e^{-k\lambda^2 t}[c_1 J_0(\lambda r) + c_2 Y_0(\lambda r)].$$

28. (a) We note that $\xi_x = \eta_x = 1$, $\xi_t = a$, and $\eta_t = -a$. Then

$$\frac{\partial u}{\partial x} = \frac{\partial u}{\partial \xi}\frac{\partial \xi}{\partial x} + \frac{\partial u}{\partial \eta}\frac{\partial \eta}{\partial x} = u_\xi + u_\eta$$

and

$$\frac{\partial^2 u}{\partial x^2} = \frac{\partial}{\partial x}(u_\xi + u_\eta) = \frac{\partial u_\xi}{\partial \xi}\frac{\partial \xi}{\partial x} + \frac{\partial u_\xi}{\partial \eta}\frac{\partial \eta}{\partial x} + \frac{\partial u_\eta}{\partial \xi}\frac{\partial \xi}{\partial x} + \frac{\partial u_\eta}{\partial \eta}\frac{\partial \eta}{\partial x}$$

$$= u_{\xi\xi} + 2u_{\xi\eta} + u_{\eta\eta}.$$

Similarly

$$\frac{\partial^2 u}{\partial t^2} = a^2(u_{\xi\xi} - 2u_{\xi\eta} + u_{\eta\eta}).$$

Thus

$$a^2\frac{\partial^2 u}{\partial x^2} = \frac{\partial^2 u}{\partial t^2} \quad \text{becomes} \quad \frac{\partial^2 u}{\partial \xi \partial \eta} = 0.$$

(b) Integrating

$$\frac{\partial^2 u}{\partial \xi \partial \eta} = \frac{\partial}{\partial \eta} u_\xi = 0$$

we obtain

$$\int \frac{\partial}{\partial \eta} u_\xi \, d\eta = \int 0 \, d\eta$$

$$u_\xi = f(\xi).$$

Integrating this result with respect to ξ we obtain

$$\int \frac{\partial u}{\partial \xi} \, d\xi = \int f(\xi) \, d\xi$$

$$u = F(\xi) + G(\eta).$$

Since $\xi = x + at$ and $\eta = x - at$, we then have

$$u = F(\xi) + G(\eta) = F(x + at) + G(x - at).$$

29. We compute

$$\frac{\partial^2 u}{\partial x^2} = m^2 e^{mx+ny}$$

$$\frac{\partial^2 u}{\partial x \partial y} = mn e^{mx+ny}$$

$$\frac{\partial^2 u}{\partial y^2} = n^2 e^{mx+ny}.$$

Substituting into the differential equation we obtain

$$m^2 e^{mx+ny} + mn e^{mx+ny} - 6n^2 e^{mx+ny} = 0$$

or

$$m^2 + mn - 6n^2 = (m + 3n)(m - 2n) = 0.$$

Then $m = -3n$ or $m = 2n$ and

$$u = e^{-3nx+ny} = e^{n(-3x+y)} \quad \text{or} \quad u = e^{2nx+ny} = e^{n(2x+y)}.$$

481

30. For $u = A_1 e^{\lambda^2 y} \cosh 2\lambda x + B_1 e^{\lambda^2 y} \sinh 2\lambda x$ we compute

$$\frac{\partial^2 u}{\partial x^2} = 4\lambda^2 A_1 e^{\lambda^2 y} \cosh 2\lambda x + 4\lambda^2 B_1 e^{\lambda^2 y} \sinh 2\lambda x$$

and

$$\frac{\partial u}{\partial y} = \lambda^2 A_1 e^{\lambda^2 y} \cosh 2\lambda x + \lambda^2 B_1 e^{\lambda^2 y} \sinh 2\lambda x.$$

Then $\partial^2 u / \partial x^2 = 4 \partial u / \partial y$.

For $u = A_2 e^{-\lambda^2 y} \cos 2\lambda x + B_2 e^{-\lambda^2 y} \sin 2\lambda x$ we compute

$$\frac{\partial^2 u}{\partial x^2} = -4\lambda^2 A_2 e^{-\lambda^2 y} \cos 2\lambda x - 4\lambda^2 B_2 e^{-\lambda^2 y} \sin 2\lambda x$$

and

$$\frac{\partial u}{\partial y} = -\lambda^2 A_2 \cos 2\lambda x - \lambda^2 B_2 \sin 2\lambda x.$$

Then $\partial^2 u / \partial x^2 = 4 \partial u / \partial y$.

For $u = A_3 x + B_3$ we compute $\partial^2 u / \partial x^2 = \partial u / \partial y = 0$. Then $\partial^2 u / \partial x^2 = 4 \partial u / \partial y$.

31. We identify $A = xy + 1$, $B = x + 2y$, and $C = 1$. Then $B^2 - 4AC = x^2 + 4y^2 - 4$. The equation $x^2 + 4y^2 = 4$ defines an ellipse. The partial differential equation is hyperbolic outside the ellipse, parabolic on the ellipse, and elliptic inside the ellipse.

Exercises 12.2

1. $k \dfrac{\partial^2 u}{\partial x^2} = \dfrac{\partial u}{\partial t}$, $\quad 0 < x < L$, $t > 0$

$u(0, t) = 0$, $\quad \dfrac{\partial u}{\partial x}\bigg|_{x=L} = 0$, $\quad t > 0$

$u(x, 0) = f(x)$, $\quad 0 < x < L$

2. $k \dfrac{\partial^2 u}{\partial x^2} = \dfrac{\partial u}{\partial t}$, $\quad 0 < x < L$, $t > 0$

$u(0, t) = u_0$, $\quad u(L, t) = u_1$, $\quad t > 0$

$u(x, 0) = 0$, $\quad 0 < x < L$

3. $k \dfrac{\partial^2 u}{\partial x^2} = \dfrac{\partial u}{\partial t}$, $\quad 0 < x < L$, $t > 0$

$u(0, t) = 100$, $\quad \dfrac{\partial u}{\partial x}\bigg|_{x=L} = -hu(L, t)$, $\quad t > 0$

$u(x, 0) = f(x)$, $\quad 0 < x < L$

4. $k\dfrac{\partial^2 u}{\partial x^2} + h(u - 50) = \dfrac{\partial u}{\partial t}, \quad 0 < x < L, \ t > 0$

$\dfrac{\partial u}{\partial x}\bigg|_{x=0} = 0, \quad \dfrac{\partial u}{\partial x}\bigg|_{x=L} \qquad t > 0$

$u(x, 0) = 100, \quad 0 < x < L$

5. $a^2 \dfrac{\partial^2 u}{\partial x^2} = \dfrac{\partial^2 u}{\partial t^2}, \quad 0 < x < L, \ t > 0$

$u(0, t) = 0, \quad u(L, t) = 0, \quad t > 0$

$u(x, 0) = x(L - x), \quad \dfrac{\partial u}{\partial x}\bigg|_{t=0} = 0, \quad 0 < x < L$

6. $a^2 \dfrac{\partial^2 u}{\partial x^2} = \dfrac{\partial^2 u}{\partial t^2}, \quad 0 < x < L, \ t > 0$

$u(0, t) = 0, \quad u(L, t) = 0, \quad t > 0$

$u(x, 0) = 0, \quad \dfrac{\partial u}{\partial x}\bigg|_{t=0} = \sin \dfrac{\pi x}{L}, \quad 0 < x < L$

7. $a^2 \dfrac{\partial^2 u}{\partial x^2} - 2\beta \dfrac{\partial u}{\partial t} = \dfrac{\partial^2 u}{\partial t^2}, \quad 0 < x < L, \ t > 0$

$u(0, t) = 0, \quad u(L, t) = \sin \pi t, \quad t > 0$

$u(x, 0) = f(x), \quad \dfrac{\partial u}{\partial t}\bigg|_{t=0} = 0, \quad 0 < x < L$

8. $a^2 \dfrac{\partial^2 u}{\partial x^2} + Ax = \dfrac{\partial^2 u}{\partial t^2}, \quad 0 < x < L, \ t > 0, \ A \text{ a constant}$

$u(0, t) = 0, \quad u(L, t) = 0, \quad t > 0$

$u(x, 0) = 0, \quad \dfrac{\partial u}{\partial x}\bigg|_{t=0} = 0, \quad 0 < x < L$

9. $\dfrac{\partial^2 u}{\partial x^2} + \dfrac{\partial^2 u}{\partial y^2} = 0, \quad 0 < x < 4, \ 0 < y < 2$

$\dfrac{\partial u}{\partial x}\bigg|_{x=0} = 0, \quad u(4, y) = f(y), \quad 0 < y < 2$

$\dfrac{\partial u}{\partial y}\bigg|_{y=0} = 0, \quad u(x, 2) = 0, \quad 0 < x < 4$

10. $\dfrac{\partial^2 u}{\partial x^2} + \dfrac{\partial^2 u}{\partial y^2} = 0, \quad 0 < x < \pi, \ y > 0$

$$u(0, y) = e^{-y}, \quad u(\pi, y) = \begin{cases} 100, & 0 < y \le 1 \\ 0, & y > 1 \end{cases}$$

$$u(x, 0) = f(x), \quad 0 < x < \pi$$

──────── Exercises 12.3 ────────

1. Using $u = XT$ and $-\lambda^2$ as a separation constant leads to

$$X'' + \lambda^2 X = 0,$$

$$X(0) = 0,$$

$$X(L) = 0,$$

and

$$T' + k\lambda^2 T = 0.$$

Then

$$X = c_1 \sin \frac{n\pi}{L} x \quad \text{and} \quad T = c_2 e^{-\frac{kn^2\pi^2}{L^2}t}$$

for $n = 1, 2, 3, \dots$ so that

$$u = \sum_{n=1}^{\infty} A_n e^{-\frac{kn^2\pi^2}{L^2}t} \sin \frac{n\pi}{L} x.$$

Imposing

$$u(x, 0) = \sum_{n=1}^{\infty} A_n \sin \frac{n\pi}{L} x$$

gives

$$A_n = \frac{2}{L} \int_0^{L/2} \sin \frac{n\pi}{L} x \, dx = \frac{2}{n\pi} \left(1 - \cos \frac{n\pi}{2} \right)$$

for $n = 1, 2, 3, \dots$ so that

$$u(x, t) = \frac{2}{\pi} \sum_{n=1}^{\infty} \frac{1 - \cos \frac{n\pi}{2}}{n} e^{-\frac{kn^2\pi^2}{L^2}t} \sin \frac{n\pi}{L} x.$$

2. Using $u = XT$ and $-\lambda^2$ as a separation constant leads to

$$X'' + \lambda^2 X = 0,$$

$$X(0) = 0,$$

$$X(L) = 0,$$

and

$$T' + k\lambda^2 T = 0.$$

Then

$$X = c_1 \sin \frac{n\pi}{L} x \quad \text{and} \quad T = c_2 e^{-\frac{kn^2\pi^2}{L^2}t}$$

for $n = 1, 2, 3, \ldots$ so that

$$u = \sum_{n=1}^{\infty} A_n e^{-\frac{kn^2\pi^2}{L^2}t} \sin \frac{n\pi}{L} x.$$

Imposing

$$u(x, 0) = \sum_{n=1}^{\infty} A_n \sin \frac{n\pi}{L} x$$

gives

$$A_n = \frac{2}{L} \int_0^L x(L - x) \sin \frac{n\pi}{L} x \, dx = \frac{4L^2}{n^3\pi^3}[1 - (-1)^n]$$

for $n = 1, 2, 3, \ldots$ so that

$$u(x, t) = \frac{4L^2}{\pi^3} \sum_{n=1}^{\infty} \frac{1 - (-1)^n}{n^3} e^{-\frac{kn^2\pi^2}{L^2}t} \sin \frac{n\pi}{L} x.$$

3. Using $u = XT$ and $-\lambda^2$ as a separation constant leads to

$$X'' + \lambda^2 X = 0,$$
$$X'(0) = 0,$$
$$X'(L) = 0,$$

and

$$T' + k\lambda^2 T = 0.$$

Then

$$X = c_1 \cos \frac{n\pi}{L} x \quad \text{and} \quad T = c_2 e^{-\frac{kn^2\pi^2}{L^2}t}$$

for $n = 0, 1, 2, \ldots$ so that

$$u = \sum_{n=0}^{\infty} A_n e^{-\frac{kn^2\pi^2}{L^2}t} \cos \frac{n\pi}{L} x.$$

Imposing

$$u(x, 0) = f(x) = \sum_{n=0}^{\infty} A_n \cos \frac{n\pi}{L} x$$

gives

$$u(x, t) = \frac{1}{L} \int_0^L f(x) \, dx + \frac{2}{L} \sum_{n=1}^{\infty} \left(\int_0^L f(x) \cos \frac{n\pi}{L} x \, dx \right) e^{-\frac{kn^2\pi^2}{L^2}t} \cos \frac{n\pi}{L} x.$$

4. If $L = 2$ and $f(x)$ is x for $0 < x < 1$ and $f(x)$ is 0 for $1 < x < 2$ then

$$u(x, t) = \frac{1}{4} + 4 \sum_{n=1}^{\infty} \left[\frac{1}{2n\pi} \sin \frac{n\pi}{2} + \frac{1}{n^2\pi^2} \left(\cos \frac{n\pi}{2} - 1 \right) \right] e^{-\frac{kn^2\pi^2}{4}t} \cos \frac{n\pi}{2} x.$$

5. Using $u = XT$ and $-\lambda^2$ as a separation constant leads to

$$X'' + \lambda^2 X = 0,$$
$$X'(0) = 0,$$
$$X'(L) = 0,$$

and

$$T' + (h + k\lambda^2)T = 0.$$

Then

$$X = c_1 \cos \frac{n\pi}{L} x \quad \text{and} \quad T = c_2 e^{-\left(h + \frac{kn^2\pi^2}{L^2}\right)t}$$

for $n = 0, 1, 2, \ldots$ so that

$$u = \sum_{n=0}^{\infty} A_n e^{-\left(h + \frac{kn^2\pi^2}{L^2}\right)t} \cos \frac{n\pi}{L} x.$$

Imposing

$$u(x,0) = f(x) = \sum_{n=0}^{\infty} A_n \cos \frac{n\pi}{L} x$$

gives

$$u(x,t) = \frac{e^{-ht}}{L} \int_0^L f(x)\, dx + \frac{2}{L} \sum_{n=1}^{\infty} \left(\int_0^L f(x) \cos \frac{n\pi}{L} x\, dx \right) e^{-\left(h + \frac{kn^2\pi^2}{L^2}\right)t} \cos \frac{n\pi}{L} x.$$

6. In Problem 5 we instead find that $X(0) = 0$ and $X(L) = 0$ so that

$$X = c_1 \sin \frac{n\pi}{L} x$$

and

$$u = \frac{2}{L} \sum_{n=1}^{\infty} \left(\int_0^L f(x) \sin \frac{n\pi}{L} x\, dx \right) e^{-\left(h + \frac{kn^2\pi^2}{L^2}\right)t} \sin \frac{n\pi}{L} x.$$

Exercises 12.4

1. Using $u = XT$ and $-\lambda^2$ as a separation constant leads to

$$X'' + \lambda^2 X = 0,$$
$$X(0) = 0,$$
$$X(L) = 0,$$

and

$$T'' + \lambda^2 a^2 T = 0.$$

Then

$$X = c_1 \sin \frac{n\pi}{L} x \quad \text{and} \quad T = c_2 \cos \frac{n\pi a}{L} t + c_3 \sin \frac{n\pi a}{L} t$$

for $n = 1, 2, 3, \ldots$ so that

$$u = \sum_{n=1}^{\infty} \left(A_n \cos \frac{n\pi a}{L} t + B_n \sin \frac{n\pi a}{L} t \right) \sin \frac{n\pi}{L} x.$$

Imposing

$$u(x, 0) = \frac{1}{4} x(L - x) = \sum_{n=1}^{\infty} A_n \sin \frac{n\pi}{L} x$$

and

$$u_t(x, 0) = 0 = \sum_{n=1}^{\infty} B_n \frac{n\pi a}{L} \sin \frac{n\pi}{L} x$$

gives

$$A_n = \frac{L}{n^3 \pi^3} [1 - (-1)^n] \quad \text{and} \quad B_n = 0$$

for $n = 1, 2, 3, \ldots$ so that

$$u(x, t) = \frac{L^2}{\pi^3} \sum_{n=1}^{\infty} \frac{1 - (-1)^n}{n^3} \cos \frac{n\pi a}{L} t \sin \frac{n\pi}{L} x.$$

2. Using $u = XT$ and $-\lambda^2$ as a separation constant leads to

$$X'' + \lambda^2 X = 0,$$

$$X(0) = 0,$$

$$X(L) = 0,$$

and

$$T'' + \lambda^2 a^2 T = 0.$$

Then

$$X = c_1 \sin \frac{n\pi}{L} x \quad \text{and} \quad T = c_2 \cos \frac{n\pi a}{L} t + c_3 \sin \frac{n\pi a}{L} t$$

for $n = 1, 2, 3, \ldots$ so that

$$u = \sum_{n=1}^{\infty} \left(A_n \cos \frac{n\pi a}{L} t + B_n \sin \frac{n\pi a}{L} t \right) \sin \frac{n\pi}{L} x.$$

Imposing

$$u(x, 0) = 0 = \sum_{n=1}^{\infty} A_n \sin \frac{n\pi}{L} x$$

and

$$u_t(x, 0) = x(L - x) = \sum_{n=1}^{\infty} B_n \frac{n\pi a}{L} \sin \frac{n\pi}{L} x$$

gives

$$A_n = 0 \quad \text{and} \quad B_n \frac{n\pi a}{L} = \frac{4L^2}{n^3\pi^3}[1 - (-1)^n]$$

for $n = 1, 2, 3, \ldots$ so that

$$u(x,t) = \frac{4L^3}{a\pi^4} \sum_{n=1}^{\infty} \frac{1 - (-1)^n}{n^4} \sin \frac{n\pi a}{L}t \sin \frac{n\pi}{L}x.$$

3. Using $u = XT$ and $-\lambda^2$ as a separation constant leads to

$$X'' + \lambda^2 X = 0,$$

$$X(0) = 0,$$

$$X(L) = 0,$$

and

$$T'' + \lambda^2 a^2 T = 0.$$

Then

$$X = c_1 \sin \frac{n\pi}{L}x \quad \text{and} \quad T = c_2 \cos \frac{n\pi a}{L}t + c_3 \sin \frac{n\pi a}{L}t$$

for $n = 1, 2, 3, \ldots$ so that

$$u = \sum_{n=1}^{\infty} \left(A_n \cos \frac{n\pi a}{L}t + B_n \sin \frac{n\pi a}{L}t \right) \sin \frac{n\pi}{L}x.$$

Imposing

$$u(x,0) = \sum_{n=1}^{\infty} A_n \sin \frac{n\pi}{L}x$$

gives

$$A_n = \frac{2}{L} \left(\int_0^{L/3} \frac{3}{L}x \sin \frac{n\pi}{L}x \, dx + \int_{L/3}^{2L/3} \sin \frac{n\pi}{L}x \, dx + \int_{2L/3}^{L} \left(3 - \frac{3}{L}x \right) \sin \frac{n\pi}{L}x \, dx \right)$$

so that

$$A_1 = \frac{6\sqrt{3}}{\pi^2},$$

$$A_2 = A_3 = A_4 = 0,$$

$$A_5 = -\frac{6\sqrt{3}}{5^2\pi^2},$$

$$A_6 = 0,$$

$$a_7 = \frac{6\sqrt{3}}{7^2\pi^2} \ \cdots.$$

Imposing

$$u_t(x,0) = 0 = \sum_{n=1}^{\infty} B_n \frac{n\pi a}{L} \sin \frac{n\pi}{L} x$$

gives $B_n = 0$ for $n = 1, 2, 3, \ldots$ so that

$$u(x,t) = \frac{6\sqrt{3}}{\pi^2} \left(\cos \frac{\pi a}{L} t \sin \frac{\pi}{L} x - \frac{1}{5^2} \cos \frac{5\pi a}{L} t \sin \frac{5\pi}{L} x + \frac{1}{7^2} \cos \frac{7\pi a}{L} t \sin \frac{7\pi}{L} x - \cdots \right).$$

4. Using $u = XT$ and $-\lambda^2$ as a separation constant leads to

$$X'' + \lambda^2 X = 0,$$

$$X(0) = 0,$$

$$X(\pi) = 0,$$

and

$$T'' + \lambda^2 a^2 T = 0.$$

Then

$$X = c_1 \sin nx \quad \text{and} \quad T = c_2 \cos nat + c_3 \sin nat$$

for $n = 1, 2, 3, \ldots$ so that

$$u = \sum_{n=1}^{\infty} (A_n \cos nat + B_n \sin nat) \sin nx.$$

Imposing

$$u(x,0) = \frac{1}{6} x(\pi^2 - x^2) = \sum_{n=1}^{\infty} A_n \sin nx \quad \text{and} \quad u_t(x,0) = 0$$

gives

$$B_n = 0 \quad \text{and} \quad A_n = \frac{2}{n^3} (-1)^{n+1}$$

for $n = 1, 2, 3, \ldots$ so that

$$u(x,t) = 2 \sum_{n=1}^{\infty} \frac{(-1)^{n+1}}{n^3} \cos nat \sin nx.$$

5. Using $u = XT$ and $-\lambda^2$ as a separation constant leads to

$$X'' + \lambda^2 X = 0,$$

$$X(0) = 0,$$

$$X(\pi) = 0,$$

and

$$T'' + \lambda^2 a^2 T = 0.$$

489

Then

$$X = c_1 \sin nx \quad \text{and} \quad T = c_2 \cos nat + c_3 \sin nat$$

for $n = 1, 2, 3, \ldots$ so that

$$u = \sum_{n=1}^{\infty} (A_n \cos nat + B_n \sin nat) \sin nx.$$

Imposing

$$u(x,0) = 0 = \sum_{n=1}^{\infty} A_n \sin nx \quad \text{and} \quad u_t(x,0) = \sin x = \sum_{n=1}^{\infty} B_n na \sin nx$$

gives

$$A_n = 0, \quad B_1 = \frac{1}{a^2}, \quad \text{and} \quad B_n = 0$$

for $n = 2, 3, 4, \ldots$ so that

$$u(x,t) = \frac{1}{a} \sin at \, \sin x.$$

6. Using $u = XT$ and $-\lambda^2$ as a separation constant leads to

$$X'' + \lambda^2 X = 0,$$
$$X(0) = 0,$$
$$X(1) = 0,$$

and

$$T'' + \lambda^2 a^2 T = 0.$$

Then

$$X = c_1 \sin n\pi x \quad \text{and} \quad T = c_2 \cos n\pi at + c_3 \sin n\pi at$$

for $n = 1, 2, 3, \ldots$ so that

$$u = \sum_{n=1}^{\infty} (A_n \cos n\pi at + B_n \sin n\pi at) \sin n\pi x.$$

Imposing

$$u(x,0) = 0.01 \sin 3\pi x = \sum_{n=1}^{\infty} A_n \sin n\pi x$$

and

$$u_t(x,0) = 0 = \sum_{n=1}^{\infty} B_n n\pi a \sin n\pi x$$

gives $B_n = 0$ for $n = 1, 2, 3, \ldots$, $A_3 = 0.01$, and $A_n = 0$ for $n = 1, 2, 4, 5, 6, \ldots$ so that

$$u(x,t) = 0.01 \sin 3\pi x \, \cos 3\pi at.$$

7. Using $u = XT$ and $-\lambda^2$ as a separation constant leads to

$$X'' + \lambda^2 X = 0,$$
$$X(0) = 0,$$
$$X(L) = 0,$$

and

$$T'' + \lambda^2 a^2 T = 0.$$

Then

$$X = c_1 \sin \frac{n\pi}{L} x \quad \text{and} \quad T = c_2 \cos \frac{n\pi a}{L} t + c_3 \sin \frac{n\pi a}{L} t$$

for $n = 1, 2, 3, \ldots$ so that

$$u = \sum_{n=1}^{\infty} \left(A_n \cos \frac{n\pi a}{L} t + B_n \sin \frac{n\pi a}{L} t \right) \sin \frac{n\pi}{L} x.$$

Imposing

$$u(x, 0) = \sum_{n=1}^{\infty} A_n \sin \frac{n\pi}{L} x$$

and

$$u_t(x, 0) = 0 = \sum_{n=1}^{\infty} B_n \frac{n\pi a}{L} \sin \frac{n\pi}{L} x$$

gives

$$B_n = 0 \quad \text{and} \quad A_n = \frac{8h}{n^2 \pi^2} \sin \frac{n\pi}{2}$$

for $n = 1, 2, 3, \ldots$ so that

$$u(x, t) = \frac{8h}{\pi^2} \sum_{n=1}^{\infty} \frac{1}{n^2} \sin \frac{n\pi}{2} \cos \frac{n\pi a}{L} t \sin \frac{n\pi}{L} x.$$

8. Using $u = XT$ and $-\lambda^2$ as a separation constant leads to

$$X'' + \lambda^2 X = 0,$$
$$X'(0) = 0,$$
$$X'(L) = 0,$$

and

$$T'' + a^2 \lambda^2 T = 0.$$

Then

$$X = c_1 \cos \frac{n\pi}{L} x \quad \text{and} \quad T = c_2 \cos \frac{n\pi a}{L} t$$

491

for $n = 1, 2, 3, \ldots$. Since $\lambda = 0$ is also an eigenvalue with eigenfunction $X(x) = 1$ we have

$$u = A_0 + \sum_{n=1}^{\infty} A_n \cos \frac{n\pi a}{L} t \cos \frac{n\pi}{L} x.$$

Imposing

$$u(x, 0) = x = \sum_{n=0}^{\infty} A_n \cos \frac{n\pi}{L} x$$

gives

$$A_0 = \frac{1}{L} \int_0^L x \, dx = \frac{L}{2}$$

and

$$A_n = \frac{2}{L} \int_0^L x \cos \frac{n\pi}{L} x \, dx = \frac{2L}{n^2 \pi^2}[(-1)^n - 1]$$

for $n = 1, 2, 3, \ldots$, so that

$$u(x, t) = \frac{L}{2} + \frac{2L}{\pi^2} \sum_{n=1}^{\infty} \frac{(-1)^n - 1}{n^2} \cos \frac{n\pi a}{L} t \cos \frac{n\pi}{L} x.$$

9. Using $u = XT$ and $-\lambda^2$ as a separation constant leads to

$$X'' + \lambda^2 X = 0,$$

$$X(0) = 0,$$

$$X(\pi) = 0,$$

and

$$T'' + 2\beta T' + \lambda^2 T = 0.$$

Then

$$X = c_1 \sin nx \quad \text{and} \quad T = e^{-\beta t}\left(c_2 \cos \sqrt{n^2 - \beta^2}\, t + c_3 \sin \sqrt{n^2 - \beta^2}\, t\right)$$

so that

$$u = \sum_{n=1}^{\infty} e^{-\beta t}\left(A_n \cos \sqrt{n^2 - \beta^2}\, t + B_n \sin \sqrt{n^2 - \beta^2}\, t\right) \sin nx.$$

Imposing

$$u(x, 0) = f(x) = \sum_{n=1}^{\infty} A_n \sin nx$$

and

$$u_t(x, 0) = 0 = \sum_{n=1}^{\infty} \left(B_n \sqrt{n^2 - \beta^2} - \beta A_n\right) \sin nx$$

gives

$$u(x, t) = e^{-\beta t} \sum_{n=1}^{\infty} A_n \left(\cos \sqrt{n^2 - \beta^2}\, t + \frac{\beta}{\sqrt{n^2 - \beta^2}} \sin \sqrt{n^2 - \beta^2}\, t\right) \sin nx,$$

where

$$A_n = \frac{2}{\pi} \int_0^\pi f(x) \sin nx \, dx.$$

10. Using $u = XT$ and $-\lambda^2$ as a separation constant leads to $X'' + \lambda^2 X = 0$, $X(0) = 0$, $X(\pi) = 0$ and $T'' + (1 + \lambda^2)T = 0$, $T'(0) = 0$. Then $X = c_2 \sin nx$ and $T = c_3 \cos \sqrt{n^2 + 1}\, t$ for $n = 1, 2, 3, \ldots$ so that

$$u = \sum_{n1}^\infty B_n \cos \sqrt{n^2 + 1}\, t \sin nx.$$

Imposing $u(x, 0) = \sum_{n=1}^\infty B_n \sin nx$ gives

$$B_n = \frac{2}{\pi} \int_0^{\pi/2} x \sin nx \, dx + \frac{2}{\pi} \int_{\pi/2}^\pi (\pi - x) \sin nx \, dx = \frac{4}{\pi n^2} \sin \frac{n\pi}{2}$$

$$= \begin{cases} 0, & n \text{ even} \\ \frac{4}{\pi n^2}(-1)^{(n+3)/2}, & n = 2k - 1, \ k = 1, 2, 3, \ldots. \end{cases}$$

Thus with $n = 2k - 1$,

$$u(x, t) = \frac{4}{\pi} \sum_{n=1}^\infty \frac{\sin \frac{n\pi}{2}}{n^2} \cos \sqrt{n^2 + 1}\, t \sin nx = \frac{4}{\pi} \sum_{k=1}^\infty \frac{(-1)^{k+1}}{(2k-1)^2} \cos \sqrt{(2k-1)^2 + 1}\, t \sin(2k - 1)x.$$

11. Separating variables in the partial differential equation gives

$$\frac{X^{(4)}}{X} = -\frac{T''}{a^2 T} = \lambda^4$$

so that

$$X^{(4)} - \lambda^4 X = 0$$

$$T'' + a^2 \lambda^4 T = 0$$

and

$$X = c_1 \cosh \lambda x + c_2 \sinh \lambda x + c_3 \cos \lambda x + c_4 \sin \lambda x$$

$$T = c_5 \cos a\lambda^2 t + c_6 \sin a\lambda^2 t.$$

The boundary conditions translate into $X(0) = X(L) = 0$ and $X''(0) = X''(L) = 0$. From $X(0) = X''(0) = 0$ we find $c_1 = c_3 = 0$. From

$$X(L) = c_2 \sinh \lambda L + c_4 \sin \lambda L = 0$$

$$X''(L) = \lambda^2 c_2 \sinh \lambda L - \lambda^2 c_4 \sin \lambda L = 0$$

we see by the subtraction that $c_4 \sin \lambda L = 0$. This equation yields the eigenvalues $\lambda = n\pi L$ for $n = 1, 2, 3, \ldots$. The corresponding eigenfunctions are

$$X = c_4 \sin \frac{n\pi}{L} x.$$

493

Thus

$$u(x,t) = \sum_{n=1}^{\infty} \left(A_n \cos \frac{n^2\pi^2}{L^2} at + B_n \sin \frac{n^2\pi^2}{L^2} at \right) \sin \frac{n\pi}{L} x.$$

From

$$u(x,0) = f(x) = \sum_{n=1}^{\infty} A_n \sin \frac{n\pi}{L} x$$

we obtain

$$A_n = \frac{2}{L} \int_0^L f(x) \sin \frac{n\pi}{L} x \, dx.$$

From

$$\frac{\partial u}{\partial t} = \sum_{n=1}^{\infty} \left(-A_n \frac{n^2\pi^2 a}{L^2} \sin \frac{n^2\pi^2}{L^2} at + B_n \frac{n^2\pi^2 a}{L^2} \cos \frac{n^2\pi^2}{L^2} at \right) \sin \frac{n\pi}{L} x$$

and

$$\frac{\partial u}{\partial t} \bigg|_{t=0} = g(x) = \sum_{n=1}^{\infty} B_n \frac{n^2\pi^2 a}{L^2} \sin \frac{n\pi}{L} x$$

we obtain

$$B_n \frac{n^2\pi^2 a}{L^2} = \frac{2}{L} \int_0^L g(x) \sin \frac{n\pi}{L} x \, dx$$

and

$$B_n = \frac{2L}{n^2\pi^2 a} \int_0^L g(x) \sin \frac{n\pi}{L} x \, dx.$$

12. In this case the boundary conditions become, for $t > 0$,

$$u(0,t) = 0, \qquad u(L,t) = 0,$$

$$\frac{\partial u}{\partial x} \bigg|_{x=0} = 0, \qquad \frac{\partial u}{\partial x} \bigg|_{x=L} = 0.$$

13. From (5) in the text we have

$$u(x,t) = \sum_{n=1}^{\infty} \left(A_n \cos \frac{n\pi a}{L} t + B_n \sin \frac{n\pi a}{L} t \right) \sin \frac{n\pi}{L} x.$$

Since $u_t(x,0) = g(x) = 0$ we have $B_n = 0$ and

$$u(x,t) = \sum_{n=1}^{\infty} A_n \cos \frac{n\pi a}{L} t \sin \frac{n\pi}{L} x$$

$$= \sum_{n=1}^{\infty} A_n \frac{1}{2} \left[\sin \left(\frac{n\pi}{L} x + \frac{n\pi a}{L} t \right) + \sin \left(\frac{n\pi}{L} x - \frac{n\pi a}{L} t \right) \right]$$

$$= \frac{1}{2} \sum_{n=1}^{\infty} A_n \left[\sin \frac{n\pi}{L} (x + at) + \sin \frac{n\pi}{L} (x - at) \right].$$

From

$$u(x, 0) = f(x) = \sum_{n=1}^{\infty} A_n \sin \frac{n\pi}{L} x$$

we identify

$$f(x + at) = \sum_{n=1}^{\infty} A_n \sin \frac{n\pi}{L} (x + at)$$

and

$$f(x - at) = \sum_{n=1}^{\infty} A_n \sin \frac{n\pi}{L} (x - at),$$

so that

$$u(x, t) = \frac{1}{2}[f(x + at) + f(x - at)].$$

14. **(a)** We have

$$u(x, t) = F(x + at) + G(x - at)$$

$$u(x, 0) = F(x) + G(x) = f(x)$$

$$u_t(x, 0) = aF'(x) - aG'(x) = g(x)$$

Integrating the last equation with respect to x gives

$$F(x) - G(x) = \frac{1}{a} \int_{x_0}^{x} g(s)\, ds + c_1.$$

Substituting $G(x) = f(x) - F(x)$ we obtain

$$F(x) = \frac{1}{2} f(x) + \frac{1}{2a} \int_{x_0}^{x} g(s)\, ds + C$$

where $c = c_1/2$. Thus

$$G(x) = \frac{1}{2} f(x) - \frac{1}{2a} \int_{x_0}^{x} g(s)\, ds - c.$$

(b) From the expressions for F and G,

$$F(x + at) = \frac{1}{2} f(x + at) + \frac{1}{2a} \int_{x_0}^{x+at} g(s)\, ds + c$$

$$G(x - at) = \frac{1}{2} f(x - at) - \frac{1}{2a} \int_{x_0}^{x-at} g(s)\, ds - c.$$

Thus,

$$u(x, t) = F(x + at) + G(x - at) = \frac{1}{2}[f(x + at) + f(x - at)] + \frac{1}{2a} \int_{x-at}^{x+at} g(s)\, ds.$$

Here we have used $-\int_{x_0}^{x-at} g(s)\, ds = \int_{x-at}^{x_0} g(s)\, ds.$

15. $u(x,t) = \dfrac{1}{2}[\sin(x+at) + \sin(x-at)] + \dfrac{1}{2a}\displaystyle\int_{x-at}^{x+at} ds$

$= \dfrac{1}{2}[\sin x \cos at + \cos x \sin at + \sin x \cos at - \cos x \sin at] + \dfrac{1}{2a}\, s\,\Big|_{x-at}^{x+at} = \sin x \cos at + t$

16. $u(x,t) = \dfrac{1}{2}\sin(x+at) + \sin(x-at)] + \dfrac{1}{2a}\displaystyle\int_{x-at}^{x+at} \cos s \, ds$

$= \sin x \cos at + \dfrac{1}{2a}[\sin(x+at) - \sin(x-at)] = \sin x \cos at + \dfrac{1}{a}\cos x \sin at$

17. $u(x,t) = 0 + \dfrac{1}{2a}\displaystyle\int_{x-at}^{x+at} \sin 2s \, ds = \dfrac{1}{2a}\left[\dfrac{-\cos(2x+2at) + \cos(2x-2at)}{2}\right]$

$= \dfrac{1}{4a}[-\cos 2x \cos 2at + \sin 2x \sin 2at + \cos 2x \cos 2at + \sin 2x \sin 2at] = \dfrac{1}{2a}\sin 2x \sin 2at$

18.

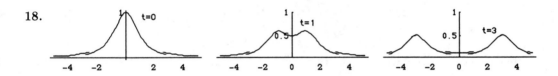

19. (a)

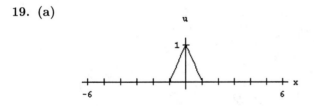

(b) Since $g(x) = 0$, d'Alembert's solution with $a = 1$ is

$$u(x,t) = \dfrac{1}{2}[f(x+t) + f(x-t)].$$

Sample plots are shown below.

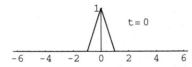 

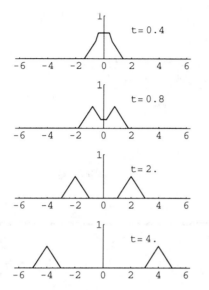

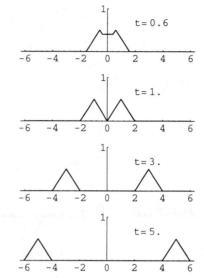

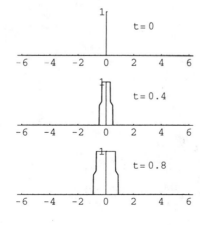

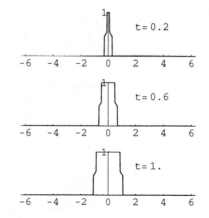

(c) The single peaked wave disolves into two peaks moving outward.

20. (a) With $a = 1$, d'Alembert's solution is

$$u(x,t) = \frac{1}{2} \int_{x-t}^{x+t} g(s)\, ds \qquad \text{where} \qquad g(s) = \begin{cases} 1, & |s| \le 0.1 \\ 0, & |s| > 0.1 \end{cases}.$$

Sample plots are shown below.

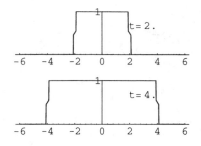 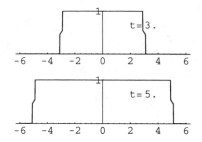

(c) The string has a roughly rectangular shape with the base on the x-axis increasing in length.

Exercises 12.5

1. Using $u = XY$ and $-\lambda^2$ as a separation constant leads to

$$X'' + \lambda^2 X = 0,$$
$$X(0) = 0,$$
$$X(a) = 0,$$

and

$$Y'' - \lambda^2 Y = 0,$$
$$Y(0) = 0.$$

Then

$$X = c_1 \sin \frac{n\pi}{a} x \quad \text{and} \quad Y = c_2 \sinh \frac{n\pi}{a} y$$

for $n = 1, 2, 3, \ldots$ so that

$$u = \sum_{n=1}^{\infty} A_n \sin \frac{n\pi}{a} x \sinh \frac{n\pi}{a} y.$$

Imposing

$$u(x, b) = f(x) = \sum_{n=1}^{\infty} A_n \sinh \frac{n\pi b}{a} \sin \frac{n\pi}{a} x$$

gives

$$A_n \sinh \frac{n\pi b}{a} = \frac{2}{a} \int_0^a f(x) \sin \frac{n\pi}{a} x \, dx$$

so that

$$u(x, y) = \sum_{n=1}^{\infty} A_n \sin \frac{n\pi}{a} x \sinh \frac{n\pi}{a} y$$

498

where

$$A_n = \frac{2}{a} \operatorname{csch} \frac{n\pi b}{a} \int_0^a f(x) \sin \frac{n\pi}{a} x \, dx.$$

2. Using $u = XY$ and $-\lambda^2$ as a separation constant leads to

$$X'' + \lambda^2 X = 0,$$

$$X(0) = 0,$$

$$X(a) = 0,$$

and

$$Y'' - \lambda^2 Y = 0,$$

$$Y'(0) = 0.$$

Then

$$X = c_1 \sin \frac{n\pi}{a} x \quad \text{and} \quad Y = c_2 \cosh \frac{n\pi}{a} y$$

for $n = 1, 2, 3, \ldots$ so that

$$u = \sum_{n=1}^{\infty} A_n \sin \frac{n\pi}{a} x \cosh \frac{n\pi}{a} y.$$

Imposing

$$u(x, b) = f(x) = \sum_{n=1}^{\infty} A_n \cosh \frac{n\pi b}{a} \sin \frac{n\pi}{a} x$$

gives

$$A_n \cosh \frac{n\pi b}{a} = \frac{2}{a} \int_0^a f(x) \sin \frac{n\pi}{a} x \, dx$$

so that

$$u(x, y) = \sum_{n=1}^{\infty} A_n \sin \frac{n\pi}{a} x \cosh \frac{n\pi}{a} y$$

where

$$A_n = \frac{2}{a} \operatorname{sech} \frac{n\pi b}{a} \int_0^a f(x) \sin \frac{n\pi}{a} x \, dx.$$

3. Using $u = XY$ and $-\lambda^2$ as a separation constant leads to

$$X'' + \lambda^2 X = 0,$$

$$X(0) = 0,$$

$$X(a) = 0,$$

and

$$Y'' - \lambda^2 Y = 0,$$

$$Y(b) = 0.$$

Then

$$X = c_1 \sin \frac{n\pi}{a} x \quad \text{and} \quad Y = c_2 \cosh \frac{n\pi}{a} y - c_2 \frac{\cosh \frac{n\pi b}{a}}{\sinh \frac{n\pi b}{a}} \sinh \frac{n\pi}{a} y$$

for $n = 1, 2, 3, \ldots$ so that

$$u = \sum_{n=1}^{\infty} A_n \sin \frac{n\pi}{a} x \left(\cosh \frac{n\pi}{a} y - \frac{\cosh \frac{n\pi b}{a}}{\sinh \frac{n\pi b}{a}} \sinh \frac{n\pi}{a} y \right).$$

Imposing

$$u(x, 0) = f(x) = \sum_{n=1}^{\infty} A_n \sin \frac{n\pi}{a} x$$

gives

$$A_n = \frac{2}{a} \int_0^a f(x) \sin \frac{n\pi}{a} x \, dx$$

so that

$$u(x, y) = \frac{2}{a} \sum_{n=1}^{\infty} \left(\int_0^a f(x) \sin \frac{n\pi}{a} x \, dx \right) \sin \frac{n\pi}{a} x \left(\cosh \frac{n\pi}{a} y - \frac{\cosh \frac{n\pi b}{a}}{\sinh \frac{n\pi b}{a}} \sinh \frac{n\pi}{a} y \right).$$

4. Using $u = XY$ and $-\lambda^2$ as a separation constant leads to

$$X'' + \lambda^2 X = 0,$$

$$X'(0) = 0,$$

$$X'(a) = 0,$$

and

$$Y'' - \lambda^2 Y = 0,$$

$$Y(b) = 0.$$

Then

$$X = c_1 \cos \frac{n\pi}{a} x$$

for $n = 0, 1, 2, \ldots$ and

$$Y = c_2(y - b) \quad \text{or} \quad Y = c_2 \cosh \frac{n\pi}{a} y - c_2 \frac{\cosh \frac{n\pi b}{a}}{\sinh \frac{n\pi b}{a}} \sinh \frac{n\pi}{a} y$$

so that

$$u = A_0(y - b) + \sum_{n=1}^{\infty} A_n \cos \frac{n\pi}{a} x \left(\cosh \frac{n\pi}{a} y - \frac{\cosh \frac{n\pi b}{a}}{\sinh \frac{n\pi b}{a}} \sinh \frac{n\pi}{a} y \right).$$

Imposing

$$u(x, 0) = x = -A_0 b + \sum_{n=1}^{\infty} A_n \cos \frac{n\pi}{a} x$$

500

gives

$$-A_0 b = \frac{1}{a} \int_0^a x \, dx = \frac{1}{2} a$$

and

$$A_n = \frac{2}{a} \int_0^a x \cos \frac{n\pi}{a} x \, dx = \frac{2a}{n^2 \pi^2}[(-1)^n - 1]$$

so that

$$u(x, y) = \frac{a}{2b}(b - y) + \frac{2a}{\pi^2} \sum_{n=1}^{\infty} \frac{(-1)^n - 1}{n^2} \cos \frac{n\pi}{a} x \left(\cosh \frac{n\pi}{a} y - \frac{\cosh \frac{n\pi b}{a}}{\sinh \frac{n\pi b}{a}} \sinh \frac{n\pi}{a} y \right).$$

5. Using $u = XY$ and λ^2 as a separation constant leads to

$$X'' + -\lambda^2 X = 0,$$

$$X(0) = 0,$$

and

$$Y'' + \lambda^2 Y = 0,$$

$$Y'(0) = 0,$$

$$Y'(1) = 0.$$

Then

$$Y = c_1 \cos n\pi y$$

for $n = 0, 1, 2, \ldots$ and

$$X = c_2 x \quad \text{or} \quad X = c_2 \sinh n\pi x$$

for $n = 1, 2, 3, \ldots$ so that

$$u = A_0 x + \sum_{n=1}^{\infty} A_n \sinh n\pi x \, \cos n\pi y.$$

Imposing

$$u(1, y) = 1 - y = A_0 + \sum_{n=1}^{\infty} A_n \sinh n\pi x \, \cos n\pi y$$

gives

$$A_0 = \int_0^1 (1 - y) \, dy$$

and

$$A_n \sinh n\pi = 2 \int_0^1 (1 - y) \cos n\pi y = \frac{2[1 - (-1)^n]}{n^2 \pi^2 \sinh n\pi}$$

for $n = 1, 2, 3, \ldots$ so that

$$u(x, y) = \frac{1}{2} x + \frac{2}{\pi^2} \sum_{n=1}^{\infty} \frac{1 - (-1)^n}{n^2 \sinh n\pi} \sinh n\pi x \, \cos n\pi y.$$

501

6. Using $u = XY$ and λ^2 as a separation constant leads to

$$X'' - \lambda^2 X = 0,$$
$$X'(1) = 0,$$

and

$$Y'' + \lambda^2 Y = 0,$$
$$Y'(0) = 0,$$
$$Y'(\pi) = 0.$$

Then

$$Y = c_1 \cos ny$$

for $n = 0, 1, 2, \ldots$ and

$$X = c_2 \cosh nx - c_2 \frac{\sinh n}{\cosh n} \sinh nx$$

for $n = 0, 1, 2, \ldots$ so that

$$u = A_0 + \sum_{n=1}^{\infty} A_n \left(\cosh nx - \frac{\sinh n}{\cosh n} \sinh nx \right) \cos ny.$$

Imposing

$$u(0, y) = g(y) = A_0 + \sum_{n=1}^{\infty} A_n \cos ny$$

gives

$$A_0 = \frac{1}{\pi} \int_0^\pi g(y) \, dy \quad \text{and} \quad A_n = \frac{2}{\pi} \int_0^\pi g(y) \cos ny \, dy$$

for $n = 1, 2, 3, \ldots$ so that

$$u(x, y) = \frac{1}{\pi} \int_0^\pi g(y) \, dy + \sum_{n=1}^{\infty} \left(\frac{2}{\pi} \int_0^\pi g(y) \cos ny \, dy \right) \left(\cosh nx - \frac{\sinh n}{\cosh n} \sinh nx \right) \cos ny.$$

7. Using $u = XY$ and λ^2 as a separation constant leads to

$$X'' - \lambda^2 X = 0,$$
$$X'(0) = X(0),$$

and

$$Y'' + \lambda^2 Y = 0,$$
$$Y(0) = 0,$$
$$Y(\pi) = 0.$$

Then

$$Y = c_1 \sin ny \quad \text{and} \quad X = c_2(n \cosh nx + \sinh nx)$$

for $n = 1, 2, 3, \ldots$ so that

$$u = \sum_{n=1}^{\infty} A_n(n \cosh nx + \sinh nx) \sin ny.$$

Imposing

$$u(\pi, y) = 1 = \sum_{n=1}^{\infty} A_n(n \cosh n\pi + \sinh n\pi) \sin ny$$

gives

$$A_n(n \cosh n\pi + \sinh n\pi) = \frac{2}{\pi} \int_0^{\pi} \sin ny \, dy = \frac{2[1 - (-1)^n]}{n\pi}$$

for $n = 1, 2, 3, \ldots$ so that

$$u(x, y) = \frac{2}{\pi} \sum_{n=1}^{\infty} \frac{1 - (-1)^n}{n} \frac{n \cosh nx + \sinh nx}{n \cosh n\pi + \sinh n\pi} \sin ny.$$

8. Using $u = XY$ and $-\lambda^2$ as a separation constant leads to

$$X'' + \lambda^2 X = 0,$$
$$X(0) = 0,$$
$$X(1) = 0$$

and

$$Y'' - \lambda^2 Y = 0,$$
$$Y'(0) = Y(0).$$

Then

$$X = c_1 \sin n\pi x \quad \text{and} \quad Y = c_2(n \cosh n\pi y + \sinh n\pi y)$$

for $n = 1, 2, 3, \ldots$ so that

$$u = \sum_{n=1}^{\infty} A_n(n \cosh n\pi y + \sinh n\pi y) \sin n\pi x.$$

Imposing

$$u(x, 1) = f(x) = \sum_{n=1}^{\infty} A_n(n \cosh n\pi + \sinh n\pi) \sin n\pi x$$

gives

$$A_n(n \cosh n\pi + \sinh n\pi) = \frac{2}{\pi} \int_0^{\pi} f(x) \sin n\pi x \, dx$$

503

for $n = 1, 2, 3, \ldots$ so that

$$u(x, y) = \sum_{n=1}^{\infty} A_n(n \cosh n\pi y + \sinh n\pi y) \sin n\pi x$$

where

$$A_n = \frac{2}{n\pi \cosh n\pi + \pi \sinh n\pi} \int_0^1 f(x) \sin n\pi x \, dx.$$

9. Using $u = XY$ and $-\lambda^2$ as a separation constant leads to

$$X'' + \lambda^2 X = 0,$$
$$X(0) = 0,$$
$$X(\pi) = 0,$$

and

$$Y'' - \lambda^2 Y = 0.$$

Then the boundedness of u as $y \to \infty$ gives $Y = c_1 e^{-ny}$ and $X = c_2 \sin nx$ for $n = 1, 2, 3, \ldots$ so that

$$u = \sum_{n=1}^{\infty} A_n e^{-ny} \sin nx.$$

Imposing

$$u(x, 0) = f(x) = \sum_{n=1}^{\infty} A_n \sin nx$$

gives

$$A_n = \frac{2}{\pi} \int_0^{\pi} f(x) \sin nx \, dx$$

so that

$$u(x, y) = \sum_{n=1}^{\infty} \left(\frac{2}{\pi} \int_0^{\pi} f(x) \sin nx \, dx \right) e^{-ny} \sin nx.$$

10. Using $u = XY$ and $-\lambda^2$ as a separation constant leads to

$$X'' + \lambda^2 X = 0,$$
$$X'(0) = 0,$$
$$X'(\pi) = 0,$$

and

$$Y'' - \lambda^2 Y = 0.$$

By the boundedness of u as $y \to \infty$ we obtain $Y = c_1 e^{-ny}$ for $n = 1, 2, 3, \ldots$ or $Y = c_1$ and $X = c_2 \cos nx$ for $n = 0, 1, 2, \ldots$ so that

$$u = A_0 + \sum_{n=1}^{\infty} A_n e^{-ny} \cos nx.$$

Imposing

$$u(x, 0) = f(x) = A_0 + \sum_{n=1}^{\infty} A_n \cos nx$$

gives

$$A_0 = \frac{1}{\pi} \int_0^{\pi} f(x)\, dx \quad \text{and} \quad A_n = \frac{2}{\pi} \int_0^{\pi} f(x) \cos nx\, dx$$

so that

$$u(x, y) = \frac{1}{\pi} \int_0^{\pi} f(x)\, dx + \sum_{n=1}^{\infty} \left(\frac{2}{\pi} \int_0^{\pi} f(x) \cos nx\, dx \right) e^{-ny} \cos nx.$$

11. Since the boundary conditions at $y = 0$ and $y = b$ are functions of x we choose to separate Laplace's equation as

$$\frac{X''}{X} = -\frac{Y''}{Y} = -\lambda^2$$

so that

$$X'' + \lambda^2 X = 0$$

$$Y'' - \lambda^2 Y = 0$$

and

$$X(x) = c_1 \cos \lambda x + c_2 \sin \lambda x$$

$$Y(y) = c_3 \cosh \lambda y + c_2 \sinh \lambda y.$$

Now $X(0) = 0$ gives $c_1 = 0$ and $X(a) = 0$ implies $\sin \lambda a = 0$ or $\lambda = n\pi/a$ for $n = 1, 2, 3, \ldots$. Thus

$$u_n(x, y) = XY = \left(A_n \cosh \frac{n\pi}{a} y + B_n \sinh \frac{n\pi}{a} y \right) \sin \frac{n\pi}{a} x$$

and

$$u(x, y) = \sum_{n=1}^{\infty} \left(A_n \cosh \frac{n\pi}{a} y + B_n \sinh \frac{n\pi}{a} y \right) \sin \frac{n\pi}{a} x. \tag{1}$$

At $y = 0$ we then have

$$f(x) = \sum_{n=1}^{\infty} A_n \sin \frac{n\pi}{a} x$$

and consequently

$$A_n = \frac{2}{a} \int_0^a f(x) \sin \frac{n\pi}{a} x\, dx. \tag{2}$$

At $y = b$,

$$g(y) = \sum_{n=1}^{\infty} \left(A_n \cosh \frac{n\pi}{a} b + B_n \sinh \frac{n\pi}{b} a \right) \sin \frac{n\pi}{a} x$$

indicates that the entire expression in the parentheses is given by

$$A_n \cosh \frac{n\pi}{a} b + B_n \sinh \frac{n\pi}{a} b = \frac{2}{a} \int_0^a g(x) \sin \frac{n\pi}{a} x \, dx.$$

We can now solve for B_n:

$$B_n \sinh \frac{n\pi}{a} b = \frac{2}{a} \int_0^a g(x) \sin \frac{n\pi}{a} x \, dx - A_n \cosh \frac{n\pi}{a} b$$

$$B_n = \frac{1}{\sinh \frac{n\pi}{a} b} \left(\frac{2}{a} \int_0^a g(x) \sin \frac{n\pi}{a} x \, dx - A_n \cosh \frac{n\pi}{a} b \right). \tag{3}$$

A solution to the given boundary-value problem consists of the series (1) with coefficients A_n and B_n given in (2) and (3), respectively.

12. Since the boundary conditions at $x = 0$ and $x = a$ are functions of y we choose to separate Laplace's equation as

$$\frac{X''}{X} = -\frac{Y''}{Y} = \lambda^2$$

so that

$$X'' - \lambda^2 X = 0$$

$$Y'' + \lambda^2 Y = 0$$

and

$$X(x) = c_1 \cosh \lambda x + c_2 \sinh \lambda x$$

$$Y(y) = c_3 \cos \lambda y + c_2 \sin \lambda y.$$

Now $Y(0) = 0$ gives $c_3 = 0$ and $Y(b) = 0$ implies $\sin \lambda b = 0$ or $\lambda = n\pi/b$ for $n = 1, 2, 3, \dots$. Thus

$$u_n(x, y) = XY = \left(A_n \cosh \frac{n\pi}{b} x + B_n \sinh \frac{n\pi}{b} x \right) \sin \frac{n\pi}{b} y$$

and

$$u(x, y) = \sum_{n=1}^{\infty} \left(A_n \cosh \frac{n\pi}{b} x + B_n \sinh \frac{n\pi}{b} x \right) \sin \frac{n\pi}{b} y. \tag{4}$$

At $x = 0$ we then have

$$F(y) = \sum_{n=1}^{\infty} A_n \sin \frac{n\pi}{b} y$$

and consequently

$$A_n = \frac{2}{b} \int_0^b F(y) \sin \frac{n\pi}{b} y \, dy. \tag{5}$$

At $x = a$,

$$G(y) = \sum_{n=1}^{\infty} \left(A_n \cosh \frac{n\pi}{b} a + B_n \sinh \frac{n\pi}{b} a \right) \sin \frac{n\pi}{b} y$$

indicates that the entire expression in the parentheses is given by

$$A_n \cosh \frac{n\pi}{b} a + B_n \sinh \frac{n\pi}{b} a = \frac{2}{b} \int_0^b G(y) \sin \frac{n\pi}{b} y \, dy.$$

We can now solve for B_n:

$$B_n \sinh \frac{n\pi}{b} a = \frac{2}{b} \int_0^b G(y) \sin \frac{n\pi}{b} y \, dy - A_n \cosh \frac{n\pi}{b} a$$

$$B_n = \frac{1}{\sinh \frac{n\pi}{b} a} \left(\frac{2}{b} \int_0^b G(y) \sin \frac{n\pi}{b} y \, dy - A_n \cosh \frac{n\pi}{b} a \right). \tag{6}$$

A solution to the given boundary-value problem consists of the series (4) with coefficients A_n and B_n given in (5) and (6), respectively.

In Problems 13 and 14 we refer to the discussion in the text under the heading **Superposition Principle**.

13. We identify $a = b = \pi$, $f(x) = 0$, $g(x) = 1$, $F(y) = 1$, and $G(y) = 1$. Then $A_n = 0$ and

$$u_1(x,y) = \sum_{n=1}^{\infty} B_n \sinh ny \sin nx$$

where

$$B_n = \frac{2}{\pi \sinh n\pi} \int_0^\pi \sin nx \, dx = \frac{2[1 - (-1)^n]}{n\pi \sinh n\pi}.$$

Next

$$u_2(x,y) = \sum_{n=1}^{\infty} (A_n \cosh nx + B_n \sinh nx) \sin ny$$

where

$$A_n = \frac{2}{\pi} \int_0^\pi \sin ny \, dy = \frac{2[1 - (-1)^n]}{n\pi}$$

and

$$B_n = \frac{1}{\sinh n\pi} \left(\frac{2}{\pi} \int_0^\pi \sin ny \, dy - A_n \cosh n\pi \right)$$

$$= \frac{1}{\sinh n\pi} \left(\frac{2[1 - (-1)^n]}{n\pi} - \frac{2[1 - (-1)^n]}{n\pi} \cosh n\pi \right)$$

$$= \frac{2[1 - (-1)^n]}{n\pi \sinh n\pi} (1 - \cosh n\pi).$$

Now

$$A_n \cosh nx + B_n \sinh nx = \frac{2[1-(-1)^n]}{n\pi}\left[\cosh nx + \frac{\sinh nx}{\sinh n\pi}(1-\cosh n\pi)\right]$$

$$= \frac{2[1-(-1)^n]}{n\pi \sinh n\pi}[\cosh nx \sinh n\pi + \sinh nx - \sinh nx \cosh n\pi]$$

$$= \frac{2[1-(-1)^n]}{n\pi \sinh n\pi}[\sinh nx + \sinh n(\pi - x)]$$

and

$$u(x,y) = u_1 + u_2 = \frac{2}{\pi}\sum_{n=1}^{\infty}\frac{1-(-1)^n}{n\sinh n\pi}\sinh ny \sin nx$$

$$+ \frac{2}{\pi}\sum_{n=1}^{\infty}\frac{[1-(-1)^n][\sinh nx + \sinh n(\pi - x)]}{n\sinh n\pi}\sin ny.$$

14. We identify $a = b = 2$, $f(x) = 0$, $g(x) = \begin{cases} x, & 0 < x < 1 \\ 2-x, & 1 < x < 2 \end{cases}$, $F(y) = 0$, and $G(y) = y(2-y)$.
Then $A_n = 0$ and

$$u_1(x,y) = \sum_{n=1}^{\infty} B_n \sinh\frac{n\pi}{2}y \sin\frac{n\pi}{2}x$$

where

$$B_n = \frac{1}{\sinh n\pi}\int_0^2 g(x)\sin\frac{n\pi}{2}x\,dx$$

$$= \frac{1}{\sinh n\pi}\left(\int_0^1 x\sin\frac{n\pi}{2}x\,dx + \int_1^2 (2-x)\sin\frac{n\pi}{2}x\,dx\right)$$

$$= \frac{8\sin\frac{n\pi}{2}}{n^2\pi^2\sinh n\pi}.$$

Next, since $A_n = 0$ in u_2, we have

$$u_2(x,y) = \sum_{n=1}^{\infty} B_n \sinh\frac{n\pi}{2}x \sin\frac{n\pi}{2}$$

where

$$B_n = \frac{1}{\sinh n\pi}\int_0^b y(2-y)\sin\frac{n\pi}{2}y\,dy = \frac{16[1-(-1)^n]}{n^3\pi^3\sinh n\pi}.$$

Thus

$$u(x,y) = u_1 + u_2 = \frac{8}{\pi^2}\sum_{n=1}^{\infty}\frac{\sin\frac{n\pi}{2}}{n^2\sinh n\pi}\sinh\frac{n\pi}{2}y \sin\frac{n\pi}{2}x$$

$$+ \frac{16}{\pi^3}\sum_{n=1}^{\infty}\frac{[1-(-1)^n]}{n^3\sinh n\pi}\sinh\frac{n\pi}{2}x \sin\frac{n\pi}{2}y.$$

1. Using $v(x, t) = u(x, t) - 100$ we wish to solve $kv_{xx} = v_t$ subject to $v(0, t) = 0$, $v(1, t) = 0$, and $v(x, 0) = -100$. Let $v = XT$ and use $-\lambda^2$ as a separation constant so that

$$X'' + \lambda^2 X = 0,$$

$$X(0) = 0,$$

$$X(1) = 0,$$

and

$$T' + \lambda^2 kT = 0.$$

Then

$$X = c_1 \sin n\pi x \quad \text{and} \quad T = c_2 e^{-kn^2\pi^2 t}$$

for $n = 1, 2, 3, \ldots$ so that

$$v = \sum_{n=1}^{\infty} A_n e^{-kn^2\pi^2 t} \sin n\pi x.$$

Imposing

$$v(x, 0) = -100 = \sum_{n=1}^{\infty} A_n \sin n\pi x$$

gives

$$A_n = 2 \int_0^1 (-100) \sin n\pi x \, dx = \frac{-200}{n\pi}[1 - (-1)^n]$$

so that

$$u(x, t) = v(x, t) + 100 = 100 + \frac{200}{\pi} \sum_{n=1}^{\infty} \frac{(-1)^n - 1}{n} e^{-kn^2\pi^2 t} \sin n\pi x.$$

2. Letting $u(x, t) = v(x, t) + \psi(x)$ and proceeding as in Example 1 in the text we find $\psi(x) = u_0 - u_0 x$. Then $v(x, t) = u(x, t) + u_0 x - u_0$ and we wish to solve $kv_{xx} = v_t$ subject to $v(0, t) = 0$, $v(1, t) = 0$, and $v(x, 0) = f(x) + u_0 x - u_0$. Let $v = XT$ and use $-\lambda^2$ as a separation constant so that

$$X'' + \lambda^2 X = 0,$$

$$X(0) = 0,$$

$$X(1) = 0,$$

and

$$T' + \lambda^2 kT = 0.$$

Then

$$X = c_1 \sin n\pi x \quad \text{and} \quad T = c_2 e^{-kn^2\pi^2 t}$$

for $n = 1, 2, 3, \ldots$ so that

$$v = \sum_{n=1}^{\infty} A_n e^{-kn^2 \pi^2 t} \sin n\pi x.$$

Imposing

$$v(x, 0) = f(x) + u_0 x - u_0 = \sum_{n=1}^{\infty} A_n \sin n\pi x$$

gives

$$A_n = 2 \int_0^1 (f(x) + u_0 x - u_0) \sin n\pi x \, dx$$

so that

$$u(x, t) = v(x, t) + u_0 - u_0 x = u_0 - u_0 x + \sum_{n=1}^{\infty} A_n e^{-kn^2 \pi^2 t} \sin n\pi x.$$

3. If we let $u(x, t) = v(x, t) + \psi(x)$, then we obtain as in Example 1 in the text

$$k\psi'' + r = 0$$

or

$$\psi(x) = -\frac{r}{2k} x^2 + c_1 x + c_2.$$

The boundary conditions become

$$u(0, t) = v(0, t) + \psi(0) = u_0$$

$$u(1, t) = v(1, t) + \psi(1) = u_0.$$

Letting $\psi(0) = \psi(1) = u_0$ we obtain homogeneous boundary conditions in v:

$$v(0, t) = 0 \quad \text{and} \quad v(1, t) = 0.$$

Now $\psi(0) = \psi(1) = u_0$ implies $c_2 = u_0$ and $c_1 = r/2k$. Thus

$$\psi(x) = -\frac{r}{2k} x^2 + \frac{r}{2k} x + u_0 = u_0 - \frac{r}{2k} x(x - 1).$$

To determine $v(x, t)$ we solve

$$k\frac{\partial^2 v}{\partial x^2} = \frac{\partial v}{\partial t}, \quad 0 < x < 1, \ t > 0$$

$$v(0, t) = 0, \quad v(1, t) = 0,$$

$$v(x, 0) = \frac{r}{2k} x(x - 1) - u_0.$$

Separating variables, we find

$$v(x, t) = \sum_{n=1}^{\infty} A_n e^{-kn^2 \pi^2 t} \sin n\pi x,$$

510

where

$$A_n = 2 \int_0^1 \left[\frac{r}{2k} x(x-1) - u_0 \right] \sin n\pi x \, dx = 2 \left[\frac{u_0}{n\pi} + \frac{r}{kn^3\pi^3} \right] [(-1)^n - 1]. \tag{1}$$

Hence, a solution of the original problem is

$$u(x,t) = \psi(x) + v(x,t)$$

$$= u_0 - \frac{r}{2k} x(x-1) + \sum_{n=1}^\infty A_n e^{-kn^2\pi^2 t} \sin n\pi x,$$

where A_n is defined in (1).

4. If we let $u(x,t) = v(x,t) + \psi(x)$, then we obtain as in Example 1 in the text

$$k\psi'' + r = 0$$

or

$$\psi(x) = -\frac{r}{2k} x^2 + c_1 x + c_2.$$

The boundary conditions become

$$u(0,t) = v(0,t) + \psi(0) = u_0$$

$$u(1,t) = v(1,t) + \psi(1) = u_1.$$

Letting $\psi(0) = u_0$ and $\psi(1) = u_1$ we obtain homogeneous boundary conditions in v:

$$v(0,t) = 0 \quad \text{and} \quad v(1,t) = 0.$$

Now $\psi(0) = u_0$ and $\psi(1) = u_1$ imply $c_2 = u_0$ and $c_1 = u_1 - u_0 + r/2k$. Thus

$$\psi(x) = -\frac{r}{2k} x^2 + \left(u_1 - u_0 + \frac{r}{2k} \right) x + u_0.$$

To determine $v(x,t)$ we solve

$$k \frac{\partial^2 v}{\partial x^2} = \frac{\partial v}{\partial t}, \quad 0 < x < 1, \ t > 0$$

$$v(0,t) = 0, \quad v(1,t) = 0,$$

$$v(x,0) = f(x) - \psi(x).$$

Separating variables, we find

$$v(x,t) = \sum_{n=1}^\infty A_n e^{-kn^2\pi^2 t} \sin n\pi x,$$

where

$$A_n = 2 \int_0^1 [f(x) - \psi(x)] \sin n\pi x \, dx. \tag{2}$$

511

Hence, a solution of the original problem is

$$u(x,t) = \psi(x) + v(x,t)$$

$$= -\frac{r}{2k}x^2 + \left(u_1 - u_0 + \frac{r}{2k}\right)x + u_0 + \sum_{n=1}^{\infty} A_n e^{-kn^2\pi^2 t} \sin n\pi x,$$

where A_n is defined in (2).

5. Substituting $u(x,t) = v(x,t) + \psi(x)$ into the partial differential equation gives

$$k\frac{\partial^2 v}{\partial x^2} + k\psi'' + Ae^{-\beta x} = \frac{\partial v}{\partial t}.$$

This equation will be homogeneous provided ψ satisfies

$$k\psi'' + Ae^{-\beta x} = 0.$$

The general solution of this differential equation is

$$\psi(x) = -\frac{A}{\beta^2 k}e^{-\beta x} + c_1 x + c_2.$$

From $\psi(0) = 0$ and $\psi(1) = 0$ we find

$$c_1 = \frac{A}{\beta^2 k}(e^{-\beta} - 1) \quad \text{and} \quad c_2 = \frac{A}{\beta^2 k}.$$

Hence

$$\psi(x) = -\frac{A}{\beta^2 k}e^{-\beta x} + \frac{A}{\beta^2 k}(e^{-\beta} - 1)x + \frac{A}{\beta^2 k}$$

$$= \frac{A}{\beta^2 k}\left[1 - e^{-\beta x} + (e^{-\beta} - 1)x\right].$$

Now the new problem is

$$k\frac{\partial^2 v}{\partial x^2} = \frac{\partial v}{\partial t} \quad, \quad 0 < x < 1, \quad t > 0,$$

$$v(0,t) = 0, \quad v(1,t) = 0, \quad t > 0,$$

$$v(x,0) = f(x) - \psi(x), \quad 0 < x < 1.$$

Identifying this as the heat equation solved in Section 12.3 in the text with $L = 1$ we obtain

$$v(x,t) = \sum_{n=1}^{\infty} A_n e^{-kn^2\pi^2 t} \sin n\pi x$$

where

$$A_n = 2\int_0^1 [f(x) - \psi(x)] \sin n\pi x \, dx.$$

Thus

$$u(x,t) = \frac{A}{\beta^2 k}\left[1 - e^{-\beta x} + (e^{-\beta} - 1)x\right] + \sum_{n=1}^{\infty} A_n e^{-kn^2\pi^2 t}\sin n\pi x.$$

6. Substituting $u(x,t) = v(x,t) + \psi(x)$ into the partial differential equation gives

$$k\frac{\partial^2 v}{\partial x^2} + k\psi'' - hv - h\psi = \frac{\partial v}{\partial t}.$$

This equation will be homogeneous provided ψ satisfies

$$k\psi'' - h\psi = 0.$$

Since k and h are positive, the general solution of this latter equation is

$$\psi(x) = c_1 \cosh\sqrt{\frac{h}{k}}\,x + c_2 \sinh\sqrt{\frac{h}{k}}\,x.$$

From $\psi(0) = 0$ and $\psi(\pi) = u_0$ we find $c_1 = 0$ and $c_2 = u_0/\sinh\sqrt{h/k}\,\pi$. Hence

$$\psi(x) = u_0 \frac{\sinh\sqrt{h/k}\,x}{\sinh\sqrt{h/k}\,\pi}.$$

Now the new problem is

$$k\frac{\partial^2 v}{\partial x^2} - hv = \frac{\partial v}{\partial t}, \quad 0 < x < \pi,\ t > 0$$

$$v(0,t) = 0, \quad v(\pi,t) = 0, \quad t > 0$$

$$v(x,0) = -\psi(x), \quad 0 < x < \pi.$$

If we let $v = XT$ then

$$\frac{X''}{X} = \frac{T' + hT}{kT} = -\lambda^2$$

gives the separated differential equations

$$X'' + \lambda^2 X = 0 \quad \text{and} \quad T' + \left(h + k\lambda^2\right)T = 0.$$

The respective solutions are

$$X(x) = c_3 \cos\lambda x + c_4 \sin\lambda x$$

$$T(t) = c_5 e^{-\left(h + k\lambda^2\right)t}.$$

From $X(0) = 0$ we get $c_3 = 0$ and from $X(\pi) = 0$ we find $\lambda = n$ for $n = 1, 2, 3, \ldots$. Consequently, it follows that

$$v(x,t) = \sum_{n=1}^{\infty} A_n e^{-\left(h + kn^2\right)t}\sin nx$$

513

where

$$A_n = -\frac{2}{\pi} \int_0^\pi \psi(x) \sin nx \, dx.$$

Hence a solution of the original problem is

$$u(x, t) = u_0 \frac{\sinh \sqrt{h/k}\, x}{\sinh \sqrt{h/k}\, \pi} + e^{-ht} \sum_{n=1}^\infty A_n e^{-kn^2 t} \sin nx$$

where

$$A_n = -\frac{2}{\pi} \int_0^\pi u_0 \frac{\sinh \sqrt{h/k}\, x}{\sinh \sqrt{h/k}\, \pi} \sin nx \, dx.$$

Using the exponential definition of the hyperbolic sine and integration by parts we find

$$A_n = \frac{2 u_0 n k (-1)^n}{\pi (h + kn^2)}.$$

7. Substituting $u(x, t) = v(x, t) + \psi(x)$ into the partial differential equation gives

$$k \frac{\partial^2 v}{\partial x^2} + k\psi'' - hv - h\psi + hu_0 = \frac{\partial v}{\partial t}.$$

This equation will be homogeneous provided ψ satisfies

$$k\psi'' - h\psi + hu_0 = 0 \quad \text{or} \quad k\psi'' - h\psi = -hu_0.$$

This second-order, linear, non-homogeneous differential equation has solution

$$\psi(x) = c_1 \cosh \sqrt{\frac{h}{k}}\, x + c_2 \sinh \sqrt{\frac{h}{k}}\, x + u_0,$$

where we assume $h > 0$ and $k > 0$. From $\psi(0) = u_0$ and $\psi(1) = 0$ we find $c_1 = 0$ and $c_2 = -u_0 / \sinh \sqrt{h/k}$. Thus, the steady-state solution is

$$\psi(x) = -\frac{u_0}{\sinh \sqrt{\frac{h}{k}}} \sinh \sqrt{\frac{h}{k}}\, x + u_0 = u_0 \left(1 - \frac{\sinh \sqrt{\frac{h}{k}}\, x}{\sinh \sqrt{\frac{h}{k}}} \right).$$

8. The partial differential equation is

$$k \frac{\partial^2 u}{\partial x^2} - hu = \frac{\partial u}{\partial t}.$$

Substituting $u(x, t) = v(x, t) + \psi(x)$ gives

$$k \frac{\partial^2 v}{\partial x^2} + k\psi'' - hv - h\psi = \frac{\partial v}{\partial t}.$$

This equation will be homogeneous provided ψ satisfies

$$k\psi'' - h\psi = 0.$$

Assuming $h > 0$ and $k > 0$, we have

$$\psi = c_1 e^{\sqrt{h/k}\,x} + c_2 e^{-\sqrt{h/k}\,x},$$

where we have used the exponential form of the solution since the rod is infinite. Now, in order that the steady-state temperature $\psi(x)$ be bounded as $x \to \infty$, we require $c_1 = 0$. Then

$$\psi(x) = c_2 e^{-\sqrt{h/k}\,x}$$

and $\psi(0) = u_0$ implies $c_2 = u_0$. Thus

$$\psi(x) = u_0 e^{-\sqrt{h/k}\,x}.$$

9. Substituting $u(x,t) = v(x,t) + \psi(x)$ into the partial differential equation gives

$$a^2 \frac{\partial^2 v}{\partial x^2} + a^2 \psi'' + Ax = \frac{\partial^2 v}{\partial t^2}.$$

This equation will be homogeneous provided ψ satisfies

$$a^2 \psi'' + Ax = 0.$$

The general solution of this differential equation is

$$\psi(x) = -\frac{A}{6a^2} x^3 + c_1 x + c_2.$$

From $\psi(0) = 0$ we obtain $c_2 = 0$, and from $\psi(1) = 0$ we obtain $c_1 = A/6a^2$. Hence

$$\psi(x) = \frac{A}{6a^2}(x - x^3).$$

Now the new problem is

$$a^2 \frac{\partial^2 v}{\partial x^2} = \frac{\partial^2 v}{\partial t^2}$$

$$v(0,t) = 0, \quad v(1,t) = 0, \quad t > 0,$$

$$v(x,0) = -\psi(x), \quad v_t(x,0) = 0, \quad 0 < x < 1.$$

Identifying this as the wave equation solved in Section 12.4 in the text with $L - 1$, $f(x) = -\psi(x)$, and $g(x) = 0$ we obtain

$$v(x,t) = \sum_{n=1}^{\infty} A_n \cos n\pi a t \sin n\pi x$$

where

$$A_n = 2 \int_0^1 [-\psi(x)] \sin n\pi x \, dx = \frac{A}{3a^2} \int_0^1 (x^3 - x) \sin n\pi x \, dx = \frac{2A(-1)^n}{a^2 \pi^3 n^3}.$$

Thus

$$u(x,t) = \frac{A}{6a^2}(x - x^3) + \frac{2A}{a^2 \pi^3} \sum_{n=1}^{\infty} \frac{(-1)^n}{n^3} \cos n\pi a t \sin n\pi x.$$

10. We solve

$$a^2 \frac{\partial^2 u}{\partial x^2} - g = \frac{\partial^2 u}{\partial t^2}, \quad 0 < x < 1, \ t > 0$$

$$u(0, t) = 0, \quad u(1, t) = 0, \quad t > 0$$

$$u(x, 0) = 0, \quad \left. \frac{\partial u}{\partial t} \right|_{t=0} = 0, \quad 0 < x < 1.$$

The partial differential equation is nonhomogeneous. The substitution $u(x, t) = v(x, t) + \psi(x)$ yields a homogeneous partial differential equation provided ψ satisfies

$$a^2 \psi'' - g = 0.$$

By integrating twice we find

$$\psi(x) = \frac{g}{2a^2} x^2 + c_1 x + c_2.$$

The imposed conditions $\psi(0) = 0$ and $\psi(1) = 0$ then lead to $c_2 = 0$ and $c_1 = -g/2a^2$. Hence

$$\psi(x) = \frac{g}{2a^2} \left(x^2 - x \right).$$

The new problem is now

$$a^2 \frac{\partial^2 v}{\partial x^2} = \frac{\partial^2 v}{\partial t^2}, \quad 0 < x < 1, \ t > 0$$

$$v(0, t) = 0, \quad v(1, t) = 0$$

$$v(x, 0) = \frac{g}{2a^2} \left(x - x^2 \right), \quad \left. \frac{\partial v}{\partial t} \right|_{t=0} = 0.$$

Substituting $v = XT$ we find in the usual manner

$$X'' + \lambda^2 X = 0$$

$$T'' + a^2 \lambda^2 T = 0$$

with solutions

$$X(x) = c_3 \cos \lambda x + c_4 \sin \lambda x$$

$$T(t) = c_5 \cos a\lambda t + c_6 \sin a\lambda t.$$

The conditions $X(0) = 0$ and $X(1) = 0$ imply in turn that $c_3 = 0$ and $\lambda = n\pi$ for $n = 1, 2, 3, \ldots$. The condition $T'(0) = 0$ implies $c_6 = 0$. Hence, by the superposition principle

$$v(x, t) = \sum_{n=1}^{\infty} A_n \cos(an\pi t) \sin(n\pi x).$$

At $t = 0$,

$$\frac{g}{2a^2}\left(x - x^2\right) = \sum_{n=1}^{\infty} A_n \sin n\pi x$$

and so

$$A_n = \frac{g}{a^2}\int_0^1 \left(x - x^2\right)\sin n\pi x \, dx = \frac{2g}{a^2 n^3 \pi^3}\left[1 - (-1)^n\right].$$

Thus the solution to the original problem is

$$u(x, t) = \psi(x) + v(x, t) = \frac{g}{2a^2}\left(x^2 - x\right) + \frac{2g}{a^2\pi^3}\sum_{n=1}^{\infty}\frac{1 - (-1)^n}{n^3}\cos(an\pi t)\sin(n\pi x).$$

11. Substituting $u(x, y) = v(x, y) + \psi(y)$ into Laplace's equation we obtain

$$\frac{\partial^2 v}{\partial x^2} + \frac{\partial^2 v}{\partial y^2} + \psi''(y) = 0.$$

This equation will be homogeneous provided ψ satisfies $\psi(y) = c_1 y + c_2$. Considering

$$u(x, 0) = v(x, 0) + \psi(0) = u_1$$

$$u(x, 1) = v(x, 1) + \psi(1) = u_0$$

$$u(0, y) = v(0, y) + \psi(y) = 0$$

we require that $\psi(0) = u_1$, $\psi_1 = u_0$ and $v(0, y) = -\psi(y)$. Then $c_1 = u_0 - u_1$ and $c_2 = u_1$. The new boundary-value problem is

$$\frac{\partial^2 v}{\partial x^2} + \frac{\partial^2 v}{\partial y^2} = 0$$

$$v(x, 0) = 0, \quad v(x, 1) = 0,$$

$$v(0, y) = -\psi(y), \quad 0 < y < 1,$$

where $v(x, y)$ is bounded at $x \to \infty$. This problem is similar to Problem 9 in Section 12.5. The solution is

$$v(x, y) = \sum_{n=1}^{\infty}\left(2\int_0^1 [-\psi(y)\sin n\pi y]\, dy\right)e^{-n\pi x}\sin n\pi y$$

$$= 2\sum_{n=1}^{\infty}\left[(u_1 - u_0)\int_0^1 y\sin n\pi y\, dy - u_1\int_0^1 \sin n\pi y\, dy\right]e^{-n\pi x}\sin n\pi y$$

$$= \frac{2}{\pi}\sum_{n=1}^{\infty}\frac{u_0(-1)^n - u_1}{n}e^{-n\pi x}\sin n\pi y.$$

517

Thus

$$u(x, y) = v(x, y) + \psi(y)$$

$$= (u_0 - u_1)y + u_1 + \frac{2}{\pi} \sum_{n=1}^{\infty} \frac{u_0(-1)^n - u_1}{n} e^{-n\pi x} \sin n\pi y.$$

12. Substituting $u(x, y) = v(x, y) + \psi(x)$ into Poisson's equation we obtain

$$\frac{\partial^2 v}{\partial x^2} + \psi''(x) + h + \frac{\partial^2 v}{\partial y^2} = 0.$$

The equation will be homogeneous provided ψ satisfies $\psi''(x) + h = 0$ or $\psi(x) = -\frac{h}{2}x^2 + c_1 x + c_2$. From $\psi(0) = 0$ we obtain $c_2 = 0$. From $\psi(\pi) = 1$ we obtain

$$c_1 = \frac{1}{\pi} + \frac{h\pi}{2}.$$

Then

$$\psi(x) = \left(\frac{1}{\pi} + \frac{h\pi}{2}\right) x - \frac{h}{2}x^2.$$

The new boundary-value problem is

$$\frac{\partial^2 v}{\partial x^2} + \frac{\partial^2 v}{\partial y^2} = 0$$

$$v(0, y) = 0, \quad v(\pi, y) = 0,$$

$$v(x, 0) = -\psi(x), \quad 0 < x < \pi.$$

This is Problem 9 in Section 12.5. The solution is

$$v(x, y) = \sum_{n=1}^{\infty} A_n e^{-ny} \sin nx$$

where

$$A_n = \frac{2}{\pi} \int_0^{\pi} [-\psi(x) \sin nx] \, dx$$

$$= \frac{2(-1)^n}{m} \left(\frac{1}{\pi} + \frac{h\pi}{2}\right) - h(-1)^n \left(\frac{\pi}{n} + \frac{2}{n^2}\right).$$

Thus

$$u(x, y) = v(x, y) + \psi(x) = \left(\frac{1}{\pi} + \frac{h\pi}{2}\right) x - \frac{h}{2}x^2 + \sum_{n=1}^{\infty} A_n e^{-ny} \sin nx.$$

_____ **Exercises 12.7** _____

1. Referring to Example 1 in the text we have

$$X(x) = c_1 \cos \lambda x + c_2 \sin \lambda x$$

and

$$T(t) = c_3 e^{-k\lambda^2 t}.$$

From $X'(0) = 0$ (since the left end of the rod is insulated), we find $c_2 = 0$. Then $X(x) = c_1 \cos \lambda x$ and the other boundary condition $X'(1) = -hX(1)$ implies

$$-\lambda \sin \lambda + h \cos \lambda = 0 \quad \text{or} \quad \cot \lambda = \frac{\lambda}{h}.$$

Denoting the consecutive positive roots of this latter equation by λ_n for $n = 1, 2, 3, \ldots$, we have

$$u(x,t) = \sum_{n=1}^{\infty} A_n e^{-k\lambda_n^2 t} \cos \lambda_n x.$$

From the initial condition $u(x,0) = 1$ we obtain

$$1 = \sum_{n=1}^{\infty} A_n \cos \lambda_n x$$

and

$$A_n = \frac{\int_0^1 \cos \lambda_n x \, dx}{\int_0^1 \cos^2 \lambda_n x \, dx} = \frac{\sin \lambda_n / \lambda_n}{\frac{1}{2} \left[1 + \frac{1}{2\lambda_n} \sin 2\lambda_n \right]}$$

$$= \frac{2 \sin \lambda_n}{\lambda_n \left[1 + \frac{1}{\lambda_n} \sin \lambda_n \cos \lambda_n \right]} = \frac{2 \sin \lambda_n}{\lambda_n \left[1 + \frac{1}{h\lambda_n} \sin \lambda_n (\lambda_n \sin \lambda_n) \right]}$$

$$= \frac{2h \sin \lambda_n}{\lambda_n [h + \sin^2 \lambda_n]}.$$

The solution is

$$u(x,t) = 2h \sum_{n=1}^{\infty} \frac{\sin \lambda_n}{\lambda_n (h + \sin^2 \lambda_n)} e^{-k\lambda_n^2 t} \cos \lambda_n x.$$

2. Substituting $u(x,t) = v(x,t) + \psi(x)$ into the partial differential equation gives

$$k \frac{\partial^2 v}{\partial x^2} + k\psi'' = \frac{\partial v}{\partial t}.$$

This equation will be homogeneous if $\psi''(x) = 0$ or $\psi(x) = c_1 x + c_2$. The boundary condition $u(0,t) = 0$ implies $\psi(0) = 0$ which implies $c_2 = 0$. Thus $\psi(x) = c_1 x$. Using the second boundary condition we obtain

$$-\left(\frac{\partial v}{\partial x} + \psi' \right) \Big|_{x=1} = -h[v(1,t) + \psi(1) - u_0],$$

which will be homogeneous when

$$-\psi'(1) = -h\psi(1) + hu_0.$$

Since $\psi(1) = \psi'(1) = c_1$ we have $-c_1 = -hc_1 + hu_0$ and $c_1 = hu_0/(h-1)$. Thus

$$\psi(x) = \frac{hu_0}{h-1}x.$$

The new boundary-value problem is

$$k\frac{\partial^2 v}{\partial x^2} = \frac{\partial v}{\partial t}, \quad 0 < x < 1, \quad t > 0$$

$$v(0,t) = 0, \quad \frac{\partial v}{\partial x}\Big|_{x=1} = -hv(1,t), \quad h > 0, \quad t > 0$$

$$v(x,0) = f(x) - \frac{hu_0}{h-1}x, \quad 0 < x < 1.$$

Referring to Example 1 in the text we see that

$$v(x,t) = \sum_{n=1}^{\infty} A_n e^{-k\lambda_n^2 t} \sin \lambda_n x$$

and

$$u(x,t) = v(x,t) + \psi(x) = \frac{hu_0}{h-1}x + \sum_{n=1}^{\infty} A_n e^{-k\lambda_n^2 t} \sin \lambda_n x$$

where

$$f(x) - \frac{hu_0}{h-1}x = \sum_{n=1}^{\infty} A_n \sin \lambda_n x$$

and λ_n is a solution of $\lambda_n \cos \lambda_n = -h \sin \lambda_n$. The coefficients are

$$A_n = \frac{\int_0^1 [f(x) - hu_0 x/(h-1)] \sin \lambda_n x\, dx}{\int_0^1 \sin^2 \lambda_n x\, dx}$$

$$= \frac{\int_0^1 [f(x) - hu_0 x/(h-1)] \sin \lambda_n x\, dx}{\frac{1}{2}\left[1 - \frac{1}{2\lambda_n}\sin 2\lambda_n\right]}$$

$$= \frac{2\int_0^1 [f(x) - hu_0 x/(h-1)] \sin \lambda_n x\, dx}{1 - \frac{1}{\lambda_n}\sin \lambda_n \cos \lambda_n}$$

$$= \frac{2\int_0^1 [f(x) - hu_0 x/(h-1)] \sin \lambda_n x\, dx}{1 - \frac{1}{h\lambda_n}(h\sin \lambda_n)\cos \lambda_n}$$

$$= \frac{2\int_0^1 [f(x) - hu_0 x/(h-1)] \sin \lambda_n x\, dx}{1 - \frac{1}{h\lambda_n}(-\lambda_n \cos \lambda_n) \cos \lambda_n}$$

$$= \frac{2h}{h + \cos^2 \lambda_n} \int_0^1 \left[f(x) - \frac{hu_0}{h-1} x \right] \sin \lambda_n x\, dx.$$

3. Separating variables in Laplace's equation gives

$$X'' + \lambda^2 X = 0$$

$$Y'' - \lambda^2 Y = 0$$

and

$$X(x) = c_1 \cos \lambda x + c_2 \sin \lambda x$$

$$Y(y) = c_3 \cosh \lambda y + c_4 \sinh \lambda y.$$

From $u(0, y) = 0$ we obtain $X(0) = 0$ and $c_1 = 0$. From $u_x(a, y) = -hu(a, y)$ we obtain $X'(a) = -hX(a)$ and

$$\lambda \cos \lambda a = -h \sin \lambda a \quad \text{or} \quad \tan \lambda a = -\frac{\lambda}{h}.$$

Let λ_n, where $n = 1, 2, 3, \ldots$, be the consecutive positive roots of this equation. From $u(x, 0) = 0$ we obtain $Y(0) = 0$ and $c_3 = 0$. Thus

$$u(x, y) = \sum_{n=1}^{\infty} A_n \sinh \lambda_n y \sin \lambda_n x.$$

Now

$$f(x) = \sum_{n=1}^{\infty} A_n \sinh \lambda_n b \sin \lambda_n x$$

and

$$A_n \sinh \lambda_n b = \frac{\int_0^a f(x) \sin \lambda_n x\, dx}{\int_0^a \sin^2 \lambda_n x\, dx}.$$

Since

$$\int_0^a \sin^2 \lambda_n x\, dx = \frac{1}{2}\left[a - \frac{1}{2\lambda_n} \sin 2\lambda_n a \right] = \frac{1}{2}\left[a - \frac{1}{\lambda_n} \sin \lambda_n a \cos \lambda_n a \right]$$

$$= \frac{1}{2}\left[a - \frac{1}{h\lambda_n}(h \sin \lambda_n a) \cos \lambda_n a \right]$$

$$= \frac{1}{2}\left[a - \frac{1}{h\lambda_n}(-\lambda_n \cos \lambda_n a) \cos \lambda_n a \right] = \frac{1}{2h}\left[ah + \cos^2 \lambda_n a \right],$$

we have

$$A_n = \frac{2h}{\sinh \lambda_n b[ah + \cos^2 \lambda_n a]} \int_0^a f(x) \sin \lambda_n x\, dx.$$

4. Letting $u(x,y) = X(x)Y(y)$ and separating variables gives

$$X''Y + XY'' = 0.$$

The boundary conditions

$$\frac{\partial u}{\partial y}\bigg|_{y=0} = 0 \quad \text{and} \quad \frac{\partial u}{\partial y}\bigg|_{y=1} = -hu(x,1)$$

correspond to

$$X(x)Y'(0) = 0 \quad \text{and} \quad X(x)Y'(1) = -hX(x)Y(1)$$

or

$$Y'(0) = 0 \quad \text{and} \quad Y'(1) = -hY(1).$$

Since these homogeneous boundary conditions are in terms of Y, we separate the differential equation as

$$\frac{X''}{X} = -\frac{Y''}{Y} = \lambda^2.$$

Then

$$Y'' + \lambda^2 Y = 0$$

and

$$X'' - \lambda^2 X = 0$$

have solutions

$$Y(y) = c_1 \cos \lambda y + c_2 \sin \lambda y$$

and

$$X(x) = c_3 e^{-\lambda x} + c_4 e^{\lambda x}.$$

We use exponential functions in the solution of $X(x)$ since $\cosh \lambda x$ and $\sinh \lambda x$ are both unbounded as $x \to \infty$. Now, $Y'(0) = 0$ implies $c_2 = 0$, so $Y(y) = c_1 \cos \lambda y$. Since $Y'(y) = -c_1 \lambda \sin \lambda y$, the boundary condition $Y'(1) = -hY(1)$ implies

$$-c_1 \lambda \sin \lambda = -hc_1 \cos \lambda \quad \text{or} \quad \cot \lambda = \frac{\lambda}{h}.$$

Consideration of the graphs of $f(\lambda) = \cot \lambda$ and $g(\lambda) = \lambda/h$ show that $\cos \lambda = \lambda h$ has an infinite number of roots. The consecutive positive roots λ_n for $n = 1, 2, 3, \ldots$, are the eigenvalues of the problem. The corresponding eigenfunctions are $Y_n(y) = c_1 \cos \lambda_n y$. The condition $\lim_{x \to \infty} u(x,y) = 0$ is equivalent to $\lim_{x \to \infty} X(x) = 0$. Thus $c_4 = 0$ and $X(x) = c_3 e^{-\lambda x}$. Therefore

$$u_n(x,y) = X_n(x)Y_n(x) = A_n e^{-\lambda_n x} \cos \lambda_n y$$

and by the superposition principle

$$u(x,y) = \sum_{n=1}^{\infty} A_n e^{-\lambda_n x} \cos \lambda_n y.$$

[It is easily shown that there are no eigenvalues corresponding to $\lambda = 0$.] Finally, the condition $u(0,y) = u_0$ implies

$$u_0 = \sum_{n=1}^{\infty} A_n \cos \lambda_n y.$$

This is not a Fourier cosine series since the coefficients λ_n of y are not integer multiples of π/p, where $p = 1$ in this problem. The functions $\cos \lambda_n y$ are however orthogonal since they are eigenfunctions of the Sturm-Lionville problem

$$Y'' + \lambda^2 Y = 0,$$

$$Y'(0) = 0$$

$$Y'(1) + hY(1) = 0,$$

with weight function $p(x) = 1$. Thus we find

$$A_n = \frac{\int_0^1 u_0 \cos \lambda_n y \, dy}{\int_0^1 \cos^2 \lambda_n y \, dy}.$$

Now

$$\int_0^1 u_0 \cos \lambda_n y \, dy = \frac{u_0}{\lambda_n} \sin \lambda_n y \bigg|_0^1 = \frac{u_0}{\lambda_n} \sin \lambda_n$$

and

$$\int_0^1 \cos^2 \lambda_n y \, dy = \frac{1}{2} \int_0^1 (1 + \cos 2\lambda_n y) \, dy = \frac{1}{2} \left[y + \frac{1}{2\lambda_n} \sin 2\lambda_n y \right]_0^1$$

$$= \frac{1}{2} \left[1 + \frac{1}{2\lambda_n} \sin 2\lambda_n \right] = \frac{1}{2} \left[1 + \frac{1}{\lambda_n} \sin \lambda_n \cos \lambda_n \right].$$

Since $\cot \lambda = \lambda/h$,

$$\frac{\cos \lambda}{\lambda} = \frac{\sin \lambda}{h}$$

and

$$\int_0^1 \cos^2 \lambda_n y \, dy = \frac{1}{2} \left[1 + \frac{\sin^2 \lambda_n}{h} \right].$$

Then

$$A_n = \frac{\frac{u_0}{\lambda_n} \sin \lambda_n}{\frac{1}{2} \left[1 + \frac{1}{h} \sin^2 \lambda_n \right]} = \frac{2hu_0 \sin \lambda_n}{\lambda_n \left(h + \sin^2 \lambda_n \right)}$$

523

and

$$u(x, y) = 2hu_0 \sum_{n=1}^{\infty} \frac{\sin \lambda_n}{\lambda_n (1 + \sin^2 \lambda_n)} e^{-\lambda_n x} \cos \lambda_n y$$

where λ_n for $n = 1, 2, 3, \ldots$ are the consecutive positive roots of $\cot \lambda = \lambda/h$.

5. The boundary-value problem is

$$k \frac{\partial^2 u}{\partial x^2} = \frac{\partial u}{\partial t}, \quad 0 < x < L, \quad t > 0,$$

$$u(0, t) = 0, \quad \frac{\partial u}{\partial x}\bigg|_{x=L} = 0, \quad t > 0,$$

$$u(x, 0) = f(x), \quad 0 < x < L.$$

Separation of variables leads to

$$X'' + \lambda^2 X = 0$$

$$T' + k\lambda^2 T = 0$$

and

$$X(x) = c_1 \cos \lambda x + c_2 \sin \lambda x$$

$$T(t) = c_3 e^{-k\lambda^2 t}.$$

From $X(0) = 0$ we find $c_1 = 0$. From $X'(L) = 0$ we obtain $\cos \lambda L = 0$ and

$$\lambda = \frac{\pi(2n - 1)}{2L}, \quad n = 1, 2, 3, \ldots .$$

Thus

$$u(x, t) = \sum_{n=1}^{\infty} A_n e^{-k(2n-1)^2 \pi^2 t / 4L^2} \sin \left(\frac{2n - 1}{2L} \right) \pi x$$

where

$$A_n = \frac{\int_0^L f(x) \sin \left(\frac{2n-1}{2L} \right) \pi x \, dx}{\int_0^L \sin^2 \left(\frac{2n-1}{2L} \right) \pi x \, dx} = \frac{2}{L} \int_0^L f(x) \sin \left(\frac{2n - 1}{2L} \right) \pi x \, dx.$$

6. Substituting $u(x, t) = v(x, t) + \psi(x)$ into the partial differential equation gives

$$a^2 \frac{\partial^2 v}{\partial x^2} + \psi''(x) = \frac{\partial^2 v}{\partial t^2}.$$

This equation will be homogeneous if $\psi''(x) = 0$ or $\psi(x) = c_1 x + c_2$. The boundary condition $u(0, t) = 0$ implies $\psi(0) = 0$ which implies $c_2 = 0$. Thus $\psi(x) = c_1 x$. Using the second boundary condition, we obtain

$$E \left(\frac{\partial v}{\partial x} + \psi' \right) \bigg|_{x=L} = F_0,$$

which will be homogeneous when

$$E\psi'(L) = F_0.$$

Since $\psi'(x) = c_1$ we conclude that $c_1 = F_0/E$ and

$$\psi(x) = \frac{F_0}{E}x.$$

The new boundary-value problem is

$$a^2 \frac{\partial^2 v}{\partial x^2} = \frac{\partial^2 v}{\partial t^2}, \quad 0 < x < L, \quad t > 0$$

$$v(0,t) = 0, \quad \frac{\partial v}{\partial x}\bigg|_{x=L} = 0, \quad t > 0,$$

$$v(x,0) = -\frac{F_0}{E}x, \quad \frac{\partial v}{\partial t}\bigg|_{t=0} = 0, \quad 0 < x < L.$$

Referring to Example 2 in the text we see that

$$v(x,t) = \sum_{n=1}^{\infty} A_n \cos a \left(\frac{2n-1}{2L}\right)\pi t \sin \left(\frac{2n-1}{2L}\right)\pi x$$

where

$$-\frac{F_0}{E}x = \sum_{n=1}^{\infty} A_n \sin \left(\frac{2n-1}{2L}\right)\pi x$$

and

$$A_n = \frac{-F_0 \int_0^L x \sin \left(\frac{2n-1}{2L}\right)\pi x \, dx}{E \int_0^L \sin^2 \left(\frac{2n-1}{2L}\right)\pi x \, dx} = \frac{8F_0 L(-1)^n}{E\pi^2(2n-1)^2}.$$

Thus

$$u(x,t) = v(x,t) + \psi(x)$$

$$= \frac{F_0}{E}x + \frac{8F_0 L}{E\pi^2}\sum_{n=1}^{\infty} \frac{(-1)^n}{(2n-1)^2}\cos a\left(\frac{2n-1}{2L}\right)\pi t \sin \left(\frac{2n-1}{2L}\right)\pi x.$$

7. Separation of variables leads to

$$Y'' + \lambda^2 Y = 0$$

$$X'' - \lambda^2 X = 0$$

and

$$Y(y) = c_1 \cos \lambda y + c_2 \sin \lambda y$$

$$X(x) = c_3 \cosh \lambda x + c_4 \sinh \lambda x.$$

From $Y(0) = 0$ we find $c_1 = 0$. From $Y'(1) = 0$ we obtain $\cos \lambda = 0$ and

$$\lambda = \frac{\pi(2n-1)}{2}, \quad n = 1, 2, 3, \ldots .$$

Thus

$$Y(y) = c_2 \sin\left(\frac{2n-1}{2}\right)\pi y.$$

From $X'(0) = 0$ we find $c_4 = 0$. Then

$$u(x,y) = \sum_{n=1}^{\infty} A_n \cosh\left(\frac{2n-1}{2}\right)\pi x \sin\left(\frac{2n-1}{2}\right)\pi y$$

where

$$u_0 = u(1,y) = \sum_{n=1}^{\infty} A_n \cosh\left(\frac{2n-1}{2}\right)\pi \sin\left(\frac{2n-1}{2}\right)\pi y$$

and

$$A_n \cosh\left(\frac{2n-1}{2}\right)\pi = \frac{\int_0^1 u_0 \sin\left(\frac{2n-1}{2}\right)\pi y\, dy}{\int_0^1 \sin^2\left(\frac{2n-1}{2}\right)\pi y\, dy} = \frac{4u_0}{(2n-1)\pi}.$$

Thus

$$u(x,y) = \frac{4u_0}{\pi}\sum_{n=1}^{\infty}\frac{1}{(2n-1)\cosh\left(\frac{2n-1}{2}\right)\pi}\cosh\left(\frac{2n-1}{2}\right)\pi x \sin\left(\frac{2n-1}{2}\right)\pi y.$$

8. The boundary-value problem is

$$k\frac{\partial^2 u}{\partial x^2} = \frac{\partial u}{\partial t}, \quad 0 < x < 1, \quad t > 0$$

$$\left.\frac{\partial u}{\partial x}\right|_{x=0} = hu(0,t), \quad \left.\frac{\partial u}{\partial x}\right|_{x=1} = -hu(1,t), \quad h > 0, \quad t > 0,$$

$$u(x,0) = f(x), \quad 0 < x < 1.$$

Referring to Example 1 in the text we have

$$X(x) = c_1 \cos \lambda x + c_2 \sin \lambda x$$

and

$$T(t) = c_3 e^{-k\lambda^2 t}.$$

Applying the boundary conditions, we obtain

$$X'(0) = hX(0)$$

$$X'(1) = -hX(1)$$

or

$$\lambda c_2 = hc_1$$

$$-\lambda c_1 \sin \lambda + \lambda c_2 \cos \lambda = -hc_1 \cos \lambda - hc_2 \sin \lambda.$$

Choosing $c_1 = \lambda$ and $c_2 = h$ (to satisfy the first equation above) we obtain

$$-\lambda^2 \sin \lambda + h\lambda \cos \lambda = -h\lambda \cos \lambda - h^2 \sin \lambda$$

$$2h\lambda \cos \lambda = (\lambda^2 - h^2) \sin \lambda.$$

The eigenvalues λ_n are the consecutive positive roots of

$$\tan \lambda = \frac{2h\lambda}{\lambda^2 - h^2}.$$

Then

$$u(x,t) = \sum_{n=1}^{\infty} A_n e^{-k\lambda_n^2 t}(\lambda_n \cos \lambda_n x + h \sin \lambda_n x)$$

where

$$f(x) = u(x,0) = \sum_{n=1}^{\infty} A_n(\lambda_n \cos \lambda_n x + h \sin \lambda_n x)$$

and

$$A_n = \frac{\int_0^1 f(x)(\lambda_n \cos \lambda_n x + h \sin \lambda_n x)dx}{\int_0^1 (\lambda_n \cos \lambda_n x + h \sin \lambda_n x)^2 dx}$$

$$= \frac{2}{\lambda_n^2 + 2h + h^2} \int_0^1 f(x)(\lambda_n \cos \lambda_n x + h \sin \lambda_n x)dx.$$

[Note: the evaluation and simplification of the integral in the denominator requires the use of the relationship $(\lambda^2 - h^2) \sin \lambda = 2h\lambda \cos \lambda$.]

9. **(a)** Using $u = XT$ and separation constant λ^4 we find

$$X^{(4)} - \lambda^4 X = 0$$

and

$$X(x) = c_1 \cos \lambda x + c_2 \sin \lambda x + c_3 \cosh \lambda x + c_4 \sinh \lambda x.$$

Since $u = XT$ the boundary conditions become

$$X(0) = 0, \quad X'(0) = 0, \quad X''(1) = 0, \quad X'''(1) = 0.$$

Now $X(0) = 0$ implies $c_1 + c_3 = 0$, while $X'(0) = 0$ implies $c_2 + c_4 = 0$. Thus

$$X(x) = c_1 \cos \lambda x + c_2 \sin \lambda x - c_1 \cosh \lambda x - c_2 \sinh \lambda x.$$

The boundary condition $X''(1) = 0$ implies

$$-c_1 \cos \lambda - c_2 \sin \lambda - c_1 \cosh \lambda - c_2 \sinh \lambda = 0$$

while the boundary condition $X'''(1) = 0$ implies

$$c_1 \sin \lambda - c_2 \cos \lambda - c_1 \sinh \lambda - c_2 \cosh \lambda = 0.$$

527

We then have the system of two equations in two unknowns

$$(\cos\lambda + \cosh\lambda)c_1 + (\sin\lambda + \sinh\lambda)c_2 = 0$$

$$(\sin\lambda - \sinh\lambda)c_1 - (\cos\lambda + \cosh\lambda)c_2 = 0.$$

This homogeneous system will have nontrivial solutions for c_1 and c_2 provided

$$\begin{vmatrix} \cos\lambda + \cosh\lambda & \sin\lambda + \sinh\lambda \\ \sin\lambda - \sinh\lambda & -\cos\lambda - \cosh\lambda \end{vmatrix} = 0$$

or

$$-2 - 2\cos\lambda\cosh\lambda = 0.$$

Thus, the eigenvalues are determined by the equation $\cos\lambda\cosh\lambda = -1$.

(b) Using a computer to graph $\cosh\lambda$ and $-1/\cos\lambda = -\sec\lambda$ we see that the first two positive eigenvalues occur near 1.9 and 4.7. Applying Newton's method with these initial values we find that the eigenvalues are $\lambda_1 = 1.8751$ and $\lambda_2 = 4.6941$.

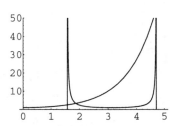

10. (a) In this case the boundary conditions are

$$u(0,t) = 0, \qquad \frac{\partial u}{\partial x}\Big|_{x=0} = 0$$

$$u(1,t) = 0, \qquad \frac{\partial u}{\partial x}\Big|_{x=1} = 0.$$

Separating variables leads to

$$X(x) = c_1\cos\lambda x + c_2\sin\lambda x + c_3\cosh\lambda x + c_4\sinh\lambda x$$

subject to

$$X(0) = 0, \quad X'(0) = 0, \quad X(1) = 0, \quad \text{and} \quad X'(1) = 0.$$

Now $X(0) = 0$ implies $c_1 + c_3 = 0$ while $X'(0) = 0$ implies $c_2 + c_4 = 0$. Thus

$$X(x) = c_1\cos\lambda x + c_2\sin\lambda x - c_1\cosh\lambda x - c_2\sinh\lambda x.$$

The boundary condition $X(1) = 0$ implies

$$c_1\cos\lambda + c_2\sin\lambda - c_1\cosh\lambda - c_2\sinh\lambda = 0$$

while the boundary condition $X'(1) = 0$ implies

$$-c_1\sin\lambda + c_2\cos\lambda - c_1\sinh\lambda - c_2\cosh\lambda = 0.$$

528

We then have the system of two equations in two unknowns

$$(\cos\lambda - \cosh\lambda)c_1 + (\sin\lambda - \sinh\lambda)c_2 = 0$$

$$-(\sin\lambda + \sinh\lambda)c_1 + (\cos\lambda - \cosh\lambda)c_2 = 0.$$

This homogeneous system will have nontrivial solutions for c_1 and c_2 provided

$$\begin{vmatrix} \cos\lambda - \cosh\lambda & \sin\lambda - \sinh\lambda \\ -\sin\lambda - \sinh\lambda & \cos\lambda - \cosh\lambda \end{vmatrix} = 0$$

or

$$2 - 2\cos\lambda\cosh\lambda = 0.$$

Thus, the eigenvalues are determined by the equation $\cos\lambda\cosh\lambda = 1$.

(b) Using a computer to graph $\cosh\lambda$ and $1/\cos\lambda = \sec\lambda$ we see that the first two positive eigenvalues occur near the vertical asymptotes of $\sec\lambda$, at $3\pi/2$ and $5\pi/2$. Applying Newton's method with these initial values we find that the eigenvalues are $\lambda_1 = 4.7300$ and $\lambda_2 = 7.8532$.

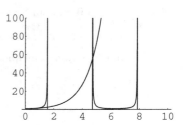

Exercises 12.8

1. This boundary-value problem was solved in Example 1 in the text. Identifying $b = c = \pi$ and $f(x, y) = u_0$ we have

$$u(x, y, t) = \sum_{m=1}^{\infty}\sum_{n=1}^{\infty} A_{mn}e^{-k(m^2+n^2)t}\sin mx \sin ny$$

where

$$A_{mn} = \frac{4}{\pi^2}\int_0^{\pi}\int_0^{\pi} u_0 \sin mx \sin ny\, dx\, dy$$

$$= \frac{4u_0}{\pi^2}\int_0^{\pi}\sin mx\, dx \int_0^{\pi}\sin ny\, dy$$

$$= \frac{4u_0}{mn\pi^2}[1 - (-1)^m][1 - (-1)^n].$$

2. As shown in Example 1 in the text, separation of variables leads to

$$X(x) = c_1 \cos\lambda x + c_2 \sin\lambda x$$

$$Y(y) = c_3 \cos\mu y + c_4 \sin\mu y$$

and

$$T(t) + c_5 e^{-k(\lambda^2 + \mu^2)t}.$$

The boundary conditions

$$\left. \begin{array}{ll} u_x(0, y, t) = 0, & u_x(1, y, t) = 0 \\ u_y(x, 0, t) = 0, & u_y(x, 1, t) = 0 \end{array} \right\} \quad \text{imply} \quad \left\{ \begin{array}{ll} X'(0) = 0, & X'(1) = 0 \\ Y'(0) = 0, & Y'(1) = 0. \end{array} \right.$$

Applying these conditions to

$$X'(x) = -\lambda c_1 \sin \lambda x + \lambda c_2 \cos \lambda x$$

and

$$Y'(y) = -\mu c_3 \sin \mu y + \mu c_4 \cos \mu y$$

gives $c_2 = c_4 = 0$ and $\sin \lambda = \sin \mu = 0$. Then

$$\lambda = m\pi, \ m = 0, 1, 2, \dots \quad \text{and} \quad \mu = n\pi, \ n = 0, 1, 2, \dots .$$

By the superposition principle

$$u(x, y, t) = A_{00} + \sum_{m=1}^{\infty} A_{m0} e^{-km^2\pi^2 t} \cos m\pi x + \sum_{n=1}^{\infty} A_{0n} e^{-kn^2\pi^2 t} \cos n\pi y$$

$$+ \sum_{m=1}^{\infty} \sum_{n=1}^{\infty} A_{mn} e^{-k(m^2 + n^2)\pi^2 t} \cos m\pi x \cos n\pi y.$$

The coefficients were computed in Problem 53, Exercises 11.3:

$$A_{00} = \frac{1}{4}$$

$$A_{m0} = \frac{(-1)^m - 1}{\pi^2 m^2}$$

$$A_{0n} = \frac{(-1)^n - 1}{\pi^2 n^2}$$

$$A_{mn} = \frac{4[(-1)^m - 1][(-1)^n - 1]}{\pi^4 m^2 n^2}.$$

In Problems 3 and 4 we need to solve the partial differential equation

$$a^2 \left(\frac{\partial^2 u}{\partial x^2} + \frac{\partial^2 u}{\partial y^2} \right) = \frac{\partial^2 u}{\partial t^2}.$$

To separate this equation we try $u(x, y, t) = X(x)Y(y)T(t)$:

$$a^2 (X''YT + XY''T) = XYT''$$

$$\frac{X''}{X} = -\frac{Y''}{Y} + \frac{T''}{a^2 T} = -\lambda^2.$$

Then

$$X'' + \lambda^2 X = 0 \tag{1}$$

$$\frac{Y''}{Y} = \frac{T''}{a^2 T} + \lambda^2 = -\mu^2$$

$$Y'' + \mu^2 Y = 0 \tag{2}$$

$$T'' + a^2 \left(\lambda^2 + \mu^2\right) T = 0. \tag{3}$$

The general solutions of equations (1), (2), and (3) are, respectively,

$$X(x) = c_1 \cos \lambda x + c_2 \sin \lambda x$$

$$Y(y) = c_3 \cos \mu y + c_4 \sin \mu y$$

$$T(t) = c_5 \cos a\sqrt{\lambda^2 + \mu^2}\, t + c_6 \sin a\sqrt{\lambda^2 + \mu^2}\, t.$$

3. The conditions $X(0) = 0$ and $Y(0) = 0$ give $c_1 = 0$ and $c_3 = 0$. The conditions $X(\pi) = 0$ and $Y(\pi) = 0$ yield two sets of eigenvalues:

$$\lambda = m, \ m = 1, 2, 3, \ldots \quad \text{and} \quad \mu = n, \ n = 1, 2, 3, \ldots .$$

A product solution of the partial differential equation that satisfies the boundary conditions is

$$u_{mn}(x, y, t) = \left(A_{mn} \cos a\sqrt{m^2 + n^2}\, t + B_{mn} \sin a\sqrt{m^2 + n^2}\, t \right) \sin mx \sin ny.$$

To satisfy the initial conditions we use the superposition principle:

$$u(x, y, t) = \sum_{m=1}^{\infty} \sum_{n=1}^{\infty} \left(A_{mn} \cos a\sqrt{m^2 + n^2}\, t + B_{mn} \sin a\sqrt{m^2 + n^2}\, t \right) \sin mx \sin ny.$$

The initial condition $u_t(x, y, 0) = 0$ implies $B_{mn} = 0$ and

$$u(x, y, t) = \sum_{m=1}^{\infty} \sum_{n=1}^{\infty} A_{mn} \cos a\sqrt{m^2 + n^2}\, t \sin mx \sin ny.$$

At $t = 0$ we have

$$xy(x - \pi)(y - \pi) = \sum_{m=1}^{\infty} \sum_{n=1}^{\infty} A_{mn} \sin mx \sin ny.$$

It follows that (See Problem 52, Exercises 11.3)

$$A_{mn} = \frac{4}{\pi^2} \int_0^\pi \int_0^\pi xy(x-\pi)(y-\pi) \sin mx \sin ny \, dx \, dy$$

$$= \frac{4}{\pi^2} \int_0^\pi x(x-\pi) \sin mx \, dx \int_0^\pi y(y-\pi) \sin ny \, dy$$

$$= \frac{16}{m^3 n^3 \pi^2} [(-1)^m - 1][(-1)^n - 1].$$

4. The conditions $X(0) = 0$ and $Y(0) = 0$ give $c_1 = 0$ and $c_3 = 0$. The conditions $X(b) = 0$ and $Y(c) = 0$ yield two sets of eigenvalues

$$\lambda = m\pi/b, \ m = 1, 2, 3, \ldots \qquad \text{and} \qquad \mu = n\pi/c, \ n = 1, 2, 3, \ldots .$$

A product solution of the partial differential that satisfies the boundary conditions is

$$u_{mn}(x, y, t) = (A_{mn} \cos a\omega_{mn}t + B_{mn} \sin a\omega_{mn}t) \sin\left(\frac{m\pi}{b}x\right) \sin\left(\frac{n\pi}{c}y\right),$$

where $\omega_{mn} = \sqrt{(m\pi/b)^2 + (n\pi/c)^2}$. To satisfy the initial conditions we use the superposition principle:

$$u(x, y, t) = \sum_{m=1}^\infty \sum_{n=1}^\infty (A_{mn} \cos a\omega_{mn}t + B_{mn} \sin a\omega_{mn}t) \sin\left(\frac{m\pi}{b}x\right) \sin\left(\frac{n\pi}{c}y\right).$$

At $t = 0$ we have

$$f(x, y) = \sum_{m=1}^\infty \sum_{n=1}^\infty A_{mn} \sin\left(\frac{m\pi}{b}x\right) \sin\left(\frac{n\pi}{c}y\right)$$

and

$$g(x, y) = \sum_{m=1}^\infty \sum_{n=1}^\infty B_{mn} a\omega_{mn} \sin\left(\frac{m\pi}{b}x\right) \sin\left(\frac{n\pi}{c}y\right).$$

It follows that (see Problem 52, Exercises 11.3)

$$A_{mn} = \frac{4}{bc} \int_0^c \int_0^b f(x, y) \sin\left(\frac{m\pi}{b}x\right) \sin\left(\frac{n\pi}{c}y\right) dx \, dy$$

$$B_{mn} = \frac{4}{abc\omega_{mn}} \int_0^c \int_0^b g(x, y) \sin\left(\frac{m\pi}{b}x\right) \sin\left(\frac{n\pi}{c}y\right) dx \, dy.$$

To separate Laplace's equation in three dimensions we try $u(x, y, z) = X(x)Y(y)Z(z)$:

$$X''YZ + XY''Z + XYZ'' = 0$$

$$\frac{X''}{X} = -\frac{Y''}{Y} - \frac{Z''}{Z} = -\lambda^2.$$

Then

$$X'' + \lambda^2 X = 0 \tag{4}$$

$$\frac{Y''}{Y} = -\frac{Z''}{Z} + \lambda^2 = -\mu^2$$

$$Y'' + \mu^2 Y = 0 \tag{5}$$

$$Z'' - (\lambda^2 + \mu^2) Z = 0. \tag{6}$$

The general solutions of equations (4), (5), and (6) are, respectively

$$X(x) = c_1 \cos \lambda x + c_2 \sin \lambda x$$

$$Y(y) = c_3 \cos \mu y + c_4 \sin \mu y$$

$$Z(z) = c_5 \cosh \sqrt{\lambda^2 + \mu^2}\, z + c_6 \sinh \sqrt{\lambda^2 + \mu^2}\, z.$$

5. The boundary and initial conditions are

$$u(0, y, z) = 0, \quad u(a, y, z) = 0$$

$$u(x, 0, z) = 0, \quad u(x, b, z) = 0$$

$$u(x, y, 0) = 0, \quad u(x, y, c) = f(x, y).$$

The conditions $X(0) = Y(0) = Z(0) = 0$ give $c_1 = c_3 = c_5 = 0$. The conditions $X(a) = 0$ and $Y(b) = 0$ yield two sets of eigenvalues:

$$\lambda = \frac{m\pi}{a}, \quad m = 1, 2, 3, \dots \quad \text{and} \quad \mu = \frac{n\pi}{b}, \quad n = 1, 2, 3, \dots.$$

By the superposition principle

$$u(x, y, t) = \sum_{m=1}^{\infty} \sum_{n=1}^{\infty} A_{mn} \sinh \omega_{mn} z \sin \frac{m\pi}{a} x \sin \frac{n\pi}{b} y$$

where

$$\omega_{mn}^2 = \frac{m^2 \pi^2}{a^2} + \frac{n^2 \pi^2}{b^2}$$

and

$$A_{mn} = \frac{4}{ab \sinh \omega_{mn} c} \int_0^b \int_0^a f(x, y) \sin \frac{m\pi}{a} x \sin \frac{n\pi}{b} y\, dx\, dy.$$

6. The boundary and initial conditions are

$$u(0, y, z) = 0, \qquad u(a, y, z) = 0,$$

$$u(x, 0, z) = 0, \qquad u(x, b, z) = 0,$$

$$u(x, y, 0) = f(x, y), \qquad u(x, y, c) = 0.$$

533

The conditions $X(0) = Y(0) = 0$ give $c_1 = c_3 = 0$. The conditions $X(a) = Y(b) = 0$ yield two sets of eigenvalues:

$$\lambda = \frac{m\pi}{a}, \quad m = 1, 2, 3, \ldots \quad \text{and} \quad \mu = \frac{n\pi}{b}, \quad n = 1, 2, 3, \ldots .$$

Let

$$\omega_{mn}^2 = \frac{m^2\pi^2}{a^2} + \frac{n^2\pi^2}{b^2}.$$

Then the boundary condition $Z(c) = 0$ gives

$$c_5 \cosh c\omega_{mn} + c_6 \sinh c\omega_{mn} = 0$$

from which we obtain

$$Z(z) = c_5 \left(\cosh \omega_{mn} z - \frac{\cosh c\omega_{mn}}{\sinh c\omega_{mn}} \sinh \omega z \right)$$

$$= \frac{c_5}{\sinh c\omega_{mn}} (\sinh c\omega_{mn} \cosh \omega_{mn} z - \cosh c\omega_{mn} \sinh \omega_{mn} z)$$

$$= c_{mn} \sinh \omega_{mn}(c - z).$$

By the superposition principle

$$u(x, y, t) = \sum_{m=1}^{\infty} \sum_{n=1}^{\infty} A_{mn} \sinh \omega_{mn}(c - z) \sin \frac{m\pi}{a} x \sin \frac{n\pi}{b} y$$

where

$$A_{mn} = \frac{4}{ab \sinh c\omega_{mn}} \int_0^b \int_0^a f(x, y) \sin \frac{m\pi}{a} x \sin \frac{n\pi}{b} y \, dx \, dy.$$

———— Chapter 12 Review Exercises ————

1. Let $u = XY$ so that

$$u_{yx} = X'Y',$$

$$X'Y' = XY,$$

and

$$\frac{X'}{X} = \frac{Y}{Y'} = \pm\lambda^2.$$

Then

$$X' \mp \lambda^2 X = 0 \quad \text{and} \quad Y \mp \lambda^2 Y' = 0.$$

For $\lambda^2 > 0$ we obtain

$$X = c_1 e^{\lambda^2 x} \quad \text{and} \quad Y = c_2 e^{y/\lambda^2}$$

For $-\lambda^2 < 0$ we obtain

$$X = c_1 e^{-\lambda^2 x} \quad \text{and} \quad Y = c_2 e^{-y/\lambda^2}.$$

For $\lambda^2 = 0$ we obtain $X = c_1$ and $Y = 0$. The general form is

$$u = XY = A_1 e^{A_2 x + y/A_2}.$$

2. If $u = XY$ then

$$u_x = X'Y,$$

$$u_y = XY',$$

$$u_{xx} = X''Y,$$

$$u_{yy} = XY'',$$

and

$$X''Y + XY'' + 2X'Y + 2XY' = 0$$

so that

$$(X'' + 2X')Y + X(Y'' + 2Y') = 0,$$

$$\frac{X'' + 2X'}{-X} = \frac{Y'' + 2Y'}{Y} = +\lambda^2,$$

$$X'' + 2X' \pm \lambda^2 X = 0,$$

and

$$Y'' + 2Y' \mp \lambda^2 Y = 0.$$

Using λ^2 as a separation constant we obtain

$$Y = c_1 e^{(-1+\sqrt{1+\lambda^2})y} + c_2 e^{(-1-\sqrt{1+\lambda^2})y}.$$

If $1 - \lambda^2 < 0$ then

$$X = e^{-x}\left(c_3 \cos\sqrt{\lambda^2 - 1}\, x + c_4 \sin\sqrt{\lambda^2 - 1}\, x\right).$$

If $1 - \lambda^2 > 0$ then

$$X = c_3 e^{(-1+\sqrt{1-\lambda^2})x} + c_4 e^{(-1-\sqrt{1-\lambda^2})x}.$$

If $1 - \lambda^2 = 0$ then

$$X = c_3 e^{-x} + c_4 x e^{-x}.$$

so that

$$u = XY = \left(A_1 e^{(-1+\sqrt{1+A_5})y} + A_2 e^{(-1-\sqrt{1+A_5})y}\right) e^{-x} \left(A_3 \cos\sqrt{A_5 - 1}\,x + A_4 \sin\sqrt{A_5 - 1}\,x\right)$$

$$= \left(A_1 e^{(-1+\sqrt{1+A_5})y} + A_2 e^{(-1-\sqrt{1+A_5})y}\right) \left(A_3 e^{(-1+\sqrt{1-A_5})x} + A_4 e^{(-1-\sqrt{1-A_5})x}\right)$$

$$= \left(A_1 e^{(-1+\sqrt{2})y} + A_2 e^{(-1-\sqrt{2})y}\right) \left(A_3 e^{-x} + A_4 x e^{-x}\right).$$

Using $-\lambda^2$ we obtain the same three solutions except that x and y are interchanged. Using $\lambda^2 = 0$ we obtain

$$u = XY = \left(A_1 + A_2 e^{-2x}\right) \left(A_3 + A_4 e^{-2y}\right).$$

3. Substituting $u(x,t) = v(x,t) + \psi(x)$ into the partial differential equation we obtain

$$k\frac{\partial^2 v}{\partial x^2} + k\psi''(x) = \frac{\partial v}{\partial t}.$$

This equation will be homogeneous provided ψ satisfies

$$k\psi'' = 0 \quad \text{or} \quad \psi = c_1 x + c_2.$$

Considering

$$u(0,t) = v(0,t) + \psi(0) = u_0$$

we set $\psi(0) = u_0$ so that $\psi(x) = c_1 x + u_0$. Now

$$-\frac{\partial u}{\partial x}\bigg|_{x=\pi} = -\frac{\partial v}{\partial x}\bigg|_{x=\pi} - \psi'(x) = v(\pi,t) + \psi(\pi) - u_1$$

is equivalent to

$$\frac{\partial v}{\partial x}\bigg|_{x=\pi} + v(\pi,t) = u_1 - \psi'(x) - \psi(\pi) = u_1 - c_1 - (c_1\pi + u_0),$$

which will be homogeneous when

$$u_1 - c_1 - c_1\pi - u_0 = 0 \quad \text{or} \quad c_1 = \frac{u_1 - u_0}{1 + \pi}.$$

The steady-state solution is

$$\psi(x) = \left(\frac{u_1 - u_0}{1 + \pi}\right) x + u_0.$$

4. The solution of the problem represents the heat of a thin rod of length π. The left boundary $x = 0$ is kept at constant temperature u_0 for $t > 0$. Heat is lost from the right end of the rod by being in contact with a medium that is held at constant temperature u_1.

5. The boundary-value problem is

$$a^2 \frac{\partial^2 u}{\partial x^2} = \frac{\partial^2 u}{\partial t^2}, \quad 0 < x < 1, \quad t > 0,$$

$$u(0, t) = 0, \quad u = (1, t) = 0, \quad t > 0,$$

$$u(x, 0) = 0, \quad \frac{\partial u}{\partial t}\bigg|_{t=0} = g(x), \quad 0 < x < 1.$$

From Section 12.4 in the text we see that $A_n = 0$,

$$B_n = \frac{2}{n\pi a} \int_0^1 g(x) \sin n\pi x \, dx = \frac{2}{n\pi a} \int_{1/4}^{3/4} h \sin n\pi x \, dx$$

$$= \frac{2h}{n\pi a} \left(-\frac{1}{n\pi} \cos n\pi x \right)\bigg|_{1/4}^{3/4} = \frac{2h}{n^2\pi^2 a} \left(\cos \frac{n\pi}{4} - \cos \frac{3n\pi}{4} \right)$$

and

$$u(x, t) = \sum_{n=1}^{\infty} B_n \sin n\pi a t \sin n\pi x.$$

6. The boundary-value problem is

$$\frac{\partial^2 u}{\partial x^2} + x^2 = \frac{\partial^2 u}{\partial t^2}, \quad 0 < x < 1, \quad t > 0,$$

$$u(0, t) = 1, \quad u(1, t) = 0, \quad t > 0,$$

$$u(x, 0) = f(x), \quad u_t(x, 0) = 0, \quad 0 < x < 1.$$

Substituting $u(x, t) = v(x, t) + \psi(x)$ into the partial differential equation gives

$$\frac{\partial^2 v}{\partial x^2} + \psi''(x) + x^2 = \frac{\partial^2 v}{\partial t^2}.$$

This equation will be homogeneous provided $\psi''(x) + x^2 = 0$ or

$$\psi(x) = -\frac{1}{12} x^4 + c_1 x + c_2.$$

From $\psi(0) = 1$ and $\psi(1) = 0$ we obtain $c_1 = -11/12$ and $c_2 = 1$. The new problem is

$$\frac{\partial^2 v}{\partial x^2} = \frac{\partial^2 v}{\partial t^2}, \quad 0 < x < 1, \quad t > 0,$$

$$v(0, t) = 0, \quad v(1, t) = 0, \quad t > 0,$$

$$v(x, 0) = f(x) - \psi(x), \quad v_t(x, 0) = 0, \quad 0 < x < 1.$$

From Section 12.4 in the text we see that $B_n = 0$,

$$A_n = 2 \int_0^1 [f(x) - \psi(x)] \sin n\pi x \, dx = 2 \int_0^1 \left[f(x) + \frac{1}{12} x^4 + \frac{11}{12} x - 1 \right] \sin n\pi x \, dx,$$

537

and

$$v(x,t) = \sum_{n=1}^{\infty} A_n \cos n\pi t \sin n\pi x.$$

Thus

$$u(x,t) = v(x,t) + \psi(x) = -\frac{1}{12}x^4 - \frac{11}{12}x + 1 + \sum_{n=1}^{\infty} A_n \cos n\pi t \sin n\pi x.$$

7. Using $u = XY$ and λ^2 as a separation constant leads to

$$X'' - \lambda^2 X = 0,$$

$$X(0) = 0,$$

and

$$Y'' + \lambda^2 Y = 0,$$

$$Y(0) = 0,$$

$$Y(\pi) = 0.$$

Then

$$Y = c_1 \sin ny \quad \text{and} \quad X = c_2 \sinh nx$$

for $n = 1, 2, 3, \ldots$ so that

$$u = \sum_{n=1}^{\infty} A_n \sinh nx \sin ny.$$

Imposing

$$u(\pi, y) = 50 = \sum_{n=1}^{\infty} A_n \sinh n\pi \sin ny$$

gives

$$A_n \sinh n\pi = \frac{100}{n\pi} \frac{1 - (-1)^n}{\sinh n\pi}$$

so that

$$u(x,y) = \frac{100}{\pi} \sum_{n=1}^{\infty} \frac{1 - (-1)^n}{n \sinh n\pi} \sinh nx \sin ny.$$

8. Using $u = XY$ and λ^2 as a separation constant leads to

$$X'' - \lambda^2 X = 0,$$

and

$$Y'' + \lambda^2 Y = 0,$$

$$Y'(0) = 0,$$

$$Y'(\pi) = 0.$$

Then

$$Y = c_1 \cos ny$$

for $n = 0, 1, 2, \ldots$ and

$$X = c_2 \quad \text{or} \quad X = c_2 e^{-nx}$$

for $n = 1, 2, 3, \ldots$ (since u must be bounded as $x \to \infty$) so that

$$u = A_0 + \sum_{n=1}^{\infty} A_n e^{-nx} \cos ny.$$

Imposing

$$u(0, y) = 50 = A_0 + \sum_{n=1}^{\infty} A_n \cos ny$$

gives

$$A_0 = \frac{1}{\pi} \int_0^{\pi} 50 \, dy = 50$$

and

$$A_n = \frac{2}{\pi} \int_0^{\pi} 50 \cos ny \, dy = 0$$

for $n = 1, 2, 3, \ldots$ so that

$$u(x, y) = 50.$$

9. Using $u = XY$ and λ^2 as a separation constant leads to

$$X'' - \lambda^2 X = 0,$$

and

$$Y'' + \lambda^2 Y = 0,$$

$$Y(0) = 0,$$

$$Y(\pi) = 0.$$

Then

$$Y = c_1 \sin ny \quad \text{and} \quad X = c_2 e^{-nx}$$

for $n = 1, 2, 3, \ldots$ (since u must be bounded as $x \to \infty$) so that

$$u = \sum_{n=1}^{\infty} A_n e^{-nx} \sin ny.$$

Imposing

$$u(0, y) = 50 = \sum_{n=1}^{\infty} A_n \sin ny$$

gives

$$A_n = \frac{2}{\pi} \int_0^\pi 50 \sin ny \, dy = \frac{100}{n\pi}[1 - (-1)^n]$$

so that

$$u(x, y) = \sum_{n=1}^\infty \frac{100}{n\pi}[1 - (-1)^n]e^{-nx} \sin ny.$$

10. The boundary-value problem is

$$k\frac{\partial^2 u}{\partial x^2} = \frac{\partial u}{\partial t}, \quad -L < x < L, \quad t > 0,$$

$$u(-L, t) = 0, \quad u(L, t) = 0, \quad t > 0,$$

$$u(x, 0) = u_0, \quad -L < x < L.$$

Referring to Section 12.3 in the text we have

$$X(x) = c_1 \cos \lambda x + c_2 \sin \lambda x$$

and

$$T(t) = c_3 e^{-k\lambda^2 t}.$$

Using the boundary conditions $u(-L, 0) = X(-L)T(0) = 0$ and $u(L, 0) = X(L)T(0) = 0$ we obtain

$$c_1 \cos(-\lambda L) + c_2 \sin(-\lambda L) = 0$$

$$c_1 \cos \lambda L + c_2 \sin \lambda L = 0$$

or

$$c_1 \cos \lambda L - c_2 \sin \lambda L = 0$$

$$c_1 \cos \lambda L + c_2 \sin \lambda L = 0.$$

Adding, we find $\cos \lambda L = 0$ which gives the eigenvalues

$$\lambda = \frac{2n - 1}{2L}\pi, \quad n = 1, 2, 3, \ldots.$$

Thus

$$u(x, t) = \sum_{n=1}^\infty A_n e^{-\left(\frac{2n-1}{2L}\pi\right)^2 kt} \cos\left(\frac{2n - 1}{2L}\right)\pi x.$$

From

$$u(x, 0) = u_0 = \sum_{n=1}^\infty A_n \cos\left(\frac{2n - 1}{2L}\right)\pi x$$

we find

$$A_n = \frac{2 \int_0^L u_0 \cos\left(\frac{2n-1}{2L}\right)\pi x \, dx}{2 \int_0^L \cos^2\left(\frac{2n-1}{2L}\right)\pi x \, dx} = \frac{u_0(-1)^{n+1}2L/\pi(2n-1)}{L/2} = \frac{4u_0(-1)^{n+1}}{\pi(2n-1)}.$$

11. The coefficients of the series

$$u(x, 0) = \sum_{n=1}^{\infty} B_n \sin nx$$

are

$$B_n = \frac{2}{\pi} \int_0^{\pi} \sin x \sin nx \, dx = \frac{2}{\pi} \int_0^{\pi} \frac{1}{2} [\cos(1 - n)x - \cos(1 + n)x] \, dx$$

$$= \frac{1}{\pi} \left[\frac{\sin(1 - n)x}{1 - n} \Big|_0^{\pi} - \frac{\sin(1 + n)x}{1 + n} \Big|_0^{\pi} \right] = 0 \text{ for } n \neq 1.$$

For $n = 1$,

$$B_1 = \frac{2}{\pi} \int_0^{\pi} \sin^2 x \, dx = \frac{1}{\pi} \int_0^{\pi} (1 - \cos 2x) \, dx = 1.$$

Thus

$$u(x, t) = \sum_{n=1}^{\infty} B_n e^{-n^2 t} \sin nx$$

reduces to $u(x, t) = e^{-t} \sin x$ for $n = 1$.

12. Substituting $u(x, t) = v(x, t) + \psi(x)$ into the partial differential equation gives

$$k \frac{\partial^2 v}{\partial x^2} + k\psi'' + \sin 2\pi x = \frac{\partial v}{\partial t}.$$

This equation will be homogeneous provided ψ satisfies

$$k\psi'' + \sin 2\pi x = 0.$$

The general solution of this equation is

$$\psi(x) = \frac{1}{4k\pi^2} \sin 2\pi x + c_1 x + c_2.$$

From $\psi(0) = \psi(1) = 0$ we find that $c_1 = c_2 = 0$ and

$$\psi(x) = \frac{1}{4k\pi^2} \sin 2\pi x.$$

Now the new problem is

$$k \frac{\partial^2 v}{\partial x^2} = \frac{\partial v}{\partial t}, \quad 0 < x < 1, \quad t > 0$$

$$v(0, t) = 0, \quad v(1, t) = 0, \quad t > 0$$

$$v(x, 0) = \sin \pi x - \psi(x), \quad 0 < x < 1.$$

If we let $v = XT$ then

$$\frac{X''}{X} = \frac{T'}{kT} = -\lambda^2$$

gives the separated differential equations

$$X'' + \lambda^2 X = 0 \quad \text{and} \quad T' + k\lambda^2 T = 0.$$

541

The respective solutions are

$$X(x) = c_3 \cos \lambda x + c_4 \sin \lambda x$$

$$T(t) = c_5 e^{-k\lambda^2 t}.$$

From $X(0) = 0$ we get $c_3 = 0$ and from $X(1) = 0$ we find $\lambda = n\pi$ for $n = 1, 2, 3, \ldots$. Consequently, it follows that

$$v(x, t) = \sum_{n=1}^{\infty} A_n e^{-kn^2\pi^2 t} \sin n\pi x$$

where

$$v(x, 0) = \sin \pi x - \frac{1}{4k\pi^2} \sin 2\pi x = 0$$

implies

$$A_n = 2 \int_0^1 \left(\sin \pi x - \frac{1}{4k\pi^2} \sin 2\pi x \right) \sin n\pi x \, dx.$$

By orthogonality $A_n = 0$ for $n = 3, 4, 5, \ldots$, and only A_1 and A_2 can be nonzero. We have

$$A_1 = 2 \left[\int_0^1 \sin^2 \pi x \, dx - \frac{1}{4k\pi^2} \int_0^1 \sin 2\pi x \sin \pi x \, dx \right] = 2 \int_0^1 \frac{1}{2} (1 - \cos 2\pi x) \, dx = 1$$

and

$$A_2 = 2 \left[\int_0^1 \sin \pi x \sin 2\pi x \, dx - \frac{1}{4k\pi^2} \int_0^1 \sin^2 2\pi x \, dx \right]$$

$$= -\frac{1}{2k\pi^2} \int_0^1 \frac{1}{2} (1 - \cos 4\pi x) \, dx = -\frac{1}{4k\pi^2}.$$

Therefore

$$v(x, t) = A_1 e^{-k\pi^2 t} \sin \pi x + A_2 e^{-k4\pi^2 t} \sin 2\pi x$$

$$= e^{-k\pi^2 t} \sin \pi x - \frac{1}{4k\pi^2} e^{-4k\pi^2 t} \sin 2\pi x$$

and

$$u(x, t) = v(x, t) + \psi(x) = e^{-k\pi^2 t} \sin \pi x + \frac{1}{4k\pi^2} (1 - e^{-4k\pi^2 t}) \sin 2\pi x.$$

13. Using $u = XT$ and $-\lambda^2$ as a separation constant we find

$$X'' + 2X' + \lambda^2 X = 0 \quad \text{and} \quad T'' + 2T' + (1 + \lambda^2)T = 0.$$

Thus for $\lambda > 1$

$$X = c_1 e^{-x} \cos \sqrt{\lambda^2 - 1}\, x + c_2 e^{-x} \sin \sqrt{\lambda^2 - 1}\, x$$

$$T = c_3 e^{-t} \cos \lambda t + c_4 e^{-t} \sin \lambda t.$$

For $0 < \lambda < 1$ we only obtain $X = 0$. Now the boundary conditions $X(0) = 0$ and $X(\pi) = 0$ give, in turn, $c_1 = 0$ and $\sqrt{\lambda^2 - 1}\,\pi = n\pi$ or $\lambda^2 = n^2 + 1$, $n = 1, 2, 3, \ldots$. The corresponding solutions are $X = c_2 e^{-x} \sin nx$. The initial condition $T'(0) = 0$ implies $c_3 = \lambda c_4$ and so

$$T = c_4 e^{-t} \left[\sqrt{n^2 + 1} \, \cos \sqrt{n^2 + 1}\,t + \sin \sqrt{n^2 + 1}\,t \right].$$

Using $u = XT$ and the superposition principle, a formal series solution is

$$u(x, t) = e^{-(x+t)} \sum_{n=1}^{\infty} A_n \left[\sqrt{n^2 + 1} \, \cos \sqrt{n^2 + 1}\,t + \sin \sqrt{n^2 + 1}\,t \right] \sin nx.$$

14. Letting $c = XT$ and separating variables we obtain

$$\frac{kX'' - hX'}{X} = \frac{T'}{T} \quad \text{or} \quad \frac{X'' - \alpha X'}{X} = \frac{T'}{kT} = -\lambda^2$$

where $\alpha = h/k$. This leads to the separated differential equations

$$X'' - \alpha X' + \lambda^2 X = 0 \quad \text{and} \quad T' + k\lambda^2 T = 0.$$

The solution of the second equation is

$$T(t) = c_3 e^{-k\lambda^2 t}.$$

For the first equation we have $m = \frac{1}{2}(\alpha \pm \sqrt{\alpha^2 - 4\lambda^2})$, and we consider three cases using the boundary conditions $X(0) = X(1) = 0$:

$\boxed{\alpha^2 > 4\lambda^2}$ The solution is $X = c_1 e^{m_1 x} + c_2 e^{m_2 x}$, where the boundary conditions imply $c_1 = c_2 = 0$, so $X = 0$. (Note in this case that if $\lambda = 0$, the solution is $X = c_1 + c_2 e^{\alpha x}$ and the boundary conditions again imply $c_1 = c_2 = 0$, so $X = 0$.)

$\boxed{\alpha^2 = 4\lambda^2}$ The solution is $X = c_1 e^{m_1 x} + c_2 x e^{m_1 x}$, where the boundary conditions imply $c_1 = c_2 = 0$, so $X = 0$.

$\boxed{\alpha^2 < 4\lambda^2}$ The solution is

$$X(x) = c_1 e^{\alpha x/2} \cos \frac{\sqrt{4\lambda^2 - \alpha^2}}{2}\,x + c_2 e^{\alpha x/2} \sin \frac{\sqrt{4\lambda^2 - \alpha^2}}{2}\,x.$$

From $X(0) = 0$ we see that $c_1 = 0$. From $X(1) = 0$ we find

$$\frac{1}{2}\sqrt{4\lambda^2 - \alpha^2} = n\pi \quad \text{or} \quad \lambda^2 = \frac{1}{4}(4n^2\pi^2 + \alpha^2).$$

Thus

$$X(x) = c_2 e^{\alpha x/2} \sin n\pi x,$$

and

$$c(x, t) = \sum_{n=1}^{\infty} A_n e^{\alpha x/2} e^{-k(4n^2\pi^2 + \alpha^2)t/4} \sin 2\pi x.$$

The initial condition $c(x, 0) = c_0$ implies

$$c_0 = \sum_{n=1}^{\infty} A_n e^{\alpha/2} \sin n\pi x. \tag{1}$$

From the self-adjoint form

$$\frac{d}{dx}[e^{-\alpha x} X'] + \lambda^2 e^{-\alpha x} X = 0$$

the eigenfunctions are orthogonal on $[0, 1]$ with weight function $e^{-\alpha x}$. That is

$$\int_0^1 e^{-\alpha x} (e^{\alpha x/2} \sin n\pi x)(e^{\alpha x/2} \sin m\pi x)\, dx = 0, \quad n \neq m.$$

Multiplying (1) by $e^{-\alpha x} e^{\alpha x/2} \sin m\pi x$ and integrating we obtain

$$\int_0^1 c_0 e^{-\alpha x} e^{\alpha x/2} \sin m\pi x\, dx = \sum_{n=1}^{\infty} A_n e^{-\alpha x} e^{\alpha x/2} (\sin m\pi x) e^{\alpha x/2} \sin n\pi x\, dx$$

$$c_0 \int_0^1 e^{-\alpha x/2} \sin n\pi x\, dx = A_n \int_0^1 \sin^2 n\pi x\, dx = \frac{1}{2} A_n$$

and

$$A_n = 2c_0 \int_0^1 e^{-\alpha x/2} \sin n\pi x\, dx = \frac{4c_0[2e^{\alpha/2} n\pi - 2n\pi(-1)^n]}{e^{\alpha/2}(\alpha^2 + 4n^2\pi^2)} = \frac{8n\pi c_0[e^{\alpha/2} - (-1)^n]}{e^{\alpha/2}(\alpha^2 + 4n^2\pi^2)}.$$

13 Boundary-Value Problems in Other Coordinate Systems

—————— **Exercises 13.1** ——————

1. We have

$$A_0 = \frac{1}{2\pi} \int_0^\pi u_0 \, d\theta = \frac{u_0}{2}$$

$$A_n = \frac{1}{\pi} \int_0^\pi u_0 \cos n\theta \, d\theta = 0$$

$$B_n = \frac{1}{\pi} \int_0^\pi u_0 \sin n\theta \, d\theta = \frac{u_0}{n\pi}[1 - (-1)^n]$$

and so

$$u(r, \theta) = \frac{u_0}{2} + \frac{u_0}{\pi} \sum_{n=1}^\infty \frac{1 - (-1)^n}{n} r^n \sin n\theta.$$

2. We have

$$A_0 = \frac{1}{2\pi} \int_0^\pi \theta \, d\theta + \frac{1}{2\pi} \int_\pi^{2\pi} (\pi - \theta) \, d\theta = 0$$

$$A_n = \frac{1}{\pi} \int_0^\pi \theta \cos n\theta \, d\theta + \frac{1}{\pi} \int_\pi^{2\pi} (\pi - \theta) \cos n\theta \, d\theta = \frac{2}{n^2\pi}[(-1)^n - 1]$$

$$B_n = \frac{1}{\pi} \int_0^\pi \theta \sin n\theta \, d\theta + \frac{1}{\pi} \int_\pi^{2\pi} (\pi - \theta) \sin n\theta \, d\theta = \frac{1}{n}[1 - (-1)^n]$$

and so

$$u(r, \theta) = \sum_{n=1}^\infty r^n \left[\frac{(-1)^n - 1}{n^2\pi} \cos n\theta + \frac{1 - (-1)^n}{n} \sin n\theta \right].$$

3. We have

$$A_0 = \frac{1}{2\pi} \int_0^{2\pi} (2\pi\theta - \theta^2) \, d\theta = \frac{2\pi^2}{3}$$

$$A_n = \frac{1}{\pi} \int_0^{2\pi} (2\pi\theta - \theta^2) \cos n\theta \, d\theta = -\frac{4}{n^2}$$

$$B_n = \frac{1}{\pi} \int_0^{2\pi} (2\pi\theta - \theta^2) \sin n\theta \, d\theta = 0$$

and so

$$u(r, \theta) = \frac{2\pi^2}{3} - 4 \sum_{n=1}^\infty \frac{r^n}{n^2} \cos n\theta.$$

4. We have

$$A_0 = \frac{1}{2\pi} \int_0^{2\pi} \theta \, d\theta = \pi$$

$$A_n = \frac{1}{\pi} \int_0^{2\pi} \theta \cos n\theta \, d\theta = 0$$

$$B_n = \frac{1}{\pi} \int_0^{2\pi} \theta \sin n\theta \, d\theta = -\frac{2}{n}$$

and so

$$u(r, \theta) = \pi - 2 \sum_{n=1}^{\infty} \frac{r^n}{n} \sin n\theta.$$

5. As in Example 1 we have $R(r) = c_3 r^n + c_4 r^{-n}$. In order that the solution be bounded as $r \to \infty$ we must define $c_3 = 0$. Hence

$$u(r, \theta) = A_0 + \sum_{n=1}^{\infty} r^{-n}(A_n \cos n\theta + B_n \sin n\theta)$$

where

$$A_0 = \frac{1}{2\pi} \int_0^{2\pi} f(\theta) \, d\theta$$

$$A_n = \frac{c^n}{\pi} \int_0^{2\pi} f(\theta) \cos n\theta \, d\theta$$

$$B_n = \frac{c^n}{\pi} \int_0^{2\pi} f(\theta) \sin n\theta \, d\theta.$$

6. We solve

$$\frac{\partial^2 u}{\partial r^2} + \frac{1}{r} \frac{\partial u}{\partial r} + \frac{1}{r^2} \frac{\partial^2 u}{\partial \theta^2} = 0, \quad 0 < \theta < \frac{\pi}{2}, \quad 0 < r < c,$$

$$u(c, \theta) = f(\theta), \quad 0 < \theta < \frac{\pi}{2},$$

$$u(r, 0) = 0, \quad u(r, \pi/2) = 0, \quad 0 < r < c.$$

Proceeding as in Example 1 of Section 13.1 in the text we obtain the separated differential equations

$$r^2 R'' + rR' - \lambda^2 R = 0$$

$$\Theta'' + \lambda^2 \Theta = 0$$

with solutions

$$\Theta(\theta) = c_1 \cos \lambda\theta + c_2 \sin \lambda\theta$$

$$R(r) = c_3 r^{\lambda} + c_4 r^{-\lambda}.$$

Since we want $R(r)$ to be bounded as $r \to 0$ we require $c_4 = 0$. Applying the boundary conditions $\Theta(0) = 0$ and $\Theta(\pi/2) = 0$ we find that $c_1 = 0$ and $\lambda = 2n$ for $n = 1, 2, 3, \ldots$. Therefore

$$u(r, \theta) = \sum_{n=1}^{\infty} A_n r^{2n} \sin 2n\theta.$$

From

$$u(c, \theta) = f(\theta) = \sum_{n=1}^{\infty} A_n c^n \sin 2n\theta$$

we find

$$A_n = \frac{4}{\pi c^{2n}} \int_0^{\pi/2} f(\theta) \sin 2n\theta \, d\theta.$$

7. Referring to the solution of Problem 6 above we have

$$\Theta(\theta) = c_1 \cos \lambda\theta + c_2 \sin \lambda\theta$$

$$R(r) = c_3 r^n.$$

Applying the boundary conditions $\Theta'(0) = 0$ and $\Theta'(\pi/2) = 0$ we find that $c_2 = 0$ and $\lambda = 2n$ for $n = 0, 1, 2, \ldots$. Therefore

$$u(r, \theta) = A_0 + \sum_{n=1}^{\infty} A_n r^{2n} \cos 2n\theta.$$

From

$$u(c, \theta) = \begin{cases} 1, & 0 < \theta < \pi/4 \\ 0, & \pi/4 < \theta < \pi/2 \end{cases} = A_0 + \sum_{n=1}^{\infty} A_n c^{2n} \cos 2n\theta$$

we find

$$A_0 = \frac{1}{\pi/2} \int_0^{\pi/4} d\theta = \frac{1}{2}$$

and

$$c^{2n} A_n = \frac{2}{\pi/2} \int_0^{\pi/4} \cos 2n\theta \, d\theta = \frac{2}{n\pi} \sin \frac{n\pi}{2}.$$

Thus

$$u(r, \theta) = \frac{1}{2} + \frac{2}{\pi} \sum_{n=1}^{\infty} \frac{1}{n} \sin \frac{n\pi}{2} \left(\frac{r}{c}\right)^{2n} \cos 2n\theta.$$

8. We solve

$$\frac{\partial^2 u}{\partial r^2} + \frac{1}{r} \frac{\partial u}{\partial r} + \frac{1}{r^2} \frac{\partial^2 u}{\partial \theta^2} = 0, \quad 0 < \theta < \pi/4, \quad r > 0$$

$$u(r, 0) = 0, \quad r > 0$$

$$u(r, \pi/4) = 30, \quad r > 0.$$

Exercises 13.1

Proceeding as in Example 1 of Section 13.1 we find the separated ordinary differential equations to be

$$r^2 R'' + r R' - \lambda^2 R = 0$$

$$\Theta'' + \lambda^2 \Theta = 0.$$

The corresponding general solutions are

$$R(r) = c_1 r^\lambda + c_2 r^{-\lambda}$$

$$\Theta(\theta) = c_3 \cos \lambda\theta + c_4 \sin \lambda\theta.$$

The condition $\Theta(0) = 0$ implies $c_3 = 0$ so that $\Theta = c_4 \sin \lambda\theta$. Now, in order that the temperature be bounded as $r \to \infty$ we define $c_1 = 0$. Similarly, in order that the temperature be bounded as $r \to 0$ we are forced to define $c_2 = 0$. Thus $R(r) = 0$ and so no nontrivial solution exists for $\lambda > 0$. For $\lambda = 0$ the separated differential equations are

$$r^2 R'' + r R' = 0 \quad \text{and} \quad \Theta'' = 0.$$

Solutions of these latter equations are

$$R(r) = c_1 + c_2 \ln r \quad \text{and} \quad \Theta(\theta) = c_2 \theta + c_3.$$

$\Theta(0) = 0$ still implies $c_3 = 0$, whereas boundedness as $r \to 0$ demands $c_2 = 0$. Thus, a product solution is

$$u = c_1 c_2 \theta = A\theta.$$

From $u(r, \pi/4) = 0$ we obtain $A = 120/\pi$. Thus, a solution to the problem is

$$u(r, \theta) = \frac{120}{\pi}\theta.$$

9. Proceeding as in Example 1 of Section 13.1 and again using the periodicity of $u(r, \theta)$, we have

$$\Theta(\theta) = c_1 \cos \lambda\theta + c_2 \sin \lambda\theta$$

where $\lambda = n$ for $n = 0, 1, 2, \ldots$. Then

$$R(r) = c_3 r^n + c_4 r^{-n}.$$

[We do not have $c_4 = 0$ in this case since $0 < a \le r$.] Since $u(b, \theta) = 0$ we have

$$u(r, \theta) = A_0 \ln \frac{r}{b} + \sum_{n=1}^{\infty} \left[\left(\frac{b}{r}\right)^n - \left(\frac{r}{b}\right)^n \right] [A_n \cos n\theta + B_n \sin n\theta].$$

From

$$u(a, \theta) = f(\theta) = A_0 \ln \frac{a}{b} + \sum_{n=1}^{\infty} \left[\left(\frac{b}{a}\right)^n - \left(\frac{a}{b}\right)^n \right] [A_n \cos n\theta + B_n \sin n\theta]$$

we find

$$A_0 \ln \frac{a}{b} = \frac{1}{2\pi} \int_0^{2\pi} f(\theta) \, d\theta,$$

$$\left[\left(\frac{b}{a}\right)^n - \left(\frac{a}{b}\right)^n \right] A_n = \frac{1}{\pi} \int_0^{2\pi} f(\theta) \cos n\theta \, d\theta,$$

and

$$\left[\left(\frac{b}{a}\right)^n - \left(\frac{a}{b}\right)^n \right] B_n = \frac{1}{\pi} \int_0^{2\pi} f(\theta) \sin n\theta \, d\theta.$$

10. Substituting $u(r, \theta) = v(r, \theta) + \psi(r)$ into the partial differential equation we obtain

$$\frac{\partial^2 v}{\partial r^2} + \psi''(r) + \frac{1}{r} \left[\frac{\partial v}{\partial r} + \psi'(r) \right] + \frac{1}{r^2} \frac{\partial^2 v}{\partial \theta^2} = 0.$$

This equation will be homogeneous provided

$$\psi''(r) + \frac{1}{r}\psi'(r) = 0 \quad \text{or} \quad r^2\psi''(r) + r\psi'(r) = 0.$$

The general solution of this Cauchy-Euler differential equation is

$$\psi(r) = c_1 + c_2 \ln r.$$

From

$$u_0 = u(a, \theta) - v(a, \theta) + \psi(a) \quad \text{and} \quad u_1 = u(b, \theta) - v(b, \theta) + \psi(b)$$

we see that in order for the boundary values $v(a, \theta)$ and $v(b, \theta)$ to be 0 we need $\psi(a) = u_0$ and $\psi(b) = u_1$. From this we have

$$\psi(a) = c_1 + c_2 \ln a = u_0$$

$$\psi(b) = c_1 + c_2 \ln b = u_1.$$

Solving for c_1 and c_2 we obtain

$$c_1 = \frac{u_1 \ln a - u_0 \ln b}{\ln a/b} \quad \text{and} \quad c_2 = \frac{u_0 - u_1}{\ln a/b}.$$

Then

$$\psi(r) = \frac{u_1 \ln a - u_0 \ln b}{\ln a/b} + \frac{u_0 - u_1}{\ln a/b} \ln r = \frac{u_0 \ln r/b - u_1 \ln r/a}{\ln a/b}.$$

From Problem 9 with $f(\theta) = 0$ we see that the solution of

$$\frac{\partial^2 v}{\partial r^2} + \frac{1}{r} \frac{\partial v}{\partial r} + \frac{1}{r^2} \frac{\partial^2 v}{\partial \theta^2} = 0, \quad 0 < \theta < 2\pi, \quad a < r < b,$$

$$v(a, \theta) = 0, \quad v(b, \theta) = 0, \quad 0 < \theta < 2\pi$$

549

is $v(r, \theta) = 0$. Thus the steady-state temperature of the ring is

$$u(r, \theta) = v(r, \theta) + \psi(r) = \frac{u_0 \ln r/b - u_1 \ln r/a}{\ln a/b}.$$

11. We solve

$$\frac{\partial^2 u}{\partial r^2} + \frac{1}{r} \frac{\partial u}{\partial r} + \frac{1}{r^2} \frac{\partial^2 u}{\partial \theta^2} = 0, \quad 0 < \theta < \pi, \quad a < r < b,$$

$$u(a, \theta) = \theta(\pi - \theta), \quad u(b, \theta) = 0, \quad 0 < \theta < \pi,$$

$$u(r, 0) = 0, \quad u(r, \pi) = 0, \quad a < r < b.$$

Proceeding as in Example 1 of this section in the text we obtain the separated differential equations

$$r^2 R'' + r R' - \lambda^2 R = 0$$

$$\Theta'' + \lambda^2 \Theta = 0$$

with solutions

$$\Theta(\theta) = c_1 \cos \lambda\theta + c_2 \sin \lambda\theta$$

$$R(r) = c_3 r^\lambda + c_4 r^{-\lambda}.$$

Applying the boundary conditions $\Theta(0) = 0$ and $\Theta(\pi) = 0$ we find that $c_1 = 0$ and $\lambda = n$ for $n = 1, 2, 3, \ldots$. The boundary condition $R(b) = 0$ gives

$$c_3 b^n + c_4 b^{-n} = 0 \quad \text{and} \quad c_4 = -c_3 b^{2n}.$$

Then
$$R(r) = c_3 \left(r^n - \frac{b^{2n}}{r^n} \right) = c_3 \left(\frac{r^{2n} - b^{2n}}{r^n} \right)$$

and
$$u(r, \theta) = \sum_{n=1}^{\infty} A_n \left(\frac{r^{2n} - b^{2n}}{r^n} \right) \sin n\theta.$$

From
$$u(a, \theta) = \theta(\pi - \theta) = \sum_{n=1}^{\infty} A_n \left(\frac{a^{2n} - b^{2n}}{a^n} \right) \sin n\theta$$

we find
$$A_n \left(\frac{a^{2n} - b^{2n}}{a^n} \right) = \frac{2}{\pi} \int_0^\pi (\theta\pi - \theta^2) \sin n\theta \, d\theta = \frac{4}{n^3 \pi} [1 - (-1)^n].$$

Thus
$$u(r, \theta) = \frac{4}{\pi} \sum_{n=1}^{\infty} \frac{1 - (-1)^n}{n^3} \frac{r^{2n} - b^{2n}}{a^{2n} - b^{2n}} \left(\frac{a}{r} \right)^n \sin n\theta.$$

12. Letting $u(r, \theta) = v(r, \theta) + \psi(\theta)$ we obtain $\psi''(\theta) = 0$ and so $\psi(\theta) = c_1\theta + c_2$. From $\psi(0) = 0$ and $\psi(\pi) = u_0$ we find, in turn, $c_2 = 0$ and $c_1 = u_0/\pi$. Therefore $\psi(\theta) = \dfrac{u_0}{\pi} \theta$.

Now $u(1,\theta) = v(1,\theta) + \psi(\theta)$ so that $v(1,\theta) = u_0 - \dfrac{u_0}{\pi}\theta$.

From
$$v(r,\theta) = \sum_{n=1}^{\infty} A_n r^n \sin n\theta \quad \text{and} \quad v(1,\theta) = \sum_{n=1}^{\infty} A_n \sin n\theta$$

we obtain
$$A_n = \frac{2}{\pi}\int_0^{\pi}\left(u_0 - \frac{u_0}{\pi}\theta\right)\sin n\theta\, d\theta = \frac{2u_0}{\pi n}.$$

Thus
$$u(r,\theta) = \frac{u_0}{\pi}\theta + \frac{2u_0}{\pi}\sum_{n=1}^{\infty}\frac{r^n}{n}\sin n\theta.$$

13. We solve

$$\frac{\partial^2 u}{\partial r^2} + \frac{1}{r}\frac{\partial u}{\partial r} + \frac{1}{r^2}\frac{\partial^2 u}{\partial \theta^2} = 0, \quad 0 < \theta < \pi, \quad 0 < r < 2,$$

$$u(2,\theta) = \begin{cases} u_0, & 0 < \theta < \pi/2 \\ 0, & \pi/2 < \theta < \pi \end{cases}$$

$$\frac{\partial u}{\partial \theta}\bigg|_{\theta=0} = 0, \quad \frac{\partial u}{\partial \theta}\bigg|_{\theta=\pi} = 0, \quad 0 < r < 2.$$

Proceeding as in Example 1 of this section in the text we obtain the separated differential equations

$$r^2 R'' + r R' - \lambda^2 R = 0$$

$$\Theta'' + \lambda^2 \Theta = 0$$

with solutions

$$\Theta(\theta) = c_1 \cos \lambda\theta + c_2 \sin \lambda\theta$$

$$R(r) = c_3 r^{\lambda} + c_4 r^{-\lambda}.$$

Applying the boundary conditions $\Theta'(0) = 0$ and $\Theta'(\pi) = 0$ we find that $c_2 = 0$ and $\lambda = n$ for $n = 0, 1, 2, \ldots$. Since we want $R(r)$ to be bounded as $r \to 0$ we require $c_4 = 0$. Thus

$$u(r,\theta) = A_0 + \sum_{n=1}^{\infty} A_n r^n \cos n\theta.$$

From
$$u(2,0) - \begin{cases} u_0, & 0 < \theta < \pi/2 \\ 0, & \pi/2 < \theta < \pi \end{cases} = A_0 + \sum_{n=1}^{\infty} A_n 2^n \cos n\theta$$

we find
$$A_0 = \frac{1}{2}\frac{2}{\pi}\int_0^{\pi/2} u_0\, d\theta = \frac{u_0}{2}$$

and
$$2^n A_n = \frac{2u_0}{\pi}\int_0^{\pi/2}\cos n\theta\, d\theta = \frac{2u_0}{\pi}\frac{\sin n\pi/2}{n}.$$

Therefore
$$u(r,\theta) = \frac{u_0}{2} + \frac{2u_0}{\pi} \sum_{n=1}^{\infty} \frac{1}{n} \left(\sin \frac{n\pi}{2} \right) \left(\frac{r}{2} \right)^n \cos n\theta.$$

14. $u(r,\theta) = \dfrac{1}{2\pi} \displaystyle\int_0^{2\pi} f(t)\, dt$

$$+ \sum_{n=1}^{\infty} r^n \left[\left(\frac{1}{c^n \pi} \int_0^{2\pi} f(t) \cos nt\, dt \right) \cos n\theta + \left(\frac{1}{c^n \pi} \int_0^{2\pi} f(t) \sin nt\, dt \right) \sin n\theta \right]$$

$$= \frac{1}{2\pi} \int_0^{2\pi} f(t) \left[1 + 2 \sum_{n=1}^{\infty} \left(\frac{r}{c} \right)^n (\cos nt \cos n\theta + \sin nt \sin n\theta) \right] dt$$

$$= \frac{1}{2\pi} \int_0^{2\pi} f(t) \left[1 + 2 \sum_{n=1}^{\infty} \left(\frac{r}{c} \right)^n \cos n(t-\theta) \right] dt$$

$$= \frac{1}{2\pi} \int_0^{2\pi} f(t) \left[1 + \sum_{n=1}^{\infty} u^n (e^{inv} + e^{-inv}) \right] dt \qquad \boxed{\text{where } u = r/c \text{ and } v = t - \theta}$$

$$= \frac{1}{2\pi} \int_0^{2\pi} f(t) \left[1 + \sum_{n=1}^{\infty} u^n e^{inv} + \sum_{n=1}^{\infty} u^n e^{-inv} \right] dt$$

$$= \frac{1}{2\pi} \int_0^{2\pi} f(t) \left[1 + \frac{ue^{iv}}{1 - ue^{iv}} + \frac{ue^{-iv}}{1 - ue^{-iv}} \right] dt \qquad \boxed{|u| < 1}$$

$$= \frac{1}{2\pi} \int_0^{2\pi} f(t) \left[\frac{1 - u^2}{1 - 2u \frac{e^{iv} + e^{-iv}}{2} + u^2} \right] dt = \frac{1}{2\pi} \int_0^{2\pi} \frac{c^2 - r^2}{c^2 - 2rc\cos(t-\theta) + r^2} f(t)\, dt$$

15. Let u_1 be the solution of the boundary-value problem

$$\frac{\partial^2 u_1}{\partial r^2} + \frac{1}{r}\frac{\partial u_1}{\partial r} + \frac{1}{r^2}\frac{\partial^2 u_1}{\partial \theta^2} = 0, \qquad 0 < \theta < \pi, \quad a < r < b$$

$$u_1(a,\theta) = f(\theta), \quad 0 < \theta < \pi$$

$$u_1(b,\theta) = 0, \quad 0 < \theta < \pi,$$

and let u^2 be the solution of the boundary-value problem

$$\frac{\partial^2 u_2}{\partial r^2} + \frac{1}{r}\frac{\partial u_2}{\partial r} + \frac{1}{r^2}\frac{\partial^2 u_1}{\partial \theta^2} = 0, \qquad 0 < \theta < \pi, \quad a < r < b$$

$$u_2(a,\theta) = 0, \quad 0 < \theta < \pi$$

$$u_2(b,\theta) = g(\theta), \quad 0 < \theta < \pi.$$

Each of these problems can be solved using the methods shown in Problem 9 of this section. Now if $u(r, \theta) = u_1(t, \theta) + u_2(r, \theta)$, then

$$u(a, \theta) = u_1(a, \theta) + u_2(a, \theta) = f(\theta)$$

$$u(b, \theta) = u_1(b, \theta) + u_2(b, \theta) = g(\theta)$$

and $u(r, \theta)$ will be the steady-state temperature of the circular ring with boundary conditions $u(a, \theta) = f(\theta)$ and $u(b, \theta) = g(\theta)$.

Exercises 13.2

1. Referring to the solution of Example 1 in the text we have

$$R(r) = c_1 J_0(\lambda_n r) \quad \text{and} \quad T(t) = c_3 \cos a\lambda_n t + c_4 \sin a\lambda_n t$$

where the eigenvalues λ_n are the positive roots of $J_0(\lambda c) = 0$. Now, the initial condition $u(r, 0) = R(r)T(0) = 0$ implies $c_3 = 0$. Thus

$$u(r, t) = \sum_{n=1}^{\infty} A_n \sin a\lambda_n t J_0(\lambda_n r) \qquad \frac{\partial u}{\partial t} = \sum_{n=1}^{\infty} a\lambda_n A_n \cos a\lambda_n t J_0(\lambda_n r).$$

From

$$\frac{\partial u}{\partial t}\bigg|_{t=0} = 1 = \sum_{n=1}^{\infty} a\lambda_n A_n J_0 \lambda_n r$$

we find

$$a\lambda_n A_n = \frac{2}{c^2 J_1^2(\lambda_n c)} \int_0^c r J_0(\lambda_n r)\, dr \qquad \boxed{x = \lambda_n r, \ dx = \lambda_n\, dr}$$

$$= \frac{2}{c^2 J_1^2(\lambda_n c)} \int_0^{\lambda_n c} \frac{1}{\lambda_n^2} x J_0(x)\, dx$$

$$= \frac{2}{c^2 J_1^2(\lambda_n c)} \int_0^{\lambda_n c} \frac{1}{\lambda_n^2} \frac{d}{dx}[x J_1(x)]\, dx \qquad \boxed{\text{see (4) of Section 11.5 in text}}$$

$$= \frac{2}{c^2 \lambda_n^2 J_1^2(\lambda_n c)} (x J_1(x)) \bigg|_0^{\lambda_n c} = \frac{2}{c\lambda_n J_1(\lambda_n c)}.$$

Then

$$A_n = \frac{2}{ac\lambda_n^2 J_1(\lambda_n c)}$$

and

$$u(r, t) = \frac{2}{ac} \sum_{n=1}^{\infty} \frac{J_0(\lambda_n r)}{\lambda_n^2 J_1(\lambda_n c)} \sin a\lambda_n t.$$

Exercises 13.2

2. From Example 1 we have $B_n = 0$ and

$$A_n = \frac{2}{J_1^2(\lambda_n)} \int_0^1 r(1 - r^2) J_0(\lambda_n r) \, dr.$$

From Problem 10, Exercises 11.5 we obtained $A_n = \dfrac{4 J_2(\lambda_n)}{\lambda_n^2 J_1^2(\lambda_n)}$. Thus

$$u(r, t) = 4 \sum_{n=1}^{\infty} \frac{J_2(\lambda_n)}{J_1^2(\lambda_n)} \cos a\lambda_n t J_0(\lambda_n r).$$

3. Referring to Example 2 in the text we have

$$R(r) = c_1 J_0(\lambda r) + c_2 Y_0(\lambda r)$$

$$Z(z) = c_3 \cosh \lambda z + c_4 \sinh \lambda z$$

where $c_2 = 0$ and $J_0(2\lambda) = 0$ defines the positive eigenvalues λ_n. From $Z(4) = 0$ we obtain

$$c_3 \cosh 4\lambda_n + c_4 \sinh 4\lambda_n = 0 \quad \text{or} \quad c_4 = -c_3 \frac{\cosh 4\lambda_n}{\sinh 4\lambda_n}.$$

Then

$$Z(z) = c_3 \left[\cosh \lambda_n z - \frac{\cosh 4\lambda_n}{\sinh 4\lambda_n} \sinh \lambda_n z \right] = c_3 \frac{\sinh 4\lambda_n \cosh \lambda_n z - \cosh 4\lambda_n \sinh \lambda_n z}{\sinh 4\lambda_n}$$

$$= c_3 \frac{\sinh \lambda_n (4 - z)}{\sinh 4\lambda_n}$$

and

$$u(r, z) = \sum_{n=1}^{\infty} A_n \frac{\sinh \lambda_n (4 - z)}{\sinh 4\lambda_n} J_0(\lambda_n r).$$

From

$$u(r, 0) = u_0 = \sum_{n=1}^{\infty} A_n J_0(\lambda_n r)$$

we obtain

$$A_n = \frac{2u_0}{4 J_1^2(2\lambda_n)} \int_0^2 r J_0(\lambda_n r) \, dr = \frac{u_0}{\lambda_n J_1(2\lambda_n)}.$$

Thus the temperature in the cylinder is

$$u(r, z) = u_0 \sum_{n=1}^{\infty} \frac{\sinh \lambda_n (4 - z) J_0(\lambda_n r)}{\lambda_n \sinh 4\lambda_n J_1(2\lambda_n)}.$$

4. (a) The boundary condition $u_r(2, z) = 0$ implies $R'(2) = 0$ or $J_0'(2\lambda) = 0$. Thus $\lambda = 0$ is also an eigenvalue and the separated equations are in this case $rR'' + R' = 0$ and $z'' = 0$. The solutions of these equations are then

$$R(r) = c_1 + c_2 \ln r, \quad Z(z) = c_3 z + c_4.$$

Now $Z(0) = 0$ yields $c_4 = 0$ and the implicit condition that the temperature is bounded as $r \to 0$ demands that we define $c_2 = 0$. Using the superposition principle then gives

$$u(r, z) = A_1 z + \sum_{n=2}^{\infty} A_n \sinh \lambda_n z J_0(\lambda_n r). \tag{1}$$

At $z = 4$ we obtain

$$f(r) = 4A_1 + \sum_{n=2}^{\infty} A_n \sinh 4\lambda_n J_0(\lambda_n r).$$

Thus from (19) and (20) of Section 11.5 in the text we can write with $b = 2$,

$$A_1 = \frac{1}{8} \int_0^2 r f(r) \, dr \tag{2}$$

$$A_n = \frac{1}{2 \sinh 4\lambda_n J_0^2(2\lambda_n)} \int_0^2 r f(r) J_0(\lambda_n r) \, dr. \tag{3}$$

A solution of the problem consists of the series (1) with coefficients A_1 and A_n defined in (2) and (3), respectively.

(b) When $f(r) = u_0$ we get $A_1 = u_0/4$ and

$$A_n = \frac{u_0 J_1(2\lambda_n)}{\lambda_n \sinh 4\lambda_n J_0^2(2\lambda_n)} = 0$$

since $J_0'(2\lambda) = 0$ is equivalent to $J_1(2\lambda) = 0$. A solution of the problem is then $u(r, z) = \dfrac{u_0}{4} z$.

5. Letting $u(r, t) = R(r)T(t)$ and separating variables we obtain

$$\frac{R'' + \frac{1}{r}R'}{R} = \frac{T'}{kT} = \mu \quad \text{and} \quad R'' + \frac{1}{r}R' - \mu R = 0, \quad T' - \mu kT = 0.$$

From the second equation we find $T(t) = e^{\mu kt}$. If $\mu > 0$, $T(t)$ increases without bound as $t \to \infty$. Thus we assume $\mu = -\lambda^2 \leq 0$. Now

$$R'' + \frac{1}{r}R' + \lambda^2 R = 0$$

is a parametric Bessel equation with solution

$$R(r) = c_1 J_0(\lambda r) + c_2 Y_0(\lambda r).$$

Since Y_0 is unbounded as $r \to 0$ we take $c_2 = 0$. Then $R(r) = c_1 J_0(\lambda r)$ and the boundary condition $u(c, t) = R(c)T(t) = 0$ implies $J_0(\lambda c) = 0$. This latter equation defines the positive eigenvalues λ_n. Thus

$$u(r, t) = \sum_{n=1}^{\infty} A_n J_0(\lambda_n r) e^{-\lambda_n^2 kt}.$$

From

$$u(r, 0) = f(r) = \sum_{n=1}^{\infty} A_n J_0(\lambda_n r)$$

we find

$$A_n = \frac{2}{c^2 J_1^2(\lambda_n c)} \int_0^c r J_0(\lambda_n r) f(r) \, dr, \quad n = 1, 2, 3, \dots .$$

6. If the edge $r = c$ is insulated we have the boundary condition $u_r(c, t) = 0$. Referring to the solution of Problem 5 above we have

$$R'(c) = \lambda c_1 J_0'(\lambda c) = 0$$

which defines an eigenvalue $\lambda = 0$ and positive eigenvalues λ_n. Thus

$$u(r, t) = A_0 + \sum_{n=1}^{\infty} A_n J_0(\lambda_n r) e^{-\lambda_n^2 k t}.$$

From

$$u(r, 0) = f(r) = A_0 + \sum_{n=1}^{\infty} A_n J_0(\lambda_n r)$$

we find

$$A_0 = \frac{2}{c^2} \int_0^c r f(r) \, dr$$

$$A_n = \frac{2}{c^2 J_0^2(\lambda_n c)} \int_0^c r J_0(\lambda_n r) f(r) \, dr.$$

7. Referring to Problem 5 above we have $T(t) = e^{-\lambda^2 k t}$ and $R(r) = c_1 J_0(\lambda r)$. The boundary condition $h u(1, t) + u_r(1, t) = 0$ implies $h J_0(\lambda) + \lambda J_0'(\lambda) = 0$ which defines positive eigenvalues λ_n. Now

$$u(r, t) = \sum_{n=1}^{\infty} A_n J_0(\lambda_n r) e^{-\lambda_n^2 k t}$$

where

$$A_n = \frac{2\lambda_n^2}{(\lambda_n^2 + h^2) J_0^2(\lambda_n)} \int_0^1 r J_0(\lambda_n r) f(r) \, dr.$$

8. We solve

$$\frac{\partial^2 u}{\partial r^2} + \frac{1}{r} \frac{\partial u}{\partial r} + \frac{\partial^2 u}{\partial z^2} = 0, \quad 0 < r < 1, \quad z > 0$$

$$\left. \frac{\partial u}{\partial r} \right|_{r=1} = -h u(1, z), \quad z > 0$$

$$u(r, 0) = u_0, \quad 0 < r < 1.$$

assuming $u = RZ$ we get

$$\frac{R'' + \frac{1}{r} R'}{R} = -\frac{Z''}{Z} = -\lambda^2$$

and so

$$rR'' + R' + \lambda^2 rR = 0 \quad \text{and} \quad Z'' - \lambda^2 Z = 0.$$

Therefore

$$R(r) = c_1 J_0(\lambda r) + c_2 Y_0(\lambda r) \quad \text{and} \quad Z(z) = c_3 e^{-\lambda z} + c_4 e^{\lambda z}.$$

We use the exponential form of the solution of $Z'' - \lambda^2 Z = 0$ since the domain of the variable z is a semi-infinite interval. As usual we define $c_2 = 0$ since the temperature is surely bounded as $r \to 0$. Hence $R(r) = c_1 J_0(\lambda r)$. Now the boundary-condition $u_r(1, z) + hu(1, z) = 0$ is equivalent to

$$\lambda J_0'(\lambda) + h J_0(\lambda) = 0. \tag{4}$$

The eigenvalues λ_n are the positive roots of (4). Finally, we must now define $c_4 = 0$ since the temperature is also expected to be bounded as $z \to \infty$. A product solution of the partial differential equation that satisfies the first boundary condition is given by

$$u_n(r, z) = A_n e^{-\lambda_n z} J_0(\lambda_n r).$$

Therefore

$$u(r, z) - \sum_{n=1}^{\infty} A_n e^{-\lambda_n z} J_0(\lambda_n r)$$

is another formal solution. At $z = 0$ we have $u_0 = A_n J_0(\lambda_n r)$. In view of (4) we use (17) and (18) of Section 11.5 in the text with the identification $b = 1$:

$$A_n = \frac{2\lambda_n^2}{(\lambda_n^2 + h^2) J_0^2(\lambda_n)} \int_0^1 r J_0(\lambda_n r) u_0 \, dr = \frac{2\lambda_n^2 u_0}{(\lambda_n^2 + h^2) J_0^2(\lambda_n)\lambda_n^2} t J_1(t) \Big|_0^{\lambda_n} = \frac{2\lambda_n u_0 J_1(\lambda_n)}{(\lambda_n^2 + h^2) J_0^2(\lambda_n)}. \tag{5}$$

Since $J_0' = -J_1$ [see (5) of Section 11.5] it follows from (4) that $\lambda_n J_1(\lambda_n) = h J_0(\lambda_n)$. Thus (5) simplifies to

$$A_n = \frac{2u_0 h}{(\lambda_n^2 + h^2) J_0^2(\lambda_n)}.$$

A solution to the boundary-value problem is then

$$u(r, z) = 2u_0 h \sum_{n=1}^{\infty} \frac{e^{-\lambda_n z}}{(\lambda_n^2 + h^2) J_0(\lambda_n)} J_0(\lambda_n r).$$

9. Substituting $u(r, t) = v(r, t) + \psi(r)$ into the partial differential equation gives

$$\frac{\partial^2 v}{\partial r^2} + \frac{1}{r} \frac{\partial v}{\partial r} + \psi'' + \frac{1}{r} \psi' = \frac{\partial v}{\partial t}.$$

This equation will be homogeneous provided $\psi'' + \frac{1}{r}\psi' = 0$ or

$$\psi(r) = c_1 \ln r + c_2.$$

Since $\ln r$ is unbounded as $r \to 0$ we take $c_1 = 0$. Then $\psi(r) = c_2$ and using

$$u(2, t) = v(2, t) + \psi(2) = 100$$

we set $c_2 = \psi(r) = 100$. Referring to Problem 5 above, the solution of the boundary-value problem

$$\frac{\partial^2 v}{\partial r^2} + \frac{1}{r}\frac{\partial v}{\partial r} = \frac{\partial v}{\partial t}, \quad 0 < r < 2, \quad t > 0,$$

$$v(2, t) = 0, \quad t > 0,$$

$$v(r, 0) = u(r, 0) - \psi(r)$$

is

$$v(r, t) = \sum_{n=1}^{\infty} A_n J_0(\lambda_n r) e^{-\lambda_n^2 t}$$

where

$$A_n = \frac{2}{2^2 J_1^2(2\lambda_n)} \int_0^2 r J_0(\lambda_n r)[u(r, 0) - \psi(r)] \, dr$$

$$= \frac{1}{2 J_1^2(2\lambda_n)} \left[\int_0^1 r J_0(\lambda_n r)[200 - 100] \, dr + \int_1^2 r J_0(\lambda_n r)[100 - 100] \, dr \right]$$

$$= \frac{50}{J_1^2(2\lambda_n)} \int_0^1 r J_0(\lambda_n r) \, dr \qquad \boxed{x = \lambda_n r, \; dx = \lambda_n \, dr}$$

$$= \frac{50}{J_1^2(2\lambda_n)} \int_0^{\lambda_n} \frac{1}{\lambda_n^2} x J_0(x) \, dx$$

$$= \frac{50}{\lambda_n^2 J_1^2(2\lambda_n)} \int_0^{\lambda_n} \frac{d}{dx}[x J_1(x)] \, dx \qquad \boxed{\text{see (4) of Section 11.5 in text}}$$

$$= \frac{50}{\lambda_n^2 J_1^2(2\lambda_n)} (x J_1(x)) \Big|_0^{\lambda_n} = \frac{50 J_1(\lambda_n)}{\lambda_n J_1^2(2\lambda_n)}.$$

Thus

$$u(r, t) = v(r, t) + \psi(r) = 100 + 50 \sum_{n=1}^{\infty} \frac{J_1(\lambda_n) J_0(\lambda_n r)}{\lambda_n J_1^2(2\lambda_n)} e^{-\lambda^2 t}.$$

10. Letting $u(r, t) = u(r, t) + \psi(r)$ we obtain $r\psi'' + \psi' = -\beta r$. The general solution of this nonhomogeneous equation is found with the aid of variation of parameters: $\psi = c_1 + c_2 \ln r - \beta \frac{r^2}{4}$. In order that this solution be bounded as $r \to 0$ we define $c_2 = 0$. Using $\psi(1) = 0$ then gives $c_1 = \frac{\beta}{4}$ and so $\psi(r) = \frac{\beta}{4}(1 - r^2)$. Using $v = RT$ we find that a solution of

$$\frac{\partial^2 v}{\partial r^2} + \frac{1}{r}\frac{\partial v}{\partial r} = \frac{\partial v}{\partial t}, \quad 0 < r < 1, \quad t > 0$$

$$v(1, t) = 0, \quad t > 0$$

$$v(r, 0) = -\psi(r), \quad 0 < r < 1$$

is

$$v(r,t) = \sum_{n=1}^{\infty} A_n e^{-\lambda_n^2 t} J_0(\lambda_n r)$$

where

$$A_n = -\frac{\beta}{4} \frac{2}{J_1^2(\lambda_n)} \int_0^1 r(1-r^2) J_0(\lambda_n r)\, dr$$

and the eigenvalues are defined by $J_0(\lambda) = 0$. From the result of Problem 10, Exercises 11.5 (see also Problem 2 of this exercise set) we get

$$A_n = -\frac{\beta J_2(\lambda_n)}{\lambda_n^2 J_1^2(\lambda_n)}.$$

Thus from $u = v + \psi(r)$ it follows that

$$u(r,t) = \frac{\beta}{4}(1-r^2) - \beta \sum_{n=1}^{\infty} \frac{J_2(\lambda_n)}{\lambda_n^2 J_1^2(\lambda_n)} e^{-\lambda_n^2 t} J_0(\lambda_n r).$$

11. (a) Writing the partial differential equation in the form

$$g\left(x \frac{\partial^2 u}{\partial x^2} + \frac{\partial u}{\partial x}\right) = \frac{\partial^2 u}{\partial t^2}$$

and separating variables we obtain

$$\frac{xX'' + X'}{X} = \frac{T''}{gT} = -\lambda^2.$$

Then

$$xX'' + X' + \lambda^2 X = 0 \quad \text{and} \quad T'' + g\lambda^2 T = 0.$$

Letting $x = \tau^2/4$ in the first equation we obtain $dx/d\tau = \tau/2$ or $d\tau/dx = 2\tau$. Then

$$\frac{dX}{dx} = \frac{dX}{d\tau} \frac{d\tau}{dx} = \frac{2}{\tau} \frac{dX}{d\tau}$$

and

$$\frac{d^2 X}{dx^2} = \frac{d}{dx}\left(\frac{2}{\tau} \frac{dX}{d\tau}\right) = \frac{2}{\tau} \frac{d}{dx}\left(\frac{dX}{d\tau}\right) + \frac{dX}{d\tau} \frac{d}{dx}\left(\frac{2}{\tau}\right)$$

$$= \frac{2}{\tau} \frac{d}{d\tau}\left(\frac{dX}{d\tau}\right) \frac{d\tau}{dx} + \frac{dX}{d\tau} \frac{d}{d\tau}\left(\frac{2}{\tau}\right) \frac{d\tau}{dx} = \frac{4}{\tau^2} \frac{d^2 X}{d\tau^2} - \frac{4}{\tau^3} \frac{dX}{d\tau}.$$

Thus

$$xX'' + X' + \lambda^2 X = \frac{\tau^2}{4}\left(\frac{4}{\tau^2} \frac{d^2 X}{d\tau^2} - \frac{4}{\tau^3} \frac{dX}{d\tau}\right) + \frac{2}{\tau} \frac{dX}{d\tau} + \lambda^2 X = \frac{d^2 X}{d\tau^2} + \frac{1}{\tau} \frac{dX}{d\tau} + \lambda^2 X = 0.$$

This is a parametric Bessel equation with solution

$$X(\tau) = c_1 J_0(\lambda\tau) + c_2 Y_0(\lambda\tau).$$

(b) To insure a finite solution at $x = 0$ (and thus $\tau = 0$) we set $c_2 = 0$. The condition $u(L, t) = X(L)T(t) = 0$ implies $X\,|_{x=L} = X\,|_{\tau=2\sqrt{L}} = c_1 J_0(2\lambda\sqrt{L}) = 0$, which defines positive eigenvalues λ_n. The solution of $T'' + g\lambda^2 T = 0$ is

$$T(t) = c_3 \cos \lambda_n \sqrt{g}\, t + c_4 \sin \lambda_n \sqrt{g}\, t.$$

The boundary condition $u_t(x, 0) = X(x)T'(0) = 0$ implies $c_4 = 0$. Thus

$$u(\tau, t) = \sum_{n=1}^{\infty} A_n \cos \lambda_n \sqrt{g}\, t\, J_0(\lambda_n \tau).$$

From

$$u(\tau, 0) = f(\tau^2/4) = \sum_{n=1}^{\infty} A_n J_0(\lambda_n \tau)$$

we find

$$A_n = \frac{2}{(2\sqrt{L})^2 J_1^2(2\lambda_n \sqrt{L})} \int_0^{2\sqrt{L}} \tau J_0(\lambda_n \tau) f(\tau^2/4)\, d\tau \qquad \boxed{v = \tau/2,\ dv = d\tau/2}$$

$$= \frac{1}{2L J_1^2(2\lambda_n \sqrt{L})} \int_0^{\sqrt{L}} 2v J_0(2\lambda_n v) f(v^2) 2\, dv$$

$$= \frac{2}{L J_1^2(2\lambda_n \sqrt{L})} \int_0^{\sqrt{L}} v J_0(2\lambda_n v) f(v^2)\, dv.$$

12. (a) First we see that

$$\frac{R''\Theta + \frac{1}{r} R'\Theta + \frac{1}{r^2} R\Theta''}{R\Theta} = \frac{T''}{a^2 T} = -\lambda^2.$$

This gives $T'' + a^2\lambda^2 T = 0$. Then from

$$\frac{R'' + \frac{1}{r} R' + \lambda^2 R}{-R/r^2} = \frac{\Theta''}{\Theta} = -\nu^2$$

we get $\Theta'' + \nu^2\Theta = 0$ and $r^2 R'' + r R' + (\lambda^2 r^2 - \nu^2)R = 0$.

(b) The general solutions of the differential equations in part (a) are

$$T = c_1 \cos a\lambda t + c_2 \sin a\lambda t$$

$$\Theta = c_3 \cos \nu\theta + c_4 \cos \nu\theta$$

$$R = c_5 J_\nu(\lambda r) + c_6 Y_\nu(\lambda r).$$

(c) Implicitly we expect $u(r, \theta, t) = u(r, \theta + 2\pi, t)$ and so Θ must be 2π-periodic. Therefore $\nu = n$, $n = 0, 1, 2, \ldots$. The corresponding eigenfunctions are 1, $\cos\theta$, $\cos 2\theta$, $\ldots$, $\sin\theta$, $\sin 2\theta$, $\ldots$. Arguing that $u(r, \theta, t)$ is bounded as $r \to 0$ we then define $c_6 = 0$ and so $R = c_3 J_n(\lambda r)$. But

$R(c) = 0$ gives $J_n(\lambda c) = 0$; this equation defines the eigenvalues λ_n. For each n, $\lambda_{ni} = x_{ni}/c$, $i = 1, 2, 3, \ldots$.

(d) $u(r, \theta, t) = \displaystyle\sum_{i=1}^{n}(A_{0i}\cos a\lambda_{0i}t + B_{0i}\sin a\lambda_{0i}t)J_0(\lambda_{0i}r)$

$$+ \sum_{n=1}^{\infty}\sum_{i=1}^{\infty}\Big[(A_{ni}\cos a\lambda_{ni}t + B_{ni}\sin a\lambda_{ni}t)\cos n\theta$$

$$+ (C_{ni}\cos a\lambda_{ni}t + D_{ni}\sin a\lambda_{ni}t)\sin n\theta\Big]J_n(\lambda_{ni}r)$$

13. (a) We need to solve $J_0(10\lambda) = 0$. From the table on page 248 in the text we see that $\lambda_1 = 0.2405$, $\lambda_2 = 0.5520$, and $\lambda_3 = 0.8654$. In order to compute A_1 we use the result of Problem 25 in Exercises 6.4:

$$A_1 = \frac{2}{10^2 J_1^2(10\lambda_1)}\int_0^{10} r J_0(\lambda_1 r)(1 - r/10)\,dr \qquad \boxed{s = \lambda_1 r}$$

$$= \frac{1}{50 J_1^2(10\lambda_1)}\int_0^{10\lambda_1}\left[\frac{1}{\lambda_1}s J_0(s) - \frac{1}{10\lambda_1^2}s^2 J_0(s)\right]\frac{1}{\lambda_1}\,ds$$

$$= \frac{1}{500\lambda_1^3 J_1^2(10\lambda_1)}\int_0^{10\lambda_1}[10\lambda_1 s J_0(s) - s^2 J_0(s)]\,ds$$

$$= \frac{1}{500\lambda_1^3 J_1^2(10\lambda_1)}\left[(10\lambda_1 s J_1(s) - s^2 J_1(s) - s J_0(s))\Big|_0^{10\lambda_1} + \int_0^{10\lambda_1}J_0(s)\,ds\right]$$

$$= \frac{1}{500\lambda_1^3 J_1^2(10\lambda_1)}\left[100\lambda_1^2 J_1(10\lambda_1) - 100\lambda_1^2 J_1(10\lambda_1) - 10\lambda_1 J_0(10\lambda_1) + \int_0^{10\lambda_1}J_0(s)\,ds\right]$$

$$= \frac{\int_0^{10\lambda_1}J_0(s)\,ds}{500\lambda_1^3 J_1^2(10\lambda_1)} = \frac{1.4703}{1.8741} = 0.7845.$$

[The value of $\int_0^{10\lambda_1}J_0(s)\,ds$ can be obtained from a table or from *Mathematica*.] In a similar fashion we compute $A_2 = 0.0687$ and $A_3 = 0.0531$. Then, from Example 1 in the text with $a = 1$ and $g(r) = 0$ [so that $B_n = 0$] we have

$$u(r, t) \approx \sum_{n=1}^{3} A_n(\cos \lambda_n t)J_0(\lambda_n r)$$

$$= 0.7845\cos(0.2405t)J_0(0.2405r) + 0.0687\cos(0.5520t)J_0(0.5520r)$$

$$+ 0.0531\cos(0.8654t)J_0(0.8654r).$$

Exercises 13.2

(b)

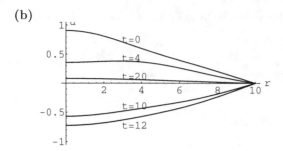

Exercises 13.3

1. To compute

$$A_n = \frac{2n+1}{2c^n} \int_0^\pi f(\theta) P_n(\cos\theta) \sin\theta \, d\theta$$

we substitute $x = \cos\theta$ and $dx = -\sin\theta \, d\theta$. Then

$$A_n = \frac{2n+1}{2c^n} \int_1^{-1} F(x) P_n(x)(-dx) = \frac{2n+1}{2c^n} \int_{-1}^1 F(x) P_n(x) \, dx$$

where

$$F(x) = \begin{cases} 0, & -1 < x < 0 \\ 50, & 0 < x < 1 \end{cases} = 50 \begin{cases} 0, & -1 < x < 0 \\ 1, & 0 < x < 1 \end{cases}.$$

The coefficients A_n are computed in Example 3 of Section 11.5. Thus

$$u(r, \theta) = \sum_{n=0}^\infty A_n r^n P_n(\cos\theta)$$

$$= 50 \left[\frac{1}{2} P_0(\cos\theta) + \frac{3}{4} \left(\frac{r}{c} \right) P_1(\cos\theta) - \frac{7}{16} \left(\frac{r}{c} \right)^3 P_3(\cos\theta) + \frac{11}{32} \left(\frac{r}{c} \right)^5 P_5(\cos\theta) + \cdots \right].$$

2. In the solution of the Cauchy-Euler equation,

$$R(r) = c_1 r^n + c_2 r^{-(n+1)},$$

we define $c_1 = 0$ since we expect the potential u to be bounded as $r \to \infty$. Hence

$$u_n(r, \theta) = A_n r^{-(n+1)} P_n(\cos\theta)$$

$$u(r, \theta) = \sum_{n=0}^\infty A_n r^{-(n+1)} P_n(\cos\theta).$$

When $r = c$ we have

$$f(\theta) = \sum_{n=0}^\infty A_n c^{-(n+1)} P_n(\cos\theta)$$

562

so that

$$A_n = c^{n+1} \frac{(2n+1)}{2} \int_0^\pi f(\theta) P_n(\cos \theta) \sin \theta \, d\theta.$$

The solution of the problem is then

$$u(r, \theta) = \sum_{n=0}^\infty \left(\frac{2n+1}{2} \int_0^\pi f(\theta) P_n(\cos \theta) \sin \theta d\theta \right) \left(\frac{c}{r} \right)^{n+1} P_n(\cos \theta).$$

3. The coefficients are given by

$$A_n = \frac{2n+1}{2c^n} \int_0^\pi \cos \theta P_n(\cos \theta) \sin \theta \, d\theta = \frac{2n+1}{2c^n} \int_0^\pi P_1(\cos \theta) P_n(\cos \theta) \sin \theta \, d\theta$$

$$\boxed{x = \cos \theta, \ dx = -\sin \theta \, d\theta}$$

$$= \frac{2n+1}{2c^n} \int_{-1}^1 P_1(x) P_n(x) \, dx.$$

Since $P_n(x)$ and $P_m(x)$ are orthogonal for $m \neq n$, $A_n = 0$ for $n \neq 1$ and

$$A_1 = \frac{2(1)+1}{2c^1} \int_{-1}^1 P_1(x) P_1(x) \, dx = \frac{3}{2c} \int_{-1}^1 x^2 dx = \frac{1}{c}.$$

Thus

$$u(r, \theta) = \frac{r}{c} P_1(\cos \theta) = \frac{r}{c} \cos \theta.$$

4. The coefficients are given by

$$A_n = \frac{2n+1}{2c^n} \int_0^\pi (1 - \cos 2\theta) P_n(\cos \theta) \sin \theta \, d\theta.$$

These were computed in Problem 16 of Section 11.5. Thus

$$u(r, \theta) = \frac{4}{3} P_0(\cos \theta) - \frac{4}{3} \left(\frac{r}{c} \right)^2 P_2(\cos \theta).$$

5. Referring to Example 1 in the text we have

$$\Theta = P_n(\cos \theta) \quad \text{and} \quad R = c_1 r^n + c_2 r^{-(n+1)}.$$

Since $u(b, \theta) = R(b)\Theta(\theta) = 0$,

$$c_1 b^n + c_2 b^{-(n+1)} = 0 \quad \text{or} \quad c_1 = -c_2 b^{-2n-1},$$

and

$$R(r) = -c_2 b^{-2n-1} r^n + c_2 r^{-(n+1)} = c_2 \left(\frac{b^{2n+1} - r^{2n+1}}{b^{2n+1} r^{n+1}} \right).$$

Then

$$u(r, \theta) = \sum_{n=0}^\infty A_n \frac{b^{2n+1} - r^{2n+1}}{b^{2n+1} r^{n+1}} P_n(\cos \theta)$$

where

$$\frac{b^{2n+1} - a^{2n+1}}{b^{2n+1}a^{n+1}} A_n = \frac{2n + 1}{2} \int_0^\pi f(\theta) P_n(\cos \theta) \sin \theta \, d\theta.$$

6. Referring to Example 1 in the text we have

$$R(r) = c_1 r^n \quad \text{and} \quad \Theta(\theta) = P_n(\cos \theta).$$

Now $\Theta(\pi/2) = 0$ implies that n is odd, so

$$u(r, \theta) = \sum_{n=0}^\infty A_{2n+1} r^{2n+1} P_{2n+1}(\cos \theta).$$

From

$$u(c, \theta) = f(\theta) = \sum_{n=0}^\infty A_{2n+1} c^{2n+1} P_{2n+1}(\cos \theta)$$

we see that

$$A_{2n+1} c^{2n+1} = (4n + 3) \int_0^{\pi/2} f(\theta) \sin \theta \, P_{2n+1}(\cos \theta) \, d\theta.$$

Thus

$$u(r, \theta) = \sum_{n=0}^\infty A_{2n+1} r^{2n+1} P_{2n+1}(\cos \theta)$$

where

$$A_{2n+1} = \frac{4n + 3}{c^{2n+1}} \int_0^{\pi/2} f(\theta) \sin \theta \, P_{2n+1}(\cos \theta) \, d\theta.$$

7. Referring to Example 1 in the text we have

$$r^2 R'' + 2r R' - \lambda^2 R = 0$$

$$\sin \theta \, \Theta'' + \cos \theta \, \Theta' + \lambda \sin \theta \, \Theta = 0.$$

Substituting $x = \cos \theta$, $0 \le \theta \le \pi/2$, the latter equation becomes

$$(1 - x^2) \frac{d^2\Theta}{dx^2} - 2x \frac{d\Theta}{dx} + \lambda^2 \Theta = 0, \quad 0 \le x \le 1.$$

Taking the solutions of this equation to be the Legendre polynomials $P_n(x)$ corresponding to $\lambda^2 = n(n + 1)$ for $n = 1, 2, 3, \ldots$, we have $\Theta = P_n(\cos \theta)$. Since

$$\frac{\partial u}{\partial \theta}\bigg|_{\theta = \pi/2} = \Theta'(\pi/2) R(r) = 0$$

we have

$$\Theta'(\pi/2) = -(\sin \pi/2) P_n'(\cos \pi/2) = -P_n'(0) = 0.$$

As noted in the hint, $P_n'(0) = 0$ only if n is even. Thus $\Theta = P_n(\cos\theta)$, $n = 0, 2, 4, \ldots$. As in Example 1, $R(r) = c_1 r^n$. Hence

$$u(r, \theta) = \sum_{n=0}^{\infty} A_{2n} r^{2n} P_{2n}(\cos\theta).$$

At $r = c$,

$$f(\theta) = \sum_{n=0}^{\infty} A_{2n} c^{2n} P_{2n}(\cos\theta).$$

Using Problem 17 in Section 11.5, we obtain

$$c^{2n} A_{2n} = (4n + 1) \int_{\pi/2}^{0} f(\theta) P_{2n}(\cos\theta)(-\sin\theta)\, d\theta$$

and

$$A_{2n} = \frac{4n + 1}{c^{2n}} \int_{0}^{\pi/2} f(\theta) \sin\theta\, P_{2n}(\cos\theta)\, d\theta.$$

8. Referring to Example 1 in the text we have

$$R(r) = c_1 r^n + c_2 r^{-(n-1)} \quad \text{and} \quad \Theta(\theta) = P_n(\cos\theta).$$

Since we expect $u(r, \theta)$ to be bounded as $r \to \infty$, we define $c_1 = 0$. Also $\Theta(\pi/2) = 0$ implies that n is odd, so

$$u(r, \theta) = \sum_{n=0}^{\infty} A_{2n+1} r^{-2(n+1)} P_{2n+1}(\cos\theta).$$

From

$$u(c, \theta) = f(\theta) = \sum_{n=0}^{\infty} A_{2n+1} c^{-2(n+1)} P_{2n+1}(\cos\theta)$$

we see that

$$A_{2n+1} c^{-2(n+1)} = (4n + 3) \int_{0}^{\pi/2} f(\theta) \sin\theta\, P_{2n+1}(\cos\theta)\, d\theta.$$

Thus

$$u(r, \theta) = \sum_{n=0}^{\infty} A_{2n+1} r^{-2(n+1)} P_{2n+1}(\cos\theta)$$

where

$$A_{2n+1} = (4n + 3) c^{2(n+1)} \int_{0}^{\pi/2} f(\theta) \sin\theta\, P_{2n+1}(\cos\theta)\, d\theta.$$

9. Checking the hint, we find

$$\frac{1}{r} \frac{\partial^2}{\partial r^2}(ru) = \frac{1}{r} \frac{\partial}{\partial r}\left[r \frac{\partial u}{\partial r} + u\right] = \frac{1}{r}\left[r \frac{\partial^2 u}{\partial r^2} + \frac{\partial u}{\partial r} + \frac{\partial u}{\partial r}\right] = \frac{\partial^2 u}{\partial r^2} + \frac{2}{r} \frac{\partial u}{\partial r}.$$

The partial differential equation then becomes

$$\frac{\partial^2}{\partial r^2}(ru) = r \frac{\partial u}{\partial t}.$$

Now, letting $ru(r,t) = v(r,t) + \psi(r)$, since the boundary condition is nonhomogeneous, we obtain

$$\frac{\partial^2}{\partial r^2}[v(r,t) + \psi(r)] = r\frac{\partial}{\partial t}\left[\frac{1}{r}v(r,t) + \psi(r)\right]$$

or

$$\frac{\partial^2 v}{\partial r^2} + \psi''(r) = \frac{\partial v}{\partial t}.$$

This differential equation will be homogeneous if $\psi''(r) = 0$ or $\psi(r) = c_1 r + c_2$. Now

$$u(r,t) = \frac{1}{r}v(r,t) + \frac{1}{r}\psi(r) \quad \text{and} \quad \frac{1}{r}\psi(r) = c_1 + \frac{c_2}{r}.$$

Since we want $u(r,t)$ to be bounded as r approaches 0, we require $c_2 = 0$. Then $\psi(r) = c_1 r$. When $r = 1$

$$u(1,t) = v(1,t) + \psi(1) = v(1,t) + c_1 = 100,$$

and we will have the homogeneous boundary condition $v(1,t) = 0$ when $c_1 = 100$. Consequently, $\psi(r) = 100r$. The initial condition

$$u(r,0) = \frac{1}{r}v(r,0) + \frac{1}{r}\psi(r) = \frac{1}{r}v(r,0) + 100 = 0$$

implies $v(r,0) = -100r$. We are thus led to solve the new boundary-value problem

$$\frac{\partial^2 v}{\partial r^2} = \frac{\partial v}{\partial t}, \quad 0 < r < 1, \quad t > 0,$$

$$v(1,t) = 0, \quad \lim_{r \to 0}\frac{1}{r}v(r,t) < \infty,$$

$$v(r,0) = -100r.$$

Letting $v(r,t) = R(r)T(t)$ and separating variables leads to

$$R'' + \lambda^2 R = 0 \quad \text{and} \quad T' + \lambda^2 T = 0$$

with solutions

$$R(r) = c_3 \cos \lambda r + c_4 \sin \lambda r \quad \text{and} \quad T(t) = c_5 e^{-\lambda^2 t}.$$

The boundary conditions are equivalent to $R(1) = 0$ and $\lim_{r \to 0}\frac{1}{r}R(r) < \infty$. Since

$$\lim_{r \to 0}\frac{1}{r}R(r) = \lim_{r \to 0}\frac{c_3 \cos \lambda r}{r} + \lim_{r \to 0}\frac{c_4 \sin \lambda r}{r} = \lim_{r \to 0}\frac{c_3 \cos \lambda r}{r} + c_4 \lambda < \infty$$

we must have $c_3 = 0$. Then $R(r) = c_4 \sin \lambda r$, and $R(1) = 0$ implies $\lambda = n\pi$ for $n = 1, 2, 3, \ldots$. Thus

$$v_n(r,t) = A_n e^{-n^2\pi^2 t}\sin n\pi r$$

566

for $n = 1, 2, 3, \ldots$. Using the condition $\lim_{r \to 0} \frac{1}{r} R(r) < \infty$ it is easily shown that there are no eigenvalues for $\lambda = 0$, nor does setting the common constant to $+\lambda^2$ when separating variables lead to any solutions. Now, by the superposition principle,

$$v(r,t) = \sum_{n=1}^{\infty} A_n e^{-n^2\pi^2 t} \sin n\pi r.$$

The initial condition $v(r,0) = -100r$ implies

$$-100r = \sum_{n=1}^{\infty} A_n \sin n\pi r.$$

This is a Fourier sine series and so

$$A_n = 2 \int_0^1 (-100r \sin n\pi r)\, dr = -200 \left[-\frac{r}{n\pi} \cos n\pi r \Big|_0^1 + \int_0^1 \frac{1}{n\pi} \cos n\pi r\, dr \right]$$

$$= -200 \left[-\frac{\cos n\pi}{n\pi} + \frac{1}{n^2\pi^2} \sin n\pi r \Big|_0^1 \right] = -200 \left[-\frac{(-1)^n}{n\pi} \right] = \frac{(-1)^n 200}{n\pi}.$$

A solution of the problem is thus

$$u(r,t) = \frac{1}{r} v(r,t) + \frac{1}{r} \psi(r) = \frac{1}{r} \sum_{n=1}^{\infty} (-1)^n \frac{20}{n\pi} e^{-n^2\pi^2 t} \sin n\pi r + \frac{1}{r}(100r)$$

$$= \frac{200}{\pi r} \sum_{n=1}^{\infty} \frac{(-1)^n}{n} e^{-n^2\pi^2 t} \sin n\pi r + 100.$$

10. Referring to Problem 9 we have

$$\frac{\partial^2 v}{\partial r^2} + \psi''(r) = \frac{\partial v}{\partial t}$$

where $\psi(r) = c_1 r$. Since

$$u(r,t) = \frac{1}{r} v(r,t) + \frac{1}{r} \psi(r) = \frac{1}{r} v(r,t) + c_1$$

we have

$$\frac{\partial u}{\partial r} = \frac{1}{r} v_r(r,t) - \frac{1}{r^2} v(r,t).$$

When $r = 1$,

$$\frac{\partial u}{\partial r} \Big|_{r=1} = v_r(1,t) - v(1,t)$$

and

$$\frac{\partial u}{\partial r} \Big|_{r=1} + hu(1,t) = v_r(1,t) - v(1,t) + h[v(1,t) + \psi(1)] = v_r(1,t) + (h-1)v(1,t) + hc_1.$$

Thus the boundary condition

$$\frac{\partial u}{\partial r} \Big|_{r=1} + hu(1,t) = hu_1$$

will be homogeneous when $hc_1 = hu_1$ or $c_1 = u_1$. Consequently $\psi(r) = u_1 r$. The initial condition

$$u(r,0) = \frac{1}{r}v(r,0) + \frac{1}{r}\psi(r) = \frac{1}{r}v(r,0) + u_1 = u_0$$

implies $v(r,0) = (u_0 - u_1)r$. We are thus led to solve the new boundary-value problem

$$\frac{\partial^2 v}{\partial r^2} = \frac{\partial v}{\partial t}, \quad 0 < r < 1, \quad t > 0,$$

$$v_r(1,t) + (h-1)v(1,t) = 0, \quad t > 0,$$

$$\lim_{r \to 0} \frac{1}{r}v(r,t) < \infty,$$

$$v(r,0) = (u_0 - u_1)r.$$

Separating variables as in Problem 9 leads to

$$R(r) = c_3 \cos \lambda r + c_4 \sin \lambda r \quad \text{and} \quad T(t) = c_5 e^{-\lambda^2 t}.$$

The boundary conditions are equivalent to

$$R'(1) + (h-1)R(1) = 0 \quad \text{and} \quad \lim_{r \to 0} \frac{1}{r}R(r) < \infty.$$

As in Problem 6 we use the second condition to determine that $c_3 = 0$ and $R(r) = c_4 \sin \lambda r$. Then

$$R'(1) + (h-1)R(1) = c_4 \lambda \cos \lambda + c_4(h-1)\sin \lambda = 0$$

and the eigenvalues λ_n are the consecutive nonnegative roots of $\tan \lambda = \lambda/(1-h)$. Now

$$v(r,t) = \sum_{n=1}^{\infty} A_n e^{-\lambda_n^2 t} \sin \lambda_n r.$$

From

$$v(r,0) = (u_0 - u_1)r = \sum_{n=1}^{\infty} A_n \sin \lambda_n r$$

we obtain

$$A_n = \frac{\displaystyle\int_0^1 (u_0 - u_1)r \sin \lambda_n r \, dr}{\displaystyle\int_0^1 \sin^2 \lambda_n r \, dr}.$$

We compute the integrals

$$\int_0^1 r \sin \lambda_n r \, dr = \left(\frac{1}{\lambda_n^2} \sin \lambda_n r - \frac{1}{\lambda_n} \cos \lambda_n r \right) \Big|_0^1 = \frac{1}{\lambda_n^2} \sin \lambda_n - \frac{1}{\lambda_n} \cos \lambda_n$$

and

$$\int_0^1 \sin^2 \lambda_n r \, dr = \left(\frac{1}{2}r - \frac{1}{4\lambda_n} \sin 2\lambda_n r \right) \Big|_0^1 = \frac{1}{2} - \frac{1}{4\lambda_n} \sin 2\lambda_n.$$

Using $\lambda_n \cos \lambda_n = -(h-1)\sin \lambda_n$ we then have

$$A_n = (u_0 - u_1)\frac{\frac{1}{\lambda_n^2}\sin \lambda_n - \frac{1}{\lambda_n}\cos \lambda_n}{\frac{1}{2} - \frac{1}{4\lambda_n}\sin 2\lambda_n} = (u_0 - u_1)\frac{4\sin \lambda_n - 4\lambda_n \cos \lambda_n}{2\lambda_n^2 - \lambda_n 2\sin \lambda_n \cos \lambda_n}$$

$$- 2(u_0 - u_1)\frac{\sin \lambda_n + (h-1)\sin \lambda_n}{\lambda_n^2 + (h-1)\sin \lambda_n \sin \lambda_n} = 2(u_0 - u_1)h\frac{\sin \lambda_n}{\lambda_n^2 + (h-1)\sin^2 \lambda_n}.$$

Therefore

$$u(r,t) = \frac{1}{r}v(r,t) + \frac{1}{r}\psi(r) = u_1 + 2(u_0 - u_1)h\sum_{n=1}^{\infty}\frac{\sin \lambda_n \sin \lambda_n r}{r[\lambda_n^2 + (h-1)\sin^2 \lambda_n]}e^{-\lambda_n^2 t}.$$

11. We write the differential equation in the form

$$a^2\frac{1}{r}\frac{\partial^2}{\partial r^2}(ru) = \frac{\partial^2 u}{\partial t^2} \quad \text{or} \quad a^2\frac{\partial^2}{\partial r^2}(ru) = r\frac{\partial^2 u}{\partial t^2},$$

and then let $v(r,t) = ru(r,t)$. The new boundary-value problem is

$$a^2\frac{\partial^2 v}{\partial r^2} = \frac{\partial^2 v}{\partial t^2}, \quad 0 < r < c, \quad t > 0$$

$$v(c,t) = 0, \quad t > 0$$

$$v(r,0) = rf(r), \quad \frac{\partial v}{\partial t}\Big|_{t=0} = rg(r).$$

Letting $v(r,t) = R(r)T(t)$ and separating variables we obtain

$$R'' + \lambda^2 R = 0$$

$$T'' + a^2\lambda^2 T = 0$$

and

$$R(r) = c_1 \cos \lambda r + c_2 \sin \lambda r$$

$$T(t) = c_3 \cos a\lambda t + c_4 \sin a\lambda t.$$

Since $u(r,t) = v(r,t)/r$, in order to insure boundedness at $r = 0$ we define $c_1 = 0$. Then $R(r) = c_2 \sin \lambda r$ and the condition $R(c) = 0$ implies $\lambda = n\pi/c$. Thus

$$v(r,t) = \sum_{n=1}^{\infty}\left(A_n \cos \frac{n\pi a}{c}t + B_n \sin \frac{n\pi a}{c}t\right)\sin \frac{n\pi}{c}r.$$

From

$$v(r,0) = rf(r) = \sum_{n=1}^{\infty}A_n \sin \frac{n\pi}{c}r$$

we see that

$$A_n = \frac{2}{c}\int_0^c rf(r)\sin \frac{n\pi}{c}r\,dr.$$

Exercises 13.3

From

$$\frac{\partial v}{\partial t}\bigg|_{t=0} = rg(r) = \sum_{n=1}^{\infty} \left(B_n \frac{n\pi a}{c}\right) \sin \frac{n\pi}{c} r$$

we see that

$$B_n = \frac{c}{n\pi a} \cdot \frac{2}{c} \int_0^c rg(r) \sin \frac{n\pi}{c} r \, dr = \frac{2}{n\pi a} \int_0^c rg(r) \sin \frac{n\pi}{c} r \, dr.$$

12. Proceeding as in Example 1 we obtain

$$\Theta(\theta) = P_n(\cos \theta) \quad \text{and} \quad R(r) = c_1 r^n + c_2 r^{-(n+1)}$$

so that

$$u(r, \theta) = \sum_{n=0}^{\infty} (A_n r^n + B_n r^{-(n+1)}) P_n(\cos \theta).$$

To satisfy $\lim_{r\to\infty} u(r, \theta) = -Er \cos \theta$ we must have $A_n = 0$ for $n = 2, 3, 4, \ldots$. Then

$$\lim_{r\to\infty} u(r, \theta) = -Er \cos \theta = A_0 \cdot 1 + A_1 r \cos \theta,$$

so $A_0 = 0$ and $A_1 = -E$. Thus

$$u(r, \theta) = -Er \cos \theta + \sum_{n=0}^{\infty} B_n r^{-(n+1)} P_n(\cos \theta).$$

Now

$$u(c, \theta) = 0 = -Ec \cos \theta + \sum_{n=0}^{\infty} B_n c^{-(n+1)} P_n(\cos \theta)$$

so

$$\sum_{n=0}^{\infty} B_n c^{-(n+1)} P_n(\cos \theta) = Ec \cos \theta$$

and

$$B_n c^{-(n+1)} = \frac{2n+1}{2} \int_0^{\pi} Ec \cos \theta \, P_n(\cos \theta) \sin \theta \, d\theta.$$

Now $\cos \theta = P_1(\cos \theta)$ so, for $n \neq 1$,

$$\int_0^{\pi} \cos \theta \, P_n(\cos \theta) \sin \theta \, d\theta = 0$$

by orthogonality. Thus $B_n = 0$ for $n \neq 1$ and

$$B_1 = \frac{3}{2} Ec^3 \int_0^{\pi} \cos^2 \theta \sin \theta \, d\theta = Ec^3.$$

Therefore,

$$u(r, \theta) = -Er \cos \theta + Ec^3 r^{-2} \cos \theta.$$

Chapter 13 Review Exercises

1. We have

$$A_0 = \frac{1}{2\pi} \int_0^\pi u_0 \, d\theta + \frac{1}{2\pi} \int_\pi^{2\pi} (-u_0) \, d\theta = 0$$

$$A_n = \frac{1}{c^n \pi} \int_0^\pi u_0 \cos n\theta \, d\theta + \frac{1}{c^n \pi} \int_\pi^{2\pi} (-u_0) \cos n\theta \, d\theta = 0$$

$$B_n = \frac{1}{c^n \pi} \int_0^\pi u_0 \sin n\theta \, d\theta + \frac{1}{c^n \pi} \int_\pi^{2\pi} (-u_0) \sin n\theta \, d\theta = \frac{2u_0}{c^n n \pi}[1 - (-1)^n]$$

and so

$$u(r, \theta) = \frac{2u_0}{\pi} \sum_{n=1}^{\infty} \frac{1 - (-1)^n}{n} \left(\frac{r}{c}\right)^n \sin n\theta.$$

2. We have

$$A_0 = \frac{1}{2\pi} \int_0^{\pi/2} d\theta + \frac{1}{2\pi} \int_{3\pi/2}^{2\pi} d\theta = \frac{1}{2}$$

$$A_n = \frac{1}{c^n \pi} \int_0^{\pi/2} \cos n\theta \, d\theta + \frac{1}{c^n \pi} \int_{3\pi/2}^{2\pi} \cos n\theta \, d\theta = \frac{1}{c^n n \pi}\left[\sin \frac{n\pi}{2} - \sin \frac{3n\pi}{2}\right]$$

$$B_n = \frac{1}{c^n \pi} \int_0^{\pi/2} \sin n\theta \, d\theta + \frac{1}{c^n \pi} \int_{3\pi/2}^{2\pi} \sin n\theta \, d\theta = \frac{1}{c^n n \pi}\left[\cos \frac{3n\pi}{2} - \cos \frac{n\pi}{2}\right]$$

and so

$$u(r, \theta) = \frac{1}{2} + \frac{1}{\pi} \sum_{n=1}^{\infty} \left(\frac{r}{c}\right)^n \left[\frac{\sin \frac{n\pi}{2} - \sin \frac{3n\pi}{2}}{n} \cos n\theta + \frac{\cos \frac{3n\pi}{2} - \cos \frac{n\pi}{2}}{n} \sin n\theta\right].$$

3. The conditions $\Theta(0) = 0$ and $\Theta(\pi) = 0$ applied to $\Theta = c_1 \cos \lambda\theta + c_2 \sin \lambda\theta$ give $c_1 = 0$ and $\lambda = n$, $n = 1, 2, 3, \ldots$, respectively. Thus we have the Fourier sine-series coefficients

$$A_n = \frac{2}{\pi} \int_0^\pi u_0(\pi\theta - \theta^2) \sin n\theta \, d\theta = \frac{4u_0}{n^3 \pi}[1 - (-1)^n].$$

Thus

$$u(r, \theta) = \frac{4u_0}{\pi} \sum_{n=1}^{\infty} \frac{1 - (-1)^n}{n^3} r^n \sin n\theta.$$

4. In this case

$$A_n = \frac{2}{\pi} \int_0^\pi \sin \theta \sin n\theta \, d\theta = \frac{1}{\pi} \int_0^\pi [\cos(1 - n)\theta - \cos(1 + n)\theta] \, d\theta = 0, \quad n \neq 1.$$

For $n = 1$,

$$A_1 = \frac{2}{\pi} \int_0^\pi \sin^2 \theta \, d\theta = \frac{1}{\pi} \int_0^\pi (1 - \cos 2\theta) \, d\theta = 1.$$

Thus

$$u(r, \theta) = \sum_{n=1}^{\infty} A_n r^n \sin n\theta$$

reduces to

$$u(r, \theta) = r \sin \theta.$$

5. We solve

$$\frac{\partial^2 u}{\partial r^2} + \frac{1}{r}\frac{\partial u}{\partial r} + \frac{1}{r^2}\frac{\partial^2 u}{\partial \theta^2} = 0, \quad 0 < \theta < \frac{\pi}{4}, \quad \frac{1}{2} < r < 1,$$

$$u(r, 0) = 0, \quad u(r, \pi/4) = 0, \quad \frac{1}{2} < r < 1,$$

$$u(1/2, \theta) = u_0, \quad u_r(1, \theta) = 0, \quad 0 < \theta < \frac{\pi}{4}.$$

Proceeding as in Example 1 in Section 13.1 in the text we obtain the separated equations

$$r^2 R'' + r R' - \lambda^2 R = 0$$

$$\Theta'' + \lambda^2 \Theta = 0$$

with solutions

$$\Theta(\theta) = c_1 \cos \lambda\theta + c_2 \sin \lambda\theta$$

$$R(r) = c_3 r^\lambda + c_4 r^{-\lambda}.$$

Applying the boundary conditions $\Theta(0) = 0$ and $\Theta(\pi/4) = 0$ gives $c_1 = 0$ and $\lambda = 4n$ for $n = 1, 2, 3, \ldots$. From $R_r(1) = 0$ we obtain $c_3 = c_4$. Therefore

$$u(r, \theta) = \sum_{n=1}^{\infty} A_n \left(r^{4n} + r^{-4n} \right) \sin 4n\theta.$$

From

$$u(1/2, \theta) = u_0 = \sum_{n=1}^{\infty} A_n \left(\frac{1}{2^{4n}} + \frac{1}{2^{-4n}} \right) \sin 4n\theta$$

we find

$$A_n \left(\frac{1}{2^{4n}} + \frac{1}{2^{-4n}} \right) = \frac{2}{\pi/4} \int_0^{\pi/4} u_0 \sin 4n\theta \, d\theta = \frac{2u_0}{n\pi}[1 - (-1)^n]$$

or

$$A_n = \frac{2u_0}{n\pi(2^{4n} + 2^{-4n})}[1 - (-1)^n].$$

Thus the steady-state temperature in the plate is

$$u(r, \theta) = \frac{2u_0}{\pi} \sum_{n=1}^{\infty} \frac{[r^{4n} + r^{-4n}][1 - (-1)^n]}{n[2^{4n} + 2^{-4n}]} \sin 4n\theta.$$

6. We solve

$$\frac{\partial^2 u}{\partial r^2} + \frac{1}{r}\frac{\partial u}{\partial r} + \frac{1}{r^2}\frac{\partial^2 u}{\partial \theta^2} = 0, \quad r > 1, \quad 0 < \theta < \pi,$$

$$u(r,0) = 0, \quad u(r,\pi) = 0, \quad r > 1,$$

$$u(1,\theta) = f(\theta), \quad 0 < \theta < \pi.$$

Separating variables we obtain

$$\Theta(\theta) = c_1 \cos \lambda\theta + c_2 \sin \lambda\theta$$

$$R(r) = c_3 r^\lambda + c_4 r^{-\lambda}.$$

Applying the boundary conditions $\Theta(0) = 0$, and $\Theta(\pi) = 0$ gives $c_1 = 0$ and $\lambda = n$ for $n = 1, 2, 3, \ldots$. Assuming $f(\theta)$ to be bounded, we expect the solution $u(r,\theta)$ to also be bounded as $r \to \infty$. This requires that $c_3 = 0$. Therefore

$$u(r,\theta) = \sum_{n=1}^{\infty} A_n r^{-n} \sin n\theta.$$

From

$$u(1,\theta) = f(\theta) = \sum_{n=1}^{\infty} A_n \sin n\theta$$

we obtain

$$A_n = \frac{2}{\pi} \int_0^\pi f(\theta) \sin n\theta \, d\theta.$$

7. Letting $u(r,t) = R(r)T(t)$ and separating variables we obtain

$$\frac{R'' + \frac{1}{r}R' - hR}{R} = \frac{T'}{T} = \mu$$

so

$$R'' + \frac{1}{r}R' - (\mu + h)R = 0 \quad \text{and} \quad T' - \mu T = 0.$$

From the second equation we find $T(t) = c_1 e^{\mu t}$. If $\mu > 0$, $T(t)$ increases without bound as $t \to \infty$. Thus we assume $\mu \leq 0$. Since $h > 0$ we can take $\mu = -\lambda^2 - h$. Then

$$R'' + \frac{1}{r}R' + \lambda^2 R = 0$$

is a parametric Bessel equation with solution

$$R(r) = c_1 J_0(\lambda r) + c_2 Y_0(\lambda r).$$

Since Y_0 is unbounded as $r \to 0$ we take $c_2 = 0$. Then $R(r) = c_1 J_0(\lambda r)$ and the boundary condition $u(1,t) = R(1)T(t) = 0$ implies $J_0(\lambda) = 0$. This latter equation defines the positive eigenvalues λ_n. Thus

$$u(r,t) = \sum_{n=1}^{\infty} A_n J_0(\lambda_n r) e^{(-\lambda_n^2 - h)t}.$$

From

$$u(r,0) = 1 = \sum_{n=1}^{\infty} A_n J_0(\lambda_n r)$$

we find

$$A_n = \frac{2}{J_1^2(\lambda_n)} \int_0^1 r J_0(\lambda_n r)\, dr \qquad \boxed{x = \lambda_n r, \ dx = \lambda_n, dr}$$

$$= \frac{2}{J_1^2(\lambda_n)} \int_0^{\lambda_n} \frac{1}{\lambda_n^2} x J_0(x)\, dx.$$

From recurrence relation (4) in Section 11.5 of the text we have

$$x J_0(x) = \frac{d}{dx} [x J_1(x)].$$

Then

$$A_n = \frac{2}{\lambda_n^2 J_1^2(\lambda_n)} \int_0^{\lambda_n} \frac{d}{dx} [x J_1(x)]\, dx = \frac{2}{\lambda_n^2 J_1^2(\lambda_n)} \left(x J_1(x) \right) \Big|_0^{\lambda_n} = \frac{2\lambda_1 J_1(\lambda_n)}{\lambda_n^2 J_1^2(\lambda_n)} = \frac{2}{\lambda_n J_1(\lambda_n)}$$

and

$$u(r,t) = 2e^{-ht} \sum_{n=1}^{\infty} \frac{J_0(\lambda_n r)}{\lambda_n J_1(\lambda_n)} e^{-\lambda_n^2 t}$$

8. Proceeding in the usual manner we find

$$u(r,t) = \sum_{n=1}^{\infty} A_n \cos a\lambda_n t J_0(\lambda_n r)$$

where the eigenvalues are defined by $J_0(\lambda) = 0$. Thus the initial condition gives

$$u_0 J_0(x_k r) = \sum_{n=1}^{\infty} A_n J_0(\lambda_n r)$$

and so

$$A_n = \frac{2}{J_1^2(\lambda_n)} \int_0^1 r \left(u_0 J_0(x_k r) \right) J_0(\lambda_n r)\, dr.$$

But $J_0(\lambda) = 0$ implies that the eigenvalues are the positive zeros of J_0, that is, $\lambda_n = x_n$, $n = 1, 2, 3, \ldots$. Therefore

$$A_n = \frac{2u_0}{J_1^2(\lambda_n)} \int_0^1 r J_0(\lambda_k r) J_0(\lambda_n r)\, dr = 0, \quad n \neq k$$

by orthogonality. For $n = k$,

$$A_k = \frac{2u_0}{J_1^2(\lambda_k)} \int_0^1 r J_0^2(\lambda_k) \, dr = u_0$$

by (6) of Section 11.5. Thus the solution $u(r, t)$ reduces to one term when $n = k$, and

$$u(r, t) = u_0 \cos a\lambda_k t J_0(\lambda_k r) = u_0 \cos a x_k t J_0(x_k r).$$

9. Referring to Example 2 in Section 13.2 we have

$$R(r) = c_1 J_0(\lambda r) + c_2 Y_0(\lambda r)$$

$$Z(z) = c_3 \cosh \lambda z + c_4 \sinh \lambda z$$

where $c_2 = 0$ and $J_0(2\lambda) = 0$ defines the positive eigenvalues λ_n. From $Z'(0) = 0$ we obtain $c_4 = 0$. Then

$$u(r, z) = \sum_{n=1}^{\infty} A_n \cosh \lambda_n z J_0(\lambda_n r).$$

From

$$u(r, 4) = 50 = \sum_{n=1}^{\infty} A_n \cosh 4\lambda_n J_0(\lambda_n r)$$

we obtain (as in Example 1 of Section 13.1)

$$A_n \cosh 4\lambda_n = \frac{2(50)}{4 J_1^2(2\lambda_n)} \int_0^2 r J_0(\lambda_n r) \, dr = \frac{50}{\lambda_n J_1(2\lambda_n)}.$$

Thus the temperature in the cylinder is

$$u(r, z) = 50 \sum_{n=1}^{\infty} \frac{\cosh \lambda_n z J_0(\lambda_n r)}{\lambda_n \cosh 4\lambda_n J_1(2\lambda_n)}.$$

10. Using $u = RZ$ and $-\lambda^2$ as a separation constant leads to

$$r^2 R'' + r R' + \lambda^2 r^2 R = 0, \quad R'(1) = 0, \quad \text{and} \quad Z'' - \lambda^2 Z = 0.$$

Thus

$$R(r) = c_1 J_0(\lambda r) + c_2 Y_0(\lambda r)$$

$$Z(z) = c_3 \cosh \lambda z + c_4 \sinh \lambda z$$

for $\lambda > 0$. Arguing that $u(r, z)$ is bounded as $r \to 0$ we defined $c_2 = 0$. Since the eigenvalues are defined by $J_0'(\lambda) = 0$ we know that $\lambda = 0$ is an eigenvalue. The solutions are then

$$R(r) = c_1 + c_2 \ln r \quad \text{and} \quad Z(z) = c_3 z + c_4$$

where $c_2 = 0$. Thus a formal solution is

$$u(r, z) = A_0 z + B_0 + \sum_{n=1}^{\infty} (A_n \sinh \lambda_n z + B_n \cosh \lambda_n z) J_0(\lambda_n r).$$

Finally, the specified conditions $z = 0$ and $z = 1$ give, in turn,

$$B_0 = 2 \int_0^1 r f(r)\, dr$$

$$B_n = \frac{2}{J_0^2(\lambda_n)} \int_0^1 r f(r) J_0(\lambda_n r)\, dr$$

$$A_0 = -B_0 + 2 \int_0^1 r g(r)\, dr$$

$$A_n = \frac{1}{\sinh \lambda_n} \left[-B_n \cosh \lambda_n + \frac{2}{J_0^2(\lambda_n)} \int_0^1 r g(r) J_0(\lambda_n r)\, dr \right].$$

11. Referring to Example 1 in Section 13.3 of the text we have

$$u(r, \theta) = \sum_{n=0}^{\infty} A_n r^n P_n(\cos\theta).$$

For $x = \cos\theta$

$$u(1, \theta) = \begin{cases} 100 & 0 < \theta < \pi/2 \\ -100 & \pi/2 < \theta < \pi \end{cases} = 100 \begin{cases} -1, & -1 < x < 0 \\ 1, & 0 < x < 1 \end{cases} = g(x).$$

From Problem 20 in Exercise 11.5 we have

$$u(r, \theta) = 100 \left[\frac{3}{2} r P_1(\cos\theta) - \frac{7}{8} r^3 P_3(\cos\theta) + \frac{11}{16} r^5 P_5(\cos\theta) + \cdots \right].$$

12. Since

$$\frac{1}{r}\frac{\partial^2}{\partial r^2}(ru) = \frac{1}{r}\frac{\partial}{\partial r}\left[r\frac{\partial u}{\partial r} + u \right] = \frac{1}{r}\left[r\frac{\partial^2 u}{\partial r^2} + \frac{\partial u}{\partial r} + \frac{\partial u}{\partial r} \right] = \frac{\partial^2 u}{\partial r^2} + \frac{2}{r}\frac{\partial u}{\partial r}$$

the differential equation becomes

$$\frac{1}{r}\frac{\partial^2}{\partial r^2}(ru) = \frac{\partial^2 u}{\partial t^2} \quad \text{or} \quad \frac{\partial^2}{\partial r^2}(ru) = r\frac{\partial^2 u}{\partial t^2}.$$

Letting $v(r, t) = ru(r, t)$ we obtain the boundary-value problem

$$\frac{\partial^2 v}{\partial r^2} = \frac{\partial^2 v}{\partial t^2}, \quad 0 < r < 1, \quad t > 0$$

$$\left.\frac{\partial v}{\partial r}\right|_{r=1} - v(1, t) = 0, \quad t > 0$$

$$v(r, 0) = r f(r), \quad \left.\frac{\partial v}{\partial t}\right|_{t=0} = r g(r), \quad 0 < r < 1.$$

If we separate variables using $v(r, t) = R(r)T(t)$ then we obtain

$$R(r) = c_1 \cos \lambda r + c_2 \sin \lambda r$$

$$T(t) = c_3 \cos \lambda t + c_4 \sin \lambda t.$$

Since $u(r,t) = v(r,t)/r$, in order to insure boundedness at $r = 0$ we define $c_1 = 0$. Then $R(r) = c_2 \sin \lambda r$. Now the boundary condition $R'(1) - R(1) = 0$ implies $\lambda \cos \lambda - \sin \lambda = 0$. Thus, the eigenvalues λ_n are the positive solutions of $\tan \lambda = \lambda$. We now have

$$v_n(r,t) = (A_n \cos \lambda_n t + B_n \sin \lambda_n t) \sin \lambda_n r.$$

For the eigenvalue $\lambda = 0$,

$$R(r) = c_1 r + c_2 \quad \text{and} \quad T(t) = c_3 t + c_4,$$

and boundedness at $r = 0$ implies $c_2 = 0$. We then take

$$v_0(r,t) = A_0 tr + B_0 r$$

so that

$$v(r,t) = A_0 tr + B_0 r + \sum_{n=1}^{\infty} (a_n \cos \lambda_n t + B_n \sin \lambda_n t) \sin \lambda_n r.$$

Now

$$v(r,0) = rf(r) = B_0 r + \sum_{n=1}^{\infty} A_n \sin \lambda_n r.$$

Since $\{r, \sin \lambda_n r\}$ is an orthogonal set on $[0,1]$,

$$\int_0^1 r \sin \lambda_n r \, dr = 0 \quad \text{and} \quad \int_0^1 \sin \lambda_n r \sin \lambda_n r \, dr = 0$$

for $m \neq n$. Therefore

$$\int_0^1 r^2 f(r) \, dr = B_0 \int_0^1 r^2 \, dr = \frac{1}{3} B_0$$

and

$$B_0 = 3 \int_0^1 r^2 f(r) \, dr.$$

Also

$$\int_0^1 rf(r) \sin \lambda_n r \, dr = A_n \int_0^1 \sin^2 \lambda_n r \, dr$$

and

$$A_n = \frac{\int_0^1 rf(r) \sin \lambda_n r \, dr}{\int_0^1 \sin^2 \lambda_n r \, dr}.$$

Now

$$\int_0^1 \sin^2 \lambda_n r \, dr = \frac{1}{2} \int_0^1 (1 - \cos 2\lambda_n r) \, dr = \frac{1}{2} \left[1 - \frac{\sin 2\lambda_n}{2\lambda_n} \right] = \frac{1}{2} [1 - \cos^2 \lambda_n].$$

Since $\tan \lambda_n = \lambda_n$,

$$1 + \lambda_n^2 = 1 + \tan^2 \lambda_n = \sec^2 \lambda_n = \frac{1}{\cos^2 \lambda_n}$$

and

$$\cos^2 \lambda_n = \frac{1}{1 + \lambda_n^2} \, .$$

Then

$$\int_0^1 \sin^2 \lambda_n r \, dr = \frac{1}{2}\left[1 - \frac{1}{1 + \lambda_n^2}\right] = \frac{\lambda_n^2}{2(1 + \lambda_n^2)}$$

and

$$A_n = \frac{2(1 + \lambda_n^2)}{\lambda_n^2} \int_0^1 r f(r) \sin \lambda_n r \, dr.$$

Similarly, setting

$$\frac{\partial v}{\partial t}\bigg|_{t=0} = rg(r) = A_0 r + \sum_{n=1}^{\infty} B_n \lambda_n \sin \lambda_n r$$

we obtain

$$A_0 = 3 \int_0^1 r^2 g(r) \, dr$$

and

$$B_n = \frac{2(1 + \lambda_n^2)}{\lambda_n^3} \int_0^1 r g(r) \sin \lambda_n r \, dr.$$

Therefore, since $v(r, t) = r u(r, t)$ we have

$$u(r, t) = A_0 t + B_0 + \sum_{n=1}^{\infty} (A_n \cos \lambda_n t + B_n \sin \lambda_n t)\frac{\sin \lambda_n r}{r} \, ,$$

where the λ_n are solutions of $\tan \lambda = \lambda$ and

$$A_0 = 3 \int_0^1 r^2 g(r) \, dr$$

$$B_0 = 3 \int_0^1 r^2 f(r) \, dr$$

$$A_n = \frac{2(1 + \lambda_n^2)}{\lambda_n^2} \int_0^1 r f(r) \sin \lambda_n r \, dr$$

$$B_n = \frac{2(1 + \lambda_n^2)}{\lambda_n^3} \int_0^1 r g(r) \sin \lambda_n r \, dr$$

for $n = 1, 2, 3, \ldots$.

13. We note that the differential equation can be expressed in the form

$$\frac{d}{dx}[xu'] = -\lambda^2 x u.$$

Thus

$$u_n \frac{d}{dx}[xu'_m] = -\lambda_m^2 x u_m u_n$$

578

and

$$u_m \frac{d}{dx}[xu_n'] = -\lambda_n^2 x u_n u_m.$$

Subtracting we obtain

$$u_n \frac{d}{dx}[xu_m'] - u_m \frac{d}{dx}[xu_n'] = (\lambda_n^2 - \lambda_m^2) x u_m u_n$$

and

$$\int_a^b u_n \frac{d}{dx}[xu_m']\,dx - \int_a^b u_m \frac{d}{dx}[xu_n'] = (\lambda_n^2 - \lambda_m^2)\int_a^b x u_m u_n\,dx.$$

Using integration by parts this becomes

$$u_n x u_m' \Big|_a^b - \int_a^b x u_m' u_n'\,dx - u_m x u_n' \Big|_a^b + \int_a^b x u_n' u_m'\,dx$$

$$= b[u_n(b)u_m'(b) - u_m(b)u_n'(b)] - a[u_n(a)u_m'(a) - u_m(a)u_n'(a)]$$

$$= (\lambda_n^2 - \lambda_m^2)\int_a^b x u_m u_n\,dx.$$

Since

$$u(x) = Y_0(\lambda a)J_0(\lambda x) - J_0(\lambda a)Y_0(\lambda x)$$

we have

$$u_n(b) = Y_0(\lambda_n a)J_0(\lambda_n b) - J_0(\lambda_n a)Y(\lambda_n b) = 0$$

by the definition of the λ_n. Similarly $u_m(b) = 0$. Also

$$u_n(a) = Y_0(\lambda a)J_0(\lambda_n a) - J_0(\lambda_n a)Y_0(\lambda_n a) = 0$$

and $u_m(a) = 0$. Therefore

$$\int_a^b x u_m u_n\,dx = \frac{1}{\lambda_n^2 - \lambda_m^2}\left(b[u_n(b)u_m'(b) - u_m(b)u_n'(b)] - a[u_n(a)u_m'(a) - u_m(a)u_n'(a)]\right) = 0$$

and the $u_n(x)$ are orthogonal with respect to the weight function x.

14. Letting $u(r,t) = R(r)T(t)$ and separating variables we obtain

$$rR'' + R' + \lambda^2 rR = 0$$

$$T' + \lambda^2 T = 0,$$

with solutions

$$R(r) = c_1 J_0(\lambda r) + c_2 Y_0(\lambda r)$$

$$T(t) = c_3 e^{-\lambda^2 t}.$$

Now the boundary conditions imply

$$R(a) = 0 = c_1 J_0(\lambda a) + c_2 Y_0(\lambda a)$$

$$R(b) = 0 = c_1 J_0(\lambda b) + c_2 Y_0(\lambda b)$$

so that

$$c_2 = -\frac{c_1 J_0(\lambda a)}{Y_0(\lambda a)}$$

and

$$c_1 J_0(\lambda b) - \frac{c_1 J_0(\lambda a)}{Y(\lambda a)} Y_0(\lambda b) = 0$$

or

$$Y_0(\lambda a) J_0(\lambda b) - J_0(\lambda a) Y_0(\lambda b) = 0.$$

This equation defines λ_n for $n = 1, 2, 3, \ldots$. Now

$$R(r) = c_1 J_0(\lambda r) - c_1 \frac{J_0(\lambda a)}{Y_0(\lambda a)} Y_0(\lambda r) = \frac{c_1}{Y_0(\lambda a)} \Big[Y_0(\lambda a) J_0(\lambda r) - J_0(\lambda a) Y_0(\lambda r) \Big]$$

and

$$u_n(r, t) = A_n \Big[Y_0(\lambda_n a) J_0(\lambda_n r) - J_0(\lambda_n a) Y_0(\lambda_n r) \Big] e^{-\lambda_n^2 t} = A_n u_n(r) e^{-\lambda_n^2 t}.$$

Thus

$$u(r, t) = \sum_{n=1}^{\infty} A_n u_n(r) e^{-\lambda_n^2 t}.$$

From the initial condition

$$u(r, 0) = f(r) = \sum_{n=1}^{\infty} A_n u_n(r)$$

we obtain

$$A_n = \frac{\int_a^b r f(r) u_n(r)\, dr}{\int_a^b r u_n^2(r)\, dr}.$$

14 Integral Transform Method

___ **Exercises 14.1** _____

1. **(a)** The result follows by letting $\tau = u^2$ or $u = \sqrt{\tau}$ in $\operatorname{erf}(\sqrt{t}) = \dfrac{2}{\sqrt{\pi}} \displaystyle\int_0^{\sqrt{t}} e^{-u^2}\, du$.

   (b) Using $\mathscr{L}\{t^{-1/2}\} = \dfrac{\sqrt{\pi}}{s^{1/2}}$ and the first translation theorem, it follows from the convolution theorem that

   $$\mathscr{L}\{\operatorname{erf}(\sqrt{t})\} = \frac{1}{\sqrt{\pi}}\,\mathscr{L}\left\{\int_0^t \frac{e^{-\tau}}{\sqrt{\tau}}\, d\tau\right\} = \frac{1}{\sqrt{\pi}}\,\mathscr{L}\{1\}\,\mathscr{L}\{t^{-1/2}e^{-t}\} = \frac{1}{\sqrt{\pi}}\,\frac{1}{s}\,\mathscr{L}\{t^{-1/2}\}\Big|_{s\to s+1}$$

   $$= \frac{1}{\sqrt{\pi}}\,\frac{1}{s}\,\frac{\sqrt{\pi}}{\sqrt{s+1}} = \frac{1}{s\sqrt{s+1}}.$$

2. Since $\operatorname{erfc}(\sqrt{t}) = 1 - \operatorname{erf}(\sqrt{t})$ we have

   $$\mathscr{L}\{\operatorname{erfc}(\sqrt{t})\} = \mathscr{L}\{1\} - \mathscr{L}\{\operatorname{erf}(\sqrt{t})\} = \frac{1}{s} - \frac{1}{s\sqrt{s+1}} = \frac{1}{s}\left[1 - \frac{1}{\sqrt{s+1}}\right].$$

3. By the first translation theorem,

   $$\mathscr{L}\{e^t\operatorname{erf}(\sqrt{t})\} = \mathscr{L}\{\operatorname{erf}(\sqrt{t})\}\Big|_{s\to s-1} = \frac{1}{s\sqrt{s+1}}\Big|_{s\to s-1} = \frac{1}{\sqrt{s}\,(s-1)}.$$

4. By the first translation theorem and the result of Problem 2,

   $$\mathscr{L}\{e^t\operatorname{erfc}(\sqrt{t})\} = \mathscr{L}\{\operatorname{erfc}(\sqrt{t})\}\Big|_{s\to s-1} = \left(\frac{1}{s} - \frac{1}{s\sqrt{s+1}}\right)\Big|_{s\to s-1} = \frac{1}{s-1} - \frac{1}{\sqrt{s}\,(s-1)}$$

   $$= \frac{\sqrt{s}-1}{\sqrt{s}\,(s-1)} = \frac{\sqrt{s}-1}{\sqrt{s}\,(\sqrt{s}+1)(\sqrt{s}-1)} = \frac{1}{\sqrt{s}\,(\sqrt{s}+1)}.$$

5. From table entry 3 and the first translation theorem we have

   $$\mathscr{L}\left\{e^{-Gt/C}\operatorname{erf}\left(\frac{x}{2}\sqrt{\frac{RC}{t}}\right)\right\} = \mathscr{L}\left\{e^{-Gt/C}\left[1 - \operatorname{erfc}\left(\frac{x}{2}\sqrt{\frac{RC}{t}}\right)\right]\right\}$$

   $$= \mathscr{L}\left\{e^{-Gt/C}\right\} - \mathscr{L}\left\{e^{-Gt/C}\operatorname{erfc}\left(\frac{x}{2}\sqrt{\frac{RC}{t}}\right)\right\}$$

   $$= \frac{1}{s+G/C} - \frac{e^{-x\sqrt{RC}\sqrt{s}}}{s}\Big|_{s\to s+G/C}.$$

581

$$= \frac{1}{s + G/C} - \frac{e^{-x\sqrt{RC}\sqrt{s+G/C}}}{s + G/C} = \frac{C}{Cs + G}\left(1 - e^{x\sqrt{RCs+RG}}\right).$$

6. We first compute

$$\frac{\sinh a\sqrt{s}}{s \sinh \sqrt{s}} = \frac{e^{a\sqrt{s}} - e^{-a\sqrt{s}}}{s(e^{\sqrt{s}} - e^{-\sqrt{s}})} = \frac{e^{(a-1)\sqrt{s}} - e^{-(a+1)\sqrt{s}}}{s(1 - e^{-2\sqrt{s}})}$$

$$= \frac{e^{(a-1)\sqrt{s}}}{s}\left[1 + e^{-2\sqrt{s}} + e^{-4\sqrt{s}} + \cdots\right] - \frac{e^{-(a+1)\sqrt{s}}}{s}\left[1 + e^{-2\sqrt{s}} + e^{-4\sqrt{s}} + \cdots\right]$$

$$= \left[\frac{e^{-(1-a)\sqrt{s}}}{s} + \frac{e^{-(3-a)\sqrt{s}}}{s} + \frac{e^{-(5-a)\sqrt{s}}}{s} + \cdots\right]$$

$$\quad - \left[\frac{e^{-(1+a)\sqrt{s}}}{s} + \frac{e^{-(3+a)\sqrt{s}}}{s} + \frac{e^{-(5+a)\sqrt{s}}}{s} + \cdots\right]$$

$$= \sum_{n=0}^{\infty}\left[\frac{e^{-(2n+1-a)\sqrt{s}}}{s} - \frac{e^{-(2n+1+a)\sqrt{s}}}{s}\right].$$

Then

$$\mathscr{L}\left\{\frac{\sinh a\sqrt{s}}{s \sinh \sqrt{s}}\right\} = \sum_{n=0}^{\infty}\left[\mathscr{L}\left\{\frac{e^{-(2n+1-a)\sqrt{s}}}{s}\right\} - \mathscr{L}\left\{-\frac{e^{-(2n+1+a)\sqrt{s}}}{s}\right\}\right]$$

$$= \sum_{n=0}^{\infty}\left[\text{erfc}\left(\frac{2n+1-a}{2\sqrt{t}}\right) - \text{erfc}\left(\frac{2n+1+a}{2\sqrt{t}}\right)\right]$$

$$= \sum_{n=0}^{\infty}\left(\left[1 - \text{erf}\left(\frac{2n+1-a}{2\sqrt{t}}\right)\right] - \left[1 - \text{erf}\left(\frac{2n+1+a}{2\sqrt{t}}\right)\right]\right)$$

$$= \sum_{n=0}^{\infty}\left[\text{erf}\left(\frac{2n+1+a}{2\sqrt{t}}\right) - \text{erf}\left(\frac{2n+1-a}{2\sqrt{t}}\right)\right].$$

7. Taking the Laplace transform of both sides of the equation we obtain

$$\mathscr{L}\{y(t)\} = \mathscr{L}\{1\} - \mathscr{L}\left\{\int_0^t \frac{y(\tau)}{\sqrt{t-\tau}}\,d\tau\right\}$$

$$Y(s) = \frac{1}{s} - Y(s)\frac{\sqrt{\pi}}{\sqrt{s}}$$

$$\frac{\sqrt{s} + \sqrt{\pi}}{\sqrt{s}}Y(s) = \frac{1}{s}$$

$$Y(s) = \frac{1}{\sqrt{s}\,(\sqrt{s} + \sqrt{\pi})}.$$

Thus

$$y(t) = \mathcal{L}^{-1} \left\{ \frac{1}{\sqrt{s}\,(\sqrt{s} + \sqrt{\pi}\,)} \right\} = e^{\pi t} \operatorname{erfc}(\sqrt{\pi t}\,). \qquad \boxed{\text{By entry 5 in the table}}$$

8. Using entries 3 and 5 in the table, we have

$$\mathcal{L} \left\{ -e^{ab} e^{b^2 t} \operatorname{erfc} \left(b\sqrt{t} + \frac{a}{2\sqrt{t}} \right) + \operatorname{erfc} \left(\frac{a}{2\sqrt{t}} \right) \right\}$$

$$= -\mathcal{L} \left\{ e^{ab} e^{b^2 t} \operatorname{erfc} \left(b\sqrt{t} + \frac{a}{2\sqrt{t}} \right) \right\} + \mathcal{L} \left\{ \frac{a}{2\sqrt{t}} \right\}$$

$$= -\frac{e^{-a\sqrt{s}}}{\sqrt{s}\,(\sqrt{s} + b)} + \frac{e^{-a\sqrt{s}}}{s}$$

$$= e^{-a\sqrt{s}} \left[\frac{1}{s} - \frac{1}{\sqrt{s}\,(\sqrt{s} + b)} \right]$$

$$= e^{-a\sqrt{s}} \left[\frac{1}{s} - \frac{\sqrt{s}}{s\,(\sqrt{s} + b)} \right]$$

$$= e^{-a\sqrt{s}} \left[\frac{\sqrt{s} + b - \sqrt{s}}{s\,(\sqrt{s} + b)} \right]$$

$$= \frac{b e^{-a\sqrt{s}}}{s\,(\sqrt{s} + b)}.$$

9. $$\int_a^b e^{-u^2}\, du = \int_a^0 e^{-u^2}\, du + \int_0^b e^{-u^2}\, du = \int_0^b e^{-u^2}\, du - \int_0^a e^{-u^2}\, du$$

$$= \frac{\sqrt{\pi}}{2} \operatorname{erf}(b) - \frac{\sqrt{\pi}}{2} \operatorname{erf}(a) = \frac{\sqrt{\pi}}{2} [\operatorname{erf}(b) - \operatorname{erf}(a)]$$

10. Since $f(x) = e^{-x^2}$ is an even function,

$$\int_{-a}^a e^{-u^2}\, du = 2 \int_0^a e^{-u^2}\, du.$$

Therefore,

$$\int_{-a}^a e^{-u^2}\, du = \sqrt{\pi} \operatorname{erf}(a).$$

Exercises 14.2

1. The boundary-value problem is

$$a^2 \frac{\partial^2 u}{\partial x^2} = \frac{\partial^2 u}{\partial t^2}, \quad 0 < x < L, \quad t > 0,$$

$$u(0, t) = 0, \quad u(L, t) = 0, \quad t > 0,$$

$$u(x, 0) = A \sin \frac{\pi}{L} x, \quad \frac{\partial u}{\partial t}\bigg|_{t=0} = 0.$$

Transforming the partial differential equation gives

$$\frac{d^2 U}{dx^2} - \left(\frac{s}{a}\right)^2 U = -\frac{s}{a^2} A \sin \frac{\pi}{L} x.$$

Using undetermined coefficients we obtain

$$U(x, s) = c_1 \cosh \frac{s}{a} x + c_2 \sinh \frac{s}{a} x + \frac{As}{s^2 + a^2 \pi^2 / L^2} \sin \frac{\pi}{L} x.$$

The transformed boundary conditions, $U(0, s) = 0$, $U(L, s) = 0$ give in turn $c_1 = 0$ and $c_2 = 0$. Therefore

$$U(x, s) = \frac{As}{s^2 + a^2 \pi^2 / L^2} \sin \frac{\pi}{L} x$$

and

$$u(x, t) = A \mathscr{L}^{-1} \left\{ \frac{s}{s^2 + a^2 \pi^2 / L^2} \right\} \sin \frac{\pi}{L} x = A \cos \frac{a\pi}{L} t \sin \frac{\pi}{L} x.$$

2. The transformed equation is

$$\frac{d^2 U}{dx^2} - s^2 U = -2 \sin \pi x - 4 \sin 3\pi x$$

and so

$$U(x, s) = c_1 \cosh sx + c_2 \sinh sx + \frac{2}{s^2 + \pi^2} \sin \pi x + \frac{4}{s^2 + 9\pi^2} \sin 3\pi x.$$

The transformed boundary conditions, $U(0, s) = 0$ and $U(1, s) = 0$ give $c_1 = 0$ and $c_2 = 0$. Thus

$$U(x, s) = \frac{2}{s^2 + \pi^2} \sin \pi x + \frac{4}{s^2 + 9\pi^2} \sin 3\pi x$$

and

$$u(x, t) = 2 \mathscr{L}^{-1} \left\{ \frac{1}{s^2 + \pi^2} \right\} \sin \pi x + 4 \mathscr{L}^{-1} \left\{ \frac{1}{s^2 + 9\pi^2} \right\} \sin 3\pi x$$

$$= \frac{2}{\pi} \sin \pi t \sin \pi x + \frac{4}{3\pi} \sin 3\pi t \sin 3\pi x.$$

3. The solution of

$$a^2 \frac{d^2 U}{dx^2} - s^2 U = 0$$

is in this case

$$U(x, s) = c_1 e^{-(x/a)s} + c_2 e^{(x/a)s}.$$

Since $\lim_{x \to \infty} u(x, t) = 0$ we have $\lim_{x \to \infty} U(x, s) = 0$. Thus $c_2 = 0$ and

$$U(x, s) = c_1 e^{-(x/a)s}.$$

If $\mathcal{L}\{u(0, t)\} = \mathcal{L}\{f(t)\} = F(s)$ then $U(0, s) = F(s)$. From this we have $c_1 = F(s)$ and

$$U(x, s) = F(s) e^{-(x/a)s}.$$

Hence, by the second translation theorem,

$$u(x, t) = f\left(t - \frac{x}{a}\right) \mathcal{U}\left(t - \frac{x}{a}\right).$$

4. Expressing $f(t)$ in the form $(\sin \pi t)[1 - \mathcal{U}(t - 1)]$ and using the result of Problem 3 we find

$$u(x, t) = f\left(t - \frac{x}{a}\right) \mathcal{U}\left(t - \frac{x}{a}\right)$$

$$= \sin \pi \left(t - \frac{x}{a}\right)\left[1 - \mathcal{U}\left(t - \frac{x}{a} - 1\right)\right] \mathcal{U}\left(t - \frac{x}{a}\right)$$

$$= \sin \pi \left(t - \frac{x}{a}\right)\left[\mathcal{U}\left(t - \frac{x}{a}\right) - \mathcal{U}\left(t - \frac{x}{a}\right)\mathcal{U}\left(t - \frac{x}{a} - 1\right)\right]$$

$$= \sin \pi \left(t - \frac{x}{a}\right)\left[\mathcal{U}\left(t - \frac{x}{a}\right) - \mathcal{U}\left(t - \frac{x}{a} - 1\right)\right]$$

Now

$$\mathcal{U}\left(t - \frac{x}{a}\right) - \mathcal{U}\left(t - \frac{x}{a} - 1\right) = \begin{cases} 0, & 0 \le t < x/a \\ 1, & x/a \le t \le x/a + 1 \\ 0, & t > x/a + 1 \end{cases}$$

$$= \begin{cases} 0, & x < a(t - 1) \text{ or } x > at \\ 1, & a(t - 1) \le x \le at \end{cases}$$

so

$$u(x, t) = \begin{cases} 0, & x < a(t - 1) \text{ or } x > at \\ \sin \pi(t - x/a), & a(t - 1) \le x \le at. \end{cases}$$

The graph is shown for $t > 1$.

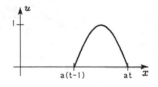

5. We use

$$U(x, s) = c_1 e^{-(x/a)s} - \frac{g}{s^3}.$$

Now

$$\mathcal{L}\{u(0, t)\} = U(0, s) = \frac{A\omega}{s^2 + \omega^2}$$

and so

$$U(0, s) = c_1 - \frac{g}{s^3} = \frac{A\omega}{s^2 + \omega^2} \quad \text{or} \quad c_1 = \frac{g}{s^3} + \frac{A\omega}{s^2 + \omega^2}.$$

Therefore

$$U(x, s) = \frac{A\omega}{s^2 + \omega^2} e^{-(x/a)s} + \frac{g}{s^3} e^{-(x/a)s} - \frac{g}{s^3}$$

and

$$u(x, t) = A\mathscr{L}^{-1}\left\{ \frac{\omega e^{-(x/a)s}}{s^2 + \omega^2} \right\} + g\mathscr{L}^{-1}\left\{ \frac{e^{-(x/a)s}}{s^3} \right\} - g\mathscr{L}^{-1}\left\{ \frac{1}{s^3} \right\}$$

$$= A \sin \omega \left(t - \frac{x}{a} \right) \mathscr{U}\left(t - \frac{x}{a} \right) + \frac{1}{2}g \left(t - \frac{x}{a} \right)^2 \mathscr{U}\left(t - \frac{x}{a} \right) - \frac{1}{2}gt^2.$$

6. Transforming the partial differential equation gives

$$\frac{d^2 U}{dx^2} - s^2 U = -\frac{\omega}{s^2 + \omega^2} \sin \pi x.$$

Using undetermined coefficients we obtain

$$U(x, s) = c_1 \cosh sx + c_2 \sinh sx + \frac{\omega}{(s^2 + \pi^2)(s^2 + \omega^2)} \sin \pi x.$$

The transformed boundary conditions $U(0, s) = 0$ and $U(1, s) = 0$ give, in turn, $c_1 = 0$ and $c_2 = 0$. Therefore

$$U(x, s) = \frac{\omega}{(s^2 + \pi^2)(s^2 + \omega^2)} \sin \pi x$$

and

$$u(x, t) = \omega \sin \pi x \, \mathscr{L}^{-1}\left\{ \frac{1}{(s^2 + \pi^2)(s^2 + \omega^2)} \right\}$$

$$= \frac{\omega}{\omega^2 - \pi^2} \sin \pi x \, \mathscr{L}^{-1}\left\{ \frac{1}{\pi} \frac{\pi}{s^2 + \pi^2} - \frac{1}{\omega} \frac{\omega}{s^2 + \omega^2} \right\}$$

$$= \frac{\omega}{\pi(\omega^2 - \pi^2)} \sin \pi t \sin \pi x - \frac{1}{\omega^2 - \pi^2} \sin \omega t \sin \pi x.$$

7. We use

$$U(x, s) = c_1 \cosh \frac{s}{a} x + c_2 \sinh \frac{s}{a} x.$$

Now $U(0, s) = 0$ implies $c_1 = 0$, so $U(x, s) = c_2 \sinh(s/a)x$. The condition $E \dfrac{dU}{dx} \bigg|_{x=L} = F_0$ then yields $c_2 = F_0 a / Es \cosh(s/a)L$ and so

$$U(x, s) = \frac{aF_0}{Es} \frac{\sinh(s/a)x}{\cosh(s/a)L}$$

$$= \frac{aF_0}{Es} \frac{e^{(s/a)x} - e^{-(s/a)x}}{e^{(s/a)L} + e^{-(s/a)L}}$$

$$= \frac{aF_0}{Es} \frac{e^{(s/a)(x-L)} - e^{-(s/a)(x+L)}}{1 + e^{-2sL/a}}$$

$$= \frac{aF_0}{E} \left[\frac{e^{-(s/a)(L-x)}}{s} - \frac{e^{-(s/a)(3L-x)}}{s} + \frac{e^{-(s/a)(5L-x)}}{s} - \cdots \right]$$

$$\quad - \frac{aF_0}{E} \left[\frac{e^{-(s/a)(L+x)}}{s} - \frac{e^{-(s/a)(3L+x)}}{s} + \frac{e^{-(s/a)(5L+x)}}{s} - \cdots \right]$$

$$= \frac{aF_0}{E} \sum_{n=0}^{\infty} (-1)^n \left[\frac{e^{-(s/a)(2nL+L-x)}}{s} - \frac{e^{-(s/a)(2nL+L+x)}}{s} \right]$$

and

$$u(x, t) = \frac{aF_0}{E} \sum_{n=0}^{\infty} (-1)^n \left[\mathcal{L}^{-1} \left\{ \frac{e^{-(s/a)(2nL+L-x)}}{s} \right\} - \mathcal{L}^{-1} \left\{ \frac{e^{-(s/a)(2nL+L+x)}}{s} \right\} \right]$$

$$= \frac{aF_0}{E} \sum_{n=0}^{\infty} (-1)^n \left[\left(t - \frac{2nL+L-x}{a} \right) \mathcal{U} \left(t - \frac{2nL+L-x}{a} \right) \right.$$

$$\left. - \left(t - \frac{2nL+L+x}{a} \right) \mathcal{U} \left(t - \frac{2nL+L+x}{a} \right) \right].$$

8. We use

$$U(x, s) = c_1 e^{-(x/a)s} + c_2 e^{(x/a)s} - \frac{v_0}{s^2}.$$

Now $\lim_{x \to \infty} \dfrac{dU}{dx} = 0$ implies $c_2 = 0$, and $U(0, s) = 0$ then gives $c_1 = v_0/s^2$. Hence

$$U(x, s) = \frac{v_0}{s^2} e^{-(x/a)s} - \frac{v_0}{s^2}$$

and

$$u(x, t) = v_0 \left(t - \frac{x}{a} \right) \mathcal{U} \left(t - \frac{x}{a} \right) - v_0 t.$$

9. Transforming the partial differential equation gives

$$\frac{d^2 U}{dx^2} - s^2 U = -sxe^{-x}.$$

Using undetermined coefficients we obtain

$$U(x,s) = c_1 e^{-sx} + c_2 e^{sx} - \frac{2s}{(s^2-1)^2} e^{-x} + \frac{s}{s^2-1} xe^{-x}.$$

The transformed boundary conditions $\lim_{x\to\infty} U(x,s) = 0$ and $U(0,s) = 0$ give, in turn, $c_2 = 0$ and $c_1 = 2s/(s^2-1)^2$. Therefore

$$U(x,s) = \frac{2s}{(s^2-1)^2} e^{-sx} - \frac{2s}{(s^2-1)^2} e^{-x} + \frac{s}{s^2-1} xe^{-x}.$$

From entries (13) and (26) in the table we obtain

$$u(x,t) = \mathscr{L}^{-1}\left\{ \frac{2s}{(s^2-1)^2} e^{-sx} - \frac{2s}{(s^2-1)^2} e^{-x} + \frac{s}{s^2-1} xe^{-x} \right\}$$

$$= 2(t-x)\sinh(t-x)\,\mathscr{U}(t-x) - te^{-x}\sinh t + xe^{-x}\cosh t.$$

10. We use

$$U(x,s) = c_1 e^{-xs} + c_2 e^{xs} + \frac{s}{s^2-1} e^{-x}.$$

Now $\lim_{x\to\infty} u(x,t) = 0$ implies $\lim_{x\to\infty} U(x,s) = 0$, so we define $c_2 = 0$. Then

$$U(x,s) = c_1 e^{-xs} + \frac{s}{s^2-1} e^{-x}.$$

Finally, $U(0,s) = 1/s$ gives $c_1 = 1/s - s/(s^2-1)$. Thus

$$U(x,s) = \frac{1}{s} - \frac{s}{s^2-1} e^{-xs} + \frac{s}{s^2-1} e^{-x}$$

and

$$u(x,t) = -\mathscr{L}^{-1}\left\{ \frac{s}{s^2-1} e^{-(x/a)s} \right\} + \mathscr{L}^{-1}\left\{ \frac{s}{s^2-1} \right\} e^{-x}$$

$$= -\cosh\left(t - \frac{x}{a}\right)\mathscr{U}\left(t - \frac{x}{a}\right) + e^{-x}\cosh t.$$

11. We use

$$U(x,s) = c_1 e^{-\sqrt{s/k}\,x} + c_2 e^{\sqrt{s/k}\,x}.$$

Now $\lim_{x\to\infty} u(x,t) = 0$ implies $\lim_{x\to\infty} U(x,s) = 0$, so we define $c_2 = 0$. Then

$$U(x,s) = c_1 e^{-\sqrt{s/k}\,x}.$$

Finally, from $U(0,s) = u_0/s$ we obtain $c_1 = u_0/s$. Thus

$$U(x,s) = u_0 \frac{e^{-\sqrt{s/k}\,x}}{s}$$

and

$$u(x,t) = u_0\,\mathscr{L}^{-1}\left\{ \frac{e^{-\sqrt{s/k}\,x}}{s} \right\} = u_0\,\mathscr{L}^{-1}\left\{ \frac{e^{-(x/\sqrt{k})\sqrt{s}}}{s} \right\} = u_0\,\mathrm{erfc}\left(\frac{x}{2\sqrt{kt}} \right).$$

12. Transforming the partial differential equation and using the initial condition gives

$$k\frac{d^2U}{dx^2} - sU = 0.$$

Since the domain of the variable x is an infinite interval we write the general solution of this differential equation as

$$U(x, s) = c_1 e^{-\sqrt{s/k}\,x} + c_2 e^{-\sqrt{s/k}\,x}.$$

Transforming the boundary conditions gives $U'(0, s) = -A/s$ and $\lim_{x\to\infty} U(x, s) = 0$. Hence we find $c_2 = 0$ and $c_1 = A\sqrt{k}/s\sqrt{s}$. From

$$U(x, s) = A\sqrt{k}\,\frac{e^{-\sqrt{s/k}\,x}}{s\sqrt{s}}$$

we see that

$$u(x, t) = A\sqrt{k}\,\mathscr{L}^{-1}\left\{\frac{e^{-\sqrt{s/k}\,x}}{s\sqrt{s}}\right\}.$$

With the identification $a = x/\sqrt{k}$ it follows from entry 55 of the table in Appendix III that

$$u(x, t) = A\sqrt{k}\left\{2\sqrt{\frac{t}{\pi}}\,e^{-x^2/4kt} - \frac{x}{\sqrt{k}}\,\text{erfc}\left(x/2\sqrt{kt}\right)\right\}$$

$$= 2A\sqrt{\frac{kt}{\pi}}\,e^{-x^2/4kt} - Ax\,\text{erfc}\left(x/2\sqrt{kt}\right).$$

13. We use

$$U(x, s) = c_1 e^{-\sqrt{s}\,x} + c_2 e^{\sqrt{s}\,x} + \frac{u_1}{s}.$$

The condition $\lim_{x\to\infty} u(x, t) = u_1$ implies $\lim_{x\to\infty} U(x, s) = u_1/s$, so we define $c_2 = 0$. Then

$$U(x, s) = c_1 e^{-\sqrt{s}\,x} + \frac{u_1}{s}.$$

From $U(0, s) = u_0/s$ we obtain $c_1 = (u_0 - u_1)/s$. Thus

$$U(x, s) = (u_0 - u_1)\frac{e^{-\sqrt{s}\,x}}{s} + \frac{u_1}{s}$$

and

$$u(x, t) = (u_0 - u_1)\mathscr{L}^{-1}\left\{\frac{e^{-x\sqrt{s}}}{s}\right\} + u_1\mathscr{L}^{-1}\left\{\frac{1}{s}\right\} = (u_0 - u_1)\,\text{erfc}\left(\frac{x}{2\sqrt{t}}\right) + u_1.$$

14. We use

$$U(x, s) = c_1 e^{-\sqrt{s}\,x} + c_2 e^{\sqrt{s}\,x} + \frac{u_1 x}{s}.$$

The condition $\lim_{x \to \infty} u(x, t)/x = u_1$ implies $\lim_{x \to \infty} U(x, s)/x = u_1/s$, so we define $c_2 = 0$. Then

$$U(x, s) = c_1 e^{-\sqrt{s}\,x} + \frac{u_1 x}{s}.$$

From $U(0, s) = u_0/s$ we obtain $c_1 = u_0/s$. Hence

$$U(x, s) = u_0 \frac{e^{-\sqrt{s}\,x}}{s} + \frac{u_1 x}{s}$$

and

$$u(x, t) = u_0 \mathcal{L}^{-1}\left\{ \frac{e^{-x\sqrt{s}}}{s} \right\} + u_1 x\, \mathcal{L}^{-1}\left\{ \frac{1}{s} \right\} = u_0 \operatorname{erfc}\left(\frac{x}{2\sqrt{t}} \right) + u_1 x.$$

15. We use

$$U(x, s) = c_1 e^{-\sqrt{s}\,x} + c_2 e^{\sqrt{s}\,x} + \frac{u_0}{s}.$$

The condition $\lim_{x \to \infty} u(x, t) = u_0$ implies $\lim_{x \to \infty} U(x, s) = u_0/s$, so we define $c_2 = 0$. Then

$$U(x, s) = c_1 e^{-\sqrt{s}\,x} + \frac{u_0}{s}.$$

The transform of the remaining boundary conditions gives

$$\frac{dU}{dx}\bigg|_{x=0} = U(0, s).$$

This condition yields $c_1 = -u_0/s(\sqrt{s} + 1)$. Thus

$$U(x, s) = -u_0 \frac{e^{-\sqrt{s}\,x}}{s(\sqrt{s} + 1)} + \frac{u_0}{s}$$

and

$$u(x, t) = -u_0 \mathcal{L}^{-1}\left\{ \frac{e^{-x\sqrt{s}}}{s(\sqrt{s} + 1)} \right\} + u_0 \mathcal{L}^{-1}\left\{ \frac{1}{s} \right\}$$

$$= u_0 e^{x+t} \operatorname{erfc}\left(\sqrt{t} + \frac{x}{2\sqrt{t}} \right) - u_0 \operatorname{erfc}\left(\frac{x}{2\sqrt{t}} \right) + u_0 \qquad \boxed{\text{By (5) in the table in 14.1.}}$$

16. We use

$$U(x, s) = c_1 e^{-\sqrt{s}\,x} + c_2 e^{\sqrt{s}\,x}.$$

The condition $\lim_{x \to \infty} u(x, t) = 0$ implies $\lim_{x \to \infty} U(x, s) = 0$, so we define $c_2 = 0$. Hence

$$U(x, s) = c_1 e^{-\sqrt{s}\,x}.$$

The remaining boundary condition transforms into

$$\frac{dU}{dx}\bigg|_{x=0} = U(0, s) - \frac{50}{s}.$$

This condition gives $c_1 = 50/s(\sqrt{s}+1)$. Therefore

$$U(x,s) = 50\,\frac{e^{-\sqrt{s}\,x}}{s(\sqrt{s}+1)}$$

and

$$u(x,t) = 50\,\mathscr{L}^{-1}\left\{\frac{e^{-x\sqrt{s}}}{s(\sqrt{s}+1)}\right\} = -50e^{x+t}\operatorname{erfc}\left(\sqrt{t}+\frac{x}{2\sqrt{t}}\right) + 50\operatorname{erfc}\left(\frac{x}{2\sqrt{t}}\right).$$

17. We use

$$U(x,s) = c_1 e^{-\sqrt{s}\,x} + c_2 e^{\sqrt{s}\,x}.$$

The condition $\lim_{x\to\infty} u(x,t) = 0$ implies $\lim_{x\to\infty} U(x,s) = 0$, so we define $c_2 = 0$. Hence

$$U(x,s) = c_1 e^{-\sqrt{s}\,x}.$$

The transform of $u(0,t) = f(t)$ is $U(0,s) = F(s)$. Therefore

$$U(x,s) = F(s)e^{-\sqrt{s}\,x}$$

and

$$u(x,t) = \mathscr{L}^{-1}\left\{F(s)e^{-x\sqrt{s}}\right\} = \frac{x}{2\sqrt{\pi}}\int_0^t \frac{f(t-\tau)e^{-x^2/4\tau}}{\tau^{3/2}}\,d\tau.$$

18. We use

$$U(x,s) = c_1 e^{-\sqrt{s}\,x} + c_2 e^{\sqrt{s}\,x}.$$

The condition $\lim_{x\to\infty} u(x,t) = 0$ implies $\lim_{x\to\infty} U(x,s) = 0$, so we define $c_2 = 0$. Then $U(x,s) = c_1 e^{-\sqrt{s}\,x}$. The transform of the remaining boundary condition gives

$$\left.\frac{dU}{dx}\right|_{x=0} = -F(s)$$

where $F(s) = \mathscr{L}\{f(t)\}$. This condition yields $c_1 = F(s)/\sqrt{s}$. Thus

$$U(x,s) = F(s)\,\frac{e^{-\sqrt{s}\,x}}{\sqrt{s}}.$$

Using entry (44) of the table and the convolution theorem we obtain

$$u(x,t) = \mathscr{L}^{-1}\left\{F(s)\cdot\frac{e^{-\sqrt{s}\,x}}{\sqrt{s}}\right\} = \frac{1}{\sqrt{\pi}}\int_0^t f(\tau)\,\frac{e^{-x^2/4(t-\tau)}}{\sqrt{t-\tau}}\,d\tau.$$

19. Transforming the partial differential equation gives

$$\frac{d^2U}{dx^2} - sU = -60.$$

Using undetermied coefficients we obtain

$$U(x,s) = c_1 e^{-\sqrt{s}\,x} + c_2 e^{\sqrt{s}\,x} + \frac{60}{s}.$$

The condition $\lim_{x\to\infty} u(x, t) = 60$ implies $\lim_{x\to\infty} U(x, s) = 60/s$, so we define $c_2 = 0$. The transform of the remaining boundary condition gives

$$U(0, s) = \frac{60}{s} + \frac{40}{s} e^{-2s}.$$

This condition yields $c_1 = \frac{40}{s} e^{-2s}$. Thus

$$U(x, s) = \frac{60}{s} + 40e^{-2s} \frac{e^{-\sqrt{s}\,x}}{s}.$$

Using entry (54) of the table in Appendix III and the second translation theorem we obtain

$$u(x, t) = \mathscr{L}^{-1}\left\{\frac{60}{s} + 40e^{-2s} \frac{e^{-\sqrt{s}\,x}}{s}\right\} = 60 + 40\,\text{erfc}\left(\frac{x}{2\sqrt{t-2}}\right)\mathscr{U}(t-2).$$

20. The solution of the transformed equation

$$\frac{d^2U}{dx^2} - sU = -100$$

by undetermined coefficients is

$$U(x, s) = c_1 e^{\sqrt{s}\,x} + c_2 e^{-\sqrt{s}\,x} + \frac{100}{s}.$$

From the fact that $\lim_{x\to\infty} U(x, s) = 100/s$ we see that $c_1 = 0$. Thus

$$U(x, s) = c_2 e^{-\sqrt{s}\,x} + \frac{100}{s}. \tag{1}$$

Now the transform of the boundary condition at $x = 0$ is

$$U(0, s) = 20\left[\frac{1}{s} - \frac{1}{s}e^{-s}\right].$$

It follows from (1) that

$$\frac{20}{s} - \frac{20}{s}e^{-s} = c_2 + \frac{100}{s} \quad \text{or} \quad c_2 = -\frac{80}{s} - \frac{20}{s}e^{-s}$$

and so

$$U(x, s) = \left(-\frac{80}{s} - \frac{20}{s}e^{-s}\right)e^{-\sqrt{s}\,x} + \frac{100}{s}$$

$$= \frac{100}{s} - \frac{80}{s}e^{-\sqrt{s}\,x} - \frac{20}{s}e^{-\sqrt{s}\,x}e^{-s}.$$

Thus

$$u(x, t) = 100\,\mathscr{L}^{-1}\left\{\frac{1}{s}\right\} - 80\,\mathscr{L}^{-1}\left\{\frac{e^{-\sqrt{s}\,x}}{s}\right\} - 20\,\mathscr{L}^{-1}\left\{\frac{e^{-\sqrt{s}\,x}}{s}e^{-s}\right\}$$

$$= 100 - 80\,\text{erfc}\left(x/2\sqrt{t}\right) - 20\,\text{erfc}\left(x/2\sqrt{t-1}\right)\mathscr{U}(t-1).$$

21. Transforming the partial differential equation gives

$$\frac{d^2U}{dx^2} - sU = 0$$

and so

$$U(x, s) = c_1 e^{-\sqrt{s}\,x} + c_2 e^{\sqrt{s}\,x}.$$

The condition $\lim_{x \to -\infty} u(x, t) = 0$ implies $\lim_{x \to -\infty} U(x, s) = 0$, so we define $c_1 = 0$. The transform of the remaining boundary condition gives

$$\frac{dU}{dx}\bigg|_{x=1} = \frac{100}{s} - U(1, s).$$

This condition yields

$$c_2 \sqrt{s}\, e^{\sqrt{s}} = \frac{100}{s} - c_2 e^{\sqrt{s}}$$

from which it follows that

$$c_2 = \frac{100}{s(\sqrt{s}+1)}\, e^{-\sqrt{s}}.$$

Thus

$$U(x, s) = 100\, \frac{e^{-(1-x)\sqrt{s}}}{s(\sqrt{s}+1)}.$$

Using entry (57) of the table in Appendix III we obtain

$$u(x, t) = 100\, \mathcal{L}^{-1}\left\{\frac{e^{-(1-x)\sqrt{s}}}{s(\sqrt{s}+1)}\right\} = 100\left[-e^{1-x+t}\operatorname{erfc}\left(\sqrt{t}+\frac{1-x}{\sqrt{t}}\right) + \operatorname{erfc}\left(\frac{1-x}{2\sqrt{t}}\right)\right].$$

22. Transforming the partial differential equation gives

$$k\frac{d^2U}{dx^2} - sU = -\frac{r}{s}.$$

Using undetermined coefficients we obtian

$$U(x, s) = c_1 e^{-\sqrt{s/k}\,x} + c_2 e^{\sqrt{s/k}\,x} + \frac{r}{s^2}.$$

The condition $\lim_{x \to \infty}\dfrac{\partial u}{\partial x} = 0$ implies $\lim_{x \to \infty}\dfrac{dU}{dx} = 0$, so we define $c_2 = 0$. The transform of the remaining boundary condition gives $U(0, s) = 0$. This condition yields $c_1 = -r/s^2$. Thus

$$U(x, s) = r\left[\frac{1}{s^2} - \frac{e^{-\sqrt{s/k}\,x}}{s^2}\right].$$

Using entries (3) and (54) of the table in Appendix III and the convolution theorem we obtain

$$u(x, t) = r\, \mathcal{L}^{-1}\left\{\frac{1}{s^2} - \frac{1}{s}\cdot\frac{e^{-\sqrt{s/k}\,x}}{s}\right\} = rt - r\int_0^t \operatorname{erfc}\left(\frac{x}{2\sqrt{k\tau}}\right) d\tau.$$

23. The solution of

$$\frac{d^2U}{dx^2} - sU = -u_0 - u_0 \sin\frac{\pi}{L}x$$

is

$$U(x,s) = c_1 \cosh(\sqrt{s}\,x) + c_2 \sinh(\sqrt{s}\,x) + \frac{u_0}{s} + \frac{u_0}{s+\pi^2/L^2}\sin\frac{\pi}{L}x.$$

The transformed boundary conditions $U(0,s) = u_0/s$ and $U(L,s) = u_0/s$ give, in turn, $c_1 = 0$ and $c_2 = 0$. Therefore

$$U(x,s) = \frac{u_0}{s} + \frac{u_0}{s+\pi^2/L^2}\sin\frac{\pi}{L}x$$

and

$$u(x,t) = u_0\,\mathscr{L}^{-1}\left\{\frac{1}{s}\right\} + u_0\,\mathscr{L}^{-1}\left\{\frac{1}{s+\pi^2/L^2}\right\}\sin\frac{\pi}{L}x = u_0 + u_0 e^{-\pi^2 t/L^2}\sin\frac{\pi}{L}x.$$

24. The transform of the partial differential equation is

$$k\frac{d^2U}{dx^2} - hU + h\frac{u_m}{s} = sU - u_0$$

or

$$k\frac{d^2U}{dx^2} - (h+s)U = -h\frac{u_m}{s} - u_0.$$

By undetermined coefficients we find

$$U(x,s) = c_1 e^{\sqrt{(h+s)/k}\,x} + c_2 e^{-\sqrt{(h+s)/k}\,x} + \frac{hu_m + u_0 s}{s(s+h)}.$$

The transformed boundary conditions are $U'(0,s) = 0$ and $U'(L,s) = 0$. These conditions imply $c_1 = 0$ and $c_2 = 0$. By partial fractions we then get

$$U(x,s) = \frac{hu_m + u_0 s}{s(s+h)} = \frac{u_m}{s} - \frac{u_m}{s+h} + \frac{u_0}{s+h}.$$

Therefore,

$$u(x,t) = u_m\,\mathscr{L}^{-1}\left\{\frac{1}{s}\right\} - u_m\,\mathscr{L}^{-1}\left\{\frac{1}{s+h}\right\} + u_0\,\mathscr{L}^{-1}\left\{\frac{1}{s+h}\right\} = u_m - u_m e^{-ht} + u_0 e^{-ht}.$$

25. We use

$$U(x,s) = c_1 \cosh\sqrt{\frac{s}{k}}\,x + c_2 \sinh\sqrt{\frac{s}{k}}\,x + \frac{u_0}{s}.$$

The transformed boundary conditions $\left.\dfrac{dU}{dx}\right|_{x=0} = 0$ and $U(1,s) = 0$ give, in turn, $c_2 = 0$ and

594

$c_1 = -u_0/s \cosh \sqrt{s/k}$. Therefore

$$U(x,s) = \frac{u_0}{s} - \frac{u_0 \cosh \sqrt{s/k}\, x}{s \cosh \sqrt{s/k}} = \frac{u_0}{s} - u_0 \frac{e^{\sqrt{s/k}\, x} + e^{-\sqrt{s/k}\, x}}{s(e^{\sqrt{s/k}} + e^{-\sqrt{s/k}})}$$

$$= \frac{u_0}{s} - u_0 \frac{e^{\sqrt{s/k}\,(x-1)} + e^{-\sqrt{s/k}\,(x+1)}}{s(1 + e^{-2\sqrt{s/k}})}$$

$$= \frac{u_0}{s} - u_0 \left[\frac{e^{-\sqrt{s/k}\,(1-x)}}{s} - \frac{e^{-\sqrt{s/k}\,(3-x)}}{s} + \frac{e^{-\sqrt{s/k}\,(5-x)}}{s} - \cdots \right]$$

$$\quad - u_0 \left[\frac{e^{-\sqrt{s/k}\,(1+x)}}{s} - \frac{e^{-\sqrt{s/k}\,(3+x)}}{s} + \frac{e^{-\sqrt{s/k}\,(5+x)}}{s} - \cdots \right]$$

$$= \frac{u_0}{s} - u_0 \sum_{n=0}^{\infty} (-1)^n \left[\frac{e^{-(2n+1-x)\sqrt{s}/\sqrt{k}}}{s} + \frac{e^{-(2n+1+x)\sqrt{s}/\sqrt{k}}}{s} \right]$$

and

$$u(x,t) = u_0 \mathscr{L}\left\{ \frac{1}{s} \right\} - u_0 \sum_{n=0}^{\infty} (-1)^n \left[\mathscr{L}^{-1}\left\{ \frac{e^{-(2n+1-x)\sqrt{s}/\sqrt{k}}}{s} \right\} - \mathscr{L}^{-1}\left\{ \frac{e^{-(2n+1+x)\sqrt{s}/\sqrt{k}}}{s} \right\} \right]$$

$$= u_0 - u_0 \sum_{n=0}^{\infty} (-1)^n \left[\operatorname{erfc}\left(\frac{2n+1-x}{2\sqrt{kt}} \right) - \operatorname{erfc}\left(\frac{2n+1+x}{2\sqrt{kt}} \right) \right].$$

26. We use

$$c(x,s) = c_1 \cosh \sqrt{\frac{s}{D}}\, x + c_2 \sinh \sqrt{\frac{s}{D}}\, x.$$

The transform of the two boundary conditions are $c(0,s) = c_0/s$ and $c(1,s) = c_0/s$. From these conditions we obtain $c_1 = c_0/s$ and

$$c_2 = c_0(1 - \cosh \sqrt{s/D})/s \sinh \sqrt{s/D}.$$

Therefore

$$c(x,s) = c_0 \left[\frac{\cosh \sqrt{s/D}\, x}{s} + \frac{(1 - \cosh \sqrt{s/D})}{s \sinh \sqrt{s/D}} \sinh \sqrt{s/D}\, x \right]$$

$$= c_0 \left[\frac{\sinh \sqrt{s/D}\,(1-x)}{s \sinh \sqrt{s/D}} + \frac{\sin \sqrt{s/D}\, x}{s \sinh \sqrt{s/D}} \right]$$

$$= c_0 \left[\frac{e^{\sqrt{s/D}\,(1-x)} - e^{-\sqrt{s/D}\,(1-x)}}{s(e^{\sqrt{s/D}} - e^{-\sqrt{s/D}})} + \frac{e^{\sqrt{s/D}\,x} - e^{-\sqrt{s/D}\,x}}{s(e^{\sqrt{s/D}} - e^{-\sqrt{s/D}})} \right]$$

$$= c_0 \left[\frac{e^{-\sqrt{s/D}\,x} - e^{-\sqrt{s/D}\,(2-x)}}{s(1 - e^{-2\sqrt{s/D}})} + \frac{e^{\sqrt{s/D}\,(x-1)} - e^{-\sqrt{s/D}\,(x+1)}}{s(1 - e^{-2\sqrt{s/D}})} \right]$$

$$= c_0 \frac{(e^{-\sqrt{s/D}\,x} - e^{-\sqrt{s/D}\,(2-x)})}{s} \left(1 + e^{-2\sqrt{s/D}} + e^{-4\sqrt{s/D}} + \cdots \right)$$

$$+ c_0 \frac{(e^{\sqrt{s/D}\,(x-1)} - e^{-\sqrt{s/D}\,(x+1)})}{s} \left(1 + e^{-2\sqrt{s/D}} + e^{-4\sqrt{s/D}} + \cdots \right)$$

$$= c_0 \sum_{n=0}^{\infty} \left[\frac{e^{-(2n+x)\sqrt{s/D}}}{s} - \frac{e^{-(2n+2-x)\sqrt{s/D}}}{s} \right]$$

$$+ c_0 \sum_{n=0}^{\infty} \left[\frac{e^{-(2n+1-x)\sqrt{s/D}}}{s} - \frac{e^{-(2n+1+x)\sqrt{s/D}}}{s} \right]$$

and

$$c(x,t) = c_0 \sum_{n=0}^{\infty} \left[\mathcal{L}^{-1} \left\{ \frac{e^{-\frac{(2n+x)}{\sqrt{D}}\sqrt{s}}}{s} \right\} - \mathcal{L}^{-1} \left\{ \frac{e^{-\frac{(2n+2-x)}{\sqrt{D}}\sqrt{s}}}{s} \right\} \right]$$

$$+ c_0 \sum_{n=0}^{\infty} \left[\mathcal{L}^{-1} \left\{ \frac{e^{-\frac{(2n+1-x)}{\sqrt{D}}\sqrt{s}}}{s} \right\} - \mathcal{L}^{-1} \left\{ \frac{e^{-\frac{(2n+1+x)}{\sqrt{D}}\sqrt{s}}}{s} \right\} \right]$$

$$= c_0 \sum_{n=0}^{\infty} \left[\mathrm{erfc}\left(\frac{2n+x}{2\sqrt{Dt}} \right) - \mathrm{erfc}\left(\frac{2n+2-x}{2\sqrt{Dt}} \right) \right]$$

$$+ c_0 \sum_{n=0}^{\infty} \left[\mathrm{erfc}\left(\frac{2n+1-x}{2\sqrt{Dt}} \right) - \mathrm{erfc}\left(\frac{2n+1+x}{2\sqrt{Dt}} \right) \right].$$

Now using $\mathrm{erfc}(x) = 1 - \mathrm{erf}(x)$ we get

$$c(x,t) = c_0 \sum_{n=0}^{\infty} \left[\mathrm{erf}\left(\frac{2n+2-x}{2\sqrt{Dt}} \right) - \mathrm{erf}\left(\frac{2n+x}{2\sqrt{Dt}} \right) \right]$$

$$+ c_0 \sum_{n=0}^{\infty} \left[\mathrm{erf}\left(\frac{2n+1+x}{2\sqrt{Dt}} \right) - \mathrm{erf}\left(\frac{2n+1-x}{2\sqrt{Dt}} \right) \right].$$

27. We use

$$U(x,s) = c_1 e^{-\sqrt{RCs+RG}\,x} + c_2 e^{\sqrt{RCs+RG}} + \frac{Cu_0}{Cs+G}.$$

The condition $\lim_{x \to \infty} \partial u / \partial x = 0$ implies $\lim_{x \to \infty} dU/dx = 0$, so we define $c_2 = 0$. Applying $U(0, s) = 0$ to

$$U(x, s) = c_1 e^{-\sqrt{RCsRG}\, x} + \frac{Cu_0}{Cs + G}$$

gives $c_1 = -Cu_0/(Cs + G)$. Therefore

$$U(x, s) = -Cu_0 \frac{e^{-\sqrt{RCs+RG}\, x}}{Cs + G} + \frac{Cu_0}{Cs + G}$$

and

$$u(x, t) = u_0 \mathscr{L}^{-1} \left\{ \frac{1}{s + G/C} \right\} - u_0 \mathscr{L}^{-1} \left\{ \frac{e^{-x\sqrt{RC}\sqrt{s+G/C}}}{s + G/C} \right\}$$

$$= u_0 e^{-Gt/C} - u_0 e^{-Gt/C} \operatorname{erfc} \left(\frac{x\sqrt{RC}}{2\sqrt{t}} \right)$$

$$= u_0 e^{-Gt/C} \left[1 - \operatorname{erfc} \left(\frac{x}{2} \sqrt{\frac{RC}{t}} \right) \right]$$

$$= u_0 e^{-Gt/C} \operatorname{erf} \left(\frac{x}{2} \sqrt{\frac{RC}{t}} \right).$$

28. We use

$$U(x, s) = c_1 e^{-\sqrt{s+h}\, x} + c_2 e^{\sqrt{s+h}\, x}.$$

The condition $\lim_{x \to \infty} u(x, t) = 0$ implies $\lim_{x \to \infty} U(x, s) = 0$, so we take $c_2 = 0$. Therefore

$$U(x, s) = c_1 e^{-\sqrt{s+h}\, x}.$$

The Laplace transform of $u(0, t) = u_0$ is $U(0, s) = u_0/s$ and so

$$U(x, s) = u_0 \frac{e^{-\sqrt{s+h}\, x}}{s}$$

and

$$u(x, t) = u_0 \mathscr{L}^{-1} \left\{ \frac{e^{-\sqrt{s+h}\, x}}{s} \right\} = u_0 \mathscr{L}^{-1} \left\{ \frac{1}{s} e^{-\sqrt{s+h}\, x} \right\}.$$

From the first translation theorem,

$$\mathscr{L}^{-1} \left\{ e^{-\sqrt{s+h}\, x} \right\} = e^{-ht} \mathscr{L}^{-1} \left\{ e^{-x\sqrt{s}} \right\} = e^{-ht} \frac{x}{2\sqrt{\pi t^3}} e^{-x^2/4t}.$$

Thus, from the convolution theorem we obtain

$$u(x, s) = \frac{u_0 x}{2\sqrt{\pi}} \int_0^t \frac{e^{-h\tau - x^2/4\tau}}{\tau^{3/2}} \, d\tau.$$

29. We use

$$U(x,s) = c_1 e^{-\sqrt{s/k}\,x} + c_2 e^{-\sqrt{s/k}\,x}.$$

The condition $\lim_{x\to\infty} u(x,t) = 0$ implies $\lim_{x\to\infty} U(x,s) = 0$, so we define $c_2 = 0$. Now the transform of the remaining boundary condition is

$$\left.\frac{dU}{dx}\right|_{x=0} = -A.$$

From this we obtain $c_1 = A\sqrt{k/s}$. Thus

$$U(x,s) = A\sqrt{k}\,\frac{e^{-\frac{x}{\sqrt{k}}\sqrt{s}}}{\sqrt{s}}$$

and

$$u(x,t) = A\sqrt{k}\,\mathscr{L}^{-1}\left\{\frac{e^{-\frac{x}{\sqrt{k}}\sqrt{s}}}{\sqrt{s}}\right\}$$

$$= A\sqrt{k}\,\frac{1}{\sqrt{\pi t}}\,e^{-x^2/4kt} \qquad \boxed{\text{By entry 1 of the table in Section 14.1}}$$

$$= A\sqrt{\frac{k}{\pi t}}\,e^{-x^2/4kt}.$$

An impulse of heat, or flash burn, takes place at $x = 0$.

30. (a) We use

$$U(x,s) = c_1 e^{-(s/a)x} + c_2 e^{(s/a)x} + \frac{v_0^2 F_0}{(a^2 - v_0^2)s^2}\,e^{-(s/v_0)x}.$$

The condition $\lim_{x\to\infty} u(x,t) = 0$ implies $\lim_{x\to\infty} U(x,s) = 0$, so we must define $c_2 = 0$. Consequently

$$U(x,s) = c_1 e^{-(s/a)x} + \frac{v_0^2 F_0}{(a^2 - v_0^2)s^2}\,e^{-(s/v_0)x}.$$

The remaining boundary condition transforms into $U(0,s) = 0$. From this we find

$$c_1 = -v_0^2 F_0/(a^2 - v_0^2)s^2.$$

Therefore, by the second translation theorem

$$U(x,s) = -\frac{v_0^2 F_0}{(a^2 - v_0^2)s^2}\,e^{-(s/a)x} + \frac{v_0^2 F_0}{(a^2 - v_0^2)s^2}\,e^{-(s/v_0)x}$$

and

$$u(x,t) = \frac{v_0^2 F_0}{a^2 - v_0^2}\left[\mathscr{L}^{-1}\left\{\frac{e^{-(x/v_0)s}}{s^2}\right\} - \mathscr{L}^{-1}\left\{\frac{e^{-(x/a)s}}{s^2}\right\}\right]$$

$$= \frac{v_0^2 F_0}{a^2 - v_0^2}\left[\left(t - \frac{x}{v_0}\right)\mathscr{U}\left(t - \frac{x}{v_0}\right) - \left(t - \frac{x}{a}\right)\mathscr{U}\left(t - \frac{x}{a}\right)\right].$$

(b) In the case when $v_0 = a$ the solution of the transformed equation is

$$U(x, s) = c_1 e^{-(s/a)x} + c_2 e^{(s/a)x} - \frac{F_0}{2as} x e^{-(s/a)x}.$$

The usual analysis then leads to $c_1 = 0$ and $c_2 = 0$. Therefore

$$U(x, s) = -\frac{F_0}{2as} x e^{-(s/a)x}$$

and

$$u(x, t) = -\frac{x F_0}{2a} \mathcal{L}^{-1} \left\{ \frac{e^{-(x/a)s}}{s} \right\} = -\frac{x F_0}{2a} \mathcal{U} \left(t - \frac{x}{a} \right).$$

———— **Exercises 14.3** ————

1. From formulas (5) and (6) in the text,

$$A(\alpha) = \int_{-1}^{0} (-1) \cos \alpha x \, dx + \int_{0}^{1} (2) \cos \alpha x \, dx = -\frac{\sin \alpha}{\alpha} + 2\frac{\sin \alpha}{\alpha} = \frac{\sin \alpha}{\alpha}$$

and

$$B(\alpha) = \int_{-1}^{0} (-1) \sin \alpha x \, dx + \int_{0}^{1} (2) \sin \alpha x \, dx$$

$$= \frac{1 - \cos \alpha}{\alpha} - 2\frac{\cos \alpha - 1}{\alpha}$$

$$= \frac{3(1 - \cos \alpha)}{\alpha}.$$

Hence

$$f(x) = \frac{1}{\pi} \int_{0}^{\infty} \frac{\sin \alpha \cos \alpha x + 3(1 - \cos \alpha) \sin \alpha x}{\alpha} \, d\alpha.$$

2. From formulas (5) and (6) in the text,

$$A(\alpha) = \int_{\pi}^{2\pi} 4 \cos \alpha x \, dx = 4\frac{\sin 2\pi\alpha - \sin \pi\alpha}{\alpha}$$

and

$$B(\alpha) = \int_{\pi}^{2\pi} 4 \sin \alpha x \, dx = 4\frac{\cos \pi\alpha - \cos 2\pi\alpha}{\alpha}.$$

Hence

$$f(x) = \frac{4}{\pi} \int_{0}^{\infty} \frac{(\sin 2\pi\alpha - \sin \pi\alpha) \cos \alpha x + (\cos \pi\alpha - \cos 2\pi\alpha) \sin \alpha x}{\alpha} \, d\alpha$$

$$= \frac{4}{\pi} \int_{0}^{\infty} \frac{\sin 2\pi\alpha \cos \alpha x - \cos 2\pi\alpha \sin \alpha x - \sin \pi\alpha \cos \alpha x + \cos \pi\alpha \sin \alpha x}{\alpha} \, d\alpha$$

$$= \frac{4}{\pi} \int_{0}^{\infty} \frac{\sin \alpha(2\pi - x) - \sin \alpha(\pi - x)}{\alpha} \, d\alpha.$$

3. From formulas (5) and (6) in the text,

$$A(\alpha) = \int_0^3 x \cos \alpha x \, dx$$

$$= \frac{x \sin \alpha x}{\alpha} \bigg|_0^3 - \frac{1}{\alpha} \int_0^3 \sin \alpha x \, dx$$

$$= \frac{3 \sin 3\alpha}{\alpha} + \frac{\cos \alpha x}{\alpha^2} \bigg|_0^3$$

$$= \frac{3\alpha \sin 3\alpha + \cos 3\alpha - 1}{\alpha^2}$$

and

$$B(\alpha) = \int_0^3 x \sin \alpha x \, dx$$

$$= -\frac{x \cos \alpha x}{\alpha} \bigg|_0^3 + \frac{1}{\alpha} \int_0^3 \cos \alpha x \, dx$$

$$= -\frac{3 \cos 3\alpha}{\alpha} + \frac{\sin \alpha x}{\alpha^2} \bigg|_0^3$$

$$= \frac{\sin 3\alpha - 3\alpha \cos 3\alpha}{\alpha^2}.$$

Hence

$$f(x) = \frac{1}{\pi} \int_0^\infty \frac{(3\alpha \sin 3\alpha + \cos 3\alpha - 1) \cos \alpha x + (\sin 3\alpha - 3\alpha \cos 3\alpha) \sin \alpha x}{\alpha^2} \, d\alpha$$

$$= \frac{1}{\pi} \int_0^\infty \frac{3\alpha(\sin 3\alpha \cos \alpha x - \cos 3\alpha \sin \alpha x) + \cos 3\alpha \cos \alpha x + \sin 3\alpha \sin \alpha x - \cos \alpha x}{\alpha^2} \, d\alpha$$

$$= \frac{1}{\pi} \int_0^\infty \frac{3\alpha \sin \alpha(3 - x) + \cos \alpha(3 - x) - \cos \alpha x}{\alpha^2} \, d\alpha.$$

4. From formulas (5) and (6) in the text,

$$A(\alpha) = \int_{-\infty}^\infty f(x) \cos \alpha x \, dx$$

$$= \int_{-\infty}^0 0 \cdot \cos \alpha x \, dx + \int_0^\pi \sin x \cos \alpha x \, dx + \int_\pi^\infty 0 \cdot \cos \alpha x \, dx$$

$$= \frac{1}{2} \int_0^\pi [\sin(1 + \alpha)x + \sin(1 - \alpha)x] \, dx$$

$$= \frac{1}{2} \left[-\frac{\cos(1 + \alpha)x}{1 + \alpha} - \frac{\cos(1 - \alpha)x}{1 - \alpha} \right]_0^\pi$$

600

$$= -\frac{1}{2}\left[\frac{\cos(1+\alpha)\pi - 1}{1+\alpha} + \frac{\cos(1-\alpha)\pi - 1}{1-\alpha}\right]$$

$$= -\frac{1}{2}\left[\frac{\cos(1+\alpha)\pi - \alpha\cos(1+\alpha)\pi + \cos(1-\alpha)\pi + \alpha\cos(1-\alpha)\pi - 2}{1-\alpha^2}\right]$$

$$= \frac{1+\cos\alpha\pi}{1-\alpha^2},$$

and

$$B(\alpha) = \int_0^\pi \sin x \sin \alpha x \, dx$$

$$= \frac{1}{2}\int_0^\pi [\cos(1-\alpha)x - \cos(1+\alpha)] \, dx$$

$$= \frac{1}{2}\left[\frac{\sin(1-\alpha)\pi}{1-\alpha} - \frac{\sin(1+\alpha)\pi}{1+\alpha}\right]$$

$$= \frac{\sin\alpha\pi}{1-\alpha^2}.$$

Hence

$$f(x) = \frac{1}{\pi}\int_0^\infty \frac{\cos\alpha x + \cos\alpha x \cos\alpha\pi + \sin\alpha x \sin\alpha\pi}{1-\alpha^2} \, d\alpha$$

$$= \frac{1}{\pi}\int_0^\infty \frac{\cos\alpha x + \cos\alpha(x-\pi)}{1-\alpha^2} \, d\alpha.$$

5. From formula (5) in the text,

$$A(\alpha) = \int_0^\infty e^{-x}\cos\alpha x \, dx.$$

Recall $\mathscr{L}\{\cos kt\} = s/(s^2 + k^2)$. If we set $s = 1$ and $k = \alpha$ we obtain

$$A(\alpha) = \frac{1}{1+\alpha^2}.$$

Now

$$B(\alpha) = \int_0^\infty e^{-x}\sin\alpha x \, dx.$$

Recall $\mathscr{L}\{\sin kt\} = k/(s^2 + k^2)$. If we set $s = 1$ and $k = \alpha$ we obtain

$$B(\alpha) = \frac{\alpha}{1+\alpha^2}.$$

Hence

$$f(x) = \frac{1}{\pi}\int_0^\infty \frac{\cos\alpha x + \alpha\sin\alpha x}{1+\alpha^2} \, d\alpha.$$

601

6. From formulas (5) and (6) in the text,

$$A(\alpha) = \int_{-1}^{1} e^x \cos \alpha x \, dx$$

$$= \frac{e(\cos \alpha + \alpha \sin \alpha) - e^{-1}(\cos \alpha - \alpha \sin \alpha)}{1 + \alpha^2}$$

$$= \frac{2(\sinh 1)\cos \alpha - 2\alpha(\cosh 1)\sin \alpha}{1 + \alpha^2}$$

and

$$B(\alpha) = \int_{-1}^{1} e^x \sin \alpha x \, dx$$

$$= \frac{e(\sin \alpha - \alpha \cos \alpha) - e^{-1}(-\sin \alpha - \alpha \cos \alpha)}{1 + \alpha^2}$$

$$= \frac{2(\cosh 1)\sin \alpha - 2\alpha(\sinh 1)\cos \alpha}{1 + \alpha^2} .$$

Hence

$$f(x) = \frac{1}{\pi} \int_{0}^{\infty} [A(\alpha) \cos \alpha x + B(\alpha) \sin \alpha x] \, d\alpha.$$

7. The function is odd. Thus from formula (11) in the text

$$B(\alpha) = 5 \int_{0}^{1} \sin \alpha x \, dx = \frac{5(1 - \cos \alpha)}{\alpha} .$$

Hence from formula (10) in the text,

$$f(x) = \frac{10}{\pi} \int_{0}^{\infty} \frac{(1 - \cos \alpha) \sin \alpha x}{\alpha} \, d\alpha.$$

8. The function is even. Thus from formula (9) in the text

$$A(\alpha) = \pi \int_{1}^{2} \cos \alpha x \, dx = \pi \left(\frac{\sin 2\alpha - \sin \alpha}{\alpha} \right).$$

Hence from formula (8) in the text,

$$f(x) = 2 \int_{0}^{\infty} \frac{(\sin 2\alpha - \sin \alpha) \cos \alpha x}{\alpha} \, d\alpha.$$

9. The function is even. Thus from formula (9) in the text

$$A(\alpha) = \int_{0}^{\pi} x \cos \alpha x \, dx = \frac{x \sin \alpha x}{\alpha} \bigg|_{0}^{\pi} - \frac{1}{\alpha} \int_{0}^{\pi} \sin \alpha x \, dx$$

$$= \frac{\pi \alpha \sin \pi \alpha}{\alpha} + \frac{1}{\alpha^2} \cos \alpha x \bigg|_{0}^{\pi} = \frac{\pi \alpha \sin \pi \alpha + \cos \pi \alpha - 1}{\alpha^2} .$$

Hence from formula (8) in the text

$$f(x) = \frac{2}{\pi} \int_0^\infty \frac{(\pi\alpha \sin \pi\alpha + \cos \pi\alpha - 1) \cos \alpha x}{\alpha^2} \, d\alpha.$$

10. The function is odd. Thus from formula (11) in the text

$$B(\alpha) = \int_0^\pi x \sin \alpha x \, dx = -\frac{x \cos \alpha x}{\alpha} \Big|_0^\pi + \frac{1}{\alpha} \int_0^\pi \cos \alpha x \, dx$$

$$= -\frac{\pi \cos \pi\alpha}{\alpha} + \frac{1}{\alpha^2} \sin \alpha x \Big|_0^\pi = \frac{-\pi\alpha \cos \pi\alpha + \sin \pi\alpha}{\alpha^2}.$$

Hence from formula (10) in the text,

$$f(x) = \frac{2}{\pi} \int_0^\infty \frac{(-\pi\alpha \cos \pi\alpha + \sin \pi\alpha) \sin \alpha x}{\alpha^2} \, d\alpha.$$

11. The function is odd. Thus from formula (11) in the text

$$B(\alpha) = \int_0^\infty (e^{-x} \sin x) \sin \alpha x \, dx$$

$$= \frac{1}{2} \int_0^\infty e^{-x} [\cos(1 - \alpha)x - \cos(1 + \alpha)x] \, dx$$

$$= \frac{1}{2} \int_0^\infty e^{-x} \cos(1 - \alpha)x \, dx - \frac{1}{2} \int_0^\infty e^{-x} \cos(1 + \alpha)x, dx.$$

Now recall

$$\mathscr{L}\{\cos kt\} = \int_0^\infty e^{-st} \cos kt \, dt = s/(s^2 + k^2).$$

If we set $s = 1$, and in turn, $k = 1 - \alpha$ and then $k = 1 + \alpha$, we obtain

$$B(\alpha) = \frac{1}{2} \frac{1}{1 + (1 - \alpha)^2} - \frac{1}{2} \frac{1}{1 + (1 + \alpha)^2} = \frac{1}{2} \frac{(1 + \alpha)^2 - (1 - \alpha)^2}{[1 + (1 - \alpha)^2][1 + (1 + \alpha)^2]}.$$

Simplifying the last expression gives

$$B(\alpha) = \frac{2\alpha}{4 + \alpha^4}.$$

Hence from formula (10) in the text

$$f(x) = \frac{4}{\pi} \int_0^\infty \frac{\alpha \sin \alpha x}{4 + \alpha^4} \, d\alpha.$$

12. The function is odd. Thus from formula (11) in the text

$$B(\alpha) = \int_0^\infty x e^{-x} \sin \alpha x \, dx.$$

Now recall

$$\mathscr{L}\{t \sin kt\} = -\frac{d}{ds} \mathscr{L}\{\sin kt\} = 2ks/(s^2 + k^2)^2.$$

If we set $s = 1$ and $k = \alpha$ we obtain

$$B(\alpha) = \frac{2\alpha}{(1+\alpha^2)^2}.$$

Hence from formula (10) in the text

$$f(x) = \frac{4}{\pi} \int_0^\infty \frac{\alpha \sin \alpha x}{(1+\alpha^2)^2} \, d\alpha.$$

13. For the cosine integral,

$$A(\alpha) = \int_0^\infty e^{-kx} \cos \alpha x \, dx = \frac{k}{k^2 + \alpha^2}.$$

Hence

$$f(x) = \frac{2}{\pi} \int_0^\infty \frac{k \cos \alpha x}{k^2 + \alpha^2} = \frac{2k}{\pi} \int_0^\infty \frac{\cos \alpha x}{k^2 + \alpha^2} \, d\alpha.$$

For the sine integral,

$$B(\alpha) = \int_0^\infty e^{-kx} \sin \alpha x \, dx = \frac{\alpha}{k^2 + \alpha^2}.$$

Hence

$$f(x) = \frac{2}{\pi} \int_0^\infty \frac{\alpha \sin \alpha x}{k^2 + \alpha^2} \, d\alpha.$$

14. From Problem 13 the cosine and sine integral representations of e^{-kx}, $k > 0$, are respectively,

$$e^{-kx} = \frac{2k}{\pi} \int_0^\infty \frac{\cos \alpha x}{k^2 + \alpha^2} \, d\alpha \quad \text{and} \quad e^{-kx} = \frac{2}{\pi} \int_0^\infty \frac{\alpha \sin \alpha x}{k^2 + \alpha^2} \, d\alpha.$$

Hence, the cosine integral representation of $f(x) = e^{-x} - e^{-3x}$ is

$$e^{-x} - e^{-3x} = \frac{2}{\pi} \int_0^\infty \frac{\cos \alpha x}{1 + \alpha^2} \, d\alpha - \frac{2(3)}{\pi} \int_0^\infty \frac{\cos \alpha x}{9 + \alpha^2} \, d\alpha = \frac{4}{\pi} \int_0^\infty \frac{3 - \alpha^2}{(1 + \alpha^2)(9 + \alpha^2)} \cos \alpha x \, d\alpha.$$

The sine integral representation of f is

$$e^{-x} - e^{-3x} = \frac{2}{\pi} \int_0^\infty \frac{\alpha \sin \alpha x}{1 + \alpha^2} \, d\alpha - \frac{2}{\pi} \int_0^\infty \frac{\alpha \sin \alpha x}{9 + \alpha^2} \, d\alpha = \frac{16}{\pi} \int_0^\infty \frac{\alpha \sin \alpha x}{(1 + \alpha^2)(9 + \alpha^2)} \, d\alpha.$$

15. For the cosine integral,

$$A(\alpha) = \int_0^\infty x e^{-2x} \cos \alpha x \, dx.$$

But we know

$$\mathscr{L}\{t \cos kt\} = -\frac{d}{ds} \frac{s}{(s^2 + k^2)} = \frac{(s^2 - k^2)}{(s^2 + k^2)^2}.$$

If we set $s = 2$ and $k = \alpha$ we obtain

$$A(\alpha) = \frac{4 - \alpha^2}{(4 + \alpha^2)^2}.$$

Hence

$$f(x) = \frac{2}{\pi} \int_0^\infty \frac{(4 - \alpha^2)\cos\alpha x}{(4 + \alpha^2)^2}\, d\alpha.$$

For the sine integral,

$$B(\alpha) = \int_0^\infty xe^{-2x}\sin\alpha x\, dx.$$

From Problem 12, we know

$$\mathscr{L}\{t\sin kt\} = \frac{2ks}{(s^2 + k^2)^2}.$$

If we set $s = 2$ and $k = \alpha$ we obtain

$$B(\alpha) = \frac{4\alpha}{(4 + \alpha^2)^2}.$$

Hence

$$f(x) = \frac{8}{\pi} \int_0^\infty \frac{\alpha\sin\alpha x}{(4 + \alpha^2)^2}\, d\alpha.$$

16. For the cosine integral,

$$A(\alpha) = \int_0^\infty e^{-x}\cos x \cos\alpha x\, dx$$

$$= \frac{1}{2} \int_0^\infty e^{-x}[\cos(1 + \alpha)x + \cos(1 - \alpha)x]\, dx$$

$$= \frac{1}{2}\frac{1}{1 + (1 + \alpha)^2} + \frac{1}{2}\frac{1}{1 + (1 - \alpha)^2}$$

$$= \frac{1}{2}\frac{1 + (1 - \alpha)^2 + 1 + (1 + \alpha)^2}{[1 + (1 + \alpha)^2][1 + (1 - \alpha)^2]}$$

$$= \frac{2 + \alpha^2}{4 + \alpha^4}.$$

Hence

$$f(x) = \frac{2}{\pi} \int_0^\infty \frac{(2 + \alpha^2)\cos\alpha x}{4 + \alpha^4}\, d\alpha.$$

For the sine integral,

$$B(\alpha) = \int_0^\infty e^{-x}\cos x \sin\alpha x\, dx$$

$$= \frac{1}{2} \int_0^\infty e^{-x}[\sin(1 + \alpha)x - \sin(1 - \alpha)x]\, dx$$

$$= \frac{1}{2}\frac{1 + \alpha}{1 + (1 + \alpha)^2} - \frac{1}{2}\frac{1 - \alpha}{1 + (1 - \alpha)^2}$$

605

$$= \frac{1}{2} \left[\frac{(1+\alpha)[1 + (1-\alpha)^2] - (1-\alpha)[1 + (1+\alpha)^2]}{[1 + (1+\alpha)^2][1 + (1-\alpha)^2]} \right]$$

$$= \frac{\alpha^3}{4 + \alpha^4} \cdot$$

Hence

$$f(x) = \frac{2}{\pi} \int_0^\infty \frac{\alpha^3 \sin \alpha x}{4 + \alpha^4} \, d\alpha.$$

17. By formula (8) in the text

$$f(x) = 2\pi \int_0^\infty e^{-\alpha} \cos \alpha x \, d\alpha = \frac{2}{\pi} \frac{1}{1 + x^2}, \quad x > 0.$$

18. From the formula for sine integral of $f(x)$ we have

$$f(x) = \frac{2}{\pi} \int_0^\infty \left(\int_0^\infty f(x) \sin \alpha x \, dx \right) \sin \alpha x \, dx$$

$$= \frac{2}{\pi} \left[\int_0^1 1 \cdot \sin \alpha x \, d\alpha + \int_1^\infty 0 \cdot \sin \alpha x \, d\alpha \right]$$

$$= \frac{2}{\pi} \frac{(-\cos \alpha x)}{x} \Big|_0^1$$

$$= \frac{2}{\pi} \frac{1 - \cos x}{x} \cdot$$

19. (a) From formula (7) in the text with $x = 2$, we have

$$\frac{1}{2} = \frac{2}{\pi} \int_0^\infty \frac{\sin \alpha \cos \alpha}{\alpha} \, d\alpha = \frac{1}{\pi} \int_0^\infty \frac{\sin 2\alpha}{\alpha} \, d\alpha.$$

If we let $\alpha = x$ we obtain

$$\int_0^\infty \frac{\sin 2x}{x} \, dx = \frac{\pi}{2} \cdot$$

(b) If we now let $2x = kt$ where $k > 0$, then $dx = (k/2)dt$ and the integral in part (a) becomes

$$\int_0^\infty \frac{\sin kt}{kt/2} (k/2) \, dt = \int_0^\infty \frac{\sin kt}{t} \, dt = \frac{\pi}{2} \cdot$$

20. With $f(x) = e^{-|x|}$, formula (16) in the text is

$$C(\alpha) = \int_{-\infty}^\infty e^{-|x|} e^{i\alpha x} dx = \int_{-\infty}^\infty e^{-|x|} \cos \alpha x \, dx + i \int_{-\infty}^\infty e^{-|x|} \sin \alpha x \, dx.$$

The imaginary part in the last line is zero since the integrand is an odd function of x. Therefore,

$$C(\alpha) = \int_{-\infty}^\infty e^{-|x|} \cos \alpha x \, dx = 2 \int_0^\infty e^{-x} \cos \alpha x \, dx = \frac{2}{1 + \alpha^2}$$

and so from formula (15) in the text,

$$f(x) = \frac{1}{\pi} \int_{-\infty}^{\infty} \frac{\cos \alpha x}{1+\alpha^2} \, d\alpha = \frac{2}{\pi} \int_0^{\infty} \frac{\cos \alpha x}{1+\alpha^2} \, d\alpha.$$

This is the same result obtained from formulas (8) and (9) in the text.

————— **Exercises 14.4** —————————————

For the boundary-value problems in this section it is sometimes useful to note that the identities

$$e^{i\alpha} = \cos \alpha + i \sin \alpha \quad \text{and} \quad e^{-i\alpha} = \cos \alpha - i \sin \alpha$$

imply

$$e^{i\alpha} + e^{-i\alpha} = 2 \cos \alpha \quad \text{and} \quad e^{i\alpha} - e^{-i\alpha} = 2i \sin \alpha$$

1. Using the Fourier transform, the partial differential equation becomes

$$\frac{dU}{dt} + k\alpha^2 U = 0 \qquad \text{and so} \qquad U(\alpha, t) = ce^{-k\alpha^2 t}.$$

Now

$$\mathscr{F}\{u(x,0)\} = U(\alpha, 0) = \mathscr{F}\left\{e^{-|x|}\right\}.$$

We have

$$\mathscr{F}\left\{e^{-|x|}\right\} = \int_{-\infty}^{\infty} e^{-|x|}e^{i\alpha x} dx = \int_{-\infty}^{\infty} e^{-|x|}(\cos \alpha x + i \sin \alpha x) \, dx = \int_{-\infty}^{\infty} e^{-|x|} \cos \alpha x \, dx.$$

The integral

$$\int_{-\infty}^{\infty} e^{-|x|} \sin \alpha x \, dx = 0$$

since the integrand is an odd function of x. Continuing we obtain

$$\mathscr{F}\left\{e^{-|x|}\right\} = 2 \int_0^{\infty} e^{-x} \cos \alpha x \, dx = \frac{2}{1+\alpha^2}.$$

But $U(\alpha, 0) = c = \dfrac{2}{1+\alpha^2}$ gives

$$U(\alpha, t) = \frac{2e^{-k\alpha^2 t}}{1+\alpha^2}$$

and so

$$u(x,t) = \frac{2}{2\pi} \int_{-\infty}^{\infty} \frac{e^{-k\alpha^2 t} e^{-i\alpha x}}{1+\alpha^2}\, d\alpha$$

$$= \frac{1}{\pi} \int_{-\infty}^{\infty} \frac{e^{-k\alpha^2 t}}{1+\alpha^2} (\cos \alpha x - i \sin \alpha x)\, d\alpha$$

$$= \frac{1}{\pi} \int_{-\infty}^{\infty} \frac{e^{-k\alpha^2 t} \cos \alpha x}{1+\alpha^2}\, d\alpha$$

$$= \frac{2}{\pi} \int_{0}^{\infty} \frac{e^{-k\alpha^2 t} \cos \alpha x}{1+\alpha^2}\, d\alpha.$$

2. Since the domain of x is $(-\infty, \infty)$ we transform the differential equation using the Fourier transform:

$$-k\alpha^2 U(\alpha, t) = \frac{du}{dt}$$

$$\frac{du}{dt} + k\alpha^2 U(\alpha, t) = 0$$

$$U(\alpha, t) = ce^{-k\alpha^2 t}. \tag{1}$$

The transform of the initial condition is

$$\mathscr{F}\{u(x,0)\} = \int_{-\infty}^{\infty} u(x,0) e^{i\alpha x}\, dx$$

$$= \int_{-1}^{0} (-100 e^{i\alpha x})\, dx + \int_{0}^{1} 100 e^{i\alpha x}\, dx$$

$$= -100 \frac{1 - e^{-i\alpha}}{i\alpha} + 100 \frac{e^{i\alpha} - 1}{i\alpha}$$

$$= 100 \frac{e^{i\alpha} + e^{-i\alpha} - 2}{i\alpha}$$

$$= 100 \frac{2\cos\alpha - 2}{i\alpha}$$

$$= 200 \frac{\cos\alpha - 1}{i\alpha}.$$

Thus

$$U(\alpha, 0) = 200 \frac{\cos\alpha - 1}{i\alpha},$$

and since $c = U(\alpha, 0)$ in (1) we have

$$U(\alpha, t) = 200 \frac{\cos\alpha - 1}{i\alpha} e^{-k\alpha^2 t}.$$

Applying the inverse Fourier transform we obtain

$$u(x,t) = \mathscr{F}^{-1}\{U(\alpha,t)\}$$

$$= \frac{1}{2\pi}\int_{-\infty}^{\infty} 200\frac{\cos\alpha - 1}{i\alpha}e^{-k\alpha^2 t}e^{-i\alpha x}\,d\alpha$$

$$= \frac{100}{\pi}\int_{-\infty}^{\infty} 200\frac{\cos\alpha - 1}{i\alpha}e^{-k\alpha^2 t}(\cos\alpha x - i\sin\alpha x)\,d\alpha$$

$$= \frac{100}{\pi}\int_{-\infty}^{\infty} \underbrace{\frac{\cos\alpha x(\cos\alpha - 1)}{i\alpha}e^{-k\alpha^2 t}}_{\text{odd function}}\,d\alpha - \frac{100}{\pi}\int_{-\infty}^{\infty} \underbrace{\frac{\sin\alpha x(\cos\alpha - 1)}{\alpha}e^{-k\alpha^2 t}}_{\text{even function}}\,d\alpha$$

$$= \frac{200}{\pi}\int_{0}^{\infty} \frac{\sin\alpha x(1 - \cos\alpha)}{\alpha}e^{-k\alpha^2 t}\,d\alpha.$$

3. Using the Fourier transform, the partial differential equation equation becomes

$$\frac{dU}{dt} + k\alpha^2 U = 0 \qquad \text{and so} \qquad U(\alpha,t) = ce^{-k\alpha^2 t}.$$

Now

$$\mathscr{F}\{u(x,0)\} = U(\alpha,0) = \sqrt{\pi}\,e^{-\alpha^2/4}$$

by the given result. This gives $c = \sqrt{\pi}\,e^{-\alpha^2/4}$ and so

$$U(\alpha,t) = \sqrt{\pi}\,e^{-(\frac{1}{4}+kt)\alpha^2}.$$

Using the given Fourier transform again we obtain

$$u(x,t) = \sqrt{\pi}\,\mathscr{F}^{-1}\{e^{-(\frac{1+4kt}{4})\alpha^2}\} = \frac{1}{\sqrt{1+4kt}}e^{-x^2/(1+4kt)}.$$

4. **(a)** We use $U(\alpha,t) = ce^{-k\alpha^2 t}$. The Fourier transform of the boundary condition is $U(\alpha,0) = F(\alpha)$. This gives $c = F(\alpha)$ and so $U(\alpha,t) = F(\alpha)e^{-k\alpha^2 t}$. By the convolution theorem and the given result, we obtain

$$u(x,t) = \mathscr{F}^{-1}\{F(\alpha)\cdot e^{-k\alpha^2 t}\} = \frac{1}{2\sqrt{k\pi t}}\int_{-\infty}^{\infty} f(\tau)e^{-(x-\tau)^2/4kt}\,d\tau.$$

(b) Using the definition of f and the solution is part (a) we obtain

$$u(x,t) = \frac{u_0}{2\sqrt{k\pi t}}\int_{-1}^{1} e^{-(x-\tau)^2/4kt}\,d\tau.$$

.If $u = \dfrac{x - \tau}{2\sqrt{kt}}$, then $d\tau = -2\sqrt{kt}\,du$ and the integral becomes

$$u(x,t) = \frac{u_0}{\sqrt{\pi}}\int_{(x-1)/2\sqrt{kt}}^{(x+1)/2\sqrt{\pi t}} e^{-u^2}\,du.$$

Using the result in Problem 9, Exercises 14.1, we have

$$u(x,t) = \frac{u_0}{2}\left[\operatorname{erf}\left(\frac{x+1}{2\sqrt{kt}}\right) - \operatorname{erf}\left(\frac{x-1}{2\sqrt{kt}}\right)\right].$$

5. Using the Fourier sine transform, the partial differential equation becomes

$$\frac{dU}{dt} + k\alpha^2 U = k\alpha u_0.$$

The general solution of this linear equation is

$$U(\alpha,t) = ce^{-k\alpha^2 t} + \frac{u_0}{\alpha}.$$

But $U(\alpha, 0) = 0$ implies $c = -u_0/\alpha$ and so

$$U(\alpha,t) = u_0\frac{1 - e^{-k\alpha^2 t}}{\alpha}$$

and

$$u(x,t) = \frac{2u_0}{\pi}\int_0^\infty \frac{1 - e^{-k\alpha^2 t}}{\alpha}\sin\alpha x\, d\alpha.$$

6. The solution of Problem 5 can be written

$$u(x,t) = \frac{2u_0}{\pi}\int_0^\infty \frac{\sin\alpha x}{\alpha}\, d\alpha - \frac{2u_0}{\pi}\int_0^\infty \frac{\sin\alpha x}{\alpha}e^{-k\alpha^2 t}\, d\alpha.$$

Using $\int_0^\infty \frac{\sin\alpha x}{\alpha}\, d\alpha = \pi/2$ the last line becomes

$$u(x,t) = u_0 - \frac{2u_0}{\pi}\int_0^\infty \frac{\sin\alpha x}{\alpha}e^{-k\alpha^2 t}\, d\alpha.$$

7. Using the Fourier sine transform we find

$$U(\alpha,t) = ce^{-k\alpha^2 t}.$$

Now

$$\mathscr{F}_S\{u(x,0)\} = U(\alpha,0) = \int_0^1 \sin\alpha x\, dx = \frac{1 - \cos\alpha}{\alpha}.$$

From this we find $c = (1 - \cos\alpha)/\alpha$ and so

$$U(\alpha,t) = \frac{1 - \cos\alpha}{\alpha}e^{-k\alpha^2 t}$$

and

$$u(x,t) = \frac{2}{\pi}\int_0^\infty \frac{1 - \cos\alpha}{\alpha}e^{-k\alpha^2 t}\sin\alpha x\, d\alpha.$$

8. Since the domain of x is $(0, \infty)$ and the condition at $x = 0$ involves $\partial u/\partial x$ we use the Fourier cosine transform:

$$-k\alpha^2 U(\alpha,t) - ku_x(0,t) = \frac{dU}{dt}$$

$$\frac{dU}{dt} + ka^2 U(\alpha, t) = k\Lambda$$

$$U(\alpha, t) = ce^{-ka^2 t} + \frac{A}{\alpha^2}.$$

Since

$$\mathscr{F}\{u(x, 0)\} = U(\alpha, 0) = 0$$

we find $c = -A/\alpha^2$, so that

$$U(\alpha, t) = A\frac{1 - e^{-ka^2 t}}{\alpha^2}.$$

Applying the inverse Fourier cosine transform we obtain

$$u(x, t) = \mathscr{F}_C^{-1}\{U(\alpha, t)\} = \frac{2A}{\pi}\int_0^\infty \frac{1 - e^{-ka^2 t}}{\alpha^2} \cos \alpha x \, d\alpha.$$

9. Using the Fourier cosine transform we find

$$U(\alpha, t) = ce^{-ka^2 t}.$$

Now

$$\mathscr{F}_C\{u(x, 0)\} = \int_0^1 \cos \alpha x \, dx = \frac{\sin \alpha}{\alpha} = U(\alpha, 0).$$

From this we obtain $c = (\sin \alpha)/\alpha$ and so

$$U(\alpha, t) = \frac{\sin \alpha}{\alpha} e^{-ka^2 t}$$

and

$$u(x, t) = \frac{2}{\pi}\int_0^\infty \frac{\sin \alpha}{\alpha} e^{-ka^2 t} \cos \alpha x \, d\alpha.$$

10. Using the Fourier sine transform we find

$$U(\alpha, t) = ce^{-ka^2 t} + \frac{1}{\alpha}.$$

Now

$$\mathscr{F}_S\{u(x, 0)\} = \mathscr{F}_S\{e^{-x}\} = \int_0^\infty e^{-x} \sin \alpha x \, dx = \frac{\alpha}{1 + \alpha^2} = U(\alpha, 0).$$

From this we obtain $c = \alpha/(1 + \alpha^2) - 1/\alpha$. Therefore

$$U(\alpha, t) = \left(\frac{\alpha}{1 + \alpha^2} - \frac{1}{\alpha}\right) e^{-ka^2 t} + \frac{1}{\alpha} = \frac{1}{\alpha} - \frac{e^{-ka^2 t}}{\alpha(1 + \alpha^2)}$$

and

$$u(x, t) = \frac{2}{\pi}\int_0^\infty \left(\frac{1}{\alpha} - \frac{e^{-ka^2 t}}{\alpha(1 + \alpha^2)}\right) \sin \alpha x \, d\alpha.$$

611

11. (a) Using the Fourier transform we obtain

$$U(\alpha, t) = c_1 \cos \alpha a t + c_2 \sin \alpha a t.$$

If we write

$$\mathscr{F}\{u(x,0)\} = \mathscr{F}\{f(x)\} = F(\alpha)$$

and

$$\mathscr{F}\{u_t(x,0)\} = \mathscr{F}\{g(x)\} = G(\alpha)$$

we first obtain

$c_1 = F(\alpha)$ from $U(\alpha, 0) = F(\alpha)$ and then $c_2 = G(\alpha)/\alpha a$ from $\dfrac{dU}{dt}\Big|_{t=0} = G(\alpha)$. Thus

$$U(\alpha, t) = F(\alpha) \cos \alpha a t + \frac{G(\alpha)}{\alpha a} \sin \alpha a t$$

and

$$u(x,t) = \frac{1}{2\pi} \int_{-\infty}^{\infty} \left(F(\alpha) \cos \alpha a t + \frac{G(\alpha)}{\alpha a} \sin \alpha a t \right) e^{-i\alpha x} d\alpha.$$

(b) If $g(x) = 0$ then $c_2 = 0$ and

$$u(x,t) = \frac{1}{2\pi} \int_{-\infty}^{\infty} F(\alpha) \cos \alpha a t e^{-i\alpha x} d\alpha$$

$$= \frac{1}{2\pi} \int_{-\infty}^{\infty} F(\alpha) \left(\frac{e^{\alpha a t i} + e^{-\alpha a t i}}{2} \right) e^{-i\alpha x} d\alpha$$

$$= \frac{1}{2} \left[\frac{1}{2\pi} \int_{-\infty}^{\infty} F(\alpha) e^{-i(x-at)\alpha} d\alpha + \frac{1}{2\pi} \int_{-\infty}^{\infty} F(\alpha) e^{-i(x+at)\alpha} d\alpha \right]$$

$$= \frac{1}{2} \left[f(x - at) + f(x + at) \right].$$

12. Using the Fourier sine transform we obtain

$$U(\alpha, t) = c_1 \cos \alpha a t + c_2 \sin \alpha a t.$$

Now

$$\mathscr{F}_S\{u(x,0)\} = \mathscr{F}\left\{ xe^{-x} \right\} = \int_0^{\infty} xe^{-x} \sin \alpha x \, dx = \frac{2\alpha}{(1+\alpha^2)^2} = U(\alpha, 0).$$

Also,

$$\mathscr{F}_S\{u_t(x,0)\} = \frac{dU}{dt}\Big|_{t=0} = 0.$$

This last condition gives $c_2 = 0$. Then $U(\alpha, 0) = 2\alpha/(1+\alpha^2)^2$ yields $c_1 = 2\alpha/(1+\alpha^2)^2$. Therefore

$$U(\alpha, t) = \frac{2\alpha}{(1+\alpha^2)^2} \cos \alpha a t$$

and

$$u(x,t) = \frac{4}{\pi} \int_0^\infty \frac{\alpha \cos \alpha at}{(1+\alpha^2)^2} \sin \alpha x \, d\alpha.$$

13. Using the Fourier cosine transform we obtain

$$U(x, \alpha) = c_1 \cosh \alpha x + c_2 \sinh \alpha x.$$

Now the Fourier cosine transforms of $u(0, y) = e^{-y}$ and $u(\pi, y) = 0$ are, respectively, $U(0, \alpha) = 1/(1 + \alpha^2)$ and $U(\pi, \alpha) = 0$. The first of these conditions gives $c_1 = 1/(1 + \alpha^2)$. The second condition gives

$$c_2 = -\frac{\cosh \alpha \pi}{(1 + \alpha^2) \sinh \alpha \pi}.$$

Hence

$$U(x, \alpha) = \frac{\cosh \alpha x}{1 + \alpha^2} - \frac{\cosh \alpha \pi \sinh \alpha x}{(1 + \alpha^2) \sinh \alpha \pi} = \frac{\sinh \alpha \pi \cosh \alpha \pi - \cosh \alpha \pi \sinh \alpha x}{(1 + \alpha^2) \sinh \alpha \pi} = \frac{\sinh \alpha(\pi - x)}{(1 + \alpha^2) \sinh \alpha \pi}$$

and

$$u(x,t) = \frac{2}{\pi} \int_0^\infty \frac{\sinh \alpha(\pi - x)}{(1 + \alpha^2) \sinh \alpha \pi} \cos \alpha y \, d\alpha.$$

14. Since the boundary condition at $y = 0$ now involves $u(x, 0)$ rather than $u'(x, 0)$, we use the Fourier sine transform. The transform of the partial differential equation is then

$$\frac{d^2 U}{dx^2} - \alpha^2 U + \alpha u(x, 0) = 0 \quad \text{or} \quad \frac{d^2 U}{dx^2} - \alpha^2 U = -\alpha.$$

The solution of this differential equation is

$$U(x, \alpha) = c_1 \cosh \alpha x + c_2 \sinh \alpha x + \frac{1}{\alpha}.$$

The transforms of the boundary conditions at $x = 0$ and $x = \pi$ in turn imply that $c_1 = 1/\alpha$ and

$$c_2 = \frac{\cosh \alpha \pi}{\alpha \sinh \alpha \pi} - \frac{1}{\alpha \sinh \alpha \pi} + \frac{\alpha}{(1 + \alpha^2) \sinh \alpha \pi}.$$

Hence

$$U(\alpha, x) = \frac{1}{\alpha} - \frac{\cosh \alpha x}{\alpha} + \frac{\cosh \alpha \pi}{\alpha \sinh \alpha \pi} \sinh \alpha x - \frac{\sinh \alpha x}{\alpha \sinh \alpha \pi} + \frac{\alpha \sinh \alpha x}{(1 + \alpha^2) \sinh \alpha \pi}$$

$$= \frac{1}{\alpha} - \frac{\sinh \alpha(\pi - x)}{\alpha \sinh \alpha \pi} - \frac{\sinh \alpha x}{\alpha(1 + \alpha^2) \sinh \alpha \pi}.$$

Taking the inverse transform it follows that

$$u(x, y) = \frac{2}{\pi} \int_0^\infty \left(\frac{1}{\alpha} - \frac{\sinh \alpha(\pi - x)}{\alpha \sinh \alpha \pi} - \frac{\sinh \alpha x}{\alpha(1 + \alpha^2) \sinh \alpha \pi} \right) \sin \alpha y \, d\alpha.$$

15. Using the Fourier cosine transform with respect to x gives

$$U(\alpha, y) = c_1 e^{-\alpha y} + c_2 e^{\alpha y}.$$

Since we expect $u(x, y)$ to be bounded as $y \to \infty$ we define $c_2 = 0$. Thus

$$U(\alpha, y) = c_1 e^{-\alpha y}.$$

Now

$$\mathscr{F}_C\{u(x, 0)\} = \int_0^1 50 \cos \alpha x \, dx = 50 \frac{\sin \alpha}{\alpha}$$

and so

$$U(\alpha, y) = 50 \frac{\sin \alpha}{\alpha} e^{-\alpha y}$$

and

$$u(x, y) = \frac{100}{\pi} \int_0^\infty \frac{\sin \alpha}{\alpha} e^{-\alpha y} \cos \alpha x \, d\alpha.$$

16. The boundary condition $u(0, y) = 0$ indicates that we now use the Fourier sine transform. We still have $U(\alpha, y) = c_1 e^{-\alpha y}$, but

$$\mathscr{F}_S\{u(x, 0)\} = \int_0^1 50 \sin \alpha x \, dx = 50(1 - \cos \alpha)/\alpha = U(\alpha, 0).$$

This gives $c_1 = 50(1 - \cos \alpha)/\alpha$ and so

$$U(\alpha, y) = 50 \frac{1 - \cos \alpha}{\alpha} e^{-\alpha y}$$

and

$$u(x, y) = \frac{100}{\pi} \int_0^\infty \frac{1 - \cos \alpha}{\alpha} e^{-\alpha y} \sin \alpha x \, d\alpha.$$

17. We use the Fourier sine transform with respect to x to obtain

$$U(\alpha, y) = c_1 \cosh \alpha y + c_2 \sinh \alpha y.$$

The transforms of $u(x, 0) = f(x)$ and $u(x, 2) = 0$ give, in turn, $U(\alpha, 0) = F(\alpha)$ and $U(\alpha, 2) = 0$. The first condition gives $c_1 = F(\alpha)$ and the second condition then yields

$$c_2 = -\frac{F(\alpha) \cosh 2\alpha}{\sinh 2\alpha}.$$

Hence

$$U(\alpha, y) = F(\alpha) \cosh \alpha y - \frac{F(\alpha) \cosh 2\alpha \sinh \alpha y}{\sinh 2\alpha}$$

$$= F(\alpha) \frac{\sinh 2\alpha \cosh \alpha y - \cosh 2\alpha \sinh \alpha y}{\sinh 2\alpha}$$

$$= F(\alpha) \frac{\sinh \alpha(2 - y)}{\sinh 2\alpha}.$$

614

and

$$u(x, y) = \frac{2}{\pi} \int_0^\infty F(\alpha) \frac{\sinh \alpha(2 - y)}{\sinh 2\alpha} \sin \alpha x \, d\alpha.$$

18. The domain of y and the boundary condition at $y = 0$ suggest that we use a Fourier cosine transform. The transformed equation is

$$\frac{d^2 U}{dx^2} - \alpha^2 U - u_y(x, 0) = 0 \quad \text{or} \quad \frac{d^2 U}{dx^2} - \alpha^2 U = 0.$$

Because the domain of the variable x is a finite interval we choose to write the general solution of the latter equation as

$$U(x, \alpha) = c_1 \cosh \alpha x + c_2 \sinh \alpha x.$$

Now $U(0, \alpha) = F(\alpha)$, where $F(\alpha)$ is the Fourier cosine transform of $f(y)$, and $U'(\pi, \alpha) = 0$ imply $c_1 = F(\alpha)$ and $c_2 = -F(\alpha) \sinh \alpha \pi / \cosh \alpha \pi$. Thus

$$U(x, \alpha) = F(\alpha) \cosh \alpha x - F(\alpha) \frac{\sinh \alpha \pi}{\cosh \alpha \pi} \sinh \alpha x = F(\alpha) \frac{\cosh \alpha(\pi - x)}{\cosh \alpha \pi}.$$

Using the inverse transform we find that a solution to the problem is

$$u(x, y) = \frac{2}{\pi} \int_0^\infty F(\alpha) \frac{\cosh \alpha(\pi - x)}{\cosh \alpha \pi} \cos \alpha y \, d\alpha.$$

19. We solve two boundary-value problems:

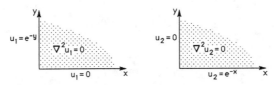

Using the Fourier sine transform with respect to y gives

$$u_1(x, y) = \frac{2}{\pi} \int_0^\infty \frac{\alpha e^{-\alpha x}}{1 + \alpha^2} \sin \alpha y \, d\alpha.$$

The Fourier sine transform with respect to x yields the solution to the second problem:

$$u_2(x, y) = \frac{2}{\pi} \int_0^\infty \frac{\alpha e^{-\alpha y}}{1 + \alpha^2} \sin \alpha x \, d\alpha.$$

We define the solution of the original problem to be

$$u(x, y) = u_1(x, y) + u_2(x, y) = \frac{2}{\pi} \int_0^\infty \frac{\alpha}{1 + \alpha^2} \left[e^{-\alpha x} \sin \alpha y + e^{-\alpha y} \sin \alpha x \right] d\alpha.$$

20. We solve the three boundary-value problems:

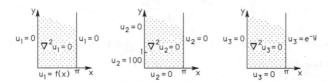

Using separation of variables we find the solution of the first problem is

$$u_1(x, y) = \sum_{n=1}^{\infty} A_n e^{-ny} \sin nx \quad \text{where} \quad A_n = \frac{2}{\pi} \int_0^{\pi} f(x) \sin nx \, dx.$$

Using the Fourier sine transform with respect to y gives the solution of the second problem:

$$u_2(x, y) = \frac{200}{\pi} \int_0^{\infty} \frac{(1 - \cos \alpha) \sinh \alpha(\pi - x)}{\alpha \sinh \alpha \pi} \sin \alpha y \, d\alpha.$$

Also, the Fourier sine transform with respect to y gives the solution of the third problem:

$$u_3(x, y) = \frac{2}{\pi} \int_0^{\infty} \frac{\alpha \sinh \alpha x}{(1 + \alpha^2) \sinh \alpha \pi} \sin \alpha y \, d\alpha.$$

The solution of the original problem is

$$u(x, y) = u_1(x, y) + u_2(x, y) + u_3(x, y).$$

21. Using the Fourier transform with respect to x gives

$$U(\alpha, y) = c_1 \cosh \alpha y + c_2 \sinh \alpha y.$$

The transform of the boundary condition $\dfrac{\partial u}{\partial y}\Big|_{y=0} = 0$ is $\dfrac{dU}{dy}\Big|_{y=0} = 0$. This condition gives $c_2 = 0$.

Hence

$$U(\alpha, y) = c_1 \cosh \alpha y.$$

Now by the given information the transform of the boundary condition $u(x, 1) = e^{-x^2}$ is $U(\alpha, 1) = \sqrt{\pi} \, e^{-\alpha^2/4}$. This condition then gives $c_1 = \sqrt{\pi} \, e^{-\alpha^2/4} \cosh \alpha$. Therefore

$$U(\alpha, y) = \sqrt{\pi} \, \frac{e^{-\alpha^2/4} \cosh \alpha y}{\cosh \alpha}$$

and

$$U(x, y) = \frac{1}{2\sqrt{\pi}} \int_{-\infty}^{\infty} \frac{e^{-\alpha^2/4} \cosh \alpha y}{\cosh \alpha} e^{-i\alpha x} \, d\alpha$$

$$= \frac{1}{2\sqrt{\pi}} \int_{-\infty}^{\infty} \frac{e^{-\alpha^2/4} \cosh \alpha y}{\cosh \alpha} \cos \alpha x \, d\alpha$$

$$= \frac{1}{\sqrt{\pi}} \int_0^{\infty} \frac{e^{-\alpha^2/4} \cosh \alpha y}{\cosh \alpha} \cos \alpha x \, d\alpha.$$

22. Entries 42 and 43 of the table in Appendix III imply

$$\int_0^\infty e^{-st}\frac{\sin at}{t}\,dt = \arctan\frac{a}{s}$$

and

$$\int_0^\infty e^{-st}\frac{\sin at \cos bt}{t}\,dt = \frac{1}{2}\arctan\frac{a+b}{s} + \frac{1}{2}\arctan\frac{a-b}{s}.$$

Identifying $\alpha = t, x = a$, and $y = s$, the solution of Problem 16 is

$$u(x,y) = \frac{100}{\pi}\int_0^\infty \frac{1-\cos\alpha}{\alpha}e^{-\alpha y}\sin\alpha x\,d\alpha$$

$$= \frac{100}{\pi}\left[\int_0^\infty \frac{\sin\alpha x}{\alpha}e^{-\alpha y}d\alpha - \int_0^\infty \frac{\sin\alpha x\cos\alpha}{\alpha}e^{-\alpha y}d\alpha\right]$$

$$= \frac{100}{\pi}\left[\arctan\frac{x}{y} - \frac{1}{2}\arctan\frac{x+1}{y} - \frac{1}{2}\arctan\frac{x-1}{y}\right].$$

23. Using integration by parts,

$$\mathscr{F}_S\{f'(x)\} = \int_0^\infty f'(x)\sin\alpha x\,dx = f(x)\sin\alpha x\,\Big|_0^\infty - \alpha\int_0^\infty f(x)\cos\alpha x\,dx.$$

If we assume that $f(x)\sin\alpha x \to 0$ as $x \to \infty$ and that f is bounded at $x = 0$, we obtain

$$\mathscr{F}_S\{f'(x)\} = -\alpha\int_0^\infty f(x)\cos\alpha x\,dx = -\alpha\mathscr{F}_C\{f(x)\}.$$

24. Using integration by parts,

$$\mathscr{F}_C\{f'(x)\} = \int_0^\infty f'(x)\cos\alpha x\,dx = f(x)\cos\alpha x\,\Big|_0^\infty + \alpha\int_0^\infty f(x)\sin\alpha x\,dx.$$

If we assume that $f(x)\cos\alpha x \to 0$ as $x \to \infty$ and that f is bounded at $x = 0$, we obtain

$$\mathscr{F}_C\{f'(x)\} = -f(0) + \alpha\mathscr{F}_S\{f(x)\}.$$

Chapter 14 Review Exercises

1. The partial differential equation and the boundary conditions indicate that the Fourier cosine transform is appropriate for the problem. We find in this case

$$u(x,y) = \frac{2}{\pi}\int_0^\infty \frac{\sinh\alpha y}{\alpha(1+\alpha^2)\cosh\alpha\pi}\cos\alpha x\,d\alpha.$$

2. We use the Laplace transform and undetermined coefficients to obtain

$$U(x,s) = c_1\cosh\sqrt{s}\,x + c_2\sinh\sqrt{s}\,x + \frac{50}{s+4\pi^2}\sin 2\pi x.$$

The transformed boundary conditions $U(0, s) = 0$ and $U(1, s) = 0$ give, in turn, $c_1 = 0$ and $c_2 = 0$. Hence

$$U(x, s) = \frac{50}{s + 4\pi^2} \sin 2\pi x$$

and

$$u(x, t) = 50 \sin 2\pi x \, \mathcal{L}^{-1}\left\{\frac{1}{s + 4\pi^2}\right\} = 50 e^{-4\pi^2 t} \sin 2\pi x.$$

3. The Laplace transform gives

$$U(x, s) = c_1 e^{-\sqrt{s+h}\,x} + c_2 e^{\sqrt{s+h}\,x} + \frac{u_0}{s + h}.$$

The condition $\lim_{x \to \infty} \partial u / \partial x = 0$ implies $\lim_{x \to \infty} dU/dx = 0$ and so we define $c_2 = 0$. Thus

$$U(x, s) = c_1 e^{-\sqrt{s+h}\,x} + \frac{u_0}{s + h}.$$

The condition $U(0, s) = 0$ then gives $c_1 = -u_0/(s + h)$ and so

$$U(x, s) = \frac{u_0}{s + h} - u_0 \frac{e^{-\sqrt{s+h}\,x}}{s + h}.$$

With the help of the first translation theorem we then obtain

$$u(x, t) = u_0 \mathcal{L}^{-1}\left\{\frac{1}{s + h}\right\} - u_0 \mathcal{L}^{-1}\left\{\frac{e^{-\sqrt{s+h}\,x}}{s + h}\right\} = u_0 e^{-ht} - u_0 e^{-ht} \operatorname{erfc}\left(\frac{x}{2\sqrt{t}}\right)$$

$$= u_0 e^{-ht}\left[1 - \operatorname{erfc}\left(\frac{x}{2\sqrt{t}}\right)\right] = u_0 e^{-ht} \operatorname{erf}\left(\frac{x}{2\sqrt{t}}\right).$$

4. Using the Fourier transform and the result $\mathscr{F}\left\{e^{-|x|}\right\} = 1/(1 + \alpha^2)$ we find

$$u(x, t) = \frac{1}{2\pi} \int_{-\infty}^{\infty} \frac{1 - e^{-\alpha^2 t}}{\alpha^2(1 + \alpha^2)} e^{-i\alpha x} d\alpha$$

$$= \frac{1}{2\pi} \int_{-\infty}^{\infty} \frac{1 - e^{-\alpha^2 t}}{\alpha^2(1 + \alpha^2)} \cos \alpha x \, d\alpha$$

$$= \frac{1}{\pi} \int_{0}^{\infty} \frac{1 - e^{-\alpha^2 t}}{\alpha^2(1 + \alpha^2)} \cos \alpha x \, d\alpha.$$

5. The Laplace transform gives

$$U(x, s) = c_1 e^{-\sqrt{s}\,x} + c_2 e^{\sqrt{s}\,x}.$$

The condition $\lim_{x \to \infty} u(x, t) = 0$ implies $\lim_{x \to \infty} U(x, s) = 0$ and so we define $c_2 = 0$. Thus

$$U(x, s) = c_1 e^{-\sqrt{s}\,x}.$$

The transform of the remaining boundary condition is $U(0, s) = 1/s^2$. This gives $c_1 = 1/s^2$. Hence

$$U(x, s) = \frac{e^{-\sqrt{s}\,x}}{s^2} \quad \text{and} \quad u(x, t) = \mathcal{L}^{-1}\left\{\frac{1}{s}\frac{e^{-\sqrt{s}\,x}}{s}\right\}.$$

Using

$$\mathcal{L}^{-1}\left\{\frac{1}{s}\right\} = 1 \quad \text{and} \quad \mathcal{L}^{-1}\left\{\frac{e^{-\sqrt{s}\,x}}{s}\right\} = \text{erfc}\left(\frac{x}{2\sqrt{t}}\right),$$

it follows from the convolution theorem that

$$u(x, t) = \int_0^t \text{erfc}\left(\frac{x}{2\sqrt{\tau}}\right) d\tau.$$

6. The Laplace transform and undetermined coefficients gives

$$U(x, s) = c_1 \cosh sx + c_2 \sinh sx + \frac{s-1}{s^2 + \pi^2}\sin \pi x.$$

The conditions $U(0, s) = 0$ and $U(1, s) = 0$ give, in turn, $c_1 = 0$ and $c_2 = 0$. Thus

$$U(x, s) = \frac{s-1}{s^2 + \pi^2}\sin \pi x$$

and

$$u(x, t) = \sin \pi x\, \mathcal{L}^{-1}\left\{\frac{s}{s^2 + \pi^2}\right\} - \frac{1}{\pi}\sin \pi x\, \mathcal{L}^{-1}\left\{\frac{\pi}{s^2 + \pi^2}\right\}$$

$$= (\sin \pi x)\cos \pi t - \frac{1}{\pi}(\sin \pi x)\sin \pi t.$$

7. The Fourier transform gives the solution

$$u(x, t) = \frac{u_0}{2\pi}\int_{-\infty}^{\infty}\left(\frac{e^{i\alpha\pi} - 1}{i\alpha}\right)e^{-i\alpha x}e^{-k\alpha^2 t}d\alpha$$

$$= \frac{u_0}{2\pi}\int_{-\infty}^{\infty}\frac{e^{i\alpha(\pi-x)} - e^{-i\alpha x}}{i\alpha}e^{-k\alpha^2 t}d\alpha$$

$$= \frac{u_0}{2\pi}\int_{-\infty}^{\infty}\frac{\cos\alpha(\pi - x) + i\sin\alpha(\pi - x) - \cos\alpha x + i\sin\alpha x}{i\alpha}e^{-k\alpha^2 t}d\alpha.$$

Since the imaginary part of the integrand of the last integral is an odd function of α, we obtain

$$u(x, t) = \frac{u_0}{2\pi}\int_{-\infty}^{\infty}\frac{\sin\alpha(\pi - x) + \sin\alpha x}{\alpha}e^{-k\alpha^2 t}d\alpha.$$

8. Using the Fourier cosine transform we obtain

$$U(x, \alpha) = c_1 \cosh \alpha x + c_2 \sinh \alpha x.$$

The condition $U(0, \alpha) = 0$ gives $c_1 = 0$. Thus

$$U(x, \alpha) = c_2 \sinh \alpha x.$$

Now

$$\mathscr{F}_C\{u(\pi,y)\} = \int_1^2 \cos\alpha y\, dy = \frac{\sin 2\alpha - \sin\alpha}{\alpha} = U(\pi,\alpha).$$

This last condition gives $c_2 = (\sin 2\alpha - \sin\alpha)/\alpha \sinh\alpha\pi$. Hence

$$U(x,\alpha) = \frac{\sin 2\alpha - \sin\alpha}{\alpha \sinh\alpha\pi}\sinh\alpha x$$

and

$$u(x,y) = \frac{2}{\pi}\int_0^\infty \frac{\sin 2\alpha - \sin\alpha}{\alpha \sinh\alpha\pi}\sinh\alpha x \cos\alpha y\, d\alpha.$$

9. We solve the two problems

$$\frac{\partial^2 u_1}{\partial x^2} + \frac{\partial^2 u_1}{\partial y^2} = 0, \quad x > 0, \quad y > 0,$$

$$u_1(0,y) = 0, \quad y > 0,$$

$$u_1(x,0) = \begin{cases} 100, & 0 < x < 1 \\ 0, & x > 1 \end{cases}$$

and

$$\frac{\partial^2 u_2}{\partial x^2} + \frac{\partial^2 u_2}{\partial y^2} = 0, \quad x > 0, \quad y > 0,$$

$$u_2(0,y) = \begin{cases} 50, & 0 < y < 1 \\ 0, & y > 1 \end{cases}$$

$$u_2(x,0) = 0.$$

Using the Fourier sine transform with respect to x we find

$$u_1(x,y) = \frac{200}{\pi}\int_0^\infty \left(\frac{1 - \cos\alpha}{\alpha}\right)e^{-\alpha y}\sin\alpha x\, d\alpha.$$

Using the Fourier sine transform with respect to y we find

$$u_2(x,y) = \frac{100}{\pi}\int_0^\infty \left(\frac{1 - \cos\alpha}{\alpha}\right)e^{-\alpha x}\sin\alpha y\, d\alpha.$$

The solution of the problem is then

$$u(x,y) = u_1(x,y) + u_2(x,y).$$

10. The Laplace transform gives

$$U(x,s) = c_1\cosh\sqrt{s}\,x + c_2\sinh\sqrt{s}\,x + \frac{r}{s^2}.$$

The condition $\dfrac{\partial u}{\partial x}\Big|_{x=0} = 0$ transforms into $\dfrac{dU}{dx}\Big|_{x=0} = 0$. This gives $c_2 = 0$. The remaining condition

$u(1,t) = 0$ transforms into $U(1,s) = 0$. This condition then implies $c_1 = -r/s^2 \cosh\sqrt{s}$. Hence

$$U(x,s) = \frac{r}{s^2} - r\,\frac{\cosh\sqrt{s}\,x}{s^2\cosh\sqrt{s}}.$$

Using geometric series and the convolution theorem we obtain

$$u(x,t) = r\,\mathcal{L}^{-1}\left\{\frac{1}{s^2}\right\} - r\,\mathcal{L}^{-1}\left\{\frac{\cosh\sqrt{s}\,x}{s^2\cosh\sqrt{s}}\right\}$$

$$= rt - r\sum_{n=0}^{\infty}(-1)^n\left[\int_0^t \operatorname{erfc}\left(\frac{2n+1-x}{2\sqrt{\tau}}\right)d\tau + \int_0^t \operatorname{erfc}\left(\frac{2n+1+x}{2\sqrt{\tau}}\right)d\tau\right].$$

11. The Fourier sine transform with respect to x and undetermined coefficients give

$$U(\alpha,y) = c_1\cosh\alpha y + c_2\sinh\alpha y + \frac{A}{\alpha}.$$

The transforms of the boundary conditions are

$$\frac{dU}{dy}\Big|_{y=0} = 0 \quad\text{and}\quad \frac{dU}{dy}\Big|_{y=\pi} = \frac{B\alpha}{1+\alpha^2}.$$

The first of these conditions gives $c_2 = 0$ and so

$$U(\alpha,y) = c_1\cosh\alpha y + \frac{A}{\alpha}.$$

The second transformed boundary condition yields $c_1 = B/(1+\alpha^2)\sinh\alpha\pi$. Therefore

$$U(\alpha,y) = \frac{B\cosh\alpha y}{(1+\alpha^2)\sinh\alpha\pi} + \frac{A}{\alpha}$$

and

$$u(x,y) = \frac{2}{\pi}\int_0^\infty\left(\frac{B\cosh\alpha y}{(1+\alpha^2)\sinh\alpha\pi} + \frac{A}{\alpha}\right)\sin\alpha x\, d\alpha.$$

12. Using the Laplace transform gives

$$U(x,s) = c_1\cosh\sqrt{s}\,x + c_2\sinh\sqrt{s}\,x.$$

The condition $u(0,t) = u_0$ transforms into $U(0,s) = u_0/s$. This gives $c_1 = u_0/s$. The condition
$u(1,t) = u_0$ transforms into $U(1,s) = u_0/s$. This implies that $c_2 = u_0(1 - \cosh\sqrt{s})/s\sinh\sqrt{s}$.

Hence

$$U(x, s) = \frac{u_0}{s} \cosh \sqrt{s}\, x + u_0 \left[\frac{1 - \cosh \sqrt{s}}{s \sinh \sqrt{s}} \right] \sinh \sqrt{s}\, x$$

$$= u_0 \left[\frac{\sinh \sqrt{s} \cosh \sqrt{s}\, x - \cosh \sinh \sqrt{s} \sinh \sqrt{s}\, x + \sinh \sqrt{s}\, x}{s \sinh \sqrt{s}} \right]$$

$$= u_0 \left[\frac{\sinh \sqrt{s}\, (1 - x) + \sinh \sqrt{s}\, x}{s \sinh \sqrt{s}} \right]$$

$$= u_0 \left[\frac{\sinh \sqrt{s}\, (1 - x)}{s \sinh \sqrt{s}} + \frac{\sinh \sqrt{s}\, x}{s \sinh \sqrt{s}} \right]$$

and

$$u(x, t) = u_0 \left[\mathscr{L}^{-1}\left\{ \frac{\sinh \sqrt{s}\,(1 - x)}{s \sinh \sqrt{s}} \right\} + \mathscr{L}^{-1}\left\{ \frac{\sinh \sqrt{s}\, x}{s \sinh \sqrt{s}} \right\} \right]$$

$$= u_0 \sum_{n=0}^{\infty} \left[\operatorname{erf}\left(\frac{2n + 2 - x}{2\sqrt{t}} \right) - \operatorname{erf}\left(\frac{2n + x}{2\sqrt{t}} \right) \right]$$

$$+ u_0 \sum_{n=0}^{\infty} \left[\operatorname{erf}\left(\frac{2n + 1 + x}{2\sqrt{t}} \right) - \operatorname{erf}\left(\frac{2n + 1 - x}{2\sqrt{t}} \right) \right].$$

13. Using the Fourier transform gives

$$U(\alpha, t) = c_1 e^{-k\alpha^2 t}.$$

Now

$$u(\alpha, 0) = \int_0^{\infty} e^{-x} e^{i\alpha x}\, dx = \frac{e^{(i\alpha - 1)x}}{i\alpha - 1} \Big|_0^{\infty} = 0 - \frac{1}{i\alpha - 1} = \frac{1}{1 - i\alpha} = c_1$$

so

$$U(\alpha, t) = \frac{1 + i\alpha}{1 + \alpha^2} e^{-k\alpha^2 t}$$

and

$$u(x, t) = \frac{1}{2\pi} \int_{-\infty}^{\infty} \frac{1 + i\alpha}{1 + \alpha^2} e^{-k\alpha^2 t} e^{-i\alpha x}\, d\alpha.$$

Since

$$\frac{1 + i\alpha}{1 + \alpha^2} (\cos \alpha x - i \sin \alpha x) = \frac{\cos \alpha x + \alpha \sin \alpha x}{1 + \alpha^2} + \frac{i(\alpha \cos \alpha x - \sin \alpha x)}{1 + \alpha^2}$$

and the integral of the product of the second term with $e^{-k\alpha^2 t}$ is 0 (it is an odd function), we have

$$u(x, t) = \frac{1}{2\pi} \int_{-\infty}^{\infty} \frac{\cos \alpha x + \alpha \sin \alpha x}{1 + \alpha^2} e^{-k\alpha^2 t}\, d\alpha.$$

15 Numerical Methods for Partial Differential Equations

Exercises 15.1

1. The figure shows the values of $u(x, y)$ along the boundary. We need to determine u_{11} and u_{21}. The system is

$$u_{21} + 2 + 0 + 0 - 4u_{11} = 0$$

or

$$1 + 2 + u_{11} + 0 - 4u_{21} = 0$$

$$-4u_{11} + u_{21} = -2$$

$$u_{11} - 4u_{21} = -3.$$

Solving we obtain $u_{11} = 11/15$ and $u_{21} = 14/15$.

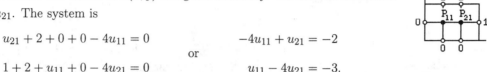

2. The figure shows the values of $u(x, y)$ along the boundary. We need to determine u_{11}, u_{21}, and u_{31}. By symmetry $u_{11} = u_{31}$ and the system is

$$u_{21} + 0 + 0 + 100 - 4u_{11} = 0$$

$$u_{31} + 0 + u_{11} + 100 - 4u_{21} = 0 \qquad \text{or}$$

$$0 + 0 + u_{21} + 100 - 4u_{31} = 0$$

$$-4u_{11} + u_{21} = -100$$

$$2u_{11} - 4u_{21} = -100.$$

Solving we obtain $u_{11} = u_{31} = 250/7$ and $u_{21} = 300/7$.

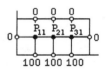

3. The figure shows the values of $u(x, y)$ along the boundary. We need to determine u_{11}, u_{21}, u_{12}, and u_{22}. By symmetry $u_{11} = u_{21}$ and $u_{12} = u_{22}$. The system is

$$u_{21} + u_{12} + 0 + 0 - 4u_{11} = 0$$

$$0 + u_{22} + u_{11} + 0 - 4u_{21} = 0$$

$$u_{22} + \sqrt{3}/2 + 0 + u_{11} - 4u_{12} = 0$$

$$0 + \sqrt{3}/2 + u_{12} + u_{21} - 4u_{22} = 0$$

or

$$3u_{11} + u_{12} = 0$$

$$u_{11} - 3u_{12} = -\frac{\sqrt{3}}{2}.$$

Solving we obtain $u_{11} = u_{21} = \sqrt{3}/16$ and $u_{12} = u_{22} = 3\sqrt{3}/16$.

4. The figure shows the values of $u(x,y)$ along the boundary. We need to determine u_{11}, u_{21}, u_{12}, and u_{22}. The system is

$$u_{21} + u_{12} + 8 + 0 - 4u_{11} = 0 \qquad -4u_{11} + u_{21} + u_{12} = -8$$

$$0 + u_{22} + u_{11} + 0 - 4u_{21} = 0 \qquad u_{11} - 4u_{21} + u_{22} = 0$$

$$\text{or}$$

$$u_{22} + 0 + 16 + u_{11} - 4u_{12} = 0 \qquad u_{11} - 4u_{12} + u_{22} = -16$$

$$0 + 0 + u_{12} + u_{21} - 4u_{22} = 0 \qquad u_{21} + u_{12} - 4u_{22} = 0.$$

Solving we obtain $u_{11} = 11/3$, $u_{21} = 4/3$, $u_{12} = 16/3$, and $u_{22} = 5/3$.

5. The figure shows the values of $u(x,y)$ along the boundary. For Gauss-Seidel the coefficients of the unknowns u_{11}, u_{21}, u_{31}, u_{12}, u_{22}, u_{32}, u_{13}, u_{23}, u_{33} are shown in the matrix

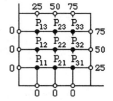

$$\begin{bmatrix} 0 & .25 & 0 & .25 & 0 & 0 & 0 & 0 & 0 \\ .25 & 0 & .25 & 0 & .25 & 0 & 0 & 0 & 0 \\ 0 & .25 & 0 & 0 & 0 & .25 & 0 & 0 & 0 \\ .25 & 0 & 0 & 0 & .25 & 0 & .25 & 0 & 0 \\ 0 & .25 & 0 & .25 & 0 & .25 & 0 & .25 & 0 \\ 0 & 0 & .25 & 0 & .25 & 0 & 0 & 0 & .25 \\ 0 & 0 & 0 & .25 & 0 & 0 & 0 & .25 & 0 \\ 0 & 0 & 0 & 0 & .25 & 0 & .25 & 0 & .25 \\ 0 & 0 & 0 & 0 & 0 & .25 & 0 & .25 & 0 \end{bmatrix}$$

The constant terms in the equations are 0, 0, 6.25, 0, 0, 12.5, 6.25, 12.5, 37.5. We use 25 as the initial guess for each variable. Then $u_{11} = 6.25$, $u_{21} = u_{12} = 12.5$, $u_{31} = u_{13} = 18.75$, $u_{22} = 25$, $u_{32} = u_{23} = 37.5$, and $u_{33} = 56.25$

6. The coefficients of the unknowns are the same as shown above in Problem 5. The constant terms are 7.5, 5, 20, 10, 0, 15, 17.5, 5, 27.5. We use 32.5 as the initial guess for each variable. Then $u_{11} = 21.92$, $u_{21} = 28.30$, $u_{31} = 38.17$, $u_{12} = 29.38$, $u_{22} = 33.13$, $u_{32} = 44.38$, $u_{13} = 22.46$, $u_{23} = 30.45$, and $u_{33} = 46.21$.

7. **(a)** Using the difference approximations for u_{xx} and u_{yy} we obtain

$$u_{xx} + u_{yy} = \frac{1}{h^2}(u_{i+1,j} + u_{i,j+1} + u_{i-1,j} + u_{i,j-1} - 4u_{ij}) = f(x,y)$$

so that

$$u_{i+1,j} + u_{i,j+1} + u_{i-1,j} + u_{i,j-1} - 4u_{ij} = h^2 f(x,y).$$

624

(b) By symmetry, as shown in the figure, we need only solve for u_1, u_2, u_3, u_4, and u_5. The difference equations are

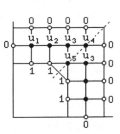

$$u_2 + 0 + 0 + 1 - 4u_1 = \frac{1}{4}(-2)$$

$$u_3 + 0 + u_1 + 1 - 4u_2 = \frac{1}{4}(-2)$$

$$u_4 + 0 + u_2 + u_5 - 4u_3 = \frac{1}{4}(-2)$$

$$0 + 0 + u_3 + u_3 - 4u_4 = \frac{1}{4}(-2)$$

$$u_3 + u_3 + 1 + 1 - 4u_5 = \frac{1}{4}(-2)$$

or

$$u_1 = 0.25u_2 + 0.375$$

$$u_2 = 0.25u_1 + 0.25u_3 + 0.375$$

$$u_3 = 0.25u_2 + 0.25u_4 + 0.25u_5 + 0.125$$

$$u_4 = 0.5u_3 + 0.125$$

$$u_5 = 0.5u_3 + 0.625.$$

Using Gauss-Seidel iteration we find $u_1 = 0.5427$, $u_2 = 0.6707$, $u_3 = 0.6402$, $u_4 = 0.4451$, and $u_5 = 0.9451$.

8. By symmetry, as shown in the figure, we need only solve for u_1, u_2, u_3, u_4, and u_5. The difference equations are

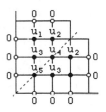

$$u_2 + 0 + 0 + u_3 - 4u_1 = -1 \qquad u_1 = 0.25u_2 + 0.25u_3 + 0.25$$

$$0 + 0 + u_1 + u_4 - 4u_2 = -1 \qquad u_2 = 0.25u_1 + 0.25u_4 + 0.25$$

$$u_4 + u_1 + 0 + u_5 - 4u_3 = -1 \quad \text{or} \quad u_3 = 0.25u_1 + 0.25u_4 + 0.25u_5 + 0.25$$

$$u_2 + u_2 + u_3 + u_3 - 4u_4 = -1 \qquad u_4 = 0.5u_2 + 0.5u_3 + 0.25$$

$$u_3 + u_3 + 0 + 0 - 4u_5 = -1 \qquad u_5 = 0.5u_3 + 0.25.$$

Using Gauss-Seidel iteration we find $u_1 = 0.6157$, $u_2 = 0.6493$, $u_3 = 0.8134$, $u_4 = 0.9813$, and $u_5 = 0.6567$.

9. Identifying $u_{ij} = u(x, t)$ the difference equation is given by

$$\frac{1}{h^2}(u_{i+1,j} - 2u_{ij} + u_{i-1,j}) = \frac{1}{k}(u_{i,j+1} - u_{ij})$$

or

$$u_{i,j+1} = \left(1 - \frac{2k}{h^2}\right)u_{ij} + \frac{k}{h^2}(u_{i+1,j} + u_{i-1,j}).$$

10. Identifying $u_{ij} = u(x, t)$ the difference equation is given by

$$\frac{1}{h^2}(u_{i+1,j} - 2u_{ij} + u_{i-1,j}) = \frac{1}{k^2}(u_{i,j+1} - 2u_{ij} + u_{i,j-1})$$

or

$$u_{i,j+1} = 2\left(1 - \frac{k^2}{h^2}\right)u_{ij} + \frac{k^2}{h^2}(u_{i+1,j} + u_{i-1,j}) - u_{i,j-1}.$$

Exercises 15.2

1. We identify $c = 1$, $a = 2$, $T = 1$, $n = 8$, and $m = 40$. Then $h = 2/8 = 0.25$, $k = 1/40 = 0.025$, and $\lambda = 2/5 = 0.4$.

TIME	X=0.25	X=0.50	X=0.75	X=1.00	X=1.25	X=1.50	X=1.75
0.000	1.0000	1.0000	1.0000	1.0000	0.0000	0.0000	0.0000
0.025	0.6000	1.0000	1.0000	0.6000	0.4000	0.0000	0.0000
0.050	0.5200	0.8400	0.8400	0.6800	0.3200	0.1600	0.0000
0.075	0.4400	0.7120	0.7760	0.6000	0.4000	0.1600	0.0640
0.100	0.3728	0.6288	0.6800	0.5904	0.3840	0.2176	0.0768
0.125	0.3261	0.5469	0.6237	0.5437	0.4000	0.2278	0.1024
0.150	0.2840	0.4893	0.5610	0.5182	0.3886	0.2465	0.1116
0.175	0.2525	0.4358	0.5152	0.4835	0.3836	0.2494	0.1209
0.200	0.2248	0.3942	0.4708	0.4562	0.3699	0.2517	0.1239
0.225	0.2027	0.3571	0.4343	0.4275	0.3571	0.2479	0.1255
0.250	0.1834	0.3262	0.4007	0.4021	0.3416	0.2426	0.1242
0.275	0.1672	0.2989	0.3715	0.3773	0.3262	0.2348	0.1219
0.300	0.1530	0.2752	0.3448	0.3545	0.3101	0.2262	0.1183
0.325	0.1407	0.2541	0.3209	0.3329	0.2943	0.2166	0.1141
0.350	0.1298	0.2354	0.2990	0.3126	0.2787	0.2067	0.1095
0.375	0.1201	0.2186	0.2790	0.2936	0.2635	0.1966	0.1046
0.400	0.1115	0.2034	0.2607	0.2757	0.2488	0.1865	0.0996
0.425	0.1036	0.1895	0.2438	0.2589	0.2347	0.1766	0.0945
0.450	0.0965	0.1769	0.2281	0.2432	0.2211	0.1670	0.0896
0.475	0.0901	0.1652	0.2136	0.2283	0.2083	0.1577	0.0847
0.500	0.0841	0.1545	0.2002	0.2144	0.1961	0.1487	0.0800

(continued)

(continued)

TIME	X=0.25	X=0.50	X=0.75	X=1.00	X=1.25	X=1.50	X=1.75
0.525	0.0786	0.1446	0.1876	0.2014	0.1845	0.1402	0.0755
0.550	0.0736	0.1354	0.1759	0.1891	0.1735	0.1320	0.0712
0.575	0.0689	0.1269	0.1650	0.1776	0.1632	0.1243	0.0670
0.600	0.0645	0.1189	0.1548	0.1668	0.1534	0.1169	0.0631
0.625	0.0605	0.1115	0.1452	0.1566	0.1442	0.1100	0.0594
0.650	0.0567	0.1046	0.1363	0.1471	0.1355	0.1034	0.0559
0.675	0.0532	0.0981	0.1279	0.1381	0.1273	0.0972	0.0525
0.700	0.0499	0.0921	0.1201	0.1297	0.1196	0.0914	0.0494
0.725	0.0468	0.0864	0.1127	0.1218	0.1124	0.0859	0.0464
0.750	0.0439	0.0811	0.1058	0.1144	0.1056	0.0807	0.0436
0.775	0.0412	0.0761	0.0994	0.1074	0.0992	0.0758	0.0410
0.800	0.0387	0.0715	0.0933	0.1009	0.0931	0.0712	0.0385
0.825	0.0363	0.0671	0.0876	0.0948	0.0875	0.0669	0.0362
0.850	0.0341	0.0630	0.0823	0.0890	0.0822	0.0628	0.0340
0.875	0.0320	0.0591	0.0772	0.0836	0.0772	0.0590	0.0319
0.900	0.0301	0.0555	0.0725	0.0785	0.0725	0.0554	0.0300
0.925	0.0282	0.0521	0.0681	0.0737	0.0681	0.0521	0.0282
0.950	0.0265	0.0490	0.0640	0.0692	0.0639	0.0489	0.0265
0.975	0.0249	0.0460	0.0601	0.0650	0.0600	0.0459	0.0249
1.000	0.0234	0.0432	0.0564	0.0610	0.0564	0.0431	0.0233

2.

(x,y)	exact	approx	abs error
(0.25,0.1)	0.3794	0.3728	0.0066
(1,0.5)	0.1854	0.2144	0.0290
(1.5,0.8)	0.0623	0.0712	0.0089

3. We identify $c = 1$, $a = 2$, $T = 1$, $n = 8$, and $m = 40$. Then $h = 2/8 = 0.25$, $k = 1/40 = 0.025$, and $\lambda = 2/5 = 0.4$.

TIME	X=0.25	X=0.50	X=0.75	X=1.00	X=1.25	X=1.50	X=1.75
0.000	1.0000	1.0000	1.0000	1.0000	0.0000	0.0000	0.0000
0.025	0.7074	0.9520	0.9566	0.7444	0.2545	0.0371	0.0053
0.050	0.5606	0.8499	0.8685	0.6633	0.3303	0.1034	0.0223
0.075	0.4684	0.7473	0.7836	0.6191	0.3614	0.1529	0.0462
0.100	0.4015	0.6577	0.7084	0.5837	0.3753	0.1871	0.0684
0.125	0.3492	0.5821	0.6428	0.5510	0.3797	0.2101	0.0861
0.150	0.3069	0.5187	0.5857	0.5199	0.3778	0.2247	0.0990
0.175	0.2721	0.4652	0.5359	0.4901	0.3716	0.2329	0.1078
0.200	0.2430	0.4198	0.4921	0.4617	0.3622	0.2362	0.1132
0.225	0.2186	0.3809	0.4533	0.4348	0.3507	0.2358	0.1160
0.250	0.1977	0.3473	0.4189	0.4093	0.3378	0.2327	0.1166
0.275	0.1798	0.3181	0.3881	0.3853	0.3240	0.2275	0.1157
0.300	0.1643	0.2924	0.3604	0.3626	0.3097	0.2208	0.1136
0.325	0.1507	0.2697	0.3353	0.3412	0.2953	0.2131	0.1107
0.350	0.1387	0.2495	0.3125	0.3211	0.2808	0.2047	0.1071
0.375	0.1281	0.2313	0.2916	0.3021	0.2666	0.1960	0.1032
0.400	0.1187	0.2150	0.2725	0.2843	0.2528	0.1871	0.0989
0.425	0.1102	0.2002	0.2549	0.2675	0.2393	0.1781	0.0946
0.450	0.1025	0.1867	0.2387	0.2517	0.2263	0.1692	0.0902
0.475	0.0955	0.1743	0.2236	0.2368	0.2139	0.1606	0.0858
0.500	0.0891	0.1630	0.2097	0.2228	0.2020	0.1521	0.0814
0.525	0.0833	0.1525	0.1967	0.2096	0.1906	0.1439	0.0772
0.550	0.0779	0.1429	0.1846	0.1973	0.1798	0.1361	0.0731
0.575	0.0729	0.1339	0.1734	0.1856	0.1696	0.1285	0.0691
0.600	0.0683	0.1256	0.1628	0.1746	0.1598	0.1214	0.0653
0.625	0.0641	0.1179	0.1530	0.1643	0.1506	0.1145	0.0617
0.650	0.0601	0.1106	0.1438	0.1546	0.1419	0.1080	0.0582
0.675	0.0564	0.1039	0.1351	0.1455	0.1336	0.1018	0.0549
0.700	0.0530	0.0976	0.1270	0.1369	0.1259	0.0959	0.0518
0.725	0.0497	0.0917	0.1194	0.1288	0.1185	0.0904	0.0488
0.750	0.0467	0.0862	0.1123	0.1212	0.1116	0.0852	0.0460
0.775	0.0439	0.0810	0.1056	0.1140	0.1050	0.0802	0.0433
0.800	0.0413	0.0762	0.0993	0.1073	0.0989	0.0755	0.0408
0.825	0.0388	0.0716	0.0934	0.1009	0.0931	0.0711	0.0384
0.850	0.0365	0.0674	0.0879	0.0950	0.0876	0.0669	0.0362
0.875	0.0343	0.0633	0.0827	0.0894	0.0824	0.0630	0.0341
0.900	0.0323	0.0596	0.0778	0.0841	0.0776	0.0593	0.0321
0.925	0.0303	0.0560	0.0732	0.0791	0.0730	0.0558	0.0302
0.950	0.0285	0.0527	0.0688	0.0744	0.0687	0.0526	0.0284
0.975	0.0268	0.0496	0.0647	0.0700	0.0647	0.0495	0.0268
1.000	0.0253	0.0466	0.0609	0.0659	0.0608	0.0465	0.0252

(x,y)	exact	approx	abs error
(0.25,0.1)	0.3794	0.4015	0.0221
(1,0.5)	0.1854	0.2228	0.0374
(1.5,0.8)	0.0623	0.0755	0.0132

4. We identify $c = 1$, $a = 2$, $T = 1$, $n = 8$, and $m = 20$. Then $h = 2/8 = 0.25$, $h = 1/20 = 0.05$, and $\lambda = 4/5 = 0.8$.

TIME	X=0.25	X=0.50	X=0.75	X=1.00	X=1.25	X=1.50	X=1.75
0.00	1.00	1.00	1.00	1.00	0.00	0.00	0.00
0.05	0.20	1.00	1.00	0.20	0.80	0.00	0.00
0.10	0.68	0.36	0.36	1.32	-0.32	0.64	0.00
0.15	-0.12	0.62	1.13	-0.76	1.76	-0.64	0.51
0.20	0.56	0.44	-0.79	2.77	-2.18	2.20	-0.82
0.25	0.01	-0.44	3.04	-4.03	5.28	-3.72	2.25
0.30	-0.36	2.70	-5.41	9.07	-9.37	8.26	-4.33
0.35	2.38	-6.24	12.67	-17.26	19.49	-15.91	9.20
0.40	-6.42	15.78	-26.40	36.08	-38.23	32.50	-18.25
0.45	16.47	-35.72	57.33	-73.35	77.80	-64.68	36.94
0.50	-38.46	80.48	-121.66	152.12	-157.11	130.60	-73.91
0.55	87.46	-176.38	259.07	-314.28	320.44	-263.18	148.83
0.60	-193.58	383.05	-547.97	652.17	-654.23	533.32	-299.84
0.65	422.59	-823.07	1156.96	-1353.07	1340.93	-1083.25	606.56
0.70	-912.01	1757.48	-2435.09	2810.16	-2753.61	2207.94	-1230.53
0.75	1953.19	-3732.17	5115.16	-5837.05	5666.65	-4512.08	2504.67
0.80	-4157.65	7893.99	-10724.47	12127.68	-11679.29	9244.30	-5112.47
0.85	8809.78	-16642.09	22452.02	-25199.62	24105.16	-18979.99	10462.92
0.90	-18599.54	34994.69	-46944.58	52365.51	-49806.79	39042.46	-21461.75
0.95	39155.48	-73432.11	98054.91	-108820.40	103010.45	-80440.31	44111.02
1.00	-82238.97	153827.58	-204634.95	226144.53	-213214.84	165961.36	-90818.86

(x,y)	exact	approx	abs error
(0.25,0.1)	0.3794	0.6800	0.3006
(1,0.5)	0.1854	152.1152	151.9298
(1.5,0.8)	0.0623	9244.3042	9244.2419

In this case $\lambda = 0.8$ is greater than 0.5 and the procedure is unstable.

5. We identify $c = 1$, $a = 2$, $T = 1$, $n = 8$, and $m = 20$. Then $h = 2/8 = 0.25$, $k = 1/20 = 0.05$, and $\lambda = 4/5 = 0.8$.

TIME	X=0.25	X=0.50	X=0.75	X=1.00	X=1.25	X=1.50	X=1.75
0.00	1.0000	1.0000	1.0000	1.0000	0.0000	0.0000	0.0000
0.05	0.5265	0.8693	0.8852	0.6141	0.3783	0.0884	0.0197
0.10	0.3972	0.6551	0.7043	0.5883	0.3723	0.1955	0.0653
0.15	0.3042	0.5150	0.5844	0.5192	0.3812	0.2261	0.1010
0.20	0.2409	0.4171	0.4901	0.4620	0.3636	0.2385	0.1145
0.25	0.1962	0.3452	0.4174	0.4092	0.3391	0.2343	0.1178
0.30	0.1631	0.2908	0.3592	0.3624	0.3105	0.2220	0.1145
0.35	0.1379	0.2482	0.3115	0.3208	0.2813	0.2056	0.1077
0.40	0.1181	0.2141	0.2718	0.2840	0.2530	0.1876	0.0993
0.45	0.1020	0.1860	0.2381	0.2514	0.2265	0.1696	0.0904
0.50	0.0888	0.1625	0.2092	0.2226	0.2020	0.1523	0.0816
0.55	0.0776	0.1425	0.1842	0.1970	0.1798	0.1361	0.0732
0.60	0.0681	0.1253	0.1625	0.1744	0.1597	0.1214	0.0654
0.65	0.0599	0.1104	0.1435	0.1544	0.1418	0.1079	0.0582
0.70	0.0528	0.0974	0.1268	0.1366	0.1257	0.0959	0.0518
0.75	0.0466	0.0860	0.1121	0.1210	0.1114	0.0851	0.0460
0.80	0.0412	0.0760	0.0991	0.1071	0.0987	0.0754	0.0408
0.85	0.0364	0.0672	0.0877	0.0948	0.0874	0.0668	0.0361
0.90	0.0322	0.0594	0.0776	0.0839	0.0774	0.0592	0.0320
0.95	0.0285	0.0526	0.0687	0.0743	0.0686	0.0524	0.0284
1.00	0.0252	0.0465	0.0608	0.0657	0.0607	0.0464	0.0251

(x,y)	exact	approx	abs error
(0.25,0.1)	0.3794	0.3972	0.0178
(1,0.5)	0.1854	0.2226	0.0372
(1.5,0.8)	0.0623	0.0754	0.0131

6. (a) We identify $c = 15/88 \approx 0.1705$, $a = 20$, $T = 10$, $n = 10$, and $m = 10$. Then $h = 2$, $k = 1$, and $\lambda = 15/352 \approx 0.0426$.

TIME	X=2	X=4	X=6	X=8	X=10	X=12	X=14	X=16	X=18
0	30.0000	30.0000	30.0000	30.0000	30.0000	30.0000	30.0000	30.0000	30.0000
1	28.7216	30.0000	30.0000	30.0000	30.0000	30.0000	30.0000	30.0000	28.7216
2	27.5521	29.9455	30.0000	30.0000	30.0000	30.0000	30.0000	29.9455	27.5521
3	26.4800	29.8459	29.9977	30.0000	30.0000	30.0000	29.9977	29.8459	26.4800
4	25.4951	29.7089	29.9913	29.9999	30.0000	29.9999	29.9913	29.7089	25.4951
5	24.5882	29.5414	29.9796	29.9995	30.0000	29.9995	29.9796	29.5414	24.5882
6	23.7515	29.3490	29.9618	29.9987	30.0000	29.9987	29.9618	29.3490	23.7515
7	22.9779	29.1365	29.9373	29.9972	29.9998	29.9972	29.9373	29.1365	22.9779
8	22.2611	28.9082	29.9057	29.9948	29.9996	29.9948	29.9057	28.9082	22.2611
9	21.5958	28.6675	29.8670	29.9912	29.9992	29.9912	29.8670	28.6675	21.5958
10	20.9768	28.4172	29.8212	29.9862	29.9985	29.9862	29.8212	28.4172	20.9768

(b) We identify $c = 15/88 \approx 0.1705$, $a = 50$, $T = 10$, $n = 10$, and $m = 10$. Then $h = 5$, $k = 1$, and $\lambda = 3/440 \approx 0.0068$.

TIME	X=5	X=10	X=15	X=20	X=25	X=30	X=35	X=40	X=45
0	30.0000	30.0000	30.0000	30.0000	30.0000	30.0000	30.0000	30.0000	30.0000
1	29.7955	30.0000	30.0000	30.0000	30.0000	30.0000	30.0000	30.0000	29.7955
2	29.5937	29.9986	30.0000	30.0000	30.0000	30.0000	30.0000	29.9986	29.5937
3	29.3947	29.9959	30.0000	30.0000	30.0000	30.0000	30.0000	29.9959	29.3947
4	29.1984	29.9918	30.0000	30.0000	30.0000	30.0000	30.0000	29.9918	29.1984
5	29.0047	29.9864	29.9999	30.0000	30.0000	30.0000	29.9999	29.9864	29.0047
6	28.8136	29.9798	29.9998	30.0000	30.0000	30.0000	29.9998	29.9798	28.8136
7	28.6251	29.9720	29.9997	30.0000	30.0000	30.0000	29.9997	29.9720	28.6251
8	28.4391	29.9630	29.9995	30.0000	30.0000	30.0000	29.9995	29.9630	28.4391
9	28.2556	29.9529	29.9992	30.0000	30.0000	30.0000	29.9992	29.9529	28.2556
10	28.0745	29.9416	29.9989	30.0000	30.0000	30.0000	29.9989	29.9416	28.0745

(c) We identify $c = 50/27 \approx 1.8519$, $a = 20$, $T = 10$, $n = 10$, and $m = 10$. Then $h = 2$, $k = 1$, and $\lambda = 25/54 \approx 0.4630$.

TIME	X=2	X=4	X=6	X=8	X=10	X=12	X=14	X=16	X=18
0	18.0000	32.0000	42.0000	48.0000	50.0000	48.0000	42.0000	32.0000	18.0000
1	16.1481	30.1481	40.1481	46.1481	48.1481	46.1481	40.1481	30.1481	16.1481
2	15.1536	28.2963	38.2963	44.2963	46.2963	44.2963	38.2963	28.2963	15.1536
3	14.2226	26.8414	36.4444	42.4444	44.4444	42.4444	36.4444	26.8414	14.2226
4	13.4801	25.4452	34.7764	40.5926	42.5926	40.5926	34.7764	25.4452	13.4801
5	12.7787	24.2258	33.1491	38.8258	40.7407	38.8258	33.1491	24.2258	12.7787
6	12.1622	23.0574	31.6460	37.0842	38.9677	37.0842	31.6460	23.0574	12.1622
7	11.5756	21.9895	30.1875	35.4385	37.2238	35.4385	30.1875	21.9895	11.5756
8	11.0378	20.9636	28.8232	33.8340	35.5707	33.8340	28.8232	20.9636	11.0378
9	10.5230	20.0070	27.5043	32.3182	33.9626	32.3182	27.5043	20.0070	10.5230
10	10.0420	19.0872	26.2620	30.8509	32.4400	30.8509	26.2620	19.0872	10.0420

(d) We identify $c = 260/159 \approx 1.6352$, $a = 100$, $T = 10$, $n = 10$, and $m = 10$. Then $h = 10$, $k = 1$, and $\lambda = 13/795 \approx 00164$.

TIME	X=10	X=20	X=30	X=40	X=50	X=60	X=70	X=80	X=90
0	8.0000	16.0000	24.0000	32.0000	40.0000	32.0000	24.0000	16.0000	8.0000
1	8.0000	16.0000	24.0000	32.0000	39.7384	32.0000	24.0000	16.0000	8.0000
2	8.0000	16.0000	24.0000	31.9957	39.4853	31.9957	24.0000	16.0000	8.0000
3	8.0000	16.0000	23.9999	31.9874	39.2403	31.9874	23.9999	16.0000	8.0000
4	8.0000	16.0000	23.9997	31.9754	39.0031	31.9754	23.9997	16.0000	8.0000
5	8.0000	16.0000	23.9993	31.9599	30.7733	31.9599	23.9993	16.0000	8.0000
6	8.0000	16.0000	23.9987	31.9412	38.5505	31.9412	23.9987	16.0000	8.0000
7	8.0000	16.0000	23.9978	31.9194	38.3343	31.9194	23.9978	16.0000	8.0000
8	8.0000	15.9999	23.9965	31.8947	38.1245	31.8947	23.9965	15.9999	8.0000
9	8.0000	15.9999	23.9949	31.8675	37.9208	31.8675	23.9949	15.9999	8.0000
10	8.0000	15.9998	23.9929	31.8377	37.7228	31.8377	23.9929	15.9998	8.0000

7. (a) We identify $c = 15/88 \approx 0.1705$, $a = 20$, $T = 10$, $n = 10$, and $m = 10$. Then $h = 2$, $k = 1$, and $\lambda = 15/352 \approx 0.0426$.

TIME	X=2.00	X=4.00	X=6.00	X=8.00	X=10.00	X=12.00	X=14.00	X=16.00	X=18.00
0.00	30.0000	30.0000	30.0000	30.0000	30.0000	30.0000	30.0000	30.0000	30.0000
1.00	28.7733	29.9749	29.9995	30.0000	30.0000	30.0000	29.9995	29.9749	28.7733
2.00	27.6450	29.9037	29.9970	29.9999	30.0000	29.9999	29.9970	29.9037	27.6450
3.00	26.6051	29.7938	29.9911	29.9997	30.0000	29.9997	29.9911	29.7938	26.6051
4.00	25.6452	29.6517	29.9805	29.9991	29.9999	29.9991	29.9805	29.6517	25.6452
5.00	24.7573	29.4829	29.9643	29.9981	29.9998	29.9981	29.9643	29.4829	24.7573
6.00	23.9347	29.2922	29.9421	29.9963	29.9996	29.9963	29.9421	29.2922	23.9347
7.00	23.1711	29.0836	29.9134	29.9936	29.9992	29.9936	29.9134	29.0836	23.1711
8.00	22.4612	28.8606	29.8782	29.9898	29.9986	29.9898	29.8782	28.8606	22.4612
9.00	21.7999	28.6263	29.8362	29.9848	29.9977	29.9848	29.8362	28.6263	21.7999
10.00	21.1829	28.3831	29.7878	29.9782	29.9964	29.9782	29.7878	28.3831	21.1829

(b) We identify $c = 15/88 \approx 0.1705$, $a = 50$, $T = 10$, $n = 10$, and $m = 10$. Then $h = 5$, $k = 1$, and $\lambda = 3/440 \approx 0.0068$.

TIME	X=5.00	X=10.00	X=15.00	X=20.00	X=25.00	X=30.00	X=35.00	X=40.00	X=45.00
0.00	30.0000	30.0000	30.0000	30.0000	30.0000	30.0000	30.0000	30.0000	30.0000
1.00	29.7968	29.9993	30.0000	30.0000	30.0000	30.0000	30.0000	29.9993	29.7968
2.00	29.5964	29.9973	30.0000	30.0000	30.0000	30.0000	30.0000	29.9973	29.5964
3.00	29.3987	29.9939	30.0000	30.0000	30.0000	30.0000	30.0000	29.9939	29.3987
4.00	29.2036	29.9893	29.9999	30.0000	30.0000	30.0000	29.9999	29.9893	29.2036
5.00	29.0112	29.9834	29.9998	30.0000	30.0000	30.0000	29.9998	29.9834	29.0112
6.00	28.8212	29.9762	29.9997	30.0000	30.0000	30.0000	29.9997	29.9762	28.8213
7.00	28.6339	29.9679	29.9995	30.0000	30.0000	30.0000	29.9995	29.9679	28.6339
8.00	28.4490	29.9585	29.9992	30.0000	30.0000	30.0000	29.9993	29.9585	28.4490
9.00	28.2665	29.9479	29.9989	30.0000	30.0000	30.0000	29.9989	29.9479	28.2665
10.00	28.0864	29.9363	29.9986	30.0000	30.0000	30.0000	29.9986	29.9363	28.0864

(c) We identify $c = 50/27 \approx 1.8519$, $a = 20$, $T = 10$, $n = 10$, and $m = 10$. Then $h = 2$, $k = 1$, and $\lambda = 25/54 \approx 0.4630$.

TIME	X=2.00	X=4.00	X=6.00	X=8.00	X=10.00	X=12.00	X=14.00	X=16.00	X=18.00
0.00	18.0000	32.0000	42.0000	48.0000	50.0000	48.0000	42.0000	32.0000	18.0000
1.00	16.4489	30.1970	40.1561	46.1495	48.1486	46.1495	40.1561	30.1970	16.4489
2.00	15.3312	28.5348	38.3465	44.3067	46.3001	44.3067	38.3465	28.5348	15.3312
3.00	14.4216	27.0416	36.6031	42.4847	44.4619	42.4847	36.6031	27.0416	14.4216
4.00	13.6371	25.6867	34.9416	40.6988	42.6453	40.6988	34.9416	25.6867	13.6371
5.00	12.9378	24.4419	33.3628	38.9611	40.8634	38.9611	33.3628	24.4419	12.9378
6.00	12.3012	23.2863	31.8624	37.2794	39.1273	37.2794	31.8624	23.2863	12.3012
7.00	11.7137	22.2051	30.4350	35.6578	37.4446	35.6578	30.4350	22.2051	11.7137
8.00	11.1659	21.1877	29.0757	34.0984	35.8202	34.0984	29.0757	21.1877	11.1659
9.00	10.6517	20.2261	27.7799	32.6014	34.2567	32.6014	27.7799	20.2261	10.6517
10.00	10.1665	19.3143	26.5439	31.1662	32.7549	31.1662	26.5439	19.3143	10.1665

(d) We identify $c = 260/159 \approx 1.6352$, $a = 100$, $T = 10$, $n = 10$, and $m = 10$. Then $h = 10$, $k = 1$, and $\lambda = 13/795 \approx 00164$.

TIME	X=10.00	X=20.00	X=30.00	X=40.00	X=50.00	X=60.00	X=70.00	X=80.00	X=90.00
0.00	8.0000	16.0000	24.0000	32.0000	40.0000	32.0000	24.0000	16.0000	8.0000
1.00	8.0000	16.0000	24.0000	31.9979	39.7425	31.9979	24.0000	16.0000	8.0000
2.00	8.0000	16.0000	23.9999	31.9918	39.4932	31.9918	23.9999	16.0000	8.0000
3.00	8.0000	16.0000	23.9997	31.9820	39.2517	31.9820	23.9997	16.0000	8.0000
4.00	8.0000	16.0000	23.9993	31.9686	39.0175	31.9686	23.9993	16.0000	8.0000
5.00	8.0000	16.0000	23.9987	31.9520	38.7905	31.9520	23.9987	16.0000	8.0000
6.00	8.0000	15.9999	23.9978	31.9323	38.5701	31.9323	23.9978	15.9999	8.0000
7.00	8.0000	15.9999	23.9966	31.9097	38.3561	31.9097	23.9966	15.9999	8.0000
8.00	8.0000	15.9998	23.9950	31.8844	38.1483	31.8844	23.9950	15.9998	8.0000
9.00	8.0000	15.9997	23.9931	31.8566	37.9463	31.8566	23.9931	15.9997	8.0000
10.00	8.0000	15.9996	23.9908	31.8265	37.7498	31.8265	23.9908	15.9996	8.0000

8. (a) We identify $c = 15/88 \approx 0.1705$, $a = 20$, $T = 10$, $n = 10$, and $m = 10$. Then $h = 2$, $k = 1$, and $\lambda = 15/352 \approx 0.0426$.

TIME	X=2	X=4	X=6	X=8	X=10	X=12	X=14	X=16	X=18
0	30.0000	30.0000	30.0000	30.0000	30.0000	30.0000	30.0000	30.0000	30.0000
1	28.7216	30.0000	30.0000	30.0000	30.0000	30.0000	30.0000	30.0000	29.5739
2	27.5521	29.9455	30.0000	30.0000	30.0000	30.0000	30.0000	29.9818	29.1840
3	26.4800	29.8459	29.9977	30.0000	30.0000	30.0000	29.9992	29.9486	28.8267
4	25.4951	29.7089	29.9913	29.9999	30.0000	30.0000	29.9971	29.9030	28.4984
5	24.5882	29.5414	29.9796	29.9995	30.0000	29.9998	29.9932	29.8471	28.1961
6	23.7515	29.3490	29.9618	29.9987	30.0000	29.9996	29.9873	29.7830	27.9172
7	22.9779	29.1365	29.9373	29.9972	29.9999	29.9991	29.9791	29.7122	27.6593
8	22.2611	28.9082	29.9057	29.9948	29.9997	29.9982	29.9686	29.6361	27.4204
9	21.5958	28.6675	29.8670	29.9912	29.9995	29.9970	29.9557	29.5558	27.1986
10	20.9768	28.4172	29.8212	29.9862	29.9990	29.9954	29.9404	29.4724	26.9923

(b) We identify $c = 15/88 \approx 0.1705$, $a = 50$, $T = 10$, $n = 10$, and $m = 10$. Then $h = 5$, $k = 1$, and $\lambda = 3/440 \approx 0.0068$.

TIME	X=5	X=10	X=15	X=20	X=25	X=30	X=35	X=40	X=45
0	30.0000	30.0000	30.0000	30.0000	30.0000	30.0000	30.0000	30.0000	30.0000
1	29.7955	30.0000	30.0000	30.0000	30.0000	30.0000	30.0000	30.0000	29.9318
2	29.5937	29.9986	30.0000	30.0000	30.0000	30.0000	30.0000	29.9995	29.8646
3	29.3947	29.9959	30.0000	30.0000	30.0000	30.0000	30.0000	29.9986	29.7982
4	29.1984	29.9918	30.0000	30.0000	30.0000	30.0000	30.0000	29.9973	29.7328
5	29.0047	29.9864	29.9999	30.0000	30.0000	30.0000	29.9955	29.9955	29.6682
6	28.8136	29.9798	29.9998	30.0000	30.0000	30.0000	29.9999	29.9933	29.6045
7	28.6251	29.9720	29.9997	30.0000	30.0000	30.0000	29.9999	29.9907	29.5417
8	28.4391	29.9630	29.9995	30.0000	30.0000	30.0000	29.9998	29.9877	29.4797
9	28.2556	29.9529	29.9992	30.0000	30.0000	30.0000	29.9997	29.9843	29.4185
10	28.0745	29.9416	29.9989	30.0000	30.0000	30.0000	29.9996	29.9805	29.3582

(c) We identify $c = 50/27 \approx 1.8519$, $a = 20$, $T = 10$, $n = 10$, and $m = 10$. Then $h = 2$, $k = 1$, and $\lambda = 25/54 \approx 0.4630$.

TIME	X=2	X=4	X=6	X=8	X=10	X=12	X=14	X=16	X=18
0	18.0000	32.0000	42.0000	48.0000	50.0000	48.0000	42.0000	32.0000	18.0000
1	16.1481	30.1481	40.1481	46.1481	48.1481	46.1481	40.1481	30.1481	25.4074
2	15.1536	28.2963	38.2963	44.2963	46.2963	44.2963	38.2963	32.5830	25.0988
3	14.2226	26.8414	36.4444	42.4444	44.4444	42.4444	38.4290	31.7631	26.2031
4	13.4801	25.4452	34.7764	40.5926	42.5926	41.5114	37.2019	32.2751	25.9054
5	12.7787	24.2258	33.1491	38.8258	41.1661	40.0168	36.9161	31.6071	26.1204
6	12.1622	23.0574	31.6460	37.2812	39.5506	39.1134	35.8938	31.5248	25.8270
7	11.5756	21.9895	30.2787	35.7230	38.2975	37.8252	35.3617	30.9096	25.7672
8	11.0378	21.0058	28.9616	34.3944	36.8869	36.9033	34.4411	30.5900	25.4779
9	10.5425	20.0742	27.7936	33.0332	35.7406	35.7558	33.7981	30.0062	25.3086
10	10.0746	19.2352	26.6455	31.8608	34.4942	34.8424	32.9489	29.5869	25.0257

(d) We identify $c = 260/159 \approx 1.6352$, $a = 100$, $T = 10$, $n = 10$, and $m = 10$. Then $h = 10$, $k = 1$, and $\lambda = 13/795 \approx 00164$.

TIME	X=10	X=20	X=30	X=40	X=50	X=60	X=70	X=80	X=90
0	8.0000	16.0000	24.0000	32.0000	40.0000	32.0000	24.0000	16.0000	8.0000
1	8.0000	16.0000	24.0000	32.0000	39.7384	32.0000	24.0000	16.0000	8.3270
2	8.0000	16.0000	24.0000	31.9957	39.4853	31.9957	24.0000	16.0053	8.6434
3	8.0000	16.0000	23.9999	31.9874	39.2403	31.9874	24.0000	16.0157	8.9495
4	8.0000	16.0000	23.9997	31.9754	39.0031	31.9754	24.0001	16.0307	9.2457
5	8.0000	16.0000	23.9993	31.9599	38.7733	31.9599	24.0002	16.0501	9.5325
6	8.0000	16.0000	23.9987	31.9412	38.5505	31.9412	24.0003	16.0735	9.8103
7	8.0000	16.0000	23.9978	31.9194	38.3343	31.9194	24.0006	16.1007	10.0793
8	8.0000	15.9999	23.9965	31.8947	38.1245	31.8948	24.0009	16.1314	10.3400
9	8.0000	15.9999	23.9949	31.8675	37.9208	31.8676	24.0013	16.1654	10.5927
10	8.0000	15.9998	23.9929	31.8377	37.7228	31.8380	24.0018	16.2024	10.8376

9. (a) We identify $c = 15/88 \approx 0.1705$, $a = 20$, $T = 10$, $n = 10$, and $m = 10$. Then $h = 2$, $k = 1$, and $\lambda = 15/352 \approx 0.0426$.

TIME	X=2.00	X=4.00	X=6.00	X=8.00	X=10.00	X=12.00	X=14.00	X=16.00	X=18.00
0.00	30.0000	30.0000	30.0000	30.0000	30.0000	30.0000	30.0000	30.0000	30.0000
1.00	28.7733	29.9749	29.9995	30.0000	30.0000	30.0000	29.9998	29.9916	29.5911
2.00	27.6450	29.9037	29.9970	29.9999	30.0000	30.0000	29.9990	29.9679	29.2150
3.00	26.6051	29.7938	29.9911	29.9997	30.0000	29.9999	29.9970	29.9313	28.8684
4.00	25.6452	29.6517	29.9805	29.9991	30.0000	29.9997	29.9935	29.8839	28.5484
5.00	24.7573	29.4829	29.9643	29.9981	29.9999	29.9994	29.9881	29.8276	28.2524
6.00	23.9347	29.2922	29.9421	29.9963	29.9997	29.9988	29.9807	29.7641	27.9782
7.00	23.1711	29.0836	29.9134	29.9936	29.9995	29.9979	29.9711	29.6945	27.7237
8.00	22.4612	28.8606	29.8782	29.9899	29.9991	29.9966	29.9594	29.6202	27.4870
9.00	21.7999	28.6263	29.8362	29.9848	29.9985	29.9949	29.9454	29.5421	27.2666
10.00	21.1829	28.3831	29.7878	29.9783	29.9976	29.9927	29.9293	29.4610	27.0610

(b) We identify $c = 15/88 \approx 0.1705$, $a = 50$, $T = 10$, $n = 10$, and $m = 10$. Then $h = 5$, $k = 1$, and $\lambda = 3/440 \approx 0.0068$.

TIME	X=5.00	X=10.00	X=15.00	X=20.00	X=25.00	X=30.00	X=35.00	X=40.00	X=45.00
0.00	30.0000	30.0000	30.0000	30.0000	30.0000	30.0000	30.0000	30.0000	30.0000
1.00	29.7968	29.9993	30.0000	30.0000	30.0000	30.0000	30.0000	29.9998	29.9323
2.00	29.5964	29.9973	30.0000	30.0000	30.0000	30.0000	30.0000	29.9991	29.8655
3.00	29.3987	29.9939	30.0000	30.0000	30.0000	30.0000	30.0000	29.9980	29.7996
4.00	29.2036	29.9893	29.9999	30.0000	30.0000	30.0000	30.0000	29.9964	29.7345
5.00	29.0112	29.9834	29.9998	30.0000	30.0000	30.0000	29.9999	29.9945	29.6704
6.00	28.8212	29.9762	29.9997	30.0000	30.0000	30.0000	29.9999	29.9921	29.6071
7.00	28.6339	29.9679	29.9995	30.0000	30.0000	30.0000	29.9998	29.9893	29.5446
8.00	28.4490	29.9585	29.9992	30.0820	30.0000	30.0000	29.9997	29.9862	29.4830
9.00	28.2665	29.9479	29.9989	30.0000	30.0000	30.0000	29.9996	29.9827	29.4222
10.00	28.0864	29.9363	29.9986	30.0000	30.0000	30.0000	29.9995	29.9788	29.3621

(c) We identify $c = 50/27 \approx 1.8519$, $a = 20$, $T = 10$, $n = 10$, and $m = 10$. Then $h = 2$, $k = 1$, and $\lambda = 25/54 \approx 0.4630$.

TIME	X=2.00	X=4.00	X=6.00	X=8.00	X=10.00	X=12.00	X=14.00	X=16.00	X=18.00
0.00	18.0000	32.0000	42.0000	48.0000	50.0000	48.0000	42.0000	32.0000	18.0000
1.00	16.4489	30.1970	40.1562	46.1502	48.1531	46.1773	40.3274	31.2520	22.9449
2.00	15.3312	28.5350	38.3477	44.3130	46.3327	44.4671	39.0872	31.5755	24.6930
3.00	14.4219	27.0429	36.6090	42.5113	44.5759	42.9362	38.1976	31.7478	25.4131
4.00	13.6381	25.6913	34.9606	40.7728	42.9127	41.5716	37.4340	31.7086	25.6986
5.00	12.9409	24.4545	33.4091	39.1182	41.3519	40.3240	36.7033	31.5136	25.7663
6.00	12.3088	23.3146	31.9546	37.5566	39.8880	39.1565	35.9745	31.2134	25.7128
7.00	11.7294	22.2589	30.5939	36.0884	38.5109	38.0470	35.2407	30.8434	25.5871
8.00	11.1946	21.2785	29.3217	34.7092	37.2109	36.9834	34.5032	30.4279	25.4167
9.00	10.6987	20.3660	28.1318	33.4130	35.9801	35.9591	33.7660	29.9836	25.2181
10.00	10.2377	19.5150	27.0178	32.1929	34.8117	34.9710	33.0338	29.5224	25.0019

(d) We identify $c = 260/159 \approx 1.6352$, $a = 100$, $T = 10$, $n = 10$, and $m = 10$. Then $h = 10$, $k = 1$, and $\lambda = 13/795 \approx 00164$.

TIME	X=10.00	X=20.00	X=30.00	X=40.00	X=50.00	X=60.00	X=70.00	X=80.00	X=90.00
0.00	8.0000	16.0000	24.0000	32.0000	40.0000	32.0000	24.0000	16.0000	8.0000
1.00	8.0000	16.0000	24.0000	31.9979	39.7425	31.9979	24.0000	16.0026	8.3218
2.00	8.0000	16.0000	23.9999	31.9918	39.4932	31.9918	24.0000	16.0102	8.6333
3.00	8.0000	16.0000	23.9997	31.9820	39.2517	31.9820	24.0001	16.0225	8.9350
4.00	8.0000	16.0000	23.9993	31.9686	39.0175	31.9687	24.0002	16.0391	9.2272
5.00	8.0000	16.0000	23.9987	31.9520	38.7905	31.9520	24.0003	16.0599	9.5103
6.00	8.0000	15.9999	23.9978	31.9323	38.5701	31.9324	24.0005	16.0845	9.7846
7.00	8.0000	15.9999	23.9966	31.9097	38.3561	31.9098	24.0008	16.1126	10.0506
8.00	8.0000	15.9998	23.9950	31.8844	38.1483	31.8846	24.0012	16.1441	10.3084
9.00	8.0000	15.9997	23.9931	31.8566	37.9463	31.8569	24.0017	16.1786	10.5585
10.00	8.0000	15.9996	23.9908	31.8265	37.7499	31.8269	24.0023	16.2160	10.8012

10. (a) With $n = 4$ we have $h = 1/2$ so that $\lambda = 1/100 = 0.01$.

(b) We observe that $\alpha = 2(1 + 1/\lambda) = 202$ and $\beta = 2(1 - 1/\lambda) = -198$. The system of equations is

$$-u_{01} + \alpha u_{11} - u_{21} = u_{20} - \beta u_{10} + u_{00}$$

$$-u_{11} + \alpha u_{21} - u_{31} = u_{30} - \beta u_{20} + u_{10}$$

$$-u_{21} + \alpha u_{31} - u_{41} = u_{40} - \beta u_{30} + u_{20}.$$

Now $u_{00} = u_{01} = u_{40} = u_{41} = 0$, so the system is

$$\alpha u_{11} - u_{21} = u_{20} - \beta u_{10}$$

$$-u_{11} + \alpha u_{21} - u_{31} = u_{30} - \beta u_{20} + u_{10}$$

$$-u_{21} + \alpha u_{31} = -\beta u_{30} + u_{20}$$

or

$$202u_{11} - u_{21} = \sin \pi + 198 \sin \frac{\pi}{2} = 198$$

$$-u_{11} + 202u_{21} - u_{31} = \sin \frac{3\pi}{2} + 198 \sin \pi + \sin \frac{\pi}{2} = 0$$

$$-u_{21} + 202u_{31} = 198 \sin \frac{3\pi}{2} + \sin \pi = -198.$$

(c) The solution of this system is $u_{11} \approx 0.9802$, $u_{21} = 0$, $u_{31} \approx -0.9802$. The corresponding entries in Table 15.4 in the text are 0.9768, 0, and -0.9768.

11. (a) The differential equation is $k \dfrac{\partial^2 u}{\partial x^2} = \dfrac{\partial u}{\partial t}$ where $k = K/\gamma\rho$. If we let $u(x,t) = v(x,t) + \psi(x)$, then

$$\frac{\partial^2 u}{\partial x^2} = \frac{\partial^2 v}{\partial x^2} + \psi'' \quad \text{and} \quad \frac{\partial u}{\partial t} = \frac{\partial v}{\partial t}.$$

Substituting into the differential equation gives

$$k \frac{\partial^2 v}{\partial x^2} + k\psi'' = \frac{\partial v}{\partial t}.$$

Requiring $k\psi'' = 0$ we have $\psi(x) = c_1 x + c_2$. The boundary conditions become

$$u(0,t) = v(0,t) + \psi(0) = 20 \quad \text{and} \quad u(20,t) = v(20,t) + \psi(20) = 30.$$

Letting $\psi(0) = 20$ and $\psi(20) = 30$ we obtain the homogeneous boundary conditions in v: $v(0,t) = v(20,t) = 0$. Now $\psi(0) = 20$ and $\psi(20) = 30$ imply that $c_1 = 1/2$ and $c_2 = 20$. The steady-state solution is $\psi(x) = \frac{1}{2} x + 20$.

(b) To use the Crank-Nicholson method we identify $c = 375/212 \approx 1.7689$, $a = 20$, $T = 400$, $n = 5$, and $m = 40$. Then $h = 4$, $k = 10$, and $\lambda = 1875/1696 \approx 1.1055$.

TIME	X=4.00	X=8.00	X=12.00	X=16.00
0.00	50.0000	50.0000	50.0000	50.0000
10.00	32.7433	44.2679	45.4228	38.2971
20.00	29.9946	36.2354	38.3148	35.8160
30.00	26.9487	32.1409	34.0874	32.9644
40.00	25.2691	29.2562	31.2704	31.2580
50.00	24.1178	27.4348	29.4296	30.1207
60.00	23.3821	26.2339	28.2356	29.3810
70.00	22.8995	25.4560	27.4554	28.8998
80.00	22.5861	24.9481	26.9482	28.5859
90.00	22.3817	24.6176	26.6175	28.3817
100.00	22.2486	24.4022	26.4023	28.2486
110.00	22.1619	24.2620	26.2620	28.1619
120.00	22.1055	24.1707	26.1707	28.1055
130.00	22.0687	24.1112	26.1112	28.0687
140.00	22.0447	24.0724	26.0724	28.0447
150.00	22.0291	24.0472	26.0472	28.0291
160.00	22.0190	24.0307	26.0307	28.0190
170.00	22.0124	24.0200	26.0200	28.0124
180.00	22.0081	24.0130	26.0130	28.0081
190.00	22.0052	24.0085	26.0085	28.0052
200.00	22.0034	24.0055	26.0055	28.0034
210.00	22.0022	24.0036	26.0036	28.0022
220.00	22.0015	24.0023	26.0023	28.0015
230.00	22.0009	24.0015	26.0015	28.0009
240.00	22.0006	24.0010	26.0010	28.0006
250.00	22.0004	24.0007	26.0007	28.0004
260.00	22.0003	24.0004	26.0004	28.0003
270.00	22.0002	24.0003	26.0003	28.0002
280.00	22.0001	24.0002	26.0002	28.0001
290.00	22.0001	24.0001	26.0001	28.0001
300.00	22.0000	24.0001	26.0001	28.0000
310.00	22.0000	24.0001	26.0001	28.0000
320.00	22.0000	24.0000	26.0000	28.0000
330.00	22.0000	24.0000	26.0000	28.0000
340.00	22.0000	24.0000	26.0000	28.0000
350.00	22.0000	24.0000	26.0000	28.0000

We observe that the approximate steady-state temperatures agree exactly with the corresponding values of $\psi(x)$.

Exercises 15.3

1. (a) Identifying $h = 1/4$ and $k = 1/10$ we see that $\lambda = 2/5$.

TIME	X=0.25	X=0.5	X=0.75
0.00	0.1875	0.2500	0.1875
0.10	0.1775	0.2400	0.1775
0.20	0.1491	0.2100	0.1491
0.30	0.1066	0.1605	0.1066
0.40	0.0556	0.0938	0.0556
0.50	0.0019	0.0148	0.0019
0.60	-0.0501	-0.0682	-0.0501
0.70	-0.0970	-0.1455	-0.0970
0.80	-0.1361	-0.2072	-0.1361
0.90	-0.1648	-0.2462	-0.1648
1.00	-0.1802	-0.2591	-0.1802

(b) Identifying $h = 2/5$ and $k = 1/10$ we see that $\lambda = 1/4$.

TIME	X=0.4	X=0.8	X=1.2	X=1.6
0.00	0.0032	0.5273	0.5273	0.0032
0.10	0.0194	0.5109	0.5109	0.0194
0.20	0.0652	0.4638	0.4638	0.0652
0.30	0.1318	0.3918	0.3918	0.1318
0.40	0.2065	0.3035	0.3035	0.2065
0.50	0.2743	0.2092	0.2092	0.2743
0.60	0.3208	0.1190	0.1190	0.3208
0.70	0.3348	0.0413	0.0413	0.3348
0.80	0.3094	-0.0180	-0.0180	0.3094
0.90	0.2443	-0.0568	-0.0568	0.2443
1.00	0.1450	-0.0768	-0.0768	0.1450

(c) Identifying $h = 1/10$ and $k = 1/25$ we see that $\lambda = 2\sqrt{2}/5$.

TIME	X=0.1	X=0.2	X=0.3	X=0.4	X=0.5	X=0.6	X=0.7	X=0.8	X=0.9
0.00	0.0000	0.0000	0.0000	0.0000	0.0000	0.5000	0.5000	0.5000	0.5000
0.04	0.0000	0.0000	0.0000	0.0000	0.0800	0.4200	0.5000	0.5000	0.4200
0.08	0.0000	0.0000	0.0000	0.0256	0.2432	0.2568	0.4744	0.4744	0.2312
0.12	0.0000	0.0000	0.0082	0.1126	0.3411	0.1589	0.3792	0.3710	0.0462
0.16	0.0000	0.0026	0.0472	0.2394	0.3076	0.1898	0.2108	0.1663	-0.0496
0.20	0.0008	0.0187	0.1334	0.3264	0.2146	0.2651	0.0215	-0.0933	-0.0605
0.24	0.0071	0.0657	0.2447	0.3159	0.1735	0.2463	-0.1266	-0.3056	-0.0625
0.28	0.0299	0.1513	0.3215	0.2371	0.2013	0.0849	-0.2127	-0.3829	-0.1223
0.32	0.0819	0.2525	0.3168	0.1737	0.2033	-0.1345	-0.2580	-0.3223	-0.2264
0.36	0.1623	0.3197	0.2458	0.1657	0.0877	-0.2853	-0.2843	-0.2104	-0.2887
0.40	0.2412	0.3129	0.1727	0.1583	-0.1223	-0.3164	-0.2874	-0.1473	-0.2336
0.44	0.2657	0.2383	0.1399	0.0658	-0.3046	-0.2761	-0.2549	-0.1565	-0.0761
0.48	0.1965	0.1410	0.1149	-0.1216	-0.3593	-0.2381	-0.1977	-0.1715	0.0800
0.52	0.0466	0.0531	0.0225	-0.3093	-0.2992	-0.2260	-0.1451	-0.1144	0.1300
0.56	-0.1161	-0.0466	-0.1662	-0.3876	-0.2188	-0.2114	-0.1085	0.0111	0.0602
0.60	-0.2194	-0.2069	-0.3875	-0.3411	-0.1901	-0.1662	-0.0666	0.1140	-0.0446
0.64	-0.2485	-0.4290	-0.5362	-0.2611	-0.2021	-0.0969	0.0012	0.1084	-0.0843
0.68	-0.2559	-0.6276	-0.5625	-0.2503	-0.1993	-0.0298	0.0720	0.0068	-0.0354
0.72	-0.3003	-0.6865	-0.5097	-0.3230	-0.1585	0.0156	0.0893	-0.0874	0.0384
0.76	-0.3722	-0.5652	-0.4538	-0.4029	-0.1147	0.0289	0.0265	-0.0849	0.0596
0.80	-0.3867	-0.3464	-0.4172	-0.4068	-0.1172	-0.0046	-0.0712	-0.0005	0.0155
0.84	-0.2647	-0.1633	-0.3546	-0.3214	-0.1763	-0.0954	-0.1249	0.0665	-0.0386
0.88	-0.0254	-0.0738	-0.2202	-0.2002	-0.2559	-0.2215	-0.1079	0.0385	-0.0468
0.92	0.2064	-0.0157	-0.0325	-0.1032	-0.3067	-0.3223	-0.0804	-0.0636	-0.0127
0.96	0.3012	0.1081	0.1380	-0.0487	-0.2974	-0.3407	-0.1250	-0.1548	0.0092
1.00	0.2378	0.3032	0.2392	-0.0141	-0.2223	-0.2762	-0.2481	-0.1840	-0.0244

2. (a) In Section 12.4 the solution of the wave equation is shown to be

$$u(x,t) = \sum_{n=1}^{\infty} (A_n \cos n\pi t + B_n \sin n\pi t) \sin n\pi x$$

where

$$A_n = 2 \int_0^1 \sin \pi x \sin n\pi x \, dx = \begin{cases} 1, & n = 1 \\ 0, & n = 2, 3, 4, \ldots \end{cases}$$

and

$$B_n = \frac{2}{n\pi} \int_0^1 0 \, dx = 0.$$

Thus $u(x,t) = \cos \pi t \sin \pi x$.

(b) We have $h = 1/4$, $k = 0.5/5 = 0.1$ and $\lambda = 0.4$. Now $u_{0,j} = u_{4,j} = 0$ or $j = 0, 1, \ldots, 5$, and the initial values of u are $u_{1,0} = u(1/4, 0) = \sin \pi/4 \approx 0.7071$, $u_{2,0} = u(1/2, 0) = \sin \pi/2 = 1$, $u_{3,0} = u(3/4, 0) = \sin 3\pi/4 \approx 0.7071$. From equation (6) in the text we have

$$u_{i,1} = 0.8(u_{i+1,0} + u_{i-1,0}) + 0.84u_{i,0} + 0.1(0).$$

Then $u_{1,1} \approx 0.6740$, $u_{2,1} = 0.9531$, $u_{3,1} = 0.6740$. From equation (3) in the text we have for $j = 1, 2, 3, \ldots$

$$u_{i,j+1} = 0.16u_{i+1,j} + 2(0.84)u_{i,j} + 0.16u_{i-1,j} - u_{i,j-1}.$$

The results of the calculations are given in the table.

TIME	x=0.25	x=0.50	x=0.75
0.0	0.7071	1.0000	0.7071
0.1	0.6740	0.9531	0.6740
0.2	0.5777	0.8169	0.5777
0.3	0.4272	0.6042	0.4272
0.4	0.2367	0.3348	0.2367
0.5	0.0241	0.0340	0.0241

(c)

i,j	approx	exact	error
1,1	0.6740	0.6725	0.0015
1,2	0.5777	0.5721	0.0056
1,3	0.4272	0.4156	0.0116
1,4	0.2367	0.2185	0.0182
1,5	0.0241	0.0000	0.0241
2,1	0.9531	0.9511	0.0021
2,2	0.8169	0.8090	0.0079
2,3	0.6042	0.5878	0.0164
2,4	0.3348	0.3090	0.0258
2,5	0.0340	0.0000	0.0340
3,1	0.6740	0.6725	0.0015
3,2	0.5777	0.5721	0.0056
3,3	0.4272	0.4156	0.0116
3,4	0.2367	0.2185	0.0182
3,5	0.0241	0.0000	0.0241

3. (a) Identifying $h = 1/5$ and $k = 0.5/10 = 0.05$ we see that $\lambda = 0.25$.

TIME	X=0.2	X=0.4	X=0.6	X=0.8
0.00	0.5878	0.9511	0.9511	0.5878
0.05	0.5808	0.9397	0.9397	0.5808
0.10	0.5599	0.9059	0.9059	0.5599
0.15	0.5256	0.8505	0.8505	0.5256
0.20	0.4788	0.7748	0.7748	0.4788
0.25	0.4206	0.6806	0.6806	0.4206
0.30	0.3524	0.5701	0.5701	0.3524
0.35	0.2757	0.4460	0.4460	0.2757
0.40	0.1924	0.3113	0.3113	0.1924
0.45	0.1046	0.1692	0.1692	0.1046
0.50	0.0142	0.0230	0.0230	0.0142

(b) Identifying $h = 1/5$ and $k = 0.5/20 = 0.025$ we see that $\lambda = 0.125$.

TIME	X=0.2	X=0.4	X=0.6	X=0.8
0.00	0.5878	0.9511	0.9511	0.5878
0.03	0.5860	0.9482	0.9482	0.5860
0.05	0.5808	0.9397	0.9397	0.5808
0.08	0.5721	0.9256	0.9256	0.5721
0.10	0.5599	0.9060	0.9060	0.5599
0.13	0.5445	0.8809	0.8809	0.5445
0.15	0.5257	0.8507	0.8507	0.5257
0.18	0.5039	0.8153	0.8153	0.5039
0.20	0.4790	0.7750	0.7750	0.4790
0.23	0.4513	0.7302	0.7302	0.4513
0.25	0.4209	0.6810	0.6810	0.4209
0.28	0.3879	0.6277	0.6277	0.3879
0.30	0.3527	0.5706	0.5706	0.3527
0.33	0.3153	0.5102	0.5102	0.3153
0.35	0.2761	0.4467	0.4467	0.2761
0.38	0.2352	0.3806	0.3806	0.2352
0.40	0.1929	0.3122	0.3122	0.1929
0.43	0.1495	0.2419	0.2419	0.1495
0.45	0.1052	0.1701	0.1701	0.1052
0.48	0.0602	0.0974	0.0974	0.0602
0.50	0.0149	0.0241	0.0241	0.0149

4. We have $\lambda = 1$. The initial values of n are $u_{1,0} = u(0.2, 0) = 0.16$, $u_{2,0} = u(0.4) = 0.24$, $u_{3,0} = 0.24$, and $u_{4,0} = 0.16$. From equation (6) in the text we have

$$u_{i,1} = \frac{1}{2}(u_{i+1,0} + u_{i-1,0}) + 0u_{i,0} + k \cdot 0 = \frac{1}{2}(u_{i+1,0} + u_{i-1,0}).$$

Then, using $u_{0,0} = u_{5,0} = 0$, we find $u_{1,1} - 0.12$, $u_{2,1} = 0.2$, $u_{3,1} = 0.2$, and $u_{4,1} = 0.12$.

5. We identify $c = 24944.4$, $k = 0.00020045$ seconds $= 0.20045$ milliseconds, and $\lambda = 0.5$. Time in the table is expressed in milliseconds.

TIME	X=10	X=20	X=30	X=40	X=50
0.00000	0.1000	0.2000	0.3000	0.2000	0.1000
0.20045	0.1000	0.2000	0.2750	0.2000	0.1000
0.40089	0.1000	0.1938	0.2125	0.1938	0.1000
0.60134	0.0984	0.1688	0.1406	0.1688	0.0984
0.80178	0.0898	0.1191	0.0828	0.1191	0.0898
1.00223	0.0661	0.0531	0.0432	0.0531	0.0661
1.20268	0.0226	-0.0121	0.0085	-0.0121	0.0226
1.40312	-0.0352	-0.0635	-0.0365	-0.0635	-0.0352
1.60357	-0.0913	-0.1011	-0.0950	-0.1011	-0.0913
1.80401	-0.1271	-0.1347	-0.1566	-0.1347	-0.1271
2.00446	-0.1329	-0.1719	-0.2072	-0.1719	-0.1329
2.20491	-0.1153	-0.2081	-0.2402	-0.2081	-0.1153
2.40535	-0.0920	-0.2292	-0.2571	-0.2292	-0.0920
2.60580	-0.0801	-0.2230	-0.2601	-0.2230	-0.0801
2.80624	-0.0838	-0.1903	-0.2445	-0.1903	-0.0838
3.00669	-0.0932	-0.1445	-0.2018	-0.1445	-0.0932
3.20713	-0.0921	-0.1003	-0.1305	-0.1003	-0.0921
3.40758	-0.0701	-0.0615	-0.0440	-0.0615	-0.0701
3.60803	-0.0284	-0.0205	0.0336	-0.0205	-0.0284
3.80847	0.0224	0.0321	0.0842	0.0321	0.0224
4.00892	0.0700	0.0953	0.1087	0.0953	0.0700
4.20936	0.1064	0.1555	0.1265	0.1555	0.1064
4.40981	0.1285	0.1962	0.1588	0.1962	0.1285
4.61026	0.1354	0.2106	0.2098	0.2106	0.1354
4.81070	0.1273	0.2060	0.2612	0.2060	0.1273
5.01115	0.1070	0.1955	0.2851	0.1955	0.1070
5.21159	0.0821	0.1853	0.2641	0.1853	0.0821
5.41204	0.0625	0.1689	0.2038	0.1689	0.0625
5.61249	0.0539	0.1347	0.1260	0.1347	0.0539
5.81293	0.0520	0.0781	0.0526	0.0781	0.0520
6.01338	0.0436	0.0086	-0.0080	0.0086	0.0436
6.21382	0.0156	-0.0564	-0.0604	-0.0564	0.0156
6.41427	-0.0343	-0.1043	-0.1107	-0.1043	-0.0343
6.61472	-0.0931	-0.1364	-0.1578	-0.1364	-0.0931
6.81516	-0.1395	-0.1630	-0.1942	-0.1630	-0.1395
7.01561	-0.1568	-0.1915	-0.2150	-0.1915	-0.1568
7.21605	-0.1436	-0.2173	-0.2240	-0.2173	-0.1436
7.41650	-0.1129	-0.2263	-0.2297	-0.2263	-0.1129
7.61695	-0.0824	-0.2078	-0.2336	-0.2078	-0.0824
7.81739	-0.0625	-0.1644	-0.2247	-0.1644	-0.0625
8.01784	-0.0526	-0.1106	-0.1856	-0.1106	-0.0526
8.21828	-0.0440	-0.0611	-0.1091	-0.0611	-0.0440
8.41873	-0.0287	-0.0192	-0.0085	-0.0192	-0.0287
8.61918	-0.0038	0.0229	0.0867	0.0229	-0.0038
8.81962	0.0287	0.0743	0.1500	0.0743	0.0287
9.02007	0.0654	0.1332	0.1755	0.1332	0.0654
9.22051	0.1027	0.1858	0.1799	0.1858	0.1027
9.42096	0.1352	0.2160	0.1872	0.2160	0.1352
9.62140	0.1540	0.2189	0.2089	0.2189	0.1540
9.82185	0.1506	0.2030	0.2356	0.2030	0.1506
10.02230	0.1226	0.1822	0.2461	0.1822	0.1226

6. We identify $c = 24944.4$, $k = 0.00010022$ seconds $= 0.10022$ milliseconds, and $\lambda = 0.25$. Time in the table is expressed in milliseconds.

TIME	X=10	X=20	X=30	X=40	X=50
0.00000	0.2000	0.2667	0.2000	0.1333	0.0667
0.10022	0.1958	0.2625	0.2000	0.1333	0.0667
0.20045	0.1836	0.2503	0.1997	0.1333	0.0667
0.30067	0.1640	0.2307	0.1985	0.1333	0.0667
0.40089	0.1384	0.2050	0.1952	0.1332	0.0667
0.50111	0.1083	0.1744	0.1886	0.1328	0.0667
0.60134	0.0755	0.1407	0.1777	0.1318	0.0666
0.70156	0.0421	0.1052	0.1615	0.1295	0.0665
0.80178	0.0100	0.0692	0.1399	0.1253	0.0661
0.90201	-0.0190	0.0340	0.1129	0.1184	0.0654
1.00223	-0.0435	0.0004	0.0813	0.1077	0.0638
1.10245	-0.0626	-0.0309	0.0464	0.0927	0.0610
1.20268	-0.0758	-0.0593	0.0095	0.0728	0.0564
1.30290	-0.0832	-0.0845	-0.0278	0.0479	0.0493
1.40312	-0.0855	-0.1060	-0.0639	0.0184	0.0390
1.50334	-0.0837	-0.1237	-0.0974	-0.0150	0.0250
1.60357	-0.0792	-0.1371	-0.1275	-0.0511	0.0069
1.70379	-0.0734	-0.1464	-0.1533	-0.0882	-0.0152
1.80401	-0.0675	-0.1515	-0.1747	-0.1249	-0.0410
1.90424	-0.0627	-0.1528	-0.1915	-0.1595	-0.0694
2.00446	-0.0596	-0.1509	-0.2039	-0.1904	-0.0991
2.10468	-0.0585	-0.1467	-0.2122	-0.2165	-0.1283
2.20491	-0.0592	-0.1410	-0.2166	-0.2368	-0.1551
2.30513	-0.0614	-0.1349	-0.2175	-0.2507	-0.1772
2.40535	-0.0643	-0.1294	-0.2154	-0.2579	-0.1929
2.50557	-0.0672	-0.1251	-0.2105	-0.2585	-0.2005
2.60580	-0.0696	-0.1227	-0.2033	-0.2524	-0.1993
2.70602	-0.0709	-0.1219	-0.1942	-0.2399	-0.1889
2.80624	-0.0710	-0.1225	-0.1833	-0.2214	-0.1699
2.90647	-0.0699	-0.1236	-0.1711	-0.1972	-0.1435
3.00669	-0.0678	-0.1244	-0.1575	-0.1681	-0.1115
3.10691	-0.0649	-0.1237	-0.1425	-0.1348	-0.0761
3.20713	-0.0617	-0.1205	-0.1258	-0.0983	-0.0395
3.30736	-0.0583	-0.1139	-0.1071	-0.0598	-0.0042
3.40758	-0.0547	-0.1035	-0.0859	-0.0209	0.0279
3.50780	-0.0508	-0.0889	-0.0617	0.0171	0.0552
3.60803	-0.0460	-0.0702	-0.0343	0.0525	0.0767
3.70825	-0.0399	-0.0478	-0.0037	0.0840	0.0919
3.80847	-0.0318	-0.0221	0.0297	0.1106	0.1008
3.90870	0.0211	0.0062	0.0640	0.1314	0.1041
4.00892	-0.0074	0.0365	0.1005	0.1464	0.1025
4.10914	0.0095	0.0680	0.1350	0.1558	0.0973
4.20936	0.0295	0.1000	0.1666	0.1602	0.0897
4.30959	0.0521	0.1318	0.1937	0.1606	0.0808
4.40981	0.0764	0.1625	0.2148	0.1581	0.0719
4.51003	0.1013	0.1911	0.2291	0.1538	0.0639
4.61026	0.1254	0.2164	0.2364	0.1485	0.0575
4.71048	0.1475	0.2373	0.2369	0.1431	0.0532
4.81070	0.1659	0.2526	0.2315	0.1379	0.0512
4.91093	0.1794	0.2611	0.2217	0.1331	0.0514
5.01115	0.1867	0.2620	0.2087	0.1288	0.0535

Chapter 15 Review Exercises

1. Using the figure we obtain the system

$$u_{21} + 0 + 0 + 0 - 4u_{11} = 0$$

$$u_{31} + 0 + u_{11} + 0 - 4u_{21} = 0$$

$$50 + 0 + u_{21} + 0 - 4u_{31} = 0.$$

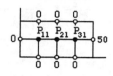

By Gauss-Elimination then,

$$\begin{bmatrix} -4 & 1 & 0 & | & 0 \\ 1 & -4 & 1 & | & 0 \\ 0 & 1 & -4 & | & -50 \end{bmatrix} \xrightarrow[\text{operations}]{\text{row}} \begin{bmatrix} 1 & -4 & 1 & | & 0 \\ 0 & 1 & -4 & | & -50 \\ 0 & 0 & 1 & | & 13.3928 \end{bmatrix}.$$

The solution is $u_{11} = 0.8929$, $u_{21} = 3.5714$, $u_{31} = 13.3928$.

2. By symmetry we observe that $u_{i,1} = u_{i,3}$ for $i = 1, 2, \ldots, 7$. We then use Gauss-Seidel iteration with an initial guess of 7.5 for all variables to solve the system

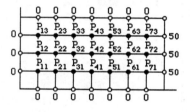

$$u_{11} = 0.25u_{21} + 0.25u_{12}$$

$$u_{21} = 0.25u_{31} + 0.25u_{22} + 0.25u_{11}$$

$$u_{31} = 0.25u_{41} + 0.25u_{32} + 0.25u_{21}$$

$$u_{41} = 0.25u_{51} + 0.25u_{42} + 0.25u_{31}$$

$$u_{51} = 0.25u_{61} + 0.25u_{52} + 0.25u_{41}$$

$$u_{61} = 0.25u_{71} + 0.25u_{62} + 0.25u_{51}$$

$$u_{71} = 12.5 + 0.25u_{72} + 0.25u_{61}$$

$$u_{12} = 0.25u_{22} + 0.5u_{11}$$

$$u_{22} = 0.25u_{32} + 0.5u_{21} + 0.25u_{12}$$

$$u_{32} = 0.25u_{42} + 0.5u_{31} + 0.25u_{22}$$

$$u_{42} = 0.25u_{52} + 0.5u_{41} + 0.25u_{32}$$

$$u_{52} = 0.25u_{62} + 0.5u_{51} + 0.25u_{42}$$

$$u_{62} = 0.25u_{72} + 0.5u_{61} + 0.25u_{52}$$

$$u_{72} = 12.5 + 0.5u_{71} + 0.25u_{62}.$$

After 30 iterations we obtain $u_{11} = u_{13} = 0.1765$, $u_{21} = u_{23} = 0.4566$, $u_{31} = u_{33} = 1.0051$, $u_{41} = u_{43} = 2.1479$, $u_{51} = u_{53} = 4.5766$, $u_{61} = u_{63} = 9.8316$, $u_{71} = u_{73} = 21.6051$, $u_{12} = 0.2494$, $u_{22} = 0.6447$, $u_{32} = 1.4162$, $u_{42} = 3.0097$, $u_{52} = 6.3269$, $u_{62} = 13.1447$, $u_{72} = 26.5887$.

3. (a)

TIME	X=0.0	X=0.2	X=0.4	X=0.6	X=0.8	X=1.0
0.00	0.0000	0.2000	0.4000	0.6000	0.8000	0.0000
0.01	0.0000	0.2000	0.4000	0.6000	0.5500	0.0000
0.02	0.0000	0.2000	0.4000	0.5375	0.4250	0.0000
0.03	0.0000	0.2000	0.3844	0.4750	0.3469	0.0000
0.04	0.0000	0.1961	0.3609	0.4203	0.2922	0.0000
0.05	0.0000	0.1883	0.3346	0.3734	0.2512	0.0000

(b)

TIME	X=0.0	X=0.2	X=0.4	X=0.6	X=0.8	X=1.0
0.00	0.0000	0.2000	0.4000	0.6000	0.8000	0.0000
0.01	0.0000	0.2000	0.4000	0.6000	0.8000	0.0000
0.02	0.0000	0.2000	0.4000	0.6000	0.5500	0.0000
0.03	0.0000	0.2000	0.4000	0.5375	0.4250	0.0000
0.04	0.0000	0.2000	0.3844	0.4750	0.3469	0.0000
0.05	0.0000	0.1961	0.3609	0.4203	0.2922	0.0000

(c) The table in part (b) is the same as the table in part (a) shifted downward one row.

Appendix

─────────── **Appendix I** ────────────────────────

1. (a) $\Gamma(5) = \Gamma(4+1) = 4! = 24$

   (b) $\Gamma(7) = \Gamma(6+1) = 6! = 720$

   (c) Using Example 1 in the text,

   $$-2\sqrt{\pi} = \Gamma\left(-\frac{1}{2}\right) = \Gamma\left(-\frac{3}{2}+1\right) = -\frac{3}{2}\Gamma\left(-\frac{3}{2}\right).$$

   Thus, $\Gamma(-3/2) = 4\sqrt{\pi}/3$.

   (d) Using (c)

   $$\frac{4\sqrt{\pi}}{3} = \Gamma\left(-\frac{3}{2}\right) = \Gamma\left(-\frac{5}{2}+1\right) = -\frac{5}{2}\Gamma\left(-\frac{5}{2}\right).$$

   Thus $\Gamma(-5/2) = -8\sqrt{\pi}/15$.

2. If $t = x^5$, then $dt = 5x^4\,dx$ and $x^5\,dx = \frac{1}{5}t^{1/5}\,dt$. Now

   $$\int_0^\infty x^5 e^{-x^5}\,dx = \int_0^\infty \frac{1}{5}t^{1/5}e^{-t}\,dt = \frac{1}{5}\int_0^\infty t^{1/5}e^{-t}\,dt$$

   $$= \frac{1}{5}\Gamma\left(\frac{6}{5}\right) = \frac{1}{5}(0.92) = 0.184.$$

3. If $t = x^3$, then $dt = 3x^2\,dx$ and $x^4\,dx = \frac{1}{3}t^{2/3}\,dt$. Now

   $$\int_0^\infty x^4 e^{-x^3}\,dx = \int_0^\infty \frac{1}{3}t^{2/3}e^{-t}\,dt = \frac{1}{3}\int_0^\infty t^{2/3}e^{-t}\,dt$$

   $$= \frac{1}{3}\Gamma\left(\frac{5}{3}\right) = \frac{1}{3}(0.89) \approx 0.297.$$

4. If $t = -\ln x = \ln\frac{1}{x}$ then $dt = -\frac{1}{x}\,dx$. Also $e^t = \frac{1}{x}$, so $x = e^{-t}$ and $dx = -x\,dt = -e^{-t}\,dt$. Thus

   $$\int_0^1 x^3\left(\ln\frac{1}{x}\right)^3\,dx = \int_\infty^0 (e^{-t})^3 t^3(-e^{-t})\,dt$$

   $$= \int_0^\infty t^3 e^{-4t}\,dt$$

   $$= \int_0^\infty \left(\frac{1}{4}u\right)^3 e^{-u}\left(\frac{1}{4}du\right) \qquad [u = 4t]$$

646

$$= \frac{1}{256} \int_0^\infty u^3 e^{-u} du = \frac{1}{256} \Gamma(4)$$

$$= \frac{1}{256} (3!) = \frac{3}{128} \, .$$

5. Since $e^{-t} \geq e^{-1}$ for $0 \leq t \leq 1$,

$$\Gamma(x) = \int_0^\infty t^{x-1} e^{-t} dt > \int_0^1 t^{x-1} e^{-t} dt \geq e^{-1} \int_0^1 t^{x-1} dt$$

$$= \frac{1}{e} \left(\frac{1}{x} t^x \right) \Big|_0^1 = \frac{1}{ex}$$

for $x > 0$. As $x \to 0^+$, we see that $\Gamma(x) \to \infty$.

6. For $x > 0$

$$\Gamma(x + 1) = \int_0^\infty t^x e^{-t} dt$$

$u = t^x$	$dv = e^{-t} \, dt$
$du = xt^{x-1} \, dt$	$v = -e^{-t}$

$$= -t^x e^{-t} \Big|_0^\infty - \int_0^\infty xt^{x-1}(-e^{-t}) \, dt$$

$$= x \int_0^\infty t^{x-1} e^{-t} dt = x\Gamma(x).$$

———— Appendix II ————

1. (a) $\mathbf{A} + \mathbf{B} = \begin{pmatrix} 4-2 & 5+6 \\ -6+8 & 9-10 \end{pmatrix} = \begin{pmatrix} 2 & 11 \\ 2 & -1 \end{pmatrix}$

(b) $\mathbf{B} - \mathbf{A} = \begin{pmatrix} -2-4 & 6-5 \\ 8+6 & -10-9 \end{pmatrix} = \begin{pmatrix} -6 & 1 \\ 14 & -19 \end{pmatrix}$

(c) $2\mathbf{A} + 3\mathbf{B} = \begin{pmatrix} 8 & 10 \\ -12 & 18 \end{pmatrix} + \begin{pmatrix} -6 & 18 \\ 24 & -30 \end{pmatrix} = \begin{pmatrix} 2 & 28 \\ 12 & -12 \end{pmatrix}$

2. (a) $\mathbf{A} - \mathbf{B} = \begin{pmatrix} -2-3 & 0+1 \\ 4-0 & 1-2 \\ 7+4 & 3+2 \end{pmatrix} = \begin{pmatrix} -5 & 1 \\ 4 & -1 \\ 11 & 5 \end{pmatrix}$

(b) $\mathbf{B} - \mathbf{A} = \begin{pmatrix} 3+2 & -1-0 \\ 0-4 & 2-1 \\ -4-7 & -2-3 \end{pmatrix} = \begin{pmatrix} 5 & -1 \\ -4 & 1 \\ -11 & -5 \end{pmatrix}$

(c) $2(\mathbf{A} + \mathbf{B}) = 2 \begin{pmatrix} 1 & -1 \\ 4 & 3 \\ 3 & 1 \end{pmatrix} = \begin{pmatrix} 2 & -2 \\ 8 & 6 \\ 6 & 2 \end{pmatrix}$

3. (a) $\mathbf{AB} = \begin{pmatrix} -2-9 & 12-6 \\ 5+12 & -30+8 \end{pmatrix} = \begin{pmatrix} -11 & 6 \\ 17 & -22 \end{pmatrix}$

(b) $\mathbf{BA} = \begin{pmatrix} -2-30 & 3+24 \\ 6-10 & -9+8 \end{pmatrix} = \begin{pmatrix} -32 & 27 \\ -4 & -1 \end{pmatrix}$

(c) $\mathbf{A}^2 = \begin{pmatrix} 4+15 & -6-12 \\ -10-20 & 15+16 \end{pmatrix} = \begin{pmatrix} 19 & -18 \\ -30 & 31 \end{pmatrix}$

(d) $\mathbf{B}^2 = \begin{pmatrix} 1+18 & -6+12 \\ -3+6 & 18+4 \end{pmatrix} = \begin{pmatrix} 19 & 6 \\ 3 & 22 \end{pmatrix}$

4. (a) $\mathbf{AB} = \begin{pmatrix} -4+4 & 6-12 & -3+8 \\ -20+10 & 30-30 & -15+20 \\ -32+12 & 48-36 & -24+24 \end{pmatrix} = \begin{pmatrix} 0 & -6 & 5 \\ -10 & 0 & 5 \\ -20 & 12 & 0 \end{pmatrix}$

(b) $\mathbf{BA} = \begin{pmatrix} -4+30-24 & -16+60-36 \\ 1-15+16 & 4-30+24 \end{pmatrix} = \begin{pmatrix} 2 & 8 \\ 2 & -2 \end{pmatrix}$

5. (a) $\mathbf{BC} = \begin{pmatrix} 9 & 24 \\ 3 & 8 \end{pmatrix}$

(b) $\mathbf{A(BC)} = \begin{pmatrix} 1 & -2 \\ -2 & 4 \end{pmatrix} \begin{pmatrix} 9 & 24 \\ 3 & 8 \end{pmatrix} = \begin{pmatrix} 3 & 8 \\ -6 & -16 \end{pmatrix}$

(c) $\mathbf{C(BA)} = \begin{pmatrix} 0 & 2 \\ 3 & 4 \end{pmatrix} \begin{pmatrix} 0 & 0 \\ 0 & 0 \end{pmatrix} = \begin{pmatrix} 0 & 0 \\ 0 & 0 \end{pmatrix}$

(d) $\mathbf{A(B+C)} = \begin{pmatrix} 1 & -2 \\ -2 & 4 \end{pmatrix} \begin{pmatrix} 6 & 5 \\ 5 & 5 \end{pmatrix} = \begin{pmatrix} -4 & -5 \\ 8 & 10 \end{pmatrix}$

6. (a) $\mathbf{AB} = \begin{pmatrix} 5 & -6 & 7 \end{pmatrix} \begin{pmatrix} 3 \\ 4 \\ -1 \end{pmatrix} = (-16)$

(b) $\mathbf{BA} = \begin{pmatrix} 3 \\ 4 \\ -1 \end{pmatrix} \begin{pmatrix} 5 & -6 & 7 \end{pmatrix} = \begin{pmatrix} 15 & -18 & 21 \\ 20 & -24 & 28 \\ -5 & 6 & -7 \end{pmatrix}$

(c) $(\mathbf{BA})\mathbf{C} = \begin{pmatrix} 15 & -18 & 21 \\ 20 & -24 & 28 \\ -5 & 6 & -7 \end{pmatrix} \begin{pmatrix} 1 & 2 & 4 \\ 0 & 1 & -1 \\ 3 & 2 & 1 \end{pmatrix} = \begin{pmatrix} 78 & 54 & 99 \\ 104 & 72 & 132 \\ -26 & -18 & -33 \end{pmatrix}$

(d) Since $\mathbf{AB}$ is 1×1 and $\mathbf{C}$ is 3×3 the product $(\mathbf{AB})\mathbf{C}$ is not defined.

7. (a) $\mathbf{A}^T\mathbf{A} = \begin{pmatrix} 4 & 8 & -10 \end{pmatrix} \begin{pmatrix} 4 \\ 8 \\ -10 \end{pmatrix} = (180)$

(b) $\mathbf{B}^T\mathbf{B} = \begin{pmatrix} 2 \\ 4 \\ 5 \end{pmatrix} \begin{pmatrix} 2 & 4 & 5 \end{pmatrix} = \begin{pmatrix} 4 & 8 & 10 \\ 8 & 16 & 20 \\ 10 & 20 & 25 \end{pmatrix}$

(c) $\mathbf{A} + \mathbf{B}^T = \begin{pmatrix} 4 \\ 8 \\ -10 \end{pmatrix} + \begin{pmatrix} 2 \\ 4 \\ 5 \end{pmatrix} = \begin{pmatrix} 6 \\ 12 \\ -5 \end{pmatrix}$

8. (a) $\mathbf{A} + \mathbf{B}^T = \begin{pmatrix} 1 & 2 \\ 2 & 4 \end{pmatrix} + \begin{pmatrix} -2 & 5 \\ 3 & 7 \end{pmatrix} = \begin{pmatrix} -1 & 7 \\ 5 & 11 \end{pmatrix}$

(b) $2\mathbf{A}^T - \mathbf{B}^T = \begin{pmatrix} 2 & 4 \\ 4 & 8 \end{pmatrix} - \begin{pmatrix} -2 & 5 \\ 3 & 7 \end{pmatrix} = \begin{pmatrix} 4 & -1 \\ 1 & 1 \end{pmatrix}$

(c) $\mathbf{A}^T(\mathbf{A} - \mathbf{B}) = \begin{pmatrix} 1 & 2 \\ 2 & 4 \end{pmatrix} \begin{pmatrix} 3 & -1 \\ -3 & -3 \end{pmatrix} = \begin{pmatrix} -3 & -7 \\ -6 & -14 \end{pmatrix}$

9. (a) $(\mathbf{AB})^T = \begin{pmatrix} 7 & 10 \\ 38 & 75 \end{pmatrix}^T = \begin{pmatrix} 7 & 38 \\ 10 & 75 \end{pmatrix}$

(b) $\mathbf{B}^T\mathbf{A}^T = \begin{pmatrix} 5 & -2 \\ 10 & -5 \end{pmatrix} \begin{pmatrix} 3 & 8 \\ 4 & 1 \end{pmatrix} = \begin{pmatrix} 7 & 38 \\ 10 & 75 \end{pmatrix}$

10. (a) $\mathbf{A}^T + \mathbf{B}^T = \begin{pmatrix} 5 & -4 \\ 9 & 6 \end{pmatrix} + \begin{pmatrix} -3 & -7 \\ 11 & 2 \end{pmatrix} = \begin{pmatrix} 2 & -11 \\ 20 & 8 \end{pmatrix}$

(b) $(\mathbf{A} + \mathbf{B})^T = \begin{pmatrix} 2 & 20 \\ -11 & 8 \end{pmatrix}^T = \begin{pmatrix} 2 & -11 \\ 20 & 8 \end{pmatrix}$

11. $\begin{pmatrix} -4 \\ 8 \end{pmatrix} - \begin{pmatrix} 4 \\ 16 \end{pmatrix} + \begin{pmatrix} -6 \\ 9 \end{pmatrix} = \begin{pmatrix} -14 \\ 1 \end{pmatrix}$

12. $\begin{pmatrix} 6t \\ 3t^2 \\ -3t \end{pmatrix} + \begin{pmatrix} -t+1 \\ -t^2+t \\ 3t-3 \end{pmatrix} - \begin{pmatrix} 6t \\ 8 \\ -10t \end{pmatrix} = \begin{pmatrix} -t+1 \\ 2t^2+t-8 \\ 10t-3 \end{pmatrix}$

13. $\begin{pmatrix} -19 \\ 18 \end{pmatrix} - \begin{pmatrix} 19 \\ 20 \end{pmatrix} = \begin{pmatrix} -38 \\ -2 \end{pmatrix}$

14. $\begin{pmatrix} -9t+3 \\ 13t-5 \\ -6t+4 \end{pmatrix} + \begin{pmatrix} -t \\ 1 \\ 4 \end{pmatrix} - \begin{pmatrix} 2 \\ 8 \\ -6 \end{pmatrix} = \begin{pmatrix} -10t+1 \\ 13t-12 \\ -6t+14 \end{pmatrix}$

15. Since $\det \mathbf{A} = 0$, $\mathbf{A}$ is singular.

16. Since $\det \mathbf{A} = 3$, $\mathbf{A}$ is nonsingular.

$$\mathbf{A}^{-1} = \frac{1}{3} \begin{pmatrix} 4 & -5 \\ -1 & 2 \end{pmatrix}$$

17. Since $\det \mathbf{A} = 4$, $\mathbf{A}$ is nonsingular.

$$\mathbf{A}^{-1} = \frac{1}{4} \begin{pmatrix} -5 & -8 \\ 3 & 4 \end{pmatrix}$$

18. Since $\det \mathbf{A} = -6$, $\mathbf{A}$ is nonsingular.

$$\mathbf{A}^{-1} = -\frac{1}{6} \begin{pmatrix} 2 & -10 \\ -2 & 7 \end{pmatrix}$$

19. Since $\det \mathbf{A} = 2$, $\mathbf{A}$ is nonsingular. The cofactors are

$$\begin{array}{lll} A_{11} = 0 & A_{12} = 2 & A_{13} = -4 \\ A_{21} = -1 & A_{22} = 2 & A_{23} = -3 \\ A_{31} = 1 & A_{32} = -2 & A_{33} = 5. \end{array}$$

Then

$$\mathbf{A}^{-1} = \frac{1}{2} \begin{pmatrix} 0 & 2 & -4 \\ -1 & 2 & -3 \\ 1 & -2 & 5 \end{pmatrix}^{T} = \frac{1}{2} \begin{pmatrix} 0 & -1 & 1 \\ 2 & 2 & -2 \\ -4 & -3 & 5 \end{pmatrix}.$$

20. Since det $\mathbf{A} = 27$, $\mathbf{A}$ is nonsingular. The cofactors are

$$\begin{array}{ccc} A_{11} = -1 & A_{12} = 4 & A_{13} = 22 \\ A_{21} = 7 & A_{22} = -1 & A_{23} = -19 \\ A_{31} = -1 & A_{32} = 4 & A_{33} = -5. \end{array}$$

Then

$$\mathbf{A}^{-1} = \frac{1}{27} \begin{pmatrix} -1 & 4 & 22 \\ 7 & -1 & -19 \\ -1 & 4 & -5 \end{pmatrix}^{T} = \frac{1}{27} \begin{pmatrix} -1 & 7 & -1 \\ 4 & -1 & 4 \\ 22 & -19 & -5 \end{pmatrix}.$$

21. Since det $\mathbf{A} = -9$, $\mathbf{A}$ is nonsingular. The cofactors are

$$\begin{array}{ccc} A_{11} = -2 & A_{12} = -13 & A_{13} = 8 \\ A_{21} = -2 & A_{22} = 5 & A_{23} = -1 \\ A_{31} = -1 & A_{32} = 7 & A_{33} = -5. \end{array}$$

Then

$$\mathbf{A}^{-1} = -\frac{1}{9} \begin{pmatrix} -2 & -13 & 8 \\ -2 & 5 & -1 \\ -1 & 7 & -5 \end{pmatrix}^{T} = -\frac{1}{9} \begin{pmatrix} -2 & -2 & -1 \\ -13 & 5 & 7 \\ 8 & -1 & -5 \end{pmatrix}.$$

22. Since det $\mathbf{A} = 0$, $\mathbf{A}$ is singular.

23. Since det $\mathbf{A}(t) = 2e^{3t} \neq 0$, $\mathbf{A}$ is nonsingular.

$$\mathbf{A}^{-1} = \frac{1}{2} e^{-3t} \begin{pmatrix} 3e^{4t} & -e^{4t} \\ -4e^{-t} & 2e^{-t} \end{pmatrix}$$

24. Since det $\mathbf{A}(t) = 2e^{2t} \neq 0$, $\mathbf{A}$ is nonsingular.

$$\mathbf{A}^{-1} = \frac{1}{2} e^{-2t} \begin{pmatrix} e^{t} \sin t & 2e^{t} \cos t \\ -e^{t} \cos t & 2e^{t} \sin t \end{pmatrix}$$

25. $\dfrac{d\mathbf{X}}{dt} = \begin{pmatrix} -5e^{-t} \\ -2e^{-t} \\ 7e^{-t} \end{pmatrix}$

26. $\dfrac{d\mathbf{X}}{dt} = \begin{pmatrix} \cos 2t + 8 \sin 2t \\ -6 \cos 2t - 10 \sin 2t \end{pmatrix}$

27. $\mathbf{X} = \begin{pmatrix} 2e^{2t} + 8e^{-3t} \\ -2e^{2t} + 4e^{-3t} \end{pmatrix}$ so that $\dfrac{d\mathbf{X}}{dt} = \begin{pmatrix} 4e^{2t} - 24e^{-3t} \\ -4e^{2t} - 12e^{-3t} \end{pmatrix}$.

28. $\dfrac{d\mathbf{X}}{dt} = \begin{pmatrix} 10te^{2t} + 5e^{2t} \\ 3t\cos 3t + \sin 3t \end{pmatrix}$

29. (a) $\dfrac{d\mathbf{A}}{dt} = \begin{pmatrix} 4e^{4t} & -\pi \sin \pi t \\ 2 & 6t \end{pmatrix}$

(b) $\displaystyle\int_0^2 \mathbf{A}(t)\, dt = \begin{pmatrix} \frac{1}{4}e^{4t} & \frac{1}{\pi}\sin \pi t \\ t^2 & t^3 - t \end{pmatrix}\Bigg|_{t=0}^{t=2} = \begin{pmatrix} \frac{1}{4}e^8 - \frac{1}{4} & 0 \\ 4 & 6 \end{pmatrix}$

(c) $\displaystyle\int_0^t \mathbf{A}(s)\, ds = \begin{pmatrix} \frac{1}{4}e^{4s} & \frac{1}{\pi}\sin \pi s \\ s^2 & s^3 - s \end{pmatrix}\Bigg|_{s=0}^{s=t} = \begin{pmatrix} \frac{1}{4}e^{4t} - \frac{1}{4} & \frac{1}{\pi}\sin \pi t \\ t^2 & t^3 - t \end{pmatrix}$

30. (a) $\dfrac{d\mathbf{A}}{dt} = \begin{pmatrix} -2t/(t^2 + 1)^2 & 3 \\ 2t & 1 \end{pmatrix}$

(b) $\dfrac{d\mathbf{B}}{dt} = \begin{pmatrix} 6 & 0 \\ -1/t^2 & 4 \end{pmatrix}$

(c) $\displaystyle\int_0^1 \mathbf{A}(t)\, dt = \begin{pmatrix} \tan^{-1} t & \frac{3}{2}t^2 \\ \frac{1}{3}t^3 & \frac{1}{2}t^2 \end{pmatrix}\Bigg|_{t=0}^{t=1} = \begin{pmatrix} \frac{\pi}{4} & \frac{3}{2} \\ \frac{1}{3} & \frac{1}{2} \end{pmatrix}$

(d) $\displaystyle\int_1^2 \mathbf{B}(t)\, dt = \begin{pmatrix} 3t^2 & 2t \\ \ln t & 2t^2 \end{pmatrix}\Bigg|_{t=1}^{t=2} = \begin{pmatrix} 9 & 2 \\ \ln 2 & 6 \end{pmatrix}$

(e) $\mathbf{A}(t)\mathbf{B}(t) = \begin{pmatrix} 6t/(t^2 + 1) + 3 & 2/(t^2 + 1) + 12t^2 \\ 6t^3 + 1 & 2t^2 + 4t^2 \end{pmatrix}$

(f) $\dfrac{d}{dt}\mathbf{A}(t)\mathbf{B}(t) = \begin{pmatrix} (6 - 6t^2)/(t^2 + 1)^2 & -4t/(t^2 + 1)^2 + 24t \\ 18t^2 & 12t \end{pmatrix}$

(g) $\displaystyle\int_1^t \mathbf{A}(s)\mathbf{B}(s)\, ds = \begin{pmatrix} 6s/(s^2 + 1) + 3 & 2/(s^2 + 1) + 12s^2 \\ 6s^3 + 1 & 6s^2 \end{pmatrix}\Bigg|_{s=1}^{s=t}$

$= \begin{pmatrix} 3t + 3\ln(t^2 + 1) - 3 - 3\ln 2 & 4t^3 + 2\tan^{-1} t - 4 - \pi/2 \\ (3/2)t^4 + t - (5/2) & 2t^3 - 2 \end{pmatrix}$

31.
$$\begin{pmatrix} 1 & 1 & -2 & | & 14 \\ 2 & -1 & 1 & | & 0 \\ 6 & 3 & 4 & | & 1 \end{pmatrix} \Longrightarrow \begin{pmatrix} 1 & 1 & -2 & | & 14 \\ 0 & -3 & 5 & | & -28 \\ 0 & 6 & 1 & | & 1 \end{pmatrix} \Longrightarrow \begin{pmatrix} 1 & 0 & -1/3 & | & 14/3 \\ 0 & 1 & -5/3 & | & 28/3 \\ 0 & 0 & 11 & | & -55 \end{pmatrix}$$

$$\Longrightarrow \begin{pmatrix} 1 & 0 & 0 & | & 3 \\ 0 & 1 & 0 & | & 1 \\ 0 & 0 & 1 & | & -5 \end{pmatrix}$$

Thus $x = 3$, $y = 1$, and $z = -5$.

32.
$$\begin{pmatrix} 5 & -2 & 4 & | & 10 \\ 1 & 1 & 1 & | & 9 \\ 4 & -3 & 3 & | & 1 \end{pmatrix} \Longrightarrow \begin{pmatrix} 1 & 1 & 1 & | & 9 \\ 0 & -7 & -1 & | & -35 \\ 0 & -7 & -1 & | & -35 \end{pmatrix} \Longrightarrow \begin{pmatrix} 1 & 0 & 6/7 & | & 4 \\ 0 & 1 & 1/7 & | & 5 \\ 0 & 0 & 0 & | & 0 \end{pmatrix}$$

Letting $z = t$ we find $y = 5 - \frac{1}{7}t$, and $x = 4 - \frac{6}{7}t$.

33.
$$\begin{pmatrix} 1 & -1 & -5 & | & 7 \\ 5 & 4 & -16 & | & -10 \\ 0 & 1 & 1 & | & -5 \end{pmatrix} \Longrightarrow \begin{pmatrix} 1 & -1 & -5 & | & 7 \\ 0 & 1 & 1 & | & -5 \\ 0 & 9 & 9 & | & -45 \end{pmatrix} \Longrightarrow \begin{pmatrix} 1 & 0 & -4 & | & 2 \\ 0 & 1 & 1 & | & -5 \\ 0 & 0 & 0 & | & 0 \end{pmatrix}$$

Letting $z = t$ we find $y = -5 - t$, and $x = 2 + 4t$.

34.
$$\begin{pmatrix} 1 & 1 & -3 & | & 6 \\ 4 & 2 & -1 & | & 7 \\ 3 & 1 & 1 & | & 4 \end{pmatrix} \Longrightarrow \begin{pmatrix} 1 & 1 & -3 & | & 6 \\ 0 & -2 & 11 & | & -17 \\ 0 & -2 & 10 & | & -14 \end{pmatrix} \Longrightarrow \begin{pmatrix} 1 & 0 & 5/2 & | & -5/2 \\ 0 & 1 & -11/2 & | & 17/2 \\ 0 & 0 & -1 & | & 3 \end{pmatrix}$$

$$\Longrightarrow \begin{pmatrix} 1 & 0 & 0 & | & 5 \\ 0 & 1 & 0 & | & -8 \\ 0 & 0 & 1 & | & -3 \end{pmatrix}$$

Thus $x = 5$, $y = -8$, and $z = -3$.

35.
$$\begin{pmatrix} 2 & 1 & 1 & | & 4 \\ 10 & -2 & 2 & | & -1 \\ 6 & -2 & 4 & | & 8 \end{pmatrix} \Longrightarrow \begin{pmatrix} 1 & 1/2 & 1/2 & | & 2 \\ 0 & -7 & -3 & | & -21 \\ 0 & -5 & 1 & | & 4 \end{pmatrix} \Longrightarrow \begin{pmatrix} 1 & 0 & 2/7 & | & 1/2 \\ 0 & 1 & 3/7 & | & 3 \\ 0 & 0 & 22/7 & | & 11 \end{pmatrix}$$

$$\Longrightarrow \begin{pmatrix} 1 & 0 & 0 & | & -1/2 \\ 0 & 1 & 0 & | & 3/2 \\ 0 & 0 & 1 & | & 7/2 \end{pmatrix}$$

Thus $x = -1/2$, $y = 3/2$, and $z = 7/2$.

36. $\begin{pmatrix} 1 & 0 & 2 & | & 8 \\ 1 & 2 & -2 & | & 4 \\ 2 & 5 & -6 & | & 6 \end{pmatrix} \Longrightarrow \begin{pmatrix} 1 & 0 & 2 & | & 8 \\ 0 & 2 & -4 & | & -4 \\ 0 & 5 & -10 & | & -10 \end{pmatrix} \Longrightarrow \begin{pmatrix} 1 & 0 & 2 & | & 8 \\ 0 & 1 & -2 & | & -2 \\ 0 & 0 & 0 & | & 0 \end{pmatrix}$

Letting $z = t$ we find $y = -2 + 2t$, and $x = 8 - 2t$.

37. $\begin{pmatrix} 1 & 1 & -1 & -1 & | & -1 \\ 1 & 1 & 1 & 1 & | & 3 \\ 1 & -1 & 1 & -1 & | & 3 \\ 4 & 1 & -2 & 1 & | & 0 \end{pmatrix} \Longrightarrow \begin{pmatrix} 1 & 1 & -1 & -1 & | & -1 \\ 0 & 0 & 2 & 2 & | & 4 \\ 0 & -2 & 2 & 0 & | & 4 \\ 0 & -3 & 2 & 5 & | & 4 \end{pmatrix} \Longrightarrow \begin{pmatrix} 1 & 0 & 0 & -1 & | & 1 \\ 0 & 1 & -1 & 0 & | & -2 \\ 0 & 0 & 2 & 2 & | & 4 \\ 0 & 0 & -1 & 5 & | & -2 \end{pmatrix}$

$$\Longrightarrow \begin{pmatrix} 1 & 0 & 0 & -1 & | & 1 \\ 0 & 1 & 0 & 1 & | & 0 \\ 0 & 0 & 1 & 1 & | & 2 \\ 0 & 0 & 0 & 6 & | & 0 \end{pmatrix} \Longrightarrow \begin{pmatrix} 1 & 0 & 0 & 0 & | & 1 \\ 0 & 1 & 0 & 0 & | & 0 \\ 0 & 0 & 1 & 0 & | & 2 \\ 0 & 0 & 0 & 1 & | & 0 \end{pmatrix}$$

Thus $x_1 = 1$, $x_2 = 0$, $x_3 = 2$, and $x_4 = 0$.

38. $\begin{pmatrix} 1 & 3 & 1 & | & 0 \\ 2 & 1 & 1 & | & 0 \\ 7 & 1 & 3 & | & 0 \end{pmatrix} \Longrightarrow \begin{pmatrix} 1 & 3 & 1 & | & 0 \\ 0 & -5 & -1 & | & 0 \\ 0 & -20 & -4 & | & 0 \end{pmatrix} \Longrightarrow \begin{pmatrix} 1 & 0 & 2/5 & | & 0 \\ 0 & 1 & 1/5 & | & 0 \\ 0 & 0 & 0 & | & 0 \end{pmatrix}$

Letting $x_3 = t$, we find $x_2 = -\frac{1}{5}t$ and $x_1 = -\frac{2}{5}t$.

39. $\begin{pmatrix} 1 & 2 & 4 & | & 2 \\ 2 & 4 & 3 & | & 1 \\ 1 & 2 & -1 & | & 7 \end{pmatrix} \Longrightarrow \begin{pmatrix} 1 & 2 & 4 & | & 2 \\ 0 & 0 & -5 & | & -3 \\ 0 & 0 & -5 & | & 5 \end{pmatrix} \Longrightarrow \begin{pmatrix} 1 & 2 & 0 & | & -2/5 \\ 0 & 0 & 1 & | & 3/5 \\ 0 & 0 & 0 & | & 8 \end{pmatrix}$

There is no solution.

40.
$$\begin{pmatrix} 1 & 1 & -1 & 3 & | & 1 \\ 0 & 1 & -1 & -4 & | & 0 \\ 1 & 2 & -2 & -1 & | & 6 \\ 4 & 7 & -7 & 0 & | & 9 \end{pmatrix} \implies \begin{pmatrix} 1 & 1 & -1 & 3 & | & 1 \\ 0 & 1 & -1 & -4 & | & 0 \\ 0 & 1 & -1 & -4 & | & 5 \\ 0 & 3 & -3 & -12 & | & 5 \end{pmatrix} \implies \begin{pmatrix} 1 & 0 & 0 & 7 & | & 1 \\ 0 & 1 & -1 & -4 & | & 0 \\ 0 & 0 & 0 & 0 & | & 5 \\ 0 & 0 & 0 & 0 & | & 5 \end{pmatrix}$$

There is no solution.

41. We solve
$$\det(\mathbf{A} - \lambda\mathbf{I}) = \begin{vmatrix} -1-\lambda & 2 \\ -7 & 8-\lambda \end{vmatrix} = (\lambda - 6)(\lambda - 1) = 0.$$

For $\lambda_1 = 6$ we have
$$\begin{pmatrix} -7 & 2 & | & 0 \\ -7 & 2 & | & 0 \end{pmatrix} \implies \begin{pmatrix} 1 & -2/7 & | & 0 \\ 0 & 0 & | & 0 \end{pmatrix}$$

so that $k_1 = \frac{2}{7}k_2$. If $k_2 = 7$ then
$$\mathbf{K}_1 = \begin{pmatrix} 2 \\ 7 \end{pmatrix}.$$

For $\lambda_2 = 1$ we have
$$\begin{pmatrix} -2 & 2 & | & 0 \\ -7 & 7 & | & 0 \end{pmatrix} \implies \begin{pmatrix} 1 & -1 & | & 0 \\ 0 & 0 & | & 0 \end{pmatrix}$$

so that $k_1 = k_2$. If $k_2 = 1$ then
$$\mathbf{K}_2 = \begin{pmatrix} 1 \\ 1 \end{pmatrix}.$$

42. We solve
$$\det(\mathbf{A} - \lambda\mathbf{I}) = \begin{vmatrix} 2-\lambda & 1 \\ 2 & 1-\lambda \end{vmatrix} = \lambda(\lambda - 3) = 0.$$

For $\lambda_1 = 0$ we have
$$\begin{pmatrix} 2 & 1 & | & 0 \\ 2 & 1 & | & 0 \end{pmatrix} \implies \begin{pmatrix} 1 & 1/2 & | & 0 \\ 0 & 0 & | & 0 \end{pmatrix}$$

so that $k_1 = -\frac{1}{2}k_2$. If $k_2 = 2$ then
$$\mathbf{K}_1 = \begin{pmatrix} -1 \\ 2 \end{pmatrix}.$$

For $\lambda_2 = 3$ we have
$$\begin{pmatrix} -1 & 1 & | & 0 \\ 2 & -2 & | & 0 \end{pmatrix} \implies \begin{pmatrix} 1 & -1 & | & 0 \\ 0 & 0 & | & 0 \end{pmatrix}$$

so that $k_1 = k_2$. If $k_2 = 1$ then

$$\mathbf{K}_2 = \begin{pmatrix} 1 \\ 1 \end{pmatrix}.$$

43. We solve

$$\det(\mathbf{A} - \lambda\mathbf{I}) = \begin{vmatrix} -8 - \lambda & -1 \\ 16 & -\lambda \end{vmatrix} = (\lambda + 4)^2 = 0.$$

For $\lambda_1 = \lambda_2 = -4$ we have

$$\begin{pmatrix} -4 & -1 & | & 0 \\ 16 & 4 & | & 0 \end{pmatrix} \Longrightarrow \begin{pmatrix} 1 & 1/4 & | & 0 \\ 0 & 0 & | & 0 \end{pmatrix}$$

so that $k_1 = -\frac{1}{4}k_2$. If $k_2 = 4$ then

$$\mathbf{K}_1 = \begin{pmatrix} -1 \\ 4 \end{pmatrix}.$$

44. We solve

$$\det(\mathbf{A} - \lambda\mathbf{I}) = \begin{vmatrix} 1 - \lambda & 1 \\ 1/4 & 1 - \lambda \end{vmatrix} = (\lambda - 3/2)(\lambda - 1/2) = 0.$$

For $\lambda_1 = 3/2$ we have

$$\begin{pmatrix} -1/2 & 1 & | & 0 \\ 1/4 & -1/2 & | & 0 \end{pmatrix} \Longrightarrow \begin{pmatrix} 1 & -2 & | & 0 \\ 0 & 0 & | & 0 \end{pmatrix}$$

so that $k_1 = 2k_2$. If $k_2 = 1$ then

$$\mathbf{K}_1 = \begin{pmatrix} 2 \\ 1 \end{pmatrix}.$$

If $\lambda_2 = 1/2$ then

$$\begin{pmatrix} 1/2 & 1 & | & 0 \\ 1/4 & 1/2 & | & 0 \end{pmatrix} \Longrightarrow \begin{pmatrix} 1 & 2 & | & 0 \\ 0 & 0 & | & 0 \end{pmatrix}$$

so that $k_1 = -2k_2$. If $k_2 = 1$ then

$$\mathbf{K}_2 = \begin{pmatrix} -2 \\ 1 \end{pmatrix}.$$

45. We solve

$$\det(\mathbf{A} - \lambda\mathbf{I}) = \begin{vmatrix} 5 - \lambda & -1 & 0 \\ 0 & -5 - \lambda & 9 \\ 5 & -1 & -\lambda \end{vmatrix} = \begin{vmatrix} 4 - \lambda & -1 & 0 \\ 4 - \lambda & -5 - \lambda & 9 \\ 4 - \lambda & -1 & -\lambda \end{vmatrix} = \lambda(4 - \lambda)(\lambda + 4) = 0.$$

If $\lambda_1 = 0$ then

$$\begin{pmatrix} 5 & -1 & 0 & | & 0 \\ 0 & -5 & 9 & | & 0 \\ 5 & -1 & 0 & | & 0 \end{pmatrix} \Longrightarrow \begin{pmatrix} 1 & 0 & -9/25 & | & 0 \\ 0 & 1 & -9/5 & | & 0 \\ 0 & 0 & 0 & | & 0 \end{pmatrix}$$

so that $k_1 = \frac{9}{25}k_3$ and $k_2 = \frac{9}{5}k_3$. If $k_3 = 25$ then

$$\mathbf{K}_1 = \begin{pmatrix} 9 \\ 45 \\ 25 \end{pmatrix}.$$

If $\lambda_2 = 4$ then

$$\begin{pmatrix} 1 & -1 & 0 & | & 0 \\ 0 & -9 & 9 & | & 0 \\ 5 & -1 & -4 & | & 0 \end{pmatrix} \Longrightarrow \begin{pmatrix} 1 & 0 & -1 & | & 0 \\ 0 & 1 & -1 & | & 0 \\ 0 & 0 & 0 & | & 0 \end{pmatrix}$$

so that $k_1 = k_3$ and $k_2 = k_3$. If $k_3 = 1$ then

$$\mathbf{K}_2 = \begin{pmatrix} 1 \\ 1 \\ 1 \end{pmatrix}.$$

If $\lambda_3 = -4$ then

$$\begin{pmatrix} 9 & -1 & 0 & | & 0 \\ 0 & -1 & 9 & | & 0 \\ 5 & -1 & 4 & | & 0 \end{pmatrix} \Longrightarrow \begin{pmatrix} 1 & 0 & -1 & | & 0 \\ 0 & 1 & -9 & | & 0 \\ 0 & 0 & 0 & | & 0 \end{pmatrix}$$

so that $k_1 = k_3$ and $k_2 = 9k_3$. If $k_3 = 1$ then

$$\mathbf{K}_3 = \begin{pmatrix} 1 \\ 9 \\ 1 \end{pmatrix}.$$

46. We solve

$$\det(\mathbf{A} - \lambda\mathbf{I}) = \begin{vmatrix} 3-\lambda & 0 & 0 \\ 0 & 2-\lambda & 0 \\ 4 & 0 & 1-\lambda \end{vmatrix} = (3-\lambda)(2-\lambda)(1-\lambda) = 0.$$

If $\lambda_1 = 1$ then

$$\begin{pmatrix} 2 & 0 & 0 & | & 0 \\ 0 & 1 & 0 & | & 0 \\ 4 & 0 & 0 & | & 0 \end{pmatrix} \Longrightarrow \begin{pmatrix} 1 & 0 & 0 & | & 0 \\ 0 & 1 & 0 & | & 0 \\ 0 & 0 & 0 & | & 0 \end{pmatrix}$$

so that $k_1 = 0$ and $k_2 = 0$. If $k_3 = 1$ then

$$\mathbf{K}_1 = \begin{pmatrix} 0 \\ 0 \\ 1 \end{pmatrix}.$$

If $\lambda_2 = 2$ then

$$\begin{pmatrix} 1 & 0 & 0 & | & 0 \\ 0 & 0 & 0 & | & 0 \\ 4 & 0 & -1 & | & 0 \end{pmatrix} \implies \begin{pmatrix} 1 & 0 & 0 & | & 0 \\ 0 & 0 & 1 & | & 0 \\ 0 & 0 & 0 & | & 0 \end{pmatrix}$$

so that $k_1 = 0$ and $k_3 = 0$. If $k_2 = 1$ then

$$\mathbf{K}_2 = \begin{pmatrix} 0 \\ 1 \\ 0 \end{pmatrix}.$$

If $\lambda_3 = 3$ then

$$\begin{pmatrix} 0 & 0 & 0 & | & 0 \\ 0 & -1 & 0 & | & 0 \\ 4 & 0 & -2 & | & 0 \end{pmatrix} \implies \begin{pmatrix} 1 & 0 & -1/2 & | & 0 \\ 0 & 1 & 0 & | & 0 \\ 0 & 0 & 0 & | & 0 \end{pmatrix}$$

so that $k_1 = \frac{1}{2}k_3$ and $k_2 = 0$. If $k_3 = 2$ then

$$\mathbf{K}_3 = \begin{pmatrix} 1 \\ 0 \\ 2 \end{pmatrix}.$$

47. We solve

$$\det(\mathbf{A} - \lambda\mathbf{I}) = \begin{vmatrix} -\lambda & 4 & 0 \\ -1 & -4-\lambda & 0 \\ 0 & 0 & -2-\lambda \end{vmatrix} = -(\lambda+2)^3 = 0.$$

For $\lambda_1 = \lambda_2 = \lambda_3 = -2$ we have

$$\begin{pmatrix} 2 & 4 & 0 & | & 0 \\ -1 & -2 & 0 & | & 0 \\ 0 & 0 & 0 & | & 0 \end{pmatrix} \implies \begin{pmatrix} 1 & 2 & 0 & | & 0 \\ 0 & 0 & 0 & | & 0 \\ 0 & 0 & 0 & | & 0 \end{pmatrix}$$

so that $k_1 = -2k_2$. If $k_2 = 1$ and $k_3 = 1$ then

$$\mathbf{K}_1 = \begin{pmatrix} -2 \\ 1 \\ 0 \end{pmatrix} \quad \text{and} \quad \mathbf{K}_2 = \begin{pmatrix} 0 \\ 0 \\ 1 \end{pmatrix}.$$

48. We solve

$$\det(\mathbf{A} - \lambda\mathbf{I}) = \begin{vmatrix} 1-\lambda & 6 & 0 \\ 0 & 2-\lambda & 1 \\ 0 & 1 & 2-\lambda \end{vmatrix} = \begin{vmatrix} 1-\lambda & 6 & 0 \\ 0 & 3-\lambda & 3-\lambda \\ 0 & 1 & 2-\lambda \end{vmatrix} = (3-\lambda)(1-\lambda)^2 = 0.$$

For $\lambda = 3$ we have

$$\begin{pmatrix} -2 & 6 & 0 & | & 0 \\ 0 & 0 & 0 & | & 0 \\ 0 & 1 & -1 & | & 0 \end{pmatrix} \implies \begin{pmatrix} 1 & 0 & -3 & | & 0 \\ 0 & 1 & -1 & | & 0 \\ 0 & 0 & 0 & | & 0 \end{pmatrix}$$

so that $k_1 = 3k_3$ and $k_2 = k_3$. If $k_3 = 1$ then

$$\mathbf{K}_1 = \begin{pmatrix} 3 \\ 1 \\ 1 \end{pmatrix}.$$

For $\lambda_2 = \lambda_3 = 1$ we have

$$\begin{pmatrix} 0 & 6 & 0 & | & 0 \\ 0 & 1 & 1 & | & 0 \\ 0 & 1 & 1 & | & 0 \end{pmatrix} \implies \begin{pmatrix} 0 & 1 & 0 & | & 0 \\ 0 & 0 & 1 & | & 0 \\ 0 & 0 & 0 & | & 0 \end{pmatrix}$$

so that $k_2 = 0$ and $k_3 = 0$. If $k_1 = 1$ then

$$\mathbf{K}_2 = \begin{pmatrix} 1 \\ 0 \\ 0 \end{pmatrix}.$$

49. We solve

$$\det(\mathbf{A} - \lambda\mathbf{I}) = \begin{vmatrix} -1-\lambda & 2 \\ -5 & 1-\lambda \end{vmatrix} = \lambda^2 + 9 = (\lambda - 3i)(\lambda + 3i) = 0.$$

For $\lambda_1 = 3i$ we have

$$\begin{pmatrix} -1-3i & 2 & | & 0 \\ -5 & 1-3i & | & 0 \end{pmatrix} \implies \begin{pmatrix} 1 & -(1/5)+(3/5)i & | & 0 \\ 0 & 0 & | & 0 \end{pmatrix}$$

so that $k_1 = \left(\frac{1}{5} - \frac{3}{5}i\right)k_2$. If $k_2 = 5$ then

$$\mathbf{K}_1 = \begin{pmatrix} 1 - 3i \\ 5 \end{pmatrix}.$$

For $\lambda_2 = -3i$ we have

$$\begin{pmatrix} -1+3i & 2 & | & 0 \\ -5 & 1+3i & | & 0 \end{pmatrix} \implies \begin{pmatrix} 1 & -\frac{1}{5}-\frac{3}{5}i & | & 0 \\ 0 & 0 & | & 0 \end{pmatrix}$$

so that $k_1 = \left(\frac{1}{5} + \frac{3}{5}i\right) k_2$. If $k_2 = 5$ then

$$\mathbf{K}_2 = \begin{pmatrix} 1 + 3i \\ 5 \end{pmatrix}.$$

50. We solve

$$\det(\mathbf{A} - \lambda\mathbf{I}) = \begin{vmatrix} 2-\lambda & -1 & 0 \\ 5 & 2-\lambda & 4 \\ 0 & 1 & 2-\lambda \end{vmatrix} = -\lambda^3 + 6\lambda^2 - 13\lambda + 10 = (\lambda - 2)(-\lambda^2 + 4\lambda - 5)$$

$$= (\lambda - 2)(\lambda - (2 + i))(\lambda - (2 - i)) = 0.$$

For $\lambda_1 = 2$ we have

$$\begin{pmatrix} 0 & -1 & 0 & | & 0 \\ 5 & 0 & 4 & | & 0 \\ 0 & 1 & 0 & | & 0 \end{pmatrix} \implies \begin{pmatrix} 1 & 0 & 4/5 & | & 0 \\ 0 & 1 & 0 & | & 0 \\ 0 & 0 & 0 & | & 0 \end{pmatrix}$$

so that $k_1 = -\frac{4}{5}k_3$ and $k_2 = 0$. If $k_3 = 5$ then

$$\mathbf{K}_1 = \begin{pmatrix} -4 \\ 0 \\ 5 \end{pmatrix}.$$

For $\lambda_2 = 2 + i$ we have

$$\begin{pmatrix} -i & -1 & 0 & | & 0 \\ 5 & -i & 4 & | & 0 \\ 0 & 1 & -i & | & 0 \end{pmatrix} \implies \begin{pmatrix} 1 & -i & 0 & | & 0 \\ 0 & 1 & -i & | & 0 \\ 0 & 0 & 0 & | & 0 \end{pmatrix}$$

so that $k_1 = ik_2$ and $k_2 = ik_3$. If $k_3 = i$ then

$$\mathbf{K}_2 = \begin{pmatrix} -i \\ -1 \\ i \end{pmatrix}.$$

For $\lambda_3 = 2 - i$ we have

$$\begin{pmatrix} i & -1 & 0 & | & 0 \\ 5 & i & 4 & | & 0 \\ 0 & 1 & i & | & 0 \end{pmatrix} \implies \begin{pmatrix} 1 & i & 0 & | & 0 \\ 0 & 1 & i & | & 0 \\ 0 & 0 & 0 & | & 0 \end{pmatrix}$$

so that $k_1 = -ik_2$ and $k_2 = -ik_3$. If $k_3 = i$ then

$$\mathbf{K}_3 = \begin{pmatrix} -1 \\ 1 \\ i \end{pmatrix}.$$

51. Let

$$\mathbf{A} = \begin{pmatrix} a_{11} & a_{12} \\ a_{21} & a_{22} \end{pmatrix}.$$

Then

$$\frac{d}{dt}[\mathbf{A}(t)\mathbf{X}(t)] = \frac{d}{dt}\begin{pmatrix} a_1 & a_2 \\ a_3 & a_4 \end{pmatrix}\begin{pmatrix} x_1 \\ x_2 \end{pmatrix} = \frac{d}{dt}\begin{pmatrix} a_1 x_1 + a_2 x_2 \\ a_3 x_1 + a_4 x_2 \end{pmatrix} = \begin{pmatrix} a_1 x_1' + a_1' x_1 + a_2 x_2' + a_2' x_2 \\ a_3 x_1' + a_3' x_1 + a_4 x_2' + a_4' x_2 \end{pmatrix}$$

$$= \begin{pmatrix} a_1 & a_2 \\ a_3 & a_4 \end{pmatrix}\begin{pmatrix} x_1' \\ x_2' \end{pmatrix} + \begin{pmatrix} a_1' & a_2' \\ a_3' & a_4' \end{pmatrix}\begin{pmatrix} x_1 \\ x_2 \end{pmatrix} = \mathbf{A}(t)\mathbf{X}'(t) + \mathbf{A}'(t)\mathbf{X}(t).$$

52. Assume $\det \mathbf{A} \neq 0$ and $\mathbf{AB} = \mathbf{I}$, so that

$$\begin{pmatrix} a_{11} & a_{12} \\ a_{21} & a_{22} \end{pmatrix}\begin{pmatrix} b_{11} & b_{12} \\ b_{21} & b_{22} \end{pmatrix} = \begin{pmatrix} 1 & 0 \\ 0 & 1 \end{pmatrix}.$$

Then

$$a_{11}b_{11} + a_{12}b_{21} = 1 \qquad a_{11}b_{12} + a_{12}b_{22} = 0$$

$$\text{and}$$

$$a_{21}b_{11} + a_{21}b_{21} = 0 \qquad a_{21}b_{12} + a_{21}b_{22} = 1$$

and by Cramer's rule

$$b_{11} = \frac{a_{22}}{\det \mathbf{A}} \qquad b_{12} = \frac{-a_{12}}{\det \mathbf{A}}$$

$$b_{21} = \frac{-a_{21}}{\det \mathbf{A}} \qquad b_{22} = \frac{a_{11}}{\det \mathbf{A}}.$$

Thus

$$\mathbf{A}^{-1} = \mathbf{B} = \frac{1}{\det \mathbf{A}}\begin{pmatrix} a_{22} & -a_{12} \\ -a_{21} & a_{11} \end{pmatrix}.$$

53. Since $\mathbf{A}$ is nonsingular, $\mathbf{AB} = \mathbf{AC}$ implies $\mathbf{A}^{-1}\mathbf{AB} = \mathbf{A}^{-1}\mathbf{AC}$. Then $\mathbf{IB} = \mathbf{IC}$ and $\mathbf{B} = \mathbf{C}$.

54. Since

$$(\mathbf{AB})(\mathbf{B}^{-1}\mathbf{A}^{-1}) = \mathbf{A}(\mathbf{BB}^{-1})\mathbf{A}^{-1} = \mathbf{AIA}^{-1} = \mathbf{AA}^{-1} = \mathbf{I}$$

and

$$(\mathbf{B}^{-1}\mathbf{A}^{-1})(\mathbf{AB}) = \mathbf{B}^{-1}(\mathbf{A}^{-1}\mathbf{A})\mathbf{B} = \mathbf{B}^{-1}\mathbf{IB} = \mathbf{B}^{-1}\mathbf{B} = \mathbf{I}$$

we have

$$(\mathbf{AB})^{-1} = \mathbf{B}^{-1}\mathbf{A}^{-1}.$$

55. No; consider

$$\mathbf{A} = \begin{pmatrix} 1 & 0 \\ 0 & 0 \end{pmatrix} \quad \text{and} \quad \mathbf{B} = \begin{pmatrix} 0 & 0 \\ 1 & 0 \end{pmatrix}.$$